HOLT
ALGEBRA 2
WITH TRIGONOMETRY

About the Authors

Eugene D. Nichols
Robert O. Lawton Distinguished Professor
of Mathematics Education
Florida State University
Tallahassee, Florida

Mervine L. Edwards
Chairman of Mathematics Department
Shore Regional High School
West Long Branch, New Jersey

E. Henry Garland
Head of the Mathematics Department
Developmental Research School
DRS Professor
Florida State University
Tallahassee, Florida

Sylvia A. Hoffman
Resource Consultant in Mathematics
Illinois State Board of Education
State of Illinois

Albert Mamary
Superintendent of Schools for Instruction
Johnson City Central School District
Johnson City, New York

William F. Palmer
Professor of Education and Director
Center for Mathematics and Science Education
Catawba College
Salisbury, North Carolina

Computer Consultant

Michael Murphy
Mathematics Department Chairman
Bonny Eagle High School
West Buxton, Maine

HOLT
ALGEBRA 2
WITH TRIGONOMETRY

Eugene D. Nichols
Mervine L. Edwards
E. Henry Garland
Sylvia A. Hoffman
Albert Mamary
William F. Palmer

HOLT, RINEHART AND WINSTON, PUBLISHERS
New York · Toronto · Mexico City · London · Sydney · Tokyo

Acknowledgments for Photographs

Cover: Radio telescope from Very Large Array, Dennis Brack/Black Star.

Page x, Michael Melford/Peter Arnold; *30*, Tom Tracy/Medichrome/Stock Shop; *56*, HRW photo by Richard Haynes; *88*, Geoffrey Gove/Image Bank; *114*, HRW photo by Richard Haynes; *138*, S.L. Craig, Jr./Bruce Coleman; *164*, John Iacono/*Sports Illustrated; 186*, Bettmann; *208*, Nancy J. Pierce/Photo Researchers; *234*, Georg Gerster/Photo Researchers; *239*, Harvey Olsen; *275*, Larry Mulvehill/Photo Researchers; *297*, Erwin & Peggy Bauer/Bruce Coleman; *302*, Allan Seiden/Image Bank; *337,* Heinz Kluetmeier/*Sports Illustrated; 345*, Judi Buie/Bruce Coleman; *356*, Bettmann; *389*, John Moss/Medichrome/Stock Shop; *423*, Focus on Sports; *454*, Bettmann; *496*, *FDA Consumer; 538*, William James Warren/West Light.

Insert 1:
Page 1, Harvey Olsen; *2, tl*, Robert Harding Picture Library, London; *tr*, Courtesy of the Wellcome Trustees/Robert Harding Picture Library; *b*, Harald Sund/Image Bank; *3, tl*, Robert Farber/Image Bank; *r*, British Museum of Ethnography/NYPL Picture Collection; *bl*, Raoul Hackel/Stock, Boston; *4, tl*, Granger; *tr*, Granger; *br*, Courtesy of the Trustees, National Gallery, London; *5, tc*, Granger; *br*, Earth Satellite Corporation; *bl*, Brown Brothers; *tl*, Harvey Olsen; *r*, Chuck O'Rear/Woodfin Camp & Assoc.; *bl*, Harvey Olsen; *7, tl*, Landsat/Rainbow; *tr*, Harvey Olsen; *b*, John Running/After Image; *8, t,* James A. Cook/Stock Broker; *b,* Bill Gallery/Stock, Boston.

Insert 2:
Page 1, Harvey Olsen; *2, t*, Granger; *br*, Robert Harding Picture Library, London; *bl*, illustration, *London News*/Robert Harding Picture Library; *3, tl*, Robert Harding Picture Library; *tr*, Mark Sherman/Bruce Coleman; *br*, Cameron Davidson/Bruce Coleman; *4, tl*, Granger; *r*, Granger; *bl*, Chuck O'Rear/West Light; *5, tl*, Mark Tuschman/Discover Magazine; *tr*, Jehangir Gazdar/Woodfin Camp & Assoc.; *b*, Nicholas Foster/Image Bank; *6, tl*, Terry A. Renna; *tr*, Keith Brofsky/Stock Broker; *bl,*Granger; *7, tc*, Werner H. Muller/Peter Arnold; *br*, Harvey Olsen; *bl*, NASA/Grant Heilman; *8, t*, Dr. E.P. Miles/Miles Color Art; *b*, E.R. Degginger/Bruce Coleman.

Insert 3:
Page 1, Ron Church/Photo Researchers; *2, l*, Bob Evans/Peter Arnold; *tr*, Granger; *br*, Culver; *3, t*, Jeff Simon/Bruce Coleman; *br*, Culver; *bl*, Jack Fields/Photo Researchers; *4, l*, Culver; *tr*, Granger; *br*, NASA/Photo Researchers; *5, tl*, Granger; *r*, Culver; *bl*, Granger; *6, l*, Jeff Simon/Bruce Coleman; *br*, Bettmann; *7, tl*, Keith Bradley/Woods Hole Oceanographic Institute; *r*, Soames Summerhays/Photo Researchers; *bl*, Shelly Lauzon/Woods Hole Oceanographic Institute; *8*, Granger.

ISBN: 0-03-002173-1

5678901234   032   98765432

Contents

1 Linear Equations and Inequalities x

1 Linear Equations 1.1
4 Linear Inequalities 1.2
7 Compound Linear Inequalities 1.3
10 Absolute Value 1.4
13 Problem Solving: Number Problems 1.5
18 Problem Solving: Perimeter Problems 1.6

20 Problem Solving: Age Problems 1.7
22 Problem Solving: Using Formulas 1.8
26 Applications: Properties of a Field
27 Chapter One Review
28 Chapter One Test
29 College Prep Test

2 Polynomials and Problem Solving 30

31 Properties of Exponents 2.1
34 Zero and Negative Integral Exponents 2.2
37 Polynomials and Synthetic Substitution 2.3
41 Multiplying Polynomials 2.4
43 Problem Solving: Mixture Problems 2.5

47 Problem Solving: Motion Problems 2.6
50 Introduction to Vectors 2.7
53 Chapter Two Review
54 Chapter Two Test
55 College Prep Test

3 Factoring and Special Products 56

57 Factoring Trinomials 3.1
60 Special Products 3.2
62 Special Factors 3.3
65 Combined Types of Factoring 3.4
68 Quadratic Equations 3.5
71 Quadratic Inequalities 3.6
74 Problem Solving: Number Problems 3.7

78 Problem Solving: Geometric Problems 3.8
83 Applications
84 Chapter Three Review
85 Chapter Three Test
86 Computer Activities
87 College Prep Test

4 Rational Expressions 88

89 Rational Numbers and Decimals 4.1
91 Simplifying Products and Quotients 4.2
94 Dimensional Analysis 4.3
96 Sums and Differences 4.4
99 Complex Rational Expressions 4.5
102 Statistics: Averages and Variability 4.6

106 Applications: Proofs with Rational Numbers
108 Chapter Four Review
109 Chapter Four Test
110 Computer Activities
111 College Prep Test
112 Cumulative Review (Chapters 1–4)

5 Using Rational Expressions   114

115 Fractional Equations   5.1
119 Problem Solving: Work Problems   5.2
122 Problem Solving: Round Trips   5.3
125 Literal Equations and Formulas   5.4
128 Dividing Polynomials   5.5

130 Remainder Theorem   5.6
134 Chapter Five Review
135 Chapter Five Test
136 Computer Activities
137 College Prep Test

6 Radicals and Rational Number Exponents   138

139 Square Roots   6.1
142 Simplifying Radicals   6.2
145 Sums, Differences, and Products of Radicals   6.3
148 Quotients of Radicals   6.4
152 Other Radicals   6.5
155 Rational Number Exponents   6.6

157 Expressions with Rational Number Exponents   6.7
160 Chapter Six Review
161 Chapter Six Test
162 Computer Activities
163 College Prep Test

7 Quadratic Formula   164

165 Completing the Square   7.1
167 Maximum Value   7.2
169 The Quadratic Formula   7.3
172 Problem Solving: The Quadratic Formula   7.4
175 Problem Solving: Irrational Measurements   7.5

178 Radical Equations   7.6
181 Applications
182 Chapter Seven Review
183 Chapter Seven Test
184 Computer Activities
185 College Prep Test

8 Complex Numbers   186

187 Sums and Differences of Complex Numbers   8.1
190 Products and Quotients of Complex Numbers   8.2
193 Equations with Imaginary Number Solutions   8.3
195 The Discriminant   8.4

198 Sum and Product of Solutions   8.5
201 Applications: Proofs with Complex Numbers
203 Chapter Eight Review
204 Chapter Eight Test
205 College Prep Test
206 Cumulative Review (Chapters 1–8)

9 Coordinate Geometry 208

209 The Coordinate Plane 9.1
212 Slope of a Line 9.2
216 Equation of a Line 9.3
220 Slope-Intercept Method of Drawing Graphs 9.4
223 The Distance Between Two Points 9.5
226 The Midpoint of a Segment 9.6

228 Parallel and Perpendicular Lines 9.7
231 Using Coordinate Geometry 9.8
234 Applications
235 Computer Activities
236 Chapter Nine Review
237 Chapter Nine Test
238 College Prep Test

10 Linear Systems, Matrices, and Determinants 239

240 Solving Linear Systems Graphically 10.1
243 Solving Linear Systems Algebraically 10.2
247 Solving Systems of Three Linear Equations 10.3
250 Two-by-Two Determinants 10.4
253 Three-by-Three Determinants 10.5
257 Matrix Addition 10.6

260 Matrix Multiplication 10.7
264 Solving Linear Systems Using Matrices 10.8
268 Problem Solving Using Linear Systems 10.9
272 Chapter Ten Review
273 Chapter Ten Test
274 College Prep Test

11 Functions 275

276 Relations and Functions 11.1
279 Values and Composition of Functions 11.2
282 Linear and Constant Functions 11.3
285 Inverse Relations and Functions 11.4
288 Direct Variation 11.5

291 Inverse Variation 11.6
294 Joint and Combined Variation 11.7
297 Applications
298 Chapter Eleven Review
299 Chapter Eleven Test
300 Computer Activities
301 College Prep Test

12 Conic Sections 302

303 Symmetric Points in a Coordinate Plane 12.1
306 The Circle 12.2
310 The Ellipse 12.3
314 The Parabola 12.4
319 The Hyperbola 12.5
324 Identifying Conic Sections 12.6

327 Equations and Properties of Conic Sections 12.7
331 Computer Activities
332 Chapter Twelve Review
333 Chapter Twelve Test
334 College Prep Test
335 Cumulative Review (Chapters 1–12)

13 Linear-Quadratic Systems 337

338 Solving Linear-Quadratic Systems of Equations 13.1
341 Solving Quadratic Systems of Equations 13.2
344 Applications
346 Solving Linear Inequality Systems 13.3

349 Solving Linear-Quadratic Inequality Systems 13.4
352 Chapter Thirteen Review
353 Chapter Thirteen Test
354 Computer Activities
355 College Prep Test

14 Progressions, Series, and Binomial Expansions 356

357 Arithmetic Progressions 14.1
361 The Arithmetic Mean 14.2
364 Arithmetic Series and Σ Notation 14.3
368 Geometric Progressions 14.4
372 The Geometric Mean 14.5
375 Geometric Series 14.6

379 Infinite Geometric Series 14.7
382 Binomial Expansions 14.8
384 Special Infinite Series 14.9
386 Chapter Fourteen Review
387 Chapter Fourteen Test
388 College Prep Test

15 Exponential, Logarithmic, and Polynomial Functions 389

390 Graphing Exponential Functions 15.1
392 Logarithmic Functions 15.2
394 Graphing Logarithmic Functions 15.3
397 Logs of Products and Quotients 15.4
400 Logs of Powers and Radicals 15.5
403 Common Logarithms and Antilogarithms 15.6
406 Using Exponential Equations 15.7

409 Higher-Degree Polynomial Equations 15.8
412 Rational Zero Theorem 15.9
416 Graphing Polynomial Functions 15.10
418 Applications: Descartes' Rule of Signs
420 Chapter Fifteen Review
421 Chapter Fifteen Test
422 College Prep Test

16 Permutations, Combinations, and Probability 423

424 Fundamental Counting Principle 16.1
428 Conditional Permutations 16.2
431 Distinguishable Permutations 16.3
433 Circular Permutations 16.4
435 Combinations 16.5
438 Probability: Simple Events 16.6

441 Probability: Compound Events 16.7
445 Probability and Arrangements 16.8
448 Chapter Sixteen Review
449 Chapter Sixteen Test
450 Computer Activities
451 College Prep Test
452 Cumulative Review (Chapters 1–16)

17 Trigonometric Functions 454

455 Angles of Rotation 17.1
458 Special Right Triangles 17.2
462 Sine and Cosine Functions 17.3
466 Tangent and Cotangent Functions 17.4
470 Secant and Cosecant Functions 17.5
473 Functions of Negative Angles 17.6
475 Values of Trigonometric Functions 17.7
477 Using Trigonometric Tables 17.8

481 Linear Trigonometric Equations 17.9
483 Interpolation with Trigonometry 17.10
486 Trigonometry of a Right Triangle 17.11
489 Using Trigonometry of a Right Triangle 17.12
492 Chapter Seventeen Review
493 Chapter Seventeen Test
494 Computer Activities
495 College Prep Test

18 Trigonometric Graphs and Identities 496

497 Graphing $y = \sin x$ and $y = \cos x$ 18.1
500 Determining Amplitude 18.2
503 Determining Period 18.3
506 Change of Period and Amplitude 18.4
509 Graphing $y = \tan x$ and $y = \cot x$ 18.5
512 Finding Logarithms of Trigonometric Functions 18.6
515 Quadratic Trigonometric Equations 18.7

518 Basic Trigonometric Identities 18.8
521 Proving Trigonometric Identities 18.9
523 Radian Measure 18.10
526 Applying Radian Measure 18.11
529 Inverses of Trigonometric Functions 18.12
533 Complex Numbers and De Moivre's Theorem 18.13
535 Chapter Eighteen Review
536 Chapter Eighteen Test
537 College Prep Test

19 Trigonometric Laws and Formulas 538

539 Law of Cosines 19.1
542 Using the Law of Cosines to Find Angle Measures 19.2
544 Area of a Triangle 19.3
546 Law of Sines 19.4
549 The Ambiguous Case 19.5
552 Sin $(A \pm B)$ and Sin $2A$ 19.6
556 Cos $(A \pm B)$ and Cos $2A$ 19.7
560 Tan $(A \pm B)$ and Tan $2A$ 19.8

563 Half-Angle Formulas 19.9
566 Double and Half-Angle Identities 19.10
568 Chapter Nineteen Review
569 Chapter Nineteen Test
570 Computer Activities
571 College Prep Test
572 Cumulative Review (Chapters 17–19)

574 **Computer Basic Section**
588 **Computations Using Logarithms**
590 **Tables**

597 **Glossary and Symbol List**
601 **Answers to Selected Exercises**
623 **Index**

1 LINEAR EQUATIONS AND INEQUALITIES

Problem Solving

Problem

Solve the equation $x + y + z = 74$ for integers x, y, and z with the following ratios: $x:y = 2:3$ and $y:z = 5:4$.

Four Steps in Problem Solving

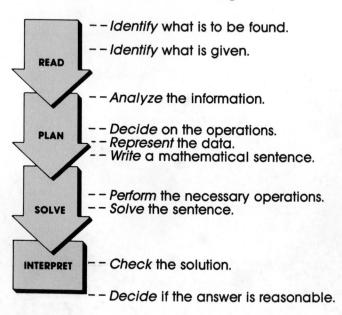

READ

– – *Identify* what is to be found.

– – *Identify* what is given.

PLAN

– – *Analyze* the information.

– – *Decide* on the operations.
– – *Represent* the data.
– – *Write* a mathematical sentence.

SOLVE

– – *Perform* the necessary operations.
– – *Solve* the sentence.

INTERPRET

– – *Check* the solution.

– – *Decide* if the answer is reasonable.

You are to find three numbers such that

(1) they are integers, (2) their sum is 74, and (3) the two ratios are true.

$x:y = 2:3 = 4:6 = 6:9 = 8:12 = 10:15$
$y:z = 5:4 = 10:8 = 15:12$

$x:y:z = 10:15:12 = 10n:15n:12n$
$$10n + 15n + 12n = 74$$

$$37n = 74$$
$$n = 2$$
$x = 10n = 20, \ y = 15n = 30, \ z = 12n = 24$

$$20 + 30 + 24 = 74$$
$$20:30 = 2:3 \text{ and } 30:24 = 5:4$$

LINEAR EQUATIONS

Objective   **To solve a linear equation in one variable**

Each of the equations below is a **linear equation** in one variable.

$$6x + 32 - 10x = 7 + 8x - 39 \qquad \frac{3}{5}y - 3 = 8 - \frac{1}{5}y \qquad 3.8n + 2.3 = 7.58 - 0.6n$$

You can solve a linear equation in one variable by
(1) combining the like terms, if any, on each side of the equation, and
(2) using one or more of the following equation properties.

Equation properties		For all numbers a, b, and c,
	Addition:	if $a = b$, then $a + c = b + c$.
	Subtraction:	if $a = b$, then $a - c = b - c$.
	Multiplication:	if $a = b$, then $a \cdot c = b \cdot c$.
	Division:	if $a = b$, then $\dfrac{a}{c} = \dfrac{b}{c}$; $c \neq 0$.

Example 1   **Find the root of $\frac{2}{5}n + 3 = 7 - \frac{1}{5}n$.**

$$\frac{2}{5}n + 3 = 7 - \frac{1}{5}n$$

$$\frac{3}{5}n + 3 = 7 \qquad \blacktriangleleft \; Add \; \frac{1}{5}n \; to \; each \; side.$$

$$\frac{3}{5}n = 4 \qquad \blacktriangleleft \; Subtract \; 3 \; from \; each \; side.$$

$$\frac{5}{3} \cdot \frac{3}{5}n = \frac{5}{3} \cdot 4 \qquad \blacktriangleleft \; Multiply \; each \; side \; by \; \frac{5}{3},$$

$$1n = 6\frac{2}{3} \qquad \qquad the \; reciprocal \; of \; \frac{3}{5}.$$

Thus, the root is $6\frac{2}{3}$.

Sometimes the variable appears on the right side of the equation but not on the left side. In that case, it may be more convenient to get the variable alone on the right side of the equation as shown below.

Example 2   **Solve $78 = 22 - 8t$. Check the solution by substitution.**

$$78 = 22 - 8t$$
$$56 = -8t$$
$$\frac{56}{-8} = \frac{-8t}{-8}$$
$$-7 = t$$

Check:

78	$22 - 8t$
78	$22 - 8(-7)$
	$22 + 56$
	78

True: $78 = 78$

Thus, the solution is -7.

A linear equation may contain parentheses. You can remove parentheses by using the **Distributive Property.**

Distributive property	For all numbers a, b, and c, $a(b + c) = ab + ac$ and $(b + c)a = ba + ca$.

Even though the Distributive Property is stated for multiplication over addition, the property remains true for multiplication over subtraction as well.

Example 3 **Solve $2(3x + 8) - 3(x - 2) = 9x - (4x - 6)$.**

$$2(3x + 8) + (-3)(x - 2) = 9x + (-1)(4x - 6) \quad \blacktriangleleft \; -(4x - 6) = -1(4x - 6)$$
$$6x + 16 - 3x + 6 = 9x - 4x + 6 \quad \blacktriangleleft \; \textit{Use the Distributive}$$
$$3x + 22 = 5x + 6 \qquad\qquad \textit{Property three times.}$$
$$-2x = -16$$
$$x = 8$$

Thus, the solution is 8.

Example 4 **Solve $\frac{1}{3}(6y - 15) - \frac{3}{4}(12y + 16) = 11$.**

$$\frac{1}{3}(6y) + \frac{1}{3}(-15) - \frac{3}{4}(12y) - \frac{3}{4}(16) = 11$$
$$2y - 5 - 9y - 12 = 11$$
$$-7y - 17 = 11$$
$$-7y = 28$$
$$y = -4$$

Thus, the solution is -4.

Example 5 **Solve $0.1 + 1.2(4n - 0.3) = 5(0.3n + 0.74)$.**

$$0.1 + 1.2(4n) + 1.2(-0.3) = 5(0.3n) + 5(0.74)$$
$$0.1 + 4.8n - 0.36 = 1.5n + 3.70$$
$$3.3n = 3.96$$
$$n = 1.2$$

Thus, the solution is 1.2.

Oral Exercises

Simplify each expression.

1. $3(5x - 8)$ 2. $-4(3y + 11)$ 3. $-(5n - 7)$ 4. $-(8 + 6t)$

5. $\frac{1}{2}(6a + 20)$ 6. $\frac{1}{3}(12x - 27)$ 7. $0.4(0.3t - 11)$ 8. $1.32(10y + 0.1)$

Written Exercises

Solve each equation.

(A) 1. $10 + \frac{2}{3}t = 6$ 2. $7 = \frac{3}{4}n - 24$

3. $-27 = 12w + 27$ 4. $126 - a = -9a$

5. $3(2n + 4) + 5(3 - 2n) = 7 - 2n$ 6. $2(7x - 4) - 4(2x - 6) = 3x + 31$

7. $5y - (2 - 3y) = 54$ 8. $6(3 + 2c) - (7c - 2) = 3c - 4$

9. $\frac{1}{4}(12x - 24) - 4x = \frac{1}{3}(18 + 15x) + 24$ 10. $\frac{1}{5}(30n + 25) - \frac{1}{8}(32n + 16) = 14n - 33$

11. $0.4t + 3(1.2t + 0.9) = 21.1$ 12. $0.7(5a - 1.2) = 2a - 0.39$

13. $6(3.2y - 2.3) = 7.2y - 13.32$ 14. $2.6(4x + 2.3) = 8.2x + 49.98$

(B) 15. $\frac{2}{7}x + 18 = 8 - \frac{3}{7}x$ 16. $\frac{4}{5}y - 22 = \frac{1}{5}y + 14$

17. $\frac{2}{9}n - 31 = 16 + \frac{7}{9}n$ 18. $18 - \frac{2}{11}a = \frac{5}{11}a - 19$

19. $3.2n - 0.05 = n + 0.06$ 20. $x - 0.01 = 12.2 - 2.3x$

21. $x - (9x - 5) = -(3x + 7)$ 22. $-4(2y - 5) - (7 - 3y) = 4(y + 3)$

23. $\frac{3}{5}(15n + 20) + 3n = \frac{2}{7}(14n - 28) - 12$ 24. $21 + \frac{5}{6}(30x + 18) = 19x - \frac{2}{5}(35x - 15)$

25. $2.5(1.2x + 6) = 1.92 + 8(4.3x + 2.42)$

26. $7(3.2n + 9.4) = 4.4(3.5n + 4.5) - 0.2$

(C) 27. $5(2n + 4) - 3(4n - 6) = 2(20 - n)$

28. $4(x - 3) - (x + 6) = 3(x - 6)$

29. $5[3(5y - 4) - 2(7y + 6)] + 3y - 8 = 0$

30. $10 - 4[6(2a + 5) - (20 + 10a)] - 2a = 0$

NON-ROUTINE PROBLEMS

1. Solve the equation $x - y + z - w = 6$ for integers x, y, z, and w with the following ratios: $x:y = 3:2$, $y:z = 2:5$, and $x:w = 5:8$.

2. Mrs. Jones invested \$12,000 in three different parts with the ratio $1:2:3$. The first part was invested at 11% and the second part at 9%. At what rate did she invest the third part if the yearly income was \$1,000?

LINEAR INEQUALITIES

Objectives
To find the solution set of a linear inequality in one variable
To graph the solutions of a linear inequality in one variable

A mathematical sentence such as $5 + 8x > 16$ or $7 - 5x < 27$ is called a **linear inequality** in one variable. The procedures used to solve a linear inequality are similar to those you have been using to solve a linear equation, with one major exception. This exception occurs when you need to multiply or divide by a negative number.

$$-8 \quad < \quad 6$$
$$-8(-2) \ \Big| \ 6(-2)$$
$$16 \quad > \quad -12$$

$$-8 < 6$$
$$\frac{-8}{-2} \ \Big| \ \frac{6}{-2}$$
$$4 \ > \ -3$$

$$-8 \quad < \quad 6$$
$$-8(2) \ \Big| \ 6(2)$$
$$-16 \ < \ 12$$

$$-8 < 6$$
$$\frac{-8}{2} \ \Big| \ \frac{6}{2}$$
$$-4 \ < \ 3$$

Notice that the *order* changes from $<$ to $>$ when each side of $-8 < 6$ is multiplied or divided by -2. Multiplying or dividing each side by 2 does not reverse the order. Also, adding the same number to each side of an inequality or subtracting the same number from each side does not reverse the order.

Inequality properties	For all numbers a, b, and c,	
	Addition:	if $a < b$, then $a + c < b + c$.
	Subtraction:	if $a < b$, then $a - c < b - c$.
	Multiplication:	if $a < b$, $c > 0$, then $a \cdot c < b \cdot c$.
		if $a < b$, $c < 0$, then $a \cdot c > b \cdot c$.
	Division:	if $a < b$, $c > 0$, then $\dfrac{a}{c} < \dfrac{b}{c}$.
		if $a < b$, $c < 0$, then $\dfrac{a}{c} > \dfrac{b}{c}$.

Example 1
Find the solution set of $8 - 3x < 2x + 28$.
Graph the solutions on a number line.

$$8 - 3x < 2x + 28$$
$$8 - 5x < 28$$
$$-5x < 20$$
$$\frac{-5x}{-5} \ \Big| \ \frac{20}{-5} \quad \blacktriangleleft \text{ Divide each side by } -5.$$
$$x > -4 \quad \blacktriangleleft \text{ Reverse the order from } < \text{ to } >.$$

Thus, $\{x \mid x > -4\}$ **is the solution set.** ◀ Read: "The set of all x such that x is greater than -4."
To graph the solutions, plot all numbers greater than -4.

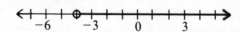

You will find it easier to read and graph an inequality like $5 > x$ if you write the variable first. Thus, $5 > x$ can be replaced by $x < 5$, and read as *x is less than 5*.

Even though the inequality properties on page 5 are stated for $<$, they are true for $>$ as well. In Example 2, the properties are applied to the $>$ relation.

Example 2   **Find the solution set of $3x - 2 > 5x + 7$.**

$$3x - 2 > 5x + 7$$
$$-2 > 2x + 7 \qquad \blacktriangleleft \text{ Subtract } 3x \text{ from each side.}$$
$$-9 > 2x \qquad \blacktriangleleft \text{ Subtract } 7 \text{ from each side.}$$
$$-4\frac{1}{2} > x \qquad \blacktriangleleft \text{ Divide each side by 2.}$$

$$x < -4\frac{1}{2} \qquad \blacktriangleleft \text{ Write the variable first.}$$

Thus, the solution set is $\left\{ x \mid x < -4\frac{1}{2} \right\}$.

Some linear inequalities contain the relation $\geq$, *is greater than or equal to*. You can solve such an inequality by using both equation and inequality properties. This is shown in Example 3.

Example 3   **Find the solution set of $3(4 - 2x) \geq 4x - 23$.**
Graph the solutions on a number line.

$$3(4 - 2x) \geq 4x - 23$$
$$12 - 6x \geq 4x - 23 \qquad \text{Combined division properties:}$$
$$-10x \geq -35 \qquad \qquad \text{If } a = b, \text{ then } \frac{a}{c} = \frac{b}{c}, c \neq 0.$$
$$\Big\downarrow \qquad \blacktriangleleft$$
$$x \leq 3\frac{1}{2} \qquad \qquad \text{If } a > b, c < 0, \text{ then } \frac{a}{c} < \frac{b}{c}.$$

Thus, the solution set is $\left\{ x \mid x \leq 3\frac{1}{2} \right\}$.

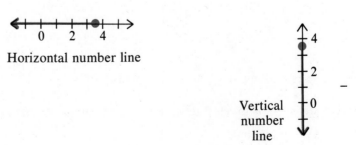

Horizontal number line

Vertical
number
line

Reading in Algebra

Match each inequality to exactly one graph.

1. $5 < x$
2. $5 \le x$
3. $x < 5$
4. $x \le 5$
5. $x < 2$
6. $x > 2$

A ●→→ 5 6

B ←+● 4 5

E ○ 2 ↓ 1 ↓

F ↑ 3 ○ 2

C ○→→ 5 6

D ←+○ 4 5

Oral Exercises

True or false?

1. $-7 < 3$
2. $-4 \le -6$
3. $5 > -9$
4. $8 \ge 8$
5. If $5a + 2 < 12$, then $5a > 10$.
6. If $8 \le y$, then $y \ge 8$.
7. If $-a < -8$, then $a < 8$.
8. If $-\dfrac{1}{2}x < 6$, then $x > -12$.

Written Exercises

Find the solution set of each inequality. Graph the solutions on a number line.

(A)
1. $-5y \le 15$
2. $-7a > -28$
3. $32 < 8x$
4. $3 > x - 2$
5. $4 \le n + 6$
6. $-4 \ge y - 6$

Find the solution set of each inequality.

7. $6 - 4x < 30$
8. $3y - 6 > 9y$
9. $5n + 27 \le -4n$
10. $8n - 2 > 6n + 12$
11. $4 - 2x \ge 3x + 19$
12. $7y - 6 > 3(y - 6)$
13. $3(2y + 6) < 8y + 11$
14. $4(2n - 3) > 5(n + 4)$
15. $-2(3x - 2) \ge 7(2 - x)$

Find the solution set of each inequality. Graph the solutions on a number line.

(B)
16. $-x < 0$
17. $22 - 3a > 29$
18. $\dfrac{1}{2}y \le 4$
19. $18 - 5y \ge 7 - 3y$
20. $\dfrac{2}{3}n < 12$
21. $8x - 7 - 10x \le 7 + 2x$

Find the solution set of each inequality.

22. $8 - 2(2x + 1) < 2x - 3$
23. $2n + 3(n - 2) \ge 2(3 - 2n)$
24. $5(x - 4) - (2 - 3x) < 0$
25. $\dfrac{1}{4}(8y - 12) > \dfrac{1}{3}(9 - 12y)$
26. $\dfrac{3}{5}(10a + 25) < \dfrac{2}{3}(6a - 9)$
27. $8n - \dfrac{3}{4}(12n + 28) \le 0$

(C)
28. $\dfrac{2x - 1}{-3} < 5$
29. $\dfrac{2(2y + 3)}{-5} > 6 - 2y$
30. $\dfrac{3(5a + 6)}{-4} < \dfrac{8 - 5a}{-4}$

31. Use algebraic properties to prove: If $x < y$, then $-y < -x$.

Show that each statement is false by giving a numerical counterexample.

32. If $b \ne 0$ and $a < b$, then $a^2 < b^2$.
33. If $a \ne 0$, $b \ne 0$, and $a < b$, then $\dfrac{1}{a} > \dfrac{1}{b}$.

COMPOUND LINEAR INEQUALITIES 1.3

Objectives

To find the solution set of a disjunction or a conjunction of two linear inequalities in one variable
To graph the solutions of a disjunction or a conjunction of two linear inequalities

A first linear inequality $x < 3$ and a second linear inequality $x < 5$ may be joined by the connective *or* or the connective *and* to form a compound inequality.

1. [$x < 3$ *or* $x < 5$] is a *disjunction* of two inequalities. One solution is 4 since at least one part of the disjunction [$4 < 3$ *or* $4 < 5$] is true.
2. [$x < 3$ *and* $x < 5$] is a *conjunction* of two inequalities. Four is *not* a solution since the two parts of the conjunction [$4 < 3$ *and* $4 < 5$] are not both true. Two is a solution because both parts of the conjunction [$2 < 3$ *and* $2 < 5$] are true.

You can find the solution set of a compound inequality and graph its solutions on a number line as shown below.

First, draw the graphs of the parts. Then,

1. For a disjunction (*or*), draw the graph of *both parts* (their union) on one number line.
2. For a conjunction (*and*), draw the graph of the *overlap* (intersection) of both parts on one number line.

Finally, use this graph of the solutions to describe the solution set.

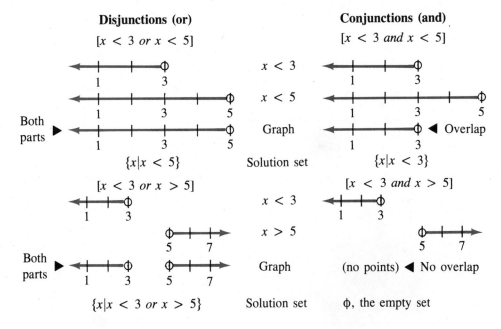

Disjunction (or)		Conjunction (and)

Disjunction (or) **Conjunction (and)**

$[x > 3$ or $x < 5]$ $[x > 3$ and $x < 5]$

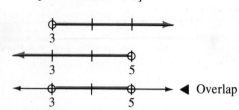

$x > 3$

$x < 5$

Graph

Solution set

Both parts ▶

{all numbers} $\{x|\ 3 < x < 5\}$

◀ Overlap

In the last conjunction, $3 < x < 5$ means $[3 < x$ *and* $x < 5]$.
You can read $3 < x < 5$ as "x is *between* 3 and 5."

Definition: $a < x < b$	$a < x < b$ means $[a < x$ *and* $x < b]$. The solutions of $a < x < b$ are the numbers *between* a and b if $a < b$.

Example 1   **Find the solution set of $[2x + 9 \leq 3$ or $3x - 7 > 5]$.**
Graph the solutions on a number line.

$$2x + 9 \leq 3 \quad \text{or} \quad 3x - 7 > 5$$
$$2x \leq -6 \quad \text{or} \quad 3x > 12$$
$$x \leq -3 \quad \text{or} \quad x > 4$$

Thus, the solution set is $\{x|x \leq -3$ or $x > 4\}$.

Example 2   **Find the solution set of $-3 < 2y - 7 \leq 1$.**
Graph the solutions on a number line.

Method 1   **Method 2**

$-3 < 2y - 7$ and $2y - 7 \leq 1$ $-3 < 2y - 7 \leq 1$
$\quad 4 < 2y \quad$ and $\quad 2y \leq 8$ ◀ *Add 7.* ▶ $4 < \quad 2y \leq 8$
$\quad 2 < y \quad$ and $\quad y \leq 4$ ◀ *Divide* ▶ $2 < \quad y \leq 4$
                                    by 2.

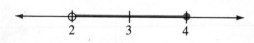

Thus, the solution set is $\{y|2 < y \leq 4\}$.

Reading in Algebra

Find the solution set.

1. $x < -2$ or $x < 3$
2. $x < -2$ and $x < 3$
3. $x < -2$ or $x > 3$
4. $x < -2$ and $x > 3$
5. $x > -2$ or $x > 3$
6. $x > -2$ and $x > 3$
7. $x > -2$ or $x < 3$
8. $x > -2$ and $x < 3$
9. $x \le -2$ or $x > -2$
10. $-6 < 3x < 12$
11. $-3 < x - 1 < 5$
12. $-6 < x + 3 < 10$

Written Exercises

Find the solution set of each compound inequality. Graph the solutions on a number line.

Ⓐ
1. $x + 7 < 3$ or $x - 2 > 2$
2. $x + 5 \le 7$ or $x - 3 \le 2$
3. $3x \ge -6$ or $5x \ge 15$
4. $4x > -8$ or $3x < 12$
5. $x + 3 < 9$ and $2x > 6$
6. $5x < -10$ and $x + 4 > 7$
7. $-2 < x - 3 < 2$
8. $4 \le x + 6 \le 8$
9. $-9 < 3x < 12$
10. $8 \le 4x < 16$

Ⓑ
11. $-11 \le 2x - 7 < 11$
12. $6 < 3x + 12 \le 21$
13. $4x + 9 < 1$ or $4x - 9 < 3$
14. $2x - 5 < -3$ and $2x + 5 < 9$
15. $5 < 4x + 5 < 25$
16. $-12 \le 5x - 2 \le -2$
17. $3 > 5x - 2$ or $4 < 3x - 2$
18. $-2 > 3x - 5$ and $7 < 4x - 5$
19. $4 - 3x \le 19$ or $2 - 5x \le 17$
20. $15 > 7 - 2x$ and $11 > 4x + 3$
21. $8 - 6x < 23$ and $15 < 4x + 1$
22. $-2 - 7x < -16$ or $-5 > 3 - 2x$

Ⓒ
23. $-6 < -3x < 9$
24. $-4 \le 1 - 5x \le 11$
25. $3x - 7 \le 5x + 6 \le 3x + 9$
26. $2 - 3x < -4$ or $2 - 3x > 14$
27. $-4 < 3x - 1 < 5$ or $3x - 1 > 11$
28. $-7 < 2x + 3 < -1$ or $1 < 2x + 3 < 7$

29. Let p and q represent any pair of statements such that each can be assigned the truth-value T for true or F for false. Complete the following truth table for the disjunction [p or q] and the conjunction [p and q] by writing either T or F in each empty cell.

p	q	p or q	p and q
T	T		
T	F		
F	T		
F	F		

Compound Linear Inequalities

ABSOLUTE VALUE

Objectives **To solve an equation involving absolute value**
To find the solution set of an inequality involving absolute value
To graph the solutions of an inequality involving absolute value

The **absolute value** of any number x is the distance between x and the origin, 0, on a number line. Therefore, the absolute value of -5 is 5 and the absolute value of 5 is 5.

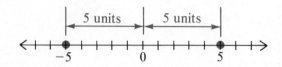

The symbol $|x|$ means the absolute value of x, so $|-5| = |5| = 5$.

Definition:
Absolute
value

$|x| = -x$, if x is a negative number.

$|x| = x$, if x is a positive number or 0.

From the definition of absolute value, the solutions of the equation $|x| = 12$ are -12 and 12 since $|-12| = 12$ and $|12| = 12$. This leads to the following equation property.

Equation
property for
absolute
value

For each number x and each number $k \geq 0$,
if $|x| = k$, then $x = -k$ or $x = k$.

Example 1 **Solve $|3n - 5| = 7$.**

$$|3n - 5| = 7$$

$$3n - 5 = -7 \quad \text{or} \quad 3n - 5 = 7 \qquad \blacktriangleleft \text{ If } |x| = k, \text{ then}$$
$$3n = -2 \qquad\qquad 3n = 12 \qquad\qquad x = -k \text{ or } x = k.$$
$$n = -\frac{2}{3} \qquad\qquad n = 4$$

Check: $n = -\dfrac{2}{3}$

$|3n - 5|$	7
$\left| 3 \cdot \dfrac{-2}{3} - 5 \right|$	7
$|-2 - 5|$	
$|-7|$	
7	

True: $7 = 7$

Check: $n = 4$

$|3n - 5|$	7
$|3 \cdot 4 - 5|$	7
$|12 - 5|$	
$|7|$	
7	

True: $7 = 7$

Thus, the solutions are $-\dfrac{2}{3}$ and 4.

The graph for $|x| < 4$ is shown below on the left. The graph for $|x| > 4$ is shown below on the right.

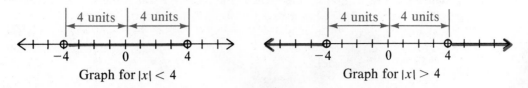

Graph for $|x| < 4$ Graph for $|x| > 4$

All numbers between -4 and 4 are solutions of the inequality. In symbols, $-4 < x < 4$. The compound inequality $-4 < x < 4$ means $-4 < x$ *and* $x < 4$.

All numbers less than -4 and all numbers greater than 4 are solutions of the inequality. In symbols, $x < -4$ *or* $x > 4$.

The graphs suggest the following inequality properties:

Inequality properties for absolute value	For all numbers x and k, if $k > 0$ and $|x| < k$, then $-k < x < k$. if $k \geq 0$ and $|x| > k$, then $x < -k$ or $x > k$.

The inequality properties for absolute value are applied in the following examples. Notice that Example 3 involves the $\geq$ relation, so combined equation and inequality properties are used.

Example 2   **Find the solution set of $|4y + 2| < 10$. Graph the solutions on a number line.**

$$|4y + 2| < 10$$
$$-10 < 4y + 2 < 10 \quad \blacktriangleleft \text{ If } |x| < k, \text{ then } -k < x < k.$$
$$-12 < 4y < 8 \quad \blacktriangleleft \text{ Subtract 2 from each side.}$$
$$-3 < y < 2 \quad \blacktriangleleft \text{ Divide each side by 4.}$$

The solution set is $\{y | -3 < y < 2\}$.

Example 3   **Find the solution set of $|2n - 1| \geq 5$. Graph the solutions on a number line.**

$$|2n - 1| \geq 5$$

		Combined properties:
$2n - 1 \leq -5$ or $2n - 1 \geq 5$	$\blacktriangleleft$	If $|x| = k$, then $x = -k$, or $x = k$.
$2n \leq -4$ $2n \geq 6$		If $|x| > k$, then $x < -k$, or $x > k$.
$n \leq -2$ $n \geq 3$		

The solution set is $\{n | n \leq -2 \text{ or } n \geq 3\}$.

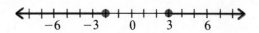

Reading in Algebra

True or false? If a statement is false, give a numerical counterexample which shows that it is false.

1. $|-x| = x$, if $x > 0$.   **2.** $|-x| = -x$, if $x < 0$.   **3.** $|x - 2| = x - 2$, if $x > 0$.

4. $|x - 2| = 2 - x$, if $x \leq 0$.   **5.** $|4 - x| = 4 - x$, if $x \geq 0$.   **6.** $|-x - 6| = x + 6$, if $x < 0$.

7. $|5 - x| = |x - 5|$, for each x.   **8.** $|-x - 8| = |x + 8|$, for each x.   **9.** $|x| = |-x|$, for each x.

Oral Exercises

Find the value of each expression.

1. $|-9 + 4|$ **2.** $|-5 - 8|$ **3.** $|-6 + 9|$ **4.** $|7 - 7|$

5. $|-3 \cdot 5|$ **6.** $|-2(-4)|$ **7.** $|2(-5) - 10|$ **8.** $|-5(-4) - 2|$

Written Exercises

Solve each equation.

Ⓐ **1.** $|x - 3| = 7$ **2.** $|a + 6| = 2$ **3.** $|4 - y| = 6$ **4.** $|2 - x| = 3$

5. $|2x + 15| = 5$ **6.** $|3y - 9| = 18$ **7.** $|4n + 6| = 2$ **8.** $|5 - 2a| = 9$

Find the solution set of each inequality. Graph the solutions on a number line.

9. $|x| < 3$ **10.** $|2n| \geq 12$ **11.** $|y| > 8$ **12.** $|5a| \leq 20$

13. $|x - 3| \geq 1$ **14.** $|y + 4| < 7$ **15.** $|a - 2| \leq 6$ **16.** $|n + 5| > 3$

Solve each equation.

Ⓑ **17.** $|4y + 3| = 5$ **18.** $|3x - 5| = 10$ **19.** $|5n + 12| - 2 = 0$ **20.** $|14 - 6z| = 10$

21. $|2x - 3.5| = 10.5$ **22.** $|4y + 6.8| = 9.2$ **23.** $2.7 - |3z + 9.3| = 0$ **24.** $|14.2 - 5n| = 0.8$

Find the solution set of each inequality. Graph the solutions on a number line.

25. $|2x - 5| < 17$ **26.** $|3y + 6| > 21$ **27.** $|3z - 15| \geq 6$ **28.** $|2n + 17| \leq 1$

29. $|4y + 6| \geq 2$ **30.** $10 \geq |5a - 15|$ **31.** $|5x + 10| - 35 < 0$ **32.** $20 < |4z - 4|$

Ⓒ **33.** $|18 - 3x| < 6$ **34.** $|75 - 2y| > 5$

35. $|x - 2| < 6$ or $|x - 2| > 10$ **36.** $|y + 5| \leq 2$ or $|y - 5| \leq 2$

Solve each equation.

37. $|x - 2| + |2 - x| = 10$ **38.** $|2y + 4| + |6 - 2y| = 22$

39. $|n + 2| + |-n - 6| = 18$ **40.** $|1 - 2x| + |-x - 7| = 21$

Find the solution set of each inequality.

41. $|x - 4| < 0.1$ **42.** $|y + 6| < 0.001$

43. $|x - 2.5| < 0.01$ **44.** $|x + 3.4| < 0.0001$

CUMULATIVE REVIEW

1. Solve $\frac{2}{3}(12y - 15) - \frac{3}{5}(15y + 20) = 3y - 2$ **2.** Solve $0.9n + 2.2(0.5n - 3) = 0.4(10n - 1.5)$.

PROBLEM SOLVING: NUMBER PROBLEMS 1.5

Objectives **To write a word phrase in algebraic terms**
To solve a word problem about one or more numbers

Many mathematical problems are presented using a word phrase or several word phrases. To solve such problems, you begin by writing the word phrases in algebraic terms. Some typical word phrases and their algebraic translations are listed below.

Word Phrase	Written in Algebraic Terms
Ten decreased by a number	$10 - n$
Twice a number, increased by 12	$2x + 12$
Six less than a number	$y - 6$
Seven more than a number	$n + 7$ or $7 + n$
Eight decreased by 5 times the sum of a number and 14	$8 - 5(x + 14)$
Is, or is the same as	$=$

Notice that two translations are given for the word phrase "Seven more than a number." The two translations are equivalent since addition is commutative. However, there is only one correct translation for the word phrase "Six less than a number." Subtraction is not commutative. Remember, for *4 less than x* write $x - 4$, not $4 - x$.

You can solve a word problem about one number if you are given enough information about the number. Example 1 illustrates this situation.

Example 1 **Seven less than twice a number is 20 more than 5 times the number. What is the number?**

Represent the data. Let x be the number.
7 less than twice x is   20 more than $5x$.

Write an equation. $2x - 7 = 5x + 20$

$-27 = 3x$

Solve the equation. $-9 = x$

Check in the problem.

Seven less than twice x	20 more than $5x$
$2(-9) - 7$	$5(-9) + 20$
$-18 - 7$	$-45 + 20$
-25	-25

Answer the question. **Thus,** the number is -9.

In some word problems you will be asked to find more than one number. To do this, you begin by representing each of the numbers in terms of one variable, as shown in Example 2.

Example 2

The greater of two numbers is 8 less than twice the sum of the smaller number and 5. Represent each number in terms of one variable.

The *greater* number is *described in terms of* the *smaller* number. Represent the *smaller* number by a variable.

Let s = smaller number.

Twice the sum of s and 5   8 less than

$$2(s + 5) \quad - \quad 8$$

Thus, s = smaller number and $2(s + 5) - 8$ = greater number.

The four steps you should use in solving any word problem are given at the beginning of this chapter. They are illustrated again in Example 3 below.

Example 3

The second of three numbers is 3 times the first. The third number is 2 more than the second number. Seven less than twice the second is the same as 12 more than the third. What are the three numbers?

READ ▶

You are asked to find three numbers. The third number is described *in terms of* the second number. The second number is described *in terms of* the first number.

PLAN ▶

Let f = 1st number
$3f$ = 2nd number
$3f + 2$ = 3rd number

7 less than twice the 2nd   is   12 more than the 3rd.
$2(3f) - 7$ = $(3f + 2) + 12$

SOLVE ▶

$$
\begin{aligned}
6f - 7 &= 3f + 14 \\
3f &= 21 \\
f &= 7
\end{aligned}
$$

$f = 7$ $3f = 21$ $3f + 2 = 23$

INTERPRET ▶

7 less than twice the 2nd	12 more than the 3rd
$2(21) - 7$	$23 + 12$
$42 - 7$	35
35	

Thus, the three numbers are 7, 21, and 23.

In Example 3, you saw the four major steps in solving a word problem.

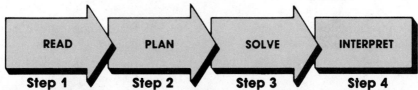

READ | PLAN | SOLVE | INTERPRET
Step 1 | Step 2 | Step 3 | Step 4

Step 2 may require some information that is not stated directly in the problem. For example, some problems require a knowledge of *consecutive integers*. Four such typical word phrases and their algebraic translations are shown in the table below, along with a numerical example of each.

Word Phrase	Example	Written in Algebraic Terms	
Consecutive integers	$-1, 0, 1, 2, 3$	$x, x + 1, x + 2, \ldots$	(x is an integer.)
Consecutive odd integers	$5, 7, 9, 11, 13$	$y, y + 2, y + 4, \ldots$	(y is an odd integer.)
Consecutive even integers	$-2, 0, 2, 4, 6$	$n, n + 2, n + 4, \ldots$	(n is an even integer.)
Consecutive multiples of 4	$12, 16, 20, 24$	$a, a + 4, a + 8, \ldots$	(a is a multiple of 4.)

A word problem about consecutive odd integers is solved in Example 4.

Example 4 **The sum of three consecutive odd integers is 69 more than twice the third odd integer. What are the three odd integers?**

Let $y =$ 1st odd integer, $y + 2 =$ 2nd odd integer, and $y + 4 =$ 3rd odd integer.

Their sum is 69 more than twice the third.
$$y + (y + 2) + (y + 4) = 2(y + 4) + 69$$
$$3y + 6 = 2y + 77$$
$$y = 71$$

$y = 71 \qquad y + 2 = 73 \qquad y + 4 = 75$

Check:

The sum of the 3 integers	69 more than twice the 3rd
$71 + 73 + 75$	$2 \cdot 75 + 69$
219	$150 + 69$
	219

Thus, the three consecutive odd integers are 71, 73, and 75.

Reading in Algebra

State each word phrase in algebraic terms.
1. Twelve less than a number
2. The sum of a number and 14
3. Twenty increased by a number
4. A number decreased by 15
5. Eight more than twice a number
6. Five less than 3 times a number
7. Four times the sum of 8 and n
8. Ten times the difference of n and 2

Oral Exercises

1. State 4 consecutive even integers, beginning with 6.
2. State 4 consecutive multiples of 10, beginning with 30.
3. State 4 consecutive integers, beginning with −10.
4. State 4 consecutive odd integers, beginning with −3.

Written Exercises

Solve each problem.

(A)
1. Nine less than 5 times a number is 1 less than the number. What is the number?
2. Sixteen more than a number is 20 more than twice the number. Find the number.
3. Five times the sum of 6 and a number is 2 more than the number. Find the number.
4. Fourteen more than twice a number is 6 times the difference of the number and 7. What is the number?

Represent each number in terms of one variable.

5. The greater of two numbers is 7 more than 5 times the smaller number.
6. The second of two numbers is 6 less than 9 times the first number.
7. The second of three numbers is 7 times the first number. The third number is 6 less than the second number.
8. A second number is 5 times a first number. A third number is 8 more than twice the second number.

Solve each problem.

9. The greater of two numbers is 3 less than twice the smaller number. The sum of the two numbers is 18. What are the two numbers?
10. The smaller of two numbers is 10 less than 4 times the greater number. The greater number is 19 more than the smaller. Find the two numbers.

11. The second of three numbers is 5 times the first number. The third number is 12 less than the first. Find the three numbers if their sum is 51.
12. The second of three numbers is 6 more than the first and the third is 7 times the second. If the third number is 12 more than 10 times the first number, what are the three numbers?

13. The sum of three consecutive integers is 12 less than 4 times the first integer. What are the three integers?
14. There are three consecutive even integers such that the sum of the first two is 22 more than the third. Find the three even integers.

15. The population of city A is 4 times the population of city B. City C has 670 fewer people than city A. What is the population of each city if their combined population is 8,330?
16. A company's total sales were $150,000 for a 3-year period. Sales for the second year were $20,000 more than the first year's sales. The third year's sales were 3 times those of the second year. Find the sales amount for each year.

17. There are four consecutive multiples of 3 such that the sum of the first three multiples is twice the fourth. Find the four multiples of 3.

18. The sum of three consecutive multiples of 6 is 30 more than the fifth consecutive multiple of 6. Find the fourth of these five consecutive multiples of 6.

19. Twice the sum of three consecutive multiples of 15 is 270. What are the three multiples of 15?

20. One-half of the sum of three consecutive multiples of 10 is 75. Find the three multiples of 10.

21. The third of three numbers is 7 times the first and the second is 8 more than the first. Six more than the third, decreased by 3 times the second, is 9 less than the first. Find the three numbers.

22. The second of three numbers is twice the first and the third is 4 less than the second. If 9 less than the first is decreased by the third, the result is −12. What are the three numbers?

23. A size A box holds twice as many jars as a size B box and a size C box holds one dozen fewer jars than a size A box. Ten size A boxes, 5 size B boxes, and 3 size C boxes will hold a total of 522 jars. Find the number of jars that each type of box will hold.

24. A company's sales for February increased $400 over its January sales figure. The March sales were twice the February figure and the sales for April doubled the figure for March. If the sales total was $18,000 for the four months, find the sales figure for each month.

25. Find three consecutive integers so that 8 times the first, decreased by the third, is 488.

26. Find three consecutive odd integers so that 6 times the second, decreased by the third, is 363.

27. Find three consecutive even integers so that 7 times the third is equal to 20 less than the first.

28. Find three consecutive integers so that 25 more than the second is equal to 5 times the third.

29. Find three consecutive integers so that 3 less than twice the third integer is the sum of the first and second integers.

30. Find three consecutive multiples of 5 for which the sum of the first and third multiples is twice the second multiple of 5.

31. Six decreased by 3 times a number is less than 12. Find the set of all such numbers.

32. Seven more than 4 times a number is greater than 3. Find the set of all these numbers.

33. Eight less than 5 times a number is greater than or equal to 8 more than the number. Find the set of all these numbers.

34. Twenty increased by twice a number is less than or equal to 6 times the number. Find the set of all such numbers.

PROBLEM SOLVING: PERIMETER PROBLEMS

Objective

To solve a word problem involving the perimeter of a geometric figure

The **perimeter** of a geometric figure is the sum of the lengths of its sides. To solve a problem that involves perimeter, you may find it helpful to draw the geometric figure and label the lengths of the sides. If the measures of any of the sides are unknown, represent these lengths in terms of one variable.

Example 1

The length of a rectangle is 4 meters (m) more than twice the width. The perimeter is 26 m. Find the length and the width. Find the area of the rectangle in square meters (m^2).

READ ▶

PLAN ▶

Let w = width of rectangle and $2w + 4$ = length of rectangle. Draw the figure and label each side.

$$2w + 4$$

w | | w

$$2w + 4$$

SOLVE ▶

$w + (2w + 4) + w + (2w + 4) = 26$ or $2(w) + 2(2w + 4) = 26$

$$6w + 8 = 26$$
$$6w = 18$$
$$w = 3 \qquad 2w + 4 = 10$$

INTERPRET ▶

Check: $\dfrac{2(w) + 2(2w + 4)}{2(3) + 2(10)} \;\bigg|\; \dfrac{26}{26}$

              26

The area, length × width, is 10 · 3, or 30.

Thus, the length is 10 m, the width is 3 m, and the area is 30 m^2.

Example 2

Side a of a triangle is 3 times as long as side b. Side c is 2 cm longer than side a. Find the length of each side of the triangle if its perimeter is 72 cm.

Let x = length of side b, $3x$ = length of side a, and $3x + 2$ = length of side c.

$3x$ x

$3x + 2$

$$x + 3x + (3x + 2) = 72$$
$$7x = 70$$
$$x = 10 \qquad 3x = 30 \qquad 3x + 2 = 32$$

Check: $\dfrac{x + 3x + (3x + 2)}{10 + 30 + 32} \;\bigg|\; \dfrac{72}{72}$

        72

Thus, the lengths of the three sides are 10 cm, 30 cm, and 32 cm.

Written Exercises

Solve each problem.

(A) 1. The length of a rectangle is 12 cm more than the width. The perimeter is 44 cm. Find the length and the width. Find the area of the rectangle in cm^2.

2. The width of a rectangle is 4 m less than the length and the perimeter is 40 m. Find the width, the length, and the area of the rectangle.

3. Side *a* of a triangle is 4 cm longer than side *b*. Side *c* is twice as long as side *b*. What is the length of each side of the triangle if its perimeter is 28 cm?

4. Side *c* of a triangle is 6 m longer than side *a*. Side *b* is 8 m shorter than side *a*. The perimeter is 52 m. Find the length of each of the three sides.

5. The length of a rectangle is 5 dm more than 3 times the width. Find the length, the width, and the area of the rectangle if its perimeter is 42 dm.

6. The length of side *b* of a triangle is 6 m less than twice the length of side *a*. Sides *b* and *c* have the same length. If the perimeter is 28 m, find the lengths of the three sides.

7. The two longest sides of a *pentagon* (5-sided figure) are each 3 times as long as the shortest side. The other sides are each 8 m longer than the shortest side. Find the length of each side if the perimeter is 79 m.

8. Each of the three longest sides of a *hexagon* (6-sided figure) is 3 cm less than 4 times the length of each of the three shortest sides. If the perimeter is 111 cm, find the length of each side.

(B) 9. A rectangle and a square have the same width. The rectangle is 6 cm longer than the square. One perimeter is twice the other. Find the dimensions of the rectangle.

10. One side of an equilateral triangle is 7 dm shorter than one side of a square. The sum of the perimeters of the two figures is 49 dm. Find the perimeter of each figure.

11. The hypotenuse of a right triangle is 10 m long and one leg is 2 m longer than the other leg. If the perimeter is 24 m, find the area of the right triangle in square meters.

12. One leg of a right triangle is 2 cm longer than twice the length of the other leg. The hypotenuse is 1 cm longer than the longer leg. Find the area if the perimeter is 30 cm.

13. A longer side of a parallelogram is 4 dm shorter than 5 times the length of a shorter side. One side of a *rhombus* (equilateral quadrilateral) and a longer side of the parallelogram are equal in length. If the difference in the two perimeters is 48 dm, find the perimeter of the rhombus.

14. The lengths of the sides of a quadrilateral are in the ratio of 2 to 3 to 5 to 6. What is the length of each side if the perimeter is 104 m?

(C) 15. The length of a rectangle is 4 cm more than 5 times the width. The perimeter is less than 44 cm. Find the set of all such possible widths.

16. The length of a rectangle is 7 dm less than 3 times the width. Find the set of all possible lengths if the perimeter must be greater than 18 dm.

PROBLEM SOLVING: AGE PROBLEMS   1.7

Objective   **To solve a word problem about ages**

Word problems about people's ages may include information about their present ages, past ages, and future ages. If you are x years old now, then 3 years ago you were $x - 3$ years old. Ten years from now, you will be $x + 10$ years old.

To solve a word problem about ages, it may be helpful to use a table to organize the information.

Example   **Carol is 6 times as old as her nephew Howard. Betty is 22 years younger than her aunt Carol. In 4 years, Carol's age will be twice the sum of Howard's and Betty's ages. How old is each person now?**

In a table, represent their ages *now* and their ages *in 4 years*. Let $h =$ Howard's age now.

	Howard	Carol	Betty
Ages now	h	$6h$	$6h - 22$
Ages in 4 yr	$h + 4$	$6h + 4$	$(6h - 22) + 4$, or $6h - 18$

In 4 years, Carol's age will be 2 times the sum of Howard's and Betty's ages.
$$6h + 4 = 2[(h + 4) + (6h - 18)]$$
$$6h + 4 = 2(7h - 14)$$
$$6h + 4 = 14h - 28$$
$$32 = 8h$$
$$4 = h$$
$$h = 4 \qquad 6h = 24 \qquad 6h - 22 = 2$$

Check:

Carol's age in 4 years	twice the sum of Howard's and Betty's ages in 4 years
$24 + 4$	$2[(4 + 4) + (2 + 4)]$
28	$2(14)$
	28

Thus, Howard is now 4 years old, Carol is 24, and Betty is 2.

Reading in Algebra

Select the correct equation for each situation.
  1. He is x years old and she is y years old. She is 4 times as old as he is.
     (a) $y = 4x$ **(b)** $4y = x$ **(c)** $y = x + 4$ **(d)** $x + 4 = y + 4$
  2. She is $x - 10$ years old and he is $x - 2$ years old. He is 3 times as old as she is.
     (a) $3(x - 2) = x - 10$ **(b)** $x - 2 = (x - 10) + 3$ **(c)** $x - 2 = 3(x - 10)$ **(d)** $x = 3(x - 10)$

Oral Exercises

Represent each age now in terms of one variable.

1. Karen is 7 years older than Jane.
2. Ed is 5 years younger than Frank.
3. Eloise is 5 times as old as Robert.
4. Sarah is half as old as Richard.
5. A horse is 10 years older than twice the age of a colt.
6. A redwood tree is 80 years older than 4 times the age of an oak tree.
7. A second city was chartered 20 years before a first city. A third city is 4 times as old as the first city.
8. An eagle is 3 times as old as a hawk. A robin was hatched 15 years after the eagle was hatched.

Written Exercises

Solve each problem.

(A) 1. Fred is 4 times as old as his niece, Selma. Ten years from now, he will be twice as old as she will be. How old is each now?

2. Raymond is 12 years younger than Susan. Four years ago, she was 4 times as old as he was. Find their present ages.

3. Byron is 2 years younger than Cindy. Eight years ago, the sum of their ages was 14. Find their present ages.

4. A fir tree is 5 times as old as a pine tree. Seven years from now, the sum of their ages will be 32. How old is the fir tree now?

5. Brenda is 4 years older than Walter and Carol is twice as old as Brenda. Three years ago, the sum of their ages was 35. How old is each now?

6. Denise is 3 times as old as Conrad and Billy is 8 years younger than Denise. Five years from now, the sum of their ages will be 49. How old is Billy?

(B) 7. An eagle is 4 times as old as a falcon. Three years ago, the eagle was 7 times as old as the falcon. Find the present age of each bird.

8. An oak tree is 20 years older than a pine tree. In eight years, the oak will be 3 times as old as the pine will be. How old is each tree now?

9. Adam is 3 times as old as Cynthia and Fred is 16 years younger than Adam. One year ago, Adam's age was twice the sum of Cynthia's and Fred's ages. Find their present ages.

10. Building A was built 20 years before building B and 30 years after building C. In 10 years, building C will be 20 years younger than the combined ages of buildings A and B. What is the present age of each building?

11. Phyllis is 6 years older than Keith and Manuel is 10 years older than twice Keith's age. If Manuel's age is added to 5 years less than twice Phyllis' age, the result is the same as 7 years more than 5 times Keith's age. How old is each now?

12. The present ages of three portraits are in the ratio of 1 to 3 to 4. In 3 years, the sum of their ages will be 49 years. Find the present ages of the portraits.

(C) 13. A person is x years old now and 10 years ago was t years old. Represent the person's age f years from now in terms of t and f.

14. In x years, a person will be y years old. Represent the person's age z years ago in terms of x, y, and z.

PROBLEM SOLVING: USING FORMULAS 1.8

Objective   **To solve a word problem by using a given formula**

Many applications of mathematics involve the use of **formulas.** One such application is the formula for finding *simple interest* on money borrowed or loaned.

If you borrow $2,500 at $8\frac{1}{2}\%$ per year for 3 years (yr) 6 months (mo), you will pay $743.75 in simple interest, since $2,500 × 0.085 × 3.5 = $743.75. At the end of 3 yr 6 mo, you will pay $3,243.75, since $2,500 + $743.75 = $3,243.75.

In general, a principal p borrowed at a rate of $r\%$ per year for t years is charged a simple interest i, where $i = prt$.
The simple interest i, or prt, added to the principal p gives the total amount A that is owed, or $A = p + prt$.

To find the value of one variable in a formula, you substitute given values for the other variables and then solve the resulting equation. This procedure is shown in the following examples, using the formulas $i = prt$ and $A = p + prt$.

Example 1   **Jack borrowed $4,000 and paid $585 in simple interest after 1 yr 6 mo. What was the annual rate of interest?**

Use the formula $i = prt$. Find r.
$$p = \$4,000 \qquad i = \$585 \qquad t = 1 \text{ yr } 6 \text{ mo, or } 1.5 \text{ yr}$$

$$i = p \cdot r \cdot t$$

Substitute for p, i, and t. ▶ $585 = 4,000 \cdot r \cdot 1.5$
Solve the equation for r. ▶ $585 = 6,000r$
$$0.0975 = r \qquad\qquad 0.0975 = 9.75\%, \text{ or } 9\tfrac{3}{4}\%$$

Thus, the rate of interest was $9\frac{3}{4}\%$ per year.

Example 2   **At the end of 2 yr, Maria paid $3,780 to cover a loan for her junior college tuition. If she was charged 4% in simple interest per year, how much money did Maria borrow?**

Use the formula $A = p + prt$. Find p.
$$A = \$3,780 \qquad r = 4\%, \text{ or } 0.04 \qquad t = 2 \text{ yr}$$

$$A = p + prt$$
$$3,780 = p + p(0.04)(2)$$
$$3,780 = 1.08\,p$$
$$3,500 = p$$

Thus, Maria borrowed $3,500.

A parallel electric circuit with two resistances is shown at the right. The bulbs have resistances of r_1 and r_2 (r-sub-one and r-sub-two) ohms. The circuit has a total resistance of R ohms, where
$$R(r_1 + r_2) = r_1 \cdot r_2.$$

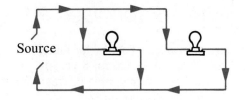

Source

Example 3

Find r_2 in ohms if $R = 4$ ohms and $r_1 = 6$ ohms. Use the formula $R(r_1 + r_2) = r_1 \cdot r_2$.

$$R = 4 \text{ ohms} \qquad r_1 = 6 \text{ ohms}$$

$$R(r_1 + r_2) = r_1 \cdot r_2$$
$$4(6 + r_2) = 6 \cdot r_2 \qquad \blacktriangleleft \textit{ Substitute for R and } r_1.$$
$$24 + 4r_2 = 6r_2 \qquad \blacktriangleleft \textit{ Solve the equation for } r_2.$$
$$24 = 2r_2$$
$$12 = r_2$$

Thus, r_2 is 12 ohms.

An object is launched into the air with an initial velocity of 50 meters per second (m/s). The figure at the right shows the height, in meters, of the object at the end of t seconds. The height h is determined by the formula

$$h = 50t - 5t^2.$$

Use this formula to verify the heights for the times given in the figure.

If an object is shot upward from the surface of the earth with an initial velocity of v meters per second, then at the end of t seconds its height h in meters is given by the formula below.

$$h = vt - 5t^2$$

(figure, right:)
$t=5\,s$ —— 125 m —— $t=5\,s$
$t=2\,s$ —— 80 m $t=8\,s$
$t=1\,s$ —— 45 m $t=9\,s$
$t=0\,s$ | 0 meters | $t=10\,s$
Earth's Surface

Example 4

Find v in meters per second if $h = 120$ m and $t = 3$ s. Use the formula $h = vt - 5t^2$.

$$h = vt - 5t^2$$
$$120 = v \cdot 3 - 5 \cdot 3^2$$
$$165 = 3v$$
$$55 = v$$

Thus, v is 55 m/s.

Written Exercises

Solve each problem. Use the formula $i = prt$.

(A) **1.** For a loan of $6,500 for 8 yr, Jose was charged $3,640 in simple interest. What was the annual interest rate?

2. For a loan of $2,500 for 5 yr 9 mo, Muna was charged $1,150 in simple interest. What was the annual interest rate?

3. When would a borrower owe $6,528 in simple interest on a loan of $6,400 at $8\frac{1}{2}\%$ per year?

4. When would a borrower owe $1,152 in simple interest on a loan of $3,200 at 8% per year?

5. Ms. Brown borrowed some money on February 1, 1972, at 6% per year. How much did she borrow if Ms. Brown paid off the debt, including $3,000 in simple interest, on February 1, 1982?

6. Mr. Adamshick borrowed some money on July 1, 1974, at $7\frac{1}{2}\%$ per year. How much did he borrow if Mr. Adamshick paid off the debt, including $1,485 in simple interest, on October 1, 1982?

Solve each problem. Use the formula $A = p + prt$.

7. Find p in dollars if $A = \$9,780$, $r = 9\%$ and $t = 7$ yr.

8. Find p in dollars if $r = 8\frac{1}{2}\%$, $t = 6$ yr, and $A = \$5,285$.

Solve each problem. Use the formula $R(r_1 + r_2) = r_1 \cdot r_2$.

9. Find R in ohms if $r_1 = 50$ ohms and $r_2 = 30$ ohms.

10. Find r_2 in ohms if $R = 19.2$ ohms and $r_1 = 32$ ohms.

11. Find r_2 in ohms if $R = 24$ ohms and $r_1 = 60$ ohms.

12. Find r_1 in ohms if $R = 21$ ohms and $r_2 = 70$ ohms.

Use the formula $h = vt - 5t^2$ for Exercises 13–20.

13. Find v in meters per second if $h = 150$ m and $t = 5$ s.

14. Find h in meters if $v = 60$ m/s and $t = 4$ s.

15. At the end of 3 seconds, what will be the height of an arrow if an archer shoots it upward with an initial velocity of 120 meters per second?

16. A certain baseball pitcher can throw a ball with an initial velocity of 35 m/s. If the pitcher throws the ball straight up, what will its height be at the end of 2 seconds and at the end of 5 seconds?

17. Explain how the two answers in Exercise 16 can be equal.

(B) **18.** If an object is sent upward, what initial velocity is required to reach a height of 132 m at the end of 4.4 s?

19. Find the initial velocity needed for a missile, launched upward, to achieve an altitude of 3,000 m at the end of 12 s.

20. Find v in meters per second if $h = 26.25$ m and $t = 3.5$ s.

21. If $h = vt - 5t^2$, the maximum value of h is $\dfrac{v^2}{20}$. What initial velocity is required to launch an object to a maximum height of 8,000 meters?

22. Find the maximum height achieved by a missile launched with an initial velocity of 250 m/s. Use the maximum value of h as $\dfrac{v^2}{20}$ from Ex. 21.

23. Find r_2 in ohms if $R = 3.2$ ohms and $r_1 = 4.8$ ohms.

24. Find r_1 in ohms if $R = 4.2$ ohms and $r_2 = 8.7$ ohms.

Solve each problem.

25. Find p in dollars if $A = \$8,295$, $r = 8\frac{1}{2}\%$, and $t = 4$ yr 6 mo.

26. Find p in dollars if $A = \$4,650$, $r = 7\frac{1}{4}\%$, and $t = 6$ yr 3 mo.

27. Find t in years and months if $A = \$12,160$, $p = \$8,000$, and $r = 8\%$.

28. Find the annual interest rate r if $A = \$697$, $p = \$400$, and $t = 9$ yr.

29. When will an investment of $1,000 at 8% per year in simple interest be doubled in total value?

30. When will an investment of p dollars at 12.5% per year in simple interest be doubled in total value?

31. You buy a truck and pay $1,000 down. The finance charge is 8% per year on the original unpaid balance. If you must pay the finance company $9,900 in 48 equal monthly payments, what is the original price of the truck?

32. Susan earned some money last summer, spent $700 of it, and put the remainder in an account that paid 6% per year. At the end of 6 mo, she had $2,575 in her account. Find the total amount she earned last summer.

Ⓒ **33.** When will an investment of p dollars at $r\%$ per year in simple interest be doubled in total value?

34. Prove: If $R(r_1 + r_2) = r_1 \cdot r_2$, then $\dfrac{1}{R} = \dfrac{1}{r_1} + \dfrac{1}{r_2}$.

The diagram shows an object launched from a platform 80 m above the earth's surface. If an object is sent upward from d meters above the surface of the earth with an initial velocity of v meters per second, then its height h in meters at the end of t seconds is determined by the formula
$$h = d + vt - 5t^2.$$

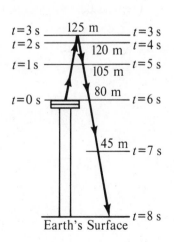

Use the formula $h = d + vt - 5t^2$ for Exercises 35–39.

35. Find h if $d = 120$ m, $v = 90$ m/s, and $t = 5$ s.

36. Find v if $h = 195$ m, $d = 75$ m, and $t = 3$ s.

37. Find d if $h = 765$ m, $v = 150$ m/s, and $t = 6$ s.

38. Find v for the diagram.

39. At the end of 3 s, what is the height of an object shot upward with an initial velocity of 75 m/s from the *bottom* of a shaft that extends 60 m below the earth's surface?

APPLICATIONS: PROPERTIES OF A FIELD

A set F is a *field* for the operations of addition and multiplication if the eleven properties below exist for all numbers x, y, and z in F.

1–A.	*Closure* for addition	$x + y$ is in F.
1–M.	*Closure* for multiplication	$x \cdot y$ is in F.
2–A.	*Commutative* for addition	$x + y = y + x$
2–M.	*Commutative* for multiplication	$x \cdot y = y \cdot x$
3–A.	*Associative* for addition	$(x + y) + z = x + (y + z)$
3–M.	*Associative* for multiplication	$(x \cdot y) \cdot z = x \cdot (y \cdot z)$
4–A.	*Zero* (additive identity)	There exists a unique number 0 in F such that $x + 0 = x$.
4–M.	*One* (multiplicative identity)	There exists a unique number 1 in F such that $x \cdot 1 = x$.
5–A.	*Opposite* (additive inverse)	For each x, there exists a unique y such that $x + y = 0$. y is the additive inverse of x, or $y = -x$.
5–M.	*Reciprocal* (multiplicative inverse)	For each $x \neq 0$, there exists a unique y such that $x \cdot y = 1$. y is the multiplicative inverse of x, or $y = \dfrac{1}{x}$.
6.	*Distributive* for multiplication over addition	$x \cdot (y + z) = x \cdot y + x \cdot z$

Written Exercises

Each set of numbers below is *not* a field for addition and multiplication. Identify the first property in the list above that does not exist for the set. (Ex. 1–4)

1. {all whole numbers: 0, 1, 2, 3, 4, . . .}
2. {all odd integers: . . . , −5, −3, −1, 1, 3, 5, . . .}
3. {all even integers: . . . , −4, −2, 0, 2, 4, . . .}
4. {all integers: . . . , −3, −2, −1, 0, 1, 2, 3, . . .}

5. Does the set of all rational numbers $\dfrac{a}{b}$ where a and b are integers, $b \neq 0$, form a field for addition and multiplication?

6. Does {0, 1, 2}, along with the addition and multiplication tables at the right, form a field?

+	0	1	2
0	0	1	2
1	1	2	0
2	2	0	1

·	0	1	2
0	0	0	0
1	0	1	2
2	0	2	1

7. Construct an addition (+) table and a multiplication (·) table for {0, 1, 2, 3} using a "four-hour" clock. Does {0, 1, 2, 3} along with the operations + and · form a field?

CHAPTER ONE REVIEW

Vocabulary
absolute value [1.4]
conjunction [1.3]
consecutive integers [1.5]
disjunction [1.3]
linear equation [1.1]
linear inequality [1.2]
perimeter [1.6]

Solve each equation.

1. $5a + 8 - 3a = 4a - 5 - 6a$ [1.1]

2. $12 + \frac{3}{5}x = \frac{2}{5}x - 3$

3. $7.2c + 2.3 = 1.4c - 0.6$

4. $0.05 + 2.7(3n - 0.3) = 4.3n$

5. $-2x - (4 - 5x) = x - 3$

6. $4c - 2(4c - 3) = 3(2c - 5)$

7. $\frac{2}{3}(6x - 9) + \frac{3}{4}(8x + 4) = 28 - 4x$

★ 8. $12 - 7[6(2x + 5) - 3(7x + 8)] = 0$

9. $|3x - 15| = 6$ [1.4]

10. $|4y + 3| = 9$

★11. $|n + 4| + |3 - 2n| = 16$

Find the solution set of each inequality. Graph the solutions on a number line.

12. $-5n - 3 \le 17$ [1.2]

13. $3x < x - 4$ or $5x - 3 > 7$ [1.3]

14. $|y + 4| < 6$ [1.4]

15. $|5z - 15| \ge 30$

★16. $|x - 4| < 6$ or $|x - 4| > 12$

Find the solution set of each inequality.

17. $2(3a + 1) > 3(4a - 5) - 2a$

★18. $\frac{4(y - 6)}{-3} > 12 - 2y$

19. $-2x < 6$ and $3x > -15$ [1.3]

Represent each of the three numbers in terms of one variable. [1.5]

20. A second number is 8 times a first number. A third number is 5 more than 3 times the second number.

Solve each problem.

21. Twice the sum of a number and 3 is the same as 24 less than 7 times the number. What is the number? [1.5]

22. The sum of three consecutive even integers is 20 more than twice the second integer. Find the three integers. [1.5]

23. Side a of a triangle is 3 times as long as side b and side c is 2 cm shorter than side a. Find the length of each side if the perimeter is 26 cm. [1.6]

24. Paul is 7 years younger than Carol. Three years from now, she will be twice as old as he will be. How old is each now? [1.7]

25. An oak tree is 6 times as old as a pine tree and a spruce tree is 20 years younger than the oak tree. Three years ago, the age of the oak was 3 times the sum of the ages of the pine and the spruce. Find the present ages of the trees. [1.7]

26. Find p in dollars if $A = \$7,623$, $r = 8\frac{1}{2}\%$, and $t = 4$ yr 3 mo. Use the formula $A = p + prt$. [1.8]

27. Find r_2 in ohms if $R = 8.4$ ohms and $r_1 = 12$ ohms. Use the formula $R(r_1 + r_2) = r_1 \cdot r_2$. [1.8]

28. Find the initial velocity needed for a missile, launched upward, to reach a height of 4,875 m at the end of 15 s. Use the formula $h = vt - 5t^2$. [1.8]

★29. The length of a rectangle is 6 cm more than the width. Find the set of all possible lengths if the perimeter must be less than 54 cm. [1.6]

★30. In $x + 4$ years, a tree will be y years old. Represent the age of the tree z years ago in terms of x, y, and z. [1.7]

CHAPTER ONE TEST

Solve each equation.

1. $8 + 5c - 6 = 9c - 22 - 12c$

2. $6 + \frac{3}{5}n = 14 - \frac{1}{5}n$

3. $3 + 5(2y - 1) = 2y - 4(y + 1)$

4. $2.5(6y - 1.2) = 22.8 - 2.2y$

5. $\frac{1}{2}(16x + 24) + \frac{2}{3}(6x - 9) = 42$

6. $|2a - 7| = 15$

Find the solution set of each inequality. Graph the solutions on a number line.

7. $3 - 4n \geq 19$

8. $|y + 2| < 5$

Find the solution set of each inequality.

9. $4(y - 2) + 1 \leq 3(2y + 1)$

10. $|2a + 3| > 11$

11. $5x > x - 12$ or $3x > x + 12$

12. $-8 \leq 3x - 2 \leq 16$

Represent each number in terms of one variable.

13. A third number is 7 less than a first number. A second number is 5 times the first number.

Solve each problem.

14. The sum of three consecutive integers is 40 less than 4 times the third integer. Find the three integers.

15. Room A has twice as many chairs as room B. Room C has 20 more chairs than room A. If the number of chairs in C is 10 less than the total number of chairs in A and B, find the number of chairs in each room.

16. Gloria is 4 times as old as Theotis. Three years ago, she was 7 times as old as he was. How old is each now?

17. Sides a and b of a triangle have the same length. Side c is 12 cm less than 3 times the length of b. Find the length of each side if the perimeter is 23 cm.

18. Find p in dollars if $A = \$5,620$, $r = 9\%$, and $t = 4$ yr 6 mo. Use the formula $A = p + prt$.

19. Find r_2 in ohms if $R = 4.8$ ohms and $r_1 = 8$ ohms. Use the formula $R(r_1 + r_2) = r_1 \cdot r_2$.

20. Building A was built 10 years before building B and 5 years after building C. In 20 years, the combined ages of buildings A and C will be 3 times that of building B. What is the present age of each building?

21. If an object is sent upward, what initial velocity is required to reach a height of 200 m at the end of 3.2 s? Use the formula $h = vt - 5t^2$

★ 22. Solve. $5[4(6x - 3) - 7(5x - 2)] = 12(1 - 5x)$

★ 23. Find and graph the solution set of $|x + 7| \leq 3$ or $|x - 7| \leq 3$.

★ 24. The length of a rectangle is 10 m less than 3 times the width. Find the set of all possible widths if the perimeter must be less than 36 m.

★ 25. Solve $|n - 5| + |-n - 3| = 12$.

COLLEGE PREP TEST

DIRECTIONS: Choose the *one* best answer to each question or problem.

1. Which investment earns the most simple interest?

    (A) \$4,444 at $8\frac{1}{2}\%$ per year for 2yr 3mo
    (B) \$4,444 at $4\frac{1}{4}\%$ per year for 4 yr 6 mo
    (C) \$2,222 at $8\frac{1}{2}\%$ per year for 4 yr 6 mo
    (D) \$1,111 at $8\frac{1}{2}\%$ per year for 9 yr
    (E) They all earn the same amount of interest.

2. If 2 more than $\frac{2}{3}$ of a certain number is 8, then the number is

    (A) 6   (B) 8   (C) 9
    (D) 15   (E) 18

3. In the equation $4x + 6x = 3(8 + tx)$, for what value of t is x equal to 6?

    (A) 2   (B) 4   (C) 6
    (D) 8   (E) 10

4. If $ax < b$, where a, x, and b are integers, then

    (A) $x < \dfrac{b}{a}$ (B) $x < b - a$

    (C) $x > \dfrac{b}{a}$ (D) $x > b - a$

    (E) None of these

5. Which fraction is greater than $\frac{1}{5}$ but less than $\frac{1}{4}$?

    (A) $\dfrac{9}{20}$ (B) $\dfrac{2}{9}$ (C) $\dfrac{1}{9}$ (D) $\dfrac{2}{7}$ (E) $\dfrac{1}{6}$

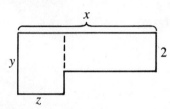

6. The figure above contains two rectangular regions and has the dimensions shown. The area of the figure in square units is

    (A) $2xyz$ (B) $xy - yz - 2x + 2z$
    (C) $2x + yz$ (D) $2x + yz - 2z$
    (E) None of these

7. If $0 < x < 1$, then which one of the following is false?

    (A) $x^2 < x$ (B) $2x < 2$

    (C) $\dfrac{2}{x} < 2$ (D) $\dfrac{x}{2} < \dfrac{1}{2}$

    (E) None of these

8. If $5(3x - 4) - 6x = 4(2x + 6) + x$, then

    (A) $x = -20$
    (B) $x = 24$
    (C) $x = -20$ or $x = 24$
    (D) There is no solution to the equation.
    (E) Every number is a solution to the equation.

9. The hypotenuse of a certain right triangle is 5 cm long and one leg is 3 cm long. Find the area of the triangle.

    (A) 3 cm^2 (B) 4 cm^2 (C) 5 cm^2
    (D) 6 cm^2 (E) 12 cm^2

2 POLYNOMIALS AND PROBLEM SOLVING

Plan How to Solve a Problem

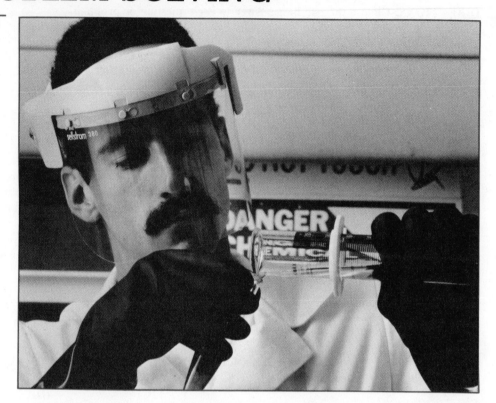

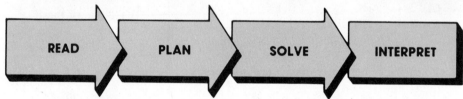

READ → PLAN → SOLVE → INTERPRET

The second step in problem solving is to *plan* how you will solve the problem. Consider this problem and the analysis that must be done before an equation is written to solve it.

How much water must be added to 250 mL of a 15% sugar solution to dilute it to a 12% solution?

Suppose x mL of water is added.

	Total	Sugar	Water
15%	250	0.15(250)	0.85(250)
12%	250 + x	0.12(250 + x)	0.88(250 + x)

To write an equation, you need to understand the following:
(1) A sugar solution consists of (a) sugar and (b) water.
(2) If water alone is added to a sugar solution, the *amount of sugar* in the total solution *remains the same.*

Using this information, an equation can be written for the problem.

Sugar in 15% solution = sugar in 12% solution
$$0.15(250) = 0.12(250 + x)$$

PROPERTIES OF EXPONENTS

Objectives
To simplify expressions containing positive integral exponents
To solve an exponential equation

The *third power* of x, or the *cube* of x, is written as x^3. In the expression x^3, x is the **base** and 3 is the **exponent.** Exponents that are positive integers indicate the number of times the base is used as a factor. Thus, x^3 means $x \cdot x \cdot x$. When no exponent is written for a base, the exponent is understood to be 1. For example, the first power of y may be written as either y^1 or y.

You can use the meaning of exponents to discover how to multiply powers of the same base.

$$x^3 \cdot x \cdot x^4 = (xxx)(x)(xxxx) = x^8$$

Notice that the sum of the exponents in $x^3 \cdot x^1 \cdot x^4$ is $3 + 1 + 4$, or 8. This suggests the following property of exponents.

Product of powers	For each number x and all positive integers m and n, $$x^m \cdot x^n = x^{m+n}.$$

Example 1
Simplify $-5a^4b \cdot 6a^3b^2$.

$-5 \cdot a^4 \cdot b^1 \cdot 6 \cdot a^3 \cdot b^2$
$(-5 \cdot 6)(a^4 \cdot a^3)(b^1 \cdot b^2)$ ◀ *Group powers with the same base.*
$-30a^7b^3$

You can use the meaning of exponents to discover how to divide powers of the same base.

$$\frac{x^5}{x^2} = \frac{xx \cdot xxx}{xx} = x^3 \qquad \frac{y^3}{y^6} = \frac{yyy}{yyy \cdot yyy} = \frac{1}{y^3} \qquad \frac{z^3}{z^3} = \frac{zzz}{zzz} = 1$$

Notice that $\frac{x^5}{x^2} = x^{5-2}$ and that $\frac{y^3}{y^6} = \frac{1}{y^{6-3}}$. This leads to the following *Quotient of Powers* properties.

Quotient of powers	For each number $x \neq 0$ and all positive integers m and n, $$\frac{x^m}{x^n} = x^{m-n} \text{ if } m > n, \quad \frac{x^m}{x^n} = \frac{1}{x^{n-m}} \text{ if } m < n, \text{ and}$$ $$\frac{x^m}{x^n} = 1 \text{ if } m = n.$$

Example 2
Simplify $\dfrac{-9a^7b^5c}{12a^4b^5c^6}$.

$$\frac{-9}{12} \cdot \frac{a^7}{a^4} \cdot \frac{b^5}{b^5} \cdot \frac{c^1}{c^6} = \frac{-3}{4} \cdot a^{7-4} \cdot 1 \cdot \frac{1}{c^{6-1}} = \frac{-3a^3}{4c^5}$$

The Product of Powers and the Quotient of Powers properties on page 31 apply only when the bases are the same. Neither property can be used with $a^7 \cdot b^3$ in Example 1 or with $\dfrac{a^3}{c^5}$ in Example 2.

An expression such as $(x^4)^3$ is called a *power of a power*. For the exponent 3, the base is x^4. From the meaning of exponents and the Product of Powers property, it follows that

$$(x^4)^3 = x^4 \cdot x^4 \cdot x^4 = x^{4+4+4} = x^{12}.$$

Notice that the product of the two exponents in $(x^4)^3$ is $4 \cdot 3$, or 12. This leads to the following property.

Power of a power	For each number x and all positive integers m and n, $$(x^m)^n = x^{m \cdot n}.$$

Example 3 **Simplify $6(c^4)^5$.**

$$6(c^4)^5 = 6 \cdot c^{4 \cdot 5} = 6c^{20}$$

An expression such as $(2a)^3$ or $\left(\dfrac{a}{b}\right)^4$ can be simplified using the following Power of a Product or Power of a Quotient property, respectively.

Power of a product **Power of a quotient**	For all numbers x and y and each positive integer n, (1) $(xy)^n = x^n \cdot y^n$; and (2) $\left(\dfrac{x}{y}\right)^n = \dfrac{x^n}{y^n}$, $(y \neq 0)$.

Using these properties, $(2a)^3 = 2^3 \cdot a^3$, or $8a^3$ and $\left(\dfrac{a}{b}\right)^4 = \dfrac{a^4}{b^4}$.

Sometimes you must use more than one property of exponents to simplify an expression. This is shown in Examples 4 and 5.

Example 4 **Simplify $4a^2(5a^4b^2c)^3$.**

$$\begin{aligned}
4a^2(5a^4b^2c)^3 &= 4a^2 \cdot 5^3(a^4)^3(b^2)^3(c^1)^3 &&\blacktriangleleft \text{ Power of a Product} \\
&= 4a^2 \cdot 125a^{12}b^6c^3 &&\blacktriangleleft \text{ Power of a Power} \\
&= 500a^{14}b^6c^3 &&\blacktriangleleft \text{ Product of Powers}
\end{aligned}$$

Example 5 **Simplify $\left(\dfrac{-5a^3b}{2c^2}\right)^4$.**

$$\left(\frac{-5a^3b}{2c^2}\right)^4 = \frac{(-5a^3b)^4}{(2c^2)^4} = \frac{(-5)^4(a^3)^4b^4}{2^4(c^2)^4} = \frac{625a^{12}b^4}{16c^8}$$

Equations such as $2^x = 16$, $4^{5x} = 64$, and $5^{4x+1} = 125$ are called **exponential equations** since the variable appears in the exponent. To solve an exponential equation, you must be familiar with powers of numbers like 2^4, 3^4, 4^3, and 5^3.

Example 6   **Solve $5^{4x+1} = 125$.**

$$5^{4x+1} = 125$$
$$5^{4x+1} = 5^3 \quad \blacktriangleleft \quad 125 = 5^3$$
$$4x + 1 = 3 \quad \blacktriangleleft \quad \text{If } 5^a = 5^b, \text{ then } a = b.$$
$$4x = 2$$
$$x = \frac{1}{2}$$

Check:
$$5^{4 \cdot \frac{1}{2} + 1} = 5^3 = 125$$

Thus, the value of x is $\frac{1}{2}$.

Reading in Algebra

Supply the missing words and expressions in the following paragraph.

In the expression $9y^4$, the exponent is _____ and its base is _____, but in the expression $(9y)^4$, the _____ is 4 and its _____ is _____. Notice that in the expression _____, the base is $7c^3$, the _____ is 2, and the second power of $7c^3$ is multiplied by $5c$.

Written Exercises

Simplify each expression.

(A)
1. $5a^2b^3 \cdot 4ab^5$
2. $-12mn^7 \cdot 6m^3n^2$
3. $-4x^5y^4(-9x^4y)$
4. $7c^3d^2 \cdot 4c^2d \cdot 10c^4d^6$
5. $\dfrac{10a^4b^7}{-5a^9b^2}$
6. $\dfrac{8x^9y^3z^2}{12x^3y^3z^8}$
7. $\dfrac{-4m^3np^5}{6m^{12}np^{10}q^2}$
8. $\dfrac{-9a^9bc^2d^7}{-6ac^{12}d^7}$
9. $(y^6)^5$
10. $(n^4)^7$
11. $(5x)^3$
12. $(-3c)^4$
13. $7(10n^5)^3$
14. $3(-3x^2)^4$
15. $4y(11y)^2$
16. $2a(5a^3)^4$
17. $\left(\dfrac{-5a}{4}\right)^3$
18. $\left(\dfrac{x^2}{y^6}\right)^4$
19. $\left(\dfrac{-7c^5}{10d^3}\right)^2$
20. $\left(\dfrac{2ab^2}{3c^3d}\right)^5$

Solve each equation.

21. $3^{x-2} = 81$
22. $10^{n+1} = 1{,}000$
23. $2^{3y} = 64$
24. $6^{-x} = 36$

(B)
25. $5^{3x+1} = 625$
26. $4^{6-y} = 256$
27. $3^{1-2n} = 243$
28. $2^{8x+6} = 1{,}024$

Simplify each expression.

29. $-4a^6b^4c^3 \cdot 9a^3b^7c$
30. $5m^7n^2p^6 \cdot 14mn^4p^3$
31. $10(-5x^2y^3)^3$
32. $3(5m^3n^2p^4)^4$
33. $6a(4a^2b^3)^3$
34. $-5x(3x^4y^3)^4$
35. $4d^3(-10c^2d^5)^3$
36. $2a^2b(-3ab^5)^4$

(C)
37. $x^ay^{3b} \cdot x^{2c}y^{4d}$
38. $x^{2a+3}y^{2b} \cdot x^{5a}y^{3b-1}$
39. $x^a(x^by^c)^d$
40. $x^{2m}y^n(x^{m+3}y^n)^3$
41. $\dfrac{x^{6a}y^{3b}}{x^{2a}y^b}$
42. $\dfrac{x^{4a+3b}y^{2a-4b}}{x^{a+b}y^{a-3b}}$
43. $\dfrac{x^{3c+2}y^{5d-1}}{x^{c-1}y^{2d+3}}$
44. $\dfrac{x^{6n-2}y^{3n+2}}{x^{2n-8}y^{4-n}}$

Properties of Exponents

ZERO AND NEGATIVE INTEGRAL EXPONENTS

Objectives   **To evaluate expressions containing zero and negative integral exponents**
To simplify expressions containing zero and negative integral exponents

In general, the Quotient of Powers properties indicate that to divide powers of the same base, you should subtract the exponents of the powers. Observe below what happens when this is applied to the quotient $\frac{5^3}{5^3}$.

$$\frac{5^3}{5^3} = 5^{3-3} = 5^0$$

But you know that $\frac{5^3}{5^3}$ is 1: $\frac{5^3}{5^3} = \frac{125}{125} = 1$

This implies that $5^0 = 1$, which leads to the following definition.

Definition: Zero exponent	For each number $x \neq 0$, $x^0 = 1$.

You can define 5^{-3} to mean $\frac{1}{5^3}$ and define x^{-n} to mean $\frac{1}{x^n}$, for each $x \neq 0$, by using the definition of zero exponent and extending the Quotient of Powers properties to include negative exponents.

$$\frac{5^0}{5^3} = 5^{0-3} = 5^{-3} \text{ and } \frac{5^0}{5^3} = \frac{1}{5^3}. \text{ Thus, } 5^{-3} = \frac{1}{5^3}.$$

This view of zero and negative exponents may be seen in the following pattern where any term is divided by 5 to obtain the next term. Notice that $5^0 = 1$ and $5^{-3} = \frac{1}{5^3}$.

... 5^3	5^2	5^1	5^0	5^{-1}	5^{-2}	5^{-3} ...
... 125	25	5	1	$\frac{1}{5}$	$\frac{1}{25}$, or $\frac{1}{5^2}$	$\frac{1}{125}$, or $\frac{1}{5^3}$ ...

Sometimes a negative exponent appears in a denominator, as in $\frac{1}{2^{-4}}$. Study the steps below which show that $\frac{1}{2^{-4}} = 2^4$.

$$\frac{1}{2^{-4}} = 1 \div (2^{-4}) = 1 \div \frac{1}{2^4} = 1 \div \frac{1}{16} = 1 \times 16 = 16 = 2^4$$

Definition: Negative exponent	For each number $x \neq 0$ and each positive integer n, $$x^{-n} = \frac{1}{x^n} \text{ and } \frac{1}{x^{-n}} = x^n.$$

In expressions like $6c^{-2}$ and $(6c)^{-2}$, you should be careful in determining the base for the exponent -2. The base for the exponent -2 in $6c^{-2}$ is c and the base in $(6c)^{-2}$ is $(6c)$.

Example 1 **Find the value of $2 \cdot 5^{-3}$ and $\dfrac{7}{(2 \cdot 5)^{-3}}$.**

$$2 \cdot 5^{-3} = \frac{2}{1} \cdot \frac{1}{5^3} = \frac{2}{125} \qquad \frac{7}{(2 \cdot 5)^{-3}} = 7(2 \cdot 5)^3 = 7 \cdot 10^3 = 7000$$

A number is written in **scientific notation** when it is in the form

$$a \times 10^c, \text{ where } 1 \le a < 10 \text{ and } c \text{ is an } \textit{integer.}$$

Notice below that the integral exponent c indicates the number of places and the direction to shift the decimal point in a to obtain **ordinary notation.**

$$8.64 \times 10^3 = 8640.0 \qquad \text{and} \qquad 8.64 \times 10^{-3} = 0.00864$$

Example 2 **Find the value of $\dfrac{6.6 \times 10^{-1}}{8.8 \times 10^2}$ in ordinary notation.**

$$\frac{6.6 \times 10^{-1}}{8.8 \times 10^2} = \frac{6.6}{8.8} \times 10^{-1} \times 10^{-2} = 0.75 \times 10^{-3} = 0.00075$$

Each of the five properties of exponents can now be extended to, and used with, expressions containing zero and negative integral exponents.

Example 3 **Simplify and write each expression with positive exponents.**

$$-5x^4 \cdot 3y^{-2} \cdot x^{-7} \cdot y^5 \qquad \qquad \frac{-8a^8 b^{-3} c^{-4}}{12a^{-2} b^{-5} c^5}$$

$$
\begin{aligned}
&-5 \cdot x^4 \cdot 3 \cdot y^{-2} \cdot x^{-7} \cdot y^5 \\
&(-5 \cdot 3)(x^4 \cdot x^{-7})(y^{-2} \cdot y^5) \\
&-15x^{4+(-7)} y^{-2+5} \\
&-15x^{-3} y^3 \\
&\frac{-15y^3}{x^3} \quad \blacktriangleleft \; x^{-3} = \frac{1}{x^3}
\end{aligned}
\qquad
\begin{aligned}
&\frac{-8}{12} \cdot \frac{a^8}{a^{-2}} \cdot \frac{b^{-3}}{b^{-5}} \cdot \frac{c^{-4}}{c^5} \\
&\frac{-2}{3} \cdot \frac{a^8 \cdot a^2}{1} \cdot \frac{b^5}{b^3} \cdot \frac{1}{c^5 \cdot c^4} \quad \blacktriangleleft \; \frac{b^{-m}}{b^{-n}} = \frac{b^n}{b^m} \\
&\frac{-2a^{10} b^2}{3c^9}
\end{aligned}
$$

Example 4 **Simplify $(5a^{-3} b^4)^{-2}$ and $\left(\dfrac{-4x^{-2}}{y^4 z^{-3}}\right)^3$.**

$$
\begin{aligned}
(5a^{-3} b^4)^{-2} &= 5^{-2}(a^{-3})^{-2}(b^4)^{-2} \\
&= 5^{-2} a^6 b^{-8} \\
&= \frac{a^6}{25b^8}
\end{aligned}
\qquad
\begin{aligned}
\left(\frac{-4x^{-2}}{y^4 z^{-3}}\right)^3 &= \frac{(-4x^{-2})^3}{(y^4 z^{-3})^3} = \frac{(-4)^3 (x^{-2})^3}{(y^4)^3 (z^{-3})^3} = \frac{-64x^{-6}}{y^{12} z^{-9}} \\
&= \frac{-64z^9}{y^{12} x^6}
\end{aligned}
$$

Reading in Algebra

For each phrase at the left, find its one correct match at the right.

1. The base of -3 in $5 \cdot 2^{-3}$
2. The base of -3 in $(5 \cdot 2)^{-3}$
3. An exponent in $5 \cdot 2^{-3}$
4. An exponent in $(5 \cdot 2)^{-3}$
5. The value of $5 \cdot 2^{-3}$
6. The value of $(5 \cdot 2)^{-3}$

A 3   **B** -3 **C** 5
D 2   **E** $5 \cdot 2$ **F** 1,000
G $\dfrac{1}{1,000}$ **H** $\dfrac{5}{8}$

Oral Exercises

Evaluate each expression if $x = 2$.

1. x^0 2. $4x^0$ 3. $(4x)^0$ 4. x^{-1} 5. $5x^{-1}$ 6. $(5x)^{-1}$

Written Exercises

Find the value of each expression.

Ⓐ 1. 8^0 2. $6 \cdot 4^0$ 3. $(6 \cdot 4)^0$ 4. 4^{-3} 5. 2^{-4} 6. $6^2 \cdot 6^0$

7. $4 \cdot 3^{-2}$ 8. $(4 \cdot 3)^{-2}$ 9. $\dfrac{2}{5^{-3}}$ 10. $\dfrac{10}{3^{-4}}$ 11. $\dfrac{2}{3 \cdot 4^{-2}}$ 12. $\dfrac{2}{(3 \cdot 4)^{-2}}$

13. $\dfrac{3^2}{2^{-3}}$ 14. $\dfrac{3^{-2}}{2^3}$ 15. $\dfrac{3^{-2}}{2^{-3}}$ 16. $\dfrac{5^{-4}}{2^{-4}}$ 17. $7^4 \cdot 7^{-6}$ 18. $7^{-4} \cdot 7^6$

19. 2.7×10^{-2} 20. $\dfrac{6.5}{10^{-3}}$ 21. 1.23×10^{-4} 22. $\dfrac{3.57}{10^{-2}}$ 23. 46.8×10^{-3} 24. $\dfrac{0.097}{10^0}$

Simplify and write each expression with positive exponents.

25. $9x^{-3}$ 26. $(3y)^{-4}$ 27. $-10a^{-3}$ 28. $(-2c)^{-5}$

29. $x^{-4} \cdot x^7$ 30. $4y^3 \cdot y^{-7}$ 31. $n^{-5} \cdot 7n^{-3}$ 32. $6a^{-4} \cdot 2a^0$

33. $\dfrac{1}{10x^{-3}}$ 34. $\dfrac{1}{(10x)^{-3}}$ 35. $\dfrac{5}{6a^{-2}}$ 36. $\dfrac{5}{(6a)^{-2}}$

37. $\dfrac{x^6}{x^{-3}}$ 38. $\dfrac{a^{-4}}{a^2}$ 39. $\dfrac{10n^{-5}}{15n^{-8}}$ 40. $\dfrac{-7c^{-6}}{14c^{-4}}$

41. $(x^{-2})^5$ 42. $(4a^3)^{-2}$ 43. $(2n^{-4})^{-5}$ 44. $5(-5y^{-2})^4$

45. $\left(\dfrac{y^{-4}}{-5}\right)^3$ 46. $\left(\dfrac{x^3}{y^{-2}}\right)^5$ 47. $\left(\dfrac{c^{-3}}{d^5}\right)^4$ 48. $\left(\dfrac{a^{-3}}{b^{-2}}\right)^6$

Find the value of each expression in ordinary notation.

Ⓑ 49. $\dfrac{8.4 \times 10^2}{4.2 \times 10^{-2}}$ 50. $\dfrac{6.3 \times 10^{-7}}{0.9 \times 10^{-9}}$ 51. $\dfrac{9.63 \times 10^{-1}}{3 \times 10^0}$ 52. $\dfrac{6.6 \times 10^{-20}}{8.8 \times 10^{-18}}$

Simplify and write each expression with positive exponents.

53. $\dfrac{a^5 b^{-4}}{c^3 d^{-6}}$ 54. $\dfrac{x^3 y^{-4}}{z^{-5}}$ 55. $\dfrac{4a^{-2} b^5}{9c^3 d^{-4}}$ 56. $\dfrac{6x^{-5}}{8y^{-2} z^{-1}}$

57. $\dfrac{x^8 y^{-14}}{x^{-2} y^{12}}$ 58. $\dfrac{a^6 b^{-6}}{a^{10} b^{-8}}$ 59. $\dfrac{6c^{-6} d^8}{9c^{-3} d^{-2}}$ 60. $\dfrac{25m^{-12} n^{-10}}{45m^{-15} n^5}$

61. $2x^{-8} \cdot 5y^{-2} \cdot x^3$ 62. $10a^7 \cdot b^{-6} \cdot 3a^{-3}$ 63. $(-3c^{-3} d^4)^{-3}$ 64. $(-2c^2 d^{-6})^4$

65. $\left(\dfrac{12x^{-3}}{y^3 z^{-5}}\right)^2$ 66. $\left(\dfrac{5a^{-2} b^3}{3c^{-1}}\right)^4$ 67. $\left(\dfrac{-2x^{-4} y}{3z^3 w^{-2}}\right)^3$ 68. $\left(\dfrac{ab^{-2} c^3}{d^{-3} ef^2}\right)^4$

Ⓒ 69. x^{a-b} if $0 < a < b$ 70. y^{c-d} if $c < d < 0$ 71. z^{3m-3n} if $m < 0 < n$

POLYNOMIALS AND SYNTHETIC SUBSTITUTION

Objectives

To evaluate a polynomial for given values of its variables
To add polynomials
To subtract polynomials

Five examples of algebraic **terms** are shown below.

$$-7 \quad z \quad 3t \quad x^2y \quad 2.3mn^2$$

Each such term is a numeral, a variable, or a product of a numeral and one or more variables.

An algebraic expression consisting of either one term or a sum of several terms is called a **polynomial**. Two examples of polynomials are shown below.

$$-3a^2b \quad 5x^2y - 6x + 4y - 20$$

The first polynomial consists of one term, and the second is the sum of four terms. Expressions like $\frac{3}{x}$ (dividing by a variable) and $\sqrt{x}$ (square root of a variable) are not classified as polynomials.

Some polynomials are classified by the number of terms they contain. For example, $5ab^3$ is a **monomial** (one term), $9n^4 - 1$ is a **binomial** (two terms), and $x^2 + xy - 5$ is a **trinomial** (three terms).

The **degree of a term** is the sum of the exponents of its variables. For example, $5a^2b^4c$ is of degree 7. The sum of the exponents is $2 + 4 + 1$, or 7. The degree of the number 5 is 0, since $5 = 5 \cdot 1 = 5x^0$.

The **degree of a polynomial** is the same as the degree of its term that has the greatest degree. For example, $4xy^2 + 7x^2y^3 - 3x$ is of degree 5. Its terms are of degrees 3, 5, and 1, respectively, and 5 is the greatest of these degrees.

Example 1

Give the degree of each polynomial.

Polynomial	Degree		Polynomial	Degree
$5x + 7$	1		$3x + 2y - 6$	1
$3x^2 - 2x - 1$	2		$x^2 - 4xy + 4y^2$	2
$x^3 + 9x^2 - 6x + 4$	3		$a^2 + 3ab^3 + a^3bc$	5

First-degree polynomials, such as $5x + 7$ and $3x + 2y - 6$, are called **linear polynomials**. Second-degree polynomials, such as $3x^2 - 2x - 1$ and $x^2 - 4xy + 4y^2$, are called **quadratic polynomials**.

You can evaluate a polynomial for given values of its variables by **direct substitution** as shown below.

If $\quad P = 5x^3y^2 - xy + 14$ and $x = -2$ and $y = 3$,
then, $P = 5(-2)^3 \cdot 3^2 - (-2) \cdot 3 + 14 = 5 \cdot -8 \cdot 9 - (-6) + 14 = -340.$

Given a polynomial $P = 3x^3 + 10x^2 - 5x - 4$ *with one variable*, you can evaluate P for $x = 2$ by **synthetic substitution**. First, notice that:

$$P = 3x^3 + 10x^2 - 5x - 4 = (3x^2 + 10x - 5) \cdot x - 4$$

$$\downarrow \qquad \downarrow \qquad \downarrow \qquad \downarrow$$

$$3 \qquad 10 \qquad -5 \qquad -4 = ([3 \cdot x + 10] \cdot x - 5) \cdot x - 4$$
$$= ([3 \cdot 2 + 10] \cdot 2 - 5) \cdot 2 - 4 = \boxed{50}\ \text{if } x = 2.$$

Write the coefficients
of P.
Repeat the cycle:
Multiply by 2 and add.

$$\downarrow \qquad \downarrow \qquad \downarrow \qquad \downarrow$$

$$
\begin{array}{c|cccc}
 & 3 & 10 & -5 & -4 \\
 & & 6 & 32 & 54 \\
\hline
2\,| & 3 & 16 & 27 & \boxed{50}
\end{array}
$$

By **synthetic substitution**, $P = 50$ if $x = 2$.

There is a **shorter form** of synthetic substitution.

$$
\begin{array}{c|cccc}
 & 3 & 10 & -5 & -4 \\
\hline
2\,| & 3 & 16 & 27 & \boxed{50}
\end{array}
$$
◀ *Do not write the multiples of 2.*

Example 2

Evaluate $2x^4 - x^3 + 5x + 3$ if $x = -2$ and again if $x = 3$. Use the shorter form of synthetic substitution.

$$
\begin{array}{c|ccccc}
 & 2 & -1 & 0 & 5 & 3 \\
\hline
-2\,| & 2 & -5 & 10 & -15 & \boxed{33} \\
\hline
3\,| & 2 & 5 & 15 & 50 & \boxed{153}
\end{array}
$$
◀ *Write 0 for the coefficient of the missing x^2 term.*

Thus, the values are 33 if $x = -2$ and 153 if $x = 3$.

You can add and subtract polynomials by combining the like (similar) terms. For example, $3a^2b$ and $4a^2b$ are like terms, but $5a^2b^3$ and $7a^3b^2$ are not like terms. To simplify the **difference** of two polynomials, use the rule:
$x - y = x + (-y)$.

Example 3

Simplify $(7a^2b - 3ab + b^2) - (8ab - 9a^2b + 2b^2)$.

$(7a^2b - 3ab + b^2) + (-8ab + 9a^2b - 2b^2)$ ◀ *Add the opposite of the second polynomial.*

$(7a^2b + 9a^2b) + (-3ab - 8ab) + (1b^2 - 2b^2)$ ◀ *Group the like terms.*
$16a^2b - 11ab - b^2$ ◀ *Combine the like terms.*

Example 4   **Subtract:** $-8x^3 - 2x^2 - 4x$
$$-8x^3 + 7x^2 \qquad - 10$$

Multiply the second $-8x^3 - 2x^2 - 4x$
polynomial by -1. ▶ $\underline{8x^3 - 7x^2 \qquad + 10}$
Then add. ▶ $-9x^2 - 4x + 10$

It is often useful to arrange the terms of a polynomial, with one variable, in *descending order* of the exponents. In the polynomial $8x^2 - 5x - 7x^4 + 6 + 2x^5$, the exponents in order from left to right are 2, 1, 4, 0, 5. This polynomial can be rewritten as $2x^5 - 7x^4 + 8x^2 - 5x + 6$, with its terms in descending (or decreasing) order of the exponents.

Reading in Algebra

Determine whether each expression is a polynomial (Yes or No).
1. $5a - 3b + 6$
2. 9
3. $2y + \sqrt{y}$
4. $\frac{1}{4}a + b^2 + \frac{2}{3}$
5. $7m + n^4 + \frac{4}{p^2}$
6. $\frac{1}{3}x^2y$

Determine whether each polynomial is linear (L), quadratic (Q), or neither (N).
7. $8y^2 - 6y - 2$
8. $-6x + 12$
9. $3n^3 + 2n^2 - 1$
10. $10x + 5y + 15$
11. $3x^2y + 4xy + 5y$
12. $xy - x^2 + y^2$

Evaluate each polynomial if $x = 2$ and $y = -1$.
13. $y - x$
14. $xy - y$
15. $-xy$
16. $7x - 5y$
17. $-10x - 2y$
18. $x^3 - x^2 + y$

Write the next row to complete each synthetic substitution.
19. $\underline{\quad 2 \quad 3 \quad -1 \quad 1 \quad}$
    $-1\rfloor$
20. $\underline{\quad 3 \quad -1 \quad 0 \quad -2 \quad}$
    $2\rfloor$
21. $\underline{\quad 2 \quad 0 \quad 0 \quad 16 \quad}$
    $-2\rfloor$

Oral Exercises

Classify each polynomial by the number of terms.
1. $y^2 - 9$
2. $4a^2x^2$
3. $4a^2 - 4a + 1$
4. $8 - c^3$

Give the degree of each polynomial.
5. $5x - 8y + 9$
6. $4xy$
7. $a^2 - 5a^2b^3c$
8. 7

Arrange the terms of each polynomial in descending order of the exponents.
9. $x^2 - x + x^4 - 5$
10. $8 + 3y + 4y^2 + y^3$
11. $-n^2 + n^6 - n^4$
12. $-8 - x^3$

Written Exercises

Evaluate each polynomial for the given values of its variables.

Ⓐ 1. $3x^2y - xy + 5y$ if $x = 3$ and $y = -2$ 2. $8x^2y + 5xy + 2xy^2$ if $x = -4$ and $y = 5$

Use synthetic substitution to evaluate each polynomial for $x = 3$ and again for $x = -2$.

3. $x^4 + x^3 - x^2 - x + 1$ 4. $2x^4 - 8x^2 - 5x - 5$

5. $2x^5 - 8x^3 + 10x - 20$ 6. $3x^5 - 10x^4 + x^3 - 4x^2 + 5x + 5$

Add. Arrange the terms of the sum in descending order of the exponents.

7. $(3y - 12y^2 + 24) + (17 - 8y + 12y^2)$ 8. $(16 + n^2 - 12n) + (13n + 15n^2 - 19)$

9. $(-15a^4 + a - 25a^3) + (6a^2 + 25a^3 - 9a)$ 10. $(12x^3 - 6x^2 - 6x) + (5x^2 + 6x + x^5)$

11. $6 + 2a + a^2$
$\underline{3 - 2a + a^2}$

12. $8 - 2n - 4n^2$
$\underline{- 6n + n^2 - 8n^3}$

Subtract. Arrange the terms of the difference in descending order of the exponents.

13. $(5x^2 + 3x - 4) - (2x^2 - 4x - 5)$ 14. $(2n - 3n^2 + 9) - (5 + n^2 - 8n)$

15. $(12y - 12y^2 + 12) - (-8y^2 + 14y + 15)$ 16. $(-x^4 + x^6 - x^2 + 10) - (5x^2 + 6x^6 - x^4)$

17. $5a^2 + 9a - 2$
$\underline{5a^2 - 3a + 7}$

18. $-4n^3 + 8n - 2$
$\underline{-4n^3 - 5n^2 - n}$

Evaluate each polynomial for the given values of its variables.

Ⓑ 19. $10xy - 5x^2$ if $x = 0.4$ and $y = 0.5$ 20. $a^3 + 100ab$ if $a = 0.2$ and $b = 1.25$

21. $1.4xy + 2.2y^2$ if $x = 2.5$ and $y = 4$ 22. $3.6x^2 - 0.4xy$ if $x = 0.5$ and $y = 10$

Use synthetic substitution to evaluate each polynomial for each of the given values of x.

23. $2x^4 + x^3 - 11x^2 - 4x + 12$; $-1, -3, 2$, and 3 for x

24. $x^6 - x^4 - 9x^3 + 8x + 8$; $1, -1, 2, -2$, and 3 for x

25. $3x^4 - 26x^3 + 72x^2 - 54x - 27$; $0, 1, -1, 2$, and 3 for x

Subtract.

26. $(6x^2y^2 - 2xy + 4) - (5xy - 8 - 8x^2y^2)$ 27. $(29x^2 - y^2 + 28xy) - (xy - x^2 - y^2)$

28. $12n^3 - n^2 + 6n$
$\underline{8n - 5n^2 - 3n^3}$

29. $6n^4 + 5n^2 - 4n - 30$
$\underline{6n^4 + 2n^3 - 9n + 20}$

Evaluate each polynomial for $x = 10$.

30. $9x^4 + 8x^3 + 7x^2 + 6x + 5$ 31. $4x^5 + 4x^3 + 4x$

32. $-x^5 - 2x^4 - 3x^3 - 4x^2 - 5x - 6$ 33. $-7x^6 - 7x^4 - 7x^2 - 7$

Evaluate each polynomial for the given values of its variables.

Ⓒ 34. $5xy^3 + 5x^2y^2 + 10x^3y$
if $x = 2a$ and $y = -3a$

35. $10x^4y^2 - 2x^3y^3 - x^2y^4 + 5xy^5$
if $x = c^2$ and $y = -2c^2$

CUMULATIVE REVIEW

1. Multiply: $3n^2(5n^2 - n - 4)$.

2. Simplify $3a \cdot 5a^4b \cdot -4a^3b^3$.

3. Simplify $10c(5c^4d)^2$.

4. Solve $|x - 9| = 3$.

5. Find the solution set of $|x - 9| > 3$.

6. Solve $5 + 0.75x = 23$.

MULTIPLYING POLYNOMIALS   2.4

Objective   **To multiply two or more polynomials**

To multiply a monomial and a trinomial, you use the Distributive Property.

$$3x(2x^2 - 4x + 6) = 3x \cdot 2x^2 + 3x(-4x) + 3x \cdot 6 = 6x^3 - 12x^2 + 18x.$$

To simplify $(a + b)(c + d)$, the indicated product of any two binomials, you can use the Distributive Property three times as follows.

$$(a + b)[c + d] = a[c + d] + b[c + d] = ac + ad + bc + bd$$

The result, $ac + ad + bc + bd$, is the sum of the products of the: **First** terms (ac), **Outer** terms (ad), **Inner** terms (bc), **Last** terms (bd). This suggests the **FOIL** method, an efficient method for multiplying any pair of binomials.

Example 1   **Simplify $(3x + 4)(2x - 5)$. Use the FOIL method.**

$$
\begin{aligned}
(3x + 4)(2x - 5) &= 3x \cdot 2x + 3x(-5) + 4 \cdot 2x + 4(-5) \\
&= 6x^2 - 15x + 8x - 20 \\
&= 6x^2 - 7x - 20
\end{aligned}
$$

To simplify $3n(2n^2 - 6)(3n^2 - 5)$, it is easier to find the product of the binomial factors first and then multiply the result by the monomial factor.

Example 2   **Simplify $3n(2n^2 - 6)(3n^2 - 5)$.**

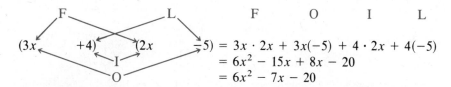

$$
\begin{aligned}
3n[(2n^2 - 6)(3n^2 - 5)] &= 3n[2n^2 \cdot 3n^2 + 2n^2 \cdot -5 - 6 \cdot 3n^2 - 6 \cdot -5] \\
&= 3n[6n^4 - 10n^2 - 18n^2 + 30] \\
&= 3n[6n^4 - 28n^2 + 30] \\
&= 18n^5 - 84n^3 + 90n
\end{aligned}
$$

In Example 3, you are to multiply a binomial and a trinomial. To do this, distribute each term of the binomial to each term of the trinomial.

Example 3   **Simplify $(3x - 2y)(5x + y - 4)$.**

$$
\begin{aligned}
&3x(5x + y - 4) - 2y(5x + y - 4) \\
&15x^2 + 3xy - 12x - 10xy - 2y^2 + 8y \\
&15x^2 - 7xy - 12x - 2y^2 + 8y
\end{aligned}
$$

Oral Exercises

Simplify.

1. $-2(8a^2 + 4a - 5)$ **2.** $5x^2(2x^2 - x + 3)$ **3.** $-3x(4x - 5y - 1)$

Written Exercises

Simplify.

Ⓐ **1.** $(4t + 5)(2t - 4)$ **2.** $(6n - 3)(2n - 1)$
3. $(2n + 8)(2n - 8)$ **4.** $(3y - 4)(3y + 4)$
5. $(3x - 2y)(2x + 4y)$ **6.** $(5m + 4n)(3m + 2n)$
7. $(x^2 + 10)(x^2 + 15)$ **8.** $(n^3 - 8)(n^3 - 9)$
9. $(2y^3 - 3)(5y^3 - 7)$ **10.** $(8x^2 + 5)(3x^2 + 4)$
11. $(7x + 0.4)(3x + 0.8)$ **12.** $(5n - 0.6)(4n - 0.7)$
13. $(0.8a + 6)(0.4a - 9)$ **14.** $(0.7x - 3)(0.6x + 5)$
15. $(5x + 4)(x^2 - 3x + 2)$ **16.** $(4y - 3)(2y^2 + 6y - 1)$
17. $(3y - 2)(3y^2 - y + 4)$ **18.** $(2x + 4)(4x^2 + 2x - 3)$
19. $4(2x - 5)(x + 3)$ **20.** $3(4a + 5)(2a - 3)$

Ⓑ **21.** $-2(3y - 4)(3y - 4)$ **22.** $-4(2x + 10)(2x - 10)$
23. $(2x^2 + 3y)(4x^2 - 8y)$ **24.** $(8a - 5b^2)(3a + 5b^2)$
25. $(3x^2 + 4y^2)(3x^2 - 5y^2)$ **26.** $(2c^3 - 3d^2)(2c^3 + 6d^2)$
27. $(3a - 2b)(4a + 5b + 3)$ **28.** $(2x + 5y)(x - y + 3)$
29. $(n^2 - 4)(3n^2 + 4n - 5)$ **30.** $(2a^2 + 3)(4a^2 - a + 6)$
31. $(3.2x + 0.4)(0.7x + 0.3)$ **32.** $(5.3n - 0.5)(0.5n - 0.7)$
33. $(3.5x + 0.9)(3.5x - 0.9)$ **34.** $(5.4n - 0.7)(5.4n + 0.7)$
35. $4x(3x + 2)(2x + 5)$ **36.** $-2y(4y - 3)(5y + 2)$
37. $2y(5y^2 - 2)(5y^2 - 2)$ **38.** $-3x(10x^3 - 5)(10x^3 + 5)$

Ⓒ **39.** Prove that $(a + b + c)(a + b + c) = a^2 + b^2 + c^2 + 2ab + 2ac + 2bc$.

Simplify.

40. $(2x + y + 5)(2x + y + 5)$ **41.** $(8x - y - 4z)(8x - y - 4z)$
42. $(x^n + 3)(x^n - 5)$ **43.** $(3x^{2a} + 4)(x^{2a} + 1)$
44. $(x^c + y^{2c})(x^c + 4y^{2c})$ **45.** $(4x^{2a} - 3y^b)(5x^{2a} + 2y^b)$

NON-ROUTINE PROBLEMS

1. If $16 \cdot 2^x = 6^{y-8}$ and $y = 8$, then $x =$
    A. -8.   **B.** -4.   **C.** -3.   **D.** 0.   **E.** 8.

2. $[7 - 2(5 - 7)^{-1}]^{-1} =$
    A. 8 **B.** $\frac{1}{8}$ **C.** 11 **D.** $\frac{1}{11}$ **E.** -8

PROBLEM SOLVING: MIXTURE PROBLEMS   2.5

Objective   **To solve a word problem involving a collection or mixture of items**

To solve certain problems that involve a *collection* or *mixture* of items, it is helpful to convert the values of the items to a common unit. For example, a collection of x nickels, $4x$ dimes, and $x + 6$ quarters is worth

$$5 \cdot x + 10 \cdot 4x + 25(x + 6) \text{ cents.}$$

You will use this technique in solving the problem below.

Example 1

The number of dimes in Martha's coin collection is 4 less than the number of nickels and the number of quarters is 3 times the number of dimes. If the value of her collection is $7.40, how many nickels, dimes, and quarters does she have?

READ ▶

PLAN ▶
Represent
the data.

Let $\quad n$ = number of nickels.
$n - 4$ = number of dimes
$3(n - 4)$ = number of quarters

Write an
equation.

$$\left(\begin{array}{c}\text{Value of} \\ \text{nickels in } \cancel{c}\end{array}\right) + \left(\begin{array}{c}\text{Value of} \\ \text{dimes in } \cancel{c}\end{array}\right) + \left(\begin{array}{c}\text{Value of} \\ \text{quarters in } \cancel{c}\end{array}\right) = \left(\begin{array}{c}\text{Value of} \\ \text{collection in } \cancel{c}\end{array}\right)$$

$$5n \quad + \quad 10(n - 4) \quad + \quad 25 \cdot 3(n - 4) \quad = \quad 100(7.40)$$

SOLVE ▶
Solve the
equation.

$$5n + 10n - 40 + 75n - 300 = 740$$
$$90n - 340 = 740$$
$$90n = 1{,}080$$
$$n = 12$$

$$n = 12, \ n - 4 = 8, \ 3(n - 4) = 24$$

INTERPRET ▶
Check the
solution.

12 nickels + 8 dimes + 24 quarters	$7.40
$12 \cdot 5\cancel{c}$ + $8 \cdot 10\cancel{c}$ + $24 \cdot 25\cancel{c}$	$7.40 \cdot 100\cancel{c}$
$60\cancel{c}$ + $80\cancel{c}$ + $600\cancel{c}$	$740\cancel{c}$
$740\cancel{c}$	

Answer the
question.

Thus, there are 12 nickels, 8 dimes, and 24 quarters.

Example 2 on page 44 is an example of a *dry mixture* problem. In such situations, a mixture is obtained by combining two types of items. The intent is to gain the same amount of money for the mixture as would be realized if the items to be mixed were sold individually. In other words, (Value of 1st Item) + (Value of 2nd Item) = (Value of Mixture).

Example 2

Peanuts worth $2.10/kg are mixed with almonds worth $3.60/kg to make 20 kg of a mixture worth $2.40/kg. How many kilograms of peanuts and how many kilograms of almonds are there in the mixture?

Let x = number of kilograms (kg) of peanuts.

	No. of kg	Value/kg	Value in ¢
Peanuts	x	210¢	$210x$
Almonds	$20 - x$	360¢	$360(20 - x)$
Mixture	20	240¢	$240 \cdot 20$

(Value of peanuts) + (Value of almonds) = (Value of mixture)
$$210x \quad + \quad 360(20 - x) \quad = \quad 240 \cdot 20$$
$$210x + 7{,}200 - 360x = 4{,}800$$
$$-150x = -2{,}400$$
$$x = 16 \qquad 20 - x = 4$$

Check:

(16 kg at $2.10/kg) + (4 kg at $3.60/kg)	20 kg at $2.40/kg
$(16 \cdot \$2.10) \quad + \quad (4 \cdot \$3.60)$	$20 \cdot \$2.40$
$\$33.60 \quad + \quad \$14.40$	$\$48.00$
$\$48.00$	

Thus, there are 16 kg of peanuts and 4 kg of almonds in the mixture.

If a given liquid is classified as a 25% iodine solution, this means that 25% of the solution is iodine and 75% is water. Thus, 6 liters (L) of a 25% iodine solution is made up of 1.5 L of iodine and 4.5 L of water: $0.25 \cdot 6$ L = 1.5 L and $0.75 \cdot 6$ L = 4.5 L. Example 3 presents a situation in which a solution is obtained that contains a given percentage of a particular ingredient.

Example 3

How many liters of water must be evaporated from 60 L of a 12% vinegar solution to make it a 36% vinegar solution?

Let x = number of liters (L) of water to be evaporated.

	Total L	L of vinegar
12% solution	60	0.12(60)
36% solution	$60 - x$	$0.36(60 - x)$

Vinegar in 12% solution = Vinegar in 36% solution
$$0.12(60) = 0.36(60 - x)$$
$$7.20 = 21.60 - 0.36x$$
$$0.36x = 14.40$$
$$x = 40 \qquad \text{(Check in the problem.)}$$

Thus, 40 L of water must be evaporated.

A pharmacist can mix a 90% iodine solution with a 20% iodine solution to obtain a 45% iodine solution. Notice that the

$$\begin{pmatrix} \text{Iodine in the} \\ \text{20\% solution} \end{pmatrix} + \begin{pmatrix} \text{Iodine in the} \\ \text{90\% solution} \end{pmatrix} = \begin{pmatrix} \text{Iodine in the} \\ \text{45\% solution} \end{pmatrix}.$$

This concept is used in Example 4.

Example 4

How many milliliters (mL) of an 80% iodine solution should be added to 20 mL of a 16% iodine solution to obtain a 30% iodine solution?

Let x = number of milliliters of 80% solution added.

	Total milliliters	Milliliters of iodine
16% solution	20	0.16(20)
80% solution	x	0.80(x)
Mixture ▶ 30% solution	20 + x	0.30(20 + x)

$$\begin{pmatrix} \text{Iodine in} \\ \text{16\% sol.} \end{pmatrix} + \begin{pmatrix} \text{Iodine in} \\ \text{80\% sol.} \end{pmatrix} = \begin{pmatrix} \text{Iodine in} \\ \text{30\% sol.} \end{pmatrix}$$

$$0.16(20) + 0.80x = 0.30(20 + x)$$
$$3.20 + 0.80x = 6.00 + 0.30x$$
$$0.50x = 2.80$$
$$x = 5.6$$

Check:

(16% of 20 mL) + (80% of 5.6 mL)	30% of (20 + 5.6) mL
0.16(20 mL) + 0.80(5.6 mL)	0.30(25.6 mL)
3.20 mL + 4.48 mL	7.68 mL
7.68 mL	

Thus, 5.6 mL of the 80% solution should be added.

Written Exercises

(A) **1.** A parking meter takes only dimes and quarters. At the end of one day, the meter held $7.55 and the number of dimes was 8 more than 20 times the number of quarters. What was the number of dimes and quarters?

2. Some brand A dog food worth 70¢/kg is to be added to 9 kg of brand B dog food worth 50¢/kg. The mixture will be worth 58¢/kg. How many kilograms of brand A should be added?

3. Some salt tablets worth $3.50/kg were mixed with some cheaper tablets worth $2.00/kg to make 12 kg of a mixture worth $2.50/kg. How many kilograms of each kind were mixed?

4. How many milliliters of water must be evaporated from 160 mL of a 20% iodine solution to make it a 40% iodine solution?

5. How many liters of water must be added to 15 L of a 28% salt solution to dilute it to a 12% solution?

6. How many liters of an 80% sulfuric acid solution must be added to 60 L of a 30% sulfuric acid solution to obtain a 50% acid solution?

7. How many liters of a 52% citrus solution must be added to 30 L of a 24% citrus solution to make a 40% citrus solution?

8. A vending machine contains 12 more dimes than quarters and twice as many nickels as dimes. Find the number of each kind of coin if their total value is $6.00.

Ⓑ **9.** Gerry paid $11.92 to buy some 10¢, 13¢, and 50¢ stamps. If the number of 13¢ stamps was twice the number of 50¢ stamps, and there were 4 more 10¢ stamps that 13¢ stamps, how many of each kind did she buy?

10. Some brand A bulk tea worth $1.40/kg and some brand B tea worth $3.00/kg are mixed to make 16 kg of a brand C worth $2.00/kg. How many kilograms of each brand are used?

11. Fifty kilograms of oats worth 90¢/kg are added to some corn worth 60¢/kg to make some animal feed worth 75¢/kg. How many kilograms of corn should be used?

12. Skimmed milk contains no butterfat. How many liters of skimmed milk must be added to 600 L of milk that is 3.5% butterfat to obtain milk with 2% butterfat?

13. How many liters of a 72% alcohol solution must be added to 15 L of an 18% alcohol solution to obtain a 45% alcohol solution?

14. A store sells the Red Dot golf ball for 60¢, the Black Dot ball for 96¢, and the Gold Dot ball for $1.08. A golf pro bought two dozen more Red Dot than Gold Dot balls and three times as many Black Dot as Red Dot balls. If the golf balls and one dollar's worth of tees cost $130.12, how many golf balls of each kind were bought?

15. A package of paper tablets sells for $16.00 and contains twice as many 60¢ tablets as 80¢ tablets. How many paper tablets of each kind are in the package?

16. Some 8¢ slices of cheese and some 16¢ slices of meat make up a package of 40 slices worth $5.20. How many slices of each kind are in the package?

17. A chemist added 9 mL of a 40% citrus solution and some milliliters of a 60% citrus solution to 210 mL of a 20% citrus solution to make a 30% solution. How many milliliters of the 60% solution were added?

18. Chemicals A, B, and C are worth 60¢, 40¢, and 80¢/g, respectively. They are mixed so that the mass of B is twice that of A and is 3 g less than the mass of C. How many grams of each chemical should be used if the mixture is worth $11.40?

NON-ROUTINE PROBLEMS

A grocer bought some apples at 3 for 26¢. He bought the same number of apples at 5 for 40¢. To "break even" (no profit and no loss), the grocer must sell all of the apples at what rate?

A. 4 for 35¢ **B.** 5 for 42¢ **C.** 6 for 50¢
D. 8 for 66¢ **E.** 11 for 92¢

CUMULATIVE REVIEW

Find the solution set. Graph the solutions on a number line.

1. $7 - 2x > 15$ or $3x - 8 > 7$

2. $|4x - 2| > 14$

3. $9 - 2x < 17$ and $3x - 7 < 8$

4. $|2x + 3| < 9$

PROBLEM SOLVING: MOTION PROBLEMS   2.6

Objective   **To solve a word problem involving two objects in motion**

An automobile that is driven at an average rate of 90 kilometers per hour (km/h) for 3 hours travels $90 \cdot 3$, or 270 km. An Olympic sprinter who runs at an average rate of 8 meters per second (m/s) for 50 seconds travels $8 \cdot 50$, or 400 m. The relationship between distance, rate, and time of an object in motion is stated below.

Distance of object in motion	The distance (d) traveled by a moving object is equal to its rate of speed (r) times the time in motion (t): $d = rt$.

The formula $d = rt$ is a mathematical *model* for solving problems involving motion.

Example 1

READ ▶

Two trains started toward each other at the same time from stations 732 km apart. One train traveled at 148 km/h and the other train traveled at 96 km/h. In how many hours did they meet?

PLAN ▶

Represent the data in a diagram and in a table.
Let t = number of hours for each train.

Faster train→   Meet   ◄Slower train
| 148 km/h for t hours | | 96 km/h for t hours |
                            732 km

	rate (r)	time (t)	distance (d)	
Faster train	148	t	$148t$	◀ Use $d = rt$.
Slower train	96	t	$96t$	

(Distance of faster train) + (Distance of slower train) = Total distance
         $148t$ + $96t$ =   732

SOLVE ▶

$$244t = 732$$
$$t = 3$$

INTERPRET ▶

Check:

(Distance of faster train) + (Distance of slower train)	Total distance
$148 \text{ km/h} \cdot 3 \text{ h} + 96 \text{ km/h} \cdot 3 \text{ h}$	732 km
$444 \text{ km} + 288 \text{ km}$	
732 km	

Thus, the trains met in 3 hours.

Suppose a car leaves a certain city at 2:00 P.M., traveling at 60 km/h. Then, at 4:00 P.M. a second car leaves the same city, traveling the same route at 90 km/h. The second car will overtake the first car as indicated below.

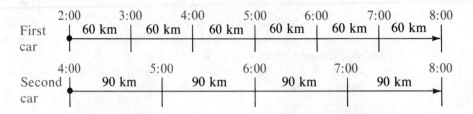

The second car will overtake the first car at 8:00 P.M. Notice that the two cars travel the same distance: 60 km/h · 6 h = 90 km/h · 4 h = 360 km.

Motion problems of this type can also be solved using the formula $d = rt$ along with appropriate equation properties. The procedure is shown in the next example.

Example 2

An aircraft carrier, traveling at 30 km/h, left a port at 7:00 A.M. At 9:00 A.M., a destroyer left the same port, traveling the same route at 40 km/h. At what time did the destroyer overtake the carrier?

Let t = number of hours the carrier traveled.

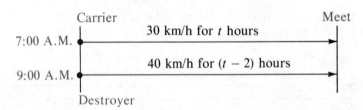

	rate	time	distance
Carrier	30	t	$30t$
Destroyer	40	$t - 2$	$40(t - 2)$

◄ Use $d = rt$.

(Distance traveled by carrier) = (Distance traveled by destroyer)
$$30t = 40(t - 2)$$
$$30t = 40t - 80$$
$$-10t = -80$$
$$t = 8$$

Check: | (Distance traveled by carrier) | (Distance traveled by destroyer) |
30 km/h · 8 h | 40 km/h · 6 h
240 km | 240 km

The time, 8 hours after 7:00 A.M., is 3:00 P.M.
Thus, the destroyer overtook the carrier at 3:00 P.M.

Written Exercises

(A)

1. A passenger train and a freight train start toward each other at the same time from towns 870 km apart. The passenger train travels 80 km/h and the freight train travels 65 km/h. In how many hours will they meet?

2. Two hockey players start toward each other at the same time from opposite ends of an ice rink that is 60 m long. In how many seconds will they meet if one player skates at a rate of 9.5 m/s and the other skates at a rate of 10.5 m/s?

3. A ship left port at 2:00 P.M. and traveled at 30 km/h. At 6:00 P.M., a helicopter left the same port and flew at 120 km/h in the same direction as the ship. At what time did the helicopter reach the ship?

4. John began the marathon run at 8:30 A.M. and averaged 12 km/h. Mary began at 9:00 A.M. and averaged 14 km/h. At what time did Mary pass John on the marathon course?

5. Two cars start from the same point at the same time but travel in opposite directions. One car averages 80 km/h and the other averages 70 km/h. In how many hours will they be 600 km apart?

6. From a dock in a river, Cindy swam downstream at 4.4 m/s while George swam upstream at 2.8 m/s. In how many seconds were they 108 m apart?

(B)

7. At 7:00 A.M., an armored van left city A and traveled at 70 km/h toward city B. At 11:00 A.M. a helicopter, headed toward the van, left city B and traveled at 110 km/h. At approximately what time did the copter reach the van if the cities are 550 km apart?

8. One bus left the city at 2:00 P.M. and averaged 75 km/h. Another bus, traveling in the opposite direction, left the city at 5:00 P.M. and averaged 85 km/h. At what time were the buses 705 km apart?

9. A plane left an airfield and averaged 600 km/h. Two hours later, a pursuit plane left the same field and traveled at 900 km/h in the same direction. How long did it take the pursuit plane to catch the slower plane?

10. A thief left a bank by car at 8:45 A.M. and averaged 110 km/h. The police left the scene at 8:48 A.M. and pursued the thief at 140 km/h. At what time was the thief caught?

11. A carrier and a destroyer left the same port at 8:00 A.M. and sailed in the same direction. The carrier averaged 30 km/h and the destroyer traveled at 75 km/h. At what time will they be 375 km apart?

12. A jet plane and a cargo plane left the same terminal at the same time and flew in the same direction. The jet flew at 1,000 km/h. After 6 hours, the planes were 2,100 km apart. Find the rate of the cargo plane.

(C)

13. An Olympic cyclist rode the first 6 hours of a race with a planned strategy. The cyclist pedaled at a constant rate for the first 3 hours and then reduced his speed by one-half for 1 hour. The last 2 hours, the rate was 3 times that of the fourth hour. Find the three different speeds if 78 km were traveled in the first 6 hours.

14. An object moved in one direction for 6.25 seconds at a constant speed. It then reversed direction for 2.50 seconds with its rate doubled, and then reversed direction again for 0.75 seconds at 40% of its initial speed. Find the three speeds if the object was 5.89 m from its start at the end of 9.50 seconds.

INTRODUCTION TO VECTORS

Objectives   **To identify the magnitude and direction of a vector**
To add and subtract vectors by drawing
To multiply a scalar and a vector by drawing

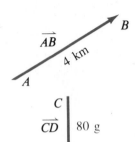

The **displacement** (movement) of an object a distance of 4 km to the northeast is shown by the **directed** line segment $\overrightarrow{AB}$ at the left. Such directed line segments are called **vectors**. The vector shown is written as: $\overrightarrow{AB}$ (with a half-arrow: $\rightarrow$) and read as "vector AB."
1. $\overrightarrow{AB}$ has a **magnitude** of 4 km and its **direction** is northeast.
2. The **initial point** of $\overrightarrow{AB}$ is A. The **terminal point** is B.

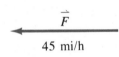

The vector $\overrightarrow{CD}$ represents the *force* exerted by an 80-g weight (magnitude) suspended on a coiled spring, stretching it downward (direction).

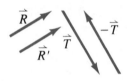

Sometimes, a single letter is used to name a vector. The **velocity** vector $\vec{F}$ shows a magnitude of 45 mi/h in the westward direction.

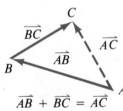

Four vectors are shown at the left. $\vec{R}$ and $\vec{R}'$ are **equivalent** vectors since they have the same magnitude and the same direction. $\vec{T}$ and $-\vec{T}$ are **opposite** vectors. They have the same magnitude and opposite directions.

The addition of vectors can be interpreted as follows. If you move from A to B and then from B to C, your *total* displacement is from A to C. Thus, $\overrightarrow{AB} + \overrightarrow{BC} = \overrightarrow{AC}$. This is called the **triangle** method of addition.

Notice that the **tail** of $\overrightarrow{BC}$ is at the **head** of $\overrightarrow{AB}$ and that the sum $\overrightarrow{AC}$ is drawn *from* the tail of $\overrightarrow{AB}$ *to* the head of $\overrightarrow{BC}$.

Given $\vec{A}$ and $\vec{B}$ as shown below, you can use equivalent vectors to draw $\vec{A} + \vec{B}$. Move the tail of $\vec{B}$ to the head of $\vec{A}$. Draw the sum from the tail of $\vec{A}$ to the head of $\vec{B}'$.

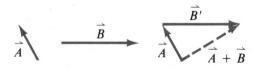

You can add three vectors by extending the method above.

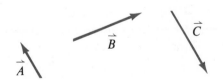

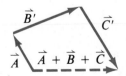

If $\vec{F}$ and $\vec{G}$ have the *same* initial point, you can use the **parallelogram** method to draw the sum $\vec{F} + \vec{G}$. Draw $\vec{F'}$ and $\vec{G'}$ to form a parallelogram. Draw the sum $\vec{F} + \vec{G}$ from the tail of $\vec{F}$ to the head of $\vec{G'}$. The sum is a **diagonal** of the parallelogram.

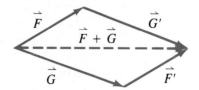

To add two *collinear* vectors $\vec{X}$ and $\vec{Y}$, move the tail of $\vec{Y}$ to the head of $\vec{X}$. Draw $\vec{X} + \vec{Y}$ from the tail of $\vec{X}$ to the head of $\vec{Y'}$. Three cases are shown below.

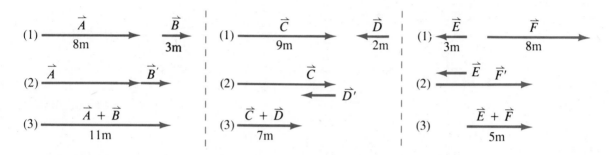

(1) $\vec{A}$ 8m $\vec{B}$ 3m   (1) $\vec{C}$ 9m $\vec{D}$ 2m   (1) $\vec{E}$ 3m $\vec{F}$ 8m

(2) $\vec{A}$ $\vec{B'}$ (2) $\vec{C}$ (2) $\vec{E}$ $\vec{F'}$

(3) $\vec{A} + \vec{B}$ 11m   (3) $\vec{C} + \vec{D}$ 7m   (3) $\vec{E} + \vec{F}$ 5m

Example 1

Use the figure at the right for the exercises below.

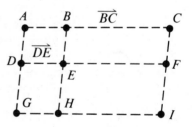

1. Name two vectors equivalent to $\overrightarrow{BC}$.
2. Name three vectors that are opposites of $\overrightarrow{BC}$.
3. Name two vectors that are collinear with $\overrightarrow{DE}$ and have the same direction as $\overrightarrow{DE}$.

Find the following sums.

4. $\overrightarrow{GB} + \overrightarrow{BF}$ **5.** $\overrightarrow{AE} + \overrightarrow{EG}$ **6.** $\overrightarrow{AD} + \overrightarrow{DG}$ **7.** $\overrightarrow{AG} + \overrightarrow{GD}$
8. $\overrightarrow{DE} + \overrightarrow{FI}$ **9.** $\overrightarrow{IE} + \overrightarrow{HG}$ **10.** $(\overrightarrow{DE} + \overrightarrow{EB}) + \overrightarrow{BC}$ **11.** $(\overrightarrow{GB} + \overrightarrow{EF}) + \overrightarrow{CF}$

Subtraction with vectors is similar to subtraction with real numbers.

$$a - b = a + (-b) \text{ for all real numbers } a \text{ and } b$$
$$\vec{X} - \vec{Y} = \vec{X} + (-\vec{Y}) \text{ for all vectors } \vec{X} \text{ and } \vec{Y}$$

To draw $\vec{A} - \vec{B}$, you draw $-\vec{B}$ and then draw $\vec{A} + (-\vec{B})$.

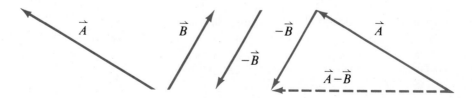

If a displacement of 2 m to the east occurs each hour for 3 h, the result is a displacement of 6 m to the east.

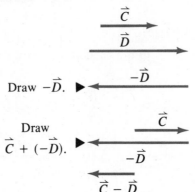

$\vec{D}$ was multiplied by the real number 3 to give the vector $3 \cdot \vec{D}$. The number 3 is called a **scalar**. The vector $3 \cdot \vec{D}$ is a **scalar multiple** of the vector $\vec{D}$.

Example 2   **Draw $\vec{C} - \vec{D}$.**   **Draw $-2 \cdot \vec{A}$.**

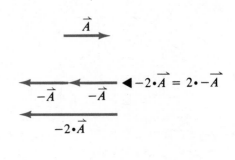

Written Exercises

Draw and label each vector. (Ex. 1–3)

Ⓐ 1. $\vec{A}$, a displacement vector for a move of 10 km to the northwest
   2. $\vec{B}$, a force vector for the pull exerted by a 20-lb weight suspended by a rope
   3. $\vec{C}$, a velocity vector for a speed of 70 km/h to the southeast

Use the figure at the right to find the following. (Ex. 4–10)

4. $\vec{AF} + \vec{DG}$	5. $\vec{HE} + \vec{FC}$	6. $\vec{AG} - \vec{DG}$
7. $\vec{BH} - \vec{FI}$	8. $\vec{CD} + \vec{DB} + \vec{BG}$	
9. $\vec{BF} + \vec{EH} + \vec{CB}$	10. $\vec{DC} - \vec{HI} - \vec{IC}$	

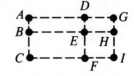

Draw *larger* copies of the vectors below. Use them for Exercises 11–29.

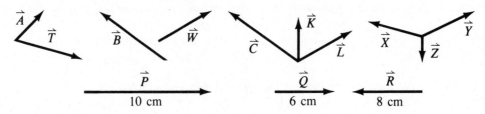

Draw each of the following. Use the vectors shown above.

Ⓑ 11. $\vec{A} + \vec{T}$ 12. $\vec{T} - \vec{A}$ 13. $\vec{B} + \vec{W}$ 14. $\vec{B} - \vec{W}$ 15. $\vec{W} - \vec{B}$
   16. $\vec{C} + \vec{K} + \vec{L}$ 17. $\vec{X} + \vec{Y} + \vec{Z}$ 18. $\vec{P} + \vec{R}$ 19. $\vec{Q} + \vec{R}$ 20. $\vec{R} - \vec{Q}$
   21. $3 \cdot \vec{Q}$ 22. $\frac{1}{2} \cdot \vec{P}$ 23. $-2 \cdot \vec{R}$ 24. $-1.5 \cdot \vec{P}$ 25. $4(\vec{P} + \vec{R})$
   26. $\vec{C} + \vec{K} - \vec{L}$ 27. $\vec{X} - \vec{Y} + \vec{Z}$ 28. $\vec{C} - (\vec{K} + \vec{L})$ 29. $3 \cdot \vec{A} + 2 \cdot \vec{T}$

CHAPTER TWO REVIEW

Vocabulary
base, exponent [2.1]
exponential equation [2.1]
negative exponent, zero exponent [2.2]
polynomial [2.3]
   binomial   linear   quadratic
   degree of   monomial   trinomial
scientific notation [2.2]
synthetic substitution [2.3]

Simplify and write each expression with positive exponents.
1. $-5a^4bc^5 \cdot 8a^2b^6c$ 2. $2m(4mn^5)^3$ [2.1]

3. $\dfrac{10x^3y^7z^8}{15x^7y^7z^7}$ 4. $\left(\dfrac{-5c^4}{2d^3}\right)^3$

★ 5. $x^m y^n (x^{2m} y^{n+3})^2$

6. $7x^{-3} \cdot 2y^0 \cdot x^{-4} \cdot 4y^2$

7. $(2a^{-2}b^3)^{-4}$ [2.2]

8. $\dfrac{6c^{-5}d^{-3}}{8c^4d^{-7}}$ 9. $\left(\dfrac{-3xy^{-2}}{2z^3w^{-3}}\right)^4$

★ 10. x^{2c-2d} if $c < d < 0$

Find the value of each expression. [2.2]
11. $\dfrac{3.45}{10^{-3}}$ 12. $\dfrac{6.9 \times 10^{-5}}{0.3 \times 10^{-2}}$

Solve each equation. [2.1]
13. $5^{x-3} = 625$ 14. $3^{4x-3} = 81$

Evaluate each polynomial for the given values of its variables. (Ex. 15–17) [2.3]
15. $3xy - 2x^2y^3 - 4$ if $x = 3$ and $y = -2$
16. $3.2c^2 + 5.4d^2$ if $c = -0.5$ and $d = -5$
★ 17. $10x^3 + 5x^2y - 5xy^2$ if $x = a^2$ and $y = 2a^2$

18. Use synthetic substitution to evaluate $3x^4 - 2x^3 - 3x - 180$ for $x = 3$ and again for $x = -2$. [2.3]
19. Subtract: $(9xy^2 - 5x^3y) - (3xy - 5x^3y)$. [2.3]

Simplify. [2.4]
20. $(5x - 2)(3x - 1)$
21. $(4x^2 + 3y^2)(4x^2 + 3y^2)$
22. $(3a - 1)(2a^2 - a - 1)$
23. $7c(5c^3 - 2)(4c^3 + 3)$

Solve each problem.
24. The number of nickels in a coin box is 8 more than the number of quarters and the number of dimes is 3 times the number of nickels. Find the number of each kind of coin if their total value is $7.00. [2.5]

25. Some caramels worth $2.60/kg are mixed with some peppermints worth $3.00/kg to make 8 kg of a mixture worth $2.85/kg. How many kilograms of each kind are there in the mixture? [2.5]

26. How many liters of water must be evaporated from 90 L of a 15% saline solution to make it a 45% saline solution? [2.5]

27. How many milliliters of a 25% iodine solution should be added to 90 mL of a 65% iodine solution to obtain a 55% iodine solution? [2.5]

28. At 6:00 A.M., a car left city A and headed for city B at 80 km/h. At 9:00 A.M., another car left city B and headed for city A at 90 km/h. At what time did the cars meet if the two cities are 580 km apart? [2.6]

29. A ship left a port at 25 km/h, and 4 hours later, a helicopter left the same port in the same direction at 100 km/h. How long did it take the helicopter to overtake the ship? [2.6]

30. In problem 29 above, how long did it take the helicopter to reach a distance of 10 km behind the ship? [2.6]

31. Steve left the gymnasium at 7:45 A.M. and jogged in one direction at 10 km/h. At 8:15 A.M., Karla left the gym and jogged in the opposite direction at 12 km/h. At what time were they 10.5 km apart? [2.6]

CHAPTER TWO TEST

Simplify and write each expression with positive exponents.

1. $-6x^2yz^4 \cdot 3xy^5z^3$
2. $8a^{-5} \cdot 5b^0 \cdot a^8 \cdot b^{-3}$
3. $\dfrac{5m^8n^3p^2}{15m^4n^3p^{10}}$
4. $\dfrac{4x^{-6}y^{-2}}{6x^{-4}y^{-4}}$
5. $(-4c^2)^3$
6. $10m(2m^3n)^4$
7. $\left(\dfrac{3x^2}{-2y}\right)^4$
8. $(3c^{-4}d^5)^{-2}$
9. $4a^2(-5a^3b)^3$
10. $\left(\dfrac{6x^{-4}}{x^4y^{-3}}\right)^2$

Find the value of each expression.

11. $\dfrac{7^{-2}}{2^{-4}}$
12. $\dfrac{4.8 \times 10^{-1}}{1.2 \times 10^{-3}}$

Solve each equation.

13. $2^{x-3} = 64$
14. $6^{3x+1} = 36$

Evaluate each polynomial for the given values of its variables. (Ex. 15–16)

15. $8x^2y - 2xy^2$ if $x = 2$ and $y = -2$
16. $3cd^3 - cd$ if $c = -2.4$ and $d = -5$
17. Use synthetic substitution to evaluate $3x^4 - 3x^3 + 4x - 10$ for $x = 2$ and again for $x = -2$.
18. Subtract: $(5x^5 - x^3 + 2x) - (x^3 - 6x - 8x^5)$.

Simplify.

19. $(5x - 3)(2x + 1)$
20. $(7m^3 + 5n)(7m^3 + 5n)$
21. $(3.5a - 0.5)(4a + 6)$
22. $4n(3n + 7)(3n - 7)$
23. $(2x + 3)(x^2 - 4x - 5)$

Solve each problem.

24. A collection of dimes and quarters is worth $2.85. If the number of dimes is 3 less than twice the number of quarters, find the number of dimes and quarters.

25. How many liters of water must be added to 30 L of a 20% chlorine solution to dilute it to a 15% chlorine solution?

26. One train left a depot at 3:00 P.M. and traveled at 80 km/h. Another train left the depot at 6:00 P.M. and traveled in the opposite direction at 150 km/h. At what time were the trains 1,620 km apart?

27. Some Brand A cleanser worth 80¢/kg is to be mixed with some Brand B cleanser worth 60¢/kg to make 15 kg of a mixture worth 68¢/kg. How many kilograms of each brand should be used?

28. John and Everett ran in the local marathon. If John averaged 17 km/h and Everett averaged 14 km/h, how long did it take John to be 4.5 km ahead of Everett?

29. Wheat flour worth 40¢/kg was mixed with some bran flour worth 60¢/kg. The amount of bran flour used in the mixture was 10 kg less than the amount of wheat flour used. If the mixture was worth $14.00, how many kilograms of each kind of flour were in the mixture?

★ 30. An automobile dealer purchased some model A, B, and C automobiles worth $3,000, $6,000, and $4,000 each, respectively. The ratio of model A to model B to model C automobiles purchased was 8 to 3 to 4. If their total value was $116,000, how many automobiles of each model were purchased?

★ 31. **Simplify and write with positive exponents.**
$$\dfrac{x^{6a+4}y^{3b-2}}{x^{2a-2}y^{b+3}}$$

★ 32. **Evaluate the polynomial for the given values of its variables.**
$4x^3y + 10x^2y^2 - 10xy^3$ if $x = a$ and $y = 2a$

★ 33. Multiply: $(3c + 2d - 5)(2c - d + 4)$.

COLLEGE PREP TEST

DIRECTIONS: In each item, you are to compare a quantity in Column 1 with a quantity in Column 2. Write the letter of the correct answer from these choices:

A   The quantity in Column 1 is greater than the quantity in Column 2.
B   The quantity in Column 2 is greater than the quantity in Column 1.
C   The quantity in Column 1 is equal to the quantity in Column 2.
D   The relationship cannot be determined from the given information.

Notes: Information centered over both columns refers to one or both of the quantities to be compared.
A symbol that appears in both columns has the same meaning in each column.

SAMPLE ITEMS AND ANSWERS

Column 1	Column 2

$$x = 7$$

S1. $x + x$ x^2

The answer is B, since $7^2 > 7 + 7$.

S2. $2x$ $3x$

The answer is D, since $2x > 3x$ if $x = -1$; $3x > 2x$ if $x = 1$; and $2x = 3x$ if $x = 0$.

Column 1	Column 2

1. $21 \cdot 1{,}234 \cdot 20$ $20 \cdot 1{,}234 \cdot 22$

$$x = -3$$
2. $x^3 - 2x$ $x^2 + 10x$

3. $5 + 2^4 + 7$ $6 + 4^2 + 5$

4. $(y + 2)(y + 3)$ $(y + 4)(y + 5)$

$$x = 10$$
5. $2x^4 + 3x^2 + 4$ $9x^3 + 8x^2 + 7x + 6$

Column 1	Column 2

6. The greater solution of $|x + 5| = 2$ The smaller solution of $|x - 5| = 7$

$$y = \frac{4}{3}x + 15$$
7. x, if $y = -45$ y, if $x = -45$

$$x^y = 16$$
8. x y

9. Number of liters of iodine in 60 L of a 25% iodine solution   Number of liters of water in 40 L of a 20% salt solution

10. Area of a rectangle, 1.5 m by 0.3 m   Area of a rectangle, 9 cm by 5 cm

11. $(x + 2)^2$ $(x + 2)(x - 2)$

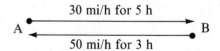

12. Average speed for round trip from A to B to A   40 mph

3 FACTORING AND SPECIAL PRODUCTS

Formulating a Problem-Solving Situation

Problem 1. How many cans of liquid floor wax must you buy to cover the floor of your recreation room?

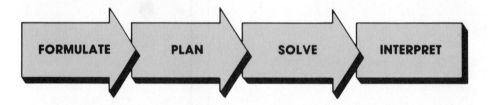

Your first step in solving a problem of this type is to *formulate* the problem-solving situation. That is, *analyze* the situation with particular *focus* on the question to be answered.

To answer the question, you must obtain the following information:
    (a)   Dimensions of floor
    (b)   Sizes of cans of floor wax available
    (c)   Surface area covered by one unit of wax of your choice

Suppose your analysis of the situation reveals these facts:
    (a)   The dimensions of the floor are 700 cm by 600 cm.
    (b)   The floor wax brand of your choice comes only in 1-L cans.
    (c)   One liter of floor wax covers an area of 20 m^2.

Problem 2. How many 1-L cans of liquid floor wax must you buy to cover the floor of your recreation room which measures 700 cm by 600 cm, if one liter of wax covers an area of 20 m^2?

FACTORING TRINOMIALS

Objective **To factor a trinomial of degree 2 or greater into two binomials**

You have multiplied two binomials, such as $(3x - 5)(2x + 1)$, by the FOIL method as shown below.

$$(3x - 5)(2x + 1) = 6x^2 - 7x - 5$$

Notice that the middle term of the trinomial is $3x - 10x$, or $-7x$.

A trinomial of the form $ax^2 + bx + c(a \neq 0)$ is called a **quadratic trinomial**. You can *factor* a quadratic trinomial into two binomials by reversing the multiplication process above. Begin by listing the possible pairs of binomial factors. Then check each pair until you find the factorization that gives the correct middle term.

Example 1 **Factor $3x^2 - 8x + 5$ into two binomials.**

$$3x^2 - 8x + 5 = (3x \underline{\ ?\ })(x \underline{\ ?\ })$$

The first terms of the binomial factors are $3x$ and x. The last terms are either $+1$ and $+5$ or -1 and -5. Since the middle term of the trinomial is negative, $(-8x)$, try -1 and -5.

Possible Factors	Middle Term	
$(3x - 1)(x - 5)$	$-15x - x = -16x$	
$(3x - 5)(x - 1)$	$-3x - 5x = -8x$	◀ *Correct middle term: $-8x$*

Thus, $3x^2 - 8x + 5 = (3x - 5)(x - 1)$.

Example 2 **Factor $3n^2 + 26n + 16$ into two binomials.**

$$3n^2 + 26n + 16 = (3n + \underline{\ ?\ })(n + \underline{\ ?\ })$$

	Possible Factors	Middle Term	
	$(3n + 1)(n + 16)$	$48n + n = 49n$	
Try 1 and 16,	$(3n + 16)(n + 1)$	$3n + 16n = 19n$	
2 and 8, ▶	$(3n + 8)(n + 2)$	$6n + 8n = 14n$	
4 and 4	$(3n + 2)(n + 8)$	$24n + 2n = 26n$	◀ *Correct middle term: $26n$*
for last terms	$(3n + 4)(n + 4)$	$12n + 4n = 16n$	
of the factors.			

Thus, $3n^2 + 26n + 16 = (3n + 2)(n + 8)$.

To factor a polynomial of any type, express it as a product of other polynomials of a lower degree. Using integers only, some polynomials cannot be factored in this manner.

Example 3

Factor $y^2 + y + 9$ into two binomials.

$$y^2 + y + 9 = (y + \underline{\ ?\ })(y + \underline{\ ?\ })$$

Possible Factors	Middle Term
$(y + 3)(y + 3)$	$3y + 3y = 6y$
$(y + 9)(y + 1)$	$y + 9y = 10y$

Neither pair of factors gives the correct middle term, y.
Thus, $y^2 + y + 9$ cannot be factored using integers.

Similar procedures should be followed when factoring quadratic trinomials that contain more than one variable or trinomials of a degree greater than 2. This is shown in Examples 4 and 5, respectively.

Example 4

Factor $2x^2 - 11xy + 5y^2$ into two binomials.

$$2x^2 - 11xy + 5y^2 = (2x - \underline{\ ?\ })(x - \underline{\ ?\ })$$

Possible Factors	Middle Term	
$(2x - 5y)(x - y)$	$-2xy - 5xy = -7xy$	
$(2x - y)(x - 5y)$	$-10xy - xy = -11xy$	◄ *Correct middle term: $-11xy$*

Thus, $2x^2 - 11xy + 5y^2 = (2x - y)(x - 5y)$.

Example 5

Factor $2x^4 - 9x^2 - 5$ into two binomials.

The first terms of the binomial factors are $2x^2$ and x^2.
The last terms are either -5 and $+1$ or $+5$ and -1.

$$2x^4 - 9x^2 - 5 = (2x^2 \ \underline{\ ?\ })(x^2 \ \underline{\ ?\ })$$

Possible Factors	Middle Term	
$(2x^2 - 5)(x^2 + 1)$	$2x^2 - 5x^2 = -3x^2$	
$(2x^2 + 1)(x^2 - 5)$	$-10x^2 + x^2 = -9x^2$	◄ *Correct middle term: $-9x^2$*
$(2x^2 - 1)(x^2 + 5)$	$10x^2 - x^2 = 9x^2$	
$(2x^2 + 5)(x^2 - 1)$	$-2x^2 + 5x^2 = 3x^2$	

Thus, $2x^4 - 9x^2 - 5 = (2x^2 + 1)(x^2 - 5)$.

Oral Exercises

Factor each trinomial into two binomials.

1. $x^2 - 7x + 12$
2. $x^2 - 13x + 12$
3. $x^2 + x - 12$
4. $x^2 - x - 12$
5. $x^2 + 8x + 12$
6. $x^2 - 4x - 12$
7. $x^2 + x + 3$
8. $x^2 - 5x + 6$
9. $x^2 - 5x - 6$

Written Exercises

Factor each trinomial into two binomials.

(A)
1. $7x^2 - 12x + 5$
2. $5y^2 - 36y + 7$
3. $5y^2 + 14y - 3$
4. $7n^2 + 4n - 11$
5. $5a^2 - 54a - 11$
6. $3x^2 - 10x - 13$
7. $6x^2 - 7x - 2$
8. $8n^2 + 14n + 3$
9. $2a^2 + a - 10$
10. $2x^2 - 11x + 14$
11. $5x^2 - 28x + 15$
12. $3y^2 + 11y + 6$
13. $9a^2 + 18a + 8$
14. $6t^2 - 11t - 10$
15. $4c^2 - 15c + 9$
16. $15n^2 + 7n - 4$
17. $x^2 + 4xy - 12y^2$
18. $x^2 + 13xy + 36y^2$
19. $x^2 - xy - 12y^2$
20. $8a^2 - 2ab - b^2$
21. $15c^2 - 8cd + d^2$
22. $21m^2 + 10mn + n^2$
23. $x^4 - 9x^2 + 14$
24. $n^4 + 8n^2 + 15$
25. $a^4 - 3a^2 - 18$
26. $10c^4 + 3c^2 - 1$
27. $15a^4 + 16a^2 - 7$
28. $21d^4 - 8d^2 - 5$
29. $x^2 - 14x - 72$
30. $72x^2 + 21x - 1$
31. $64y^2 - 12y - 1$
32. $8a^2 - 5a - 10$
33. $y^2 + 30y - 64$

(B)
34. $10x^2 + 23x + 12$
35. $12n^2 - 17n + 6$
36. $12y^2 - 8y - 15$
37. $18a^2 - 35a + 12$
38. $24x^2 - 46x - 35$
39. $16n^2 + 10n - 21$
40. $20c^2 - 48c + 27$
41. $24a^2 - 5a - 36$
42. $6y^2 + yz - 12z^2$
43. $5x^2 + 4xy - 12y^2$
44. $36m^2 - 7mn - 15n^2$
45. $24c^2 + 17cd - 20d^2$
46. $15x^4 - 8x^2 - 12$
47. $12x^4 - 23x^2 + 10$
48. $18y^4 + 9y^2 - 20$
49. $6y^4 - 23y^2 - 18$
50. $10n^4 - 31n^2 + 24$
51. $9a^4 + 24a^2 - 20$
52. $2x^4y^4 - 7x^2y^2 + 6$
53. $3a^4b^4 - a^2b^2 - 10$
54. $5m^4n^4 + 4m^2n^2 - 12$
55. $6c^4d^4 + 23c^2d^2 + 20$
56. $5x^4 + 23x^2y^2 + 24y^4$
57. $14c^4 - 9c^2d^2 - 18d^4$
58. $2x^8 + 2x^4 - 1$
59. $x^6 - 7x^3y^2 + 6y^4$
60. $5m^4 - 4m^2n^3 - n^6$

(C)
61. $x^{6m} - 7x^{3m} + 12$
62. $6x^{2m} + 19x^m + 15$
63. $9x^{4m} + 24x^{2m} - 20$
64. $8x^{6m} - 2x^{3m} - 21$
65. $5x^{2a} + 23x^a y^b + 12y^{2b}$
66. $8x^{4a} - 26x^{2a}y^{2b} + 15y^{4b}$
67. $12x^{6a} - 8x^{3a}y^{2b} - 15y^{4b}$
68. $x^{2n+4} + 7x^{n+2} + 12$
69. $x^{2n-2} + 3x^{n-1} - 10$
70. $x^{8n+6} - 2x^{4n+3} - 3$
71. $x^{2a+6} + 2x^{a+3}y^{a-2} + y^{2a-4}$
72. $2x^{2a+2} + 5x^{a+1}y^a + 3y^{2a}$
73. $x^2 + (m + n)x + mn$
74. $acx^2 + (ad + bc)x + bd$

CUMULATIVE REVIEW

Multiply.

1. $(x + 6)(x + 6)$
2. $(x + 10)(2x^2 - 3x - 4)$
3. $(3x^2 - 2y)(3x^2 + 2y)$

Factoring Trinomials

SPECIAL PRODUCTS

Objective — **To multiply a pair of related polynomials**

When working with polynomials, one special product that often occurs is the *product of the sum and the difference of two terms*. The general form is $(a + b)(a - b)$. As shown below, the FOIL method can be used to complete the multiplication.

$$(4x + 5)(4x - 5) = 16x^2 - 20x + 20x - 25 = 16x^2 - 25$$

Notice that there is no middle term in the product. Since $16x^2$ and 25 are both squares, you can rewrite $16x^2 - 25$ as $(4x)^2 - (5)^2$.

Product of sum and difference of two terms	The product of the sum and difference of two terms is the difference of the squares of the two terms. $(a + b)(a - b) = a^2 - b^2$, for all numbers a and b.

Example 1

Multiply: $(5x + 3y)(5x - 3y)$.

$$(5x + 3y)(5x - 3y) = (5x)^2 - (3y)^2$$
$$= 25x^2 - 9y^2$$

Multiply: $(6n^2 - 10)(6n^2 + 10)$.

$$(6n^2 - 10)(6n^2 + 10) = (6n^2)^2 - 10^2$$
$$= 36n^4 - 100$$

The *square of a binomial* is another special product. The general form is $(a + b)^2$ or $(a - b)^2$. [Recall that $a - b$ means $a + (-b)$.] You can simplify $(3n + 5)^2$ by using the FOIL method.

$$(3n + 5)^2 = (3n + 5)(3n + 5) = 9n^2 + 15n + 15n + 25 = 9n^2 + 30n + 25$$

Notice that the product $9n^2 + 30n + 25$ can be written as $(3n)^2 + 2 \cdot 3n \cdot 5 + 5^2$. This leads to the following statement and corresponding formulas.

Square of a binomial	The square of a binomial is the sum of the square of the first term, twice the product of the terms, and the square of the last term. $(a + b)^2 = a^2 + 2ab + b^2$ and $(a - b)^2 = a^2 - 2ab + b^2$, for all numbers a and b.

Example 2

Multiply: $(6c + 5d)^2$.

$$(a + b)^2 = a^2 + 2ab + b^2$$
$$(6c + 5d)^2 = (6c)^2 + 2(6c \cdot 5d) + (5d)^2$$
$$= 36c^2 + 60cd + 25d^2$$

Multiply: $(4x^2 - 3)^2$.

$$(a - b)^2 = a^2 - 2ab + b^2$$
$$(4x^2 - 3)^2 = (4x^2)^2 - 2(4x^2 \cdot 3) + 3^2$$
$$= 16x^4 - 24x^2 + 9$$

The trinomials, $36c^2 + 60cd + 25d^2$ and $16x^4 - 24x^2 + 9$, are *perfect square trinomials*. A perfect square trinomial is the square of a binomial.

Expressions such as $(x + 3)(x^2 - 3x + 9)$ and $(x - 5)(x^2 + 5x + 25)$ represent two other special products. You can use the distributive property to expand such products, as shown below.

$(x + 3)(x^2 - 3x + 9)$
$= x(x^2 - 3x + 9) + 3(x^2 - 3x + 9)$
$= x^3 - 3x^2 + 9x + 3x^2 - 9x + 27$
$= x^3 + 27$, or $x^3 + 3^3$

$(x - 5)(x^2 + 5x + 25)$
$= x(x^2 + 5x + 25) - 5(x^2 + 5x + 25)$
$= x^3 + 5x^2 + 25x - 5x^2 - 25x - 125$
$= x^3 - 125$, or $x^3 - 5^3$

Notice that each product is either a *sum* or a *difference of two cubes*. This suggests the special product formulas stated below.

Two special products	$(a + b)(a^2 - ab + b^2) = a^3 + b^3$ and $(a - b)(a^2 + ab + b^2) = a^3 - b^3$, for all numbers a and b.

Example 3   **Multiply: $(4x + 5)(16x^2 - 20x + 25)$.**

$$\text{Use } (a + b)[\ a^2 - ab + b^2] = a^3 + b^3.$$
$$(4x + 5)(16x^2 - 20x + 25) = (4x + 5)[(4x)^2 - 4x \cdot 5 + 5^2] = (4x)^3 + 5^3$$
$$= 64x^3 + 125$$

Example 4   **Multiply: $(2x - 3y)(4x^2 + 6xy + 9y^2)$.**

$$\text{Use } (a - b)[\ a^2 + ab + b^2] = a^3 - b^3.$$
$$(2x - 3y)(4x^2 + 6xy + 9y^2) = (2x - 3y)[(2x)^2 + 2x \cdot 3y + (3y)^2] = (2x)^3 - (3y)^3$$
$$= 8x^3 - 27y^3$$

Written Exercises

Multiply each pair of polynomials

(A)
1. $(n + 12)(n - 12)$
2. $(10y - 7)(10y + 7)$
3. $(9x + 6)(9x - 6)$
4. $(2x + 5)^2$
5. $(1 - 6n)^2$
6. $(8y + 10)^2$
7. $(10 - 6a)(10 + 6a)$
8. $(12x + 11)(12x - 11)$
9. $(25 - 3y)(25 + 3y)$
10. $(1 - 9n)^2$
11. $(7c + 10)^2$
12. $(15 - 2x)^2$
13. $(x + 2)(x^2 - 2x + 4)$
14. $(3y - 1)(9y^2 + 3y + 1)$
15. $(4 - 5a)(16 + 20a + 25a^2)$
16. $(4c + 10)(16c^2 - 40c + 100)$

(B)
17. $(5x + 8y)(5x - 8y)$
18. $(a - 6b)(a + 6b)$
19. $(4n^2 + 1)(4n^2 - 1)$
20. $(a + 7b)^2$
21. $(3x - 10y)^2$
22. $(c^2 + 5d)^2$
23. $(7y^2 - 4)(7y^2 + 4)$
24. $(12c + 3d^2)(12c - 3d^2)$
25. $(10y^2 - 5z^2)(10y^2 + 5z^2)$
26. $(5x - 4y^2)^2$
27. $(4m^2 + 3n^2)^2$
28. $(10x^3 - 2y^3)^2$
29. $(8x + y)(64x^2 - 8xy + y^2)$
30. $(5c - 6d)(25c^2 + 30cd + 36d^2)$
31. $(m^2 - 3n)(m^4 + 3m^2n + 9n^2)$
32. $(3x + 4y^2)(9x^2 - 12xy^2 + 16y^4)$

SPECIAL FACTORS

Objectives   **To factor a difference of two squares**
To factor a perfect square trinomial
To factor a sum or a difference of two cubes

Many polynomials have the form of one of the following special products that were discussed in the previous lesson.

$$a^2 + 2ab + b^2 \qquad a^2 - b^2 \qquad a^3 + b^3 \qquad a^3 - b^3$$

You can factor polynomials of these types by reversing the appropriate special product formulas. For example, since $(y + 6)(y - 6) = y^2 - 6^2 = y^2 - 36$, you can factor $y^2 - 36$ as follows:

$$y^2 - 36 = y^2 - 6^2 = (y + 6)(y - 6).$$

The expression $y^2 - 36$ is called a **difference of two squares.**

Factoring a difference of two squares	$a^2 - b^2 = (a + b)(a - b)$, for all numbers a and b.

Example 1

Factor $49c^2 - 16$.

$$\begin{aligned} 49c^2 - 16 &= (7c)^2 - 4^2 \\ &= (7c + 4)(7c - 4) \end{aligned}$$

Factor $9x^4 - 25y^2$.

$$\begin{aligned} 9x^4 - 25y^2 &= (3x^2)^2 - (5y)^2 \\ &= (3x^2 + 5y)(3x^2 - 5y) \end{aligned}$$

Trinomials of the form $a^2 + 2ab + b^2$ and $a^2 - 2ab + b^2$ are **perfect square trinomials.** When a given trinomial is to be factored, you should first check to see if it is a perfect square trinomial. For example, $x^2 + 10x + 25$ is a perfect square trinomial since it can be written in the form $a^2 + 2ab + b^2$ as $x^2 + 2 \cdot x \cdot 5 + 5^2$.
You can factor $x^2 + 10x + 25$ by reversing one of the special product formulas.

$$(x + 5)^2 = x^2 + 2 \cdot x \cdot 5 + 5^2 = x^2 + 10x + 25$$
$$x^2 + 10x + 25 = x^2 + 2 \cdot x \cdot 5 + 5^2 = (x + 5)^2$$

Factoring a perfect square trinomial	$a^2 + 2ab + b^2 = (a + b)^2$ and $a^2 - 2ab + b^2 = (a - b)^2$, for all numbers a and b.

Example 2

Factor $16n^2 + 40n + 25$.

$$\begin{aligned} &16n^2 + 40n + 25 \\ &= (4n)^2 + 2 \cdot 4n \cdot 5 + 5^2 \\ &= (4n + 5)^2 \end{aligned}$$

Factor $36c^4 - 12c^2d + d^2$.

$$\begin{aligned} &36c^4 - 12c^2d + d^2 \\ &= (6c^2)^2 - 2 \cdot 6c^2 \cdot d + d^2 \\ &= (6c^2 - d)^2 \end{aligned}$$

The factorization of $16n^2 + 40n + 25$ in Example 2 can be checked by squaring $4n + 5$: $(4n + 5)^2 = (4n)^2 + 2 \cdot 4n \cdot 5 + 5^2 = 16n^2 + 40n + 25$.

You can factor a **sum or** a **difference of two cubes** by using the reverse of the two special product formulas below.

$$(a + b)(a^2 - ab + b^2) = a^3 + b^3$$
$$(a - b)(a^2 + ab + b^2) = a^3 - b^3$$

Factoring a sum or difference of two cubes	Sum: $a^3 + b^3 = (a + b)(a^2 - ab + b^2)$ and Difference: $a^3 - b^3 = (a - b)(a^2 + ab + b^2)$, for all numbers a and b.

Example 3

Factor $8x^3 + 27$.

$8x^3 + 27$, or $(2x)^3 + 3^3$, is a sum of two cubes.

$$a^3 + b^3 = (a + b)(a^2 - ab + b^2)$$

$$8x^3 + 27 = (2x)^3 + 3^3 = (2x + 3)[(2x)^2 - 2x \cdot 3 + 3^2]$$
$$= (2x + 3)(4x^2 - 6x + 9)$$

Example 4

Factor $125x^3 - 64y^3$.

$125x^3 - 64y^3$, or $(5x)^3 - (4y)^3$, is a difference of two cubes.

$$a^3 - b^3 = (a - b)(a^2 + ab + b^2)$$

$$125x^3 - 64y^3 = (5x)^3 - (4y)^3 = (5x - 4y)[(5x)^2 + 5x \cdot 4y + (4y)^2]$$
$$= (5x - 4y)(25x^2 + 20xy + 16y^2)$$

Sometimes a polynomial can be factored after its terms are grouped appropriately.

$$m^2 - 2mn + n^2 - c^2 - 2cd - d^2 = (m^2 - 2mn + n^2) - (c^2 + 2cd + d^2)$$
$$= (m - n)^2 - (c + d)^2$$
$$= [(m - n) + (c + d)][(m - n) - (c + d)]$$
$$= (m - n + c + d)(m - n - c - d)$$

Example 5

Factor each polynomial.

$m^2 + 2mn + n^2 - 6m - 6n + 9$

$(m^2 + 2mn + n^2) - 6(m + n) + 9$
$(m + n)^2 - 2(m + n) \cdot 3 + 3^2$
$$a^2 \quad -2 \cdot a \cdot b \quad + b^2$$

$(a - b)^2$
$(m + n - 3)^2$

$9x^2 - c^2 + 8c - 16$

$9x^2 - (c^2 - 8c + 16)$
$(3x)^2 - (c - 4)^2$
$$a^2 - b^2$$

$(a + b)(a - b)$
$(3x + c - 4)(3x - [c - 4])$
$(3x + c - 4)(3x - c + 4)$

Special Factors

63

Reading in Algebra

Determine whether each polynomial is a difference of two squares (DS), a perfect square trinomial (PS), a sum of two cubes (SC), a difference of two cubes (DC), or none of these (N).

1. $x^3 - 8$　　　　2. $y^2 + 16$　　　　3. $x^2 + 6x + 9$　　　　4. $9c^2 - 4$
5. $n^2 - 10n - 25$　　6. $27x^3 + 1$　　　7. $(a + b)^2 - 4$　　8. $x^6 - 1$
9. $z^2 + 2z - 1$　　　10. $b^3 - 15c^3$　　11. $x^2y^2 - 1$　　　12. $49 - 14y + y^2$

Oral Exercises

State each binomial as a difference of two squares, a sum of two cubes, or a difference of two cubes.

1. $100n^2 - 9$　　　　2. $125n^3 + 1$　　　3. $25 - 4y^2$　　　　4. $1,000 - 27n^3$
5. $x^6 + 8z^3$　　　　6. $64x^2 - 81y^2$　　7. $x^9 - 125y^3$　　8. $x^4 - 36y^2$

Determine whether each trinomial is a perfect square trinomial.

9. $25x^2 + 50x + 16$　　10. $4y^2 - 20y + 25$　　11. $9n^2 - 13n + 4$　　12. $9m^4 + 24m^2 + 16$

Written Exercises

Factor each polynomial.

(A) 1. $4b^2 - 49$
3. $y^2 + 8y + 16$
5. $x^3 + 27$
7. $c^4 - 25$
9. $9n^2 - 24n + 16$
11. $y^4 + 18y^2 + 81$
13. $125c^3 - 1$
15. $x^2 + 6x + 9 - y^2$

(B) 17. $x^2 + 2xy + y^2 + 4x + 4y + 4$
19. $x^2 - y^2 - 10y - 25$

21. $49c^4d^4 - 100$
23. $25c^2 + 20cd + 4d^2$
25. $64m^3 + n^3$
27. $125t^3 - 8v^3$
29. $16c^6 - 40c^3d + 25d^2$
31. $25x^6 - 36d^4$
33. $0.25x^2 - 1.21y^2$

(C) 35. $m^2 - 2mn + n^2 - x^2 - 2xy - y^2$
37. $x^2 + 2cx + 2dx + c^2 + 2cd + d^2$
39. $25x^2 - 30x + 9 - 4a^2 + 4ab - b^2$

41. $9x^{4m+6} + 12x^{2m+3}y^n + 4y^{2n}$
43. $x^{3c} + y^{12d}$

2. $121 - 36n^2$
4. $1 - 20a + 100a^2$
6. $y^3 - 8$
8. $49 - d^4$
10. $25a^2 + 20a + 4$
12. $16d^4 - 40d^2 + 25$
14. $1,000x^3 + 1$
16. $m^2 - 2mn + n^2 - 25$
18. $c^2 - 2cd + d^2 + 10c - 10d + 25$
20. $t^2 + 2tv + 12t + v^2 + 12v + 36$

22. $36x^6 - y^2$
24. $36x^4 - 60x^2y + 25y^2$
26. $c^3 - 125d^3$
28. $1,000x^3 + 27y^3$
30. $4a^6b^4 + 28a^3b^2 + 49$
32. $27x^3y^3 - 125z^6$
34. $1.44a^2 - 6ab + 6.25b^2$
36. $4x^2 + 4xy + y^2 - c^2 + 4c - 4$
38. $4x^2 - 12xy + 9y^2 - 20x + 30y + 25$
40. $25x^2 + 40xy - 30x + 16y^2 - 24y + 9$

42. $x^{4n} - y^{6n+2}$
44. $8x^{6a} - y^{3b+9}$

COMBINED TYPES OF FACTORING

Objective

To factor a polynomial completely

Some numbers can be factored into more than two factors. For example,

$$30 = 3 \cdot 10 = 3 \cdot 2 \cdot 5 \qquad \text{and} \qquad 36 = 4 \cdot 9 = 2 \cdot 2 \cdot 3 \cdot 3.$$

Likewise, some polynomials can be factored into more than two factors. To factor a polynomial *completely,* you may have to use several types of factoring. The following steps show you the order in which the factoring should be accomplished.

Factoring a polynomial completely	**Step 1** Factor out the *greatest common factor* (GCF), if any.
	Step 2 Factor the resulting polynomial, if possible.
	Step 3 Factor each polynomial factor where possible.

Example 1

Factor $12x^4y^3 + 18x^2y^2 - 24x^2y$ completely.

Step 1 The GCF of 12, 18, and 24 is 6. The GCF of x^4 and x^2 is x^2. The GCF of y^3, y^2, and y is y. Factor out $6x^2y$.

$$12x^4y^3 + 18x^2y^2 - 24x^2y = 6x^2y(2x^2y^2 + 3y - 4)$$

Step 2 The resulting polynomial, $2x^2y^2 + 3y - 4$, cannot be factored.

Thus, $12x^4y^3 + 18x^2y^2 - 24x^2y = 6x^2y(2x^2y^2 + 3y - 4)$.

In Example 1, the monomial $6x^2y$ is called a **common monomial factor.** It is a factor of each term of the original polynomial.

Example 2

Factor $15x^2 + 10x - 40$ completely.

$$15x^2 + 10x - 40$$

Step 1 Factor out 5, the GCF. $\quad = 5(3x^2 + 2x - 8)$

Step 2 Factor $3x^2 + 2x - 8$. $\quad = 5(3x - 4)(x + 2)$

Step 3 $3x - 4$ and $x + 2$ cannot be factored.

Thus, $15x^2 + 10x - 40 = 5(3x - 4)(x + 2)$.

Example 3

Factor $27y^4 - 75y^2$ completely.

$$27y^4 - 75y^2$$

Step 1 Factor out $3y^2$, the GCF. $\quad = 3y^2(9y^2 - 25)$ ◀ *Difference of*

Step 2 Factor $9y^2 - 25$. $\quad = 3y^2(3y + 5)(3y - 5)$ *two squares*

Step 3 $3y + 5$ and $3y - 5$ cannot be factored.

Thus, $27y^4 - 75y^2 = 3y^2(3y + 5)(3y - 5)$.

Combined Types of Factoring   **65**

A complete factorization may contain more than two binomials, as shown in the next example.

Example 4   **Factor $8x^4 + 10x^2 - 3$ completely.**

Step 1   There is no common monomial factor other than 1.

$$8x^4 + 10x^2 - 3$$

Step 2   Factor $8x^4 + 10x^2 - 3$. $= (2x^2 + 3)(4x^2 - 1)$
Step 3   Factor $4x^2 - 1$. $= (2x^2 + 3)(2x + 1)(2x - 1)$

Thus, $8x^4 + 10x^2 - 3 = (2x^2 + 3)(2x + 1)(2x - 1)$.

To factor a polynomial in which the coefficient of the first term is negative, begin by factoring out -1 as a common monomial factor.

Example 5   **Factor $-9x^2 + 12x - 4$ completely.**

$$-9x^2 + 12x - 4$$

Step 1   Factor out -1. $= -1(9x^2 - 12x + 4)$ ◀ *Perfect square*
Step 2   Factor $9x^2 - 12x + 4$. $= -1(3x - 2)^2$ *trinomial*

Thus, $-9x^2 + 12x - 4 = -1(3x - 2)^2$.

Factoring completely may involve factoring a sum or a difference of two cubes. This is shown in Example 6.

Example 6   **Factor $24x^3 - 81$ completely.**

$$24x^3 - 81$$

Step 1   Factor out 3, the GCF. $= 3(8x^3 - 27)$
Step 2   Factor $8x^3 - 27$. $= 3(2x - 3)(4x^2 + 6x + 9)$

Thus, $24x^3 - 81 = 3(2x - 3)(4x^2 + 6x + 9)$.

Some polynomials with an even number of terms may be factored by "grouping pairs of terms" and factoring out a **common binomial factor.**

Example 7   **Factor $5xy - 15x - 2yz + 6z$ completely.**

$5xy - 15x - 2yz + 6z = (5xy - 15x) + (-2yz + 6z) = 5x(y - 3) - 2z(y - 3)$
$= (5x - 2z)(y - 3)$

Thus, $5xy - 15x - 2yz + 6z = (5x - 2z)(y - 3)$.

Reading in Algebra

Determine whether each expression is factored completely. If not, state why not.

1. $17 \cdot 21 \cdot 59$ **2.** $19 \cdot 23 \cdot 37$ **3.** $8(2x + 6y)$ **4.** $4a(5a + 4b)$

5. $-7(x^2 + xy)$ **6.** $5(x^2 - 2x + 3)$ **7.** $2y(y^2 - 7y + 12)$ **8.** $-3(9n^2 - 1)$

Oral Exercises

Find the GCF of each group of expressions.

1. $18x^2$; $36x$; 12 **2.** $5x^4$; $-2x^2$; $7x$ **3.** x^4; x^2; x^3

4. $12y^2$; $6y$; $24y^3$ **5.** $-10x^2$; $15x^3$ **6.** $x(5a - 3b)$; $-4(5a - 3b)$

Factor -1 out of each polynomial.

7. $-5a + 3b$ **8.** $-2x^2 - 7y$ **9.** $-y^2 + y - 12$

Written Exercises

Factor each polynomial completely.

Ⓐ
1. $12n^2 - 15n - 3$ **2.** $3a^4 - 8a^3 + a^2$

3. $3x^2 - 21x + 36$ **4.** $2n^2 + 4n - 30$

5. $16y^2 - 4$ **6.** $18c^2 - 50$

7. $4n^2 + 40n + 100$ **8.** $5x^2 - 40x + 80$

9. $y^4 - 7y^2 - 18$ **10.** $2a^4 - 9a^2 + 4$

11. $-x^2 + 8x - 16$ **12.** $-y^2 + 36$

13. $3y^3 + 81$ **14.** $5n^3 - 625$

15. $8xy + 20x + 6y + 15$ **16.** $6cd - 21c - 10d + 35$

17. $-4n^2 + 4n + 3$ **18.** $-9t^2 - 30t - 25$

Ⓑ
19. $x^3 + 4x^2 + 6x + 24$ **20.** $6a^3 + 20a^2 - 21a - 70$

21. $75x^2y^5 - 30x^3y^4 + 45x^4y^3$ **22.** $44x^3y^2z^4 - 100x^2y^3z^5 - 64x^3y^3z^3$

23. $8a^3 - 4a^2 - 40a$ **24.** $3a^3 - 75ab^2$

25. $2x^3y - 4x^2y + 2xy$ **26.** $36a^3b + 120a^2b^2 + 100ab^3$

27. $9x^4 - 7x^2 - 16$ **28.** $4n^4 - 17n^2 + 18$

29. $n^4 - 13n^2 + 36$ **30.** $25y^4 - 101y^2 + 4$

31. $-3y^2 + 27$ **32.** $-18x^2 + 32$

33. $-4ac^2 - 4ac - a$ **34.** $-x^3 + 6x^2y - 9xy^2$

35. $24x^3 - 375$ **36.** $-54y^3 - 128$

37. $y^4 + 64y$ **38.** $3x^4 - 3{,}000x$

39. $12ab - 9ac - 28bd + 21cd$ **40.** $8x^3 - 15y - 20xy + 6x^2$

41. $a^3 - a^2b - a^2b^2 + ab^3$ **42.** $4c^3 + 8c^2d - 4cd^2 - 8d^3$

43. $4y^8 - 13y^4 + 9$ **44.** $36x^8 - 13x^4 + 1$

Ⓒ
45. $x^{a+6} + x^4$ **46.** $x^{5c} - x^{4c}$ **47.** $y^{n+5} + y^{n+4}$

48. $x^{5c} - 9x^{3c}$ **49.** $x^{4n+2} + 6x^{2n+2} + 9x^2$ **50.** $y^{4n} - y^n$

51. Factor $x^6 - y^6$ completely as a difference of two squares

52. Factor $x^6 - y^6$ completely as a difference of two cubes.

53. Use factoring to show that the answers in Exercises 51 and 52 are equivalent.

Combined Types of Factoring

67

QUADRATIC EQUATIONS

Objective   **To solve a quadratic equation by factoring**

Equations such as $5x^2 + 17x - 12 = 0$, $9x^2 - 25 = 0$, and $7x^2 + 14x = 0$ are called **quadratic equations.** Each equation contains a polynomial of the second degree

Standard form of a quadratic equation   The *standard form of a quadratic equation* is $ax^2 + bx + c = 0$ where a, b, and c are numbers, and $a \neq 0$.

You can solve some quadratic equations by writing the equation in standard form, factoring, and then setting each factor equal to 0. The method is based upon the following property.

Zero-product property   For all numbers m and n, if $mn = 0$, then $m = 0$ or $n = 0$.

Example 1   **Find the roots of $3c^2 - 10c - 8 = 0$.**

The equation is in standard form.	$3c^2 - 10c - 8 = 0$
Factor $3c^2 - 10c - 8$.	$(3c + 2)(c - 4) = 0$
Use the Zero-Product Property.	$3c + 2 = 0$ or $c - 4 = 0$
Solve both linear equations.	$3c = -2$ $c = 4$
	$c = -\dfrac{2}{3}$

Check both solutions:

$$\begin{array}{c|c}
3c^2 - 10c - 8 & 0 \\
\hline
3\left(-\dfrac{2}{3}\right)^2 - 10\left(-\dfrac{2}{3}\right) - 8 & 0 \\
\dfrac{4}{3} + \dfrac{20}{3} - 8, \text{ or } 0 &
\end{array}
\qquad
\begin{array}{c|c}
3c^2 - 10c - 8 & 0 \\
\hline
3 \cdot 4^2 - 10 \cdot 4 - 8 & 0 \\
48 - 40 - 8 & \\
0 &
\end{array}$$

Thus, the roots are $-\frac{2}{3}$ and 4.

Example 2   **Solve $5x = 6 - 4x^2$.**

Step 1	Write in standard form.	$4x^2 + 5x - 6 = 0$
Step 2	Factor.	$(4x - 3)(x + 2) = 0$
Step 3	Set each factor equal to 0.	$4x - 3 = 0$ or $x + 2 = 0$
Step 4	Solve the linear equations.	$4x = 3$ $x = -2$
		$x = \dfrac{3}{4}$

Thus, $\frac{3}{4}$ and -2 are the solutions.

To solve a quadratic equation by factoring:

Step 1 Write the equation in standard form.
Step 2 Factor.
Step 3 Set each factor equal to 0.
Step 4 Solve the linear equations.

Sometimes it is more convenient to use the standard form, $0 = ax^2 + bx + c$, so that the coefficient of x^2 is a positive number.

Example 3

Solve $-7x^2 = 21x$.

$$0 = 7x^2 + 21x$$
$$0 = 7x(x + 3) \qquad \blacktriangleleft \text{ *The GCF is 7x.*}$$

$7x = 0 \qquad$ or $\qquad x + 3 = 0$
$\;\;x = 0 \qquad\qquad\qquad\; x = -3$

Thus, the solutions are 0 and -3.

Solve $25 = 9n^2$.

$$0 = 9n^2 - 25$$
$$0 = (3n + 5)(3n - 5)$$

$3n + 5 = 0 \qquad$ or $\qquad 3n - 5 = 0$
$\qquad 3n = -5 \qquad\qquad\qquad 3n = 5$
$\qquad\; n = -\dfrac{5}{3} \qquad\qquad\qquad n = \dfrac{5}{3}$

Thus, the solutions are $-\dfrac{5}{3}$ and $\dfrac{5}{3}$.

The Zero-Product Property can be extended to any number of factors. In the next example, the property is used to solve the *fourth degree equation* $x^4 - 13x^2 + 36 = 0$. The fourth degree polynomial is factored completely, and then each of the four factors is set equal to 0.

Example 4

Solve $x^4 - 13x^2 + 36 = 0$.

$$x^4 - 13x^2 + 36 = 0$$
$$(x^2 - 4)(x^2 - 9) = 0$$
$$(x + 2)(x - 2)(x + 3)(x - 3) = 0 \qquad \blacktriangleleft \text{ *Factor completely.*}$$

$x + 2 = 0 \qquad$ or $\qquad x - 2 = 0 \qquad$ or $\qquad x + 3 = 0 \qquad$ or $\qquad x - 3 = 0$
$\quad x = -2 \qquad\qquad\qquad x = 2 \qquad\qquad\qquad x = -3 \qquad\qquad\qquad x = 3$

Thus, the solutions are -2, 2, -3, and 3.

In some quadratic equations, the terms contain a common numerical factor.

Example 5

Solve $3n^2 - 15n + 18 = 0$.

Divide each side by 3, the GCF.

$$n^2 - 5n + 6 = 0$$
$$(n - 2)(n - 3) = 0$$

$n - 2 = 0 \qquad$ or $\qquad n - 3 = 0$
$\quad n = 2 \qquad\qquad\qquad n = 3$

Thus, the solutions are 2 and 3.

Reading in Algebra

Determine whether each statement is true or false.

1. $4x^2 - 8x = 0$ is a quadratic equation.
2. $5x - 15 = 0$ is a quadratic equation.
3. A quadratic equation may have two solutions.
4. A fourth degree equation may have four solutions.
5. If $x(x - 5) = 0$, then $x = 0$ or $x - 5 = 0$.
6. If $(x - 1)(x - 3) = 8$, then $x - 1 = 8$ or $x - 3 = 8$.
7. If $(2n - 5)(3n - 2) = 0$, then $n = 5$ or $n = 2$.
8. If $(y - 5)(y - 5) = 0$, then $y = 5$.

Written Exercises

Solve each equation.

(A)
1. $x^2 - 13x + 40 = 0$
2. $a^2 - a - 42 = 0$
3. $0 = x^2 + 15x + 50$
4. $y^2 = 12 - y$
5. $6 = b^2 - b$
6. $6y = 16 - y^2$
7. $2a^2 - 10a = 0$
8. $3x^2 = -12x$
9. $15a = a^2$
10. $z^2 - 64 = 0$
11. $0 = c^2 - 36$
12. $16 = z^2$
13. $x^4 - 26x^2 + 25 = 0$
14. $a^4 - 29a^2 + 100 = 0$
15. $900 + x^4 = 109x^2$
16. $3y^2 - 21y + 30 = 0$
17. $5c^2 + 20c - 60 = 0$
18. $0 = 40z + 100 + 4z^2$
19. $2c^2 - 9c + 4 = 0$
20. $3z^2 = 8z + 3$
21. $-3c^2 = c - 2$
22. $8n^2 + 2n - 1 = 0$
23. $1 = 7w - 10w^2$
24. $0 = 12t^2 + 7t + 1$
25. $9c^2 - 16 = 0$
26. $16z^2 = 1$
27. $36 = 25c^2$
28. $y^4 - y^2 - 12 = 0$
29. $b^4 = 24b^2 + 25$
30. $4y^2 = y^4 - 45$
31. $2b^2 + 12b + 18 = 0$
32. $36y^2 = 12 - 6y$
33. $60x - 180 = 5x^2$
34. $(x - 7)^2 = 29 - x^2$
35. $(8 - n)^2 = 2(24 - n^2)$
36. $(3y + 2)^2 = (2y - 5)^2$

(B)
37. $2b^2 - 7b + 6 = 0$
38. $3y^2 = 15 - 4y$
39. $0 = 7 + 9n + 2n^2$
40. $25n^2 - 4n = 0$
41. $12p^2 = 30p$
42. $-22y = 6y^2$
43. $30t^2 - 125t + 120 = 0$
44. $30y + 24y^2 = 75$
45. $70 + 2m = 24m^2$
46. $4y^4 - 37y^2 + 9 = 0$
47. $25b^4 = 34b^2 - 9$
48. $52y^2 = 9y^4 + 64$
49. $4x^4 + 31x^2 - 90 = 0$
50. $16c^4 + 23c^2 = 18$
51. $10 - 36t^2 = 16t^4$
52. $a^3 - 7a^2 + 10a = 0$
53. $5x^3 = 30x - 25x^2$
54. $4y^3 + 48y^2 = 4y^4$
55. $x(x + 5)(x - 4) = x^3$
56. $3y(y - 2)(y + 4) = 3y^3$
57. $(2x^2 + 3)^2 = 4x(x^3 + 6)$

Solve each equation for x in terms of a.

(C)
58. $3x^2 + 9a^2 = 12ax$
59. $0 = 6x^2 + 8ax - 8a^2$
60. $x^4 - 10a^2x^2 + 9a^4 = 0$
61. $42a^2 = 24x^2 - 38ax$
62. $36a^4 = 25a^2x^2 - 4x^4$
63. $a^8 + 100x^4 = 29a^4x^2$

Solve. Each equation has four solutions.

64. $|x^2 - 17| = 8$
65. $|x^2 - 10| = 6$
66. $|20 - x^2| = 16$

CUMULATIVE REVIEW

Find the solution set of each inequality. Graph the solutions on a number line.

1. $-3x < 2x + 35$
2. $|x - 2| < 5$
3. $|x + 3| > 7$

QUADRATIC INEQUALITIES

Objectives

To find the solution set of a quadratic inequality by factoring
To graph the solutions of a quadratic inequality

An inequality such as $x^2 - x - 6 > 0$ is called a **quadratic inequality.** The standard form of a quadratic inequality is $ax^2 + bx + c > 0$ (or $ax^2 + bx + c < 0$), where $a \neq 0$. Each quadratic inequality contains a polynomial of the second degree.

To solve the quadratic inequality $x^2 - x - 6 > 0$, you can use the following property for the *positive* product of two factors.

If $m \cdot n > 0$, then ($m < 0$ *and* $n < 0$) *or* ($m > 0$ *and* $n > 0$).

This property will lead to two conjunctions that must be solved.

Example 1

Find the solution set of $x^2 - x - 6 > 0$. Graph the solutions.

$$x^2 - x - 6 > 0 \quad \blacktriangleleft \textit{Factor the polynomial.}$$
$$(x + 2)(x - 3) > 0$$
$$(x + 2 < 0 \text{ and } x - 3 < 0) \text{ or } (x + 2 > 0 \quad \text{and } x - 3 > 0)$$
$$(x < -2 \text{ and } \qquad x < 3) \text{ or } \qquad (x > -2 \text{ and } \qquad x > 3)$$
$$\underbrace{\qquad\qquad}_{x < -2} \qquad \text{or} \qquad \underbrace{\qquad\qquad\qquad}_{x > 3}$$

Thus, the solution set is $\{x \mid x < -2 \text{ or } x > 3\}$.

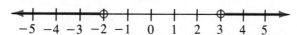

The quadratic inequality $x^2 - 16 < 0$ can be solved using the following property for the *negative* product of two factors.

If $m \cdot n < 0$, then ($m < 0$ *and* $n > 0$) *or* ($m > 0$ *and* $n < 0$).

This property produces two conjunctions that must be solved.

To solve a quadratic inequality in which the coefficient of the square term is negative, it may be helpful to multiply the inequality by -1.

Example 2

Find the solution set of $-x^2 + 16 \geq 0$. Graph the solutions.

$$-x^2 + 16 \geq 0 \quad \blacktriangleleft \textit{Multiply each side by -1 and}$$
$$x^2 - 16 \leq 0 \qquad \textit{reverse the order to} \leq.$$
$$(x + 4)(x - 4) \leq 0$$
$$(x + 4 \leq 0 \text{ and } x - 4 \geq 0) \text{ or } (x + 4 \geq 0 \quad \text{and } x - 4 \leq 0)$$
$$(x \leq -4 \text{ and } \qquad x \geq 4) \text{ or } \qquad (x \geq -4 \text{ and } \qquad x \leq 4)$$
$$\underbrace{\qquad\qquad}_{\text{no solution}} \qquad \qquad \underbrace{\qquad\qquad\qquad}_{-4 \leq x \leq 4}$$

Thus, the solution set is $\{x \mid -4 \leq x \leq 4\}$.

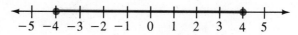

For a *polynomial inequality* of degree 3 or higher, it is more convenient to graph the solutions first and then write the solution set.

Example 3   **Graph the solutions of $x^3 - 2x^2 - 15x > 0$. Find the solution set.**

$$x^3 - 2x^2 - 15x > 0$$
$$x(x^2 - 2x - 15) > 0$$
$$x(x + 3)(x - 5) > 0$$

The solutions of $x(x + 3)(x - 5) = 0$ are 0, -3, and 5. Plot these numbers on a number line.

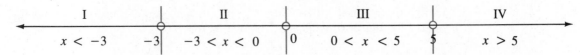

| I | II | III | IV |
| $x < -3$ | -3 $-3 < x < 0$ | 0 $0 < x < 5$ | 5 $x > 5$ |

The points at -3, 0, and 5 separate the number line into four parts, I–IV. Select the coordinate of a point in each part and evaluate $x(x + 3)(x - 5)$ to find whether the value is positive (>0) or negative (<0).

Part		x	$x(x + 3)(x - 5)$	Value
I.	$x < -3$	-4	$-4(-1)(-9)$	$-36 < 0$
II.	$-3 < x < 0$	-2	$-2(+1)(-7)$	$+14 > 0$
III.	$0 < x < 5$	1	$+1(+4)(-4)$	$-16 < 0$
IV.	$x > 5$	6	$+6(+9)(+1)$	$+54 > 0$

The graph of the solutions will be parts II and IV where the value is positive.

Thus, the solution set is $\{x \mid -3 < x < 0 \text{ or } x > 5\}$.

Example 4   **Graph the solutions of $x^4 - 10x^2 + 9 \leq 0$. Find the solution set.**

$$x^4 - 10x^2 + 9 \leq 0$$
$$(x^2 - 9)(x^2 - 1) \leq 0$$
$$(x + 3)(x - 3)(x + 1)(x - 1) \leq 0$$

Plot the solutions of $(x + 3)(x - 3)(x + 1)(x - 1) = 0$ on a number line.

| I | II | III | IV | V |
| $x < -3$ | -3 $-3 < x < -1$ | -1 $-1 < x < 1$ | 1 $1 < x < 3$ | 3 $x > 3$ |

Part		$(x + 3)(x + 1)(x - 1)(x - 3)$	Value	
I.	$x < -3$	(neg.) (neg.) (neg.) (neg.)	pos. > 0	
II.	$-3 < x < -1$	(pos.) (neg.) (neg.) (neg.)	neg. < 0	YES
III.	$-1 < x < 1$	(pos.) (pos.) (neg.) (neg.)	pos. > 0	
IV.	$1 < x < 3$	(pos.) (pos.) (pos.) (neg.)	neg. < 0	YES
V.	$x > 3$	(pos.) (pos.) (pos.) (pos.)	pos. > 0	

Draw parts II and IV and their endpoints on the number line.

Thus, the solution set is $\{x \mid -3 \leq x \leq -1 \text{ or } 1 \leq x \leq 3\}$.

Reading in Algebra

Match each inequality to exactly one graph.
1. $3 < x < 7$
2. $x < 3$ or $x > 7$
3. $x < 3$ or $7 < x < 9$
4. $3 < x < 7$ or $x > 9$
5. $x^2 \geq 0$
6. $x^2 < 0$

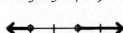

A ←———●——+——●——●——→
    3   5   7   9

B ←——●——+——+——●——●——→
    3   5   7   9

C ←——●——+——●————●——→
    3   5   7   9

D ←——●——+——●——+——→
    3   5   7   9

E ←——+——+——+——+——→
    3   5   7   9

F ←——+——+——+——+——→
    3   5   7   9

Oral Exercises

Tell whether each sentence is true or false.
1. If $x = 4$, then $(x - 2)(x - 6) < 0$.
2. If $x = -2$, then $(x + 4)(x - 3) > 0$.
3. If $x = 8$, then $(x - 2)(x - 6) < 0$.
4. If $x = -6$, then $(x + 4)(x - 3) > 0$.
5. If $x < -5$, then $(x + 5)(x - 5) > 0$.
6. If $x > 7$, then $(x + 1)(x - 7) < 0$.
7. If $-3 < x < 2$, then $(x + 3)(x - 2) > 0$.
8. If $1 < x < 5$, then $(x - 1)(x - 5) < 0$.

Written Exercises

Find the solution set of each inequality. Graph the solutions.

Ⓐ
1. $n^2 - 8n + 15 > 0$
2. $x^2 - 2x - 8 < 0$
3. $c^2 + 8c + 12 \leq 0$
4. $n^2 - 7n + 10 \geq 0$
5. $x^2 - 9 \geq 0$
6. $y^2 - 25 \leq 0$
7. $-y^2 + 4 < 0$
8. $-c^2 + 36 \geq 0$
9. $n^2 - 5n > 0$
10. $a^2 + 8a < 0$
11. $-4c^2 - 12c \geq 0$
12. $-2x^2 + 10x \leq 0$
13. $y^3 - y^2 - 6y < 0$
14. $x^3 + 6x^2 - 16x > 0$

Ⓑ
15. $x^3 - 25x \geq 0$
16. $y^3 - 100y \leq 0$
17. $2x^2 - x - 10 < 0$
18. $4a^2 + 5a \geq 9$
19. $2y^2 + 7y \geq 0$
20. $3a^2 < 10a$
21. $-c^2 - c + 12 < 0$
22. $-x^2 \geq 8 - 6x$
23. $0 < -n^2 + 4n - 3$
24. $0 \geq 35 - 2y - y^2$
25. $y^4 - 13y^2 + 36 < 0$
26. $x^4 - 26x^2 + 25 > 0$
27. $a^4 - 16 \geq 0$
28. $n^4 < 81$
29. $n^4 - 14n^2 - 32 > 0$
30. $-y^4 + 4y^2 + 45 < 0$
31. $x^2 - 4x + 4 > 0$
32. $x^2 - 6x + 9 \leq 0$
33. $3x^2 + 1 \geq 2x^2$
34. $y^3 - 3y^2 \geq 25y - 75$
35. $y^3 - 2y^2 < 16y - 32$
36. $y^2 - 5 > 2y^2 + 5$
37. $a^2 < 4$ or $a^2 > 25$
38. $a^2 \geq 4$ and $a^2 \leq 25$
39. $a^2 < 16$ and $a^3 < 4a$

Ⓒ
40. $|x^2 - 17| < 8$
41. $|x^2 - 10| > 6$
42. $|x^2| < 16$
43. $|(x - 2)^2 - 5| < 4$
44. $|(x + 3)^2 - 20| > 16$
45. $|x^2| > 9$

Quadratic Inequalities

73

PROBLEM SOLVING: NUMBER PROBLEMS   3.7

Objective   **To solve a word problem that leads to a quadratic equation**

Many word problems can be solved by using quadratic equations. Since a quadratic equation often has two solutions, a given word problem may have two sets of answers.

In Example 1, a problem about *consecutive integers* is solved. Recall that consecutive multiples of 5 can be represented by x, $x + 5$, $x + 10$, . . ., where x is a multiple of 5.

Example 1   **Find three consecutive multiples of 5 so that the square of the third number, decreased by 5 times the second number, is the same as 25 more than twice the product of the first two numbers.**

Represent
the data. ▶   Let x, $x + 5$, and $x + 10$ be the three consecutive multiples of 5, where x is a multiple of 5.

Write an
equation. ▶
$$\text{(third)}^2 - (5 \cdot \text{second}) = (2 \cdot \text{first} \cdot \text{second}) + 25$$
$$(x + 10)^2 - 5(x + 5) = 2x(x + 5) + 25$$
$$x^2 + 20x + 100 - 5x - 25 = 2x^2 + 10x + 25$$

Solve the
equation. ▶
$$0 = x^2 - 5x - 50$$
$$0 = (x + 5)(x - 10)$$
$$x + 5 = 0 \qquad \text{or} \qquad x - 10 = 0$$
$$x = -5 \qquad\qquad\qquad x = 10$$

There are two sets of answers. The first multiple, x, may be either -5 or 10. If $x = -5$, then $x + 5 = 0$ and $x + 10 = 5$. If $x = 10$, then $x + 5 = 15$ and $x + 10 = 20$.

Check in the
problem. ▶

$(-5, 0, 5)$

$\text{(third)}^2 - (5 \cdot \text{second})$	$(2 \cdot \text{first} \cdot \text{second}) + 25$
$5^2 \quad - \quad 5 \cdot 0$	$2(-5 \cdot 0) \quad + 25$
$25 \quad - \quad\quad 0$	$0 \qquad\quad + 25$
$\qquad\quad 25$	$\qquad 25$

$(10, 15, 20)$

$\text{(third)}^2 - (5 \cdot \text{second})$	$(2 \cdot \text{first} \cdot \text{second}) + 25$
$20^2 \quad - \quad 5 \cdot 15$	$2 \cdot 10 \cdot 15 \quad + 25$
$400 \quad - \quad\quad 75$	$300 \qquad\quad + 25$
$\qquad\quad 325$	$\qquad 325$

Answer the
questions. ▶   **Thus,** the three consecutive multiples are -5, 0, 5, or 10, 15, 20.

When using a quadratic equation to solve a word problem, you should determine if both solutions of the equation provide answers that are "reasonable" in the problem. In Example 1, both solutions check in the equation, and both provide logical answers to the word problem.

Example 2   **Some light bulbs are placed in boxes, and the boxes are then packed in cartons. The number of bulbs in each box is 4 less than the number of boxes in each carton. Find the number of bulbs in each box if a full carton contains 60 light bulbs.**

Let x = number of boxes in each carton.
    $x - 4$ = number of bulbs in each box

$$\binom{\text{Number of boxes}}{\text{in each carton}} \cdot \binom{\text{Number of bulbs}}{\text{in each box}} = \binom{\text{Total number of bulbs}}{\text{in each carton}}$$

$$x(x - 4) = 60$$
$$x^2 - 4x = 60$$
$$x^2 - 4x - 60 = 0$$
$$(x + 6)(x - 10) = 0$$
$$x = -6 \quad \text{or} \quad x = 10$$

The number of boxes in each carton cannot be -6. If $x = 10$, then there are 10 boxes in each carton and $10 - 4$, or 6 bulbs in each box.

Check: (No. of boxes in each carton)(No. of bulbs in each box) | 60
$$10 \cdot 6$$ | 60
$$60$$

Thus, there are 6 bulbs in each box.

To solve a word problem involving two numbers, read carefully to see if the second number is described *in terms of* the first number. If so, you should use a variable to represent the first number. This is shown in Example 3.

Example 3   **A second number is 5 less than twice a first number. If the second number is multiplied by 3 more than the first number, the result is 21. Find all such pairs of numbers.**

Let x = first number.
    $2x - 5$ = second number

Second number times 3 more than first number is 21.
$$(2x - 5)(x + 3) = 21$$
$$2x^2 + x - 15 = 21$$
$$2x^2 + x - 36 = 0$$
$$(2x + 9)(x - 4) = 0$$
$$2x + 9 = 0 \quad \text{or} \quad x - 4 = 0$$
$$2x = -9 \qquad\qquad x = 4$$
$$x = -4.5$$

There are two sets of answers. The first number may be either -4.5, or 4. If $x = -4.5$, then $2x - 5 = -14$. If $x = 4$, then $2x - 5 = 3$. Check both pairs of numbers in the problem.

Thus, the pairs of first and second numbers are $(-4.5, -14)$ or $(4, 3)$.

Oral Exercises

Beginning with 12, give four integers for each description.
1. consecutive integers
2. consecutive even integers
3. consecutive multiples of 12
4. consecutive multiples of 3

Beginning with −9, give four integers for each description.
5. consecutive integers
6. consecutive odd integers
7. consecutive multiples of 3
8. consecutive multiples of 9

Written Exercises

Solve each problem.

(A)

1. Find three consecutive even integers so that the first integer times the second integer is 24. [Hint: Use x, $x + 2$, and $x + 4$ to represent the integers.]

2. Find three consecutive integers so that the product of the second and the third integers is 42.

3. Find three consecutive integers so that the square of the first, increased by the square of the third, is 100.

4. Find three consecutive odd integers so that the sum of the squares of the first two integers is 130.

5. Forty chairs are placed in rows so that the number of chairs in each row is 3 less than the number of rows. Find the number of chairs in each row.

6. Forty-four students are seated in rows in a lecture hall. The number of students in each row is 7 more than the number of rows. Find the number of students seated in each row.

7. One-hundred forty peaches were packed in some boxes so that the number of boxes was 6 less than twice the number of peaches in each box. Find the number of boxes used.

8. An album contains 1,020 stamps. How many pages are in the album if the number of pages is 8 more than 4 times the number of stamps on each page?

9. One number is 4 less than another number. Their product is 21. Find all such pairs of numbers.

10. The product of two numbers is 33. One number is 2 more than 3 times the other number. Find both numbers.

11. The product of two numbers is 9, and one number is 4 times the other number. Find the two numbers.

12. Find two numbers whose product is 25 if one number is 9 times the other number.

(B)

13. Find three consecutive multiples of 5 so that the product of the first and the third numbers is 200.

14. Find three consecutive multiples of 10 so that 15 times the third number is the same as the product of the first two numbers.

15. For which three consecutive multiples of 4 is 400 equal to the square of the first number, increased by the square of the second number?

16. For which three consecutive multiples of 6 will the sum of their squares be 504?

17. An 8-story hotel has the same number of rooms on each of the 8 floors. The number of square meters of carpet in each room is 10 more than twice the number of rooms on each floor. Find the area of the carpet in one room if there are 8,000 m² of carpet in the hotel.

18. Twelve kilograms of flour are placed in some bags and the bags are packed in boxes. The number of kilograms of flour in each bag is 3 less than the number of bags in each box, and the number of boxes is 1 less than the number of bags in each box. Find the number of boxes used.

19. A second number is 5 more than a first number. If the second number is multiplied by 2 less than the first number, the product is 8. Find the numbers.

20. A first number is 3 less than twice a second number. If the first number is multiplied by 3 less than the second number, the result is 5. Find the numbers.

21. Find four consecutive multiples of 5 such that the product of the second and the third numbers is 50 more than 10 times the fourth number.

22. For which five consecutive multiples of 3 is the product of the second and the fourth numbers equal to 90 less than the sum of the squares of the first and the fifth numbers?

Ⓒ **23.** Find four consecutive multiples of 0.5 in which the product of the first and the third numbers is 0.25 less than 3 times the second number.

24. Find four consecutive multiples of π in which the product of the second and the third multiples is $2\pi^2$ more than 3π times the fourth multiple.

25. For which three consecutive integers is the sum of the first and the second integers the same as 3 less than twice the third integer?

26. Find three consecutive odd integers so that the square of the second integer is equal to the product of the first and the third integers.

NON-ROUTINE PROBLEMS

1. Prove that the following statement is true.

If a, b, and c are *odd integers* and if x is an *integer,* then $ax^2 + bx + c$ *cannot* be an *even* integer ($ax^2 + bx + c$ *must be* an *odd* integer).

Hint: x is either an *odd* integer *or* an *even* integer.

2. If $4^{2x} = 25$, then $4^{-x} = \underline{\ ?\ }$.

CUMULATIVE REVIEW

1. The length of a rectangle is 5 units more than its width. Represent the perimeter and the area of the rectangle in algebraic terms.

2. The base of a triangle is 8 units shorter than 4 times its height. Represent the area of the triangle in algebraic terms.

PROBLEM SOLVING: GEOMETRIC PROBLEMS

Objectives
To find the length of each side of a right triangle by using the Pythagorean relation
To solve an area problem that leads to a quadratic equation

For each right triangle ABC with right angle C as shown below, the lengths a, b, and c of the three sides are related by the **Pythagorean relation:** $a^2 + b^2 = c^2$.

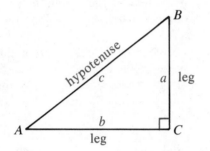

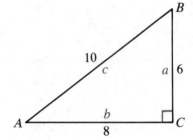

$$a^2 + b^2 = c^2$$
$$6^2 + 8^2 = 10^2$$
$$36 + 64 = 100$$
$$100 = 100$$

Example 1

One leg of a right triangle is 7 m longer than the other leg. The length of the hypotenuse is 13 m. Find the length of each leg.

Represent the data. ▶

Let x = length of one leg.
    $x + 7$ = length of other leg

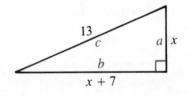

Write an equation. ▶

$$a^2 + b^2 = c^2$$
$$x^2 + (x + 7)^2 = 13^2$$
$$x^2 + (x^2 + 14x + 49) = 169$$
$$2x^2 + 14x - 120 = 0$$

Solve the equation. ▶

$$x^2 + 7x - 60 = 0 \quad ◀ \textit{Divide each side by 2, the GCF.}$$
$$(x + 12)(x - 5) = 0$$
$$x = -12 \quad \text{or} \quad x = 5$$

Check in the problem. ▶

The length of a side of a triangle cannot be -12 m.
If $x = 5$, then $x + 7 = 12$. Check to determine if $a^2 + b^2 = c^2$.

$$a = 5, b = 12, c = 13$$

$a^2 + b^2$	c^2
$5^2 + 12^2$	13^2
$25 + 144$	169
169	

Answer the question. ▶

Thus, the length of one leg is 5 m and the length of the other leg is 12 m.

The surface enclosed by a rectangle is measured in *square units*. This measure is the **area** of the rectangle. The word problem in Example 2 on page 79 involves the area of a rectangle.

Example 2

The length of a rectangle is 3 cm more than twice its width. Find the width and the length if the area is 44 cm².

Let w = width.

       $2w + 3$ = length

width $\times$ length = area of rectangle

$$w(2w + 3) = 44$$
$$2w^2 + 3w = 44$$
$$2w^2 + 3w - 44 = 0$$
$$(2w + 11)(w - 4) = 0$$

$$2w + 11 = 0 \qquad \text{or} \qquad w - 4 = 0$$
$$2w = -11 \qquad\qquad\qquad w = 4$$
$$w = -5.5$$

The width of a rectangle cannot be -5.5 cm.

If $w = 4$, then $2w + 3 = 11$.

width $\times$ length	area
4 cm $\times$ 11 cm	44 cm²
44 cm²	

Thus, the width is 4 cm and the length is 11 cm.

The word problem in Example 3 involves the area of a triangle. The area of a triangle equals $\frac{1}{2}$ times the length of the base times the height of the triangle.

Example 3

The height of a triangle is 6 m less than three times the length of its base. Find the length of the base and the height of the triangle if its area is 36 m².

Let b = length of base.

       $3b - 6$ = height

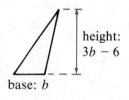

height:
$3b - 6$

base: b

$\frac{1}{2} \times$ base $\times$ height = area of triangle

$$\frac{1}{2} b(3b - 6) = 36$$

$$2 \cdot \frac{1}{2} \cdot b(3b - 6) = 36 \cdot 2 \qquad \blacktriangleleft \textit{Multiply each side by 2,}$$
$$b(3b - 6) = 72 \qquad \textit{the reciprocal of } \frac{1}{2}.$$
$$3b^2 - 6b - 72 = 0$$
$$b^2 - 2b - 24 = 0 \qquad \blacktriangleleft \textit{Divide each side by 3, the GCF.}$$
$$(b - 6)(b + 4) = 0$$
$$b = 6 \qquad \text{or} \qquad b = -4 \qquad \blacktriangleleft \textit{The length of the base cannot be } -4 \textit{ m.}$$

If $b = 6$, then $3b - 6 = 12$. Check to see that 6 m and 12 m work in the problem.

Thus, the length of the base is 6 m and the height is 12 m.

Some geometric problems involve addition or subtraction of measures. In the next example, one area is subtracted from another area.

Example 4   **The length of a rectangular floor is 4 m shorter than 3 times its width. The width of a rectangular carpet on the floor is 2 m shorter than the floor's width. The carpet is 2 m longer than twice its own width. Find the area of the floor if 44 m^2 of the floor are not covered by the carpet.**

Let x = width of floor.
    $3x - 4$ = length of floor

Sketch the two rectangles.

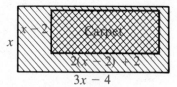

(Area of floor) − (Area of carpet) = (Area not covered)
$$x(3x - 4) - (x - 2)(2x - 2) = 44$$
$$3x^2 - 4x - (2x^2 - 6x + 4) = 44$$
$$x^2 + 2x - 48 = 0$$
$$(x - 6)(x + 8) = 0$$
$$x = 6 \quad \text{or} \quad x = -8 \quad \blacktriangleleft \quad \textit{The width cannot be } -8 \textit{ m.}$$

If $x = 6$, then $3x - 4 = 14$, $x - 2 = 4$, and $2(x - 2) + 2 = 10$.

(Area of floor) − (Area of carpet)	(Area not covered)
6 m × 14 m − 4 m × 10 m	44 m^2
84 m^2 − 40 m^2	
44 m^2	

Thus, the area of the floor is 6 m × 14 m, or 84 m^2.

Written Exercises

Solve each problem.

(A) **1.** One leg of a right triangle is 2 m longer than the other leg. The length of the hypotenuse is 10 m. Find the length of each leg.

2. One leg of a right triangle is 4 m long. The hypotenuse is 1 m shorter than twice the length of the other leg. Find the length of the other leg and the length of the hypotenuse.

3. One leg of a right triangle is 7 cm longer than the other leg. The hypotenuse is 1 cm longer than the longer leg. Find the lengths of the three sides.

4. In a right triangle, the hypotenuse is 8 dm longer than one leg and 4 dm longer than the other leg. Find the lengths of the three sides.

5. The length of a rectangle is 3 cm less than twice its width. Find the length and the width if the area is 20 cm^2.

6. The length of a rectangle is 5 m more than three times its width. Find the length and the width if the area is 42 m^2.

7. The height of a triangle is six times the length of its base. Find the length of the base and the height of the triangle if its area is 12 m^2.

8. The base of a triangle is 2 m shorter than its height and the area is 24 m^2. Find the height and the length of the base.

9. The base of a triangle is 4 dm longer than twice the height. If the area is 63 dm^2, find the length of the base.

10. The area of a triangle is 54 cm^2. Find the height of the triangle if the height is 6 cm less than 4 times the length of the base.

11. The width of one square is twice the width of another square. The sum of their areas is 125 m^2. Find the area of each square.

12. One square is 3 times as wide as another square. Find the area of each square if the difference of the two areas is 32 cm^2.

13. A square and a rectangle have the same width. The length of the rectangle is 4 times its width. If the sum of the two areas is 45 cm^2, find the area of each figure.

14. The length of a rectangle is 5 m more than twice its width. The width of a square is the same as the length of the rectangle. Find the area of each figure if one area, decreased by the other area, is 88 m^2.

(B) 15. A rectangle is 1 m narrower than it is long, and one of its diagonals is 2 m longer than the width of the rectangle. What is the length of the diagonal?

16. A diagonal of a rectangle is 5 cm longer than 4 times the width of the rectangle. Find the length of the rectangle if it is 1 cm less than the length of the diagonal.

17. The length of a rectangle is 4 cm more than its width and 4 cm less than the length of a diagonal of the rectangle. What is the area of the rectangle?

18. The length of a rectangle is 2 m more than twice its width and 1 m less than the length of a diagonal of the rectangle. Find the area of the rectangle.

19. A rectangle is twice as long as it is wide. The height of a triangle is twice the length of its base and the same as the length of the rectangle. Find the area of each figure if the sum of their areas is 108 m^2.

20. The base of a triangle is 2 cm shorter than the width of a rectangle. The rectangle's length is 4 times its width and the triangle's height is the same as the rectangle's length. Find the area of each figure if the difference of the two areas is 30 cm^2.

21. The length of a rectangular wall is 3 m more than twice its height. The width of a picture on the wall is 1 m less than the height of the wall. The picture is 2 m longer than it is wide. Find the area of the wall if 19 m^2 of the wall are not covered by the picture.

22. The length of a patio floor is 1 m less than twice its width. The floor is extended by building a second patio floor that is 3 m longer and 2 m narrower than the first floor. Find the area of the original floor if the total area of the patio is 48 m^2 after the extension is built.

© **23.** A rectangular lawn measures 26 m by 36 m and is surrounded by a uniform sidewalk. The outer edge of the sidewalk is a rectangle with an area of 1,200 m². Find the width of the sidewalk.

24. A rectangular picture measures 32 cm by 50 cm. It is surrounded by a uniform frame whose outer edge is a rectangle with an area of 2,320 cm². Find the width of the the frame.

25. A rectangular piece of cardboard is 5 cm longer than it is wide. A 3-centimeter square (3 cm by 3 cm) is cut out of each corner. Then the four flaps are turned up to form an open box with a volume of 450 cm³. Find the length and the width of the original piece of cardboard.

26. A 4-decimeter square is cut from each corner of a rectangular piece of plastic whose length is 2 dm less than twice its width. Then the four flaps are folded to form an open carton with a volume of 1,040 dm³. Find the length and the width of the original piece of plastic.

CALCULATOR ACTIVITIES

Find the area of each rectangle described below, in terms of the given unit of measure.

1. 296 in. by 225 in., in square feet

2. 17.5 ft by 38.75 ft, in square inches

3. 1.24 m by 0.35 m, in square centimeters

4. 4,225 cm by 2,416 cm, in square meters

5. 222 ft by 363 ft, in square yards

6. 497 yd by 268 yd, in square feet

The lengths of the three sides of a triangle are given. Determine whether each triangle is a right triangle by using the Pythagorean relation.

7. 98 cm, 336 cm, 350 cm

8. 0.75 yd, 0.4 yd, 0.85 yd

9. 110 in., 226 in., 286 in.

10. 0.65 m, 4.2 m, 4.25 m

NON-ROUTINE PROBLEMS

Each edge of a cube is 3 in. long. The cube is painted on each of its six faces, and then it is cut up to form 27 one-inch cubes.

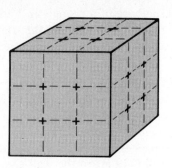

1. What is the least number of straight cuts needed to form the 27 one-inch cubes?

Find the number of one-inch cubes that are painted on the given number of faces.

2. three faces

3. exactly two faces

4. exactly one face

5. zero faces

APPLICATIONS

$$\boxed{\text{Read}} \longrightarrow \boxed{\text{Plan}} \longrightarrow \boxed{\text{Solve}} \longrightarrow \boxed{\text{Interpret}}$$

A certain investment plan requires regular payments of $30 a month with an expected earnings rate of 2% per month.

The expected value in the plan at the end of the 1st month is the payment plus the interest the payment earned for the month.

$$\underset{\downarrow}{30} \quad + \quad \overset{\downarrow}{\overbrace{30(0.02)}} = 30(1 + 0.02)$$

or **30(1.02)**

The expected value in the plan at the end of the 2nd month is the payment and the 1st month's value, plus the 2nd month's interest on both.

$$[\overset{\downarrow}{30} \quad + \quad \overbrace{30(1.02)}] \quad + \quad \overbrace{[30 + 30(1.02)]\overset{\downarrow}{(0.02)}}$$
$$= [30 + 30(1.02)](1 + 0.02)$$
$$\text{or } \mathbf{30(1.02) + 30(1.02)^2}$$

Now find the expected value in the plan at the end of the 3rd month.

Payment + Value already in the plan + 3rd month's earnings on both

$$[\overset{\downarrow}{30} \quad + \quad \overbrace{30(1.02) + 30(1.02)^2}^{\downarrow} \quad + \quad \overbrace{[30 + 30(1.02) + 30(1.02)^2]\overset{\downarrow}{(0.02)}}]$$
$$= [30 + 30(1.02) + 30(1.02)^2](1 + 0.02)$$
$$\text{or, } \mathbf{30(1.02) + 30(1.02)^2 + 30(1.02)^3}$$

Generalize the formula for r% per month. Let $x = 1 + 0.01r$.
$$30x + 30x^2 + 30x^3$$

If p dollars, rather than 30 dollars, is added monthly to the investment plan above, show the expected value at the end of n months.

$$px + px^2 + px^3 + \cdots + px^{n-1} + px^n$$

Solve each problem.
Tina's pension plan requires regular payments of $40 a month. The plan has an expected earnings rate of 3% per month.

1. What is the expected value after 6 months if she doubles the payments beginning with the fourth month?

2. What would the earnings be on Tina's plan after 6 months if she makes double payments from the first month but only receives half the monthly earnings rate?

CHAPTER THREE REVIEW

Vocabulary
area [3.8]
common binomial factor [3.4]
common monomial factor [3.4]
difference of two cubes [3.3]
difference of two squares [3.3]
perfect square trinomial [3.3]
Pythagorean relation [3.8]
quadratic trinomial [3.1]
standard form
  quadratic equation [3.5]
  quadratic inequality [3.6]
sum of two cubes [3.3]

Factor each polynomial.
1. $5c^2 - 37c + 14$ [3.1]
2. $9y^4 + 18y^2 + 8$
3. $25a^4b^4 + 5a^2b^2 - 12$
★ 4. $2x^{2n+2} - 11x^{n+1} + 5$
5. $9a^2 - 16$
6. $n^3 + 64$ [3.3]
7. $a^4 - 16a^2 + 64$
8. $100x^4 - 49y^2$
9. $8c^3 - 125d^3$
10. $25x^2 + 40xy + 16y^2$
11. $25x^2 - y^2 + 6y - 9$
★ 12. $x^{6m} + 10x^{3m}y^n + 25y^{2n}$

Multiply each pair of polynomials. [3.2]
13. $(4n + 6)(4n - 6)$
14. $(3a - 5)^2$
15. $(x + 3)(x^2 - 3x + 9)$
16. $(5x^2 + 4y)^2$
17. $(8y^2 - 6z)(8y^2 + 6z)$
18. $(2m - n)(4m^2 + 2mn + n^2)$

Factor each polynomial completely. [3.4]
19. $18x^3 - 15x^2 + 3x$
20. $-2a^2 + 3a + 14$
21. $12y^2 - 27$
22. $-y^2 + 10y - 25$
23. $2n^4 - 15n^2 - 27$
24. $4c^4 - 25c^2 + 36$
25. $2a^3b - 4a^2b - 30ab$
26. $3x^3 - 24y^3$
27. $x^3 + 4x^2 + 5x + 20$
28. $15ac - 6bc - 20ad + 8bd$
★ 29. $2x^{7n} - 18x^{3n}$

Solve each equation. [3.5]
30. $x^2 = x + 12$
31. $-6y^2 = 14y$
32. $25a^2 = 4$
33. $n^4 - 29n^2 + 100 = 0$
★ 34. $|x^2 - 5| = 4$

Find the solution set of each inequality.
Graph the solutions. [3.6]
35. $y^2 + 2y - 8 < 0$
36. $x^3 - 2x^2 - 24x > 0$
37. $2n^2 + n \geq 15$
38. $y^4 - 23y^2 - 50 > 0$
★ 39. $|x^2 - 20| > 16$

Solve each problem.
40. Find three consecutive odd integers so that the product of the second integer and the third integer is 63. [3.7]

41. Fifty-four cans are packed in some boxes so that the number of boxes is 3 more than the number of cans in each box. Find the number of boxes used. [3.7]

42. Find three consecutive multiples of 6 so that 36 less than the product of the second and third numbers is equal to the sum of the squares of the first and the second numbers. [3.7]

43. A second number is 1 less than 3 times a first number. If the second number is multiplied by 2 more than the first number, the product is 10. Find both numbers. [3.7]

44. One leg of a right triangle is 3 m longer than the other leg. The hypotenuse is 6 m longer than the shorter leg. Find the lengths of the three sides. [3.8]

45. The height of a triangle is 8 cm less than twice the length of its base. Find the length of the base and the height of the triangle if its area is 12 cm^2. [3.8]

46. The length of one rectangle is 6 m more than 2 times its width. The length and width of a second rectangle are twice the length and width, respectively, of the first rectangle. Find the area of each rectangle if the difference of the two areas is 324 m^2. [3.8]

CHAPTER THREE TEST

Factor each polynomial.
1. $3x^2 - 13x - 10$
2. $9a^2 + 30ab + 25b^2$
3. $4c^2 - 81$
4. $27n^3 - 125$
5. $9x^2 + 6xy + y^2 - 16$

Multiply each pair of polynomials.
6. $(7y + 8)(7y - 8)$
7. $(4a + 5b)^2$
8. $(2y - 3)(4y^2 + 6y + 9)$

Factor each polynomial completely.
9. $98 - 32x^2$
10. $2xy^2 - 2xy - 12x$
11. $-24x^2 - 6x + 45$
12. $4a^4 - a^2 - 18$

Solve each equation.
13. $a^2 = 11a$
14. $2x^2 = 12 - 5x$
15. $0 = a^4 - 37a^2 + 36$

Find the solution set of each inequality. Graph the solutions.
16. $x^2 + 3x - 18 \leq 0$
17. $x^3 - 64x \geq 0$
18. $a^4 - 29a^2 + 100 < 0$

Solve each problem.
19. One number is 6 less than another number. The product of the smaller number and 3 more than the larger number is 10. Find both numbers.

20. The length of the base of a triangle is 10 cm less than 4 times the height. Find the length of the base and the height of the triangle if its area is 25 cm^2.

21. Find three consecutive even integers such that the square of the second integer, increased by twice the product of the first two integers, is the same as 24 more than twice the square of the first integer.

22. The length of a rectangle is 2 m more than its width and 2 m shorter than the length of a diagonal. Find the length of the diagonal and the area of the rectangle.

23. A rectangular garden is 10 m longer than 4 times its width. The garden is to be enlarged by adding a second rectangular garden, 3 times as wide but half as long as the first garden. Find the area of the original garden if the total area of the enlarged garden will be 665 m^2.

★ 24. Factor $2x^{2n+2} - x^{n+1}y^{2n} - y^{4n}$.

★ 25. Solve $25x^2 = 40ax - 16a^2$ for x in terms of a.

★ 26. Solve $|x^2 - 20| = 16$.

Find the solution set. Graph the solutions.
★ 27. $x^2 < 25$ and $x^2 > 9$

★ 28. $|x^2 - 5| < 4$

Using String Variables

OBJECTIVE: To check factors of quadratic polynomials by using character strings

The letters A, B, C, etc., are *variables* used to represent numbers. Letters can be used to represent words or algebraic expressions by attaching a dollar sign to them: A$, B$, C$, etc. They are then called *string variables*. Consider the assignment statements: LET A$ = "DOG" and LET B$ = "2X + 3". "DOG" and "2X + 3" are called *character strings*. Character strings may be linked together by means of the + sign. PRINT A$ + B$ for A$ and B$ above would make the computer print DOG2X + 3.

Enter and run this program.

```
110 A =  INT ( RND (1) * 12) + 1
120 B =  INT ( RND (1) * 12) + 1
130 A$ = "(X+" +  STR$ (A) + ")(X+" +  STR$ (B) + ")"
140 B$ = "(X+" +  STR$ (B) + ")(X+" +  STR$ (A) + ")"
150  PRINT "FACTOR: ";
160  PRINT "X^2 + "A + B"X +"A * B
170  INPUT C$
180  IF C$ = A$ THEN 220
190  IF C$ = B$ THEN 220
200  PRINT "THOSE FACTORS ARE NOT CORRECT."
210  GOTO 150
220  PRINT "CORRECT"
230  GOTO 110
```

See the Computer Section beginning on page 574 for more information.

Lines 110 and 120 ask the computer to randomly pick a number from 1 to 12. The STR$ function in lines 130 and 140 converts the numbers represented by A and B into character strings, which are then incorporated into the larger strings assigned to A$ and B$.

Notice the way this program determines a polynomial's factored solutions in lines 110–140 before forming the polynomial itself in line 160.

EXERCISES

1. Make the changes necessary in the program above so that the factored solutions take the form (X − A) (X − B).

2. Write a program involving factoring out a common constant factor so that the results take the form A(X + B).

LAND EXPLORATION

Discovery. From ancient sailors mapping the globe to satellite photography enhanced by advanced computers, land exploration has been a challenge throughout history.

Janus display. This computer graphics display uses color to highlight topographical features for sophisticated mapping.

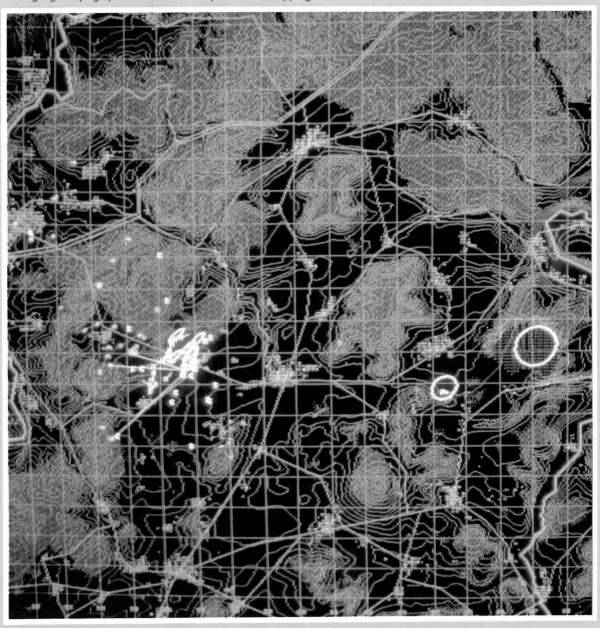

Assyrian astrolabe. Early mariners used this delicate disk to "sight" stars and observe their apparent movement across the sky.

Cross-staff being used by two people. Simpler and more sturdy than the astrolabe, a cross-staff could be aimed at the sun and stars to determine latitude.

Silhouette of First Officer with sextant. Combining the features and functions of such navigational tools as the astrolabe, cross-staff, and quadrant, sextants are used in marine navigation today.

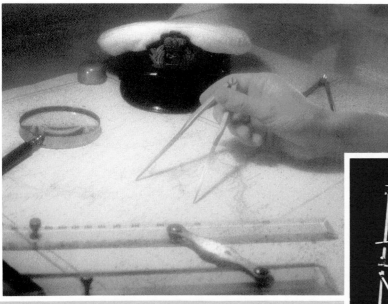

Navigator pictured at work. Once a ship's latitude and longitude have been determined, the navigator is responsible for charting the course.

Micronesian sailing chart of the Marshall Islands. Ribs of coconut palms tied together formed a map indicating winds and sea currents; shells indicated groups of islands.

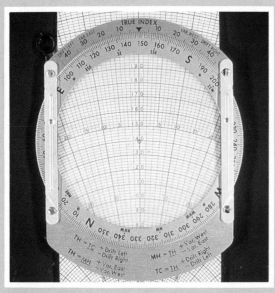

Charting-navigation tools. To chart a course accurately, navigational tools must account for the curvature of the earth on flat maps.

Keere's map of the world. Early
maps were based on information
derived from the ships' logs of the
time and were very inaccurate.

ORBIS TERRARUM TYPUS DE INTEGRO MULTIS IN LOCIS EMENDATUS auctore Petro Kerio anno 1607

Detail of Thomas Hood chart. The
carefully kept charts and logs of
ancient navigators enabled
cartographers to update their
maps with details of coastal
outlines and more accurate
scales of distance.

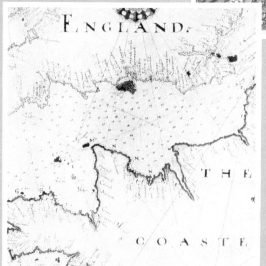

"The Cognoscenti"—painting of
English explorers. The great
explorers, anxious to find less
treacherous trade routes to Asia,
were responsible for discovering
untold riches in the New World.

Fr. Jacques Marquette and Louis Jolliet making a portage during their descent of the Mississippi River in 1673. Transporting boats over land was often necessary to chart the interior of the New World.

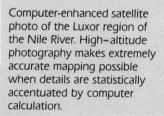

Computer-enhanced satellite photo of the Luxor region of the Nile River. High–altitude photography makes extremely accurate mapping possible when details are statistically accentuated by computer calculation.

Matt Henson, a member of Peary's expedition to the North Pole. In 1909, one of the earth's last frontiers was reached when Commander Robert Peary and his party reached the Pole.

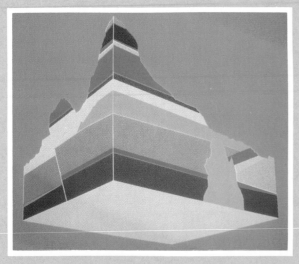

Geological cross-section. Stratigraphy is the subdivision of rock strata (layers) into mappable units to determine time relationships with similar sequences at other places.

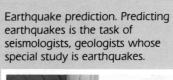

Earthquake prediction. Predicting earthquakes is the task of seismologists, geologists whose special study is earthquakes.

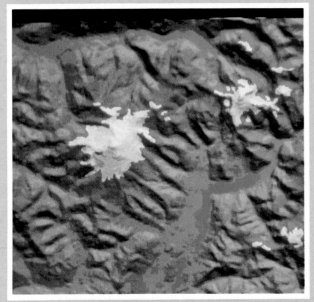

Computer graph. Data can be entered into a computer to reconstruct a topographical image in three dimensions.

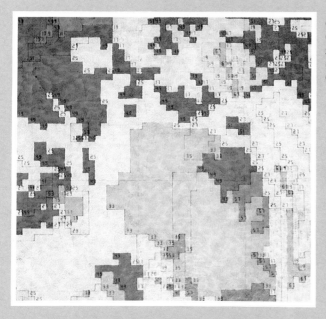

Computer-aided information. Compiling data relating to areas is simplified when digitized maps are used, as in this environmental data chart.

Computer graphics. Some computers can be programmed to rotate a reconstructed image so that it can be viewed from different angles.

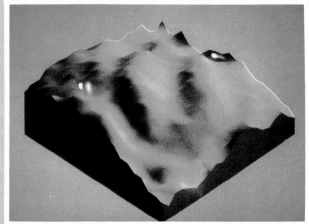

Lorrie Brow, geologist. A geologist studies not only the materials that make up our planet, but how their synthesis affects the earth's topology.

A cartographer works with the concept of distance. A chief value of maps drawn accurately to scale is that simple measurements can be converted to actual distances.

Stereoscopic plotter. A stereoscopic plotter generates a three-dimensional image on a flat surface.

COLLEGE PREP TEST

DIRECTIONS: Choose the *one* best answer to each question or problem.

1. Which value of x is not a solution of the inequality $(x + 40)(x - 20)(x + 60) \geq 0$?

   (A) -53 (B) -7 (C) 20
   (D) 36 (E) None of these

2. For which equation is the product of its solutions not a negative number?

   (A) $x^2 - 10x - 24 = 0$ (B) $x^2 - 25 = 0$
   (C) $x^2 + 7x - 18 = 0$ (D) $x^2 + 6x = 0$
   (E) None of these

3. Which figure has an area different from the other four figures?
   [1 meter (m) = 100 centimeters (cm)]

   (A) A rectangle, 1.2 m by 0.3 m
   (B) A rectangle, 2 m by 0.18 m
   (C) A rectangle, 9 cm by 4 cm
   (D) A triangle, height of 120 cm and base 60 cm long
   (E) A parallelogram, height of 60 cm and base 60 cm long

4. If x is an integer, for what value(s) of x will $x^2 + 12x + 36$ be a positive perfect square?

   (A) -6 (B) 0 (C) 6
   (D) Every integer except -6
   (E) Every integer except 0

5. If $y = \frac{5}{4}x + 20$, then $y = x$ only if

   (A) $x = 0$ (B) $y = 20$ (C) $x = 100$
   (D) $y = -80$ (E) None of these

6. Howard is n years old now. Ten years ago, he was t years old. Find his age y years from now.

   (A) $t + y + 10$ (B) $t + 10$
   (C) $t + y$ (D) $y + 10$
   (E) $n + t + y$

7. Represent the total value of x nickels, y dimes, and z quarters.

   (A) $x + y + z$ cents   (B) 40 cents
   (C) $0.05x + 0.10y + 0.25z$ dollars
   (D) $5x + 10y + 25z$ dollars
   (E) None of these

8. What is the GCF of $2(6x^2y - 9xy^3) + (15a^3x + 10ay^2)$?

   (A) 2 (B) $6xy$ (C) $10a$
   (D) $30axy$ (E) None of these

9. For which equation is the product of its solutions the greatest number?

   (A) $x^2 - 36 = 0$
   (B) $x^2 + 8x = 0$
   (C) $x^2 + 5x + 6 = 0$
   (D) $x^2 - 2x - 15 = 0$
   (E) $x^2 - 7x + 12 = 0$

10. A car traveled x miles in a hours, then y miles in b hours, and finally z miles in c hours. Find the average speed in miles per hour for the trip of $x + y + z$ miles.

    (A) $\frac{x}{a} + \frac{y}{b} + \frac{z}{c}$ (B) $\frac{x + y + z}{a + b + c}$
    (C) $\left(\frac{x}{a} + \frac{y}{b} + \frac{z}{c}\right) \div 3$
    (D) $(x + y + z)(a + b + c)$
    (E) None of these

4 RATIONAL EXPRESSIONS

Non-Routine Problem Solving

Problem

Is 2 equal to 1?

Study each step in the following "proof." Can you find the error?

1. Let $a = b$.

2. $a \cdot a = a \cdot b$ Multiply each side by a.

3. $a^2 = ab$ Simplify.

4. $a^2 - b^2 = ab - b^2$ Subtract b^2 from each side.

5. $(a + b)(a - b) = b(a - b)$ Factor each side.

6. $\dfrac{(a + b)(a - b)}{(a - b)} = \dfrac{b(a - b)}{(a - b)}$ Divide each side by $(a - b)$.

7. $a + b = b$ Divide out the common factor, $a - b$.

8. $b + b = b$ From Step 1, $a = b$. Substitute b for a.

9. $2b = 1b$ Simplify.

10. $\dfrac{2b}{b} = \dfrac{1b}{b}$ Divide each side by b.

11. Thus, $2 = 1$.   Divide out the common factor, b.

RATIONAL NUMBERS AND DECIMALS   4.1

Objective   **To show that a given number is a rational number**

Every integer, mixed number, and terminating decimal can be expressed in the form $\frac{a}{b}$ where a and b are integers. These numbers are examples of **rational numbers.**

Definition: Rational number	A *rational number* is one that can be written in the form $\frac{a}{b}$, where a and b are integers, $b \neq 0$.

Example 1   **Show that each of 18, $-7\frac{3}{4}$ and 27.89 is a rational number.**

Rewrite each number in the form $\frac{a}{b}$, where a and b are integers.

$$18 = \frac{18}{1} \qquad\qquad -7\frac{3}{4} = \frac{-31}{4} \qquad\qquad 27.89 = \frac{2{,}789}{100}$$

You can use division to express a rational number as a decimal.

$$\frac{5}{4} = 4\overline{\smash{\big)}5.00}^{\,1.25} \qquad\qquad \frac{2}{3} = 3\overline{\smash{\big)}2.000\ldots}^{\,0.666\ldots}$$

The decimal for $\frac{5}{4}$ is a **terminating decimal.** The decimal for $\frac{2}{3}$ is a **repeating decimal** and can be written either as $0.666\ldots$ or as $0.6\overline{6}$ with a bar above the repeating digit.

A repeating decimal may have more than one digit in its repeating block of digits. For example,

$$\frac{26}{11} = 11\overline{\smash{\big)}26.00\ 00\ 00\ \ldots}^{\,2.36\ 36\ 36\ \ldots} \, .$$

Notice that $2.36\overline{36}$ is a repeating decimal with a repeating block of two digits.

As shown on page 90, the procedure above can be reversed. That is, every repeating decimal can be written in the form $\frac{a}{b}$, where a and b are integers.

Summary   Every rational number has an equivalent decimal that either terminates or repeats.
Every terminating decimal is a rational number.
Every repeating decimal is a rational number.

Example 2   **Show that $2.3\overline{36}$ and $1.6\overline{2}$ are rational numbers.**

Each must be rewritten in the form $\frac{a}{b}$, where a and b are integers.

Let $n = 2.363636.\ldots$

$$100n = 236.363636\ldots$$

Subtract. $\underline{\quad n = \quad 2.363636\ldots}$

$$99n = 234.0$$

$$n = \frac{234}{99}, \text{ or } \frac{26}{11}$$

Let $n = 1.6222.\ldots$

$$10n = 16.2222\ldots$$

Subtract. $\underline{\quad n = \quad 1.6222\ldots}$

$$9n = 14.60$$

$$n = \frac{14.6}{9} \quad \blacktriangleleft \begin{array}{l} \textit{14.6 is not} \\ \textit{an integer.} \end{array}$$

$$n = \frac{146}{90}, \text{ or } \frac{73}{45}$$

Notice in Example 2 that when a two-digit block like $\overline{36}$ repeats, you multiply each side of the equation by 100. When a one-digit block like $\overline{2}$ repeats, you multiply by 10.

Written Exercises

Show that each of the following is a rational number.

Ⓐ **1.** -9

2. $-2\frac{3}{10}$

3. -34.719

4. 0

5. $0.8\overline{8}$

6. $0.25\overline{25}$

7. $0.78\overline{8}$

8. $0.70\overline{70}$

9. $0.43\overline{43}$

10. $0.624\overline{4}$

11. $0.463\overline{63}$

12. $0.517\overline{7}$

Ⓑ **13.** $4.4\overline{4}$

14. $68.68\overline{68}$

15. $5.05\overline{05}$

16. $22.2\overline{2}$

17. $5.38\overline{8}$

18. $8.741\overline{41}$

19. $3.816\overline{6}$

20. $47.239\overline{39}$

21. $0.571\overline{571}$

22. $0.743\overline{8438}$

23. $8.318\overline{318}$

24. $1.234\overline{72347}$

Ⓒ **25.** $0.999\ldots$

26. $0.4999\ldots$

27. $3.74999\ldots$

28. $-4.124999\ldots$

CALCULATOR ACTIVITIES

Three different ways of expressing the repeating block of digits in the decimal for $\frac{1810}{333}$ are given below.

$1810 \oplus 333 \ominus 5.4354354$ $5.43\overline{543}$ $5.435\overline{435}$ $5.4354\overline{354}$

Find all of the different ways to express the repeating block of digits in the decimal for each rational number.

1. $\frac{80}{33}$

2. $\frac{532}{99}$

3. $\frac{235}{111}$

4. $\frac{7700}{777}$

Express each rational number as a decimal and tell whether it terminates or repeats.

5. $\frac{19}{40}$ **6.** $\frac{37}{30}$ **7.** $\frac{5}{8}$ **8.** $\frac{4}{9}$ **9.** $\frac{119}{125}$ **10.** $\frac{22}{7}$ **11.** $\frac{234,500}{9,999}$

SIMPLIFYING PRODUCTS AND QUOTIENTS

Objectives   **To determine the values of the variables for which a rational expression is undefined**
To simplify a rational expression, a product of rational expressions, and a quotient of rational expressions

Each expression below is a **quotient of polynomials.**

$$\frac{2x + y}{3x - 2y} \qquad \frac{a^2 - 7a + 12}{3a - 9} \qquad \frac{0}{4n} \qquad \frac{x + 3}{1}, \text{ or } x + 3$$

Such expressions are called **rational expressions.** An expression like $\frac{\sqrt{x + 3}}{2y}$ is *not* a rational expression since $\sqrt{x + 3}$ is not a polynomial.

Definition: Rational expression	A *rational expression* is either a polynomial or a quotient of polynomials.

You can *evaluate* a rational expression, given values of the variables.
If $x = 4$ and $y = -5$, the value of $\frac{2x + y}{3x - 2y}$ is $\frac{2 \cdot 4 - 5}{3 \cdot 4 - 2 \cdot -5}$, or $\frac{3}{22}$.
If $x = 4$ and $y = 6$, the value of $\frac{2x + y}{3x - 2y}$ is **undefined** because $\frac{2 \cdot 4 + 6}{3 \cdot 4 - 2 \cdot 6}$, or $\frac{14}{0}$, is undefined. **Division by 0 is undefined.**

Example 1   **For what values of x is the rational expression $\dfrac{5}{x^2 - 7x + 12}$ undefined?**

The rational expression is undefined when $x^2 - 7x + 12 = 0$.
If $x^2 - 7x + 12 = 0$, then $(x - 3)(x - 4) = 0$, and $x = 3$ or $x = 4$.
Thus, $\dfrac{5}{x^2 - 7x + 12}$ is undefined if $x = 3$ or $x = 4$.

To simplify a fraction, you can factor
its terms and divide out common factors.
Rational expressions are simplified
in a similar way in Example 2.

$$\frac{30}{42} = \frac{\cancel{2} \cdot 5 \cdot \cancel{3}}{\cancel{2} \cdot \cancel{3} \cdot 7} = \frac{5}{7}$$

Example 2   **Simplify** $\dfrac{8x - 12}{12x - 18}$ **and** $\dfrac{3n^2 - 15n + 18}{5n^2 - 45}$.

$$\frac{8x - 12}{12x - 18} = \frac{4(2x - 3)}{6(2x - 3)} = \frac{4}{6} = \frac{2}{3}$$

$$\frac{3n^2 - 15n + 18}{5n^2 - 45} = \frac{3(n^2 - 5n + 6)}{5(n^2 - 9)} = \frac{3(n - 2)(n - 3)}{5(n + 3)(n - 3)} = \frac{3(n - 2)}{5(n + 3)}$$

If $\frac{a}{b}$ and $\frac{c}{d}$ are rational expressions, $b \neq 0$, $d \neq 0$, then $\frac{a}{b} \cdot \frac{c}{d} = \frac{a \cdot c}{b \cdot d}$.

To multiply either rational numbers or rational expressions, it is sometimes convenient to begin by factoring and dividing out common factors as shown.

$$\frac{2}{21} \cdot \frac{35}{11} = \frac{2 \cdot \cancel{7} \cdot 5}{\cancel{7} \cdot 3 \cdot 11} = \frac{10}{33} \qquad \frac{2}{5x + 20} \cdot \frac{3x + 12}{7} = \frac{2 \cdot 3\cancel{(x + 4)}}{5\cancel{(x + 4)} \cdot 7} = \frac{6}{35}$$

Example 3 **Simplify** $\dfrac{10c + 15}{10x - 50} \cdot \dfrac{x^2 - 25}{6c + 9}$ **and** $\dfrac{3x^8}{5x + 10} \cdot \dfrac{x^2 - 4}{6x^3}$.

$$\frac{10c + 15}{10x - 50} \cdot \frac{x^2 - 25}{6c + 9} = \frac{5(2c + 3) \cdot (x + 5)(x - 5)}{10(x - 5) \cdot 3(2c + 3)} = \frac{x + 5}{6}$$

$$\frac{3x^8}{5x + 10} \cdot \frac{x^2 - 4}{6x^3} = \frac{3 \cdot x^8(x + 2)(x - 2)}{6 \cdot x^3 \cdot 5(x + 2)} = \frac{x^5(x - 2)}{10}$$

Before identifying the common factors, be sure that each polynomial is factored *completely* as shown in Example 4.

Example 4 **Simplify** $\dfrac{2x^2 - 18}{24 - 6x} \cdot \dfrac{2x^2 - 9x + 4}{3x^2 + 24x + 45}$.

$$\frac{2(x^2 - 9)}{-6(x - 4)} \cdot \frac{(2x - 1)(x - 4)}{3(x^2 + 8x + 15)} = \frac{2(x + 3)(x - 3) \cdot (2x - 1)(x - 4)}{-6(x - 4) \cdot 3(x + 5)(x + 3)} = \frac{(x - 3)(2x - 1)}{-9(x + 5)}$$

The reciprocal of $\frac{2}{3}$ is $\frac{3}{2}$

and $\frac{8}{9} \div \frac{2}{3} = \frac{8}{9} \cdot \frac{3}{2} = \frac{4}{3}$.

To divide a rational number by a nonzero second rational number, you multiply the first number by the *reciprocal (multiplicative inverse)* of the second number.

Rational expressions are divided in a similar way.

If $\frac{a}{b}$ and $\frac{c}{d}$ are rational expressions, $b \neq 0$, $c \neq 0$, $d \neq 0$, then $\frac{a}{b} \div \frac{c}{d} = \frac{a}{b} \cdot \frac{d}{c}$.

Example 5 **Simplify** $\dfrac{4a + 8}{5a - 20} \div \dfrac{a^2 - 3a - 10}{a^2 - 4a}$.

The reciprocal of $\dfrac{a^2 - 3a - 10}{a^2 - 4a}$ is $\dfrac{a^2 - 4a}{a^2 - 3a - 10}$.

$$\frac{4a + 8}{5a - 20} \cdot \frac{a^2 - 4a}{a^2 - 3a - 10} = \frac{4(a + 2) \cdot a(a - 4)}{5(a - 4) \cdot (a - 5)(a + 2)} = \frac{4a}{5(a - 5)}$$

Reading in Algebra

Determine whether each expression is a rational expression. If it is not, then tell why not.

1. $\dfrac{3x^2 - 15x + 18}{4x - 8}$

2. $\dfrac{\sqrt{y - 3}}{2y - 6}$

3. $4a^2 - 25b^2$

4. $\dfrac{5n + 10}{10}$

5. $\dfrac{0}{x + 6}$

Evaluate each rational expression for $x = 4$ and $y = -3$, if possible.

6. $\dfrac{x + y}{3x + 4y}$

7. $\dfrac{x^2 + 2xy + y^2}{x + y}$

8. $\dfrac{y^2 + y}{x^2 - 16}$

9. $\dfrac{x^2 - y^2}{x^2 + y^2}$

10. $\dfrac{x^3}{y^3 - 9y}$

Written Exercises

For what value(s) of the variable is each rational expression undefined?

(A)

1. $\dfrac{5}{3x}$

2. $\dfrac{n - 3}{n + 4}$

3. $\dfrac{7}{3 - y}$

4. $\dfrac{x + 4}{x^2 - 25}$

5. $\dfrac{2c}{c^2 - 8c + 15}$

6. $\dfrac{a^2 - 2a - 24}{a^2 + 2a - 15}$

Simplify.

7. $\dfrac{10y - 20}{25y - 75}$

8. $\dfrac{3a - 15}{10 - 2a}$

9. $\dfrac{4 - c^2}{5c - 10}$

10. $\dfrac{3n^2 + 21n + 36}{2n^2 - 4n - 48}$

11. $\dfrac{6x^2 - 30x + 24}{10x^3 - 100x + 240}$

12. $\dfrac{5y^2 + 15y - 50}{50 - 2y^2}$

13. $\dfrac{-xy}{6a^6} \cdot \dfrac{4a^2}{-5xy^3z}$

14. $\dfrac{5bc^4}{3} \cdot \dfrac{3}{10ab^4c}$

15. $\dfrac{9c^3}{8x + 32} \cdot \dfrac{2x + 8}{-3c^4}$

16. $\dfrac{7x + 35}{15n} \div \dfrac{9x + 45}{10n^2}$

17. $\dfrac{-4a + 8}{-2x - 10} \div \dfrac{6 - 3a}{x^2 - 25}$

18. $\dfrac{5a + 25}{4a - 12} \div \dfrac{2a^2 - 2a - 60}{a^2 - 3a}$

19. $\dfrac{n^2 - 5n + 6}{3n + 12} \cdot \dfrac{4n + 16}{n^2 + 3n - 10}$

20. $\dfrac{9a + 27}{2a^2 + 8a + 8} \cdot \dfrac{4a^2 - 16}{6a + 18}$

21. $\dfrac{y^2 + 6y}{3y^2 + 6y - 24} \cdot \dfrac{10 - 5y}{-10y}$

(B)

22. $\dfrac{y^3 + 64}{y^2 - 4y + 16}$

23. $\dfrac{5x^2 + 15x + 45}{x^3 - 27}$

24. $\dfrac{15c^2d^{-6}(-5c^2 - 30c - 45)}{-10c^{-7}d^{-2}(3c^2 - 12c - 63)}$

25. $\dfrac{6a}{2n^2 + 3n - 9} \div \dfrac{8a^4}{4n^2 - 9}$

26. $\dfrac{3x^2 - 7x + 2}{4a^5b^2} \div \dfrac{2x^2 - 3x - 2}{8a^4b^2}$

27. $\dfrac{3c^2 - 5c - 2}{6c^2} \cdot \dfrac{4c^2 - 8c}{c^2 - 4c + 4}$

28. $\dfrac{2x^2 + 7x + 6}{2x^2 - 6x - 20} \cdot \dfrac{15 - 3x}{6x + 6}$

29. $\dfrac{x^2 - 5xy + 6y^2}{x + y} \div (4x - 12y)$

30. $\dfrac{c^2 + 2cd + d^2}{7a^2 - 7b^2} \div \dfrac{8c + 8d}{4b - 4a}$

31. $\dfrac{10x^3 - 4x^2 + 15x - 6}{10x^2 - 24x + 8} \cdot \dfrac{2x^3 - 4x^2 - 3x + 6}{4x^4 - 9}$

32. $\dfrac{y^3 - 8}{y^2 - 4y + 16} \cdot \dfrac{y^3 + 64}{3y - 6}$

(C)

33. $\dfrac{x^{4n+1} + x^{5n}}{x^{6n} - x^{5n+2}}$

34. $\dfrac{x^{4n} + 7x^{2n} + 12}{x^{3n} + 5x^{2n} + 4x^n + 20}$

35. $\dfrac{x^{6n} - y^{3n}}{x^{4n} - y^{2n}}$

36. $\dfrac{x^{2a} - y^{2b}}{4x^{2c} - 20x^c + 25} \cdot \dfrac{8x^c - 20}{3x^a + 3y^b}$

37. $\dfrac{x^{3a} + x^{2a+1}}{x^{b+c}} \div \dfrac{x^{3a} - x^{a+2}}{x^{a+2b} - x^{2b+1}}$

For what values of the variables is each rational expression undefined?

38. $\dfrac{5x}{x^2 - 2xy + y^2}$

39. $\dfrac{x + 1}{x^2 - y^2}$

40. $\dfrac{3x + 4}{x^3 + 4x^2y - 5xy^2}$

Simplifying Products and Quotients

DIMENSIONAL ANALYSIS

Objective

To convert a measurement in terms of one unit to an equivalent measurement in terms of a different unit using dimensional analysis

In designing a new type of tire, an engineer needed to convert a speed of 90 *miles per hour* to *yards per second*. To do this, the engineer used the following four facts:

$$1 \text{ h} = 60 \text{ min} \qquad 1 \text{ min} = 60 \text{ s} \qquad 1 \text{ mi} = 5{,}280 \text{ ft} \qquad 1 \text{ yd} = 3 \text{ ft}$$

Four **conversion factors** were set up:

$$\frac{1 \text{ h}}{60 \text{ min}} \qquad \frac{1 \text{ min}}{60 \text{ s}} \qquad \frac{5{,}280 \text{ ft}}{1 \text{ mi}} \qquad \frac{1 \text{ yd}}{3 \text{ ft}}$$

These conversion factors were used to convert 90 mi/h to yards per second (yd/s) as shown below.

$$\frac{90 \text{ mi}}{1 \text{ h}} \times \frac{1 \text{ h}}{60 \text{ min}} \times \frac{1 \text{ min}}{60 \text{ s}} \times \frac{5{,}280 \text{ ft}}{1 \text{ mi}} \times \frac{1 \text{ yd}}{3 \text{ ft}}$$

$$\frac{90 \times 5{,}280 \times 1 \text{ yd}}{60 \times 60 \text{ s} \times 3} = \frac{44 \text{ yd}}{1 \text{ s}}, \text{ or } 44 \text{ yd/s}$$

This procedure for converting from one measurement unit to another measurement unit is called **dimensional analysis.** The process is *similar* to the one you have used for *multiplying rational expressions*. You are cautioned to inspect each conversion factor to determine whether it *or* its reciprocal is needed. For example,

to convert 180 ft to inches, use $\dfrac{12 \text{ in.}}{1 \text{ ft}}$ as shown on the left below;

to convert 180 in. to feet, use $\dfrac{1 \text{ ft}}{12 \text{ in.}}$ as shown on the right below.

$$\frac{180 \text{ ft}}{1} \times \frac{12 \text{ in.}}{1 \text{ ft}} = 2{,}160 \text{ in.} \qquad \frac{180 \text{ in.}}{1} \times \frac{1 \text{ ft}}{12 \text{ in.}} = 15 \text{ ft}$$

When working with *area* or *volume*, you sometimes have to convert a measurement with *square* units or *cubic* units, respectively. To find the number of square inches in a square foot, or cubic inches in a cubic foot, you can proceed as follows.

$$\frac{12 \text{ in.}}{1 \text{ ft}} = 1 \qquad \left(\frac{12 \text{ in.}}{1 \text{ ft}}\right)^2 = \frac{144 \text{ in.}^2}{1 \text{ ft}^2} \qquad \left(\frac{12 \text{ in.}}{1 \text{ ft}}\right)^3 = \frac{1{,}728 \text{ in.}^3}{1 \text{ ft}^3}$$

Example 1

Convert 5 square yards (yd²) to square feet (ft²).

$$\frac{5 \text{ yd}^2}{1} \cdot \left(\frac{3 \text{ ft}}{1 \text{ yd}}\right)^2 = \frac{5 \text{ yd}^2}{1} \cdot \frac{9 \text{ ft}^2}{1 \text{ yd}^2} = 45 \text{ ft}^2$$

Written Exercises

Convert each measure as indicated. Use only one conversion factor in each exercise.

(A) 1. 24 pt to quarts
3. 15 yd to feet
5. 42 oz to pounds
7. 20 g to milligrams
9. 750 mL to liters
11. 24 ft^2 to square inches
13. 135 ft^3 to cubic yards

2. 12 gal to quarts
4. 144 in. to yards
6. 6 T to pounds
8. 400 g to kilograms
10. 5 m to centimeters
12. 36 ft^2 to square yards
14. 2.5 m^2 to square centimeters

Use dimensional analysis to convert each measure to the indicated unit.

(B) 15. 30 mi/h to feet per second

16. 15 T/h to pounds per minute

17. 10 ounces per pint to pounds per gallon

18. 8 ounces per square inch to pounds per square foot

Example

For 11 h a test car was driven at an average speed of 45 mi/h. The car averaged 22 mi/gal of gasohol. Determine the dollar cost of the fuel used if gasohol costs 129¢/gal.

$$\frac{11 \text{ h}}{1} \times \frac{45 \text{ mi}}{1 \text{ h}} \times \frac{1 \text{ gal}}{22 \text{ mi}} \times \frac{129¢}{1 \text{ gal}} \times \frac{\$1}{100¢}$$

$$= \frac{11 \times 45 \times 129 \times \$1}{22 \times 100}, \text{ or } \$29.03$$

Use dimensional analysis to solve each problem.

19. How many platforms can be built with 14 kegs of rivets if each keg contains 660 rivets, each platform contains 21 braces, and each brace requires 44 rivets?

20. A construction crew uses 135 ft^3 of sand per day during a 5-day work week. If sand weighs 1.215 tons per cubic yard (T/yd^3), how many pounds of sand does the crew use in one work week?

(C) 21. Each football player must have both ankles taped before each practice session and before each game. One roll of tape is needed for every 3 ankles that are taped. One carton of tape costs $12.75 and contains 6 rolls of tape. What is the cost for all the tape used by a 45-player football team during a season that consists of 60 practice sessions and 10 games?

22. A city recreation department enrolled 450 youths in its summer softball program and assigned 15 players to each team. Each team was furnished with 20 bats for the season. They were purchased at $37.80 per carton with a dozen bats in each carton. What was the cost of the bats if the department received a 5% discount for buying in quantity?

Dimensional Analysis

SUMS AND DIFFERENCES

Objective   **To add and subtract rational expressions**

Rational expressions with a *common denominator* are added in the same way that rational numbers with a common denominator are added.

$$\frac{3}{10} + \frac{1}{10} = \frac{3+1}{10} = \frac{4}{10} = \frac{2}{5} \qquad \frac{2y}{3x} + \frac{y}{3x} = \frac{2y+y}{3x} = \frac{3y}{3x} = \frac{y}{x}$$

If $\frac{a}{c}$ and $\frac{b}{c}$ are rational expressions, $c \neq 0$, then $\frac{a}{c} + \frac{b}{c} = \frac{a+b}{c}$.

Example 1   **Simplify** $\dfrac{n^2 - 6n}{n^2 + 8n + 16} + \dfrac{4n - 24}{n^2 + 8n + 16}$.

$$\frac{n^2 - 6n + 4n - 24}{n^2 + 8n + 16} = \frac{n^2 - 2n - 24}{n^2 + 8n + 16} = \frac{(n-6)(n+4)}{(n+4)(n+4)} = \frac{n-6}{n+4}$$

You can subtract rational expressions by using the definition of subtraction: $x - y = x + (-y)$. The **opposite** (*additive inverse*) of a rational number can be written in a convenient form using the property: $-\dfrac{a}{b} = \dfrac{-a}{b}$. For example,

$$-\frac{8}{2} = \frac{-8}{2} \quad \text{and} \quad -\frac{3x-2}{x^2-4} = \frac{-(3x-2)}{x^2-4} = \frac{-3x+2}{x^2-4}.$$

Example 2   **Simplify** $\dfrac{7x + 15}{6x + 12} - \dfrac{2x + 5}{6x + 12}$.

$$\frac{7x+15}{6x+12} + \frac{-(2x+5)}{6x+12} = \frac{7x+15-2x-5}{6x+12} = \frac{5x+10}{6x+12} = \frac{5(x+2)}{6(x+2)} = \frac{5}{6}$$

To add $\dfrac{7}{51}$ and $\dfrac{5}{34}$, where the denominators are different, you use the property: $\dfrac{x}{y} = \dfrac{x(c)}{y(c)}$, $y \neq 0$, $c \neq 0$.

$$\frac{7}{51} + \frac{5}{34} = \frac{7}{3 \cdot 17} + \frac{5}{2 \cdot 17} = \frac{7(2)}{3 \cdot 17(2)} + \frac{5(3)}{2 \cdot 17(3)} = \frac{14 + 15}{3 \cdot 17 \cdot 2} = \frac{29}{104}$$

To add or subtract rational expressions with different denominators, you can proceed through the following steps:

Step 1 Factor each denominator completely.
Step 2 Find the *least common denominator* (LCD) of the rational expressions. [The LCD of the rational expressions is the least common multiple (LCM) of their denominators.]
Step 3 Express each rational expression in terms of the LCD.
Step 4 Add or subtract the equivalent rational expressions with a common denominator.

For example, $\dfrac{4x + 3}{5x + 10} - \dfrac{x - 7}{3x + 6} = \dfrac{4x + 3}{5(x + 2)} + \dfrac{-x + 7}{3(x + 2)}$

The LCD is $3 \cdot 5(x + 2)$. ▶

$$= \dfrac{3(4x + 3)}{3 \cdot 5(x + 2)} + \dfrac{5(-x + 7)}{5 \cdot 3(x + 2)} = \dfrac{7x + 44}{15(x + 2)}$$

Example 3 **Simplify** $\dfrac{3}{4n - 12} - \dfrac{5 - 2n}{n^2 + 2n - 15}$.

$\dfrac{3}{4(n - 3)} + \dfrac{-5 + 2n}{(n - 3)(n + 5)} = \dfrac{3(n + 5)}{4(n - 3)(n + 5)} + \dfrac{4(-5 + 2n)}{4(n - 3)(n + 5)}$ ◀ *The LCD is $4(n - 3)(n + 5)$.*

$$= \dfrac{3n + 15 - 20 + 8n}{4(n - 3)(n + 5)} = \dfrac{11n - 5}{4(n - 3)(n + 5)}$$

Another convenient form for the opposite of a rational expression is given by the property: $-\dfrac{a}{b} = \dfrac{a}{-b}$. To simplify the *difference* $\dfrac{n + 5}{n - 7} - \dfrac{3}{7 - n}$, notice that the denominators $n - 7$ and $7 - n$ are a pair of opposites. That is, $-(7 - n) = n - 7$.

Thus, $\dfrac{n + 5}{n - 7} - \dfrac{3}{7 - n} = \dfrac{n + 5}{n - 7} + \dfrac{3}{-(7 - n)} = \dfrac{n + 5}{n - 7} + \dfrac{3}{n - 7} = \dfrac{n + 8}{n - 7}$.

You can use another property, $\dfrac{a}{b} = \dfrac{-a}{-b}$, to simplify the *sum* $\dfrac{6}{m - 5} + \dfrac{2}{5 - m}$ where the denominators are a pair of opposites.

$$\dfrac{6}{m - 5} + \dfrac{2}{5 - m} = \dfrac{6}{m - 5} + \dfrac{-2}{-(5 - m)} = \dfrac{6}{m - 5} + \dfrac{-2}{m - 5} = \dfrac{4}{m - 5}$$

Example 4 **Simplify** $\dfrac{5a}{a^2 - 9} - \dfrac{4}{a + 3} + \dfrac{2}{3 - a}$.

$\dfrac{5a}{(a + 3)(a - 3)} - \dfrac{4}{a + 3} + \dfrac{-2}{-(3 - a)} = \dfrac{5a}{(a + 3)(a - 3)} + \dfrac{-4}{a + 3} + \dfrac{-2}{a - 3}$

$$= \dfrac{5a}{(a + 3)(a - 3)} + \dfrac{-4}{a + 3} \cdot \dfrac{a - 3}{a - 3} + \dfrac{-2}{a - 3} \cdot \dfrac{a + 3}{a + 3}$$

$$= \dfrac{5a - 4a + 12 - 2a - 6}{(a + 3)(a - 3)}, \text{ or } \dfrac{-a + 6}{(a + 3)(a - 3)}$$

Oral Exercises

Determine whether each statement is true or false.

1. $\dfrac{48}{-3} = -\dfrac{48}{3}$

2. $-\dfrac{6}{9} = \dfrac{-6}{-9}$

3. $\dfrac{12}{-4} = \dfrac{-12}{4}$

4. $\dfrac{x-7}{-(x+3)} = \dfrac{-(x-7)}{x+3}$

5. $8 - x = x - 8$

6. $-(9 - x^2) = x^2 - 9$

7. $-\dfrac{2y}{-(3-y)} = \dfrac{2y}{y-3}$

8. $\dfrac{2n+3}{4-n^2} = \dfrac{-2n-3}{n^2-4}$

Written Exercises

(A) Simplify.

1. $\dfrac{2x+5}{5x+10} + \dfrac{x+1}{5x+10}$

2. $\dfrac{7a-10}{3a-12} - \dfrac{a+8}{3a-12}$

3. $\dfrac{6n+5}{2n+6} - \dfrac{2n-7}{2n+6}$

4. $\dfrac{2x^2-55}{x^2+7x} + \dfrac{6-x^2}{x^2+7x}$

5. $\dfrac{7a-15}{a^2-36} - \dfrac{2a+15}{a^2-36}$

6. $\dfrac{5c^2-8c}{c^2-9} - \dfrac{4c+9c^2}{c^2-9}$

7. $\dfrac{7x}{12} + \dfrac{3x}{8} - \dfrac{5x}{6}$

8. $\dfrac{3n+9}{10} - \dfrac{2n-1}{25} + \dfrac{n-8}{5}$

9. $\dfrac{a+2}{a} + \dfrac{a-6}{3a} + \dfrac{a-9}{5a}$

10. $\dfrac{2x}{3x+12} + \dfrac{5x}{6x+24}$

11. $\dfrac{3y-5}{2y-6} - \dfrac{4y-2}{5y-15}$

12. $\dfrac{5}{2n-4} + \dfrac{4}{3n+12}$

13. $\dfrac{2x}{x^2-4} + \dfrac{4}{x+2}$

14. $\dfrac{8}{n+5} - \dfrac{2n-15}{n^2-25}$

15. $\dfrac{3y-20}{y^2-y-20} + \dfrac{6}{y-5}$

16. $\dfrac{c-10}{c-4} - \dfrac{c+2}{4-c}$

17. $\dfrac{2a+11}{a+2} - \dfrac{2a-3}{-a-2}$

18. $\dfrac{3x}{x-5} + \dfrac{x}{5-x} - \dfrac{2x-3}{5-x}$

19. $\dfrac{4}{x+3} + \dfrac{x+2}{x^2-9} + \dfrac{3}{x-3}$

20. $\dfrac{3c-4}{c^2-25} - \dfrac{2}{c+5} + \dfrac{6}{5-c}$

21. $\dfrac{6}{x+2} - \dfrac{8x+5}{x^2-4} - \dfrac{5}{2-x}$

22. $2a + 1 + \dfrac{5a}{a-3}$

23. $3x - 4 - \dfrac{10}{x+4}$

24. $\dfrac{3y+7}{2y+5} - (5y-1)$

(B)

25. $\dfrac{y^2-8y}{y^2+10y+16} + \dfrac{3y-14}{y^2+10y+16}$

26. $\dfrac{3n^2+11n}{n^2+16n+64} - \dfrac{n^2+40}{n^2+16n+64}$

27. $\dfrac{4a^2-16}{a^2-8a+16} - \dfrac{a^2+8a}{a^2-8a+16}$

28. $\dfrac{3x+7}{2x^2+10x+12} + \dfrac{x+5}{2x^2+10x+12}$

29. $\dfrac{2y+3}{3y^2+6y-9} + \dfrac{3}{2y+6} - \dfrac{1}{3y-3}$

30. $\dfrac{7}{6x-30} - \dfrac{9x-6}{2x^2-18x+40} + \dfrac{11}{3x-12}$

31. $\dfrac{8m+2n}{m^2-2mn+n^2} + \dfrac{4m}{5am-5an}$

32. $\dfrac{2a-5b}{18a^2-8b^2} - \dfrac{3b}{6ac+4bc}$

33. $\dfrac{3x-5}{x^2-9} + \dfrac{7-x}{9-x^2} - \dfrac{1}{x+3}$

34. $\dfrac{5y+2}{2y^2-72} - \dfrac{3y-50}{72-2y^2}$

35. $\dfrac{3}{4y+12} + \dfrac{7}{6-2y} - \dfrac{5y}{9-y^2}$

36. $\dfrac{3x}{4-x^2} - \dfrac{4}{5x+10} + \dfrac{5}{6-3x}$

(C)

37. $\dfrac{y^2+y+1}{y+3} + \dfrac{y^2-3y+9}{y-1}$

38. $\dfrac{x^2-2x+4}{x-4} - \dfrac{x^2+4x+16}{x+2}$

39. $\dfrac{20-x^c}{x^{2c}+4x^c} + \dfrac{x^c+10}{x^c+4}$

40. $\dfrac{y^n}{y^n+6} - \dfrac{6-5y^n}{y^{2n}+6y^n}$

41. $\dfrac{1}{x^{3a}y^{5n}} + \dfrac{1}{x^{2a+4}y^{n+3}}$

42. $\dfrac{5y^b}{6x^c} - \dfrac{2}{9x^c y^b} + \dfrac{4x^c}{3y^{2b}}$

COMPLEX RATIONAL EXPRESSIONS   4.5

Objective   **To simplify a complex rational expression**

Each of the two expressions at the right contains at least one fraction or one rational expression in the numerator and the denominator. The first expression is a **complex fraction,** and the second is a **complex rational expression.**

$$\dfrac{\dfrac{3}{8} + \dfrac{7}{12}}{\dfrac{5}{6} - \dfrac{1}{4}} \qquad \dfrac{\dfrac{2}{x} + \dfrac{3}{xy}}{\dfrac{3}{x^2} + \dfrac{1}{5}}$$

You can simplify a complex fraction or a complex rational expression by

multiplication. For example, the LCD of the four fractions in $\dfrac{\dfrac{3}{8} + \dfrac{7}{12}}{\dfrac{5}{6} - \dfrac{1}{4}}$ is 24.

To simplify, multiply the numerator and denominator by 24.

$$\frac{24\left(\dfrac{3}{8} + \dfrac{7}{12}\right)}{24\left(\dfrac{5}{6} - \dfrac{1}{4}\right)} = \frac{\overset{3}{\cancel{24}} \cdot \dfrac{3}{\cancel{8}_1} + \overset{2}{\cancel{24}} \cdot \dfrac{7}{\cancel{12}_1}}{\overset{4}{\cancel{24}} \cdot \dfrac{5}{\cancel{6}_1} + \overset{6}{\cancel{24}} \cdot \dfrac{-1}{\cancel{4}_1}} = \frac{9 + 14}{20 - 6} = \frac{23}{14}$$

Example 1   **Simplify** $\dfrac{\dfrac{2}{x} + \dfrac{3}{xy}}{\dfrac{3}{x^2} + \dfrac{1}{5}}$.

Simplify $\dfrac{5a - \dfrac{1}{2a}}{\dfrac{1}{3b} + 4b}$.

Multiply the numerator and denominator by $5x^2y$, the LCD of the fractions.

Multiply the numerator and denominator by $6ab$, the LCD of the fractions.

$$\frac{5x^2y\left(\dfrac{2}{x} + \dfrac{3}{xy}\right)}{5x^2y\left(\dfrac{3}{x^2} + \dfrac{1}{5}\right)}$$

$$\frac{6ab\left(5a + \dfrac{-1}{2a}\right)}{6ab\left(\dfrac{1}{3b} + 4b\right)}$$

$$\frac{5x^2y \cdot \dfrac{2}{x} + 5 \cdot x^2y \cdot \dfrac{3}{xy}}{5x^2y \cdot \dfrac{3}{x^2} + 5 \cdot x^2y \cdot \dfrac{1}{5}}$$

$$\frac{6ab \cdot 5a + 6ab \cdot \dfrac{-1}{2a}}{6ab \cdot \dfrac{1}{3b} + 6ab \cdot 4b}$$

$$\frac{10xy + 15x}{15y + x^2y}$$

$$\frac{30a^2b - 3b}{2a + 24ab^2}$$

You should examine every answer to see if it can be simplified further by factoring the numerator or denominator (or both) and then dividing out any common factors.

Example 2   **Simplify** $\dfrac{1 - \dfrac{6}{x} + \dfrac{5}{x^2}}{1 - \dfrac{3}{x} - \dfrac{10}{x^2}}.$

$$\dfrac{x^2\left(1 + \dfrac{-6}{x} + \dfrac{5}{x^2}\right)}{x^2\left(1 + \dfrac{-3}{x} + \dfrac{-10}{x^2}\right)} = \dfrac{x^2 \cdot 1 + x^2 \cdot \dfrac{-6}{x} + x^2 \cdot \dfrac{5}{x^2}}{x^2 \cdot 1 + x^2 \cdot \dfrac{-3}{x} + x^2 \cdot \dfrac{-10}{x^2}}$$ ◀ *The LCD is x^2.*

$$= \dfrac{x^2 - 6x + 5}{x^2 - 3x - 10}$$ ◀ *Factor numerator and denominator.*

$$= \dfrac{(x - 1)(x - 5)}{(x + 2)(x - 5)}, \text{ or } \dfrac{x - 1}{x + 2}$$

For some complex rational expressions, you will need to factor the denominators of the individual fractions to find their LCD. This is shown in Example 3.

Example 3   **Simplify** $\dfrac{\dfrac{2x}{x^2 - 25} + \dfrac{5}{x - 5}}{\dfrac{4}{x + 5} + \dfrac{3x}{x^2 - 25}}.$

$$\dfrac{\dfrac{2x}{(x + 5)(x - 5)} + \dfrac{5}{x - 5}}{\dfrac{4}{x + 5} + \dfrac{3x}{(x + 5)(x - 5)}}$$

$$\dfrac{(x + 5)(x - 5) \cdot \dfrac{2x}{(x + 5)(x - 5)} + (x + 5)(x - 5) \cdot \dfrac{5}{x - 5}}{(x + 5)(x - 5) \cdot \dfrac{4}{x + 5} + (x + 5)(x - 5) \cdot \dfrac{3x}{(x + 5)(x - 5)}}$$ ◀ *The LCD is $(x + 5)(x - 5)$.*

$$\dfrac{2x + 5(x + 5)}{4(x - 5) + 3x}$$

$$\dfrac{2x + 5x + 25}{4x - 20 + 3x}$$

$$\dfrac{7x + 25}{7x - 20}$$

Written Exercises

Simplify each complex expression.

A

1. $\dfrac{\dfrac{3}{5} - \dfrac{1}{4}}{\dfrac{5}{2} - \dfrac{7}{10}}$

2. $\dfrac{5 + \dfrac{1}{4}}{2 + \dfrac{2}{3}}$

3. $\dfrac{3\dfrac{1}{2} + 4\dfrac{2}{3}}{5\dfrac{1}{6} + 2\dfrac{3}{4}}$

4. $\dfrac{\dfrac{4}{xy} - \dfrac{6}{y^2}}{\dfrac{8}{x^2} + \dfrac{3}{xy}}$

5. $\dfrac{\dfrac{4}{3a} + \dfrac{3}{2b}}{\dfrac{1}{6a} - \dfrac{3}{4b}}$

6. $\dfrac{\dfrac{5}{3x^2} + \dfrac{7}{2x}}{\dfrac{1}{6x} + \dfrac{10}{x^2}}$

7. $\dfrac{\dfrac{3}{5x} + 4}{2 - \dfrac{3}{10y}}$

8. $\dfrac{2a - \dfrac{1}{4c^2}}{\dfrac{5}{6c} - 3a}$

9. $\dfrac{7m - 3n + \dfrac{2}{mn}}{\dfrac{9}{2m} + \dfrac{4}{3n} - 5}$

10. $\dfrac{1 + \dfrac{8}{x} + \dfrac{12}{x^2}}{1 + \dfrac{6}{x} + \dfrac{8}{x^2}}$

11. $\dfrac{1 - \dfrac{2}{a} - \dfrac{8}{a^2}}{1 - \dfrac{7}{a} + \dfrac{12}{a^2}}$

12. $\dfrac{2 + \dfrac{5}{c} + \dfrac{3}{c^2}}{\dfrac{2}{c^2} + \dfrac{5}{c} + 3}$

13. $\dfrac{\dfrac{2y}{y^2 - 4} + \dfrac{3}{y - 2}}{\dfrac{5}{y + 2} + \dfrac{4}{y - 2}}$

14. $\dfrac{\dfrac{3a}{a^2 - 3a - 10} + \dfrac{2}{a + 2}}{\dfrac{2a}{a^2 - 3a - 10} + \dfrac{3}{a - 5}}$

15. $\dfrac{\dfrac{5n}{2n^2 - n - 1} - \dfrac{4}{n - 1}}{\dfrac{6}{2n + 1} + \dfrac{7n}{2n^2 - n - 1}}$

B

16. $\dfrac{\dfrac{4}{x - 2} + \dfrac{3}{x}}{\dfrac{5}{x} + \dfrac{3}{x - 2}}$

17. $\dfrac{\dfrac{6}{a + 3} - \dfrac{4}{a - 4}}{\dfrac{2}{a - 4} + \dfrac{5}{a + 3}}$

18. $\dfrac{1 - \dfrac{9}{x^2}}{\dfrac{1}{x} - \dfrac{3}{x^2}}$

19. $\dfrac{\dfrac{10a}{a^2 + 6a + 8}}{\dfrac{7}{a + 4} + \dfrac{3}{a + 2}}$

20. $\dfrac{\dfrac{12}{c + 5} - \dfrac{4}{c - 5}}{\dfrac{8c}{c^2 - 25}}$

21. $\dfrac{\dfrac{x - 7}{x^2 - 5x + 6}}{\dfrac{5}{x - 2} - \dfrac{4}{x - 3}}$

22. $\dfrac{\dfrac{x}{y} + 3 - \dfrac{4y}{x}}{\dfrac{x}{y} - \dfrac{y}{x}}$

23. $\dfrac{\dfrac{a}{b} - 1 - \dfrac{6b}{a}}{\dfrac{a}{b} + 4 + \dfrac{4b}{a}}$

24. $\dfrac{\dfrac{3}{y} + \dfrac{3}{x} - \dfrac{6}{xy}}{\dfrac{4}{x} + \dfrac{4}{y} - \dfrac{8}{xy}}$

25. $\dfrac{x^{-2} y^{-2}}{x^{-4} - y^{-4}}$

26. $\dfrac{x^{-2} - y^{-2}}{x^{-4} - y^{-4}}$

27. $\dfrac{x^{-2} - 1}{x - 1}$

NON-ROUTINE PROBLEMS

Simplify.

1. $5 + \dfrac{1}{4 + \dfrac{1}{3 + \dfrac{1}{2 + \dfrac{1}{1 + x}}}}$

2. $2 - \dfrac{1}{2 - \dfrac{1}{2 - \dfrac{1}{2 - x}}}$

Complex Rational Expressions

STATISTICS: AVERAGES AND VARIABILITY

Objectives

To find the mean, median and mode for a set of data
To find the mean for grouped data
To find the range, variance, and standard deviation for a set of data

One of the most frequently used statistical concepts is that of a **measure of central tendency.** Three measures of central tendency you will study are the mean, median, and mode.

Definition: Mean	The *mean* is the value obtained by adding the scores and dividing the sum by the number of scores.

Example 1

Find the mean, to the nearest hundredth, for the following scores:
88, 92, 48, 68, 71, 83, 88, 74, 64, 58, 59, 98, 96, 90, 88.

Let $\bar{x}$ represent the mean.

Add the scores.
$$\bar{x} = \frac{88 + 92 + 48 + 68 + 71 + 83 + 88 + 74 + 64 + 58 + 59 + 98 + 96 + 90 + 88}{15}$$

Divide by the number of scores.
$$\bar{x} = \frac{1,165}{15} = 77.67$$

Thus, the mean, $\bar{x}$, is 77.67.

Definition: Mode	The *mode* is the value that occurs most frequently and is obtained by listing the frequency of each score and choosing the one that occurs most frequently.

Example 2

Find the mode for the following scores:
50, 48, 51, 50, 49, 48, 46, 50, 49, 48, 44, 43, 48.

Group like scores.

51,   50 50 50   49 49   48 48 48 48   46 44 43
↑   3 times   twice   4 times   once
once

The most frequently occurring score is 48. It occurs 4 times.

Thus, the mode is 48.

Sometimes there is more than one mode for a set of data. This occurs when no one score occurs most frequently, but rather two or more scores have the same frequency.

Example 3

Find the median for the following scores:
300, 298, 301, 303, 297, 298, 300, 304, 296, 295, 296.

Arrange the scores in order.
▶ 295 296 296 297 298 298 300 300 301 303 304

Choose the middle score. ▶   5 below   5 above

Thus, the median is 298.

When many scores are given, it is often useful to make a frequency table to represent the data and to find the mean.

Example 4

Make a frequency table and determine the mean for the following scores:

95	90	95	75	90	80	95	90	95	85
90	80	75	85	70	85	90	100	90	70
85	85	80	70	75	95	100	85	100	75.

Make a frequency table.

Label the columns. ▶

Score (s)	Tally	Frequency (f)	Sum (f · s)
100					3	300
95	⊬	5	475
90	⊬		6	540
85	⊬		6	510
80					3	240
75						4	300
70					3	210
		30	2575

Count the tallies to get the frequency f.

The sum is the frequency times the score.

$$\text{Mean} = \frac{\text{sum } (f \cdot s)}{\text{frequency}} = \frac{2{,}575}{30} \doteq 85.833$$

Thus, the mean is 85.833.

Sometimes you may wish to determine whether scores cluster around the mean or are spread out from the mean. This measure is called the *variability* of the scores. The *range, variance,* and *standard deviation* are used to measure variability.

Definition: Range	The *range* is obtained by finding the difference between the highest and lowest scores.

Example 5

Find the range for the following scores:
12, 18, 16, 24, 10, 19, 22, 17, 9, 12, 16, 15.

Subtract the lowest score from the highest score.

$$24 - 9 = 15$$

Thus, the range is 15.

The indicated sum of a set of numbers can be written in a compact form using the capital Greek letter Σ, read "sigma."

Definitions: Variance Standard deviation	For a set of n measures $x_1, x_2, x_3, \ldots, x_n$ with a mean of $\bar{x}$, the *variance* s^2 of these measures is $$s^2 = \frac{(x_1 - \bar{x})^2 + (x_2 - \bar{x})^2 + (x_3 - \bar{x})^2 + \ldots + (x_n - \bar{x})^2}{n}.$$ The *standard deviation* of the measures is the square root of the variance, or s. Sometimes the lowercase Greek letter sigma σ is used to represent the standard deviation.

Example 6

Find the variance and standard deviation for the following scores:
8, 10, 12, 9, 8, 12, 11.

x_i	$\bar{x}$	$(x_i - \bar{x})$	$(x_i - \bar{x})^2$
8	10	−2	4
10	10	0	0
12	10	2	4
9	10	−1	1
8	10	−2	4
12	10	2	4
11	10	1	1
Total			18

$$s^2 = \frac{\Sigma(x_i - \bar{x})^2}{n} = \frac{18}{7} \doteq 2.571$$

$$\sigma = s = \sqrt{\frac{\Sigma(x_i - \bar{x})^2}{n}} = \sqrt{2.571} \doteq 1.60$$

Most scores cluster around the mean. Their spread is not large. If the scores had a larger spread, the standard deviation would have been larger. Approximately 68% of the scores will fall between ± one standard deviation of the mean, that is between $\bar{x} \pm 1\sigma$.

Written Exercises

Find the mean.

1. 93, 92, 88, 90, 86, 75, 68, 72, 75, 100, 86, 84, 78, 63, 48, 82

2. 14, 9, 12, 10, 9, 13, 14, 16, 19, 20, 12, 8, 9, 15, 16, 14, 19, 18, 16, 17

3. 245, 296, 205, 308, 310, 420, 495, 310, 187, 204

4. 896, 998, 1,005, 1,496, 400, 1,296, 983, 1,438

Find the mode(s).

5. 7, 8, 7, 6, 6, 6, 5, 7, 8, 7, 9

6. 48, 47, 47, 46, 48, 45, 43, 42

7. 98, 97, 98, 96, 100, 96, 95, 96

8. 101, 100, 99, 100, 99, 100, 100, 98

Find the median.

9. 100, 98, 86, 99, 100, 93, 84, 93, 79, 76, 93, 98, 100, 99, 98, 96

10. 12, 11, 13, 10, 10, 9, 8, 7, 9, 12, 9, 15, 16, 10, 9, 8, 7, 9, 8, 10, 14

11. 98, 97, 96, 98, 100, 101, 83, 96, 101

12. 53, 54, 56, 58, 93, 92, 73, 68, 93

Make a frequency table and determine the mean for the following scores.

13. 100, 101, 98, 97, 96, 99, 94, 93, 92, 88, 86, 88, 85, 92, 98, 97, 98, 88, 93, 96, 94, 96, 93, 97, 96, 100, 93, 99, 88, 86, 87, 92

14. 10, 12, 11, 10, 8, 12, 15, 16, 14, 12, 11, 8, 7, 4, 5, 9, 10, 18, 17, 16, 9, 8, 14, 16, 18, 12, 13, 14, 20, 19, 18, 16, 14, 6, 9, 17

Find the range.

15. 93, 98, 100, 98, 96, 94, 93, 92, 87

16. 10, 15, 16, 19, 21, 18, 17, 9, 28

17. 100, 103, 195, 201, 93, 200, 99

18. 48, 37, 36, 34, 33, 19, 16, 18, 19, 10

Find the variance and standard deviation for the following data.

19. 12, 11, 14, 16, 19, 8, 10, 14, 16, 10

20. 100, 98, 93, 92, 96, 93, 88, 76

21. 46, 43, 98, 38, 51, 87, 78

22. 42, 83, 82, 94, 48, 95

Solve each problem.

23. The mean for 10 scores is 94. Nine of the scores are known and are the following: 98, 99, 96, 94, 98, 92, 94, 90, 95. What is the missing score?

24. The mean for 12 scores is 48. Eleven of the scores are known and are the following: 52, 56, 54, 40, 42, 46, 48, 47, 44, 40, 46. What is the missing score?

25. The average of 20 students' test scores is 78. If the 2 lowest and the 2 highest scores are removed, the average of the remaining scores is 74. What is the average of the scores removed?

26. The average of 98 students' test scores is 92. If the 4 lowest and the 4 highest scores are removed, the average of the remaining scores is 94. What is the average of the scores removed?

APPLICATIONS: PROOFS WITH RATIONAL NUMBERS

To prove theorems involving rational numbers, you can use the properties of a field (page 26) and the definitions below. As you study each proof, you will be asked to supply reasons for some statements. Recall the following definitions:

1. $\dfrac{a}{b}$ is a **rational** number if and only if
a and b are integers and $b \neq 0$.

2. $\dfrac{a}{b} = a \cdot \dfrac{1}{b}$

Theorem 1 $\dfrac{a}{c} + \dfrac{b}{c} = \dfrac{a + b}{c}$

Proof

1. $\dfrac{a}{c} + \dfrac{b}{c} = a \cdot \dfrac{1}{c} + b \cdot \dfrac{1}{c}$	1. Definition 2
2. $\phantom{\dfrac{a}{c} + \dfrac{b}{c}} = (a + b)\dfrac{1}{c}$	2. Distributive property
3. $\phantom{\dfrac{a}{c} + \dfrac{b}{c}} = \dfrac{a + b}{c}$	3. Definition 2

The associative property allows you to regroup three factors. Sometimes, it is necessary to regroup four factors. This fact is established in Theorem 2.

Theorem 2 $(xy)(zw) = (xz)(yw)$

Proof

1. $(xy)(zw) = x[y(zw)]$	1. Associative for mult.
2. $ = x[(yz)w]$	2. $\underline{\ ?\ }$
3. $ = x[(zy)w]$	3. $\underline{\ ?\ }$
4. $ = x[z(yw)]$	4. $\underline{\ ?\ }$
5. $ = (xz)(yw)$	5. $\underline{\ ?\ }$

To prove $\dfrac{1}{a} \cdot \dfrac{1}{b} = \dfrac{1}{ab}$, use the property of reciprocals: If $x \cdot y = 1$, then $y = \dfrac{1}{x}$.

Theorem 3 $\dfrac{1}{a} \cdot \dfrac{1}{b} = \dfrac{1}{ab}$

Proof

1. $(ab)\left(\dfrac{1}{a} \cdot \dfrac{1}{b}\right) = \left(a \cdot \dfrac{1}{a}\right)\left(b \cdot \dfrac{1}{b}\right)$	1. Theorem $\underline{\ ?\ }$
2. $\phantom{(ab)\left(\dfrac{1}{a} \cdot \dfrac{1}{b}\right)} = 1 \cdot 1$	2. Prop. of reciprocals: $x \cdot \dfrac{1}{x} = 1$
3. $\phantom{(ab)\left(\dfrac{1}{a} \cdot \dfrac{1}{b}\right)} = 1$	3. Multiplicative identity
4. $(ab)\left(\dfrac{1}{a} \cdot \dfrac{1}{b}\right) = 1$	4. Statements 1–3
5. $\dfrac{1}{a} \cdot \dfrac{1}{b} = \dfrac{1}{ab}$	5. Prop. of reciprocals: If $x \cdot y = 1$, then $y = \dfrac{1}{x}$.

Theorem 4 $\dfrac{a}{b} \cdot \dfrac{c}{d} = \dfrac{ac}{bd}$

Proof

1. $\dfrac{a}{b} \cdot \dfrac{c}{d} = \left(a \cdot \dfrac{1}{b}\right)\left(c \cdot \dfrac{1}{d}\right)$	1. $\underline{\ ?\ }$
2. $\phantom{\dfrac{a}{b} \cdot \dfrac{c}{d}} = (ac)\left(\dfrac{1}{b} \cdot \dfrac{1}{d}\right)$	2. Theorem $\underline{\ ?\ }$
3. $\phantom{\dfrac{a}{b} \cdot \dfrac{c}{d}} = (ac)\left(\dfrac{1}{bd}\right)$	3. Theorem $\underline{\ ?\ }$
4. $\phantom{\dfrac{a}{b} \cdot \dfrac{c}{d}} = \dfrac{ac}{bd}$	4. $\underline{\ ?\ }$

Theorem 5 $\dfrac{a}{a} = 1$

Proof

1. $\dfrac{a}{a} = a \cdot \dfrac{1}{a}$	1. $\underline{\ ?\ }$
2. $\phantom{\dfrac{a}{a}} = 1$	2. $\underline{\ ?\ }$

Theorem 6 $\dfrac{a}{b} + \dfrac{c}{d} = \dfrac{ad + bc}{bd}$

Proof

1. $\dfrac{a}{b} + \dfrac{c}{d} = \dfrac{a}{b} \cdot 1 + \dfrac{c}{d} \cdot 1$	1. $\underline{\ ?\ }$
2. $\phantom{\dfrac{a}{b} + \dfrac{c}{d}} = \dfrac{a}{b} \cdot \dfrac{d}{d} + \dfrac{c}{d} \cdot \dfrac{b}{b}$	2. Theorem $\underline{\ ?\ }$
3. $\phantom{\dfrac{a}{b} + \dfrac{c}{d}} = \dfrac{ad}{bd} + \dfrac{cb}{db}$	3. Theorem $\underline{\ ?\ }$
4. $\phantom{\dfrac{a}{b} + \dfrac{c}{d}} = \dfrac{ad}{bd} + \dfrac{bc}{bd}$	4. $\underline{\ ?\ }$
5. $\phantom{\dfrac{a}{b} + \dfrac{c}{d}} = \dfrac{ad + bc}{bd}$	5. Theorem $\underline{\ ?\ }$

Theorem 7 The set of rational numbers is *closed* under addition.

Prove that $\dfrac{a}{b} + \dfrac{c}{d}$ is a rational number.

Proof

1. a, b, c, and d are integers.	1. Definition $\underline{\ ?\ }$
2. $\dfrac{a}{b} + \dfrac{c}{d} = \dfrac{ad + bc}{bd}$	2. Theorem $\underline{\ ?\ }$
3. ad, bc, and bd are integers.	3. Closure for $\underline{\ ?\ }$ of integers
4. $ad + bc$ is an integer.	4. Closure for $\underline{\ ?\ }$ of $\underline{\ ?\ }$
5. $\dfrac{ad + bc}{bd}$ is a rational number.	5. Definition $\underline{\ ?\ }$
6. $\dfrac{a}{b} + \dfrac{c}{d}$ is a rational number.	6. Statements 2 and 5

Prove that the set of rational numbers is closed under multiplication.

CHAPTER FOUR REVIEW

Vocabulary
complex rational expression [4.5]
conversion factor [4.3]
least common denominator (LCD) [4.4]
opposite of rational expression [4.4]
rational expression [4.2]
rational number [4.1]
reciprocal [4.2]
repeating decimal [4.1]
terminating decimal [4.1]

Evaluate each rational expression for $x = 3$ and $y = -2$, if possible. [4.2]

1. $\dfrac{x - y}{4x + 6y}$

2. $\dfrac{y^2 + 5y + 6}{x^2 + 5x + 6}$

For what values of the variable is each rational expression undefined? [4.2]

3. $\dfrac{a - 4}{5a + 10}$

4. $\dfrac{4c}{c^2 - 5c + 6}$

Simplify each rational expression. [4.2]

5. $\dfrac{9y + 18}{12y + 48}$

6. $\dfrac{x^3 + 27}{2x^2 - 4x - 30}$

7. $\dfrac{3n - 6}{8 - 4n}$

8. $\dfrac{2a^2 - 50}{30 + 9a - 3a^2}$

9. $\dfrac{4a - \dfrac{2}{3a^2}}{\dfrac{5}{6b} + 3b^2}$ [4.5]

10. $\dfrac{\dfrac{4n}{n^2 - 5n + 6} - \dfrac{2}{n - 3}}{\dfrac{6}{n - 2} + \dfrac{2n}{n^2 - 5n + 6}}$

★11. $\dfrac{1 + \dfrac{9}{x^a} + \dfrac{20}{x^{2a}}}{1 + \dfrac{7}{x^a} + \dfrac{10}{x^{2a}}}$

Simplify. [4.2]

12. $\dfrac{10a^2}{3b} \cdot \dfrac{b^4 c^2}{-15a^3}$

13. $\dfrac{x^2 - 1}{3x + 12} \cdot \dfrac{x^2 - 16}{x^2 - 5x + 4}$

14. $\dfrac{3c^2 + c - 2}{6 - 9c} \cdot \dfrac{9}{8 + 9c}$

★15. $\dfrac{x^{2a}}{y^{2a}} \cdot \dfrac{y^{5a}}{z^{8a}} \cdot \dfrac{z^{3a}}{x^a}$

Simplify. (Ex. 16–18) [4.2]

16. $\dfrac{-9x^7}{2y^5} \div \dfrac{12x^3}{y^3 z}$

17. $\dfrac{5c - 10}{3x + 3y} \div \dfrac{2c^2 - 8}{x^2 - 2xy + y^2}$

18. $\dfrac{n^2 - 2n - 8}{8n^3} \div \dfrac{n^2 - 6n + 8}{6n^2}$

19. Convert 40 ounces per square inch to pounds per square foot. (1 lb = 16 oz) [4.3]

Simplify. [4.4]

20. $\dfrac{7a + 5}{3a + 6} - \dfrac{2a - 5}{3a + 6}$

21. $\dfrac{2x + 5}{x^2 - 9} + \dfrac{3x + 10}{x^2 - 9}$

22. $\dfrac{2x + 7}{x^2 - x - 12} + \dfrac{2x + 5}{x^2 - x - 12}$

23. $\dfrac{3n + 2}{2n^2 + 3n - 5} - \dfrac{9 - 4n}{2n^2 + 3n - 5}$

24. $\dfrac{2y + 1}{5y - 10} - \dfrac{5 - y}{3y - 6}$

25. $3x + 2 + \dfrac{4x}{x + 3}$

26. $\dfrac{2y + 17}{3y - 12} - \dfrac{11 - 9y}{12 - 3y}$

27. $\dfrac{7}{x + 5} - \dfrac{x + 3}{25 - x^2} + \dfrac{1}{5 - x}$

28. $\dfrac{3n - 1}{2n^2 + 4n - 30} + \dfrac{2}{3n - 9} - \dfrac{5}{6n + 30}$

★29. $\dfrac{a^2 - a + 1}{a - 3} - \dfrac{a^2 + 3a + 9}{a + 1}$

Show that each of the following is a rational number. [4.1]

30. $-3\dfrac{4}{5}$

31. $0.58\overline{8}$

32. 18.63

33. $2.734\overline{34}$

1. Evaluate $\dfrac{4x + 3y}{x - 3y}$ for $x = 5$ and $y = -1$.

2. For what values of the variable is the rational expression $\dfrac{n + 3}{n^2 - 6n + 8}$ undefined?

Simplify each rational expression.

3. $\dfrac{4y - 12}{21 - 7y}$

4. $\dfrac{3x^2 - 75}{2x^2 - 9x - 5}$

Simplify.

5. $\dfrac{4ac}{b^2} \cdot \dfrac{3b^5}{16a^3c}$

6. $\dfrac{6n^2 - 24}{3n^2 - 5n - 2} \cdot \dfrac{9n + 3}{8n + 16}$

Simplify.

7. $\dfrac{n^2 - 4}{2x - 3y} \div \dfrac{5n + 10}{6x - 9y}$

8. $\dfrac{3a^2 - 5a - 12}{x^2 - 9y^2} \div \dfrac{4a^2 - 36}{5x + 15y}$

Simplify.

9. $\dfrac{5c}{12} + \dfrac{14c}{12} - \dfrac{c}{12}$

10. $\dfrac{4a + 21}{3a + 15} + \dfrac{2a - 4}{3a + 15} - \dfrac{a - 8}{3a + 15}$

11. $\dfrac{4y + 3}{y^2 - 3y - 10} - \dfrac{y - 3}{y^2 - 3y - 10}$

12. $2x - 3 + \dfrac{6x}{x + 2}$

13. $\dfrac{3}{4y - 12} - \dfrac{3}{15 - 5y}$

Simplify.

14. $\dfrac{2c + 1}{c^2 - 5c + 6} + \dfrac{3}{2c - 6} - \dfrac{1}{c - 2}$

15. $\dfrac{6}{x + 4} - \dfrac{3x}{x^2 - 16} + \dfrac{5}{4 - x}$

16. Convert 5 meters per second to kilometers per hour.

Simplify each complex rational expression.

17. $\dfrac{\dfrac{1}{5a^2} - \dfrac{2}{b}}{\dfrac{7}{10a} + \dfrac{3}{2b^2}}$

18. $\dfrac{\dfrac{2x}{x^2 - 9} + \dfrac{4}{x + 3}}{\dfrac{2}{x - 3} + \dfrac{4x}{x^2 - 9}}$

Show that each of the following is a rational number.

19. $-8\dfrac{2}{3}$

20. $2.47\overline{7}$

21. $2.69\overline{69}$

Simplify.

★ 22. $\dfrac{6x^{a+n}}{x^{2n} + 6x^n} \cdot \dfrac{x^{3n} - 36x^n}{x^{a+n} - 6x^a}$

★ 23. $\dfrac{c^2 - 2c + 4}{c - 5} + \dfrac{c^2 + 5c + 25}{c + 2}$

★ 24. $\dfrac{\dfrac{8y^n}{y^{2n} - 16} - \dfrac{6}{y^n - 4}}{\dfrac{10}{y^n + 4} - \dfrac{4y^n}{y^{2n} - 16}}$

COMPUTER ACTIVITIES

Decimal Forms of Fractions

OBJECTIVE: To determine whether a given fraction will yield a terminating or repeating decimal, using a loop

The program below uses long division to convert fractions into decimals. A terminating decimal results when the division yields a zero remainder, and a repeating decimal when division yields the same remainder a second time. The program first determines whether a fraction is terminating or repeating. The program will then print out the decimal at the user's request.

```
110   DIM R(40)
120    INPUT "ENTER NUMERATOR.";N
130    INPUT "ENTER DENOMINATOR.";D
140 A =   INT (N / D)
150    IF A$ <  > "YES" THEN 170          (Can be true only the 2nd time through)
160    PRINT A".";
170 R(1) = N - A * D
180    IF R(1) = 0 THEN 320               (Test for 0 for termination)
190    FOR I = 2 TO D
200 M = R(I - 1) * 10                     (Another test for termination)
210 A =   INT (M / D)                     (These lines compare remainders to see if
220    IF A$ <  > "YES" THEN 240          the same remainder will show up a second
230    PRINT A;                           time.)
240 R(I) = M - A * D
250    IF R(I) = 0 THEN 320
260    FOR J = 1 TO I - 1
270    IF R(I) = R(J) THEN 300
280    NEXT J
290    NEXT I
300    IF A$ = "YES" THEN  END
310    PRINT "THIS IS A REPEATING DECIMAL.": GOTO 340
320    IF A$ = "YES" THEN 370
330    PRINT "THIS IS A TERMINATING DECIMAL."
340    INPUT "DO YOU WANT TO SEE THE DECIMAL?";A$
350    IF A$ <  > "YES" THEN 370
360    GOTO 140
370    END
```

See the Computer Section beginning on page 574 for more information.

EXERCISES

Run the program above for the fractions $\frac{1}{2}$, $\frac{3}{10}$, $\frac{8}{2}$, $\frac{3}{8}$, $\frac{5}{16}$, $\frac{1}{3}$, $\frac{5}{6}$, $\frac{3}{7}$, and $\frac{1}{23}$.

COLLEGE PREP TEST

DIRECTIONS: Choose the *one* best answer to each question or problem.

1. Let $\frac{a}{b}$ represent a rational number where a and b are positive integers. Which statement is always true?

   (A) $3a > 2b$ (B) $b - a < a$

   (C) $\frac{a + b}{a - b} > \frac{a - b}{a + b}$ (D) $\frac{6a}{6b} = 1$

   (E) None of these

2. Let $\frac{a}{b}$ and $\frac{c}{d}$ represent any two *unequal* rational numbers, $b \neq 0$, $d \neq 0$. Which statement is always true?

   (A) $\frac{a}{b} < \frac{c}{d}$ (B) $a \cdot d = b \cdot c$

   (C) If $\frac{a}{b} < \frac{c}{d}$, then $\frac{a}{b} < \frac{a + c}{b + d} < \frac{c}{d}$.

   (D) If $\frac{a}{b} < \frac{c}{d}$, then $a \cdot b < c \cdot d$.

   (E) None of these

3. Let $\frac{a}{b}$ and $\frac{c}{d}$ represent any two nonzero rational expressions. Which statement is true?

   (A) $\frac{a}{b} \div \frac{c}{d} = \frac{d}{c} \div \frac{b}{a}$ (B) $\frac{a}{b} + \frac{c}{d} = \frac{a + c}{b + d}$

   (C) $\frac{a}{b} \cdot \frac{c}{d} = \frac{a \cdot d}{b \cdot c}$ (D) $\frac{a}{b} \div \frac{c}{d} = \frac{b}{a} \cdot \frac{c}{d}$

   (E) None of these

4. There are $x + 6$ males and $y - 3$ females in a choir. Find the ratio of the number of females to the total number of people in the choir.

   (A) $\frac{y - x - 9}{x + y + 3}$ (B) $\frac{y - 3}{x + y + 3}$

   (C) $\frac{y - 3}{x + 6}$ (D) $\frac{x + 6}{y - 3}$

   (E) $\frac{x + y + 3}{y - 3}$

5. If $\frac{x + y}{a - b} = \frac{2}{3}$, find the value of $\frac{9x + 9y}{10a - 10b}$.

   (A) $\frac{2}{3}$ (B) $\frac{3}{5}$ (C) $\frac{9}{10}$ (D) $\frac{20}{27}$

   (E) None of these

6. If $\frac{ax^2 + 2ax + a}{x^2 + 2x + 1} = 6$, find the value of a.

   (A) $x^2 + 2x + 1$ (B) $x + 1$ (C) 0

   (D) 6 (E) $\frac{1}{6}$

7. If $a = \frac{1}{3}$ and $b = \frac{1}{5}$, find the value of $\frac{a + b}{a - b}$.

   (A) $\frac{1}{4}$ (B) $\frac{2}{15}$ (C) $\frac{8}{15}$ (D) $\frac{16}{15}$ (E) 4

8. Find the arithmetic mean (average) of $\frac{x}{2}$, $\frac{x}{3}$, and $\frac{x}{6}$.

   (A) x (B) $\frac{x}{2}$ (C) $\frac{x}{3}$ (D) $\frac{x}{6}$ (E) $\frac{x}{18}$

CUMULATIVE REVIEW

DIRECTIONS: Choose the *one* best answer to each question or problem. (Exercises 1–11)

1. Solve $\frac{5}{9}x - 17 = 25 - \frac{2}{9}x$.

   (A) $10\frac{2}{7}$ (B) 24

   (C) 54   (D) 126

2. Solve $|7 - 4x| = 21$.

   (A) 7   (B) -7

   (C) $-7, 3\frac{1}{2}$ (D) $7, -3\frac{1}{2}$

3. Find two numbers such that the first number is 9 more than 4 times the second number, and 4 more than the second number is twice the first number.

   (A) $7, -\frac{1}{2}$ (B) $1, -2$

   (C) $-1, 2$ (D) $-7, \frac{1}{2}$

4. Solve $5^{3x-2} = 625$.

   (A) 2   (B) 4   (C) 125   (D) 209

5. Simplify $-3xy^3(-2x^2yz^3)^4$.

   (A) $6x^9y^7z^{12}$ (B) $16x^8y^4z^{12}$
   (C) $-48x^7y^7z^7$ (D) $-48x^9y^7z^{12}$

6. Some powdered milk worth \$3.50/kg and some cocoa worth \$4.25/kg form a 10-kg mixture that is worth \$39.20. How much cocoa is in the mixture?

   (A) 2.5 kg   (B) 4.4 kg
   (C) 5.6 kg   (D) 5.8 kg

7. Factor $-4c^3 + 20c^2 + 24c$ completely.

   (A) $-4c(c - 3)(c - 2)$
   (B) $-4c(c - 6)(c + 1)$
   (C) $-c(c - 6)(4c + 4)$
   (D) $4(6 - c)(c + 1)$

8. Solve $7x^2 - 8 = 8 - 2x^2$.

   (A) 0   (B) $\frac{4}{3}$

   (C) $-\frac{3}{4}, \frac{3}{4}$ (D) $-\frac{4}{3}, \frac{4}{3}$

9. Choose the graph of the solutions of $x^3 - 9x \le 0$.

10. If $x = 7$ and $y = -2$, find the value of $\frac{5x + 4y}{x^2 - y^2}$.

   (A) Does not exist

   (B) $\frac{43}{53}$ (C) $\frac{3}{5}$ (D) 0

11. Simplify: $\frac{5a - 10}{6a + 12} \cdot \frac{2a^2 - 8}{a^2 - 2a - 8}$.

   (A) $\frac{5a - 20}{12a + 24}$ (B) $\frac{12a + 24}{5a - 20}$

   (C) $\frac{5(a - 2)^2}{3(a + 2)(a - 4)}$ (D) $\frac{5}{12}$

Solve each equation.

12. $23 + \frac{6}{7}x = \frac{1}{7}x + 33$

13. $\frac{2}{3}(18a + 6) - (5a - 12) = 14$

14. $4.2(3n - 0.8) = 4.9n - 12.6$

15. $|3y - 8| = 7$

16. $2^{4x-3} = 32$

17. $2x^2 = 20 - 3x$

18. Use synthetic substitution to evaluate $5x^4 - 5x^2 + 10$ for $x = -3$.

Find the solution set of each inequality. Graph the solutions.

19. $4(7 - x) < 2x - 14$

20. $|y + 2| \leq 5$

21. $x^2 - 4x - 12 > 0$

22. $6x - 4 < 2x$ and $5 - 3x > -4$

Simplify each expression.

23. $-8a^3b^4c^{-2} \cdot 4a^{-5}bc^6$

24. $-10x^2y(5x^4y)^3$

25. $\dfrac{9m^{-8}n^7(5n + 25)}{10m^{-2}n^{-2}(3n^2 - 75)}$

26. $\dfrac{8.4 \times 10^5}{2.4 \times 10^7}$

Simplify.

27. $(7m^2 + 5n)(3m^2 - 5n)$

28. $6x(5x - 3)(4x + 5)$

29. $(9y^2 - 8)(9y^2 + 8)$

30. $(5c - 3d)^2$

31. $\dfrac{4a - 28}{a^2 - 36} \cdot \dfrac{a^2 + 2a - 24}{6a - 42}$

32. Convert 90 km/h to m/s.

Factor each polynomial.

33. $12x^2 + 8xy - 15y^2$

34. $3x^4 - 14x^2 - 5$

35. $36a^2 - 49b^2$

36. $8x^3 + 125$

37. $x^2 + 8xy + 16y^2 - 9$

Factor each polynomial completely.

38. $6y^2 - 45y - 24$

39. $50c^3 + 40c^2 + 8c$

40. $10x^4 - 160y^4$

41. $4a^4 - 29a^2 + 25$

42. $y^4 - 8y$

43. $x^3 + 2x^2 - 6x - 12$

Solve each problem.

44. The larger of two numbers is 4 less than 7 times the smaller number. Thirty more than the larger number is the same as 6 more than 9 times the smaller number. Find the numbers.

45. Side b of a triangle is 4 times as long as side a, and side c is 6 cm shorter than side b. Find the length of each side if the perimeter is 66 cm.

46. Find p in dollars if $A = \$4{,}832$, $r = 8\frac{1}{2}\%$, and $t = 6$ years. Use the formula $A = p + prt$.

47. How many liters of water must be added to 20 L of a 40% sugar solution to obtain a 25% solution?

48. At 8:00 A.M., a car left city A and headed for city B at 75 km/h. At 10:00 A.M., a car left city B at 90 km/h and headed for city A. At what time did the cars meet if the two cities are 645 km apart?

49. Find three consecutive even integers such that the product of the second and the third integers is 80.

50. The height of a triangle is 10 cm less than twice the length of its base. Find the length of the base and the height if the area of the triangle is 14 cm^2.

5 USING RATIONAL EXPRESSIONS

Non-Routine Problem Solving

Problem 1. You must make a choice between three successive discounts of 10%, 10%, and 5% and two successive discounts of 20% and 5% on an article priced at $1,000. How much money, if any, can you save by making the better choice?

Problem 2. If a waiter receives a 25% cut in wages, he will return to his original pay with a raise of what percent?

FRACTIONAL EQUATIONS

Objective   **To solve a fractional equation**

The equation $\frac{3}{4} + \frac{1}{6} = \frac{11}{12}$ is a true equation. If you multiply each side by 12, the LCD of all fractions in the equation, the resulting equation is also true, but it contains no fractions.

$$\frac{3}{4} + \frac{1}{6} = \frac{11}{12} : \textit{True}$$
$$12\left(\frac{3}{4} + \frac{1}{6}\right) = 12 \cdot \frac{11}{12}$$
$$12 \cdot \frac{3}{4} + 12 \cdot \frac{1}{6} = 12 \cdot \frac{11}{12}$$
$$9 + 2 = 11 : \textit{True}$$

As shown in Examples 1 and 2 below, this multiplication procedure can be used to simplify the work of solving equations that contain rational number coefficients.

Example 1   **Solve** $\dfrac{x-2}{6} + \dfrac{3x}{8} - \dfrac{3x+8}{24} = 1.$

Multiply each side by 24, the LCD. ▶

$$24\left(\frac{x-2}{6} + \frac{3x}{8} + \frac{-(3x+8)}{24}\right) = 24 \cdot 1$$
$$24 \cdot \frac{x-2}{6} + 24 \cdot \frac{3x}{8} + 24 \cdot \frac{-3x-8}{24} = 24 \cdot 1$$
$$4(x-2) + 3 \cdot 3x + 1(-3x-8) = 24$$
$$4x - 8 + 9x - 3x - 8 = 24$$
$$10x = 40$$
$$x = 4$$

Thus, the solution is 4.

Example 2   **Solve** $\dfrac{1}{3}y^2 + \dfrac{5}{12}y = \dfrac{1}{2}.$

Multiply each side by 12, the LCD. ▶

$$12 \cdot \frac{1}{3}y^2 + 12 \cdot \frac{5}{12}y = 12 \cdot \frac{1}{2}$$
$$4y^2 + 5y = 6 \quad ◀ \text{ Quadratic equation}$$
$$4y^2 + 5y - 6 = 0 \quad ◀ \text{ Standard form}$$
$$(4y - 3)(y + 2) = 0$$

$$4y - 3 = 0 \quad \text{or} \quad y + 2 = 0$$
$$y = \frac{3}{4} \qquad\qquad y = -2$$

Thus, $\frac{3}{4}$ and -2 are the solutions.

An equation such as $\dfrac{2n - 9}{n - 7} + \dfrac{n}{2} = \dfrac{5}{n - 7}$, in which a variable appears in a denominator, is called a **fractional equation.** You can use multiplication to help solve a fractional equation. However, multiplying by the LCD may produce an equation that is *not* equivalent to the original equation, as is shown in Example 3.

Example 3   **Solve** $\dfrac{2n - 9}{n - 7} + \dfrac{n}{2} = \dfrac{5}{n - 7}$.

Multiply each side
by $2(n - 7)$, the LCD. ▶
$$2(n - 7)\left[\dfrac{2n - 9}{n - 7} + \dfrac{n}{2}\right] = 2(n - 7) \cdot \dfrac{5}{n - 7}$$
$$2(2n - 9) + n(n - 7) = 2 \cdot 5$$
$$4n - 18 + n^2 - 7n = 10$$
$$n^2 - 3n - 28 = 0$$
$$(n - 7)(n + 4) = 0$$
$$n = 7 \qquad n = -4 \qquad ◀ \text{ } \textit{The ``possible''}$$

solutions are 7 and -4.

Check both numbers in the original equation.

$n = 7$ ▶

$\dfrac{2n - 9}{n - 7} + \dfrac{n}{2}$	$\dfrac{5}{n - 7}$
$\dfrac{14 - 9}{7 - 7} + \dfrac{7}{2}$	$\dfrac{5}{7 - 7}$
$\dfrac{5}{0} + \dfrac{7}{2}$	$\dfrac{5}{0}$

$n = -4$ ▶

$\dfrac{2n - 9}{n - 7} + \dfrac{n}{2}$	$\dfrac{5}{n - 7}$
$\dfrac{-8 - 9}{-4 - 7} + \dfrac{-4}{2}$	$\dfrac{5}{-4 - 7}$
$\dfrac{-17}{-11} - 2$	$\dfrac{5}{-11}$
$-\dfrac{5}{11}$	$-\dfrac{5}{11}$

7 is not a solution since $\dfrac{5}{0}$ is not defined.

Thus, -4 is the only solution of the original equation.

In Example 3, 7 is a solution of the *derived equation, $n^2 - 3n - 28 = 0$,* but it is not a solution of the original equation. In this case, 7 is called an **extraneous solution.**

Definition: Extraneous solution	An *extraneous solution* of an equation is a solution of a derived equation that is not a solution of the original equation.

Even though multiplying both sides of a fractional equation by a polynomial may produce extraneous solutions, all of the solutions of the original equation are retained. For this reason, all possible solutions must be checked in the original equation. You should notice that the extraneous solutions of a fractional equation are the numbers that make the value of at least one denominator equal to 0. In Example 3, 7 is an extraneous solution and 7 is also the number for which $n - 7$ has the value of 0.

Example 4   Solve $\dfrac{5}{2y + 6} - \dfrac{2y - 4}{y^2 - y - 12} = \dfrac{3}{y - 4}$.

$$\dfrac{5}{2(y + 3)} + \dfrac{-(2y - 4)}{(y + 3)(y - 4)} = \dfrac{3}{y - 4} \quad \blacktriangleleft \enspace \textit{The LCD is}$$
$$\textit{2(y + 3)(y - 4).}$$

$$2(y + 3)(y - 4) \cdot \dfrac{5}{2(y + 3)} + 2(y + 3)(y - 4) \cdot \dfrac{-2y + 4}{(y + 3)(y - 4)} = 2(y + 3)(y - 4) \cdot \dfrac{3}{y - 4}$$

$$5(y - 4) + 2(-2y + 4) = 3 \cdot 2(y + 3)$$
$$5y - 20 - 4y + 8 = 6(y + 3)$$
$$y - 12 = 6y + 18$$
$$-5y = 30$$
$$y = -6$$

Check in the original equation. If $y = -6$, none of the denominators has the value of 0.

Thus, -6 is the solution.

A fractional equation like $\dfrac{3}{x + 1} = \dfrac{2}{x - 4}$ is called a **proportion.** A precise definition of *proportion* follows.

Definition:
Proportion

A *proportion* is an equation of the form $\dfrac{a}{b} = \dfrac{c}{d}$, where a, b, c, and d are numbers, $b \neq 0$, $d \neq 0$.

For the proportion $\dfrac{a}{b} = \dfrac{c}{d}$, a and d are called the **extremes** and b and c are called the **means.** In the true proportion $\dfrac{2}{3} = \dfrac{10}{15}$, notice that *the product of the extremes equals the product of the means*: $2 \cdot 15 = 3 \cdot 10$. This suggests the Proportion Property below.

Proportion property

For all numbers a, b, c, and d, $b \neq 0$, $d \neq 0$, if $\dfrac{a}{b} = \dfrac{c}{d}$, then $a \cdot d = b \cdot c$.

Example 5   Solve $\dfrac{3}{x + 1} = \dfrac{2}{x - 4}$.

$$\dfrac{3}{x + 1} = \dfrac{2}{x - 4} \quad \blacktriangleleft \enspace \dfrac{a}{b} = \dfrac{c}{d}$$
$$3(x - 4) = (x + 1)2 \quad \blacktriangleleft \enspace a \cdot d = b \cdot c$$
$$3x - 12 = 2x + 2$$
$$x = 14$$

Thus, the solution is 14.

Check:

$\dfrac{3}{x + 1}$	$\dfrac{2}{x - 4}$
$\dfrac{3}{14 + 1}$	$\dfrac{2}{14 - 4}$
$\dfrac{1}{5}$	$\dfrac{1}{5}$

Reading in Algebra

Supply the missing words and expressions in the paragraph below.

$\frac{2x}{15} = \frac{6}{5x}$ is a special type of fractional equation. It is called a _____ .

The extremes are $2x$ and _____ , and the _____ are 15 and _____ . A

derived equation is $10x^2 =$ _____ , or $x^2 =$ _____ . So, the solutions are

_____ and _____ . Neither solution is an _____ solution, since the

value of $5x$ is not equal to _____ if $x = 3$ or if $x = -3$.

Written Exercises

Solve each equation.

(A) 1. $\frac{x}{10} + \frac{x}{6} + \frac{x}{15} = 1$

2. $\frac{2n - 3}{2} = \frac{3}{4} + \frac{n - 4}{8}$

3. $\frac{a - 1}{3} + \frac{a + 2}{6} = 2$

4. $\frac{2}{5} + \frac{2}{y} = 1$

5. $\frac{6}{x} + \frac{9}{2x} = 3$

6. $\frac{2}{3n^2} = \frac{1}{4n^2} + \frac{5}{6n}$

7. $\frac{1}{2}x^2 + x = 12$

8. $\frac{1}{9}y^2 + \frac{4}{3}y + 3 = 0$

9. $\frac{1}{10}x^2 - \frac{1}{2}x = 5$

10. $\frac{3n - 7}{n - 5} + \frac{n}{2} = \frac{8}{n - 5}$

11. $\frac{a - 4}{a + 3} = \frac{3a + 2}{a + 3} + \frac{a}{4}$

12. $\frac{4c - 3}{c - 4} - \frac{2c}{3} = \frac{2c + 5}{c - 4}$

13. $\frac{2}{y - 3} + \frac{2}{y} = 1$

14. $\frac{3}{x} + \frac{2}{x + 2} = 2$

15. $\frac{10}{n + 4} - \frac{1}{n} = 1$

16. $\frac{5}{x + 4} = \frac{3}{x - 2}$

17. $\frac{4a + 3}{3} = \frac{2a + 5}{4}$

18. $\frac{3}{2y + 1} = \frac{2}{3y - 2}$

19. $\frac{a + 1}{8} = \frac{2}{a + 1}$

20. $\frac{x - 3}{2} = \frac{1}{x - 4}$

21. $\frac{2z - 3}{2z + 3} = \frac{z - 2}{2z - 3}$

22. $\frac{5}{x - 3} - \frac{6}{x^2 - 9} = \frac{4}{x + 3}$

23. $\frac{5}{n + 2} = \frac{3}{n - 2} - \frac{2n}{n^2 - 4}$

24. $\frac{7y - 9}{y^2 - 25} - \frac{3}{y - 5} = \frac{2}{y + 5}$

(B) 25. $\frac{3}{4}y^2 + y = 1$

26. $\frac{1}{7}x^2 = 1 - \frac{9}{28}x$

27. $\frac{7}{4} = \frac{13}{6}y - \frac{2}{3}y^2$

28. $\frac{3}{x + 1} + \frac{x - 2}{3} = \frac{13}{3x + 3}$

29. $\frac{3}{10}a^2 = \frac{11}{20}a + \frac{1}{2}$

30. $\frac{n}{n - 3} + \frac{n - 2}{6} = \frac{5n - 1}{4n - 12}$

31. $\frac{4y + 9}{y + 1} = \frac{2y + 7}{y - 1}$

32. $\frac{2x + 2}{3x + 1} = \frac{x - 2}{x - 1}$

33. $\frac{a^2 + 5a - 2}{a + 3} = \frac{4a - 1}{3}$

34. $\frac{3}{n - 2} + \frac{6}{n^2 - 5n + 6} = \frac{4}{n - 3}$

35. $\frac{1}{x - 2} = \frac{2x + 1}{x^2 + 2x - 8} + \frac{2}{x + 4}$

36. $\frac{x}{2x - 6} - \frac{3}{x^2 - 6x + 9} = \frac{x - 2}{3x - 9}$

37. $\frac{1}{2a} - \frac{9}{a^2 + 6a} = \frac{2 - a}{2a + 12}$

(C) 38. $\frac{4}{x^2 + 2x - 15} + \frac{5}{x^2 - x - 6} = \frac{3}{x^2 + 7x + 10}$

39. $\frac{6}{y^2 - 9} + \frac{4}{6 + y - y^2} + \frac{2}{y^2 + 5y + 6} = 0$

NON-ROUTINE PROBLEMS

Solve $\frac{10}{2x^3 - x^2 - 8x + 4} + \frac{24}{2x^3 - x^2 - 2x + 1} = \frac{40}{x^4 - 5x^2 + 4}$.

PROBLEM SOLVING: WORK PROBLEMS 5.2

Objective **To solve a work problem that leads to a fractional equation**

A certain house can be painted in 6 days. The chart below indicates the part of the house that can be painted in a given number of days.

Number of days	1	2		5	x	6	
Part of job done	$\frac{1}{6}$	$\frac{2}{6}$,	or $\frac{1}{3}$	$\frac{5}{6}$	$\frac{x}{6}$	$\frac{6}{6}$,	or 1

In general, if a job can be done in 6 days, then $\frac{x}{6}$ represents the part of the job completed in x days and $\frac{6}{6}$, or 1, represents the *total job*.

Example 1 **Palmer's crew can do the cement work for a new building in 6 days. Hoffman's crew would need 8 days to complete the same job. How many days will the job take if the two crews work together?**

Let x = the number of days it will take if the crews work together.
Represent the data in a table.

Represent the data. ▶

	Palmer	Hoffman
Part done in 1 day	$\frac{1}{6}$	$\frac{1}{8}$
Part done in x days	$\frac{x}{6}$	$\frac{x}{8}$

Job takes x days. ▶

After x days ▶ (Palmer's part) + (Hoffman's part) = Total job

Write an equation. ▶
$$\frac{x}{6} + \frac{x}{8} = 1$$

The LCD is 24. ▶
$$24 \cdot \frac{x}{6} + 24 \cdot \frac{x}{8} = 24 \cdot 1$$

Solve the equation. ▶
$$4x + 3x = 24$$
$$7x = 24$$
$$x = \frac{24}{7}, \text{ or } 3\frac{3}{7}$$

Check in the problem. ▶ To check, add the parts done in $\frac{24}{7}$ days.
$$\frac{x}{6} + \frac{x}{8} = \frac{1}{6} \cdot \frac{24}{7} + \frac{1}{8} \cdot \frac{24}{7} = \frac{4}{7} + \frac{3}{7} = \frac{7}{7}, \text{ or } 1$$

Thus, the job will take $3\frac{3}{7}$ days if the two crews work together.

Answer the question. ▶

If a job requires x hours to complete, then $\frac{1}{x}$ represents the part of the job that is done in 1 hour and $\frac{n}{x}$ represents the part of the job done in n hours.

Example 2   Al, Betty, and Carl can harvest a strawberry crop in 12 hours (h) if they work together. If each person worked alone, Al could complete the job in 30 h, and Carl would take twice as long as Betty. How long would it take Carl to do the job alone?

Let x = Betty's time to do the job alone.
      $2x$ = Carl's time to do the job alone

	Al	Betty	Carl
Part done in 1 h	$\dfrac{1}{30}$	$\dfrac{1}{x}$	$\dfrac{1}{2x}$
Part done in 12 h	$\dfrac{12}{30}$	$\dfrac{12}{x}$	$\dfrac{12}{2x}$

Job takes 12 h if they work together. ▶

After 12 h ▶

$$\begin{matrix} \text{Al's} \\ \text{part} \end{matrix} + \begin{matrix} \text{Betty's} \\ \text{part} \end{matrix} + \begin{matrix} \text{Carl's} \\ \text{part} \end{matrix} = \begin{matrix} \text{Total} \\ \text{job} \end{matrix}$$

$$\frac{12}{30} + \frac{12}{x} + \frac{12}{2x} = 1$$

The LCD is 30x. ▶

$$30x \cdot \frac{12}{30} + 30x \cdot \frac{12}{x} + 30x \cdot \frac{12}{2x} = 30x \cdot 1$$
$$12x + 360 + 180 = 30x$$
$$540 = 18x$$
$$30 = x \qquad 2x = 60$$

To check, add the parts done in 12 h.
$$\frac{12}{30} + \frac{12}{x} + \frac{12}{2x} = \frac{12}{30} + \frac{12}{30} + \frac{12}{60} = \frac{2}{5} + \frac{2}{5} + \frac{1}{5} = \frac{5}{5}, \text{ or } 1$$

Thus, it would take Carl 60 h to do the job alone.

Example 3   Machine A can do a job in 12 h and machine B can do the job in 8 h. If B starts 2 h after A has started, find the total time needed for the two machines to do the complete job.

Let x = the time that A and B work together.
(Part done by A in 2 hours) + (Part done by A and B together in x hours) = Total job
$$\frac{2}{12} + \left(\frac{x}{12} + \frac{x}{8} \right) = 1$$
$$4 + 2x + 3x = 24$$
$$x = 4 \qquad 2\text{ h} + 4\text{ h} = 6\text{ h}$$

Thus, the two machines need 6 h to do the complete job.

Oral Exercises

A farmer takes 15 days to harvest a crop. Represent the part harvested in the given period of time.

1. 1 day   **2.** 4 days   **3.** x days   **4.** 15 days

A carpenter can build a cabinet in x hours. Represent the part built in the given period of time.

5. 1 hour   **6.** 3 hours   **7.** y hours   **8.** x hours

Written Exercises

Solve each problem.

(A) **1.** Mr. Adams can plant a wheat crop in 10 days and his daughter can do it in 15 days. How many days will it take if they work together?

2. Mrs. Brown can paint 3 average-sized rooms in 8 hr. Her son would need 12 h to do it. If they work together, how long will it take to paint 6 average-sized rooms?

3. Paul can put carpet on a floor in 10 h. If Irene helps him, the job is done in 6 h. How long would it take Irene if she worked alone?

4. Diane can clean the attic in 6 h. If Sam helps her, the attic can be cleaned in 4 h. How long would it take Sam working alone?

5. Machine A can do a job in 15 h. Machines B and C can do the same job in 12 h and 20 h, respectively. How many hours will the job take if the three machines operate at the same time?

6. Work crews 1, 2, 3, and 4 can load a freight train in 16 h, 10 h, 12 h, and 15 h, respectively. When will the freight train be loaded if the four crews start at 8:00 A.M. and work together?

7. Kim can complete a job in 6 weeks (wk) and the same job would take Kevin 10 wk. How long would it take Derek working alone if, working together, all three can complete the job in 2 wk?

8. Linda can keypunch 1,000 cards in 150 minutes (min). The same job would take John 3 h. How long would it take Martha to keypunch 1,000 cards if, working together, all three can keypunch 2,000 cards in 2 h?

(B) **9.** Work crew A takes 15 h to do a job and crew B can do it in 10 h. If B starts 3 h after A has begun, what is the total time needed for the two crews to do the job?

10. Ned can mow a lawn in 75 min and Pedro can do it in 50 min. If Pedro watches Ned mow for 20 min and then helps to finish the job, find the total time for the job.

11. Gene, Marie, and Merv can complete a job in 20 h if they work together. Gene can do the job alone in twice the time it would take Marie and half the time that Merv would need. Find the time each person would take to do the job if each worked alone.

12. Work crews A, B, and C can pave a road in 10 days if they work together. If they work alone, crew A would take 60 days to do the job and crew B would take 3 times as long as crew C to do the job. How long would it take crews B and C if each worked alone?

PROBLEM SOLVING: ROUND TRIPS 5.3

Objective   **To solve a motion problem that leads to a fractional equation**

An automobile that is driven a distance of 240 km at an average rate of 80 km/h will take $\frac{240 \text{ km}}{80 \text{ km/h}}$, or 3 h, for the trip. You can use the formula below to solve motion problems involving *time*.

$$\text{time}(t) = \frac{\text{distance}(d)}{\text{rate}(r)}$$

Example 1   **The Pai family drove from their home into the country at a speed of 80 km/h. They returned over the same road at 70 km/h. If the round trip took 6 hours, how far did the Pais drive into the country?**

Let x = the distance of the trip to the country.
Represent the data with a vector diagram.

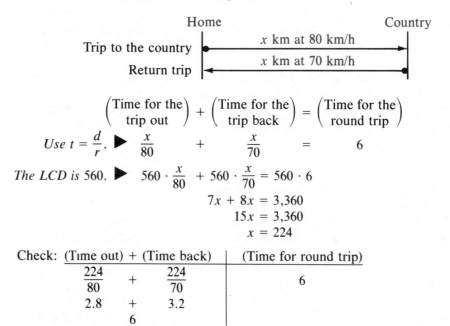

$$\left(\begin{array}{c}\text{Time for the} \\ \text{trip out}\end{array}\right) + \left(\begin{array}{c}\text{Time for the} \\ \text{trip back}\end{array}\right) = \left(\begin{array}{c}\text{Time for the} \\ \text{round trip}\end{array}\right)$$

Use $t = \frac{d}{r}$. ▶ $\dfrac{x}{80}$ + $\dfrac{x}{70}$ =   6

The LCD is 560. ▶ $560 \cdot \dfrac{x}{80} + 560 \cdot \dfrac{x}{70} = 560 \cdot 6$

$$7x + 8x = 3{,}360$$
$$15x = 3{,}360$$
$$x = 224$$

Check: (Time out) + (Time back) | (Time for round trip)

$\dfrac{224}{80}$ + $\dfrac{224}{70}$ | 6

2.8   +   3.2

6

Thus, the Pais drove 224 km into the country.

A certain falcon can fly at a speed of 40 km/h in still air. If a wind is blowing at 12 km/h, the falcon can fly at 40 + 12, or 52 km/h, with the wind and at 40 − 12, or 28 km/h, against the wind. A wind in the direction of an object's motion is called a *tailwind* and a wind in the direction against the object's motion is called a *headwind*.

If a boat can travel at 20 km/h in still water and if a river has a current of 4 km/h, then the boat can travel *downstream,* or with the current, at 24 km/h. The boat can travel *upstream,* or against the current, at 20 − 4, or 16 km/h.

Example 2 **A boat can travel 20 km/h in still water. It can travel 47 km downstream in the same time that it can travel 33 km upstream. What is the rate of the current?**

Let c = the rate of the current. Represent the data in a table.

	distance (d)	rate (r)	time (t)	
Downstream	47	$20 + c$	$\dfrac{47}{20 + c}$	◀ $t = \dfrac{d}{r}$
Upstream	33	$20 - c$	$\dfrac{33}{20 - c}$	◀ $t = \dfrac{d}{r}$

Time downstream = Time upstream

$$\frac{47}{20 + c} = \frac{33}{20 - c}$$
$$47(20 - c) = 33(20 + c)$$
$$940 - 47c = 660 + 33c$$
$$280 = 80c$$
$$3.5 = c$$

Check:

Time downstream	Time upstream
$\dfrac{47}{20 + c}$	$\dfrac{33}{20 - c}$
$\dfrac{47}{20 + 3.5}$	$\dfrac{33}{20 - 3.5}$
$\dfrac{47}{23.5}$	$\dfrac{33}{16.5}$
2	2

Thus, the rate of the current is 3.5 km/h.

Reading in Algebra

A car traveled from P to Q at 50 mi/h for 3 h and then returned from Q to P at 30 mi/h for 5 h. Choose the one correct ending for each statement.

1. The distance of the round trip was
   A   80 mi.
   B 150 mi.
   C 300 mi.

2. The time for the round trip was
   A 8 h.
   B 5 h.
   C 3 h.

3. The average speed for the round trip was
   A   40 mi/h.
   B 37.5 mi/h.
   C   80 mi/h.

4. For the trip from P to Q and for the trip from Q to P, the two
   A times were equal.   B distances were equal.   C rates were equal.

5. At the end of the first 2 h, the car was
   A 50 mi from Q.   B 50 mi from P.   C 60 mi from P.

6. The car was halfway between P and Q at the end of the first
   A $\frac{1}{2}$ h.   B $3\frac{1}{2}$ h.   C $5\frac{1}{2}$ h.

Written Exercises

Solve each problem.

(A) 1. Alice drove to the ocean at 90 km/h. She returned home on the same road at 60 km/h. If the round trip took 2 h, how far does Alice live from the ocean?

2. You left home at 9:00 A.M. and drove to the seaquarium at 75 km/h. After 2 h of sightseeing, you returned home at 50 km/h. If you arrived home at 2:00 P.M., how far do you live from the seaquarium?

3. A scout in a canoe can travel 12 km/h in still water. If the scout traveled 30 km downstream in the same time that it took to travel 15 km upstream, what was the rate of the current?

4. A plane flew 780 km with a tailwind in the same time it would have flown 720 km in the opposite direction. Find the rate of the tailwind if the plane flies at 500 km/h in still air.

5. You jog along a straight trail at 15 km/h. Then you rest for 1 hour and jog back home on that trail at 18 km/h. If you were gone from home for 3 h, how far away from home did you go?

6. A certain salmon can swim 3 km/h in still water. In a river, the salmon traveled 8 km upstream in the same time it took to swim 24 km downstream. What was the rate of the current?

(B) 7. A camper rowed 24 km downstream and then returned the same distance upstream in twice the time of the downstream trip. If the camper can row 6 km/h in still water, find the rate of the current.

8. Harvey hiked along a road at 5 km/h. Sara met him with a truck and they returned on the same road at 40 km/h. If the round trip took Harvey 4 h and 30 min, find the distance that he hiked.

9. One plane flew 390 km with the help of a tailwind in twice the time that another plane flew 180 km against the headwind. If each plane can fly 500 km/h in still air, find the rate of the tailwind.

10. With the help of a tailwind, a plane flew 378 km in $\frac{3}{4}$ of the time that a second plane flew 456 km in the opposite direction. Find the rate of the tailwind if each plane can fly 600 km/h in still air.

11. A marathon race course is in the shape of a square and each side of the square is called a leg of the race. Find the length of the course if a runner ran each leg at 12 km/h, 15 km/h, 10 km/h, and 16 km/h, respectively, in a total time of 3.75 h.

12. A gymnast climbed a vertical rope at 3.2 m/s and then descended at 4.8 m/s. Find the height of the rope if the gymnast was timed at 6.25 s for the round trip.

CUMULATIVE REVIEW

Solve each equation.

1. $8(3 - 5x) - (6x + 15) - 4(-7x - 8) = 95$

2. $10^{3x-1} = 1,000$

3. $|5x + 3| = 18$

LITERAL EQUATIONS AND FORMULAS   5.4

Objective   **To solve a formula or a literal equation for one of its variables**

A formula for the perimeter of a rectangle is
$p = 2(l + w)$. Since a formula is an equation,
$p = 2(l + w)$ can be solved for the length l in
terms of the perimeter p and the width w using
equation properties, as shown at the right.

$$p = 2(l + w)$$
$$p = 2l + 2w$$
$$p - 2w = 2l$$
$$\frac{p - 2w}{2} = l$$

A formula like $p = 2(l + w)$ and an equation like $ax + b = cx + d$ are called
literal equations, since they contain more than one letter, or variable. To solve
a literal equation for one specific variable, you should rewrite the equation so
that all of the terms containing that variable are alone on one side of the
equation.

Example 1   **Solve $ax + b = cx - d$ for x.**

$$ax + b = cx - d$$
$$ax - cx = -d - b \quad \blacktriangleleft \text{ All x-terms are alone on the left side.}$$
$$(a - c)x = -d - b \quad \blacktriangleleft \text{ Factor out x.}$$
$$x = \frac{-d - b}{a - c} \quad \blacktriangleleft \text{ Divide by $a - c$.}$$

There are several expressions that are equivalent to $\dfrac{-d - b}{a - c}$ in Example 1.

If $x = \dfrac{-d - b}{a - c}$, then $x = -\dfrac{d + b}{a - c} = -\dfrac{-d - b}{c - a} = \dfrac{d + b}{c - a}$.

To solve the equation $3(5 - 2y) = 7$, you would simplify the equation by
applying the Distributive Property. As shown in Example 2, a similar
procedure is followed in solving the literal equation $a(b - cy) = d$ for y.

Example 2   **Solve $a(b - cy) = d$ for y.**

$$a(b - cy) = d$$
$$ab - acy = d \quad \blacktriangleleft \text{ Use the Distributive Property.}$$
$$-acy = d - ab \quad \blacktriangleleft \text{ The y-term is alone on the left side.}$$
$$y = \frac{d - ab}{-ac}, \text{ or } \frac{ab - d}{ac}$$

The formula $\dfrac{1}{R} = \dfrac{1}{x} + \dfrac{1}{y}$ is a *literal fractional equation*. Recall that a fractional
equation can be simplified by multiplying each side by the LCD.

Example 3

Solve $\dfrac{1}{R} = \dfrac{1}{x} + \dfrac{1}{y}$ for x. Then find the value of x if $R = 3.2$ and $y = 4.8$.

Multiply each side by Rxy, the LCD.

$$Rxy \cdot \dfrac{1}{R} = Rxy \cdot \dfrac{1}{x} + Rxy \cdot \dfrac{1}{y}$$
$$xy = Ry + Rx$$
$$xy - Rx = Ry$$
$$x(y - R) = Ry$$
$$x = \dfrac{Ry}{y - R}$$

$R = 3.2 \qquad y = 4.8$

$$x = \dfrac{Ry}{y - R}$$
$$= \dfrac{3.2(4.8)}{4.8 - 3.2}$$
$$= \dfrac{15.36}{1.6}$$
$$x = 9.6$$

To solve a literal equation like $\dfrac{y - 3}{x + 4} = -5$ for either x or y, you can first rewrite it as a proportion. This technique is used in Example 4.

Example 4

Solve $\dfrac{y - 3}{x + 4} = -5$ for y.

$$\dfrac{y - 3}{x + 4} = \dfrac{-5}{1}$$ ◀ *Rewrite as a proportion.*
$$(y - 3)1 = (x + 4)(-5)$$ ◀ *If $\dfrac{a}{b} = \dfrac{c}{d}$, then $ad = bc$.*
$$y - 3 = -5x - 20$$
$$y = -5x - 17$$

Oral Exercises

Solve each equation for x.

1. $a + x + b = c$ **2.** $-x = a - b$ **3.** $ax = bc$ **4.** $ax = ab$

5. $\dfrac{x}{a} = b$ **6.** $ax = b + c$ **7.** $\dfrac{x}{a} = \dfrac{b}{c}$ **8.** $1 = \dfrac{a}{x}$

Written Exercises

Solve each equation for x.

Ⓐ
1. $ax - b = c$ **2.** $a = bx + c$ **3.** $ax = b + cx$

4. $5x + 4 = ax - b$ **5.** $7ax + b = 3b - ax$ **6.** $3ax - 4b = 2cx + 5d$

7. $a(x - b) = c$ **8.** $a(x + 2b) = c(x - d)$ **9.** $a(2x - b) = c(dx - 1)$

10. $\dfrac{a}{x} = b$ **11.** $\dfrac{a}{b} \cdot x = c + d$ **12.** $\dfrac{ax}{b} = \dfrac{c}{d}$

13. $\dfrac{a}{b} = \dfrac{x}{c} + d$ **14.** $\dfrac{2}{x} = \dfrac{3}{a} + \dfrac{4}{b}$ **15.** $\dfrac{8}{a} = \dfrac{6}{x} - \dfrac{4}{b}$

Solve each equation for y.

16. $\dfrac{y - 10}{x - 8} = \dfrac{1}{2}$

17. $\dfrac{y - 2}{x + 5} = \dfrac{4}{5}$

18. $\dfrac{y + 3}{x + 6} = \dfrac{-2}{3}$

19. $\dfrac{y - 6}{x + 1} = 4$

20. $\dfrac{y + 8}{x - 4} = -3$

21. $\dfrac{y - a}{x - b} = m$

Solve each formula for the specified variable. Then find the value of that variable for the given data.

22. Solve $p = 2(l + w)$ for w.
$p = 23.4$; $l = 4.8$

23. Solve $A = B + BCD$ for B.
$A = 48.35$; $C = 5.8$; $D = 16.5$

24. Solve $A = \dfrac{1}{2} bh$ for b.
$A = 30.6$; $h = 3.6$

25. Solve $\dfrac{1}{x} = \dfrac{1}{a} + \dfrac{1}{b}$ for x.
$a = 8.12$; $b = 8.7$

Solve each equation for x.

(B) 26. $\dfrac{2x}{g} = t^2$

27. $gt^2 = 2(200t - x)$

28. $ax + b(x - c) = d(x + e)$

29. $3a(2x - b) - a(b - x) = 3ab + 7c(a + b)$

30. $\dfrac{5a}{x + b} = \dfrac{a}{x - b}$

31. $\dfrac{x - a}{x - b} = \dfrac{c}{d}$

32. $\dfrac{x - a}{4} + \dfrac{x + b}{12} = \dfrac{c}{3}$

33. $\dfrac{x - 4}{a} + \dfrac{3}{b} = \dfrac{2x - 3}{c}$

34. $\dfrac{ax - b}{cx + d} = 5$

35. $\dfrac{ax + b}{c} = dx + e$

Solve each formula for the specified variable. Then find the value of that variable for the given data.

36. Solve $T = mg - mf$ for m.
$T = 912.7$; $g = 93.7$; $f = 2.43$

37. Solve $A = ah + bh + ch$ for h.
$A = 252$; $a = 2.8$; $b = 3.5$; $c = 4.2$

38. Solve $\dfrac{1}{R} = \dfrac{1}{a} + \dfrac{1}{b}$ for b.
$R = 4.8$; $a = 19.2$

39. Solve $V = \dfrac{1}{3} Bh$ for h.
$V = 2.4 \times 10^5$; $B = 4.8 \times 10^2$

40. Solve $gt^2 - h = 3(400t - h)$ for h.
$g = 32$; $t = 9$

41. Solve $\dfrac{h - 80}{16t} = 30 - t$ for h.
$t = 4$

Solve each equation for x.

(C) 42. $x^4 - 5a^2x^2 + 4a^4 = 0$

43. $16x^4 - 40a^2x^2 + 9a^4 = 0$

44. $x^3 - ax^2 - b^2x + ab^2 = 0$

45. $4x^4 - 36a^2x^2 - b^2x^2 + 9a^2b^2 = 0$

46. $|ax - b| = c$, if $c \geq 0$

47. $\dfrac{x - a}{x - b} = \dfrac{x - a}{x - c}$

CALCULATOR ACTIVITIES

Use the formula $y = \dfrac{m(ab - c) + d}{mc}$ to find the value of y for the given data.

1. $m = 28$; $a = 360$; $b = 212$; $c = 415$; $d = 198,\,660$
2. $m = 2.8$; $a = 7.2$; $b = 10.6$; $c = 8.3$; $d = 18.704$

DIVIDING POLYNOMIALS 5.5

Objectives   **To divide a polynomial of two or more terms with one variable by a binomial**
To determine if a binomial is a factor of a polynomial

Two forms for dividing 679 by 32, using the divide-multiply-subtract cycle, are shown below.

ORDINARY
FORM

$$
\begin{array}{r}
21 \\
32\,\overline{|\,679} \\
64 \\
\hline
39 \\
32 \\
\hline
7
\end{array}
$$

Divide: $600 \div 30$
Multiply: $20(30 + 2)$
Subtract: $(600 + 70) - (600 + 40)$
Divide: $30 \div 30$
Multiply: $1(30 + 2)$
Subtract: $(30 + 9) - (30 + 2)$

EXPANDED
FORM

$$
\begin{array}{r}
20 + 1 \\
30 + 2\,\overline{|\,600 + 70 + 9} \\
600 + 40 \\
\hline
30 + 9 \\
30 + 2 \\
\hline
7
\end{array}
$$

Thus, $679 \div 32 = 21$, with remainder 7. Check: $679 = 32 \times 21 + 7$.

The check shows that:
(1) Dividend = Divisor $\times$ Quotient + Remainder.
(2) 32 is *not* a factor of 679 because there is a nonzero remainder.

The *divide-multiply-subtract* cycle shown above in the expanded form can be used to divide a polynomial, with one variable, by a binomial.

Example 1   **Divide:** $(4x^2 - 5x - 30) \div (x - 3)$**. Is** $x - 3$ **a factor of** $4x^2 - 5x - 30$**?**

Step 1

Divide: $4x^2 \div x$.
Multiply: $4x(x - 3)$.
Subtract: $(4x^2 - 5x)$
$-(4x^2 - 12x)$.

$$
\begin{array}{r}
4x \\
x - 3\,\overline{|\,4x^2 - 5x - 30} \\
4x^2 - 12x \\
\hline
7x
\end{array}
$$

Step 2

Divide: $7x \div x$.
Multiply: $7(x - 3)$.
Subtract: $(7x - 30)$
$-(7x - 21)$.

$$
\begin{array}{r}
4x + 7 \\
x - 3\,\overline{|\,4x^2 - 5x - 30} \\
4x^2 - 12x \\
\hline
7x - 30 \\
7x - 21 \\
\hline
-\,9
\end{array}
$$

Thus, the polynomial quotient is $4x + 7$ and the remainder is -9.

Check: $4x^2 - 5x - 30 = (x - 3)(4x + 7) + (-9)$
So, $x - 3$ is not a factor of $4x^2 - 5x - 30$, since there is a nonzero remainder.

**Factor
Theorem**

A binomial B is a factor of a polynomial P if

$$P \div B = Q, \text{ or } P = B \cdot Q + R,$$

where Q is a polynomial and the remainder R is 0.

Example 2   **Divide: $(30 - 2x^2 - 2x + 6x^3) \div (3x + 5)$. Is $3x + 5$ a factor of the dividend?**

$$
\begin{array}{r}
2x^2 - 4x + 6 \\
3x + 5 \overline{\smash{\big)}\ 6x^3 - 2x^2 - 2x + 30} \\
\underline{6x^3 + 10x^2} \\
-12x^2 - 2x \\
\underline{-12x^2 - 20x} \\
18x + 30 \\
\underline{18x + 30} \\
0
\end{array}
$$

◀ *Arrange the terms of the dividend in descending order of the exponents.*
Check: $30 - 2x^2 - 2x + 6x^3$
  $= (3x + 5)(2x^2 - 4x + 6) + 0$
Thus, the quotient is $2x^2 - 4x + 6$, the remainder is 0, and $3x + 5$ is a factor of the dividend.

If the dividend has missing powers of the variable, you can use 0 as the coefficient to replace the missing terms.

Example 3   **Divide: $(10x^3 - 34x + 4) \div (2x - 4)$. Is the divisor a factor of the dividend?**

$$
\begin{array}{r}
5x^2 + 10x + 3 \\
2x - 4 \overline{\smash{\big)}\ 10x^3 + 0x^2 - 34x + 4} \\
\underline{10x^3 - 20x^2} \\
20x^2 - 34x \\
\underline{20x^2 - 40x} \\
6x + 4 \\
\underline{6x - 12} \\
16
\end{array}
$$

◀ *Replace the missing x^2-term with $0x^2$.*
Check: $10x^3 - 34x + 4$
  $= (2x - 4)(5x^2 + 10x + 3) + 16$
Thus, the polynomial quotient is $5x^2 + 10x + 3$, the remainder is 16, and the divisor is not a factor of the dividend.

Oral Exercises

Simplify each quotient. Assume that no divisor is equal to 0.

1. $(14x^5 - 21x^3) \div (7x)$ **2.** $(5mn^2 + 6m^2n) \div (mn)$ **3.** $(18a^2b^2 - 12a^2b) \div (3a^2b)$

4. $\dfrac{15m^4 - 25m^3 - 20m^2}{5m^2}$ **5.** $\dfrac{-6cd^4 + 10c^5d^2 - 16c^3d^3}{2cd}$ **6.** $\dfrac{24x^3y^3 + 16x^2y^2 - 28xy^3}{4xy^2}$

Written Exercises

Divide to find the polynomial quotient and remainder. Is the divisor a factor of the dividend?

(A) **1.** $(2c^2 + c - 2) \div (c + 1)$
3. $(10x^2 - 8x - 24) \div (x - 2)$
5. $(4y^2 + 7y - 4) \div (y + 3)$
7. $(3x^3 - x^2 - 17x + 9) \div (x - 3)$
9. $(15 - 14n + 8n^2) \div (4n - 5)$
11. $(a^3 - 13a - 12) \div (a + 3)$

2. $(y^2 + 3y + 1) \div (y + 2)$
4. $(2n^2 + 19n + 35) \div 2n + 5)$
6. $(6a^2 - 13a + 2) \div (3a - 2)$
8. $(3n^4 + 13n^3 + 4n^2 - 2n - 8) \div (n + 4)$
10. $(14x^2 - 3x + 3x^3 + 7) \div (x + 5)$
12. $(9y^4 + 5y^2 - 12) \div (3y - 2)$

(B) **13.** $(6x^3 + 4x^2 - 3x - 8) \div (2x^2 - 1)$
15. $(a^3 + 1) \div (a + 1)$
17. $(2n^4 - 4n^3 + 7n^2 - 12n + 3) \div (n^2 + 3)$
19. $(y^6 - y^5 - y^4 + y^2 - y - 1) \div (y^4 + 1)$

14. $(3c^4 + 2c^3 - 8c - 48) \div (c^2 - 4)$
16. $(8y^3 - 125) \div (2y - 5)$
18. $(4x^4 + 3x^3 + 13x^2 - 15x + 35) \div (x^2 - 5)$
20. $(9a^6 - 3a^5 + a^2 - 4) \div (3a^3 - 1)$

(C) **21.** $(x^{3m} - 4x^{2m} + 3x^m - 10) \div (x^m - 4)$ **22.** $(x^{3c} + 3x^{2c} - x^{2c+1} - 3x^{c+1} + 2) \div (x^c + 3)$
23. $(24x^3y + 7x^2y^2 - 6x^2y - 6xy^3 - 4xy^2) \div (3x + 2y)$

Dividing Polynomials   **129**

REMAINDER THEOREM

Objectives
To divide a polynomial by a binomial, using synthetic division
To evaluate a polynomial for a given value of its variable, using synthetic substitution
To determine if a given binomial is a factor of a given polynomial, using synthetic substitution.

The division at the right can be done in a shorter way. The shorter method eliminates repetitious rewriting of the variable and some of the coefficients.
The method is called **synthetic division** and is illustrated below.

$$
\begin{array}{r}
3x^2 - 4x + 5 \\
x - 2 \overline{\smash{\big)}\ 3x^3 - 10x^2 + 13x - 26} \\
\underline{3x^3 - 6x^2} \\
-4x^2 + 13x \\
\underline{-4x^2 + 8x} \\
5x - 26 \\
\underline{5x - 10} \\
-16
\end{array}
$$

Divisor ▶ $x - 2$ 3 -10 13 -26 ◀ *Coefficients of dividend*
                                                              Bring down the 3.
                              6   -8 10

2 | 3 -4 5 $\boxed{-16}$ ◀ *Multiply by 2*
                                          and then add.

$(x - 2) \cdot (3x^2 - 4x + 5) + (-16) = 3x^3 - 10x^2 + 13x - 26$

Check: Divisor × Quotient + Remainder = Dividend

Notice the following.
1. The remainder is -16.
2. $x - 2$ is *not* a factor of the dividend, since there is a nonzero remainder, -16.
3. If $x = 2$, then $3x^3 - 10x^2 + 13x - 26 = -16$ since
$$
\begin{aligned}
3x^3 - 10x^2 + 13x - 26 &= (x - 2)(3x^2 - 4x + 5) - 16 \\
&= (2 - 2)(3x^2 - 4x + 5) - 16 \\
&= 0 - 16, \text{ or } -16.
\end{aligned}
$$

Example 1

Given $P = 4x^4 - 14x^3 - x + 27$, use synthetic division to divide P by $x - 3$. Find the polynomial quotient Q and the remainder R.

$x - 3$ 4 -14 0 -1 27 ◀ *Replace the missing*
                                            x^2-term with $0x^2$.
                  12   -6 -18 -57

3 | 4 -2 -6 -19 $\boxed{-30}$

$P = (x - 3)(4x^3 - 2x^2 - 6x - 19) - 30$

Thus, $Q = 4x^3 - 2x^2 - 6x - 19$ and $R = -30$.

In Example 1, a polynomial P was divided by $x - 3$ to obtain a polynomial quotient Q and a remainder R of -30, so that

$$\begin{aligned} P &= 4x^4 - 14x^3 - x + 27 \\ &= (x - 3)(4x^3 - 2x^2 - 6x - 19) - 30 \\ &= (x - 3) \cdot Q - 30. \end{aligned}$$

If $x = 3$, then $P = (3 - 3) \cdot Q - 30 = 0 - 30 = -30$. The value of P is -30, the same as the remainder R, when $x = 3$. This suggests the Remainder Theorem, which is stated below.

Remainder Theorem	For the polynomials P, Q, and $x - a$, if $P = (x - a) \cdot Q + R$ for some number R, then R is the value of P when $x = a$.

The synthetic division in Example 1 can be shortened to the method of synthetic substitution studied in Lesson 2.3 (page 37).

Synthetic Division

$$\begin{array}{r|rrrrr} & 4 & -14 & 0 & -1 & 27 \\ & & 12 & -6 & -18 & -57 \\ \hline 3| & 4 & -2 & -6 & -19 & \boxed{-30} \end{array}$$

Synthetic Substitution

$$\begin{array}{r|rrrrr} & 4 & -14 & 0 & -1 & 27 \\ \hline 3| & 4 & -2 & -6 & -19 & \boxed{-30} \end{array}$$

The multiples of 3 are not written.

The Remainder Theorem states that the binomial divisor must be in the form $x - a$. A divisor like $x + 5$ can be rewritten as $x - (-5)$ where $a = -5$.

Example 2 **Given: $P = 2x^5 + 11x^4 - x^3 - 30x^2 + 6x + 30$. Use synthetic substitution to divide P by $x + 5$. Find (1) the polynomial quotient and the remainder, (2) the value of P if $x = -5$, and (3) whether $x + 5$ is a factor of P.**

Rewrite $x + 5$ as $x - (-5)$.

$$\begin{array}{r|rrrrrr} & 2 & 11 & -1 & -30 & 6 & 30 \\ \hline -5| & 2 & 1 & -6 & 0 & 6 & \boxed{0} \end{array}$$

◀ *Find and add the multiples of -5.*

$$P = (x + 5)(2x^4 + x^3 - 6x^2 + 0x + 6) + 0$$

Thus, (1) the quotient is $2x^4 + x^3 - 6x^2 + 6$ and the remainder is 0, (2) the value of P is 0 when $x = -5$, and (3) $x + 5$ is a factor of P.

Example 3 **Evaluate $2x^4 + 2x^3 - 13x^2 - x + 6$ using each of -3, -2, 2, and 3 for x. Name the binomial factors found, if any.**

$$\begin{array}{r|rrrrr} & 2 & 2 & -13 & -1 & 6 \\ \hline -3| & 2 & -4 & -1 & 2 & \boxed{0} \\ -2| & 2 & -2 & -9 & 17 & \boxed{-28} \\ 2| & 2 & 6 & -1 & -3 & \boxed{0} \\ 3| & 2 & 8 & 11 & 32 & \boxed{102} \end{array}$$

Thus, the values of P are 0, -28, 0, and 102 when the values of x are -3, -2, 2, and 3, respectively; $x + 3$ and $x - 2$ are binomial factors of the polynomial.

Reading in Algebra

An example of synthetic substitution is shown at the right. Complete the paragraph below using the information in the example.

$$-5 \rfloor \begin{array}{cccc} 2 & 11 & 0 & -10 \\ \hline 2 & 1 & -5 & 15 \end{array}$$

The dividend is _____, the divisor is _____, 15 is the

_____, and the polynomial quotient is _____. The value

of $2x^3 + 11x^2 - 10$ is _____ when the value of x is _____. The binomial

$x + 5$ is not a _____ of $2x^3 + 11x^2 - 10$.

Oral Exercises

Determine whether each statement is true or false.

1. If $P = (x + 2)(x^2 - 3x - 12) + 4$, then $x + 2$ is a factor of P.
2. If $P = (x - 4)(2x^2 + 7x - 3)$, then $x - 4$ is a factor of P.
3. If $P = (x - 3)(x^2 - 7x + 12)$, then $P = 0$ when $x = 3$.
4. If $P = (x + 6)(x^2 + 8x + 16) + 5$, then $P = 5$ when $x = -6$.

Written Exercises

Find the polynomial quotient and the remainder. Use synthetic division or synthetic substitution.

(A)
1. $(x^3 + 2x^2 - 8x - 3) \div (x - 1)$
2. $(x^3 - 13x^2 + 20x + 40) \div (x - 5)$
3. $(3y^4 + y^3 - 14y^2 + 10y - 9) \div (y + 3)$
4. $(-4y^3 + 2y^2 - 3y + 1) \div (y + 1)$
5. $(n^5 - 2n^3 + n - 2) \div (n + 2)$
6. $(2n^5 - 6n^4 - 12n^2 - 15) \div (n - 4)$

Evaluate each polynomial for the given data. Use synthetic substitution.

7. $6x^5 - 5x^4 - 4x^3 - 3x^2 - 2x + 1$ if $x = 2$
8. $2x^5 + 9x^4 + 8x^3 - 7x^2 - 6x + 8$ if $x = -3$
9. $3y^3 + 22y^2 + 10y + 20$ if $y = -10$
10. $5y^3 - 80y^2 + 80y - 5$ if $y = 15$
11. $n^4 - n^3 + n^2 + 1$ if $n = 4$
12. $n^4 + n^3 - n^2 + n - 1$ if $n = -5$
13. $6x^4 + 4x^2 - 2$ if $x = -4$
14. $4x^5 - 6x^3 + 7x$ if $x = 2$

Determine whether the binomial is a factor of the polynomial.

15. $x - 3; 2x^3 - 11x^2 + 18x - 12$
16. $x + 6; 3x^3 + 17x^2 - 8x - 12$
17. $y + 2; y^4 + 2y^3 - 3y^2 + 2y + 16$
18. $y - 5; 3y^4 - 12y^3 - 20y^2 + 30y + 5$
19. $n - 1; 4n^5 - 2n^3 + 6n^2 - 9n + 1$
20. $n + 3; 5n^5 + 10n^4 - 10n^3 - 34n + 33$

(B)
21. $y + 8; 3y^4 + 21y^3 - 31y^2 - 51y + 40$
22. $y - 9; 12y^3 - 8y^2 + 100y - 8,100$
23. $n - 10; n^3 - 9,000$
24. $n + 20; n^3 + 8,000$
25. $x + 2; x^6 + x^4 + x^2 - 84$
26. $x - 3; x^5 - x^3 - 9$
27. $y - 2; y^6 + 64$
28. $y + 1; y^7 + 1$

Evaluate each polynomial for the given data. Use synthetic substitution.

29. $n^6 - 2n^4 - 3n^2 - 4$ if $n = -3$
30. $4n^7 + 3n^5 + 2n^3 + n$ if $n = 2$
31. $3x^5 + 7x^4 + 9x^3 + 6x^2 + 4x + 2$ if $x = 10$
32. $7x^6 - 8x^4 + 9x^2$ if $x = -2$

Evaluate each polynomial using each of -3, -2, -1, 0, 1, 2, and 3 for x.
Use synthetic substitution. Name the binomial factors found, if any.

33. $x^4 - 2x^3 - 6x^2 + 6x + 9$

34. $x^4 - 3x^2 - 4$

35. $2x^4 + 2x^3 - 7x^2 - 3x + 6$

36. $4x^4 - 4x^2 - 3$

CALCULATOR ACTIVITIES

You can evaluate a polynomial with one variable for a *positive* number by using the *multiply-and-add* cycle of synthetic substitution. Evaluate $5x^4 - 17x^3 - 23x^2 + 49x + 90$ if $x = 7$.

$5 \otimes 7 \ominus 17 \otimes 7 \ominus 23 \otimes 7 \oplus 49 \otimes 7 \oplus 90 \ominus 5{,}480$,
the value of the polynomial when $x = 7$.

The sequence shown above is for a simple (nonscientific) calculator. If you use a scientific calculator, press $\ominus$ after each addition or subtraction of a number as shown below.

$5 \otimes 7 \ominus 17 \ominus \otimes 7 \ominus 23 \ominus \otimes 7 \oplus 49 \ominus \otimes 7 \oplus 90 \ominus 5{,}480$

Evaluate each polynomial for $x = 12$.

2. $24x^4 + 25x^3 - 26x^2 - 27x - 28$

1. $18x^3 - 47x^2 + 62x - 75$

3. $8x^5 - 7x^4 - 6x^3 + 5x^2 + 4x - 3$

You can evaluate a polynomial for a *negative* number as follows:

Evaluate $5x^4 - 17x^3 - 23x^2 + 49x + 90$ if $x = -6$.

Given: $x = -6$. (1) Let $x = -y$, (2) simplify, and (3) evaluate for $y = 6$.
(1) $5(-y)^4 - 17(-y)^3 - 23(-y)^2 + 49(-y) + 90$ if $-y = -6$.

(2) $5y^4 \quad + 17y^3 \quad - 23y^2 \quad - 49y \quad + 90$ if $y = 6$.
(3) $5 \otimes 6 \oplus 17 \otimes 6 \ominus 23 \otimes 6 \ominus 49 \otimes 6 \oplus 90 \ominus 9{,}120$,
the value of the polynomial when $x = -6$.

Evaluate each polynomial for $x = -15$.

5. $5x^3 - 4x^2 + 9x + 800$

4. $3x^4 - 8x^3 - 7x^2 + 6x + 9$

6. $-2x^5 - 2x^4 + 3x^3 + 3x^2 - 4x + 4$

There are no missing terms in the exercises above. If one or more terms are missing, you can evaluate as follows.

Evaluate $4x^5 - 6x^3 + 7x$ if $x = 2$.
$4x^5 + 0x^4 - 6x^3 + 0x^2 + 7x \quad + 0$

$4 \otimes 2 \underbrace{\otimes 2}_{0x^4} \ominus 6 \otimes 2 \underbrace{\otimes 2}_{0x^2} \oplus 7 \otimes 2 \oplus 0 \ominus 94$

Evaluate each polynomial for $x = 5$ and again for $x = -3$.

7. $7x^6 - 2x^4 + 6x^3 - 8x + 12$

8. $8x^6 + 4x^5 - 2x^2 - 10x$

CHAPTER FIVE REVIEW

Vocabulary

extraneous solution [5.1]
fractional equation [5.1]
literal equation [5.4]
proportion [5.1]
    extremes of [5.1]
    means of [5.1]
synthetic division [5.6]

Solve each equation. [5.1]

1. $\dfrac{n-4}{3} - \dfrac{n-2}{4} = \dfrac{n}{2}$

2. $\dfrac{1}{6} y^2 = \dfrac{3}{2} y - 3$

3. $\dfrac{8}{5n-2} = \dfrac{4}{2n+7}$

4. $\dfrac{a-1}{3a-7} = \dfrac{a-2}{a+1}$

5. $\dfrac{3}{x+2} + \dfrac{2}{x} = 3$

6. $\dfrac{2y-1}{y-3} = \dfrac{y+2}{y-3} - \dfrac{y}{4}$

7. $\dfrac{7}{n-2} = \dfrac{n-3}{n^2-7n+10} + \dfrac{1}{n-5}$

8. $\dfrac{a}{a+5} - \dfrac{5}{a^2-25} = \dfrac{a-6}{2a-10}$

★9. $\dfrac{3}{x^2+2x-24} + \dfrac{4}{x^2+8x+12} = \dfrac{5}{x^2-2x-8}$

Solve each problem.

10. Mr. Brown can plant his crops in 30 days. His son would take 40 days and his daughter 60 days. How many days will it take if they work together? [5.2]

11. Lori can put carpet on a floor in 9 h. If Brent helps her, the job is done in 6 h. How many hours would it take Brent if he works alone? [5.2]

12. A train traveled from city A to city B at 40 km/h and then returned at 60 km/h. If the round trip took 8 h, what is the rail distance between cities A and B? [5.3]

13. A family can row their boat 5 km/h in still water. They traveled 30 km downstream in the same time that they traveled 20 km upstream. What was the rate of the current? [5.3]

14. One plane flew 498 km with a tailwind in $\dfrac{3}{4}$ of the time that another plane flew 616 km against the headwind. If each plane can fly 400 km/h in still air, find the rate of the tailwind. [5.3]

Solve each equation for x. [5.4]

15. $a(x-b) = c(2x+3)$

16. $\dfrac{7a}{x+b} = \dfrac{3c}{x-b}$

★17. $x^4 - a^2 x^2 - 9b^2 x^2 + 9a^2 b^2 = 0$

Solve each formula for the specified variable. Then find the value of that variable for the given data. [5.4]

18. Solve $\dfrac{1}{a} = \dfrac{2}{b} + \dfrac{3}{c}$ for b.
    $a = 2.4; c = 9.2$

19. Solve $A = \dfrac{1}{2} h(a+b)$ for h.
    $A = 300; a = 17; b = 23$

Divide. Assume that no divisor is equal to 0. [5.5]

20. $(9x^3 - 19x - 7) \div (3x - 5)$
21. $(4y - 3y^2 + 2y^3 - 6) \div (y^2 + 2)$
★22. $(x^{2n+2} + 2x^{2n+1} - 5x^{n+2} - 10x^{n+1})$
    $\div (x^n - 5)$

23. Divide: $(n^3 + 5n^2 - 4n - 6) \div (n - 2)$. Use synthetic division. [5.6]

Determine whether the binomial is a factor of the polynomial in each case. [5.6]

24. $x + 5;\ 3x^3 + 17x^2 + 8x - 10$
25. $y - 3;\ 4y^3 - 10y^2 - 12$
26. $c + 5;\ 4c^4 - 80c^2 + 500$

Evaluate each polynomial for the given data. Use synthetic substitution. Name the binomial factors found, if any. [5.6]

27. $5n^3 - 7n^2 - 10n - 22$ if $n = 3$
28. $x^6 - 24x^4 - 75x^2 + 250$ if $x = -5$
29. $2x^4 - 7x^3 - 23x^2 + 28x + 60$ using each of $-2, -1, 2$ and 5 for x.

Solve each equation.

1. $\dfrac{y + 8}{4} - \dfrac{2y - 3}{6} = \dfrac{y}{3}$

2. $\dfrac{1}{4}n^2 = \dfrac{5}{2}n - 6$

3. $\dfrac{9}{2x - 12} = \dfrac{3}{x - 2}$

4. $\dfrac{2y + 1}{y - 1} + \dfrac{y}{5} = \dfrac{y + 2}{y - 1}$

5. $\dfrac{n - 6}{n^2 - 2n - 8} + \dfrac{3}{n - 4} = \dfrac{2}{n + 2}$

Solve each problem.

6. Sandra can decorate a room for a party in 8 h and Edward can do it in 12 h. If Lois helps, all three can do the job in 3 h, working together. How long would it take Lois if she works alone?

7. Samuel jogged along a straight trail at 18 km/h and returned on the same trail at 9 km/h. If the round trip took 2 h, how far did he jog in one direction?

8. With the help of a tailwind, a plane flew 1,230 km in twice the time that a second plane flew 585 km against the headwind. Find the rate of the wind if both planes can fly 800 km/h in still air.

Solve each equation for x.

9. $5ax + c = 4c - ax$

10. $\dfrac{x + a}{d} = \dfrac{bx + c}{4}$

11. Solve $A = \dfrac{h}{2}(b + c)$ for c. Then find c if $A = 26$, $h = 4$, and $b = 5$.

Divide. Assume that no divisor is equal to 0.

12. $(6x^3 + 4 - 2x - 17x^2) \div (3x + 2)$

13. $(27n^3 - 8) \div (3n - 2)$

14. Divide: $(y^5 - 16y^3 + 5y - 20) \div (y - 4)$. Use synthetic division.

Determine whether the binomial is a factor of the polynomial in each case.

15. $n + 3$; $4n^3 + 8n^2 - 10n + 6$

16. $x - 5$; $x^4 - 20x^2 - 25$

Evaluate each polynomial for the given data. Use synthetic substitution. Name the binomial factors found, if any.

17. $7y^3 - 18y^2 + 25y - 39$ if $y = 2$

18. $8n^4 + 50n^3 - 60n + 80$ if $n = -6$

19. $x^4 - x^3 - 22x^2 - 44x - 24$ using each of -1, -3 and 6 for x.

★ 20. Divide: $(3x^{4n} + 4x^{3n} + x^{2n}) \div (x^{2n} + x^n)$.

★ 21. Solve $\dfrac{5}{x^2 - 25} + \dfrac{4}{x^2 - 9x + 20} = \dfrac{3}{x^2 + x - 20}$.

COMPUTER ACTIVITIES

Dividing Polynomials

OBJECTIVE: To divide a polynomial by a binomial of the form X + A

The computer program below uses the concept of synthetic division to divide a polynomial by a binomial. The program also contains two loops, one to INPUT the coefficients of the polynomials, and one to carry out synthetic division.

```
110    PRINT "WHAT IS THE DEGREE OF THE POLYNOMIAL";:
         INPUT N
120    DIM C(2 * N)
130    HOME
140    FOR I = N TO 0 STEP  - 1
150    PRINT "WHAT IS THE COEFFICIENT OF THE "I"
         DEGREE TERM?";
160    INPUT C(N + I - 1)
170    NEXT I
180    PRINT "WHAT IS THE CONSTANT IN THE BINOMIAL?";:
         INPUT K
190    PRINT C(1)"X^"N - 1;
200  D(1) = C(1)
210    FOR I = 2 TO N
220  D(I) = C(I) - K * D(I - 1)
230    IF D(I) < 0 THEN 250
240    PRINT "+";
250    PRINT D(I);: IF N - I = 0 THEN 270
260    PRINT "X^"N - I;
270    NEXT I
280  D(I) = C(I) - K * D(I - 1)
290    PRINT " REM "D(I)
300    GOTO 110
310    END
```

See the Computer Section beginning on page 574 for more information.

Note the use of the STEP option in line 140's FOR statement. STEP -1 makes the computer count backward by ones. Using STEP, you can increment the variable following FOR by any value. Another example is FOR I = 2 TO 20 STEP 2. This statement will make the computer count by twos: 2, 4, 6, ..., 20.

EXERCISES

1. Write a computer program that prints the squares of the odd numbers from 15 to 39 inclusive.

2. Run the program above to do the division in Exercises 1–6 on page 132.

COLLEGE PREP TEST

DIRECTIONS: Choose the *one* best answer to each question or problem.

1. The expression $\dfrac{5}{0}$ is undefined. Which expression could be undefined if x and y are two *different positive* numbers?

   (A) $\dfrac{x + y}{x - y}$ (B) $\dfrac{x}{y}$ (C) $\dfrac{2x - y}{x + y}$

   (D) $\dfrac{x + y}{3x - 4y}$ (E) None of these

2. If 15, 22, and 30 are each factors of a certain number, then another factor of the number must be

   (A) 4 (B) 9 (C) 25
   (D) 55 (E) None of these

3. Which numeral does not represent $\dfrac{11}{4}$?

   (A) $2\dfrac{3}{4}$ (B) 2.75 (C) 2.75000 . . .
   (D) 2.74999 . . . (E) None of these

4. $\dfrac{\dfrac{1}{2} + \dfrac{1}{4} + \dfrac{1}{8}}{\dfrac{1}{16} + \dfrac{1}{32} + \dfrac{1}{64}} =$

   (A) $\dfrac{1}{24}$ (B) $\dfrac{1}{8}$ (C) 8 (D) 24
   (E) None of these

5. If $\dfrac{1}{6} + \dfrac{1}{1.5} = \dfrac{1}{x}$, then $x =$

   (A) $\dfrac{5}{6}$ (B) 1.2 (C) 7.5 (D) 12
   (E) None of these

6. Juan drove 6 mi to the airport in 12 min and returned in 18 min. Find the average speed for the round trip in miles per hour.

   (A) 24 mi/h (B) 30 mi/h (C) 35 mi/h
   (D) 40 mi/h (E) None of these

7. If $\dfrac{1}{x + y} = \dfrac{1}{x - y}$, then

   (A) $x = 0$ and $y \neq 0$ (B) $y = 0$ and $x \neq 0$
   (C) $x = y$ and $x \neq 0$ (D) $x = -y$ and $x \neq 0$
   (E) None of these

8. Albert can do a certain job in 6 days, and Brenda can do it in x days. Carol works 3 times as fast as Albert. What part of the job will be done at the end of y days if all three work together?

   (A) $\dfrac{y}{6} + \dfrac{y}{x} + \dfrac{y}{2}$ (B) $\dfrac{1}{6} + \dfrac{1}{x} + \dfrac{1}{2}$
   (C) $\dfrac{y}{6} + \dfrac{y}{x} + \dfrac{y}{18}$ (D) $\dfrac{1}{6} + \dfrac{1}{x} + \dfrac{1}{18}$
   (E) $\dfrac{6}{y} + \dfrac{x}{y} + \dfrac{2}{y}$

9. A certain plane can fly 400 mi/h in still air. The plane can go round trip between points M and N in 4 h when the wind is blowing at a rate of 20 mi/h from M toward N. Find the distance between M and N.

   (A) 780 mi (B) 798 mi
   (C) 800 mi (D) 802 mi
   (E) 805 mi

6 RADICALS AND RATIONAL NUMBER EXPONENTS

If you complete courses in chemistry and mathematics at the secondary or college level, you may qualify for an interesting and well-paid career in chemical science. The chemical industry is among the five leading manufacturing industries in the United States.

Chemical laboratory research has led to the production of textiles of high quality. More than 25% of textiles produced are made from synthetic fibers such as rayon or nylon. Chemical treatments are used to provide us with fabrics that are water-repellent, shrink-proof, and wrinkle-resistant.

Project

A chemist was asked to change 50 mL of 32% iodine solution to 28% iodine solution. He had 40% and 24% iodine solutions available. The amount of the 24% solution he used was 4 times the amount of the 40% solution. How many milliliters of each solution did he add to the 50 mL of the 32% solution?

To write the equation, you need to understand the following:
(1) An iodine solution consists of iodine and water.
(2) The amount of iodine in the separate solutions is the same as in the total solution.

SQUARE ROOTS

Objectives

To solve a quadratic equation by using the definition of square root
To approximate a square root to the nearest hundredth by using a square root table
To determine whether a given number is rational or irrational

You can solve the quadratic equation $x^2 = 16$ by factoring:

$$x^2 = 16$$
$$x^2 - 16 = 0$$
$$(x + 4)(x - 4) = 0$$

The solutions are -4 and 4.

The solutions to this equation can also be found by using the concept of **square root** as shown below.

$$x^2 = 16$$

$$x = \sqrt{16} = 4 \quad \text{or} \quad x = -\sqrt{16} = -4$$

$\sqrt{16}$ represents the *positive* square root of 16, which is 4.
$-\sqrt{16}$ represents the *negative* square root of 16, which is -4.
Notice that $(\sqrt{16})^2 = 16$ since $4^2 = 16$ and $(-\sqrt{16})^2 = 16$ since $(-4)^2 = 16$.

Definition: Square root	If $x^2 = k$, then $x = \sqrt{k}$ or $x = -\sqrt{k}$, for each number $k \geq 0$. $\sqrt{k}$ is the *principal*, or *positive*, *square root* of k and $-\sqrt{k}$ is the *negative square root* of k.

The symbol $\sqrt{k}$ (read "the square root of k") is called a **radical,** where k is the **radicand,** and $\sqrt{}$ is the **radical sign,** or **square root sign.**

The number $\sqrt{25}$ can easily be rewritten without a radical sign since the radicand, 25, is a *perfect square*. That is, 25 is the square of an integer. The number $\sqrt{10}$ cannot be simplified in this way since the radicand, 10, is not a perfect square.

Example 1

Solve each equation.
$$x^2 + 5 = 15 \qquad\qquad 3y^2 = 75$$

Subtract 5 from each side of the equation to obtain the form $x^2 = k$.

$$x^2 = 10$$
$$x = \sqrt{10} \quad \text{or} \quad x = -\sqrt{10}$$

Thus, the solutions are $\sqrt{10}$ and $-\sqrt{10}$.

Divide each side of the equation by 3 to obtain the form $y^2 = k$.

$$y^2 = 25$$
$$y = \sqrt{25} \quad \text{or} \quad y = -\sqrt{25}$$
$$y = \sqrt{5^2} = 5 \quad \text{or} \quad y = -\sqrt{5^2} = -5$$

Thus, the solutions are 5 and -5.

In Example 1, $\sqrt{10}$ and $-\sqrt{10}$ cannot be rewritten as integers, but each number can be approximated to the nearest hundredth by using the table of square roots in the back of the book. To the nearest hundredth, $\sqrt{10} \doteq 3.16$ and $-\sqrt{10} \doteq -3.16$. The symbol $\doteq$ means *is approximately equal to*.

Example 2   **Solve each equation. Give the solutions to the nearest hundredth.**

$$5y^2 - 9 = 206 \qquad\qquad 20n^2 = 4n^2 + 81$$

$$5y^2 = 215 \qquad\qquad\qquad\qquad 16n^2 = 81$$
$$y^2 = 43 \qquad\qquad\qquad\qquad\qquad n^2 = \frac{81}{16}$$
$$y = \sqrt{43} \quad \text{or} \quad y = -\sqrt{43}$$
$$n = \sqrt{\frac{81}{16}} = \frac{9}{4} \quad \text{or} \quad n = -\sqrt{\frac{81}{16}} = -\frac{9}{4}$$

Use the table of square roots.

The solutions are 6.56 and -6.56.   The solutions are 2.25 and -2.25.

The numbers $\sqrt{\dfrac{81}{16}}$ and $\sqrt{1.44}$ can be rewritten as rational numbers:

$$\sqrt{\frac{81}{16}} = \sqrt{\left(\frac{9}{4}\right)^2} = \frac{9}{4} \qquad \sqrt{1.44} = \sqrt{(1.2)^2} = 1.2$$

Notice that $\dfrac{81}{16}$ and 1.44 are *perfect squares* since they are the squares of the rational numbers $\dfrac{9}{4}$ and 1.2, respectively.

The number $\sqrt{35}$ is not a rational number. It cannot be written in the form $\dfrac{a}{b}$ where a and b are integers since 35 is not a perfect square. To 13 decimal places, $\sqrt{35} \doteq 5.9160797830996$. The decimal for $\sqrt{35}$ is *nonterminating and nonrepeating*. Numbers of this type are called **irrational numbers.** In general, for $k \geq 0$,

(1) $\sqrt{k}$ is a rational number if k is a perfect square, and
(2) $\sqrt{k}$ is an irrational number if k is not a perfect square.

Example 3   **Determine whether each number is rational or irrational.**

Number	Answer	(Reason)
$3.888\ldots$	Rational	A repeating decimal
5.37	Rational	A terminating decimal
$4.3131131113\ldots$	Irrational	A nonterminating, nonrepeating decimal
$\sqrt{\dfrac{2}{9}}$	Irrational	$\dfrac{2}{9}$ is not a perfect square.
$\sqrt{\dfrac{169}{9}}$	Rational	$\dfrac{169}{9} = \left(\dfrac{13}{3}\right)^2$
$\sqrt{99}$	Irrational	99 is not a perfect square.
$-\sqrt{6.25}$	Rational	$6.25 = 2.5^2$

Your work in mathematics often involves both rational and irrational numbers. All the rational numbers together with all the irrational numbers are called the **real numbers**.

Reading in Algebra

Determine whether each statement is true or false.
1. $\sqrt{17}$ is a radical sign.
2. The radical sign in $\sqrt{6}$ is $\sqrt{}$.
3. $\sqrt{15}$ is a radical.
4. The radicand in $\sqrt{52}$ is 52.
5. $\frac{4}{9}$ is a perfect square.
6. $\frac{25}{11}$ is not a perfect square.
7. 0.9 is a perfect square.
8. 0.49 is not a perfect square.
9. $\sqrt{2}$ is a real number.
10. $\sqrt{4}$ is a real number.
11. $\sqrt{17} \doteq 4.12$ to the nearest hundredth.
12. $\sqrt{19} \doteq 4.35$ to the nearest hundredth.

Written Exercises

Solve each equation.
(A)
1. $x^2 = 64$
2. $y^2 = 3$
3. $4n^2 = 64$
4. $c^2 + 6 = 21$
5. $a^2 - 7 = 22$
6. $900 = x^2$
7. $6y^2 + 14 = 434$
8. $16n^2 = 7n^2 + 4$

Solve each equation. Approximate the solutions to the nearest hundredth.
9. $x^2 = 87$
10. $3y^2 = 195$
11. $4n^2 = 25$
12. $25c^2 - 7 = 9$
13. $9a^2 = 2a^2 + 294$
14. $4n^2 = 20n^2 - 9$
15. $3x^2 - 86 = x^2 + 86$
16. $6n^2 + 75 = n^2 + 455$

Determine whether each number is rational or irrational.
17. $5.282282228\ldots$
18. $0.123123123\ldots$
19. $4.343343334\ldots$
20. $7.65765765\ldots$
21. $\sqrt{169}$
22. $\sqrt{47}$
23. $-\sqrt{400}$
24. $-\sqrt{75}$
25. 87.6543
26. 19.4747
27. $-\sqrt{3,000}$
28. $\sqrt{2,025}$
(B)
29. $\sqrt{\dfrac{49}{81}}$
30. $\sqrt{\dfrac{18}{100}}$
31. $\sqrt{\dfrac{18}{8}}$
32. $\sqrt{\dfrac{20}{40}}$
33. $\sqrt{2.5}$
34. $\sqrt{2.25}$
35. $\sqrt{0.36}$
36. π

Solve each equation. Approximate the solutions to the nearest hundredth.
37. $144n^2 = 64$
38. $60x^2 = 9 - 4x^2$
39. $7y^2 - 9 = 16y^2 - 58$
40. $y^2 = 0.25$
41. $n^2 - 0.0009 = 0.004$
42. $5x^2 - 1 = 4x^2 + 0.21$

(C)
43. $x^4 - 30x^2 + 125 = 0$
44. $6x^4 = 30x^2 - 36$
45. $\dfrac{x^2 + 3}{4} + \dfrac{2x^2 - 5}{3} = \dfrac{3x^2 + 2}{6}$
46. $\dfrac{2}{3}(9x^2 + 21) - \dfrac{3}{4}(12x^2 - 20) + 148 = 0$

CUMULATIVE REVIEW

Simplify each expression.
1. $(-7x^2y^5z^9)^2$
2. $(-4ab^2c^7)^3$
3. $10m^2n(3m^3np^6)^4$

SIMPLIFYING RADICALS

Objective
To simplify a radical in which the radicand contains a perfect square factor

At the right, you see that $\sqrt{100 \cdot 36} = \sqrt{100} \cdot \sqrt{36}$. In general, the square root of the product of two or more real numbers is the product of the square roots of the numbers.

$\sqrt{100 \cdot 36}$	$\sqrt{100} \cdot \sqrt{36}$
$\sqrt{3{,}600}$	$10 \cdot 6$
60	60

Square root of a product	$\sqrt{x \cdot y} = \sqrt{x} \cdot \sqrt{y}$ for all real numbers $x \geq 0$, $y \geq 0$.

Example 1
Simplify $\sqrt{189}$. Approximate the result to the nearest hundredth.

$$\begin{aligned}
\sqrt{189} &= \sqrt{9 \cdot 21} && \blacktriangleleft \text{ 9 is a perfect square factor of 189.}\\
&= \sqrt{9} \cdot \sqrt{21} && \blacktriangleleft \ \sqrt{x \cdot y} = \sqrt{x} \cdot \sqrt{y}\\
&= 3 \cdot \sqrt{21}, \text{ or } 3\sqrt{21}\\
&\doteq 3 \cdot 4.583 && \blacktriangleleft \ \sqrt{21} \doteq 4.583\\
&\doteq 13.749\\
&\doteq 13.75 && \blacktriangleleft \text{ Round to the nearest hundredth.}
\end{aligned}$$
Thus, $\sqrt{189} \doteq 13.75$.

In Example 1, the expression $3\sqrt{21}$ is the **simplest radical form** of $\sqrt{189}$. A radical or square root expression is in simplest radical form whenever the radicand is the smallest possible integer. In other words, the radicand must be an integer that does not contain a perfect square factor, other than 1. Unless otherwise indicated, all answers involving radicals should be given in simplest radical form.

Example 2
Simplify $-5\sqrt{72}$.

$$\begin{aligned}
-5\sqrt{72} &= -5\sqrt{36 \cdot 2}\\
&= -5\sqrt{36} \cdot \sqrt{2}\\
&= -5 \cdot 6 \cdot \sqrt{2}\\
&= -30\sqrt{2}
\end{aligned}$$
Thus, $-5\sqrt{72} = -30\sqrt{2}$.

When simplifying a radical, the work is easier if you identify the *greatest perfect square* factor of the radicand. In Example 2, 9 and 4 are also perfect square factors of 72. Notice what happens when these numbers are used to simplify $-5\sqrt{72}$.

$$-5\sqrt{72} = -5\sqrt{9 \cdot 4 \cdot 2} = -5\sqrt{9} \cdot \sqrt{4} \cdot \sqrt{2} = -5 \cdot 3 \cdot 2\sqrt{2} = -30\sqrt{2}$$

In this case, the Square Root of a Product property is extended to three numbers.

Notice that $\sqrt{5^2} = 5$ but $\sqrt{(-5)^2} \neq -5$ because $\sqrt{(-5)^2} = \sqrt{25} = 5$. This shows that it is *not* true that:

$$\sqrt{x^2} = x \text{ for } each \text{ real number } x.$$

You can simplify radicals like $\sqrt{x^2}$, $\sqrt{x^4}$, and $\sqrt{x^6}$ where the exponents are *even* integers and x is a *negative* number ($x < 0$). Study the statements below.

1. $\sqrt{(-6)^2} \neq -6$ but $\sqrt{(-6)^2} = \sqrt{36} = 6 = |-6|$.

         $\sqrt{x^2} \neq x^1$ but $\sqrt{x^2} = |x^1| \text{ if } x < 0$.

2. $\sqrt{(-3)^4} = \sqrt{81} = 9 = (-3)^2 \text{ and } \sqrt{x^4} = x^2 \text{ if } x < 0$.

3. $\sqrt{(-2)^6} \neq (-2)^3$ but $\sqrt{(-2)^6} = \sqrt{64} = 8 = |(-2)^3|$.

         $\sqrt{x^6} \neq x^3.$ but $\sqrt{x^6} = |x^3| \text{ if } x < 0$.

Thus, if x is *any* real number, then

$$\sqrt{x^2} = |x|, \quad \sqrt{x^4} = x^2, \quad \sqrt{x^6} = |x^3|, \quad \sqrt{x^8} = x^4, \quad \text{and so on, } or$$

$$\sqrt{x^{2n}} = |x^n| \text{ if } n \text{ is an } odd \text{ integer and}$$
$$\sqrt{x^{2n}} = x^n \text{ if } n \text{ is an } even \text{ integer.}$$

Radicals like $\sqrt{x^1}$ and $\sqrt{x^3}$, where the exponents are *odd* integers, are *not* real numbers if x is a *negative* number ($x < 0$), as shown below.

$$\sqrt{-2} \text{ and } \sqrt{(-2)^3}, \text{ or } \sqrt{-8}, \text{ are } not \text{ real numbers.}$$

To simplify $\sqrt{x^7}$ and $\sqrt{x^9}$, where the exponents are *odd* integers and x is a *non-negative* number ($x \geq 0$), rewrite the radicands as shown below.

$$\sqrt{x^7} = \sqrt{x^6 \cdot x^1} = \sqrt{x^6}\sqrt{x} = x^3\sqrt{x} \text{ if } x \geq 0.$$
$$\sqrt{x^9} = \sqrt{x^8 \cdot x^1} = \sqrt{x^8}\sqrt{x} = x^4\sqrt{x} \text{ if } x \geq 0.$$

Example 3   **Simplify $7ab\sqrt{24a^{10}b^7}$ if $b \geq 0$.**

$$7ab\sqrt{24a^{10}b^7} = 7ab\sqrt{4 \cdot 6 \cdot a^{10} \cdot b^6 \cdot b^1}$$
$$= 7ab \cdot \sqrt{4}\sqrt{a^{10}}\sqrt{b^6} \cdot \sqrt{6b}$$
$$= 7a^1b^1 \cdot 2 \cdot |a^5| \cdot b^3\sqrt{6b}$$
$$= 14a|a^5| \cdot b^4\sqrt{6b}$$

Reading in Algebra

Determine whether each statement is true or false.

1. The square root of the product of three positive real numbers is the product of their square roots.

2. The greatest perfect square factor of 200 is 25.

3. The greatest perfect square factor of 88 is 4.

4. The simplest radical form of $\sqrt{12}$ is $4\sqrt{3}$.

5. The simplest radical form of $\sqrt{32}$ is $2\sqrt{8}$.

6. There are exactly nine integers between 0 and 101 that are perfect squares.

7. $4x^{14}$ is a perfect square.

8. $9y^9$ is a perfect square.

Match each expression at the left with its *one* equivalent expression at the right.

9. $\sqrt{a^2b^5}$ if $b \geq 0$ **(A)** $ab^2\sqrt{b}$ **(B)** $|a|b^2\sqrt{b}$ **(C)** $a|b^2|\sqrt{b}$

10. $a\sqrt{a^2b^5}$ if $b \geq 0$ **(A)** $a^2b^2\sqrt{b}$ **(B)** $|a^2|b^2\sqrt{b}$ **(C)** $a|a|b^2\sqrt{b}$

11. $ab\sqrt{a^3b^6}$ if $a \geq 0$ **(A)** $a^2b^4\sqrt{a}$ **(B)** $a^2b|b^3|\sqrt{a}$ **(C)** $a^2|b^4|\sqrt{a}$

12. $ab^2\sqrt{a^3b^6}$ if $a \geq 0$ **(A)** $a^2b^5\sqrt{a}$ **(B)** $a^2b^2|b^3|\sqrt{a}$ **(C)** $|a^2|b^5\sqrt{a}$

Written Exercises

Simplify each expression. Approximate the result to the nearest hundredth.

Ⓐ   1. $\sqrt{250}$ 2. $-\sqrt{135}$ 3. $\sqrt{180}$ 4. $-\sqrt{243}$

5. $3\sqrt{120}$ 6. $-10\sqrt{490}$ 7. $2.5\sqrt{162}$ 8. $0.75\sqrt{512}$

Simplify each expression. Assume that no denominator has the value of 0.

9. $2\sqrt{27}$ 10. $5\sqrt{32}$ 11. $-6\sqrt{45}$ 12. $-10\sqrt{96}$

13. $5\sqrt{4x^6}$ 14. $\sqrt{12b^2}$ 15. $9a\sqrt{16a^{12}}$ 16. $6y^3\sqrt{24y^{10}}$

17. $\sqrt{n^9}$, $n \geq 0$ 18. $8\sqrt{y^{15}}$, $y \geq 0$ 19. $-2y^4\sqrt{25y^5}$, $y \geq 0$ 20. $-12c^2\sqrt{8c^{17}}$, $c \geq 0$

Ⓑ   21. $\sqrt{24c^{10}d^3}$, $d \geq 0$ 22. $ab^2\sqrt{a^{11}b^{16}}$, $a \geq 0$ 23. $8x^2\sqrt{9x^5y^5}$; $x \geq 0$, $y \geq 0$

24. $5mn\sqrt{8m^{20}n^7}$, $n \geq 0$ 25. $\sqrt{90a^4b^{15}c^{18}}$, $b \geq 0$ 26. $c^2d\sqrt{180c^8d^9}$, $d \geq 0$

27. $7c^2\sqrt{128c^{12}d}$, $d \geq 0$ 28. $m^2n^4\sqrt{m^{11}n^6p^3}$; $m \geq 0$, $p \geq 0$ 29. $x^{10}\sqrt{243x^{30}y^{25}}$, $y \geq 0$

30. $\dfrac{\sqrt{x^7y^8z^9}}{x^2y^3z^4}$; $x > 0$, $z > 0$ 31. $\dfrac{\sqrt{27a^8b^{11}c^{12}}}{6ab^5c^{10}}$, $b > 0$ 32. $\dfrac{3z\sqrt{108x^5y^7z^9}}{14x^2y^2}$; $x > 0$, $y > 0$, $z > 0$

Simplify. Assume that m and n are integers and that x and y are positive real numbers.

Ⓒ   33. $\sqrt{x^{4m}}$ 34. $\sqrt{x^{4m+1}}$ 35. $\sqrt{x^{2m+3}}$ 36. $x^m\sqrt{x^{6m+5}}$

37. $x^my^n\sqrt{x^{6m}y^{8n}}$ 38. $x^{2m-3}\sqrt{x^{4m+9}}$ 39. $xy\sqrt{x^{6m}y^{8n+3}}$ 40. $x^{2m}y^{3n}\sqrt{x^{2m}y^{3n}}$

NON-ROUTINE PROBLEMS

1. If n is any integer, then $n^2(n^2 - 1)$ is *always* divisible by what positive integers?

2. At a party, Mantia placed 8 saucers around a rectangular table. She put 24 peanuts in the 8 saucers so that there were 9 peanuts along each side of the table.

    Ira said, "I can pick up the 24 peanuts, add 8 more, and place the 32 peanuts in the 8 saucers so there will still be 9 peanuts along each side of the table." How can Ira do this?

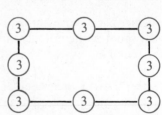

SUMS, DIFFERENCES, AND PRODUCTS OF RADICALS 6.3

Objectives   **To simplify an expression containing sums or differences of radicals**
To simplify an expression containing products of radicals

You have simplified algebraic expressions by combining (adding or subtracting) like terms, using the Distributive property. For example, $5x^2y + 3x^2y + 2xy^2 = (5 + 3)x^2y + 2xy^2 = 8x^2y + 2xy^2$. Expressions containing like radicals can be simplified in a similar way.

$$6\sqrt{10} + 8\sqrt{10} + 2\sqrt{7} = (6 + 8)\sqrt{10} + 2\sqrt{7} = 14\sqrt{10} + 2\sqrt{7}$$

The expression $14\sqrt{10} + 2\sqrt{7}$ is in its simplest radical form since $14\sqrt{10}$ and $2\sqrt{7}$ are *not* like radicals and, therefore, cannot be combined. The like radicals in this case are $6\sqrt{10}$ and $8\sqrt{10}$.

Example 1   **Simplify $8\sqrt{3} + 4\sqrt{6} - 2\sqrt{3} - 7\sqrt{6} + 9\sqrt{5}$.**

$8\sqrt{3} + 4\sqrt{6} - 2\sqrt{3} - 7\sqrt{6} + 9\sqrt{5}$
$= (8\sqrt{3} - 2\sqrt{3}) + (4\sqrt{6} - 7\sqrt{6}) + 9\sqrt{5}$ ◀ *Group the like radicals together.*
$= 6\sqrt{3} - 3\sqrt{6} + 9\sqrt{5}$ ◀ *Combine the like radicals.*

Example 2   **Simplify each expression.**
$4\sqrt{18} - \sqrt{8} + \sqrt{2}$ $\qquad\qquad$ $\sqrt{36ab^3} - \sqrt{ab^3} + 5b\sqrt{ab}\,; a \ge 0, b \ge 0$

$4\sqrt{9 \cdot 2} - \sqrt{4 \cdot 2} + \sqrt{2}$ $\qquad$ $\sqrt{36 \cdot a \cdot b^2 \cdot b} - \sqrt{a \cdot b^2 \cdot b} + 5b\sqrt{ab}$
$= 4 \cdot 3\sqrt{2} - 2\sqrt{2} + \sqrt{2}$ $\qquad\qquad$ $= 6b\sqrt{ab} - b\sqrt{ab} + 5b\sqrt{ab}$
$= 12\sqrt{2} - 2\sqrt{2} + \sqrt{2}$ $\qquad\qquad$ $= 10b\sqrt{ab}$
$= 11\sqrt{2}$

You can simplify $\sqrt{50} \cdot \sqrt{12}$ as shown at the right. In general, the product of the square roots of two or more real numbers is the square root of the product of the numbers.

$$\begin{aligned}\sqrt{50} \cdot \sqrt{12} &= \sqrt{50 \cdot 12}\\ &= \sqrt{600}\\ &= \sqrt{100 \cdot 6}\\ &= 10\sqrt{6}\end{aligned}$$

Product of square roots	For all real numbers $x \ge 0$, $y \ge 0$, $\sqrt{x} \cdot \sqrt{y} = \sqrt{x \cdot y}$ and $\sqrt{x} \cdot \sqrt{x} = x$.

Using the first product property above,
$$\sqrt{12} \cdot \sqrt{2} = \sqrt{12 \cdot 2} = \sqrt{24} = \sqrt{4 \cdot 6} = 2\sqrt{6}.$$

The second product property on page 145 involves a special case in the multiplication of radicals. Notice that $\sqrt{x} \cdot \sqrt{x} = (\sqrt{x})^2 = x$. Using this property, $\sqrt{17} \cdot \sqrt{17} = 17$ and $\sqrt{m+n} \cdot \sqrt{m+n} = m+n$.

Example 3   **Simplify each expression.**

$$-7\sqrt{30} \cdot 6\sqrt{30}$$

$$\sqrt{6a^3b^8} \cdot \sqrt{8a^5b^3}; \ a \geq 0, \ b \geq 0$$

$-7 \cdot 6 \cdot \sqrt{30} \cdot \sqrt{30}$

$= -42 \cdot 30$ ◀ $\sqrt{x} \cdot \sqrt{x} = x$

$= -1{,}260$

$\sqrt{6a^3b^8 \cdot 8a^5b^3}$ ◀ $\sqrt{x} \cdot \sqrt{y} = \sqrt{x \cdot y}$

$= \sqrt{6 \cdot 8 \cdot a^3 \cdot a^5 \cdot b^8 \cdot b^3}$

$= \sqrt{48a^8b^{11}}$

$= \sqrt{16 \cdot 3 \cdot a^8 \cdot b^{10} \cdot b}$

$= 4a^4b^5\sqrt{3b}$

You can also simplify expressions containing products of radicals in which binomials or trinomials are involved. Two such cases are shown in Example 4.

Example 4   **Simplify each expression.**

$$5\sqrt{3}(2\sqrt{6} + 4\sqrt{3} - \sqrt{7})$$

$$(3\sqrt{5} + \sqrt{2})(4\sqrt{5} - 6\sqrt{2})$$

Use the Distributive property to multiply.

$5\sqrt{3} \cdot 2\sqrt{6} + 5\sqrt{3} \cdot 4\sqrt{3} - 5\sqrt{3} \cdot \sqrt{7}$

$= 10\sqrt{18} + 20 \cdot 3 - 5\sqrt{21}$

$= 10 \cdot \sqrt{9 \cdot 2} + 60 - 5\sqrt{21}$

$= 30\sqrt{2} + 60 - 5\sqrt{21}$

Use the FOIL method to multiply.

$3\sqrt{5} \cdot 4\sqrt{5} - 3\sqrt{5} \cdot 6\sqrt{2} + \sqrt{2} \cdot 4\sqrt{5} - \sqrt{2} \cdot 6\sqrt{2}$

$= 12 \cdot 5 - 18\sqrt{10} + 4\sqrt{10} - 6 \cdot 2$

$= 60 - 18\sqrt{10} + 4\sqrt{10} - 12$

$= 48 - 14\sqrt{10}$

The special product formulas, $(a + b)(a - b) = a^2 - b^2$ and $(a - b)^2 = a^2 - 2ab + b^2$, can sometimes be used to simplify expressions containing products of radicals.

Example 5   **Simplify each expression.**

$$(2\sqrt{7} + 4\sqrt{3})(2\sqrt{7} - 4\sqrt{3}) \qquad (3\sqrt{c} - 5\sqrt{d})^2; \ c \geq 0, \ d \geq 0$$

$(a + b)(a - b) = a^2 - b^2$ ▶

$(2\sqrt{7} + 4\sqrt{3})(2\sqrt{7} - 4\sqrt{3})$

$= (2\sqrt{7})^2 - (4\sqrt{3})^2$

$= 28 - 48$, or -20

$(3\sqrt{c} - 5\sqrt{d})^2$

$= (3\sqrt{c})^2 - 2 \cdot 3\sqrt{c} \cdot 5\sqrt{d} + (5\sqrt{d})^2$

$= 9c - 30\sqrt{cd} + 25d$

In Example 5, notice how the Power of a Product property, $(xy)^n = x^n \cdot y^n$, is used with radical expressions:

$$(2\sqrt{7})^2 = (2)^2 \cdot (\sqrt{7})^2 = 4 \cdot 7 = 28 \text{ and}$$
$$(5\sqrt{d})^2 = (5)^2 \cdot (\sqrt{d})^2 = 25 \cdot d = 25d.$$

Oral Exercises

Simplify. Assume that values of the variables are nonnegative numbers.

1. $\sqrt{11} \cdot \sqrt{7}$
2. $-3\sqrt{6} \cdot \sqrt{5}$
3. $5\sqrt{c} \cdot 2\sqrt{d}$
4. $\sqrt{2a} \cdot \sqrt{3b}$
5. $(\sqrt{c})^2$
6. $\sqrt{2} \cdot \sqrt{32}$
7. $\sqrt{3} \cdot \sqrt{12}$
8. $\sqrt{5} \cdot \sqrt{2} \cdot \sqrt{10}$
9. $\sqrt{3} + \sqrt{3}$
10. $5\sqrt{2} - \sqrt{2}$

Written Exercises

Simplify. Assume that values of the variables are nonnegative numbers.

(A)
1. $3\sqrt{10} + 9\sqrt{7} - 8\sqrt{10} - 3\sqrt{7}$
2. $4\sqrt{5} - 7\sqrt{6} - \sqrt{5} + 10\sqrt{3} + \sqrt{6}$
3. $7\sqrt{45} - \sqrt{20} + \sqrt{5} - \sqrt{80}$
4. $3\sqrt{50} + 4\sqrt{18} - 2\sqrt{32} - \sqrt{8}$
5. $\sqrt{25n} - 7\sqrt{n} + \sqrt{16n}$
6. $\sqrt{x} - 6\sqrt{4x} - \sqrt{9x} + 5\sqrt{36x}$
7. $8\sqrt{6} \cdot 5\sqrt{10}$
8. $-2\sqrt{15} \cdot 4\sqrt{5} \cdot \sqrt{6}$
9. $\sqrt{6y} \cdot \sqrt{3y^9}$
10. $4\sqrt{3x^3} \cdot 5\sqrt{15x^8}$
11. $3\sqrt{5}(4\sqrt{5} - \sqrt{10} + 2\sqrt{7})$
12. $-4\sqrt{6}(\sqrt{15} + 8\sqrt{7} - 2\sqrt{6})$
13. $(4\sqrt{11} + 6)(2\sqrt{11} - 8)$
14. $(\sqrt{10} - 2\sqrt{3})(6\sqrt{10} - \sqrt{3})$
15. $(2\sqrt{7} - 4\sqrt{2})(5\sqrt{7} + 6\sqrt{2})$
16. $(8\sqrt{30} + 4\sqrt{7})(\sqrt{30} + \sqrt{7})$
17. $(-3\sqrt{11})^2$
18. $(7\sqrt{x})^2$
19. $(20\sqrt{17})^2$
20. $(-8\sqrt{y} + 2)^2$
21. $(5\sqrt{3} + 7)(5\sqrt{3} - 7)$
22. $(6\sqrt{5} + 4\sqrt{6})(6\sqrt{5} - 4\sqrt{6})$
23. $(\sqrt{7} + \sqrt{6})^2$
24. $(5\sqrt{2} - 4\sqrt{3})^2$
25. $(4\sqrt{x} + 2\sqrt{5})(4\sqrt{x} - 2\sqrt{5})$
26. $(3\sqrt{7} - 8\sqrt{y})(3\sqrt{7} + 8\sqrt{y})$
27. $(6\sqrt{n} - 6)^2$
28. $(5\sqrt{c} + 4\sqrt{10})^2$

(B)
29. $\sqrt{8} + \sqrt{12} + \sqrt{18} + \sqrt{27}$
30. $5\sqrt{2} - 4\sqrt{20} - 3\sqrt{45} + 2\sqrt{50}$
31. $2\sqrt{c^2d} + \sqrt{36c^2d} - 3c\sqrt{d}$
32. $12b^2\sqrt{ab} - \sqrt{25ab^5} + 10\sqrt{ab^5}$
33. $5\sqrt{3m^2n^3} + 2mn\sqrt{27n}$
34. $2x^2\sqrt{20x^3y} - x^2\sqrt{45x^3y} - x^3\sqrt{5xy}$
35. $\sqrt{3c^3d^5} \cdot 8\sqrt{8cd^7}$
36. $-8x\sqrt{6xy^5} \cdot 5y\sqrt{8x^5y^3} \cdot xy\sqrt{xy}$
37. $(4\sqrt{c} + \sqrt{7})(8\sqrt{c} - 3\sqrt{7})$
38. $(5\sqrt{a} - \sqrt{b})(4\sqrt{a} - \sqrt{b})$
39. $(\sqrt{x} - 3\sqrt{y})(3\sqrt{x} + \sqrt{y})$
40. $(x\sqrt{y} + y\sqrt{x})(2x\sqrt{y} + 3y\sqrt{x})$
41. $(5\sqrt{c} - 4\sqrt{d})(5\sqrt{c} + 4\sqrt{d})$
42. $(7\sqrt{a} + 2b)(7\sqrt{a} - 2b)$
43. $(\sqrt{a} + \sqrt{b})^2$
44. $(6\sqrt{c} - 5\sqrt{d})^2$

(C)
45. $(\sqrt{x + 4} + 5)(\sqrt{x + 4} + 7)$
46. $(\sqrt{3x - 7} - 4)(\sqrt{3x - 7} + 4)$, $3x - 7 \geq 0$
47. $(\sqrt{x + y} + \sqrt{y})(\sqrt{x + y} - \sqrt{y})$
48. $(6 - \sqrt{y - 3})^2$, $y \geq 3$
49. $(\sqrt{n + 5} - \sqrt{n})^2$
50. $(3\sqrt{c - 8} + 4\sqrt{c})^2$, $c \geq 8$

Factor each binomial as a difference of two squares.
[*Hint:* Rewrite $x - y$ as $(\sqrt{x})^2 - (\sqrt{y})^2$.]

51. $m - n$
52. $4x - y$
53. $c - 36d$
54. $5a - 3b$
55. $x^2 - 2y$
56. $9c - 7d^2$
57. $6m^2 - 25n$
58. $49a^2b - 10cd^2$

Sums, Differences, and Products of Radicals

QUOTIENTS OF RADICALS   6.4

Objective   **To simplify an expression containing a quotient of radicals**

The quotient of the square roots of two real numbers is the square root of the quotient of the numbers.

Quotient of square roots	For all real numbers $x \geq 0$, $y > 0$, $\dfrac{\sqrt{x}}{\sqrt{y}} = \sqrt{\dfrac{x}{y}}$.

The expression $\dfrac{\sqrt{24}}{\sqrt{3}}$ contains a quotient of two square roots in which the radicand in the numerator is exactly divisible by the radicand in the denominator. Expressions of this type can be simplified by applying the Quotient of Square Roots property, as shown in Example 1.

Example 1   **Simplify each expression.**

$$\frac{\sqrt{24}}{\sqrt{3}} \qquad\qquad \frac{\sqrt{28n^6}}{\sqrt{7n^2}}, n \neq 0 \qquad\qquad \frac{\sqrt{18c^7}}{\sqrt{6c^2}}, c > 0$$

$$= \sqrt{\frac{24}{3}} \quad \blacktriangleleft \quad \frac{\sqrt{x}}{\sqrt{y}} = \sqrt{\frac{x}{y}} \qquad = \sqrt{\frac{28n^6}{7n^2}} \qquad = \sqrt{\frac{18c^7}{6c^2}}$$

$$= \sqrt{8} \qquad\qquad\qquad\qquad\qquad = \sqrt{4n^4} \qquad\qquad = \sqrt{3c^5}$$

$$= 2\sqrt{2} \qquad\qquad\qquad\qquad\qquad = 2n^2 \qquad\qquad = c^2\sqrt{3c}$$

To simplify an expression like $\dfrac{\sqrt{5}}{3\sqrt{2}}$, in which the radicand in the numerator is *not* exactly divisible by the radicand in the denominator, a procedure called **rationalizing the denominator** is used. The denominator, $3\sqrt{2}$, is an irrational number. Notice what happens when $\dfrac{\sqrt{5}}{3\sqrt{2}}$ is multiplied by $\dfrac{\sqrt{2}}{\sqrt{2}}$, or 1.

$$\frac{\sqrt{5}}{3\sqrt{2}} = \frac{\sqrt{5}}{3\sqrt{2}} \cdot \frac{\sqrt{2}}{\sqrt{2}} = \frac{\sqrt{5} \cdot \sqrt{2}}{3\sqrt{2} \cdot \sqrt{2}} = \frac{\sqrt{10}}{3 \cdot 2} = \frac{\sqrt{10}}{6}$$

The new denominator, 6, is a rational number.

Example 2   **Simplify** $\dfrac{6\sqrt{8}}{\sqrt{6}}$.

$$\frac{6\sqrt{8}}{\sqrt{6}} = \frac{6\sqrt{8}}{\sqrt{6}} \cdot \frac{\sqrt{6}}{\sqrt{6}} = \frac{6\sqrt{8} \cdot \sqrt{6}}{\sqrt{6} \cdot \sqrt{6}} = \frac{6\sqrt{48}}{6} = \sqrt{48} = \sqrt{16 \cdot 3} = 4\sqrt{3}$$

Contrast the following two methods of rationalizing the denominator of $\dfrac{1}{\sqrt{12}}$:

Method 1: $\dfrac{1}{\sqrt{12}} = \dfrac{1}{2\sqrt{3}} \cdot \dfrac{\sqrt{3}}{\sqrt{3}} = \dfrac{\sqrt{3}}{6}$

Method 2: $\dfrac{1}{\sqrt{12}} = \dfrac{1}{\sqrt{12}} \cdot \dfrac{\sqrt{12}}{\sqrt{12}} = \dfrac{\sqrt{12}}{12} = \dfrac{2\sqrt{3}}{12} = \dfrac{\sqrt{3}}{6}$

In Method 1, simplifying the radical in the denominator before rationalizing made the work easier. The numerator and denominator of the original fraction were multiplied by a smaller number, $\sqrt{3}$, as opposed to $\sqrt{12}$, which was used in Method 2.

Example 3

Simplify $\dfrac{5\sqrt{7}}{\sqrt{12}}$ **and** $\dfrac{9\sqrt{5x}}{\sqrt{24x^4y^3}}$; $x > 0,\ y > 0.$

$$\dfrac{5\sqrt{7}}{\sqrt{12}} = \dfrac{5\sqrt{7}}{2\sqrt{3}} = \dfrac{5\sqrt{7}}{2\sqrt{3}} \cdot \dfrac{\sqrt{3}}{\sqrt{3}}$$

$$= \dfrac{5\sqrt{7} \cdot \sqrt{3}}{2\sqrt{3} \cdot \sqrt{3}}$$

$$= \dfrac{5\sqrt{21}}{6}$$

$$\dfrac{9\sqrt{5x}}{\sqrt{24x^4y^3}} = \dfrac{9\sqrt{5x}}{2x^2y\sqrt{6y}}$$

$$= \dfrac{9\sqrt{5x}}{2x^2y\sqrt{6y}} \cdot \dfrac{\sqrt{6y}}{\sqrt{6y}}$$

$$= \dfrac{9\sqrt{30xy}}{12x^2y^2},\ \text{or}\ \dfrac{3\sqrt{30xy}}{4x^2y^2}$$

The denominator of $\dfrac{7}{\sqrt{15} - \sqrt{10}}$ contains two terms. To rationalize this denominator, you use the **conjugate** of $\sqrt{15} - \sqrt{10}$, which is $\sqrt{15} + \sqrt{10}$. The product of these two irrational expressions is $(\sqrt{15} - \sqrt{10})(\sqrt{15} + \sqrt{10})$, which is equal to $15 - 10$, or the rational number 5.

Property of conjugates	The two binomials, $a + b$ and $a - b$, are a pair of *conjugates*. Their product is a difference of two squares: $(a + b)(a - b) = a^2 - b^2$.

The Property of Conjugates is used in Example 4 to simplify two radical expressions whose denominators are binomials.

Example 4

Simplify $\dfrac{7}{\sqrt{15} - \sqrt{10}}$ **and** $\dfrac{6\sqrt{y}}{5x + 2\sqrt{y}}$; $y > 0,\ 5x + 2\sqrt{y} \neq 0.$

$$\dfrac{7}{\sqrt{15} - \sqrt{10}} = \dfrac{7}{\sqrt{15} - \sqrt{10}} \cdot \dfrac{\sqrt{15} + \sqrt{10}}{\sqrt{15} + \sqrt{10}}$$

$$= \dfrac{7(\sqrt{15} + \sqrt{10})}{(\sqrt{15} - \sqrt{10})(\sqrt{15} + \sqrt{10})}$$

$(a - b)(a + b)$ ▶ $= a^2 - b^2$

$$= \dfrac{7\sqrt{15} + 7\sqrt{10}}{(\sqrt{15})^2 - (\sqrt{10})^2}$$

$$= \dfrac{7\sqrt{15} + 7\sqrt{10}}{5}$$

$$\dfrac{6\sqrt{y}}{5x + 2\sqrt{y}} = \dfrac{6\sqrt{y}}{5x + 2\sqrt{y}} \cdot \dfrac{5x - 2\sqrt{y}}{5x - 2\sqrt{y}}$$

$$= \dfrac{6\sqrt{y}(5x - 2\sqrt{y})}{(5x + 2\sqrt{y})(5x - 2\sqrt{y})}$$

$$= \dfrac{30x\sqrt{y} - 12y}{(5x)^2 - (2\sqrt{y})^2}$$

$$= \dfrac{30x\sqrt{y} - 12y}{25x^2 - 4y}$$

To simplify $\sqrt{\dfrac{7}{10}}$, you use the Quotient of Square Roots property to rewrite it as $\dfrac{\sqrt{7}}{\sqrt{10}}$ and then proceed to rationalize the denominator.

Example 5   **Simplify $\sqrt{\dfrac{7}{10}}$.**

$$\sqrt{\frac{7}{10}} = \frac{\sqrt{7}}{\sqrt{10}} = \frac{\sqrt{7}}{\sqrt{10}} \cdot \frac{\sqrt{10}}{\sqrt{10}} = \frac{\sqrt{70}}{10}$$

Example 6   **Simplify $\sqrt{\dfrac{5c}{12d^3}}$; $c > 0$, $d > 0$.**

$$\sqrt{\frac{5c}{12d^3}} = \frac{\sqrt{5c}}{\sqrt{12d^3}} = \frac{\sqrt{5c}}{2d\sqrt{3d}} = \frac{\sqrt{5c}}{2d\sqrt{3d}} \cdot \frac{\sqrt{3d}}{\sqrt{3d}} = \frac{\sqrt{15cd}}{6d^2}$$

Summary   An expression is in simplest radical form when
(1) each radicand contains no factor, other than 1, that is a perfect square,
(2) the denominator contains no radicals, and
(3) each radicand contains no fractions.

Reading in Algebra

Indicate why each expression is not in simplest radical form.

1. $\dfrac{5x^2}{\sqrt{3x}}$

2. $\dfrac{\sqrt{8y}}{5y}$

3. $\sqrt{\dfrac{7m}{3n}}$

4. $\dfrac{4\sqrt{3}}{6 - 5\sqrt{2}}$

Oral Exercises

Determine the conjugate of each expression.

1. $3x + 2y$

2. $2\sqrt{3} - 5\sqrt{2}$

3. $-2 + \sqrt{6}$

4. $-\sqrt{30} - \sqrt{10}$

Written Exercises

Simplify. Assume that no denominator has the value of 0 and that values of the variables and radicands are nonnegative numbers.

(A)

1. $\dfrac{\sqrt{21}}{\sqrt{7}}$

2. $\dfrac{\sqrt{90}}{\sqrt{10}}$

3. $\dfrac{\sqrt{88}}{\sqrt{11}}$

4. $\dfrac{\sqrt{150}}{\sqrt{2}}$

5. $\dfrac{\sqrt{a^{10}}}{\sqrt{a^6}}$

6. $\dfrac{\sqrt{x^6}}{\sqrt{x}}$

7. $\dfrac{\sqrt{45n^9}}{\sqrt{5n^3}}$

8. $\dfrac{\sqrt{24x^{12}}}{\sqrt{3x^5}}$

9. $\dfrac{1}{\sqrt{7}}$

10. $\dfrac{8}{5\sqrt{3}}$

11. $\dfrac{\sqrt{11}}{2\sqrt{6}}$

12. $\dfrac{-7\sqrt{6}}{6\sqrt{5}}$

13. $\dfrac{7}{\sqrt{12}}$

14. $\dfrac{6}{\sqrt{8}}$

15. $\dfrac{-5}{\sqrt{32}}$

16. $\dfrac{5\sqrt{44}}{\sqrt{18}}$

17. $\dfrac{5}{\sqrt{x^3}}$

18. $\dfrac{6}{\sqrt{9c^5}}$

19. $\dfrac{-10d}{\sqrt{32d^3}}$

20. $\dfrac{15x^2}{\sqrt{5x^7}}$

21. $\dfrac{7}{6 - \sqrt{30}}$

22. $\dfrac{-5}{3 + \sqrt{2}}$

23. $\dfrac{\sqrt{2}}{\sqrt{15} - 3}$

24. $\dfrac{-\sqrt{7}}{\sqrt{26} + 4}$

25. $\sqrt{\dfrac{5}{11}}$

26. $\sqrt{\dfrac{3c}{5d}}$

27. $\sqrt{\dfrac{23x}{8y}}$

28. $\sqrt{\dfrac{27m}{32n}}$

(B) 29. $\sqrt{\dfrac{4x^7}{7y^3}}$

30. $\sqrt{\dfrac{7a^3}{8m^3n^5}}$

31. $\sqrt{\dfrac{3c - 5}{6}}$

32. $\sqrt{\dfrac{5}{4x + 9}}$

33. $\dfrac{10}{\sqrt{22} + \sqrt{11}}$

34. $\dfrac{2}{5\sqrt{3} - \sqrt{15}}$

35. $\dfrac{\sqrt{5}}{2\sqrt{5} + 3\sqrt{2}}$

36. $\dfrac{2\sqrt{3}}{4\sqrt{2} - 3\sqrt{3}}$

37. $\dfrac{5 + \sqrt{10}}{8 - \sqrt{10}}$

38. $\dfrac{\sqrt{7} + 2}{3\sqrt{7} + 9}$

39. $\dfrac{3\sqrt{x}}{5\sqrt{x} - 2\sqrt{y}}$

40. $\dfrac{\sqrt{cd}}{6\sqrt{c} - 3\sqrt{d}}$

(C) 41. $\dfrac{3\sqrt{x + y}}{\sqrt{x + y} + 2}$

42. $\dfrac{\sqrt{c + d}}{5 - \sqrt{c + d}}$

43. $\dfrac{4\sqrt{5}}{(\sqrt{5} + \sqrt{3}) - 2}$

44. $\dfrac{\sqrt{2}}{(\sqrt{30} - \sqrt{2}) + 3}$

NON-ROUTINE PROBLEMS

1. Find the value of $\dfrac{x - 1}{x + 1}$ in simplest radical form if you replace

each x with $\dfrac{n - 1}{n + 1}$ and replace each n with $\sqrt{2}$.

2. $1 + \sqrt{3} + \dfrac{1}{1 - \sqrt{3}} + \dfrac{1}{1 + \sqrt{3}} =$

    (A) $2 + \sqrt{3}$ **(B)** $2 + 2\sqrt{3}$ **(C)** $\sqrt{3}$ **(D)** $1 + \sqrt{3}$ **(E)** $1 - \sqrt{3}$

Simplify. Answer in simplest radical form.

3. $\dfrac{\sqrt{\sqrt{\sqrt{16x^{16}}}}}{\sqrt{\sqrt{25y^{24}}}}$

4. $\dfrac{\sqrt{\sqrt{\sqrt{\sqrt{256x^{64}}}}}}{\sqrt{\sqrt{\sqrt{81y^{32}}}}}$

CUMULATIVE REVIEW

1. Tonya mixed some Brand A dog food worth 45¢/lb and some Brand B dog food worth 63¢/lb to make a 20-lb package that was worth $10.44. Find the amount of each brand that was used.

2. One bus left a terminal at 7:00 A.M. and headed north at 80 km/h. A second bus left the same terminal at 8:30 A.M. and traveled south at 90 km/h. At what time were the buses 460 km apart?

3. Find three consecutive multiples of 4 such that the product of the second and the third numbers is 80 more than 20 times the first number.

4. How many milliliters of water must be added to 240 mL of an $87\frac{1}{2}\%$ iodine solution to dilute it to a $62\frac{1}{2}\%$ solution?

OTHER RADICALS

Objective **To simplify expressions containing cube roots, fourth roots, or fifth roots**

You have solved equations like $x^2 = 49$ by using the definition of square root. In a similar way, you can solve equations such as $x^3 = -8$, $x^4 = 16$, and $x^5 = -32$.

If $x^3 = -8$,	If $x^4 = 16$,	If $x^5 = -32$,
then $x = \sqrt[3]{-8} = -2$	then $x = \sqrt[4]{16} = 2$ or $x = -\sqrt[4]{16} = -2$	then $x = \sqrt[5]{-32} = -2$
since $(-2)^3 = -8$.	since $2^4 = 16$ and $(-2)^4 = 16$.	since $(-2)^5 = -32$.

The three illustrations above suggest that:

(1) if $x^3 = k$, then $x = \sqrt[3]{k}$,
(2) if $x^4 = k$, and $k \geq 0$, then $x = \sqrt[4]{k}$ or $x = -\sqrt[4]{k}$, and
(3) if $x^5 = k$, then $x = \sqrt[5]{k}$.

This leads to the following definition of the **nth root** of a real number k.

Definition: nth root	For all odd integers $n > 1$ and all real numbers k, if $x^n = k$, then $x = \sqrt[n]{k}$. For all even integers $n > 1$ and all real numbers $k \geq 0$, if $x^n = k$, then $x = \sqrt[n]{k}$ or $x = -\sqrt[n]{k}$.

The symbol $\sqrt[n]{k}$ (read "the nth root of k") is called a **radical** where k is the **radicand**, $\sqrt[n]{}$ is the **radical sign**, and n is the **index**. [Note: The index 2 is usually omitted so that $\sqrt{k}$ and $\sqrt[2]{k}$ both mean "the square root of k."]
You may find the following table of powers helpful when you are required to simplify cube roots, fourth roots, or fifth roots.

Base	n	2	3	4	5	10	x^2	x^3	x^4
3rd power	n^3	8	27	64	125	1,000	x^6	x^9	x^{12}
4th power	n^4	16	81	256	625	10,000	x^8	x^{12}	x^{16}
5th power	n^5	32	243	1,024	3,125	100,000	x^{10}	x^{15}	x^{20}

All the properties involving operations with square roots can be extended to other radicals. For example, the Square Root of a Product property can be extended to cube roots as shown below.

$$\sqrt[3]{8x^{12}} = \sqrt[3]{8} \cdot \sqrt[3]{x^{12}} = 2x^4$$

Use the table above to verify that $\sqrt[3]{8} = 2$ and $\sqrt[3]{x^{12}} = x^4$.

nth root of a product	For all odd integers $n > 1$ and all real numbers x and y, $\sqrt[n]{x \cdot y} = \sqrt[n]{x} \cdot \sqrt[n]{y}$. For all even integers $n > 1$ and all real numbers $x \geq 0$, $y \geq 0$, $\sqrt[n]{x \cdot y} = \sqrt[n]{x} \cdot \sqrt[n]{y}$.

You can use the nth Root of a Product property to simplify radicals such as $\sqrt[5]{-64}$. First find the greatest 5th power that is also a factor of -64, as shown in Example 1.

Example 1

Simplify $\sqrt[5]{-64}$, $\sqrt[3]{c^8}$, and $\sqrt[4]{64a^{21}}$, $a \geq 0$.

$$\sqrt[5]{-64} = \sqrt[5]{-32 \cdot 2} = \sqrt[5]{-32} \cdot \sqrt[5]{2} = -2\sqrt[5]{2}$$

$$\sqrt[3]{c^8} = \sqrt[3]{c^6} \cdot \sqrt[3]{c^2} = c^2\sqrt[3]{c^2}$$

$$\sqrt[4]{64a^{21}} = \sqrt[4]{16 \cdot 4 \cdot a^{20} \cdot a} = \sqrt[4]{16} \cdot \sqrt[4]{4} \cdot \sqrt[4]{a^{20}} \cdot \sqrt[4]{a} = 2a^5\sqrt[4]{4a}$$

You can combine (add or subtract) like radicals and simplify products of radicals for nth roots as you have done previously with square roots.

Example 2

Simplify $5\sqrt[3]{3} + \sqrt[3]{24} + \sqrt[3]{81}$.

$$5\sqrt[3]{3} + \sqrt[3]{24} + \sqrt[3]{81} = 5\sqrt[3]{3} + \sqrt[3]{8 \cdot 3} + \sqrt[3]{27 \cdot 3} = 5\sqrt[3]{3} + 2\sqrt[3]{3} + 3\sqrt[3]{3} = 10\sqrt[3]{3}$$

Example 3

Simplify $\sqrt[4]{8d^3} \cdot \sqrt[4]{10d} \cdot \sqrt[4]{d^2}$, $d \geq 0$.

$$\sqrt[4]{8d^3} \cdot \sqrt[4]{10d} \cdot \sqrt[4]{d^2} = \sqrt[4]{8d^3 \cdot 10d \cdot d^2} = \sqrt[4]{80d^6} = \sqrt[4]{16 \cdot 5 \cdot d^4 \cdot d^2} = 2d\sqrt[4]{5d^2}$$

Recall that $\sqrt{5} \cdot \sqrt{5} = (\sqrt{5})^2 = 5$. Similarly, notice that

$$\sqrt[3]{27} \cdot \sqrt[3]{27} \cdot \sqrt[3]{27} = (\sqrt[3]{27})^3 = (3)^3 = 27 \text{ and}$$
$$\sqrt[4]{16} \cdot \sqrt[4]{16} \cdot \sqrt[4]{16} \cdot \sqrt[4]{16} = (\sqrt[4]{16})^4 = (2)^4 = 16.$$

In general, $(\sqrt[n]{k})^n = k$. This property may be used to *rationalize the denominator* of $\dfrac{5}{\sqrt[3]{36}}$.

$$\frac{5}{\sqrt[3]{36}} = \frac{5}{\sqrt[3]{6 \cdot 6}} = \frac{5}{\sqrt[3]{6} \cdot \sqrt[3]{6}} = \frac{5}{\sqrt[3]{6} \cdot \sqrt[3]{6}} \cdot \frac{\sqrt[3]{6}}{\sqrt[3]{6}} = \frac{5 \cdot \sqrt[3]{6}}{(\sqrt[3]{6})^3} = \frac{5\sqrt[3]{6}}{6}$$

nth root of a quotient	For all odd integers $n > 1$ and all real numbers x and y, $y \neq 0$, and for all even integers $n > 1$ and all real numbers $x \geq 0$, $y > 0$, $$\sqrt[n]{\frac{x}{y}} = \frac{\sqrt[n]{x}}{\sqrt[n]{y}}.$$

Example 4   Simplify $\sqrt[4]{\dfrac{7}{25a^3}}$, $a > 0$.

$$\sqrt[4]{\dfrac{7}{25a^3}} = \dfrac{\sqrt[4]{7}}{\sqrt[4]{25a^3}} = \dfrac{\sqrt[4]{7}}{\sqrt[4]{5 \cdot 5 \cdot a \cdot a \cdot a}} = \dfrac{\sqrt[4]{7}}{\sqrt[4]{5 \cdot 5 \cdot a \cdot a \cdot a}} \cdot \dfrac{\sqrt[4]{5 \cdot 5 \cdot a}}{\sqrt[4]{5 \cdot 5 \cdot a}} = \dfrac{\sqrt[4]{7 \cdot 25a}}{\sqrt[4]{5^4 a^4}} = \dfrac{\sqrt[4]{175a}}{5a}$$

If a denominator has two terms and at least one term involves a cube root, then the factors of $a^3 \pm b^3$ can be used to rationalize the denominator.

Example 5   Simplify $\dfrac{3}{2\sqrt[3]{x} + \sqrt[3]{5}}$.

$$\underbrace{\dfrac{3}{2\sqrt[3]{x} + \sqrt[3]{5}}}_{} \cdot \underbrace{\dfrac{4\sqrt[3]{x^2} - 2\sqrt[3]{5x} + \sqrt[3]{25}}{4\sqrt[3]{x^2} - 2\sqrt[3]{5x} + \sqrt[3]{25}}}_{} = \dfrac{3(4\sqrt[3]{x^2} - 2\sqrt[3]{5x} + \sqrt[3]{25})}{(2\sqrt[3]{x})^3 + (\sqrt[3]{5})^3}$$

$$\quad (a + b) \qquad\quad (a^2 - ab + b^2) \;=\; \qquad a^3 + b^3$$

$$= \dfrac{12\sqrt[3]{x^2} - 6\sqrt[3]{5x} + 3\sqrt[3]{25}}{8x + 5}$$

Written Exercises

Simplify each expression. Assume that no denominator has the value of 0.

(A) 1. $\sqrt[3]{-27}$ 2. $\sqrt[3]{-125}$ 3. $\sqrt[4]{625}$ 4. $-\sqrt[4]{81}$

5. $\sqrt[3]{x^9}$ 6. $\sqrt[4]{y^8}$ 7. $\sqrt[4]{y^{20}}$, $y \geq 0$ 8. $\sqrt[3]{x^3 y^6}$

9. $\sqrt[4]{32}$ 10. $\sqrt[3]{-40}$ 11. $\sqrt[3]{128}$ 12. $\sqrt[4]{243}$

13. $\sqrt[3]{a^5}$ 14. $\sqrt[4]{b^7}$, $b \geq 0$ 15. $\sqrt[3]{x^{10}}$ 16. $\sqrt[4]{y^9}$, $y \geq 0$

17. $\sqrt[4]{8} + \sqrt[4]{8}$ 18. $6\sqrt[3]{4} + \sqrt[3]{4}$ 19. $\sqrt[3]{32} + 3\sqrt[3]{2}$ 20. $\sqrt[3]{54} + \sqrt[3]{2}$

21. $\sqrt[3]{-64x^6}$ 22. $\sqrt[4]{81y^{16}}$ 23. $\sqrt[3]{250a^7}$ 24. $\sqrt[4]{162b^{19}}$, $b \geq 0$

25. $\sqrt[4]{7a^2} \cdot \sqrt[4]{3a^2}$, $a \geq 0$ 26. $\sqrt[3]{4a} \cdot \sqrt[3]{16a}$ 27. $\sqrt[4]{8x^3} \cdot \sqrt[4]{20x^6}$, $x \geq 0$ 28. $\sqrt[3]{25y^4} \cdot \sqrt[3]{10y}$

29. $\dfrac{2}{\sqrt[3]{7}}$ 30. $\dfrac{6x}{\sqrt[4]{3}}$ 31. $\dfrac{7\sqrt[3]{32}}{\sqrt[3]{4}}$ 32. $\dfrac{3\sqrt[4]{144}}{5\sqrt[4]{9}}$

33. $\sqrt[4]{\dfrac{1}{5}}$ 34. $\sqrt[3]{\dfrac{5}{6}}$ 35. $\sqrt[4]{\dfrac{11}{8}}$ 36. $\sqrt[3]{\dfrac{10}{9}}$

(B) 37. $\sqrt[3]{216}$ 38. $\sqrt[4]{256}$ 39. $\sqrt[5]{-243}$

40. $\sqrt[3]{2{,}000x^7 y^{22}}$, $x \geq 0$ 41. $5ab\sqrt[3]{270a^5 b^{10}}$ 42. $\sqrt[4]{64m^{20}n^{14}}$

43. $8\sqrt[3]{4} - \sqrt[3]{32} + \sqrt[3]{108}$ 44. $\sqrt[4]{2} + 3\sqrt[4]{32} - \sqrt[4]{162}$ 45. $\sqrt[5]{2} - \sqrt[5]{64} + \sqrt[5]{486}$

46. $\sqrt[4]{16c^7} + \sqrt[4]{81c^7}$, $c \geq 0$ 47. $\sqrt[4]{8a^5} + \sqrt[4]{27a^5}$ 48. $\sqrt[5]{32y^9} - \sqrt[5]{243y^9}$

49. $\sqrt[3]{2x^2} \cdot \sqrt[3]{8x^2} \cdot \sqrt[3]{4x^2}$ 50. $\sqrt[3]{3y^3} \cdot \sqrt[3]{6y^3} \cdot \sqrt[3]{9y^3}$, $y \geq 0$ 51. $\sqrt[5]{3n^4} \cdot \sqrt[5]{8n^3} \cdot \sqrt[5]{4n^2}$

52. $\sqrt[3]{\dfrac{5}{4y^2}}$ 53. $\dfrac{6a^2}{\sqrt[4]{8a^2}}$, $a > 0$ 54. $\dfrac{12n}{\sqrt[3]{9n^7}}$

55. $\dfrac{4}{\sqrt[3]{7} - \sqrt[3]{5}}$ 56. $\dfrac{-2}{\sqrt[3]{a} + 4}$ 57. $\dfrac{c}{5\sqrt[3]{c} + 2d}$ 58. $\dfrac{2t}{\sqrt[3]{6t} - 4\sqrt[3]{2v}}$

Simplify. Assume that n is an integer, $n > 1$, and that all other variables represent positive real numbers.

(C) 59. $\sqrt[3]{x^{6n}}$ 60. $\sqrt[4]{y^{4n+12}}$ 61. $\sqrt[3]{x^{2n+1}} \cdot \sqrt[3]{x^{4n+2}}$ 62. $y^n \sqrt[4]{y^{3n}} + \sqrt[4]{y^{7n}}$

63. $\sqrt[n]{x^{4n}}$ 64. $\sqrt[n]{x^{3n+2}}$ 65. $\sqrt[n]{a^n x^{5n+1}}$ 66. $\sqrt[n]{ab^{2n} x^{2n+3}}$

RATIONAL NUMBER EXPONENTS 6.6

Objectives **To write a radical expression in exponential form**
To write an exponential expression in radical form
To evaluate an expression containing rational number exponents

Each of the properties and definitions for integral exponents can be extended to rational number exponents. Using the Product of Powers property, $3^{\frac{1}{2}} \cdot 3^{\frac{1}{2}} = 3^1 = 3$. But you know that $\sqrt{3} \cdot \sqrt{3} = 3$. Thus, since $3^{\frac{1}{2}}$ and $\sqrt{3}$ are two different symbols for the same number, $3^{\frac{1}{2}} = \sqrt{3}$.
Similarly, $5^{\frac{1}{3}} = \sqrt[3]{5}$ since $5^{\frac{1}{3}} \cdot 5^{\frac{1}{3}} \cdot 5^{\frac{1}{3}} = 5$ and $\sqrt[3]{5} \cdot \sqrt[3]{5} \cdot \sqrt[3]{5} = 5$. Notice that $3^{\frac{1}{2}}$ and $5^{\frac{1}{3}}$ are written in *exponential form* and that $\sqrt{3}$ and $\sqrt[3]{5}$ are written in *radical form*.

Definition: $x^{\frac{1}{n}}$	For all integers $n > 1$ and all real numbers x, $$x^{\frac{1}{n}} = \sqrt[n]{x},$$ except when n is even and $x < 0$.

You will use the definition of $x^{\frac{1}{n}}$ to write radical expressions in exponential form, to write exponential expressions in radical form, and to evaluate expressions containing rational number exponents.

Example 1 **Write $7\sqrt[4]{a}$ and $\sqrt[4]{7a}$, $a \geq 0$, in exponential form.**

$$7\sqrt[4]{a} = 7 \cdot \sqrt[4]{a} = 7a^{\frac{1}{4}} \qquad \qquad \sqrt[4]{7a} = (7a)^{\frac{1}{4}}$$

Example 2 **Write $5y^{\frac{1}{3}}$ and $(5y)^{\frac{1}{3}}$ in radical form.**

$$5y^{\frac{1}{3}} = 5 \cdot y^{\frac{1}{3}} = 5\sqrt[3]{y} \qquad \qquad (5y)^{\frac{1}{3}} = \sqrt[3]{5y}$$

Example 3 **Find the value of $36^{\frac{1}{2}}$ and $10 \cdot 81^{\frac{1}{4}}$.**

$$36^{\frac{1}{2}} = \sqrt{36} = 6 \qquad \qquad 10 \cdot 81^{\frac{1}{4}} = 10 \cdot \sqrt[4]{81} = 10 \cdot 3 = 30$$

Since $\frac{2}{3} = 2 \cdot \frac{1}{3}$, you can rewrite $8^{\frac{2}{3}}$ as $(8^2)^{\frac{1}{3}}$ or $(8^{\frac{1}{3}})^2$, using the Power of a Power property. Then, $(8^2)^{\frac{1}{3}} = \sqrt[3]{8^2}$ and $(8^{\frac{1}{3}})^2 = (\sqrt[3]{8})^2$. Notice that $\sqrt[3]{8^2} = \sqrt[3]{64} = 4$ and $(\sqrt[3]{8})^2 = (2)^2 = 4$. Thus, $\sqrt[3]{8^2} = (\sqrt[3]{8})^2$.

Example 4   Write $\sqrt[5]{x^4}$ and $(\sqrt[4]{15})^3$ in exponential form.

$$\sqrt[5]{x^4} = x^{\frac{4}{5}} \blacktriangleleft \sqrt[n]{x^m} = x^{\frac{m}{n}} \qquad (\sqrt[4]{15})^3 = 15^{\frac{3}{4}} \blacktriangleleft (\sqrt[n]{x})^m = x^{\frac{m}{n}}$$

Example 5   Write $5y^{\frac{2}{3}}$ and $(5y)^{\frac{2}{3}}$ in radical form.

$$5y^{\frac{2}{3}} = 5 \cdot y^{\frac{2}{3}} = 5\sqrt[3]{y^2} \qquad\qquad (5y)^{\frac{2}{3}} = \sqrt[3]{(5y)^2} = \sqrt[3]{25y^2}$$

Example 6   Find the value of $2 \cdot 81^{\frac{5}{4}}$ and $8^{-\frac{2}{3}}$.

$$2 \cdot 81^{\frac{5}{4}} = 2(\sqrt[4]{81})^5 = 2 \cdot 3^5 = 2 \cdot 243 = 486$$
$$8^{-\frac{2}{3}} = \frac{1}{8^{\frac{2}{3}}} = \frac{1}{(\sqrt[3]{8})^2} = \frac{1}{2^2} = \frac{1}{4} \blacktriangleleft x^{-n} = \frac{1}{x^n}$$

Notice in Example 6 how an expression containing a negative rational number exponent is changed into an equivalent expression containing a positive rational number exponent.

Written Exercises

Write each expression in exponential form.

(A) **1.** $\sqrt{5}$ **2.** $\sqrt[3]{30}$ **3.** $\sqrt[4]{21}$ **4.** $7\sqrt[3]{a}$ **5.** $\sqrt{3b}$, $b \geq 0$ **6.** $3\sqrt[5]{9c}$

Write each expression in radical form.

7. $17^{\frac{1}{2}}$ **8.** $a^{\frac{1}{5}}$ **9.** $6c^{\frac{1}{3}}$ **10.** $(6c)^{\frac{1}{3}}$ **11.** $(7n)^{\frac{1}{4}}$, $n \geq 0$ **12.** $7n^{\frac{1}{4}}$, $n \geq 0$

Find the value of each expression.

13. $25^{\frac{1}{2}}$ **14.** $27^{\frac{1}{3}}$ **15.** $81^{\frac{1}{4}}$ **16.** $20 \cdot 32^{\frac{1}{5}}$ **17.** $-7 \cdot 64^{\frac{1}{2}}$ **18.** $-10 \cdot 625^{\frac{1}{4}}$

19. $8^{-\frac{1}{3}}$ **20.** $36^{-\frac{1}{2}}$ **21.** $16^{-\frac{1}{4}}$ **22.** $6 \cdot 9^{-\frac{1}{2}}$ **23.** $8^{\frac{1}{3}} \cdot 16^{\frac{1}{4}}$ **24.** $-8 \cdot 32^{-\frac{1}{5}}$

(B) **25.** $64^{\frac{2}{3}}$ **26.** $16^{\frac{5}{4}}$ **27.** $243^{\frac{2}{5}}$ **28.** $11 \cdot 25^{\frac{3}{2}}$ **29.** $5 \cdot 81^{\frac{3}{4}}$ **30.** $9 \cdot 100^{\frac{5}{2}}$

31. $625^{-\frac{3}{4}}$ **32.** $4^{-\frac{5}{2}}$ **33.** $27^{-\frac{5}{3}}$ **34.** $12 \cdot 8^{-\frac{4}{3}}$ **35.** $10 \cdot 16^{-\frac{5}{4}}$ **36.** $25^{-\frac{3}{2}} \cdot 16^{\frac{3}{4}}$

Write each expression in exponential form.

37. $\sqrt[5]{c^3}$ **38** $(\sqrt[3]{10})^5$ **39.** $\sqrt[6]{x^5}$, $x \geq 0$ **40.** $(\sqrt[4]{23})^3$ **41.** $\sqrt{n^7}$, $n \geq 0$ **42.** $(\sqrt{53})^9$

Write each expression in simplest radical form.

43. $5^{\frac{3}{2}}$ **44.** $2^{\frac{5}{3}}$ **45.** $10x^{\frac{2}{3}}$ **46.** $(10x)^{\frac{2}{3}}$ **47.** $(6t)^{\frac{3}{4}}$, $t \geq 0$ **48.** $6t^{\frac{3}{4}}$, $t \geq 0$

EXPRESSIONS WITH RATIONAL NUMBER EXPONENTS   6.7

Objectives
To simplify an expression containing rational number exponents
To solve an exponential equation using properties of rational number exponents

To simplify an expression containing rational number exponents, the properties for integral exponents and rationalizing denominators can be used as shown below.

$$8^{-\frac{3}{7}} = \frac{1}{8^{\frac{3}{7}}} = \frac{1}{8^{\frac{3}{7}}} \cdot \frac{8^{\frac{4}{7}}}{8^{\frac{4}{7}}} = \frac{8^{\frac{4}{7}}}{8^{\frac{7}{7}}}, \text{ or } \frac{8^{\frac{4}{7}}}{8}$$

Notice that such an expression is not in *simplest form* until the exponent is a *positive* number and the *denominator* is a rational number.

Example 1   **Simplify $5a^{-1} \cdot 2a^{\frac{1}{2}} \cdot 4a^{-\frac{3}{4}}$, $a > 0$.**

$$5a^{-1} \cdot 2a^{\frac{1}{2}} \cdot 4a^{-\frac{3}{4}} = \frac{5 \cdot 2 \cdot 4a^{\frac{1}{2}}}{a^1 \cdot a^{\frac{3}{4}}} = \frac{40a^{\frac{2}{4}}}{a^{\frac{4}{4}} a^{\frac{3}{4}}} = \frac{40}{a^{\frac{5}{4}}} \cdot \frac{a^{\frac{3}{4}}}{a^{\frac{3}{4}}} = \frac{40a^{\frac{3}{4}}}{a^2}$$

Example 2   **Simplify $(a^{\frac{1}{2}}b^{\frac{2}{3}})^6$ and $(27c^6d^{-3})^{\frac{2}{3}}$; $a \geq 0$, $d \neq 0$.**

$(xy)^n = x^n y^n$ ▶ $(a^{\frac{1}{2}}b^{\frac{2}{3}})^6 = (a^{\frac{1}{2}})^6(b^{\frac{2}{3}})^6$ | $(27c^6d^{-3})^{\frac{2}{3}} = 27^{\frac{2}{3}}(c^6)^{\frac{2}{3}}(d^{-3})^{\frac{2}{3}}$

$(x^m)^n = x^{mn}$ ▶ $= a^3b^4$ | $= (\sqrt[3]{27})^2c^4d^{-2}$ ◀ $x^{\frac{m}{n}} = (\sqrt[n]{x})^m$

| | $= \frac{9c^4}{d^2}$

The Quotient of Powers and Power of a Quotient properties are used in simplifying expressions in Examples 3 and 4, respectively.

Example 3   **Simplify $\dfrac{t^{\frac{1}{2}}v^{-\frac{1}{5}}}{t^{\frac{2}{3}}v^{-\frac{4}{5}}}$; $t > 0$, $v \neq 0$.**

$$\frac{t^{\frac{1}{2}}v^{-\frac{1}{5}}}{t^{\frac{2}{3}}v^{-\frac{4}{5}}} = \frac{t^{\frac{3}{6}}v^{\frac{4}{5}}}{t^{\frac{4}{6}}v^{\frac{1}{5}}} = \frac{v^{\frac{3}{5}}}{t^{\frac{1}{6}}} \cdot \frac{t^{\frac{5}{6}}}{t^{\frac{5}{6}}} = \frac{v^{\frac{3}{5}}t^{\frac{5}{6}}}{t}$$

Example 4   **Simplify $\left(\dfrac{8a^{12}}{27b^9}\right)^{\frac{2}{3}}$, $b \neq 0$.**

$\left(\dfrac{x}{y}\right)^n = \dfrac{x^n}{y^n}$ ▶ $\left(\dfrac{8a^{12}}{27b^9}\right)^{\frac{2}{3}} = \dfrac{(8a^{12})^{\frac{2}{3}}}{(27b^9)^{\frac{2}{3}}} = \dfrac{8^{\frac{2}{3}}(a^{12})^{\frac{2}{3}}}{27^{\frac{2}{3}}(b^9)^{\frac{2}{3}}} = \dfrac{(\sqrt[3]{8})^2a^8}{(\sqrt[3]{27})^2b^6} = \dfrac{4a^8}{9b^6}$

You have solved an exponential equation like $3^{2x-1} = 3^7$ by setting the exponents equal to each other and solving the equation $2x - 1 = 7$. This method of solving an exponential equation is based upon the property that if $a^m = a^n$, then $m = n$. As shown in Example 5, some exponential equations that involve rational number exponents can be solved in a similar way.

Example 5 **Solve each equation.**

$8^x = 16$
Write 8 and 16 as powers of 2.
$(8)^x = 16$
$(2^3)^x = 2^4$
$2^{3x} = 2^4$
$3x = 4$ ◀ *If $a^m = a^n$, then $m = n$.*
$x = \dfrac{4}{3}$

Thus, the solution is $\dfrac{4}{3}$.

$3^{8x} = \dfrac{1}{81}$

Write $\dfrac{1}{81}$ as a power of 3.

$3^{8x} = 3^{-4}$ ◀ $\dfrac{1}{81} = \dfrac{1}{3^4} = 3^{-4}$

$8x = -4$
$x = -\dfrac{1}{2}$

Thus, the solution is $-\dfrac{1}{2}$.

Written Exercises

Simplify each expression. Assume that values of the variables are positive numbers.

(A) 1. $6^{\frac{3}{8}} \cdot 6^{-\frac{1}{8}} \cdot 6^{\frac{5}{8}}$
  2. $x^{\frac{4}{3}} \cdot x^{\frac{5}{3}} \cdot x^{-\frac{7}{3}}$
  3. $y^3 \cdot y^{-\frac{1}{2}}$
  4. $5a^2 \cdot 6a^{-\frac{3}{8}}$

5. $(x^{\frac{2}{5}})^{10}$
6. $(y^{12})^{\frac{2}{3}}$
7. $(n^{10})^{\frac{3}{5}}$
8. $(b^{-\frac{2}{3}})^{-9}$

9. $(x^{\frac{1}{4}}y^{\frac{3}{4}})^8$
10. $(25a^2b^8)^{\frac{1}{2}}$
11. $(-4m^{\frac{1}{2}}n^{\frac{3}{2}})^2$
12. $(16c^{12}d^8)^{\frac{3}{4}}$

13. $\dfrac{a^{\frac{9}{7}}}{a^{\frac{3}{7}}}$
14. $\dfrac{x^{\frac{4}{5}}}{x^{-\frac{2}{5}}}$
15. $\dfrac{n^{-\frac{1}{3}}}{n^{-\frac{5}{3}}}$
16. $\dfrac{b^3}{b^{\frac{3}{4}}}$

17. $\left(\dfrac{x^{\frac{3}{4}}}{y^{\frac{1}{2}}}\right)^8$
18. $\left(\dfrac{a^{\frac{2}{3}}}{b^{\frac{3}{5}}}\right)^{15}$
19. $\left(\dfrac{25m^6}{9n^8}\right)^{\frac{1}{2}}$
20. $\left(\dfrac{27c^6}{8d^{12}}\right)^{\frac{2}{3}}$

Solve each equation.

21. $9^x = 27$
22. $3^{9x} = \dfrac{1}{27}$
23. $125^x = 25$
24. $4^{6x} = \dfrac{1}{64}$

(B) 25. $25^{2x+1} = 125$
26. $2^{2x+2} = \dfrac{1}{16}$
27. $32^{3x-2} = 16$
28. $10^{3x-1} = \dfrac{1}{10,000}$

Simplify each expression. Assume that values of the variables are positive numbers.

29. $x^{\frac{2}{5}} \cdot x^{\frac{1}{5}} \cdot x^{-\frac{4}{5}}$
30. $b^{-6} \cdot b^{\frac{5}{8}}$
31. $7a^{\frac{3}{5}} \cdot 6a^{\frac{1}{2}} \cdot 5a^{\frac{3}{4}}$
32. $-10n^{-\frac{3}{4}} \cdot 2n^{-\frac{5}{8}}$

33. $(x^{\frac{2}{3}})^{-9}$
34. $(x^{\frac{1}{2}}y^{-\frac{3}{4}})^8$
35. $(8m^9n^{-6})^{\frac{1}{3}}$
36. $(125a^{-9}b^{12})^{-\frac{2}{3}}$

37. $\dfrac{x^{\frac{1}{3}}y^{\frac{5}{4}}}{x^{\frac{2}{3}}y^{\frac{3}{4}}}$

38. $\dfrac{a^{\frac{2}{5}}b^{\frac{1}{6}}}{a^{\frac{3}{10}}b^{\frac{1}{4}}}$

39. $\dfrac{m^{-\frac{2}{3}}n^{-4}}{m^{-\frac{1}{3}}n^{-5}}$

40. $\dfrac{27^{\frac{3}{4}}x^{-\frac{1}{2}}y^{\frac{1}{4}}}{27^{\frac{1}{12}}x^{\frac{3}{4}}y^{-\frac{1}{6}}}$

41. $\left(\dfrac{x^{-\frac{3}{4}}}{y^{-\frac{1}{4}}}\right)^{8}$

42. $\left(\dfrac{-2a^{-\frac{1}{3}}}{3b^{-\frac{2}{3}}}\right)^{3}$

43. $\left(\dfrac{64m^{-3}}{27n^{-6}}\right)^{\frac{2}{3}}$

44. $\left(\dfrac{4c^{-6}}{25d^{-4}}\right)^{\frac{3}{2}}$

Simplify and write each expression with the least possible number of radical signs.

Example   Simplify $\sqrt[3]{x}\cdot\sqrt[4]{x}$ and $(\sqrt[4]{x^3}+\sqrt[6]{y})(\sqrt[4]{x^3}-\sqrt[6]{y})$.

Two radical signs ▶ $\sqrt[3]{x}\cdot\sqrt[4]{x}=x^{\frac{1}{3}}\cdot x^{\frac{1}{4}}$ ◀ $\sqrt[n]{x}=x^{\frac{1}{n}}$

$\qquad\qquad\qquad\qquad = x^{\frac{7}{12}}$ ◀ $x^m\cdot x^n = x^{m+n}$

One radical sign ▶ $\qquad = \sqrt[12]{x^7}$ ◀ $x^{\frac{m}{n}}=\sqrt[n]{x^m}$

Four radical signs ▶ $(\sqrt[4]{x^3}+\sqrt[6]{y})(\sqrt[4]{x^3}-\sqrt[6]{y})$

$\qquad\qquad\qquad (x^{\frac{3}{4}}+y^{\frac{1}{6}})(x^{\frac{3}{4}}-y^{\frac{1}{6}})$ ◀ $\sqrt[n]{x^m}=x^{\frac{m}{n}}$

$\qquad\qquad\qquad (x^{\frac{3}{4}})^2-(y^{\frac{1}{6}})^2$ ◀ $(a+b)(a-b)=a^2-b^2$

$\qquad\qquad\qquad x^{\frac{3}{2}}-y^{\frac{1}{3}}$

$\qquad\qquad\qquad \sqrt{x^3}-\sqrt[3]{y}$

Two radical signs ▶ $x\sqrt{x}-\sqrt[3]{y}$

Ⓒ **45.** $\sqrt{x}\cdot\sqrt[5]{x}$

46. $\sqrt[6]{y}\cdot\sqrt{y}$

47. $\sqrt[3]{z}\cdot\sqrt[4]{z}\cdot\sqrt[6]{z}$

48. $\sqrt[3]{x^2}\cdot\sqrt[4]{x^3}$

49. $\sqrt{a}\cdot\sqrt[3]{b}\cdot\sqrt[4]{c}$

50. $\sqrt[6]{m^5}\cdot\sqrt[3]{n^2}$

51. $(\sqrt[3]{x}+\sqrt[4]{x})(\sqrt[3]{x}-\sqrt[4]{x})$

52. $(\sqrt{x}-\sqrt[3]{y^2})(\sqrt{x}+\sqrt[3]{y^2})$

53. $(\sqrt[4]{x^3}+\sqrt[8]{y^5})(\sqrt[4]{x^3}-\sqrt[8]{y^5})$

54. $(\sqrt{x}+\sqrt[4]{x})^2$

55. $(3\sqrt{x}-2\sqrt[6]{y})^2$

56. $(\sqrt[4]{x}+\sqrt[4]{y})(\sqrt[4]{x}+\sqrt[4]{y})$

57. $\dfrac{\sqrt{x}}{\sqrt[3]{x}}$

58. $\dfrac{\sqrt[4]{x^3}}{\sqrt[5]{x^2}}$

59. $\dfrac{\sqrt[5]{x}}{\sqrt[4]{x}}$

60. $(\sqrt[3]{x}-\sqrt[3]{2})(\sqrt[3]{x^2}+\sqrt[3]{2x}+\sqrt[3]{4})$

61. $(\sqrt[3]{x}+\sqrt[3]{y})(\sqrt[3]{x^2}-\sqrt[3]{xy}+\sqrt[3]{y^2})$

Use one of the special products, $(a+b)(a^2-ab+b^2)=a^3+b^3$ or $(a-b)(a^2+ab+b^2)=a^3-b^3$, to rationalize each denominator.

62. $\dfrac{14}{5^{\frac{1}{3}}+2^{\frac{1}{3}}}$

63. $\dfrac{-2}{x^{\frac{1}{3}}-y^{\frac{1}{3}}}$

64. $\dfrac{8}{6^{\frac{2}{3}}-12^{\frac{1}{3}}+2^{\frac{2}{3}}}$

65. $\dfrac{1}{2x^{\frac{1}{3}}+3^{\frac{1}{3}}}$

66. $\dfrac{c}{3c^{\frac{1}{3}}-5d^{\frac{1}{3}}}$

67. $\dfrac{x-y}{x^{\frac{2}{3}}+x^{\frac{1}{3}}y^{\frac{1}{3}}+y^{\frac{2}{3}}}$

Expressions with Rational Number Exponents

159

CHAPTER SIX REVIEW

Vocabulary
conjugates   [6.4]   index   [6.5]
irrational number   [6.1]   like radicals   [6.3]
nth root, $\sqrt[n]{k}$ [6.5]   radical   [6.1]
radical sign   [6.1]   radicand   [6.1]

rational number exponent, $x^{\frac{1}{n}}, x^{\frac{m}{n}}$ [6.6]
real number   [6.1]
simplest radical form [6.4]
square root, $\sqrt{k}$ [6.1]

Solve each equation. Approximate the solutions to the nearest hundredth.   [6.1]
1. $5y^2 + 24 = 174$ 2. $n^2 - 7 = 18 - 3n^2$

Determine whether each number is rational or irrational. [6.1]
3. $8.43843843\ldots$ 4. $\sqrt{160}$
5. $-\sqrt{1600}$ 6. $\sqrt{0.16}$

Simplify. Assume that values of the variables are _all_ real numbers.
7. $\sqrt{20c^2}$ 8. $3m\sqrt{8m^4}$ [6.2]

In the remaining exercises, assume that (1) values of the variables are nonnegative numbers and (2) no denominator has the value of 0 (unless stated otherwise).
9. $m^2n\sqrt{32m^7n^3}$ [6.2]
10. $\sqrt{8} + \sqrt{50} - \sqrt{18}$ [6.3]
11. $(4\sqrt{3} - 2\sqrt{5})^2$ 12. $6\sqrt{c^5d} + c\sqrt{16c^3d}$
13. $\dfrac{\sqrt{27a^8}}{\sqrt{3a^3}}$ 14. $\dfrac{-5\sqrt{2}}{3\sqrt{7}}$ [6.4]
15. $\dfrac{\sqrt{2}}{2\sqrt{7} + \sqrt{10}}$
16. $\sqrt[3]{-16x^{12}}$ 17. $\sqrt[5]{2c^2} \cdot \sqrt[5]{16c^8}$ [6.5]
18. $\sqrt[4]{32x^8y^{11}}$ 19. $\sqrt[3]{27a^7} + \sqrt[3]{64a^7}$
20. $\sqrt[3]{\dfrac{7}{9}}$ 21. $\dfrac{2c^2}{\sqrt[4]{4c^3}}$
22. $\dfrac{a}{\sqrt[3]{x} + 2\sqrt[3]{y}}$

Write each expression in exponential form.   [6.6]
23. $(\sqrt[3]{17})^2$ 24. $6\sqrt[4]{3a}$
25. $\sqrt{c^5}$

Write each expression in radical form. [6.6]
26. $2a^{\frac{5}{3}}$ 27. $4y^{\frac{3}{5}}$
28. $(4y)^{\frac{3}{5}}$

Find the value of each expression. [6.6]
29. $-15 \cdot 32^{\frac{1}{5}}$ 30. $125^{\frac{4}{3}}$
31. $20 \cdot 625^{\frac{3}{4}}$ 32. $32^{-\frac{2}{5}}$

Simplify each expression. [6.7]
33. $-9a^{\frac{3}{4}} \cdot 2a^{-\frac{1}{2}}$ 34. $(49a^2b^{10})^{\frac{1}{2}}$
35. $(a^{\frac{1}{3}}b^{-\frac{2}{3}})^{-9}$ 36. $\dfrac{x^{\frac{1}{4}}y^{-\frac{1}{5}}}{x^{\frac{3}{8}}y^{-\frac{3}{5}}}$
37. $\left(\dfrac{8x^9}{27y^{-12}}\right)^{\frac{2}{3}}$

Solve each equation. (x is a real number.) [6.7]
38. $4^x = 8$ 39. $27^{2x-1} = 9$
40. $3^{10x} = \dfrac{1}{9}$ 41. $5^{4x+1} = \dfrac{1}{125}$

★42. Solve $\dfrac{2x^2 - 3}{4} + \dfrac{x^2 + 5}{6} = \dfrac{5x^2 + 4}{8}$.

Approximate the solutions to the nearest hundredth. [6.1]

Simplify each expression.
★43. $x^m y^{2n}\sqrt{x^{4m+5}y^{6n}}$ [6.3]
★44. $\dfrac{4\sqrt{x - y}}{\sqrt{x - y} + 3}$ [6.4]
★45. $\sqrt[n]{x^{2n}y^{3n+5}}$ [6.7]
★46. Factor $9x - 4y$ as a difference of two squares. [6.3]
★47. Write $(x^{2n}y^{n+2})^{\frac{a}{n}}$ in simplest radical form. [6.6]

CHAPTER SIX TEST

Solve each equation. Approximate the solutions to the nearest hundredth.

1. $7a^2 + 19 = 236$

2. $3y^2 - 11 = 7y^2 - 60$

Determine whether each number is rational or irrational.

3. $\sqrt{90}$ **4.** $-\sqrt{900}$

Simplify. Assume that values of the variables are *all* real numbers.

5. $\sqrt{27n^8}$ **6.** $5\sqrt{12a^6}$

In the remaining exercises, assume that (1) values of the variables are nonnegative numbers and (2) no denominator has the value of 0 (unless stated otherwise).

7. $5\sqrt{12x} - 6\sqrt{48x} + \sqrt{75x}$

8. $(3\sqrt{2} + \sqrt{5})(5\sqrt{2} - \sqrt{5})$

9. $\sqrt[3]{9x^4} \cdot \sqrt[3]{3x^5}$ **10.** $8\sqrt{16x^5y^9}$

11. $m\sqrt[4]{2n^5} + mn\sqrt[4]{32n}$

12. $\dfrac{4}{3\sqrt{5} + \sqrt{30}}$ **13.** $\dfrac{5}{2\sqrt[3]{x} - \sqrt[3]{y}}$

14. $\dfrac{2x}{\sqrt[3]{5x^2}}$ **15.** $\sqrt[4]{\dfrac{3}{8}}$

16. $\dfrac{8}{5\sqrt{12c^3}}$ **17.** $(27x^6y^{12})^{\frac{2}{3}}$

18. $\dfrac{x^{\frac{5}{7}}y^{-\frac{1}{4}}}{x^{-\frac{1}{7}}y^{\frac{1}{2}}}$ **19.** $\left(\dfrac{c^{\frac{2}{3}}}{d^{\frac{1}{2}}}\right)^{12}$

20. Write $\sqrt[4]{a^3}$ in exponential form.

21. Write $(2x)^{\frac{4}{5}}$ in radical form.

Find the value of each expression.

22. $25^{\frac{3}{2}}$ **23.** $6 \cdot 81^{-\frac{3}{4}}$

Solve each equation. (*x* is a real number.)

24. $25^{2x+1} = 125$ **25.** $2^{4x-1} = \dfrac{1}{32}$

★26. Solve $\dfrac{x^2 - 10}{5} + \dfrac{x^2}{10} = \dfrac{x^2 + 10}{4}$.

Approximate the solutions to the nearest hundredth.

Simplify each expression.

★27. $x^{2c}y^d\sqrt{x^{2c+3}y^{6d+5}}$

★28. $\dfrac{\sqrt{x+y}}{\sqrt{x+y}+6}$

★29. Simplify and write $(\sqrt{x} + \sqrt[6]{y})(\sqrt{x} + \sqrt[6]{y})$ with the least possible number of radical signs.

Simplifying Radicals

OBJECTIVE: To simplify radicals of any integral index

On page 582, there is another version of the program below that simplifies square roots. You may wish to scrutinize that program before undertaking the program below.

This program functions very similarly to the program on page 582. This program, however, allows you to simplify radicals for any integral index you specify. The program on page 582 tests only the squares of whole numbers 2 and greater for divisibility into the radicand. This program tests the nth power of whole numbers 2 and greater for divisibility into the radicand, for any index n you specify for the radical.

Like the program on page 582, this program tests the divisibility of one number by another by checking whether the quotient of the numbers is equal to the greatest INTeger of the quotient. When they are equal, we know that the number is divisible by the other.

```
110   PRINT "WHAT IS THE RADICAND TO BE SIMPLIFIED";:
         INPUT R
120   PRINT "WHAT IS THE INDEX";: INPUT X
130 C = 1:S = R
140   IF  ABS (S ^ (1 / X) -  INT (S ^ (1 / X))) < .05
         THEN 240
150   FOR I = 2 TO  SQR (S)
160 B = 1
170   FOR N = 1 TO X:B = B * I: NEXT N
180   IF  ABS (S / B -  INT (S / B)) > .0001 THEN 210
190 C = C * I:S = S / B: IF S < B THEN 220
200   GOTO 180
210   NEXT I
220   PRINT "RADICAL("R") = "C" * RAD ("S")"
230   GOTO 110
240   PRINT "RADICAL ("S") = S ^ (1 / X)
250   GOTO 110
```

See the Computer Section on page 574 for more information.

EXERCISES

1. Write a program that multiplies two radicals of the form $(a\sqrt{b})\,(c\sqrt{d})$ and then simplifies the product.

2. Expand the program above to multiply two radicals of the form $(a\sqrt[3]{b})$ $(c\sqrt[3]{d})$ and to simplify the result.

Directions: In each item, you are to compare a quantity in Column 1 with a quantity in Column 2. Write the letter of the correct answer from these choices:

A The quantity in Column 1 is greater than the quantity in Column 2.
B The quantity in Column 2 is greater than the quantity in Column 1.
C The quantity in Column 1 is equal to the quantity in Column 2.
D The relationship cannot be determined from the given information.

Notes: Information centered over both columns refers to one or both of the quantities to be compared.
A symbol that appears in both columns has the same meaning in each column.
All variables represent real numbers.

SAMPLE ITEMS AND ANSWERS

Column 1	Column 2
$x = 8$ and $y = 9$	
S1. $\sqrt{2x}$	$\sqrt[3]{3y}$

The answer is **A** because $\sqrt{2 \cdot 8} = \sqrt{16} = 4$, $\sqrt[3]{3 \cdot 9} = \sqrt[3]{27} = 3$, and $4 > 3$.

$0 < x < y$	
S2. $\sqrt{2x}$	$\sqrt{y}$

The answer is **D**. If $x = 8$ and $y = 9$, then $\sqrt{2x} > \sqrt{y}$. If $x = 2$ and $y = 4$, then $\sqrt{2x} = \sqrt{y}$.

	Column 1	Column 2
1.	$\sqrt[3]{65}$	4
2.	$\dfrac{1}{3}$	$\sqrt[3]{\dfrac{2}{27}}$
3.	$\sqrt[4]{x + y}$	$\sqrt[4]{x - y}$
4.	The sum of the two solutions of $x^2 = 10$	The sum of the two solutions of $x^4 = 16$

	Column 1	Column 2
	$x = 12$ and $y = 9$	
5.	$\sqrt{xy}$	$\sqrt{6x + 4y}$
	$x = y = 50$	
6.	$7 + \sqrt{x}$	$\sqrt{x} + \sqrt{y}$
	$x > 0$ and $y > 0$	
7.	$\sqrt{x^2 + y^2}$	$x + y$
	$x = -2$ and $y = 2$	
8.	$\sqrt{8x^2 y^3}$	$2xy\sqrt{2y}$

Items 9 and 10 refer to Rectangles I and II below. The rectangles are not drawn to scale.

$\sqrt{2}$ □ I $\sqrt{50}$ $\sqrt{5}$ □ II $\sqrt{20}$

9. Area of Rectangle I   Area of Rectangle II

10. Perimeter of Rectangle I   Perimeter of Rectangle II

---

7 QUADRATIC FORMULA

**Non-Routine
Problem
Solving**

Problem 1.   Automobiles A and B had equal amounts of fuel. The two cars began at the same time and were driven at the same speed. Car A ran out of fuel in 4 hours and car B ran out of fuel in 3 hours. In how many hours was A's amount of fuel twice B's amount of fuel?

Problem 2.   How many hours will it take a car averaging 80 km/h between stops to travel x kilometers if it is to make k stops of n minutes each? Simplify your answer.

COMPLETING THE SQUARE

Objective   **To solve a quadratic equation by completing the square**

As shown below, you can solve a *quadratic equation* like $y^2 = 49$ or $(n + 3)^2 = 16$ by using the definition of square root: if $x^2 = k$, then $x = \sqrt{k}$ or $x = -\sqrt{k}$.

$$y^2 = 49$$
$$y = \sqrt{49} \quad \text{or} \quad y = -\sqrt{49}$$
$$y = 7 \qquad\qquad y = -7$$
Solutions: 7, −7

$$(n + 3)^2 = 16$$
$$n + 3 = \sqrt{16} \quad \text{or} \quad n + 3 = -\sqrt{16}$$
$$n + 3 = 4 \quad \text{or} \quad n + 3 = -4$$
$$n = 1 \qquad\qquad n = -7$$
Solutions: 1, −7

The equation $y^2 + 10y = 3$ can be solved in a similar way.

Example 1   **Solve $y^2 + 10y = 3$.**

$$y^2 + 10y \qquad\; = 3$$
$$y^2 + 10y + 25 = 3 + 25 \qquad \blacktriangleleft \text{ Add 25 to each side.}$$
$$(y + 5)^2 = 28$$
$$y + 5 = \pm\sqrt{28} \qquad \blacktriangleleft \text{ If } x^2 = k, \text{ then } x = \pm\sqrt{k}.$$
$$y + 5 = \pm 2\sqrt{7}$$
$$y = -5 \pm 2\sqrt{7} \qquad \blacktriangleleft \text{ Add } -5 \text{ to each side.}$$

Thus, the solutions are $-5 + 2\sqrt{7}$ and $-5 - 2\sqrt{7}$.

The procedure used to solve the quadratic equation in Example 1 is called **completing the square.** By adding 25 to each side of the equation, the left side becomes a *perfect square trinomial, $y^2 + 10y + 25$,* that can be factored into the square of a binomial, $(y + 5)^2$.

To solve a quadratic equation of the form $x^2 + bx = c$ by completing the square, you divide b by 2, square the result, then add $\left(\frac{b}{2}\right)^2$ to each side.

Example 2   **Solve $n^2 - 3n = 10$ by completing the square.**

$$n^2 - 3n \qquad\quad = 10 \qquad \blacktriangleleft b = -3$$
$$n^2 - 3n + \frac{9}{4} = 10 + \frac{9}{4} \qquad \blacktriangleleft \text{ Add } \left(\frac{-3}{2}\right)^2, \text{ or } \frac{9}{4}, \text{ to each side.}$$
$$\left(n - \frac{3}{2}\right)^2 = \frac{49}{4}$$
$$n - \frac{3}{2} = \pm\sqrt{\frac{49}{4}} \qquad \blacktriangleleft \text{ If } x^2 = k, \text{ then } x = \pm\sqrt{k}.$$
$$n = \frac{3}{2} \pm \frac{7}{2}, \text{ or } \frac{3 \pm 7}{2}$$

Thus, the solutions are 5 and −2.

To solve $9y^2 + 12y - 8 = 0$ by completing the square, you must rewrite the equation in the general form of $y^2 + by = c$.

Example 3

Solve $9y^2 + 12y - 8 = 0$ by completing the square.

$$9y^2 + 12y - 8 = 0$$
$$9y^2 + 12y = 8 \qquad \blacktriangleleft \text{ Add 8 to each side.}$$
$$y^2 + \frac{12}{9}y = \frac{8}{9} \qquad \blacktriangleleft \text{ Divide each side by 9.}$$
$$y^2 + \frac{4}{3}y = \frac{8}{9} \qquad \blacktriangleleft \ \frac{12}{9} = \frac{4}{3}, \text{ so } b = \frac{4}{3}.$$
$$y^2 + \frac{4}{3}y + \frac{4}{9} = \frac{8}{9} + \frac{4}{9} \qquad \blacktriangleleft \text{ Add } \left(\frac{1}{2} \cdot \frac{4}{3}\right)^2, \text{ or } \frac{4}{9}, \text{ to each side.}$$
$$\left(y + \frac{2}{3}\right)^2 = \frac{12}{9}$$
$$y + \frac{2}{3} = \pm\sqrt{\frac{12}{9}}$$
$$y = -\frac{2}{3} \pm \frac{2\sqrt{3}}{3} \qquad \blacktriangleleft \ \pm\sqrt{\frac{12}{9}} = \pm\frac{\sqrt{12}}{\sqrt{9}} = \pm\frac{2\sqrt{3}}{3}$$

Thus, the solutions are $\dfrac{-2 + 2\sqrt{3}}{3}$ and $\dfrac{-2 - 2\sqrt{3}}{3}$.

Oral Exercises

Find the number to be added to each expression to complete the square.

1. $a^2 + 14a$ **2.** $n^2 - 8n$ **3.** $x^2 + 5x$ **4.** $y^2 - y$ **5.** $b^2 + \frac{3}{2}b$ **6.** $c^2 - \frac{2}{5}c$

Written Exercises

Solve each equation by completing the square.

(A)   **1.** $x^2 + 4x = 21$ **2.** $y^2 - 8y = 33$ **3.** $n^2 + 12n = -20$

4. $a^2 - 6a = 5$ **5.** $b^2 + 10b = -5$ **6.** $c^2 - 2c = 74$

7. $y^2 + 3y - 6 = 0$ **8.** $t^2 - 5t + 2 = 0$ **9.** $x^2 - x - 7 = 0$

10. $5x^2 + 3x - 2 = 0$ **11.** $2a^2 - 7a - 4 = 0$ **12.** $3y^2 + 10y + 3 = 0$

13. $a^2 - 5a - 5 = 0$ **14.** $y^2 + 7y = 0$ **15.** $n^2 = n$

(B)   **16.** $2n^2 = 5 - 5n$ **17.** $7x - 1 = 5x^2$ **18.** $3a^2 + 4a = 3$

19. $\frac{1}{3}y^2 - 2y - 1 = 0$ **20.** $\frac{1}{8}n^2 - n = \frac{1}{2}$ **21.** $\frac{5}{3}x^2 - \frac{5}{2}x - \frac{5}{6} = 0$

22. $\dfrac{x^2 + 12x}{24} = \dfrac{1}{12}$ **23.** $\dfrac{y^2 + 15}{10} = y$ **24.** $2n = \dfrac{3 - n^2}{4}$

(C)   **25.** $x^2 + (2\sqrt{3})x - 24 = 0$ **26.** $y^2 - 4\sqrt{2} \cdot y + 8 = 0$ **27.** $n^2 + 6n\sqrt{5} + 45 = 0$

28. $a^2 - 2a\sqrt{3} + 1 = 0$ **29.** $b^2 - 6b\sqrt{2} + 15 = 0$ **30.** $4x^2 + 4x\sqrt{5} + 3 = 0$

MAXIMUM VALUE

Objectives

To find the maximum value of a quadratic polynomial
To solve a real-life problem involving a maximum value

The value of $-5t^2 + 40t$ will increase from 0 to some **maximum value** and then decrease to 0, as the value of t increases from 0 to 8.

t	0	1	2	5	6	8
$-5t^2 + 40t$	0	35	60	75	60	0

Example 1

Find the maximum value of $-5t^2 + 40t$ and the corresponding value of t.

$-5t^2 + 40t$
$-5(t^2 - 8t \quad)$ ◀ *Factor out -5.*
$-5(t^2 - 8t + 16) + 5(16)$ ◀ *Complete the square. Add $-5(16) + 5(16)$, or 0.*
$-5(t - 4)^2 + 80$

If $t = 4$, then: If $t \neq 4$, then:
$\quad\quad (t - 4)^2 = 0$ $\quad\quad (t - 4)^2 > 0$
$\quad -5(t - 4)^2 = 0$ $\quad -5(t - 4)^2 < 0$
$-5(t - 4)^2 + 80 = 80.$ $-5(t - 4)^2 + 80 < 80.$

Thus, the maximum value is 80, and this occurs when $t = 4$.

Example 2

Find the maximum value of $-10x^2 + 100x + 6{,}000$ and the corresponding value of x.

$-10x^2 + 100x \quad\quad + 6{,}000$
$-10(x^2 - 10x \quad) + 6{,}000$ ◀ *Factor out -10 from $-10x^2 + 100x$.*
$-10(x^2 - 10x + 25) + 6{,}000 + 10(25)$ ◀ *Complete the square. Add $-10(25) + 10(25)$, or 0.*
$-10(x - 5)^2 + 6{,}250$

Thus, the maximum value is 6,250, and this occurs when $x = 5$.

In Chapter 1, you learned a method for finding the maximum height of an object that is shot upward from the earth's surface with a given initial velocity.

If $h = vt - 5t^2$, the maximum value of h is $\dfrac{v^2}{20}$.

The maximum height can also be found by the method of completing the square.

Example 3

Find the maximum height and the corresponding time of an object shot upward from the earth's surface at an initial velocity of 40 m/s.

$h = -5t^2 + vt = -5t^2 + 40t = -5(t - 4)^2 + 80$ ◀ *See Example 1 for the steps.*

Thus, the maximum height is 80 m at the end of 4 s.

Example 4　**An orange grove has 20 trees per acre, and the average yield is 300 oranges per tree. For each additional tree per acre, the average yield will be reduced by 10 oranges per tree. How many trees per acre will yield the maximum number of oranges per acre, and what is the maximum number of oranges per acre?**

Additional trees per acre	(Trees per acre) × (Oranges per tree)	= (Oranges per acre)
2 ▶	$(20 + 2)(300 - 10 \cdot 2)$	= 6,160
3 ▶	$(20 + 3)(300 - 10 \cdot 3)$	= 6,210
⋮	⋮　　　　⋮	
x ▶	$(20 + x)(300 - 10x)$	$= 6{,}000 + 100x - 10x^2$
		$= -10x^2 + 100x + 6{,}000$
	See Example 2 for the steps. ▶	$= -10(x - 5)^2 + 6{,}250$

The maximum value is 6,250, and this occurs when $x = 5$.
Thus, 20 + 5, or 25 trees per acre, will yield the maximum number of 6,250 oranges per acre.

Written Exercises

Find the maximum value of each polynomial and the corresponding value of x. (Ex. 1–6)

Ⓐ **1.** $-2x^2 + 12x$　　　　^　　**2.** $64x - 16x^2$　　　　**3.** $10 + 6x - x^2$
4. $-3x^2 + 6x + 25$　　　**5.** $625 + 200x - 20x^2$　　**6.** $-2x^2 + 5x + 20$

7. An arrow is shot upward at 60 m/s. Find the maximum height of the arrow and the corresponding time. Use the method of completing the square and the formula $h = -5t^2 + vt$.

Ⓑ **8.** A chartered bus ride costs $30 per ticket if there are 20 passengers. The ticket price is reduced by $1 for each additional passenger beyond 20. How many passengers will produce the maximum income, and what is the maximum income?

9. A rectangular chicken pen next to a barn wall is enclosed by 120 m of chicken wire on three sides and the barn wall on the fourth side. Find the maximum area that can be enclosed in this way. Find the dimensions of the pen.

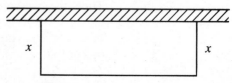

10. A rectangular field adjacent to the straight bank of a river is to be fenced, but there is to be no fencing along the riverbank. If 180 m of fencing is available, what is the maximum area that can be enclosed?

11. Four hundred people will attend a sidewalk display if tickets cost $1 each. Attendance will decrease by 20 people for each 10-cent increase in ticket price. What ticket price will yield the maximum income, and what is the maximum income from ticket sales?

12. A publishing company can get 1,000 subscribers for a new magazine if the monthly subscription rate is $5. It will get 100 more subscribers for each 10-cent decrease in the monthly rate. What monthly rate will produce the maximum monthly income, and what will that income be?

THE QUADRATIC FORMULA

Objective   **To solve a quadratic equation by using the quadratic formula**

Every quadratic equation can be written in the standard form of $ax^2 + bx + c = 0$, where $a \neq 0$. You can solve this general equation for x by the method of completing the square. The result is called the **quadratic formula.**

$$ax^2 + bx + c = 0$$
$$ax^2 + bx = -c$$
$$x^2 + \frac{b}{a}x = \frac{-c}{a} \qquad \blacktriangleleft \text{ Divide each side by } a.$$
$$x^2 + \frac{b}{a}x + \left(\frac{b}{2a}\right)^2 = \frac{-c}{a} + \left(\frac{b}{2a}\right)^2 \qquad \blacktriangleleft \text{ Add } \left(\frac{b}{2a}\right)^2 \text{ to each side.}$$
$$x^2 + \frac{b}{a}x + \frac{b^2}{4a^2} = \frac{-c}{a} + \frac{b^2}{4a^2}$$
$$\left(x + \frac{b}{2a}\right)^2 = \frac{-4ac + b^2}{4a^2}$$
$$x + \frac{b}{2a} = \pm \sqrt{\frac{b^2 - 4ac}{4a^2}} \qquad \blacktriangleleft \text{ If } x^2 = k, \text{ then } x = \pm\sqrt{k}.$$
$$x = \frac{-b}{2a} \pm \frac{\sqrt{b^2 - 4ac}}{2a}$$
$$x = \frac{-b \pm \sqrt{b^2 - 4ac}}{2a}$$

The quadratic formula	The solutions of a quadratic equation of the form $ax^2 + bx + c = 0$, $a \neq 0$, are given by the formula $x = \dfrac{-b \pm \sqrt{b^2 - 4ac}}{2a}$.

The quadratic formula can be used to solve any quadratic equation. You must first identify the values of a, b, and c. Then, you substitute these values in the formula, as shown in Example 1 below.

Example 1   **Solve $3x^2 + 5x - 4 = 0$ by using the quadratic formula.**

$$ax^2 + bx + c = 0$$
$$\downarrow \quad \downarrow \quad \downarrow$$
$$3x^2 + 5x - 4 = 0$$

$$a = 3, b = 5, c = -4$$

$$x = \frac{-b \pm \sqrt{b^2 - 4ac}}{2a}$$
$$= \frac{-(5) \pm \sqrt{5^2 - 4 \cdot 3 \cdot (-4)}}{2 \cdot 3}$$
$$= \frac{-5 \pm \sqrt{25 + 48}}{6}$$
$$= \frac{-5 \pm \sqrt{73}}{6}$$

Thus, the solutions are $\dfrac{-5 + \sqrt{73}}{6}$ and $\dfrac{-5 - \sqrt{73}}{6}$.

If a quadratic equation is not given in standard form, you should rewrite it in standard form in order to identify the values of a, b, and c.

Example 2   **Solve $4x^2 = 11 + 4x$ by using the quadratic formula.**

$$4x^2 - 4x - 11 = 0 \qquad \blacktriangleleft \text{ Write the equation in the}$$
$$\text{standard form of } ax^2 + bx + c = 0.$$

$$a = 4, \; b = -4, \; c = -11 \qquad \blacktriangleleft \text{ Identify the values of } a, \, b, \text{ and } c.$$

$$x = \frac{-b \pm \sqrt{b^2 - 4ac}}{2a}$$

$$= \frac{-(-4) \pm \sqrt{(-4)^2 - 4 \cdot 4 \cdot (-11)}}{2 \cdot 4} \qquad \blacktriangleleft \text{ Substitute.}$$

$$= \frac{4 \pm \sqrt{192}}{8}$$

$$= \frac{4 \pm 8\sqrt{3}}{8} \qquad \blacktriangleleft \; \sqrt{192} = \sqrt{64 \cdot 3}$$

$$= \frac{4(1 \pm 2\sqrt{3})}{4 \cdot 2}$$

$$= \frac{1 \pm 2\sqrt{3}}{2} \qquad \blacktriangleleft \text{ Divide numerator and}$$
$$\text{denominator by } 4.$$

Thus, the solutions are $\dfrac{1 + 2\sqrt{3}}{2}$ and $\dfrac{1 - 2\sqrt{3}}{2}$.

An equation of the form $ax^2 + bx + c = 0$, where $a = 0$, is *not* a quadratic equation since the square term is eliminated: $0 \cdot x^2 = 0$. However, there is no such restriction on the value of b or c. The standard form of the quadratic equation $5x^2 - 9x = 0$ is $5x^2 - 9x + 0 = 0$, where $c = 0$. The standard form of the quadratic equation $y^2 - 150 = 0$ is $y^2 + 0y - 150 = 0$. Equations of either type can be solved by using the quadratic formula.

Example 3   **Solve $5x^2 - 9x = 0$ and $y^2 - 150 = 0$ by using the quadratic formula.**

$$5x^2 - 9x = 0$$
$$5x^2 - 9x + 0 = 0$$
$$a = 5, \; b = -9, \; c = 0$$

$$x = \frac{-(-9) \pm \sqrt{(-9)^2 - 4 \cdot 5 \cdot 0}}{2 \cdot 5}$$

$$= \frac{9 \pm \sqrt{81 - 0}}{10}$$

$$= \frac{9 \pm 9}{10}, \text{ or } \frac{9 + 9}{10} \text{ and } \frac{9 - 9}{10}$$

The solutions are $\dfrac{9}{5}$ and 0.

$$y^2 - 150 = 0$$
$$1y^2 + 0y - 150 = 0$$
$$a = 1, \; b = 0, \; c = -150$$

$$y = \frac{-(0) \pm \sqrt{0^2 - 4 \cdot 1 \cdot (-150)}}{2 \cdot 1}$$

$$= \frac{0 \pm \sqrt{0 + 600}}{2}$$

$$= \frac{\pm 10\sqrt{6}}{2}, \text{ or } \pm 5\sqrt{6}$$

The solutions are $5\sqrt{6}$ and $-5\sqrt{6}$.

Reading in Algebra

For each equation in the left column, identify every correct description in the right column.

1. $9x + 5^2 = 0$ **A**   A quadratic equation in standard form
2. $x^2 - 14 = 0$ **B**   A quadratic equation not in standard form
3. $3x - 5 = x^2$ **C**   Not a quadratic equation
4. $5x^2 - 9x = 0$ **D** $a = 1$ **E** $b = 0$
5. $2x^2 - 7x - 6 = 0$ **F** $c = 0$

6. Identify three different methods for solving the quadratic equation $x^2 - 3x - 10 = 0$.

Oral Exercises

Identify the values of a, b, and c for each quadratic equation.

1. $6x^2 + 3x - 2 = 0$ 2. $y^2 - 8y + 7 = 0$ 3. $-2n^2 - n = 0$ 4. $z^2 - 8 = 0$

Give each quadratic equation in standard form.

5. $x^2 - 6x = 7$ 6. $4 + 3y^2 = -y$ 7. $-3 = t - 4t^2$ 8. $5x = -10x^2$

Written Exercises

Simplify each expression.

(A)

1. $\dfrac{-5 \pm \sqrt{8}}{2}$ 2. $\dfrac{6 \pm \sqrt{20}}{4}$ 3. $\dfrac{-10 \pm \sqrt{32}}{2}$

4. $\dfrac{8 \pm \sqrt{12}}{4}$ 5. $\dfrac{10 \pm \sqrt{36 - 36}}{4}$ 6. $\dfrac{0 \pm \sqrt{24}}{6}$

Solve each equation by using the quadratic formula.

7. $4x^2 + x - 2 = 0$ 8. $3y^2 + 7y + 3 = 0$ 9. $2n^2 - n - 2 = 0$
10. $2y^2 - 4y + 1 = 0$ 11. $t^2 + 4t - 1 = 0$ 12. $2x^2 + 6x + 3 = 0$
13. $4n^2 + 7 = 12n$ 14. $0 = 4x^2 + 8x + 1$ 15. $9y^2 = 6y + 7$
16. $3x^2 + 6x = 2$ 17. $2y^2 = 6y - 1$ 18. $5n^2 = 1 - 2n$
19. $y^2 - 8y + 16 = 0$ 20. $x^2 - 6x = 0$ 21. $t^2 - 11 = 0$
22. $t^2 - 20 = 0$ 23. $0 = 9y^2 + 12y + 4$ 24. $3x^2 + 2x = 0$
25. $4x^2 = 7x$ 26. $5t^2 - 11 = 0$ 27. $4y^2 = 20y - 25$

(B)

28. $\dfrac{3}{2}x^2 + \dfrac{1}{2}x - 3 = 0$ 29. $\dfrac{1}{2}y^2 - \dfrac{3}{2}y + \dfrac{5}{6} = 0$ 30. $\dfrac{1}{4}n^2 = \dfrac{1}{2}n + \dfrac{3}{8}$

31. $\dfrac{y^2}{15} + \dfrac{5}{3} = \dfrac{2y}{3}$ 32. $\dfrac{n^2}{4} = \dfrac{n}{2} - \dfrac{1}{6}$ 33. $\dfrac{x^2}{3} - \dfrac{1}{2} = \dfrac{x}{3}$

34. $6\left(2x^2 + \dfrac{4}{3}x\right) = 3(3x^2 - 1)$ 35. $\dfrac{y + 3}{2y - 1} = \dfrac{2y + 3}{y + 5}$ 36. $\dfrac{n^2 + 1}{n} + \dfrac{n - 2}{2n} = \dfrac{1}{3n}$

(C)

37. $x^2 - (2\sqrt{2})x - 6 = 0$ 38. $x^2\sqrt{6} - 4x - 2\sqrt{6} = 0$ 39. $5x^2 + 2x\sqrt{10} - 1 = 0$
40. $(y + 2)^2 + 7(y + 2) - 3$ 41. $(n + \sqrt{5})^2 - 5(n + \sqrt{5})$ 42. $(n^2)^2 - 3n^2\sqrt{2} + 4 = 0$
    $= 0$ [Hint: Let $x = y + 2$.] $- 5 = 0$ [Hint: Let $x = n^2$.]
                               [Hint: Let $x = n + \sqrt{5}$.]

43. $x^{-2} - 6x^{-1} + 4 = 0$ 44. $(x^{\frac{1}{2}})^2 - (3\sqrt{3})x^{\frac{1}{2}} + 6 = 0$ 45. $8x^{-4} - 6x^{-2} + 1 = 0$

PROBLEM SOLVING: THE QUADRATIC FORMULA

Objective **To solve a word problem by using the quadratic formula**

Sometimes a statement about the relationships between two numbers is in the form of a quadratic equation. Since a given quadratic equation may have two real number solutions, there may be two values for each of the numbers.

Example 1

A second number is 6 more than twice a first number. If the second number is multiplied by 3 more than the first number, the product is 9. Find the two numbers.

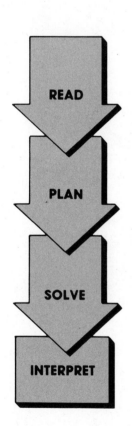

READ

PLAN

SOLVE

INTERPRET

Let f = first number.

$2f + 6$ = second number

(second number) · (3 more than first number) = 9

$$(2f + 6)(f + 3) = 9$$
$$2f^2 + 12f + 18 = 9$$
$$2f^2 + 12f + 9 = 0$$

Use the quadratic formula: $a = 2$, $b = 12$, and $c = 9$.

$$
\begin{aligned}
f &= \frac{-12 \pm \sqrt{12^2 - 4 \cdot 2 \cdot 9}}{2 \cdot 2} \\
&= \frac{-12 \pm \sqrt{72}}{4} \\
&= \frac{-12 \pm 6\sqrt{2}}{4}, \text{ or } \frac{-6 \pm 3\sqrt{2}}{2}
\end{aligned}
$$

If $f = \dfrac{-6 + 3\sqrt{2}}{2}$, then $2f + 6 = 2 \cdot \dfrac{-6 + 3\sqrt{2}}{2} + 6 = 3\sqrt{2}$.

If $f = \dfrac{-6 - 3\sqrt{2}}{2}$, then $2f + 6 = 2 \cdot \dfrac{-6 - 3\sqrt{2}}{2} + 6 = -3\sqrt{2}$.

Each pair of answers checks in the problem.

Thus, the two numbers are $\dfrac{-6 + 3\sqrt{2}}{2}$, $3\sqrt{2}$ or $\dfrac{-6 - 3\sqrt{2}}{2}$, $-3\sqrt{2}$.

Some problems involve the use of formulas so that when values are substituted for the variables, a quadratic equation is obtained. The resulting quadratic equation can then be solved using the quadratic formula. One formula of this type is $h = vt - 5t^2$, which you used earlier in this chapter. It is used again in Example 2 on page 173.

Example 2   **An object is to be shot upward from the earth's surface with an initial velocity of 25 m/s. To the nearest tenth of a second, when will the height of the object be 15 m?**

$$h = vt - 5t^2$$
$$15 = 25t - 5t^2 \quad \blacktriangleleft \; \text{Substitute 15 for } h \text{ and 25 for } v.$$
$$3 = 5t - t^2$$

$$t^2 - 5t + 3 = 0$$
$$t = \frac{5 \pm \sqrt{25 - 12}}{2} = \frac{5 \pm \sqrt{13}}{2}$$
$$= \frac{5 \pm 3.606}{2}, \text{ or } 4.303 \text{ and } 0.697$$

Check the irrational solutions, $\dfrac{5 \pm \sqrt{13}}{2}$, in the original equation.

Check: $\dfrac{5 + \sqrt{13}}{2}$

15	$25t - 5t^2$
15	$25 \cdot \dfrac{5 + \sqrt{13}}{2} - 5\left(\dfrac{5 + \sqrt{13}}{2}\right)^2$
	$\dfrac{125 + 25\sqrt{13}}{2} - 5\left(\dfrac{25 + 10\sqrt{13} + 13}{4}\right)$
	$\dfrac{125 + 25\sqrt{13}}{2} - \dfrac{95 + 25\sqrt{13}}{2}$
	$\dfrac{30}{2}, \text{ or } 15$

Check: $\dfrac{5 - \sqrt{13}}{2}$

15	$25t - 5t^2$
15	$25 \cdot \dfrac{5 - \sqrt{13}}{2} - 5\left(\dfrac{5 - \sqrt{13}}{2}\right)^2$
	$\dfrac{125 - 25\sqrt{13}}{2} - 5\left(\dfrac{25 - 10\sqrt{13} + 13}{4}\right)$
	$\dfrac{125 - 25\sqrt{13}}{2} - \dfrac{95 - 25\sqrt{13}}{2}$
	$\dfrac{30}{2}, \text{ or } 15$

Thus, the height will be 15 m at the end of 0.7 s and again at the end of 4.3 s, to the nearest 0.1 s.

Written Exercises

Solve each problem. Give the answers in simplest radical form.

(A) 1. A second number is 2 less than a first number. Their product is 4. Find the numbers.

2. The product of two numbers is 1. Find the numbers if one of them is 4 more than the other.

3. One number is 2 less than twice another number. Find these numbers if their product is 3.

4. Find two numbers such that their product is -2 and one number is 6 more than 3 times the other number.

Solve each problem using the formula $h = vt - 5t^2$. Approximate each answer to the nearest 0.1 s, unless directed otherwise.

5. An object is shot upward from the earth's surface with an initial velocity of 50 m/s. When will the height of the object be 100 m?

6. If a pellet is launched straight up from the surface of the earth with an initial velocity of 75 m/s, when will the altitude of the pellet be 225 m?

(B) 7. A certain baseball pitcher can throw a ball with an initial velocity of 40 m/s. If the ball is thrown straight up, when will its height be 15 m?

Problem Solving: The Quadratic Formula

8. An archer shot an arrow upward with an initial velocity of 125 m/s. When will its height be 755 m on its upward flight? When will its height be 755 m on its downward flight?

9. An object was shot upward from the earth with an initial velocity of 38 m/s. When did it return to earth?

10. The *maximum value* of h in the formula $h = vt - 5t^2$ is $\frac{v^2}{20}$. What initial velocity is needed for an object to reach a maximum height of 400 m.? Give the answer to the nearest meter per second.

Solve each problem. Give the answers in simplest radical form.

11. A second number is 4 more than 3 times a first number. If the second number is multiplied by 2 more than the first number, the product is 2. Find the numbers.

12. The smaller of two numbers is 3 less than twice the greater number. The product of the smaller number and 8 more than 4 times the greater number is -16. Find the numbers.

Solve each problem using the formula $h = d + vt - 5t^2$. Approximate each answer to the nearest 0.1 s. (See explanation on page 24, if necessary.)

Ⓒ 13. An object is launched upward from the top of a tower 80 m tall with an initial velocity of 40 m/s. When will the object be 130 m above the earth's surface?

14. A pellet was sent up from the top of a tower 60 m tall with an initial velocity of 35 m/s. When will it be 115 m above the earth?

15. When will the pellet in Exercise 14 return to earth?

16. When would the pellet in Exercise 14 be 35 m above the surface if it were launched from 20 m below the surface?

CALCULATOR ACTIVITIES

In Example 2, you can verify that the height of the object will be within 0.01 m of 15 m (15 ± 0.01 m) when the time $t = 0.697$ s.

First, rewrite the equation $15 = 25t - 5t^2$ as $t(25 - 5t) = 15$.
Second, show that $0.697 \times (25 - 5 \times 0.697)$ is between 14.99 and 15.01.

Compute ▶ ⊖ 5 ⊗ .697 ⊕ 25 ⊗ .697 ⊜ 14.995955, which is between 14.99 and 15.01.

1. In Example 2, verify that $h = 15 \pm 0.01$ m when $t = 4.303$ s.
2. If $h = 25t - 5t^2$, verify that $h = 22 \pm 0.01$ m if $t = 1.1399$ s or $t = 3.8601$ s.
3. If $h = 82t - 5t^2$, verify that $h = 334.6 \pm 0.01$ m when $t = 8.765$ s.
4. If $h = 647t - 5t^2$, verify that $h \neq 633.7 \pm 0.01$ m when $t = 0.987$ s.

PROBLEM SOLVING: IRRATIONAL MEASUREMENTS

Objective

To solve a word problem involving a geometric figure with irrational dimensions

You can use the Pythagorean relation, $a^2 + b^2 = c^2$, and the quadratic formula to solve some problems involving a right triangle. As shown in Example 1, measurements may be given as irrational numbers in simplest radical form.

Example 1

One leg of a right triangle is twice as long as the other leg. The hypotenuse is 3 cm longer than the longer leg. Find the length of each side of the triangle.

Let w = length of the shorter leg.
    $2w$ = length of the longer leg
   $2w + 3$ = length of the hypotenuse

$$a^2 + b^2 = c^2$$

$$w^2 + (2w)^2 = (2w + 3)^2$$
$$w^2 + 4w^2 = 4w^2 + 12w + 9$$
$$w^2 - 12w - 9 = 0 \quad \blacktriangleleft \textit{ Quadratic equation: } a = 1, b = -12, c = -9$$

$$w = \frac{-(-12) \pm \sqrt{(-12)^2 - 4 \cdot 1(-9)}}{2 \cdot 1}$$

$$= \frac{12 \pm \sqrt{180}}{2}$$

$$= \frac{12 \pm 6\sqrt{5}}{2}, \text{ or } 6 \pm 3\sqrt{5}$$

One solution of the quadratic equation is $6 - 3\sqrt{5}$, but it is *not* a solution to the problem. The length of a side of a triangle cannot be a negative number: $(6 - 3\sqrt{5}) < 0$ since $6 - 3 \cdot 2.236 < 0$.

$$w = 6 + 3\sqrt{5} \qquad 2w = 12 + 6\sqrt{5} \qquad 2w + 3 = 15 + 6\sqrt{5}$$

Check:	$a^2 + b^2$	c^2
	$(6 + 3\sqrt{5})^2 + (12 + 6\sqrt{5})^2$	$(15 + 6\sqrt{5})^2$
	$36 + 36\sqrt{5} + 45 + 144 + 144\sqrt{5} + 180$	$225 + 180\sqrt{5} + 180$
	$405 + 180\sqrt{5}$	$405 + 180\sqrt{5}$

Thus, the lengths of the two legs and the hypotenuse are $6 + 3\sqrt{5}$ cm, $12 + 6\sqrt{5}$ cm, and $15 + 6\sqrt{5}$ cm, respectively.

Sometimes measurements in the form of irrational numbers may have to be changed into decimal form. This is shown in Example 2 on page 176.

Example 2   **A 10-cm square (10 cm by 10 cm) is cut from each corner of a rectangular piece of cardboard that is 4 times as long as it is wide. The flaps are folded up to form an open carton with a volume of 7,000 cm³. Find the dimensions of the original piece of cardboard to the nearest 0.1 cm.**

Let x be the width and $4x$ the length of the original piece of cardboard.

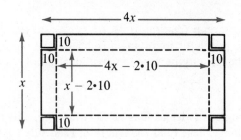
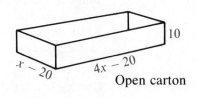

Open carton

length × width × height = volume

$(4x - 20)(x - 20) \cdot 10 = 7,000$
$(4x - 20)(x - 20) = 700$
$4x^2 - 100x + 400 = 700$
$4x^2 - 100x - 300 = 0$
$x^2 - 25x - 75 = 0$

$x = \dfrac{-(-25) \pm \sqrt{(-25)^2 - 4 \cdot 1(-75)}}{2}$

$= \dfrac{25 \pm \sqrt{925}}{2}$

$= \dfrac{25 \pm 5\sqrt{37}}{2}$

$\doteq \dfrac{25 \pm 5(6.083)}{2}$

$= 27.7075 \quad \text{or} \quad -2.7075$

The width cannot be negative, so $x = 27.7075$.
The length, $4x = 4 \cdot 27.7075$, or 110.8300.

Check: See the Non-Routine Problems on page 177.

Thus, the width of the original piece of cardboard was 27.7 cm, and the length was 110.8 cm, to the nearest 0.1 cm.

Written Exercises

Solve each problem. Give the answers in simplest radical form, unless directed otherwise.

(A) **1.** One leg of a right triangle is twice as long as the other leg and the hypotenuse is 6 m long. Find the length of each leg of the triangle.

2. In a certain right triangle, one leg is 1 cm shorter than the other leg and the hypotenuse is 3 cm longer than the shorter leg. Find the length of each side of the triangle.

3. A 2-cm square is cut from each corner of a rectangular piece of cardboard that is 20 cm longer than it is wide. The four flaps are turned up to form an open box with a volume of 50 cm³. Find the dimensions of the original piece of cardboard to the nearest 0.1 cm.

4. A rectangular piece of plastic is 3 times as long as it is wide. After a 6-cm square is cut from each corner, the flaps are pressed upward to form an open pan with a volume of 720 cm³. To the nearest 0.1 cm, find the length and the width of the original piece of plastic.

5. A diagonal of a square is 8 cm long. Find the width of the square.

6. Find the width of a square if one diagonal is $3\sqrt{22}$ mm long.

B 7. A rectangular lawn measures 20 m by 40 m and is surrounded by a sidewalk of uniform width. The outer edge of the sidewalk is a rectangle with an area of 1,000 m². Find the width of a strip of the sidewalk to the nearest 0.1 m.

8. A rectangular piece of mirror glass measures 40 cm by 60 cm. It is surrounded by a uniform frame whose outer edge is a rectangle with an area of 3,200 cm². To the nearest 0.1 cm, find the width of a strip of the frame.

9. A rectangle is 2 dm longer than it is wide. A diagonal of the rectangle is 1 dm longer than the rectangle's length. Find the length of the diagonal and the area of the rectangle.

10. A rectangle's length is 1 mm more than 3 times its width. The length of one of its diagonals is 1 mm more than the rectangles' length. Find the length, the width, and the area of the rectangle.

11. A diagonal of a square is 2 m longer than a side of the square. Find the length of the diagonal and the area of the square.

12. One side of a square is 3 dm shorter than one diagonal of the square. Find the length of the side and the area of the square.

CALCULATOR ACTIVITIES

In Example 2 on page 176, the open carton measures $4x - 20$ cm by $x - 20$ cm by 10 cm and the volume is 7,000 cm³. Determine whether the volume is within $7,000 \pm 1$ cm³ if $x = 27.7075$ cm.

If $x = 27.7075$, then $4x - 20 = 90.8300$ and $x - 20 = 7.7075$.

$90.83 \otimes 7.7075 \otimes 10 \ominus 7000.7222$, which is within 7000 ± 1.

A certain box measures $x - 10$ cm by $3x - 10$ cm by 20 cm. Determine whether the volume of the box is within 1 cm³ of the given volume for the given value of x.
1. 11,950 cm³ if $x = 21.168$ cm
2. 2,250 cm³ if $x = 13.625$ cm
3. 39,240 cm³ if $x = 32.456$ cm
4. 1,270 cm³ if $x = 12.347$ cm

NON-ROUTINE PROBLEMS

Show that $10(x - 20)(4x - 20) = 7,000$ if $x = \dfrac{25 + 5\sqrt{37}}{2}$.

RADICAL EQUATIONS

Objective   **To solve a radical equation**

Equations like $\sqrt[3]{6x + 10} = -2$ and $3\sqrt{2y + 2} = 2\sqrt{5y - 1}$ contain a variable in a radicand. They are called **radical equations.**

Example 1   **Solve $\sqrt[3]{6x + 10} = -2$.**

$$\sqrt[3]{6x + 10} = -2$$
$$(\sqrt[3]{6x + 10})^3 = (-2)^3 \quad \blacktriangleleft \text{ If } a = b, \text{ then } a^3 = b^3.$$
$$6x + 10 = -8$$
$$6x = -18$$
$$x = -3$$

Check in the original equation.

$\sqrt[3]{6x + 10}$	-2
$\sqrt[3]{6(-3) + 10}$	-2
$\sqrt[3]{-18 + 10}$	
$\sqrt[3]{-8}$	
-2	

Thus, the solution is -3.

Example 2   **Solve $3\sqrt{2y + 2} = 2\sqrt{5y - 1}$.**

$$3\sqrt{2y + 2} = 2\sqrt{5y - 1}$$
$$(3\sqrt{2y + 2})^2 = (2\sqrt{5y - 1})^2 \quad \blacktriangleleft \text{ If } a = b, \text{ then } (a)^2 = (b)^2.$$
$$9(2y + 2) = 4(5y - 1)$$
$$18y + 18 = 20y - 4$$
$$22 = 2y$$
$$11 = y$$

Check in the original equation.

$3\sqrt{2y + 2}$	$2\sqrt{5y - 1}$
$3\sqrt{2 \cdot 11 + 2}$	$2\sqrt{5 \cdot 11 - 1}$
$3\sqrt{24}$	$2\sqrt{54}$
$3 \cdot 2\sqrt{6}$	$2 \cdot 3\sqrt{6}$
$6\sqrt{6}$	$6\sqrt{6}$

Thus, the solution is 11.

It is necessary to check the apparent solutions of a radical equation since raising each side of an equation to the same power may introduce *extraneous solutions*.

Example 3   **Solve $3 + \sqrt{3x + 1} = x$.**

"Isolate" the radical. Rewrite the equation so that the term containing the radical is alone on one side.

$$\sqrt{3x + 1} = x - 3$$
$$(\sqrt{3x + 1})^2 = (x - 3)^2$$
$$3x + 1 = x^2 - 6x + 9$$

Quadratic equation $\blacktriangleright$ $\quad 0 = x^2 - 9x + 8$

$$0 = (x - 1)(x - 8)$$
$$x = 1 \text{ or } x = 8$$

Check: $x = 1$

$3 + \sqrt{3x + 1}$	x
$3 + \sqrt{3 \cdot 1 + 1}$	1
$3 + \sqrt{4}$	
$3 + 2, \text{ or } 5$	

1 does *not* check. $\blacktriangleright$

Check: $x = 8$

$3 + \sqrt{3x + 1}$	x
$3 + \sqrt{3 \cdot 8 + 1}$	8
$3 + \sqrt{25}$	
$3 + 5, \text{ or } 8$	

Thus, the solution is 8.

In some radical equations, the radicand contains an exponent.

Example 4 **Solve $\sqrt[3]{2y^{-1}} = -4$.**

$$(\sqrt[3]{2y^{-1}})^3 = (-4)^3$$
$$2y^{-1} = -64$$
$$y^{-1} = -32$$
$$\frac{1}{y} = -32$$
$$y = \frac{1}{-32}, \text{ or } -\frac{1}{32}$$

Thus, the solution is $-\frac{1}{32}$.

Check:

$\sqrt[3]{2y^{-1}}$	-4
$\sqrt[3]{2 \cdot \left(\dfrac{1}{-32}\right)^{-1}}$	-4
$\sqrt[3]{2(-32)}$	
$\sqrt[3]{-64}$, or -4	

Example 5 **Solve $\sqrt[4]{x^2 - 8} = 2$.**

$$\left(\sqrt[4]{x^2 - 8}\right)^4 = 2^4$$
$$x^2 - 8 = 16$$
$$x^2 = 24$$
$$x = \pm 2\sqrt{6}$$

Thus, the solutions are $2\sqrt{6}$ and $-2\sqrt{6}$.

Check:

$\sqrt[4]{x^2 - 8}$	2
$\sqrt[4]{(\pm 2\sqrt{6})^2 - 8}$	2
$\sqrt[4]{24 - 8}$	
$\sqrt[4]{16}$, or 2	

An equation such as $\sqrt{5x + 1} - \sqrt{3x - 5} = 2$ in Example 6 contains two radicals and a third term. To solve this equation, you need to isolate the radicals, one at a time.

Example 6 **Solve $\sqrt{5x + 1} - \sqrt{3x - 5} = 2$.**

Isolate one of the two radicals.
$$\sqrt{5x + 1} = 2 + \sqrt{3x - 5}$$
$$(\sqrt{5x + 1})^2 = (2 + \sqrt{3x - 5})^2$$
$$5x + 1 = 4 + 4\sqrt{3x - 5} + 3x - 5$$
Isolate the other radical.
$$2x + 2 = 4\sqrt{3x - 5}$$
$$x + 1 = 2\sqrt{3x - 5}$$
$$(x + 1)^2 = (2\sqrt{3x - 5})^2$$
$$x^2 + 2x + 1 = 4(3x - 5)$$
$$x^2 + 2x + 1 = 12x - 20$$
$$x^2 - 10x + 21 = 0$$
$$(x - 3)(x - 7) = 0$$
$$x = 3 \quad \text{or} \quad x = 7$$

Check:

	$\sqrt{5x + 1} - \sqrt{3x - 5}$	2
$x = 3$	$\sqrt{5 \cdot 3 + 1} - \sqrt{3 \cdot 3 - 5}$	2
	$\sqrt{16} - \sqrt{4}$	
	$4 - 2$, or 2	

Check:

	$\sqrt{5x + 1} - \sqrt{3x - 5}$	2
$x = 7$	$\sqrt{5 \cdot 7 + 1} - \sqrt{3 \cdot 7 - 5}$	2
	$\sqrt{36} - \sqrt{16}$	
	$6 - 4$, or 2	

Thus, the solutions are 3 and 7.

Radical Equations

Oral Exercises

For each equation, the only apparent solutions are 3 and 5. Check 3 and 5 in each equation to find the solution, or solutions, of the equation.

1. $\sqrt{x-1}=2$ **2.** $\sqrt{x+1}=x-1$ **3.** $\sqrt{x+1}=-2$ **4.** $\sqrt{x-1}=x-3$ **5.** $\sqrt{2x-6}=x-3$

Written Exercises

Solve each equation.

(A)

1. $\sqrt{x-2}=7$

2. $\sqrt{2y-6}=6$

3. $\sqrt{3c}=\sqrt{c+6}$

4. $\sqrt[3]{2a}=4$

5. $\sqrt[3]{7x-1}=\sqrt[3]{5x+7}$

6. $\sqrt[4]{5x+1}-\sqrt[4]{6x-2}=0$

7. $2\sqrt{x}=\sqrt{3x+5}$

8. $\sqrt{12y+3}-3\sqrt{2y}=0$

9. $3\sqrt{2a}-4=2\sqrt{a}-2$

10. $2\sqrt[3]{4y}=-1$

11. $3\sqrt[4]{2x}=2$

12. $2\sqrt[3]{3n}=3\sqrt[3]{n-1}$

13. $x-4=\sqrt{2x}$

14. $y-5=\sqrt{y-3}$

15. $\sqrt{2w-3}=w-3$

16. $\sqrt[3]{y^{-1}}=5$

17. $\sqrt[3]{c^2}=2$

18. $\sqrt[3]{x^2+9}=3$

19. $\sqrt[4]{x^3}=2$

20. $\sqrt[4]{y^{-2}}=3$

21. $\sqrt[4]{a^2+16}=4$

22. $\sqrt{8y}=\sqrt{2y}+2$

23. $\sqrt{x+7}=\sqrt{x}+1$

24. $\sqrt{3n}-2=\sqrt{n}-2$

(B)

25. $\sqrt[3]{3y^{-1}}=-6$

26. $\sqrt[4]{2y^{-2}}=\frac{1}{3}$

27. $\sqrt{(x-3)^{-2}}=\frac{1}{4}$

28. $3+\sqrt{2t+5}=0$

29. $-2+\sqrt{4x+3}=0$

30. $-5-\sqrt{7y-3}=0$

31. $6+\sqrt{3x}=x$

32. $y-\sqrt{4y-3}=2$

33. $2+\sqrt{5x+6}=3x$

34. $4-\sqrt{10-3a}=a$

35. $2\sqrt{3x-2}=x+2$

36. $3\sqrt{2n+3}=2n+5$

37. $\sqrt{x^2+x-3}=3$

38. $\sqrt[3]{y^2-12y}=4$

39. $\sqrt[4]{5x^2+3}=2\sqrt[4]{x}$

40. $\sqrt{2y+3}=1+\sqrt{y+1}$

41. $\sqrt{4x-3}=2+\sqrt{2x-5}$

42. $\sqrt{3a+1}-\sqrt{a-4}=3$

43. $\sqrt{\dfrac{x-3}{x+2}}=\dfrac{2}{3}$

44. $\sqrt{\dfrac{2}{y+3}}=\sqrt{\dfrac{3}{2y+2}}$

45. $\dfrac{\sqrt{x-2}}{x-2}=\dfrac{x-5}{\sqrt{x-2}}$

46. $\sqrt{y+15}-\sqrt{2y+7}=1$

47. $\sqrt{3x-5}+\sqrt{x-1}=2$

48. $\sqrt{3x+9}-\sqrt{2x+7}=1$

(C) **49.** The time T for one period of a simple pendulum is determined by the formula $T=2\pi\sqrt{\dfrac{l}{g}}$. Solve for l, the length of the pendulum's string.

50. The time t on an upward launch is found from the formula $t=\dfrac{v\pm\sqrt{v^2+20(d-h)}}{10}$. Solve for h, the height at the time t.

51. Solve $\sqrt{x-1}=\sqrt[4]{8x+1}$ for x.

52. Solve $\sqrt[3]{x+6}=\sqrt[6]{32x}$ for x.

53. The radius r of a sphere is given by the formula $r=\sqrt[3]{\dfrac{3V}{4\pi}}$. Solve for V, the volume of the sphere.

54. One side s of an equilateral triangle is given by the formula $s=\dfrac{2\sqrt{A}}{\sqrt[4]{3}}$. Solve for A, the area of the triangle.

55. Solve $x=\dfrac{N\pm\sqrt{N^2-4M(T-D)}}{2M}$ for T.

56. Solve $2Lx+R=\sqrt{R^2-\dfrac{4L}{C}}$ for $\dfrac{-1}{C}$.

CUMULATIVE REVIEW

Simplify each expression.

1. $5\sqrt{72}$

2. $3x^2y\sqrt{24x^7y^8}$

3. $5\sqrt{2}+2\sqrt{18}+3\sqrt{2}$

4. $(2\sqrt{7}-\sqrt{10})(2\sqrt{7}+\sqrt{10})$

APPLICATIONS

Read → Plan → Solve → Interpret

If a weight is suspended on a string, the result is a simple pendulum. Ideally; the motion of a pendulum is **periodic.** If the weight on the pendulum is pulled to one side and released, it will swing away and back. One such back-and-forth motion is called a **vibration.**

The action of some clocks is based on the periodic motion of a pendulum. Of course, some mechanical contrivance is usually employed to keep the periodic motion stable. The time, T, for one period of a simple pendulum is given by the radical equation

$$T = 2\pi \sqrt{\frac{l}{g}}$$

where l is the length of the string, and g is the acceleration due to gravity. Notice that the period, T, depends on the length, l, of the string and not upon the amount of weight that is used.

The formula above can be solved for l.

$$\left(\frac{T}{2\pi}\right)^2 = \left(\sqrt{\frac{l}{g}}\right)^2 \qquad \frac{l}{g} = \frac{T^2}{4\pi^2} \qquad \text{So, } l = \frac{gT^2}{4\pi^2}.$$

Solve each problem. Use the more appropriate of the two formulas above. Use 10 m/s² for g and 3.14 for π.

1. Find the length of the string for a simple pendulum if the time for one period is 1.57 s.
2. Find the time for one period of a simple pendulum if the length of the string is 0.225 m.
3. How should the length of the string be changed in order to double the period of a given simple pendulum? [*Hint*: Replace T by $2T$ in the second formula.]

For a sphere of radius r and volume V, $r = \sqrt[3]{\dfrac{3V}{4\pi}}$.

4. Solve the formula for V.
5. Find V if $r = 30$ cm. Use 3.14 for π.
6. Find the change in the volume of a sphere if the radius of a given sphere is tripled.

For a cylinder of radius r and volume V, $r = \sqrt{\dfrac{V}{\pi h}}$.

7. Solve the formula for V.
8. Find V if $r = 4.5$ cm and $h = 8$ cm. Use 3.14 for π.
9. What is the change in the volume of a cylinder if the radius and the height of a given cylinder are each doubled?

Vocabulary
completing the square [7.1]
quadratic formula [7.3]
radical equation [7.6]

Solve each equation by completing the square. [7.1]

1. $x^2 - 6x = 3$

2. $2n^2 = 1 - 5n$

★ 3. $x^2 + 2x\sqrt{2} - 48 = 0$

Solve each equation by using the quadratic formula. [7.3]

4. $3y^2 + 2y =$

5. $5c^2 =$

6. $\dfrac{x^2}{8} = \dfrac{x}{2} - \dfrac{1}{4}$

★ 7. $x^{-2} - 8x^{-1} + 3 = 0$

8. Find the maximum value of $-3x^2 + 24x + 152$ and the corresponding value of x. [7.2]

9. An object is sent upward at 80 m/s. Find the maximum height it reaches and the corresponding time. Use the formula $h = vt - 5t^2$. [7.2]

10. A second number is 2 more than 4 times a first number. Find the numbers if their product is 5. [7.4]

11. An object is shot upward from the earth's surface with an initial velocity of 60 m/s. To the nearest 0.1 s, when will its height be 150 m? Use the formula $h = vt - 5t^2$. [7.4]

12. A 4-cm square is cut from each corner of a rectangular piece of sheet metal that is twice as long as it is wide. Find the dimensions of the original rectangle if the flaps are turned up to form a pan with a volume of 800 cm^3. Express each dimension in simplest radical form. [7.5]

13. A rectangle is 1 m longer than it is wide. The length of one of its diagonals is 3 m more than the rectangle's width. Find the length of the diagonal and the area of the rectangle. Give the answers in simplest radical form. [7.5]

★ 14. A pellet is sent up from the top of a tower 50 m tall with an initial velocity of 45 m/s. To the nearest 0.1 s, when will its height be 125 m? Use the formula $h = d + vt - 5t^2$. [7.4]

Solve each equation. [7.6]

15. $3\sqrt{2x + 4} = 12$

16. $x - 1 = \sqrt{x + 5}$

17. $\sqrt[4]{5y^{-1}} = \dfrac{1}{2}$

18. $\sqrt[3]{n^2 - 6n} = 3$

19. $\sqrt{3x - 5} - \sqrt{x - 2} = 1$

★ 20. $\sqrt[3]{4x + 6} = 2\sqrt[6]{2x}$

CHAPTER SEVEN TEST

Solve each equation by completing the square.

1. $x^2 - 8x - 4 = 0$

2. $3c^2 = -7c - 1$

Solve each equation by using the quadratic formula.

3. $2y^2 + 3y = 3$

4. $5x^2 = 11x$

5. $\dfrac{x^2}{4} = \dfrac{x}{2} - \dfrac{1}{6}$

6. One leg of a right triangle is 3 times as long as the other leg and 2 cm shorter than the hypotenuse. Find the length of the hypotenuse. Give the answer in simplest radical form.

7. A second number is 2 less than 4 times a first number. If the second number is multiplied by 1 less than the first number, the product is 1. Find the numbers.

8. A rectangular piece of land is to be fenced on three sides with 240 m of fencing. Find the maximum area of the piece of land.

9. To the nearest 0.1 s, when will an object reach a height of 15 m if it is sent up from the earth with an initial velocity of 25 m/s? Use the formula $h = vt - 5t^2$.

10. A 3-cm square is cut from each corner of a rectangular piece of plastic that is 3 times as long as it is wide. Find the dimensions of the original rectangle if the four flaps are turned up to form an open box with a volume of 270 cm^3. Express each dimension in simplest radical form.

11. Find the maximum value of $-2x^2 + 24x + 228$ and the corresponding value of x.

Solve each equation.

12. $\sqrt[4]{7x + 4} = 3$

13. $y - \sqrt{2y - 1} = 2$

14. $\sqrt[3]{n^2 - 2n} = 2$

15. $\sqrt{3x + 1} - \sqrt{x - 1} = 2$

★ 16. Solve $x^2 + 4x\sqrt{5} + 15 = 0$ by completing the square.

★ 17. Solve $\sqrt{x - 1} = \sqrt[4]{3x + 1}$.

COMPUTER ACTIVITIES

Solutions of Quadratic Equations

OBJECTIVE: To find solutions of quadratic equations in radical form

On page 583, you find the solutions of a quadratic equation by inserting coefficient values into the quadratic formula. Unfortunately, the computer only outputs answers in decimal form. The program below will output the solutions in radical form as an alternative to decimal solutions.

In this program, line 130 tests whether or not the discriminant is a perfect square. The loop from lines 140 to 170 simplifies the radical. Lines 180 through 230 reduce the numerator and denominator to simplest terms.

Notice the use of the symbol $<>$ in lines 150, 190, 200, and 210. This combination of the greater than and less than is the BASIC symbol for "is not equal to." Notice also the use of the ABS function in line 180.

```
110   PRINT "GIVE THE VALUES OF A,B, AND C.";: INPUT
      A,B,C
120 D =  - B:E = 1:F = B * B - 4 * A * C:G = 2 * A
125   IF F < 0 THEN 270
130   IF  SQR (F) =  INT ( SQR (F)) THEN 280
140   FOR I = 2 TO  SQR (F)
150   IF F / I ^ 2 <  > INT (F / I ^ 2) THEN 170
160 E = E * I:F = F / I ^ 2
170   NEXT I
180   FOR I = 2 TO  ABS (D)
190   IF D / I <  > INT (D / I) THEN 230
200   IF E / I <  > INT (E / I) THEN 230
210   IF G / I <  > INT (G / I) THEN 230
220 D = D / I:E = E / I:G = G / I: GOTO 190
230   NEXT I
240   PRINT "THE SOLUTIONS ARE ";
250   PRINT "("D"/"G" + "E" SQR ("F" ))/"G"  AND "
260   PRINT "("D"/"G" - "E"SQR ("F"))/"G: GOTO 110
270   PRINT "NO REAL SOLUTIONS": GOTO 110
280   PRINT "THE SOLUTIONS ARE "D +  SQR (F)"/"G" AND";
290   PRINT D -  SQR (F)"/"G: GOTO 110
```

See the Computer Section beginning on page 574 for more information.

EXERCISES

1. Run the program above by inputting the values from the illustrative exercises on pages 165 and 166.

2. Write a program that solves problems similar to the ones in Exercise 2 on page 167.

COLLEGE PREP TEST

DIRECTIONS: Choose the _one_ best answer to each question or problem.

1. One diagonal of a rectangle is 10 cm long. The rectangle's length is 8 cm. Find the area of the rectangle.

   (A) 24 cm^2 (B) 36 cm^2 (C) 48 cm^2
   (D) 80 cm^2 (E) None of these

2. Which equation has two different positive integers as its solutions?

   (A) $(x - 5)^2 = 0$ (B) $(x - 1)^2 = 4$
   (C) $(x + 4)^2 = 9$ (D) $(x - 6)^2 = 16$
   (E) None of these

3. For which pair of numbers is the product of _2 more than the first_ and _3 less than the second_ greatest?

   (A) 9 and 7   (B) 8 and 8
   (C) 7 and 9   (D) -12 and -3
   (E) All products are the same.

4. If $x = -2$, then

   (A) $\sqrt{x^2} = x$ (B) $\sqrt[3]{x^3} = -x$
   (C) $\sqrt[4]{x^4} = x$ (D) $\sqrt[3]{x^3} = x$
   (E) All of these

5. For which equation is the sum of its two solutions greater than their product?

   (A) $x^2 - 7x + 12 = 0$
   (B) $x^2 + 9x - 10 = 0$
   (C) $x^2 + 8x = 0$
   (D) $x^2 + 4 = 0$
   (E) None of these

6. Find a solution of
   $\sqrt{4x} + \sqrt{2x} = \sqrt{4x - 12} + \sqrt{2x + 12}$.

   (A) 2   (B) 4   (C) 6   (D) 8
   (E) None of these

7. A rectangular piece of cardboard is 70 cm by 50 cm. A 10-cm square is cut from each corner and the four flaps are folded up to form an open box. Find the volume of the box.

   (A) 1,500 cm^3 (B) 15,000 cm^3
   (C) 24,000 cm^3 (D) 35,000 cm^3
   (E) None of these

8. Choose the number that is not equal to the other three?

   (A) $4 + 2\sqrt{3}$ (B) $\dfrac{8 + \sqrt{48}}{2}$
   (C) $6 + \sqrt{48} - 2 - \sqrt{12}$
   (D) $\sqrt{2}(\sqrt{8} + \sqrt{6})$
   (E) They are all equal.

9. The formula, $h = 5t(20 - t)$, gives the height h of an object at t seconds. The greatest height is achieved at which time?

   (A) 0 s   (B) 5 s   (C) 10 s   (D) 20 s
   (E) All heights are the same.

10. Find an equation with integral coefficients having $3 - 2\sqrt{3}$ as a solution.

    (A) $x^2 = 9x - 12$ (B) $x^2 - 6x = 3$
    (C) $x^2 - 6x + 15 = 0$ (D) $x^2 = 21$
    (E) None of these

8 COMPLEX NUMBERS

HISTORY

Sonya Kovalevsky 1850–1891

Sonya Kovalevsky was born in Moscow, Russia, on January 15, 1850. She is best known for her work on partial differential equations and elliptic functions.

At age 18, Sonya left Russia and moved to Germany. She overcame many obstacles to study mathematics. She was unable to enter the University of Berlin because she was a woman. However, after studying with Karl Weierstrass she was awarded a Ph.D. degree in absentia from the University of Gottingham. Sonya was one of the few women of her time to receive the Prix Bordin award from the Paris Academy for outstanding work in mathematics.

In 1884, she became a professor of mathematics at Stockholm University in Sweden, and in 1889 the university named her a professor for life.

Project   Write a report on Sonya Kovalevsky, emphasizing her work in elliptic functions.

SUMS AND DIFFERENCES OF COMPLEX NUMBERS

Objectives   **To simplify a square root expression where the radicand is a negative real number**
To solve a quadratic equation whose solutions are imaginary numbers
To simplify an expression containing a sum or difference of complex numbers

The equation $x^2 = -1$ does not have a real number solution since there is no real number whose square is negative. In order to solve equations of this type, mathematicians invented the numbers i and $-i$.
If $x^2 = -1$, then x would have to be $\sqrt{-1}$ or $-\sqrt{-1}$, where $(\sqrt{-1})^2 = -1$ and $(-\sqrt{-1})^2 = -1$. The number $\sqrt{-1}$ is represented by i and the number $-\sqrt{-1}$ is represented by $-i$.

Definition: i and $-i$	$i = \sqrt{-1}$ and $i^2 = -1$ $-i = -\sqrt{-1}$ and $(-i)^2 = -1$

Notice that i and $-i$ are *not* variables; each of them names one specific number. You can use the definition of i to simplify a square root expression where the radicand is a negative real number.

Example 1   **Simplify $\sqrt{-16}$, $\sqrt{-7}$, and $\sqrt{-12}$.**

$$\begin{aligned}\sqrt{-16} &= \sqrt{-1(16)} \\ &= \sqrt{-1} \cdot \sqrt{16} \\ &= 4i\end{aligned}$$
$$\begin{aligned}\sqrt{-7} &= \sqrt{-1(7)} \\ &= \sqrt{-1} \cdot \sqrt{7} \\ &= i\sqrt{7}\end{aligned}$$
$$\begin{aligned}\sqrt{-12} &= \sqrt{-1(12)} \\ &= \sqrt{-1} \cdot \sqrt{12} \\ &= i \cdot 2\sqrt{3} \\ &= 2i\sqrt{3}\end{aligned}$$

In Example 1, the property $\sqrt{xy} = \sqrt{x} \cdot \sqrt{y}$ for $x \geq 0$, $y \geq 0$, was extended to include $\sqrt{-1 \cdot y} = \sqrt{-1} \cdot \sqrt{y}$ for $y \geq 0$. In general, $\sqrt{-y} = i\sqrt{y}$ for $y \geq 0$.

Example 2   **Simplify each expression.**
$$-3 + \sqrt{-4} \qquad\qquad -5\sqrt{-8} \qquad\qquad 2 - \sqrt{-10}$$

$$\begin{aligned}&-3 + i\sqrt{4} \\ &-3 + i \cdot 2 \\ &-3 + 2i\end{aligned}$$
$$\begin{aligned}&-5 \cdot i\sqrt{8} \\ &-5 \cdot i \cdot 2\sqrt{2} \\ &-10i\sqrt{2}\end{aligned}$$
$$\begin{aligned}&2 - i\sqrt{10}\end{aligned}$$

The numbers $-3 + 2i$, $-10i\sqrt{2}$ and $2 - i\sqrt{10}$ are examples of **complex numbers**. A complex number is one that can be written in the form $a + bi$, where a and b are real numbers. Note that $-10i\sqrt{2} = 0 + (-10\sqrt{2})i$, where $a = 0$, $b = -10\sqrt{2}$, and $2 - i\sqrt{10} = 2 + (-\sqrt{10})i$, where $a = 2$, $b = -\sqrt{10}$.

Definition: Complex number	A *complex number* is a number that can be written in the form $a + bi$ where a and b are real numbers and $i = \sqrt{-1}$.

Complex numbers like $-3 + 2i$ and $-10i\sqrt{2}$, where $b \neq 0$, are called **imaginary numbers**. A complex number like $-10i\sqrt{2}$, where $a = 0$, is called a **pure imaginary number**. Notice that every *real number* is also a complex number. For example, the real number -5 can be written as $-5 + 0i$, where $b = 0$.

You can solve equations like $x^2 = -25$ and $2y^2 + 16 = 0$ if imaginary numbers are permitted as solutions.

Example 3

Solve $x^2 = -25$.

$$x^2 = -25$$
$$x = \pm\sqrt{-25}$$
$$= \pm 5i$$

The solutions are $5i$ and $-5i$.

Solve $2y^2 + 16 = 0$.

$$2y^2 = -16$$
$$y^2 = -8$$
$$y = \pm\sqrt{-8} = \pm 2i\sqrt{2}$$

The solutions are $2i\sqrt{2}$ and $-2i\sqrt{2}$.

An expression containing a sum of two complex numbers can be simplified by combining the like terms as you would with a pair of binomials. For example,

$$(3 + 5i) + (2 - 8i) = (3 + 2) + (5i - 8i) = (3 + 2) + (5 - 8)i = 5 - 3i.$$

This leads to the following property:

Sum of complex numbers	For all real numbers a, b, c, and d, $(a + bi) + (c + di) = (a + c) + (b + d)i.$

To simplify an expression containing a difference of two complex numbers, $(a + bi) - (c + di)$, add the additive inverse (opposite) of $c + di$, which is $-c - di$.

Example 4

Simplify each expression.

$$(8 - 9i) - (6 - 4i)$$

$$(8 - 9i) + (-6 + 4i)$$
$$(8 - 6) + (-9 + 4)i$$
$$2 - 5i$$

$$(-9 + \sqrt{-2}) + (6 - \sqrt{-18})$$

$$(-9 + i\sqrt{2}) + (6 - 3i\sqrt{2})$$
$$(-9 + 6) + (\sqrt{2} - 3\sqrt{2})i$$
$$-3 - 2i\sqrt{2}$$

Reading in Algebra

Which of the words at the right apply to each complex number?

1. $5i$ **2.** $8 - 4i$ **3.** -25 **A** imaginary   **B** real   **C** irrational

4. i **5.** i^2 **6.** $5 + 6\sqrt{2}$ **D** rational   **E** pure imaginary

Oral Exercises

Show that each number is a complex number $a + bi$ by identifying the real numbers a and b.

1. $6i$
2. 7
3. i
4. $-\sqrt{10}$
5. $-9i$
6. $i\sqrt{3}$
7. -2.5
8. $-i$
9. $-i\sqrt{5}$
10. $7 - 12i$

Written Exercises

Simplify each expression.

(A)
1. $\sqrt{-36}$
2. $-\sqrt{-81}$
3. $\sqrt{-64}$
4. $\sqrt{-15}$
5. $\sqrt{-22}$
6. $-\sqrt{-7}$
7. $\sqrt{-40}$
8. $-\sqrt{-27}$
9. $2\sqrt{-45}$
10. $-5\sqrt{-4}$
11. $-6\sqrt{-20}$
12. $10\sqrt{-90}$
13. $4 + \sqrt{-9}$
14. $-20 - \sqrt{-30}$
15. $8 - \sqrt{-50}$

Solve each equation.

16. $x^2 = -49$
17. $y^2 = -3$
18. $z^2 = -48$
19. $3y^2 = -30$
20. $x^2 = -100$
21. $4a^2 = -200$

Simplify each expression.

22. $(3 + 6i) + (4 + 3i)$
23. $(7 + 8i) - (3 + 6i)$
24. $(-2 - 5i) + (8 + 12i)$
25. $(6 + 12i) - (-3 + 8i)$
26. $(5 + i) + (-6 - 3i)$
27. $(-7 + i) - (5 - i)$
28. $(7 - i) + (-10 + 10i)$
29. $(-14 + 6i) - (-6 - 6i)$
30. $(-8 + 4i) + (-8 - 4i)$

(B)
31. $7 + 8\sqrt{-16}$
32. $-5 - 5\sqrt{-25}$
33. $\sqrt{60} - 4\sqrt{-15}$
34. $20 - 10\sqrt{-12}$
35. $-1 + 4\sqrt{-80}$
36. $-\sqrt{8} - 2\sqrt{-32}$

Solve each equation.

37. $x^2 + 81 = 0$
38. $y^2 + 45 = 0$
39. $0 = n^2 + 65$
40. $5t^2 + 21 = t^2 - 43$
41. $4x^2 + 21 = 3 - 2x^2$
42. $8y^2 + 12 = 8 - y^2$

Simplify each expression.

43. $(3 + \sqrt{-36}) + (-8 - \sqrt{-16})$
44. $(-9 + 2\sqrt{-25}) - (7 - 3\sqrt{-49})$
45. $(6 - 5\sqrt{-3}) - (1 - 2\sqrt{-3})$
46. $(-10 + 6\sqrt{-20}) + (-4 - 4\sqrt{-5})$
47. $(12 + \sqrt{-18}) + (-4 + 4\sqrt{-2})$
48. $(7 - 3\sqrt{-8}) - (17 - 2\sqrt{-50})$

The **absolute value** of a *complex number* $a + bi$ is defined as

$|a + bi| = \sqrt{a^2 + b^2}$, where a and b are real numbers.

Simplify. Use the definition of $|a + bi|$.

(C)
49. $|3 - 4i|$
50. $|2 + 6i|$
51. $|-3|$
52. $|5i|$
53. $|6 - 8i|$ and $|6 + 8i|$
54. $|3 + 4i|$ and $|-3 - 4i|$

Prove each theorem.

55. $|a + bi| = |a - bi|$
56. $|a + bi| = |-a - bi|$

Sums and Differences of Complex Numbers $\qquad$ **189**

PRODUCTS AND QUOTIENTS OF COMPLEX NUMBERS

Objectives

To simplify an expression containing a product of complex numbers
To simplify an expression containing a quotient of complex numbers

The fact that $i^2 = -1$ permits you to simplify products that involve imaginary numbers. Before multiplying two pure imaginary numbers, each number should be written as the product of i and a real number.

Example 1

Simplify each product.

$6i \cdot 4i$ $\qquad\qquad -\sqrt{-6} \cdot 5\sqrt{-3}$

$$6i \cdot 4i$$
$$= 24i^2$$
$i^2 = -1 \blacktriangleright \quad = 24(-1)$
$$= -24$$

$-\sqrt{-6} \cdot 5\sqrt{-3} = -i\sqrt{6} \cdot 5i\sqrt{3}$ $\qquad\blacktriangleleft \sqrt{-y} = i\sqrt{y}$
$$= -5i^2\sqrt{18}$$
$$= -5(-1) \cdot 3\sqrt{2}$$
$$= 15\sqrt{2}$$

In Example 1, notice that $\sqrt{-6} \cdot \sqrt{-3}$ is *not* equal to $\sqrt{-6(-3)}$, or $\sqrt{18}$. The correct way to simplify this product is shown below.

$$\sqrt{-6} \cdot \sqrt{-3} = i\sqrt{6} \cdot i\sqrt{3} = i^2\sqrt{18} = (-1) \cdot 3\sqrt{2} = -3\sqrt{2}$$

The Product of Square Roots property on page 145, $\sqrt{x} \cdot \sqrt{y} = \sqrt{x \cdot y}$, applies only when each radicand is a positive real number or 0. The following property is used to multiply two pure imaginary numbers.

Product of pure imaginary numbers	$\sqrt{-x} \cdot \sqrt{-y} = i\sqrt{x} \cdot i\sqrt{y} = i^2\sqrt{x \cdot y} = -\sqrt{xy}$, where $-x$ and $-y$ are negative real numbers.

The multiplication in Example 2 involves both real and imaginary numbers.

Example 2

Simplify each product.

$5\sqrt{10} \cdot 6\sqrt{-2}$ $\qquad\qquad 2\sqrt{5}(3\sqrt{10} - 4i\sqrt{15})$

$$5\sqrt{10} \cdot 6\sqrt{-2}$$
$$= 5\sqrt{10} \cdot 6i\sqrt{2}$$
$$= 30i\sqrt{20}$$
$$= 30i \cdot 2\sqrt{5}$$
$$= 60i\sqrt{5}$$

$2\sqrt{5}(3\sqrt{10} - 4i\sqrt{15})$ $\qquad\blacktriangleleft$ *Use the*
$$= 6\sqrt{50} - 8i\sqrt{75} \qquad\quad\text{\textit{Distributive Property.}}$$
$$= 6 \cdot 5\sqrt{2} - 8i \cdot 5\sqrt{3}$$
$$= 30\sqrt{2} - 40i\sqrt{3}$$

You should notice in Example 1 that the product of two pure imaginary numbers is a real number, whereas in Example 2 the product of a real number and an imaginary number is an imaginary number.

To simplify a product that involves complex numbers that are imaginary numbers, but not pure imaginary, the FOIL method of multiplying two binomials can be used.

Example 3   **Simplify $(3 + 2i)(5 + 4i)$.**

$$\overset{\quad\text{F}\qquad\qquad\text{O}\qquad\qquad\text{I}\qquad\qquad\text{L}}{(3 + 2i)(5 + 4i) = 3 \cdot 5 + 3 \cdot 4i + 2i \cdot 5 + 2i \cdot 4i}$$
$$= 15 + 12i + 10i + 8i^2$$
$$= 7 + 22i \quad \blacktriangleleft \;\; 8i^2 = 8(-1) = -8$$

In Example 3, notice that the product of the two imaginary numbers is an imaginary number. This is not always the case, however. Two numbers of the form $a + bi$ and $a - bi$ are imaginary, but $(a + bi)(a - bi)$ is a real number.

**Definition:
Conjugate
complex
numbers**

Two complex numbers of the form $a + bi$ and $a - bi$ are *conjugates*. Their product, $(a + bi)(a - bi)$, is a real number, $a^2 + b^2$.

Example 4   **Simplify $(-2 + 3i)(-2 - 3i)$.**

$$(-2 + 3i)(-2 - 3i) = (-2)^2 + (3)^2 = 4 + 9 = 13 \quad \blacktriangleleft \;\; (a + bi)(a - bi) = a^2 + b^2$$

Recall that $\dfrac{2}{3\sqrt{5}} = \dfrac{2}{3\sqrt{5}} \cdot \dfrac{\sqrt{5}}{\sqrt{5}} = \dfrac{2\sqrt{5}}{15}$ and

$$\frac{7}{4 - \sqrt{6}} = \frac{7}{4 - \sqrt{6}} \cdot \frac{4 + \sqrt{6}}{4 + \sqrt{6}} = \frac{7(4 + \sqrt{6})}{16 - 6} = \frac{28 + 7\sqrt{6}}{10}.$$

You can simplify quotients of complex numbers by rationalizing the denominators in a similar way.

Example 5   **Simplify $-5 \div (6i)$ by rationalizing the denominator.**

Write the quotient as a fraction.
$$\frac{-5}{6i} = \frac{-5}{6i} \cdot \frac{i}{i} = \frac{-5i}{6i^2} = \frac{-5i}{6(-1)} = \frac{-5i}{-6} = \frac{5i}{6}$$

Example 6   **Simplify $\dfrac{2 - 4i}{5 + 3i}$ by rationalizing the denominator.**

$$\frac{2 - 4i}{5 + 3i} = \frac{2 - 4i}{5 + 3i} \cdot \frac{5 - 3i}{5 - 3i} = \frac{10 - 6i - 20i + 12i^2}{25 + 9} = \frac{-2 - 26i}{34} = \frac{-1 - 13i}{17}$$

Powers of i can be simplified by using i^2 as a factor as many times as needed. For example,
$$i^9 = i^2 \cdot i^2 \cdot i^2 \cdot i^2 \cdot i = (-1)(-1)(-1)(-1)i = i.$$

Products and Quotients of Complex Numbers

Example 7   **Simplify $(2i)^7$.**

$$(2i)^7 = 2^7 \cdot i^7 = 2^7 \cdot i^4 \cdot i^2 \cdot i = 128 \cdot 1(-1) \cdot i = -128i \quad \blacktriangleleft \quad i^4 = i^2 \cdot i^2 = 1$$

Reading in Algebra

Determine whether each statement is always true (A), sometimes true (S), or never true (N).

1. The product of two pure imaginary numbers is an imaginary number.

2. The product of two imaginary numbers is a real number.

3. The product of a real number and an imaginary number is an imaginary number.

4. The quotient of two imaginary numbers is an imaginary number.

Written Exercises

Simplify each product.

Ⓐ **1.** $6i \cdot 7i$

2. $-i\sqrt{3} \cdot i\sqrt{5}$

3. $i\sqrt{3} \cdot i\sqrt{2} \cdot i\sqrt{6}$

4. $\sqrt{-5} \cdot \sqrt{-6}$

5. $-3\sqrt{-2} \cdot 4\sqrt{-11}$

6. $2\sqrt{3} \cdot 3\sqrt{-2}$

7. $2\sqrt{-6} \cdot 4\sqrt{-2}$

8. $-4\sqrt{-15} \cdot 2\sqrt{6}$

9. $\sqrt{-5} \cdot \sqrt{-10} \cdot \sqrt{-2}$

10. $\sqrt{3}(2\sqrt{6} + i\sqrt{15})$

11. $3\sqrt{6}(\sqrt{2} - \sqrt{-10})$

12. $\sqrt{-6}(\sqrt{11} + \sqrt{-3})$

13. $(4 + 3i)(5 + 6i)$

14. $(3 - 4i)(2 + 2i)$

15. $(6 - 5i)(-3 + i)$

16. $(9 + 11i)(9 - 11i)$

17. $(-7 - 6i)(-7 + 6i)$

18. $(-3 + 4i)(-3 - 4i)$

19. $(4 + 5i)^2$

20. $(6 - 2i)^2$

21. $(-5 - i)^2$

22. $(-4i)^3$

23. $(i\sqrt{3})^4$

24. $(-2i)^5$

Simplify each quotient by rationalizing the denominator.

25. $7 \div (4i)$

26. $-3 \div (5i)$

27. $8 \div (-6i)$

28. $-10 \div (-10i)$

29. $\dfrac{-2}{5i}$

30. $\dfrac{1}{-9i}$

31. $\dfrac{7}{i}$

32. $\dfrac{-8}{12i}$

33. $\dfrac{5}{3 + 2i}$

34. $-3 \div (4 - 2i)$

35. $\dfrac{i}{-1 + i}$

36. $5i \div (-3 - i)$

Ⓑ **37.** $\dfrac{2 + 3i}{3 - 4i}$

38. $\dfrac{5 - 3i}{1 - i}$

39. $\dfrac{1 - 3i}{-2 - 4i}$

40. $\dfrac{50}{3\sqrt{5} + \sqrt{-5}}$

Simplify each product. (Ex. 41–52)

41. $(10i)^6$

42. $(-2i)^8$

43. i^{11}

44. $(2i)^{10}$

45. $(4 + 2i\sqrt{3})^2$

46. $(6 - 2\sqrt{-6})^2$

47. $(2i\sqrt{7} + 3i\sqrt{2})^2$

48. $(\sqrt{-17} - \sqrt{-10})^2$

49. $(3i\sqrt{7})^4$

50. $(-2\sqrt{-2})^{10}$

51. $(1 - i\sqrt{3})^3$

52. $(2 + 2i\sqrt{3})^3$

Simplify. Use the definition of $|a + bi|$. (See page 189.)

Ⓒ **53.** $|4 + 3i| \cdot |2 - 2i|$ and $|(4 + 3i)(2 - 2i)|$

54. Prove that $|a + bi| \cdot |c + di| = |(a + bi)(c + di)|$.

55. Prove that the product of two conjugate complex numbers is a real number.

56. Prove that the sum of two conjugate complex numbers is a real number.

EQUATIONS WITH IMAGINARY NUMBER SOLUTIONS

Objectives   **To solve an equation whose solutions are imaginary numbers**
To find the three cube roots of a nonzero integer that is a perfect cube

You have used the quadratic formula, $x = \dfrac{-b \pm \sqrt{b^2 - 4ac}}{2a}$, to find the real number solutions of a given quadratic equation. As shown in Example 1, the formula can also be used when the solutions are imaginary numbers.

Example 1   **Solve $3x^2 + 2 = 4x$.**

Write the equation in standard form and use the quadratic formula.
$$3x^2 - 4x + 2 = 0$$

$a = 3, b = -4, c = 2$ ▶ $x = \dfrac{-(-4) \pm \sqrt{(-4)^2 - 4 \cdot 3 \cdot 2}}{2 \cdot 3} = \dfrac{4 \pm \sqrt{-8}}{6} = \dfrac{4 \pm 2i\sqrt{2}}{6} = \dfrac{2 \pm i\sqrt{2}}{3}$

Thus, the solutions are $\dfrac{2 + i\sqrt{2}}{3}$ and $\dfrac{2 - i\sqrt{2}}{3}$.

A fourth-degree equation may have imaginary number solutions. One such equation is solved by factoring in Example 2. Notice that two of the solutions of the equation are imaginary numbers and two are real numbers.

Example 2   **Solve $2x^4 + 3x^2 - 20 = 0$.**

$$2x^4 + 3x^2 - 20 = 0$$
$$(x^2 + 4)(2x^2 - 5) = 0$$

$x^2 + 4 = 0$ or $2x^2 - 5 = 0$
$x^2 = -4$ $2x^2 = 5$
$x = \pm\sqrt{-4}$ $x^2 = \dfrac{5}{2}$
$x = \pm 2i$

$$x = \pm\sqrt{\dfrac{5}{2}} = \pm\dfrac{\sqrt{5}}{\sqrt{2}} \cdot \dfrac{\sqrt{2}}{\sqrt{2}} = \pm\dfrac{\sqrt{10}}{2}$$

Thus, the four solutions are $2i$, $-2i$, $\dfrac{\sqrt{10}}{2}$, and $-\dfrac{\sqrt{10}}{2}$.

Every real number, except 0, has two different square roots and three different cube roots. However, some of the roots may be imaginary numbers. You can use a third-degree equation to find the three cube roots of a nonzero integer that is a perfect cube, such as -8.

Example 3   **Find the three cube roots of −8.**

The cube roots of −8 are the solutions of the equation $x^3 = -8$.

$$x^3 = -8$$
$$x^3 + 8 = 0 \quad \blacktriangleleft \; x^3 + 8 \text{ is a sum of two cubes.}$$
$$x^3 + 2^3 = 0$$
$$(x + 2)(x^2 - 2x + 4) = 0 \quad \blacktriangleleft \; m^3 + n^3 = (m + n)(m^2 - mn + n^2)$$

$$x + 2 = 0 \quad \text{or} \quad x^2 - 2x + 4 = 0$$

$$x = -2 \qquad x = \frac{2 \pm \sqrt{4 - 16}}{2} = \frac{2 \pm \sqrt{-12}}{2} = \frac{2 \pm 2i\sqrt{3}}{2} = 1 \pm i\sqrt{3}$$

Thus, the three cube roots of −8 are -2, $1 + i\sqrt{3}$, and $1 - i\sqrt{3}$.

In Example 3, the three cube roots of −8 were found as solutions of the equation $x^3 = -8$. You can check these roots by showing that the third power of each number is −8.

$$-2: \qquad (-2)^3 = (-2)(-2)(-2) = -8$$
$$1 + i\sqrt{3}: \; (1 + i\sqrt{3})^3 = (1 + i\sqrt{3})(1 + i\sqrt{3})^2 = (1 + i\sqrt{3})(1 + 2i\sqrt{3} + 3i^2)$$
$$= (1 + i\sqrt{3})(-2 + 2i\sqrt{3}) = -2 + 6i^2, \text{ or } -8$$
$$1 - i\sqrt{3}: \text{ See exercise 51 on page 192.}$$

Written Exercises

Solve each equation.

Ⓐ
1. $x^2 - 3x + 3 = 0$
2. $2y^2 + 3y + 3 = 0$
3. $3n^2 - 3n + 1 = 0$
4. $2y^2 + 4 = 5y$
5. $c^2 + 4 = 3c$
6. $3x^2 = x - 2$
7. $x^2 - 4x + 5 = 0$
8. $n^2 + 2n + 2 = 0$
9. $y^2 - 8y + 20 = 0$
10. $2y^2 - 2y + 13 = 0$
11. $3t^2 + 2 = 2t$
12. $4x^2 = 2x - 1$
13. $x^4 + 7x^2 - 18 = 0$
14. $x^4 - 16 = 0$
15. $y^4 = 7y^2 + 8$
16. $y^4 + 2y^2 - 15 = 0$
17. $c^4 = 10c^2 + 24$
18. $x^4 = 24 - 5x^2$

Find the three cube roots of each number.

19. -1
20. 8
21. -27

Ⓑ
22. 64
23. -125
24. $1{,}000$

Solve each equation.

25. $\frac{1}{2}x^2 = x - \frac{3}{4}$
26. $\frac{1}{3}x^2 + \frac{3}{2} = \frac{1}{3} - x$
27. $\frac{5x}{12} - \frac{x^2}{6} = \frac{1}{2}$
28. $y^4 + 13y^2 + 36 = 0$
29. $n^4 + 10n^2 + 16 = 0$
30. $c^4 + 16c^2 + 64 = 0$
31. $4x^4 + 19x^2 + 12 = 0$
32. $16x^4 + 16x^2 + 3 = 0$
33. $2x^4 + 11x^2 + 9 = 0$

Ⓒ
34. $x^2 + 2ix + 3 = 0$
35. $x^2 - 6ix + 8 = 0$
36. $2ix^2 + 5x - 2i = 0$
37. $\sqrt{x} = 3i$
38. $\sqrt{x - 1} = 2i$
39. $\sqrt{x + 8} = 3 - i$

Find the three cube roots of each number. Then raise each cube root to the third power.

40. $\frac{1}{8}$
41. $-\frac{8}{27}$
42. $\frac{125}{64}$

THE DISCRIMINANT

Objectives

To determine the nature of the solutions of a quadratic equation by examining its discriminant

To find the coefficients of a quadratic equation having exactly one solution by using the value of its discriminant

The quadratic formula, $x = \dfrac{-b \pm \sqrt{b^2 - 4ac}}{2a}$, can be used to find the solutions of any quadratic equation. However, it is not necessary to solve a given quadratic equation in order to tell the *number* and *type* of solutions that it will have. All of this information about the nature of the solutions can be obtained from the value of $b^2 - 4ac$, the radicand in the quadratic formula.

Equation	Solutions	$b^2 - 4ac$	Nature of Solutions
$6x^2 - x - 2 = 0$	$\dfrac{1 \pm \sqrt{49}}{12} = \dfrac{1 \pm 7}{12}$	49: positive perfect square	Two real number solutions: rational
$x^2 + 3x - 5 = 0$	$\dfrac{-3 \pm \sqrt{29}}{2}$	29: positive	Two real number solutions: irrational
$\dfrac{3}{2}x^2 - 2x + \dfrac{2}{3} = 0$	$\dfrac{2 \pm \sqrt{0}}{3} = \dfrac{2}{3}$	0: perfect square	One real number solution: rational
$2x^2 + 3x + 3 = 0$	$\dfrac{-3 \pm \sqrt{-15}}{4} = \dfrac{-3 \pm i\sqrt{15}}{4}$	−15: negative	Two imaginary number solutions

The expression $b^2 - 4ac$ is called the **discriminant** because its value determines the nature of the solutions of a quadratic equation.

Nature of solutions of quadratic equation

If $ax^2 + bx + c = 0$, where a, b, and c are real numbers ($a \neq 0$), then $x = \dfrac{-b \pm \sqrt{b^2 - 4ac}}{2a}$, and $b^2 - 4ac$ is the discriminant of the equation.

The nature of the solutions of the equation is determined as follows:
(1) if $b^2 - 4ac > 0$, there are two real number solutions.
(2) if $b^2 - 4ac = 0$, there is exactly one real number solution.
(3) if $b^2 - 4ac < 0$, there are two imaginary number solutions.

If a, b, and c are rational numbers and $b^2 - 4ac \geq 0$, the solutions are *rational* when $b^2 - 4ac$ is a perfect square, and the solutions are *irrational* when $b^2 - 4ac$ is not a perfect square.

The Discriminant

Example 1

Determine the nature of the solutions of $x^2 + 5x - 3 = 0$ without solving the equation.

Identify a, b, and c and compute the value of $b^2 - 4ac$.

$$a = 1, b = 5, c = -3$$
$$b^2 - 4ac = 5^2 - 4(1)(-3) = 25 + 12 = 37$$

$b^2 - 4ac = 37$; positive but not a perfect square.
Thus, $x^2 + 5x - 3 = 0$ has two real number solutions that are irrational.

Example 2

Determine the nature of the solutions of $3y^2 = 4y - 2$ without solving the equation.

$$3y^2 = 4y - 2$$
$$3y^2 - 4y + 2 = 0 \qquad \blacktriangleleft \text{ } \textit{Rewrite the equation in standard form.}$$
$$a = 3, b = -4, c = 2 \qquad \blacktriangleleft \text{ } \textit{Identify a, b, and c.}$$
$$b^2 - 4ac = (-4)^2 - 4(3)(2) = 16 - 24 = -8$$

$b^2 - 4ac = -8$; negative
Thus, $3y^2 = 4y - 2$ has two imaginary number solutions.

Example 3

Determine the nature of the solutions of $\frac{1}{9} n^2 + \frac{2}{3} n + 1 = 0$ without solving the equation.

$$\frac{1}{9} n^2 + \frac{2}{3} n + 1 = 0$$

$$a = \frac{1}{9}, b = \frac{2}{3}, c = 1$$

$$b^2 - 4ac = \left(\frac{2}{3}\right)^2 - 4\left(\frac{1}{9}\right)(1) = \frac{4}{9} - \frac{4}{9} = 0$$

$b^2 - 4ac = 0$; a perfect square
Thus, $\frac{1}{9} n^2 + \frac{2}{3} n + 1 = 0$ has exactly one real number solution that is rational.

Example 4

For what values of k will $2x^2 + (k + 2)x + 8 = 0$ have exactly one solution?

If $b^2 - 4ac = 0$, there is exactly one solution to the equation.
$$b^2 - 4ac = 0 \qquad a = 2, b = k + 2, c = 8$$
$$(k + 2)^2 - 4(2)(8) = 0$$
$$k^2 + 4k + 4 - 64 = 0$$
$$k^2 + 4k - 60 = 0$$
$$(k - 6)(k + 10) = 0$$
$$k = 6 \quad \text{or} \quad k = -10$$
Thus, the equation will have exactly one solution if $k = 6$ or $k = -10$.

Reading in Algebra

You are given a quadratic equation $ax^2 + bx + c = 0$, where a, b, and c are real numbers ($a \neq 0$). Match each expression or statement at the left to exactly one phrase at the right.

1. $b^2 - 4ac < 0$
2. $b^2 - 4ac > 0$
3. $b^2 - 4ac = 0$
4. $b^2 - 4ac$
5. $\sqrt{b^2 - 4ac}$
6. $b^2 - 4ac$ is a positive number but not a perfect square.

A the discriminant
B exactly one solution
C two imaginary solutions
D two real solutions
E two irrational solutions
F the radical in the quadratic formula

Written Exercises

Determine the nature of the solutions without solving the equation.

(A)
1. $x^2 + 6x - 16 = 0$
2. $y^2 - 8y + 2 = 0$
3. $n^2 + 2n + 3 = 0$
4. $z^2 + 16 = 8z$
5. $4y^2 + 12y + 9 = 0$
6. $3n^2 - 4n + 3 = 0$
7. $6x^2 = 4 - 5x$
8. $15d^2 - 25d + 9 = 0$
9. $9x^2 + 0x + 4 = 0$
10. $4y^2 - 5y = 0$
11. $n^2 = 15$
12. $25t^2 = 0$

For what value(s) of k will each quadratic equation have exactly one solution?

13. $x^2 - 20x + k^2 = 0$
14. $25y^2 + 10ky + 16 = 0$
15. $9ky^2 + 1 = 2ky$
(B)
16. $x^2 - 16x + (k - 1)^2 = 0$
17. $4y^2 - (k - 3)y + 9 = 0$
18. $(2k + 1)x^2 = 12x - k$

Determine the nature of the solutions without solving the equation.

19. $\frac{1}{3}x^2 - \frac{5}{6}x - 2 = 0$
20. $\frac{5}{2}y^2 + 2y + \frac{2}{5} = 0$
21. $\frac{2}{9}n^2 = \frac{4}{3}n - 2$

22. $(2y - 3)(3y + 4) = -10$
23. $(5x + 1)(2x + 1) = 2$
24. $(t - 5)(t - 3) + 3 = 0$
25. $-5n^2 + 8n - 3 = 0$
26. $-3y^2 - 3y - 108 = 0$
27. $-2x(3x + 4) + 4 = 0$

Determine the nature of the solutions without solving the equation. The coefficient of x in each equation is an irrational number.

(C)
28. $x^2 + 2\sqrt{2}\,x - 6 = 0$
29. $x^2 - 2\sqrt{3}\,x + 3 = 0$
30. $x^2 + 4\sqrt{5}\,x + 4 = 0$

31. For what values of k will $2x^2 + kx + 8 = 0$ have two real solutions?
32. For what values of k will $2x^2 + (k + 1)x + 18 = 0$ have two imaginary solutions?

NON-ROUTINE PROBLEMS

Given a pair of quadratic equations,
(1) $px^2 + qx + r = 0$ and   (2) $ptx^2 + qtx + rt = 0$,
prove that the discriminant of equation (2) is t^2 times the discriminant of equation (1).

The Discriminant

SUM AND PRODUCT OF SOLUTIONS   8.5

Objectives   **To find the sum and the product of the solutions of a quadratic equation**
To write a quadratic equation, given its solutions

The solutions of the quadratic equation $3x^2 - 4x - 20 = 0$ are -2 and $\frac{10}{3}$.

The equation is in the standard form of $ax^2 + bx + c = 0$, where $a = 3$, $b = -4$, and $c = -20$. Let r and s represent the solutions of $3x^2 - 4x - 20 = 0$. Then,

$$(1) \quad r + s = -2 + \frac{10}{3} = \frac{4}{3} = \frac{-b}{a} \text{ and}$$

$$(2) \quad r \cdot s = -2 \cdot \frac{10}{3} = \frac{-20}{3} = \frac{c}{a}.$$

The *sum*, $r + s$, and the *product*, $r \cdot s$, of the solutions of $3x^2 - 4x - 20 = 0$ are related to the value of the coefficients a, b, and c.

$$3x^2 - 4x - 20 = 0 \qquad\qquad ax^2 + bx + c = 0$$
$$x^2 - \frac{4}{3}x - \frac{20}{3} = 0 \qquad\qquad x^2 + \frac{b}{a}x + \frac{c}{a} = 0$$

Notice that $r + s = \frac{4}{3} = -\frac{b}{a}$ or $\frac{b}{a} = -(r + s)$, and

$$r \cdot s = -\frac{20}{3} = \frac{c}{a}.$$

These relationships remain true for any quadratic equation.

Sum and product of solutions of a quadratic equation

If r and s are the solutions of $ax^2 + bx + c = 0$ $(a \neq 0)$, then
$$r + s = -\frac{b}{a} \qquad \text{and} \qquad r \cdot s = \frac{c}{a}.$$

Example 1   **Find the sum and the product of the solutions of $4x^2 + 7x + 3 = 0$.**

$$4x^2 + 7x + 3 = 0$$
$$x^2 + \frac{7}{4}x + \frac{3}{4} = 0 \quad \blacktriangleleft \textit{ Divide each side by a, or 4.}$$
$$\frac{b}{a} = \frac{7}{4}, \text{ or } -\frac{b}{a} = -\frac{7}{4}, \text{ and } \frac{c}{a} = \frac{3}{4}.$$

Thus, the sum of the solutions is $-\frac{7}{4}$ and the product is $\frac{3}{4}$.

If you are given two numbers, either real or imaginary, you can write a quadratic equation that has the two numbers as its solutions.

Example 2   **Write a quadratic equation, in standard form, that has** $-\dfrac{3}{4}$ **and** $\dfrac{2}{3}$ **as its solutions.**

Let $r = -\dfrac{3}{4}$ and $s = \dfrac{2}{3}$. Find $\dfrac{b}{a}$ and $\dfrac{c}{a}$.

$$\dfrac{b}{a} = -(r + s) = -\left(-\dfrac{3}{4} + \dfrac{2}{3}\right) \qquad\qquad \dfrac{c}{a} = r \cdot s$$

$$= -\left(-\dfrac{9}{12} + \dfrac{8}{12}\right) \qquad\qquad = -\dfrac{3}{4} \cdot \dfrac{2}{3}$$

$$= \dfrac{1}{12} \qquad\qquad\qquad\qquad = -\dfrac{1}{2}$$

$$x^2 + \dfrac{b}{a}x + \dfrac{c}{a} = 0$$

Substitute. ▶ $\quad x^2 + \left(\dfrac{1}{12}\right)x + \left(-\dfrac{1}{2}\right) = 0$

Write in standard form. ▶ $\qquad 12x^2 + x - 6 = 0$

Thus, $12x^2 + x - 6 = 0$ is a quadratic equation, in standard form, whose solutions are $-\dfrac{3}{4}$ and $\dfrac{2}{3}$.

A similar procedure is used to write a quadratic equation that has exactly one given real number solution. You begin by letting $r = s$.

Example 3   **Write a quadratic equation, in standard form, that has** $\dfrac{3}{7}$ **as its only solution.**

Let $r = s = \dfrac{3}{7}$.

$$\dfrac{b}{a} = -(r + s) = -\left(\dfrac{3}{7} + \dfrac{3}{7}\right) = -\dfrac{6}{7} \qquad\qquad \dfrac{c}{a} = r \cdot s = \dfrac{3}{7} \cdot \dfrac{3}{7} = \dfrac{9}{49}$$

$$x^2 + \dfrac{b}{a}x + \dfrac{c}{a} = 0$$

Substitute. ▶ $\quad x^2 + \left(-\dfrac{6}{7}\right)x + \left(\dfrac{9}{49}\right) = 0$

Write in standard form. ▶ $\qquad 49x^2 - 42x + 9 = 0$

Thus, the required equation is $49x^2 - 42x + 9 = 0$.

The two numbers represented by $\dfrac{4 \pm 2\sqrt{3}}{3}$ are a pair of conjugate irrational numbers, $\dfrac{4 + 2\sqrt{3}}{3}$ and $\dfrac{4 - 2\sqrt{3}}{3}$. As shown in Example 4 on page 200, a quadratic equation can be written that has these two numbers as its solutions.

Sum and Product of Solutions

Example 4

Write a quadratic equation, in standard form, that has $\dfrac{4 \pm 2\sqrt{3}}{3}$ as its solutions.

Let $r = \dfrac{4 + 2\sqrt{3}}{3}$ and $s = \dfrac{4 - 2\sqrt{3}}{3}$.

$$\frac{b}{a} = -\left(\frac{4 + 2\sqrt{3}}{3} + \frac{4 - 2\sqrt{3}}{3}\right) = -\frac{4 + 2\sqrt{3} + 4 - 2\sqrt{3}}{3} = -\frac{8}{3}$$

$$\frac{c}{a} = \frac{4 + 2\sqrt{3}}{3} \cdot \frac{4 - 2\sqrt{3}}{3} = \frac{4^2 - (2\sqrt{3})^2}{3 \cdot 3} = \frac{16 - 12}{9} = \frac{4}{9}$$

$$x^2 - \frac{8}{3}x + \frac{4}{9} = 0$$

Thus, $9x^2 - 24x + 4 = 0$ is the required equation.

Written Exercises

Without solving the equation, find the sum and the product of its solutions.

(A)
1. $2x^2 + 9x - 6 = 0$ 2. $3y^2 + 4y - 1 = 0$ 3. $4n^2 - 12n - 2 = 0$ 4. $x^2 + 9x + 14 = 0$
5. $6t^2 + 5t = 0$ 6. $3x^2 - 9 = 0$ 7. $y^2 = 6y - 9$ 8. $12 = 6n + 3n^2$.

Write a quadratic equation, in standard form, that has the given solution(s).

9. 3 and 5   10. -6 and 2   11. 0 and 4   12. -9 and 0

13. $\dfrac{1}{4}$ and $\dfrac{1}{2}$ 14. $-\dfrac{3}{4}$ and $-\dfrac{5}{8}$ 15. -6 and $\dfrac{2}{3}$ 16. $\dfrac{9}{10}$ and 2

17. 7   18. -8 19. $-\dfrac{1}{2}$ 20. $\dfrac{10}{7}$

(B)
21. $-\sqrt{7}$ and $\sqrt{7}$ 22. $-4\sqrt{3}$ and $4\sqrt{3}$ 23. $-i\sqrt{10}$ and $i\sqrt{10}$ 24. $-2i\sqrt{6}$ and $2i\sqrt{6}$
25. $4 \pm \sqrt{5}$ 26. $-3 \pm 2\sqrt{10}$ 27. $5 \pm 2i\sqrt{7}$ 28. $-6 \pm 3i\sqrt{5}$
29. $\dfrac{2 \pm \sqrt{3}}{2}$ 30. $\dfrac{-2 \pm 3\sqrt{2}}{3}$ 31. $\dfrac{-3 \pm 2i\sqrt{3}}{3}$ 32. $\dfrac{4 \pm 2i\sqrt{5}}{3}$

Without solving the equation, find the sum and the product of its solutions.

33. $4x^2 + \dfrac{1}{3}x - \dfrac{1}{2} = 0$ 34. $\dfrac{1}{3}y^2 - 3y + 1 = 0$

35. $2.5x^2 + 5x - 10 = 0$ 36. $0.2y^2 - 4y + 0.6 = 0$

(C)
37. Prove that the sum and the product of $\dfrac{-b + \sqrt{b^2 - 4ac}}{2a}$ and

$\dfrac{-b - \sqrt{b^2 - 4ac}}{2a}$, $a \neq 0$, are $-\dfrac{b}{a}$ and $\dfrac{c}{a}$, respectively.

38. For what values of k will the *sum* of the solutions of $x^2 - (k^2 - 2k)x + 12 = 0$ be 8?

39. For what values of k will the *product* of the solutions of $2x^2 + x + (4k^2 - 4k - 3) = 0$ be 0?

40. Prove: If a, b, and c are odd integers, then $ax^2 + bx + c = 0$ has no integral solutions. [Hint: First prove that there are no *even* integral solutions and then that there are no *odd* integral solutions.]

APPLICATIONS: PROOFS WITH COMPLEX NUMBERS

You can use the properties of a field (page 26) for the *real* numbers and some definitions to prove theorems involving complex numbers. As you study each proof, you will be asked to supply a property, definition, or theorem as a reason for a statement in the proof.

Definitions:
1. $a + bi$ is a complex number if and only if a and b are real numbers.
2. $a + bi = c + di$ if and only if $a = c$ and $b = d$.
3. $(a + bi) + (c + di) = (a + c) + (b + d)i$
4. $x - y = x + (-y)$ for all real numbers x and y.
5. $(a + bi) - (c + di) = (a + bi) + [-(c + di)]$
6. $(a + bi)(c + di) = (ac - bd) + (ad + bc)i$

Theorem 1   The set of complex numbers is *closed* under addition.
Prove that $(a + bi) + (c + di)$ is a complex number.

Proof

1. $(a + bi) + (c + di) = (a + c) + (b + d)i$	1. Definition $\underset{?}{\_\_}$
2. $a + c$ and $b + d$ are real numbers.	2. Closure for addition of real numbers
3. $(a + c) + (b + d)i$ is a complex number.	3. Definition $\underset{?}{\_\_}$
4. $(a + bi) + (c + di)$ is a complex number.	4. Statement 1 (Substitution)

Theorem 2   Addition of complex numbers is *commutative*.
Prove that $(a + bi) + (c + di) = (c + di) + (a + bi)$.

Proof

1. $(a + bi) + (c + di) = (a + c) + (b + d)i$	1. $\underset{?}{\_\_}$
2. $ = (c + a) + (d + b)i$	2. $\underset{?}{\_\_}$
3. $(c + di) + (a + bi) = (c + a) + (d + b)i$	3. $\underset{?}{\_\_}$
4. $(a + bi) + (c + di) = (c + di) + (a + bi)$	4. Statements 2 and 3 (Transitive prop. for =)

Theorem 3 $0 + 0i$ is *the* additive identity.
Solve $(a + bi) + (x + yi) = a + bi$ for $x + yi$, the additive identity.

Proof

1. $(a + bi) + (x + yi) = a + bi$	1. Given
2. $(a + x) + (b + y)i = a + bi$	2. $\underset{?}{\_\_}$
3. $a + x = a$ and $b + y = b$	3. Definition $\underset{?}{\_\_}$
4. $x = 0$ and $y = 0$	4. Additive identity for real numbers
5. $x + yi = 0 + 0i$	5. Substitution

Theorem 4 $-a + (-bi)$ is the additive inverse of $a + bi$.
Solve $(a + bi) + (x + yi) = 0 + 0i$ for $x + yi$, or $-(a + bi)$.

Proof

1. $(a + bi) + (x + yi) = 0 + 0i$	1. $\underline{?}$
2. $(a + x) + (b + y)i = 0 + 0i$	2. $\underline{?}$
3. $a + x = 0$ and $b + y = 0$	3. $\underline{?}$
4. $\quad x = -a$ and $y = -b$	4. $\underline{?}$
5. $x + yi = -a + (-bi)$	5. $\underline{?}$

Theorem 5 $(a + bi) - (c + di) = (a - c) + (b - d)i$

Proof

1. $(a + bi) - (c + di)$	$= (a + bi) + [-(c + di)]$	1. Definition $\underline{?}$
2.	$= (a + bi) + [-c + (-di)]$	2. Theorem $\underline{?}$
3.	$= [a + (-c)] + [b + (-d)]i$	3. $\underline{?}$
4.	$= (a - c) + (b - d)i$	4. Definition $\underline{?}$

Theorem 6 $1 + 0i$ is the multiplicative identity.
Solve $(a + bi)(x + yi) = a + bi$ for $x + yi$, the
multiplicative identity.

Proof

1. $(a + bi)(x + yi) = a + bi; \, a + bi \neq 0 + 0i$ | 1. Given
2. $(ax - by) + (ay + bx)i = a + bi$ | 2. Definition $\underline{?}$
3. $ax - by = a$ and $ay + bx = b$ | 3. $\underline{?}$

$$x = \frac{by + a}{a} \text{ and } x = \frac{b - ay}{b}$$

$$\frac{by + a}{a} = \frac{b - ay}{b}$$

$$b^2y + ab = ab - a^2y$$

$a^2y + b^2y = 0$
$(a^2 + b^2)y = 0$
$\qquad y = 0 \; [a^2 + b^2 \neq 0]$
$$x = \frac{b - ay}{b} = \frac{b}{b} = 1$$

Thus, $x + yi = 1 + 0i$.
$[(0 + 0i)(1 + 0i) = 0 + 0i]$

Prove each theorem.

Theorem 7   The set of complex numbers is closed under multiplication.
(Review the proof of Theorem 1.)

Theorem 8   Multiplication of complex numbers is commutative.
(Review the proof of Theorem 2.)

Theorem 9   Addition of complex numbers is associative.
Prove: $[(a + bi) + (c + di)] + (e + fi) = (a + bi) + [(c + di) + (e + fi)]$.

Theorem 10 $\dfrac{a}{a^2 + b^2} + \dfrac{-b}{a^2 + b^2} \cdot i$ is the multiplicative inverse of $a + bi$.

Proof: Solve $(a + bi)(x + yi) = 1 + 0i$ for $x + yi$ where
$a + bi \neq 0 + 0i$. (Review the proof of Theorem 6.)

CHAPTER EIGHT REVIEW

Vocabulary
complex number [8.1]
conjugate complex numbers [8.2]
discriminant [8.4]
imaginary number [8.1]
pure imaginary number [8.1]

Simplify each expression.
1. $5\sqrt{-12}$ [8.1]
2. $(-3 + 4i) - (5 - i)$
3. $(9 - 4\sqrt{-18}) + (-2 + 3\sqrt{-50})$
4. $2\sqrt{-6} \cdot 3\sqrt{-15}$ [8.2]
5. $(2i)^{11}$
6. $(7 - 2i)(-6 + 3i)$
7. $(4 - \sqrt{-10})^2$

Simplify each quotient by rationalizing the denominator. [8.2]
8. $-3 \div (4i)$
9. $4i \div (5 - 3i)$
10. $\dfrac{3 + 2i}{-2 + i}$

Solve each equation. (Exercises 11–16)
11. $x^2 = -60$ [8.1]
12. $4y^2 + 100 = 0$
13. $x^2 + 4x + 6 = 0$ [8.3]
14. $2x^2 + 13 = 2x$
15. $y^4 + 12y^2 + 32 = 0$
★ 16. $x^2 - 2ix + 15 = 0$

17. Find the three cube roots of 27. [8.3]

Determine the nature of the solutions without solving the equation. [8.4]
18. $2x^2 + 5x + 2 = 0$
19. $4y^2 + 25 = 0$
20. $\dfrac{1}{2}n^2 - n = \dfrac{2}{3}$
★ 21. $x^2 - 6x\sqrt{2} - 14 = 0$

22. For what values of k will
$2kx^2 + 12x + (4k + 1) = 0$ have exactly
one solution? [8.4]

Without solving the equation, find the sum and the product of its solutions. [8.5]
23. $4x^2 + 3x - 2 = 0$
24. $\dfrac{1}{6}y^2 - \dfrac{4}{3}y + 2 = 0$

Write a quadratic equation, in standard form, that has the given pair of numbers as its solutions. [8.5]
25. -8 and 5
26. $\dfrac{1}{6}$ and $\dfrac{3}{4}$
27. $-5 \pm 2\sqrt{7}$
28. $\dfrac{1 \pm 2i\sqrt{5}}{2}$

★ 29. For what values of k will the sum of the
solutions of $x^2 - (k^2 - 3k)x + 24 = 0$
be 10? [8.5]

CHAPTER EIGHT TEST

Simplify each expression.

1. $-8 + 4\sqrt{-20}$

2. $(12 - 10i) - (-3 + 4i)$

3. $5\sqrt{-3} \cdot 2\sqrt{-6}$

4. $(6 - 5i)(6 + 5i)$

5. $(3 + 2i\sqrt{5})^2$

Simplify each quotient by rationalizing the denominator.

6. $-7 \div (10i)$

7. $\dfrac{3i}{4 + 3i}$

Solve each equation. (Ex. 8–11)

8. $2x^2 + 24 = 0$

9. $y^2 - 6y + 10 = 0$

10. $3x^2 + 5 = 2x$

11. $t^4 + 7t^2 + 12 = 0$

12. Find the three cube roots of 125.

Determine the nature of the solutions without solving the equation. (Ex. 13–15)

13. $3x^2 + 4x = 4$

14. $2y^2 - 3y + 4 = 0$

15. $\dfrac{5}{3}x^2 + 2x + \dfrac{3}{5} = 0$

16. For what values of k will $y^2 + (k + 2)y + 16 = 0$ have exactly one solution?

17. Without solving the equation, find the sum and the product of the solutions of $12x^2 + 2 = 3x$.

Write a quadratic equation, in standard form, that has the given pair of numbers as its solutions. (Ex. 18–19)

18. $-\dfrac{1}{3}$ and $\dfrac{2}{3}$

19. $-2 \pm i\sqrt{6}$

★ 20. Compute the value of the discriminant for $x^2 - 2x\sqrt{3} - 24 = 0$ and determine the nature of the solutions.

★ 21. For what values of k will $5x^2 - kx + 5 = 0$ have two real number solutions?

★ 22. Factor $x^2 + 9$ as a difference of two squares. [Hint: Use the fact that $i^2 = -1$.]

Use the definition $|a + bi| = \sqrt{a^2 + b^2}$ in Ex. 23–24.

★ 23. Simplify $|5 + 5i|$ and $|-3 - 3i|$.

★ 24. Prove that $|a + bi| = |a|\sqrt{2}$ if $a = b$.

COLLEGE PREP TEST

DIRECTIONS: Choose the *one* best answer to each question or problem.

1. Let x and y represent two imaginary numbers. Which statement(s) could be true?

   I xy is a real number.
   II xy is an imaginary number.
   III $x + y$ is a real number.

   (A) I only   **(B)** II only
   (C) I or II only   **(D)** II or III only
   (E) I, II, or III

2. If $i = \sqrt{-1}$ and $i^2 = -1$, which statement is *false*?

   (A) $i^4 = 1$ **(B)** $i^6 = i^2$ **(C)** $i^3 = -i$
   (D) $i^5 = i$ **(E)** None of these

3. For what value(s) of k will
   $$kx^2 + 2x + k = 0$$
   have exactly one solution?

   (A) $-1, 0,$ and 1 **(B)** 0 and 1 only
   (C) -1 and 1 only   **(D)** 1 only
   (E) None of these

4. If $n \div 7$ gives a remainder of 5, then $(n + 3) \div 7$ gives what remainder?

   (A) 1   **(B)** 2   **(C)** 3   **(D)** 4   **(E)** 5

5. For which equation(s) is the sum of the solutions greater than the product of the solutions?

   (A) $7x^2 - 28x + 28 = 0$
   (B) $8x^2 + 12x - 8 = 0$
   (C) $9x^2 - 3x - 2 = 0$
   (D) All of these
   (E) None of these

6. The conjugate of $-(-2i + 3)$ is

   (A) $3 - 2i$ **(B)** $-3 - 2i$ **(C)** $3 + 2i$
   (D) $-3 + 2i$ **(E)** None of these

7. For every complex number x, $x\star$ is defined to be x^2. Find the value of $[(-2i)\star]\star$.

   (A) $16i$ **(B)** 16   **(C)** $-16i$
   (D) -16 **(E)** None of these

8. Which equation does *not* have two imaginary number solutions?

   (A) $x^2 + 4 = 0$ **(B)** $\dfrac{x}{2} + \dfrac{2}{x} = 0$

   (C) $(x + 2)^2 = 0$ **(D)** $\dfrac{(-x)^2}{3} + \sqrt{\dfrac{16}{9}} = 0$

   (E) None of these

9. If $-2i \cdot 3i \cdot 4i \cdot N = 3i \cdot 4i$, then $N =$

   (A) $-\dfrac{i}{2}$ **(B)** $-\dfrac{1}{2}$ **(C)** $\dfrac{i}{2}$

   (D) $\dfrac{1}{2}$ **(E)** None of these

10. For what value(s) of k will
    $$x^2 + kx + 1 = 0$$
    have two imaginary number solutions?

    (A) $k < -2$ or $k > 2$ **(B)** $-2 < k < 2$
    (C) $-4 < k < 4$ **(D)** 0 only
    (E) None of these

11. If $x = a + bi$ and $y = a - bi$, where a and b are real numbers and $i = \sqrt{-1}$, which statement(s) is (are) true?

    I xy is always an imaginary number.
    II xy is always a real number.
    III $x + y$ is always a real number.

    (A) I only   **(B)** II only   **(C)** III only
    (D) I and III only   **(E)** II and III only

CUMULATIVE REVIEW

DIRECTIONS: Choose the _one_ best answer to each question or problem. (Exercises 1–12)

1. Factor $3x^3 - 4x^2 + 15x - 20$ completely by grouping pairs of terms.

(A) $(3x^2 + 5)(x - 4)$ (B) $(x^2 - 4)(3x + 5)$
(C) $(3x^2 - 4)(x + 5)$ (D) $(x^2 + 5)(3x - 4)$

2. If $x^2 - 5x - 14 > 0$, then

(A) $-2 < x < 7$ (B) $x < -2$ or $x > 7$
(C) $-7 < x < 2$ (D) $x < -7$ or $x > 2$

3. Multiply: $\dfrac{8x^2 - 72y^2}{10a^{-4}b^{-3}c^2} \cdot \dfrac{-6a^{-2}b^3c^{-6}}{6x^2 + 3xy - 45y^2}$.

(A) $-\dfrac{8a^2b^6(x - 3y)}{5c^8(2x - 5y)}$ (B) $\dfrac{5(x + 3y)}{16a^2bc^3(2x - 5y)}$
(C) $\dfrac{-8a^2b^6(x - 3y)}{5c^8(2x + 5y)}$ (D) $\dfrac{8(x + 3y)}{-5a^2c^4(2x + 5y)}$

4. Machine X can do a certain job in 4 h and machine Y can do the same job in 6 h. If machines X, Y, and Z can do the job in 2 h working together, how long would it take Z to do the job working alone?

(A) 12 h   (B) 10 h   (C) 8 h   (D) 6 h

5. Solve $\dfrac{x + 1}{3x - 1} = \dfrac{x + 4}{2x + 6}$.

(A) $\dfrac{3}{5}, \dfrac{1}{4}$ (B) 2, 5
(C) $-2, 5$ (D) $-5, 2$

6. Solve $32^{2x+1} = 8$ for x.

(A) 1   (B) $\dfrac{7}{2}$ (C) $-\dfrac{1}{5}$ (D) $\dfrac{1}{4}$

7. Solve $a(bx - c) = x - 2(x + c)$ for x.

(A) $\dfrac{ac - 2c - x}{ab}$ (B) $\dfrac{ac - 2c}{ab}$
(C) $\dfrac{ac - 2c}{ab + 1}$ (D) $\dfrac{2c - ac}{ab + 1}$

8. Use synthetic substitution to find the value of $3x^4 + 20x^3 - 2x + 20$ if $x = -6$.

(A) -400 (B) 32
(C) 104   (D) 8,216

9. Find the value of $\dfrac{81^{-\frac{5}{4}}}{9^{-\frac{3}{2}}}$.

(A) 9   (B) $\dfrac{1}{4}$ (C) $\dfrac{1}{9}$ (D) $\dfrac{5}{2}$

10. Solve $n - 1 = \sqrt{3n + 7}$.

(A) -1 (B) 6
(C) $-1, 6$ (D) None of these

11. Factor $4x^2 - 12x + 9 - 25y^2$ as a difference of squares.

(A) $(2x - 3 - 5y)^2$
(B) $(2x + 5y - 3)(2x - 5y - 3)$
(C) $(2x - 3 - 5y)(2x - 3 - 5y)$
(D) $(2x - 5y + 3)(2x - 5y - 3)$

12. Find the three cube roots of 64.

(A) $4, -2 + 2i\sqrt{3}, -2 - 2i\sqrt{3}$
(B) $4, -2 + 4i\sqrt{3}, -2 - 4i\sqrt{3}$
(C) $-4, 2 + 2i\sqrt{3}, 2 - 2i\sqrt{3}$
(D) $4, 2 + 4i\sqrt{3}, 2 - 4i\sqrt{3}$

Solve each equation.

13. $|5x - 4| = 26$

14. $\dfrac{5}{x + 4} = \dfrac{3x - 4}{x^2 + 2x - 8} + \dfrac{1}{2x - 4}$

15. $\dfrac{y + 5}{2y + 2} = \dfrac{y - 4}{y - 3}$ 16. $\dfrac{9}{n - 2} - \dfrac{6}{n} = 3$

17. $8^{3x-1} = 4$ 18. $5^{2x-3} = \dfrac{1}{25}$

19. $n + 1 = \sqrt{n + 7}$

20. $\sqrt[3]{y^2 - 3y + 10} = 4$

21. $n^4 + 16n^2 + 48 = 0$

22. $x^2 + 13 = 6x$

Solve each equation for x.

23. $a(3x + t) = b + c(4x - v)$

24. $\dfrac{6}{m} = \dfrac{4}{n} + \dfrac{2}{x}$

Factor each polynomial completely.

25. $2n^3 - 54$ 26. $6x^3 - 3x^2 + 10x - 5$

Find the solution set of each inequality. Graph the solutions on a number line.

27. $8(2x - 3) - (14x - 19) \le 3x - 2$

28. $y^2 - 2y - 24 > 0$

29. $n^4 - 26n^2 + 25 < 0$

Simplify each expression. Assume that no divisor has the value of 0.

30. $(-3a^{-5}b^2)^{-4}$ 31. $\dfrac{-14x^{-2}y^3}{28x^6y^{-6}}$

32. $\dfrac{40mn^2 - 90m}{30n^2 + 75n + 45}$

33. $\dfrac{6 - 2a}{3a^2 + 12a} \div \dfrac{a^2 - 9}{a^3 + 3a^2 - 4a}$

34. $\dfrac{4}{x + 4} - \dfrac{x + 1}{16 - x^2} + \dfrac{2}{12 - 3x}$

35. $\dfrac{4x + 6y}{\dfrac{4}{x} + \dfrac{6}{y}}$

36. $-5a^2b\sqrt{80a^7b^9c^{11}}; \ a \ge 0, \ b \ge 0, \ c \ge 0$

37. $\sqrt{8c^3d} + c\sqrt{50cd} - 4\sqrt{18c^3d}; \ c \ge 0, \ d \ge 0$

38. $\sqrt{\dfrac{25a^5}{12b^3}}; \ a \ge 0, \ b > 0$ 39. $\dfrac{6c^2}{\sqrt[4]{8c^2}}, \ c > 0$

40. $\sqrt[3]{-8x^7} + \sqrt[3]{125x^7}$

41. $(-5 + \sqrt{-27}) - (4 - 2\sqrt{-3})$ 42. $\dfrac{6i}{1 + 7i}$

Use synthetic substitution in Exercises 43–45.

43. $(5x^3 + 8 - 77x) \div (x + 4)$

44. Is $x + 4$ a factor of $5x^3 - 77x + 8$? (See Exercise 43.)

45. Evaluate $2n^3 - 15n^2 - 10n + 33$ if $n = 8$.

Determine whether each number is rational or irrational.

46. $\sqrt{40}$ 47. $83.43434\ldots$

48. $-\sqrt{400}$ 49. $9.29929992\ldots$

Find the value of each expression.

50. $-10 \cdot 32^{\frac{3}{5}} \cdot 36^{\frac{1}{2}}$ 51. $\dfrac{5^{-3}}{8^{-\frac{2}{3}}}$

52. Write $4\sqrt[3]{x^2}$ in exponential form.

53. Write $(5y)^{\frac{3}{4}}$ in radical form.

54. How many liters of a 20% salt solution must be added to 8 L of a 60% salt solution to obtain a 50% salt solution?

55. Machines A, B, and C can complete a certain job in 30 min, 40 min, and 1 h, respectively. How long will the job take if the machines work together?

56. One plane flew 390 km with a tailwind in the same time that another plane flew 360 km against the wind. Find the rate of the wind if each plane can fly at 500 km/h in still air.

57. The length of a rectangle is twice its width and 5 m less than the length of one of its diagonals. Find the length of the diagonal and the area of the rectangle.

58. Without solving $5y^2 + 3y - 10 = 0$, find the sum and the product of its solutions.

9 COORDINATE GEOMETRY

CAREERS
Computer Programming

If you are organized in your work, consider alternative ways to solve difficult problems, pay attention to details, look for shortcuts, and enjoy mathematics and problem solving, you may want to consider a career in the field of programming.

The person who writes a computer program is called a computer programmer. In addition to the high school mathematics program, a computer programmer must receive further education. This additional training may be taken at a college that offers courses in computer science or at a vocational school that specializes in computer programming.

Project

Write a program to determine if a triangle is a right triangle, given the measures of the three sides.

To write the program, you need to know the following:
  (1) If the square of the hypotenuse is equal to the sum of the squares of the two legs, then the triangle is a right triangle.
  (2) The dimensions of the three sides.

THE COORDINATE PLANE 9.1

Objectives **To graph an ordered pair of real numbers in a coordinate plane**
To give the coordinates of a point in a coordinate plane
To draw the graph of a linear equation in two variables

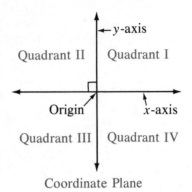

When a horizontal number line is perpendicular to a vertical number line, a **coordinate plane** is formed. The horizontal line is called the **x-axis** and the vertical line is called the **y-axis.** The two axes intersect in a point called the **origin,** and they separate the plane into four sections called **quadrants.** The four quadrants are numbered in a counterclockwise direction, as shown in the figure.

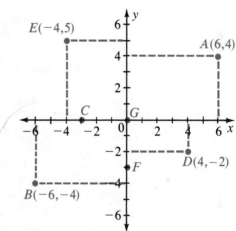

Each point in a coordinate plane is assigned a unique **ordered pair** of real numbers, (x, y). The two numbers paired with a given point are called the **coordinates** of that point. In the second figure, the coordinates of point A are written as the ordered pair of numbers, (6, 4). [The order is important; (6, 4) and (4, 6) are two different ordered pairs.] For each ordered pair of real numbers, there is a unique point in the coordinate plane. The point is called the **graph** of the ordered pair. Thus, the point A is the graph of (6, 4).

You can "graph the ordered pair of numbers" (6, 4) or "plot the point" A(6, 4) in the following way. Starting at the origin, move to the right 6 units and then up 4 units. Place a dot in the coordinate plane to represent the graph. In general, the x-coordinate, or **abscissa,** indicates the direction and distance to move horizontally from the origin along the x-axis. The y-coordinate, or **ordinate,** indicates the direction and distance to move vertically, or parallel to the y-axis.

A(6, 4)

right 6, up 4

B(−6, −4)

left 6, down 4

D(4, −2)

right 4, down 2

E(−4, 5)

left 4, up 5

In the second figure, notice that the points C, F, and G are not located in a quadrant. Each point falls on one of the axes. Point C is the graph of (−3, 0) and point F is the graph of (0, −3). Point G is at the origin, so it is the graph of (0, 0).

An equation of the form $ax + by = c$, where a, b, and c are real numbers with a and b both not equal to 0, is called a **linear equation in two variables** x and y. One solution of a linear equation in two variables is an ordered pair of real numbers that satisfies the equation. For example, $(0, 4)$ is a solution of $-2x + y = 4$ since $-2 \cdot 0 + 4 = 0 + 4 = 4$.

The graph of a given linear equation in two variables is the set of all the points whose coordinates (ordered pairs) satisfy the equation. As shown in the examples, the graph of any linear equation in two variables is a line.

Example 1　**Draw the graph of $-2x + y = 4$.**

First, solve the equation for y in terms of x.

$-2x + y = 4$
$\quad\quad y = 2x + 4$

Second, make a table by choosing values for x and finding corresponding values for y.

Third, plot the points and draw the graph.

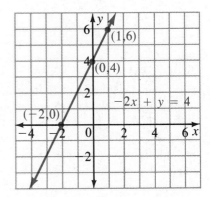

x	$2x + 4$	y	ordered pair
0	$2 \cdot 0 + 4$	4	$(0, 4)$
1	$2 \cdot 1 + 4$	6	$(1, 6)$
-2	$2 \cdot -2 + 4$	0	$(-2, 0)$

Example 2　**Draw the graph of $3x + 2y = 5$.**

$3x + 2y = 5$
$\quad\quad 2y = -3x + 5$
$\quad\quad\ \ y = -\dfrac{3}{2}x + \dfrac{5}{2}$

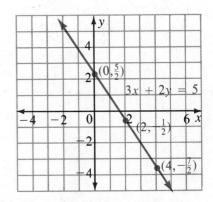

x	$-\dfrac{3}{2}x + \dfrac{5}{2}$	y	ordered pair
0	$-\dfrac{3}{2} \cdot 0 + \dfrac{5}{2}$	$\dfrac{5}{2}$	$\left(0, \dfrac{5}{2}\right)$
2	$-\dfrac{3}{2} \cdot 2 + \dfrac{5}{2}$	$-\dfrac{1}{2}$	$\left(2, -\dfrac{1}{2}\right)$
4	$-\dfrac{3}{2} \cdot 4 + \dfrac{5}{2}$	$-\dfrac{7}{2}$	$\left(4, -\dfrac{7}{2}\right)$

You could have used different values than those used for x in the tables. Although only two points are needed to determine the graph of a linear equation, it is wise to plot a third point as a check.

Reading in Algebra

Let $P(x, y)$ represent any point in a coordinate plane. Match each term at the left to exactly one description at the right.

1. Quadrant I
2. Quadrant II
3. Quadrant III
4. Quadrant IV
5. x-axis
6. y-axis
7. origin

A $x > 0$ and $y < 0$
B $x = 0$ and y is a real number
C $x < 0$ and $y > 0$
D $(0, 0)$
E $x < 0$ and $y < 0$
F $y = 0$ and x is a real number
G $x > 0$ and $y > 0$

Written Exercises

On the same coordinate plane, graph each ordered pair.

(A) 1. $(-3, 2)$ 2. $(4, -1)$ 3. $(0, 1)$ 4. $(-1, -3)$ 5. $(4, 3)$
6. $(0, 0)$ 7. $(-1, 4)$ 8. $(5, 0)$ 9. $(-3, 0)$ 10. $(0, -5)$

Use the coordinate system at the right for Exercises 11–13.

11. Give the coordinates of each of the labeled points.

12. Which of the labeled points do not lie in a quadrant?

13. In which quadrant does each of the other labeled points lie?

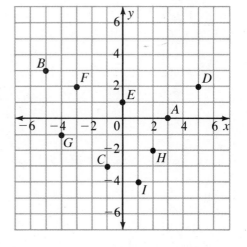

Draw the graph of each equation.

14. $y = -3x - 1$ 15. $y = -2x + 3$ 16. $y = 3x + 2$ 17. $y - 4x = -2$
18. $y = 4x + 3$ 19. $y = 2x + 4$ 20. $y = -2x - 3$ 21. $x - y = 5$
22. $3x + 3y = 9$ 23. $4x - 2y = 10$ 24. $-4x - 2y = 10$ 25. $10x - 5y = 15$
26. $x - y = 0$ 27. $6x + 3y = 9$ 28. $6x - 3y = 15$ 29. $4x + 2y = 8$

(B) 30. $2x - 4y = 12$ 31. $3x + 2y = 12$ 32. $-3x - y = 5$ 33. $5x + 3y = 9$
34. $x - 2y + 5 = 0$ 35. $-x + 2y - 3 = 0$ 36. $-4x - 3y = 9$ 37. $4x - 3y = 8$
38. $\frac{2}{3}x + \frac{1}{6}y = 4$ 39. $-\frac{1}{5}x + \frac{3}{10}y = 2$ 40. $\frac{1}{5}x + \frac{2}{3}y = -1$ 41. $\frac{1}{3}x + 3y = 8$

42. $-1.2x + 3.6y = 2.4$ 43. $0.2x + 0.3y = 1.2$ 44. $0.12x - 0.02y = 0.06$

(C) 45. $\sqrt{2}x + 3y = 9$ 46. $x - \sqrt{3}y = 10$ 47. $\sqrt{3}x + \sqrt{2}y = -2$

The Coordinate Plane

211

SLOPE OF A LINE

Objectives   **To find the length of a vertical or a horizontal segment, given the coordinates of its endpoints**
To find the slope of a line, given the coordinates of any two points on the line

If you are given the coordinates of the endpoints of a vertical or horizontal segment in a coordinate plane, you can find the *length* of the segment.

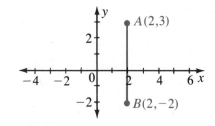

Notice that $\overline{AB}$ (read: segment AB) in the figure at the right is vertical and that the points A and B have the same x-coordinate. You can find AB (read: the length of $\overline{AB}$) by subtracting the y-coordinates of the two points A and B and then taking the absolute value of the difference.

$$AB = |y\text{-coord. of } A - y\text{-coord. of } B| = |3 - (-2)| = 5$$

It is necessary to use absolute value since a length cannot be negative.

Length of a vertical segment	For a vertical segment $\overline{AB}$ with $A(x_1, y_1)$ and $B(x_1, y_2)$, $AB = |y_2 - y_1|$, or $|y_1 - y_2|$.	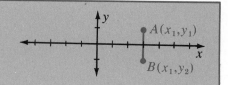

In the figure at the right, $\overline{CD}$ is horizontal and the points C and D have the same y-coordinate. You can find CD by subtraction of the x-coordinates, as shown below.

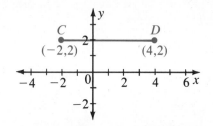

$$CD = |x\text{-coord. of } D - x\text{-coord. of } C| = |4 - (-2)| = 6$$

Length of a horizontal segment	For a horizontal segment $\overline{CD}$ with $C(x_1, y_1)$ and $D(x_2, y_1)$, $CD = |x_2 - x_1|$, or $|x_1 - x_2|$.	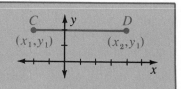

When variables are used for coordinates in this chapter, you should assume that the variables represent real numbers.

Example 1

Find the length of $\overline{BC}$ and $\overline{AC}$ in the figure.

Then find the ratio $\dfrac{BC}{AC}$.

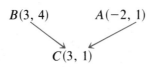

Find the coordinates of C.

$\overline{BC}$ is a vertical segment. So, C has the same x-coordinate as $B(3, 4)$.

$\overline{AC}$ is a horizontal segment. So, C has the same y-coordinate as $A(-2, 1)$.

$$B(3, 4) \qquad A(-2, 1)$$
$$\searrow \qquad \swarrow$$
$$C(3, 1)$$

$$BC = |y_2 - y_1| \qquad\qquad AC = |x_2 - x_1|$$
$$= |4 - 1|, \text{ or } 3 \qquad\qquad = |3 - (-2)|, \text{ or } 5$$

Thus, $BC = 3$ and $AC = 5$ and the ratio $\dfrac{BC}{AC} = \dfrac{3}{5}$.

Notice that for $\overleftrightarrow{AB}$ (read: line AB) in the figure above, the ratio $\frac{3}{5}$ shows how the change in the y-coordinates compares to the change in the x-coordinates as you move from point A to point B on the line:

$$\frac{\text{change in } y\text{-coordinates}}{\text{change in corresponding } x\text{-coordinates}} = \frac{3}{5}.$$

This ratio is called the **slope** of $\overleftrightarrow{AB}$. The letter m is used to represent slope.

**Definition:
Slope of
a line**

The *slope, m, of a line* that contains the points $A(x_1, y_1)$ and $B(x_2, y_2)$ is given by the following formula:

$$m = \frac{\text{change in } y\text{-coordinates}}{\text{change in corresponding } x\text{-coordinates}}, \text{ or } m = \frac{y_2 - y_1}{x_2 - x_1}.$$

Example 2

Find the slope of each line.

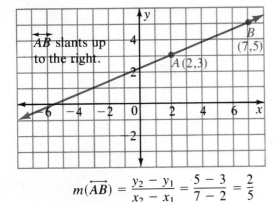

$$m(\overleftrightarrow{AB}) = \frac{y_2 - y_1}{x_2 - x_1} = \frac{5 - 3}{7 - 2} = \frac{2}{5}$$

The slope of $\overleftrightarrow{AB}$ is $\frac{2}{5}$.

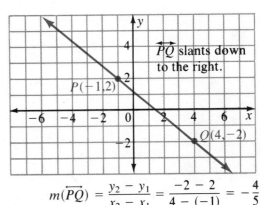

$$m(\overleftrightarrow{PQ}) = \frac{y_2 - y_1}{x_2 - x_1} = \frac{-2 - 2}{4 - (-1)} = -\frac{4}{5}$$

The slope of $\overleftrightarrow{PQ}$ is $-\frac{4}{5}$.

The slope of a line is a measure of its "steepness." In Example 2, you have seen that if a line slants up to the right it has a positive slope and that if a line slants down to the right it has a negative slope. You will now see what happens when different pairs of points on the same line are used to compute its slope.

Example 3

Find the slope of the line in the figure, using points A and B, B and C, C and D.

$$m = \frac{y_2 - y_1}{x_2 - x_1}$$

$$m(\overleftrightarrow{AB}) = \frac{0 - (-1)}{-3 - (-6)} = \frac{1}{3}$$

$$m(\overleftrightarrow{BC}) = \frac{2 - 0}{3 - (-3)} = \frac{2}{6}, \text{ or } \frac{1}{3}$$

$$m(\overleftrightarrow{CD}) = \frac{3 - 2}{6 - 3} = \frac{1}{3}$$

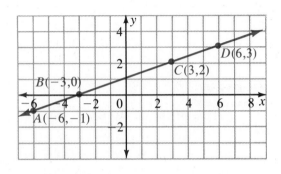

Thus, the slope of the line is $\frac{1}{3}$ using all three pairs of points.

Example 3 shows you that the slope of a given line is uniform. That is, for any two points on the line, the slope is the same.

Example 4

Find the slope of each line.

Vertical line

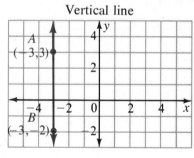

Horizontal line

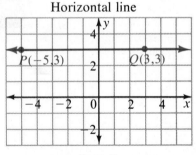

$$m = \frac{y_2 - y_1}{x_2 - x_1}$$

$$m(\overleftrightarrow{AB}) = \frac{3 - (-2)}{-3 - (-3)} = \frac{5}{0}$$

The slope is undefined.

$$m = \frac{y_2 - y_1}{x_2 - x_1}$$

$$m(\overrightarrow{PQ}) = \frac{3 - 3}{3 - (-5)} = \frac{0}{8}$$

The slope is 0.

Since all points of a vertical line have the same x-coordinates, the slope of any vertical line is *undefined*. Similarly, since all points of a horizontal line have the same y-coordinates, the slope of any horizontal line is *0*.

Reading in Algebra

Match each sentence to exactly one of the statements at the right.
1. Points A, B, C, and D are on the same line.
2. The coordinates of P and Q are $P(3, 1)$ and $Q(5, 1)$.
3. The length of segment PQ is 2.
4. The slope of line PQ is 2.
5. A line is vertical.
6. The coordinates of four points are $A(1, 5)$, $B(6, 5)$, $C(1, 1)$ and $D(4, 1)$.
7. A line slants up to the right.
8. A line slants down to the right.

  A $m(\overrightarrow{PQ}) = 2$
  B $m(\overrightarrow{AB}) = m(\overrightarrow{CD})$
  C $m(\overrightarrow{AB}) = m(\overrightarrow{BC}) = m(\overrightarrow{CD})$
  D $PQ = 2$
  E $m = 0$
  F $m < 0$
  G $m > 0$
  H m is undefined

Written Exercises

Find the length of $\overline{AB}$ for the given coordinates.

(A)
1. $A(3, 2)$, $B(3, -3)$ 2. $A(1, -4)$, $B(-2, -4)$ 3. $A(2, 3)$, $B(-3, 3)$ 4. $A(-2, 1)$, $B(-2, 3)$
5. $A(2, -4)$, $B(-2, -4)$ 6. $A(5, -1)$, $B(5, 1)$ 7. $A(-2, 5)$, $B(-2, -5)$ 8. $A(6, -4)$, $B(-6, -4)$

Find the slope of $\overleftrightarrow{AB}$ for the given coordinates. Tell whether the line slants up to the right, slants down to the right, is horizontal, or is vertical.

 9. $A(-6, 1)$, $B(4, 4)$ 10. $A(-1, -4)$, $B(-3, 2)$ 11. $A(-2, 9)$, $B(1, -3)$
12. $A(3, 8)$, $B(-3, 8)$ 13. $A(2, -3)$, $B(2, 3)$ 14. $A(6, 5)$, $B(-4, -5)$
15. $A(7, 8)$, $B(1, 11)$ 16. $A(-4, -1)$, $B(-2, 4)$ 17. $A(-4, -7)$, $B(3, -7)$
18. $A(-3, -1)$, $B(-6, -4)$ 19. $A(-9, -7)$, $B(10, 8)$ 20. $A(6, 2)$, $B(-1, -4)$
21. $A(-4, -2)$, $B(-1, 7)$ 22. $A(4, -2)$, $B(8, -6)$ 23. $A(-3, -5,)$, $B(-5, -5)$
24. $A(6, 0)$, $B(0, -8)$

Show that $m(\overrightarrow{AB}) = m(\overrightarrow{BC}) = m(\overrightarrow{AC})$ for the given coordinates.
25. $A(-2, -4)$, $B(0, 0)$, $C(4, 8)$ 26. $A(-6, -4)$, $B(-4, -3)$, $C(-2, -2)$
27. $A(3, 6)$, $B(0, 4)$, $C(-3, 2)$ 28. $A(5, -1)$, $B(0, -1)$, $C(-6, -1)$

Find the slope of $\overrightarrow{PQ}$ for the given coordinates.

(B)
29. $P(-2, -3)$, $Q(-\frac{1}{2}, -\frac{1}{3})$ 30. $P(-\frac{1}{3}, \frac{1}{2})$, $Q(\frac{5}{6}, 4)$ 31. $P(\frac{1}{5}, \frac{1}{4})$, $Q(\frac{3}{5}, \frac{3}{4})$
32. $P(-\frac{1}{6}, \frac{1}{4})$, $Q(\frac{1}{2}, \frac{1}{2})$ 33. $P(\frac{2}{3}, \frac{3}{4})$, $Q(\frac{1}{6}, -\frac{1}{2})$ 34. $P(\frac{2}{5}, -\frac{1}{3})$, $Q(-\frac{1}{3}, \frac{1}{3})$

35. $P(a, -b)$, $Q(-a, b)$ 36. $P(4x, 5y)$, $Q(-2x, -7y)$
37. $P(6a, -2b)$, $Q(7a, 5b)$ 38. $P(-10a, 12b)$, $Q(-2a, -7b)$
(C) 39. $P(x_1, y_1)$, $Q(x_1 + t, y_1 + s)$ 40. $P(2m - 5n, t + v)$, $Q(-n, 3v - t)$

Find the missing coordinate for the given data.
41. $A(4, y)$, $B(5, 0)$, $m(\overrightarrow{AB}) = -2$ 42. $M(-3, 2)$, $N(x, -4)$, $m(\overrightarrow{MN}) = -\frac{1}{2}$
43. $P(-\frac{1}{3}, -\frac{1}{2})$, $Q(\frac{4}{3}, y)$, $m(\overrightarrow{PQ}) = \frac{1}{5}$ 44. $F(x, -\frac{1}{5})$, $G(\frac{2}{5}, -\frac{4}{5})$, $m(\overrightarrow{FG}) = -\frac{3}{4}$

CUMULATIVE REVIEW

Solve each equation.

1. $3x - 2(1 - x) = 13$ 2. $\dfrac{x - 5}{2} = \dfrac{x - 3}{5}$ 3. $\dfrac{3}{x^2 - 7x + 12} - \dfrac{x}{x - 3} = \dfrac{2x - 1}{x - 4}$

EQUATION OF A LINE

Objectives

To write an equation of a line, given its slope and the coordinates of a point on the line
To write an equation of a line, given its slope and *y*-intercept
To find the slope and the *y*-intercept of a line, given an equation of the line
To write an equation of a line, given the coordinates of two points on the line

The line in the figure passes through the points $P(x_1, y_1)$ and $Q(x_2, y_2)$. Let $R(x, y)$ be any other point on $\overleftrightarrow{PQ}$. You can find an equation of this line by applying the definition of slope. Consider the points P and R.

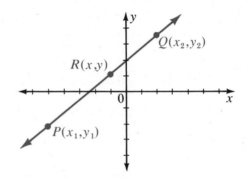

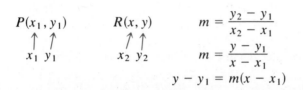

$$m = \frac{y_2 - y_1}{x_2 - x_1}$$

$$m = \frac{y - y_1}{x - x_1}$$

$$y - y_1 = m(x - x_1)$$

Point-slope form of an equation of a line	$y - y_1 = m(x - x_1)$ is an equation of a nonvertical line whose slope is a real number, m, and passes through the point $P(x_1, y_1)$.

The point-slope form enables you to write an equation of a nonvertical line when its slope and the coordinates of one point of the line are known. The process is shown in Example 1.

Example 1

Write an equation of the line that passes through the point $P(3, -4)$ and whose slope is $-\dfrac{2}{3}$.

$$y - y_1 = m(x - x_1) \qquad \blacktriangleleft \text{ Use the point-slope form.}$$

$$y - (-4) = -\frac{2}{3}(x - 3)$$

$$y + 4 = -\frac{2}{3}(x - 3)$$

$$3y + 12 = -2x + 6 \qquad \blacktriangleleft \text{ Multiply each side by 3.}$$

$$2x + 3y = -6$$

Thus, an equation of the line is $2x + 3y = -6$.

An equation of a line is a *linear equation in two variables*. The **standard form** is $ax + by = c$, where a, b, and c are integers, and a and b both not equal to 0. Notice that the equation in Example 1 is written in this standard form.

The **y-intercept** of a line is the *y*-coordinate of the point where the line intersects the *y*-axis. Generally, the letter *b* is used to represent the *y*-intercept of a line.

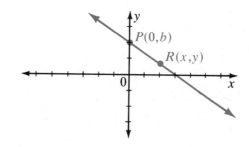

In the figure at the right, the point $P(0, b)$ is the point where the given line intersects the *y*-axis, with *b* the *y*-intercept. Let $R(x, y)$ be any other point on the line. You can find an equation of the line by using the point-slope form of an equation.

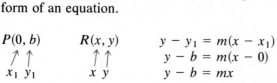

$$y - y_1 = m(x - x_1)$$
$$y - b = m(x - 0)$$
$$y - b = mx$$
$$y = mx + b$$

Slope-intercept form of an equation of a line	$y = mx + b$ is an equation of a nonvertical line whose slope is a real number, m, and whose *y*-intercept is *b*.

Example 2   **Write an equation, in standard form, of the line whose slope is $-\dfrac{5}{2}$ and whose *y*-intercept is −3.**

$$y = mx + b \qquad \blacktriangleleft \textit{ Use the slope-intercept form.}$$
$$y = -\frac{5}{2}x + (-3) \qquad \blacktriangleleft \textit{ } m = -\frac{5}{2},\ b = -3$$
$$2y = -5x - 6 \qquad \blacktriangleleft \textit{ Multiply each side by 2.}$$
$$5x + 2y = -6$$

Thus, an equation of the line is $5x + 2y = -6$.

Given an equation of a line, you can find the slope and the *y*-intercept of the line.

Example 3   **Find the slope and the *y*-intercept of the line whose equation is $3x + 2y = 9$.**

Rewrite the equation in the form $y = mx + b$.
$$3x + 2y = 9$$
$$2y = -3x + 9$$
$$y = -\frac{3}{2}x + \frac{9}{2} \qquad \blacktriangleleft \textit{ Divide each side by 2.}$$

Thus, the slope is $-\dfrac{3}{2}$ and the *y*-intercept is $\dfrac{9}{2}$.

Example 4 on page 218 shows you how to find an equation of a line when the coordinates of two points of the line are known.

Equation of a Line

Example 4   **Write an equation, in standard form, of the line that passes through the points $P(-3, 1)$ and $Q(5, -4)$.**

Find the slope of $\overleftrightarrow{PQ}$. ▶ $m = \dfrac{y_2 - y_1}{x_2 - x_1} = \dfrac{-4 - 1}{5 - (-3)} = -\dfrac{5}{8}$

Use the point-slope form. ▶ $y - y_1 = m(x - x_1)$

$$y - 1 = -\frac{5}{8}[x - (-3)]$$

$$y - 1 = -\frac{5}{8}(x + 3)$$

$$8y - 8 = -5x - 15$$

$$5x + 8y = -7$$

Thus, an equation of the line is $5x + 8y = -7$.

In the figure at the right, $\overleftrightarrow{AB}$ is a horizontal line and $\overleftrightarrow{PQ}$ is a vertical line. You can write an equation of $\overleftrightarrow{AB}$ and an equation of $\overleftrightarrow{PQ}$ as shown below. Recall that the slope of a horizontal line is 0 and the slope of a vertical line is undefined.

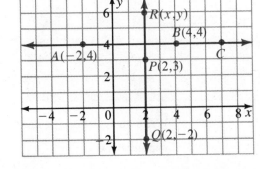

The y-intercept of $(\overleftrightarrow{AB})$ is 4 and $m(\overleftrightarrow{AB})$ is 0.

$$y = mx + b$$
$$y = 0 \cdot x + 4$$
$$y = 4$$

For each point C on $\overleftrightarrow{AB}$, the y-coordinate is 4 and the x-coordinate is a real number. Therefore, $y = 4$ is an equation of $\overleftrightarrow{AB}$.

$\overleftrightarrow{PQ}$ does not intersect the y-axis, so there is no y-intercept; $m(\overleftrightarrow{PQ})$ is undefined. Let $R(x, y)$ be any point on $\overleftrightarrow{PQ}$. Then $m(\overleftrightarrow{PQ}) = m(\overleftrightarrow{PR}) = \dfrac{y - 3}{x - 2}$, which is undefined only for $x = 2$. For each point R on $\overleftrightarrow{PQ}$, the x-coordinate is 2 and the y-coordinate is a real number. Therefore, $x = 2$ is an equation of $\overleftrightarrow{PQ}$.

Example 5   **For the given data, write an equation of each line.**

Data	Equation
Line passes through $A(7, 3)$ and $B(7, -2)$	$x = 7$
Line passes through $C(-5, -4)$ and $D(6, -4)$	$y = -4$
Line passes through $T(-10, 8)$ and m is undefined	$x = -10$
Line with $m = 0$ and $b = 6$	$y = 6$

The linear equations $y = c$ and $x = c$, where c is a constant, represent the *standard form* of the equation of a horizontal line and a vertical line, respectively.

Reading in Algebra

A description of each of five lines is given at the left. Match each description to the _one_ best form to use to find an equation of each line.

1. $P(-1, -5)$ is on the line and its slope is $-\frac{3}{5}$.
2. The line is vertical and $P(-2, 3)$ is on the line.
3. The y-intercept of the line is -4 and its slope is 3.
4. The line is horizontal and $P(-2, 3)$ is on the line.
5. $P(-3, 5)$ and $Q(2, -1)$ are on the line.

A $y = mx + b$
B $x = c$
C $y = c$
D $y - y_1 = m(x - x_1)$

Written Exercises

A line has the given slope, y-intercept, or contains the indicated point(s). Write an equation, in standard form, of each line.

(A)
1. $P(2, -5)$, $m = \frac{2}{3}$
2. $Q(2, -2)$, $m = -\frac{1}{2}$
3. $R(-1, 6)$, $m = \frac{5}{6}$
4. $A(-2, -3)$, $m = -1$
5. $M(-2, 4)$, $m = -4$
6. $R(3, -2)$, $m = -3$
7. $m = -\frac{4}{3}$, $b = 2$
8. $m = \frac{3}{5}$, $b = -10$
9. $m = -\frac{5}{8}$, $b = -\frac{3}{8}$
10. $m = 2$, $b = 3$
11. $m = -4$, $b = -4$
12. $m = -3$, $b = \frac{7}{3}$
13. $A(-1, 3)$, $B(0, 4)$
14. $A(0, -3)$, $B(-5, -6)$
15. $A(2, -3)$, $B(0, 4)$
16. $M(-1, 6)$, $N(-4, 2)$
17. $M(7, 3)$, $N(-3, 6)$
18. $M(-1, 5)$, $N(-3, -4)$
19. $P(7, 5)$, $Q(-3, 5)$
20. $P(5, -4)$, $Q(-2, -4)$
21. $P(8, 4)$, $Q(8, 9)$
22. $A(0, 7)$, $B(4, 0)$
23. $M(-5, 0)$, $N(0, 9)$
24. $R(0, -6)$, $Q(10, 0)$
25. $R(8, 12)$, $m = 0$
26. $S(-6, -4)$, m is undefined.
27. $T(7, -3)$, line is horizontal.

Find the slope and the y-intercept of the line described by the given equation.

28. $2x + 3y = 9$
29. $3x - 5y = 10$
30. $7x + 3y = -21$
31. $5x - 3y - 15 = 0$
32. $5y - x + 20 = 0$
33. $3x + 4y - 40 = 0$
34. $7x - 3(4 - y) = 0$
35. $15 - 4(-2x - 3y) = 39$
36. $3y - (18 + 6x) = 0$

(B)
37. $y - (4x + 5) = -4(3x + 5)$
38. $3x + 5(x - y) - (y - x) = 12$
39. $2(x + 3y) - (2x - y) = 14$
40. $-2(3y + 4x) - 5(2x - 3y) = 15$
41. $7y + 5(-3y - 4x) = -(7 - 3x) + 2y$
42. $\frac{4}{5}(10y - 20x) - \frac{2}{3}(6x - 15y) = 0$
43. $-\frac{1}{2}(6x - 8y) + \frac{4}{5}(10x + 15y) = 6 - y$

(C)
44. $4x - 5[2(6x - 3) - 4(6y - 3)] = 6$
45. $-8 - 3[-(4 - 3y) - 2(5x + 2y)] = x - y$
46. $0.3x - 4.3[-(x - 2) - 2.1(4 - y)] = 1.8$
47. $1.6y - 3.1[(y + 5) - 1.6(4 - x)] = 7.3$

Write an equation, in standard form, of each line.

48. The line having an x-intercept of -2 and a y-intercept of 3

49. The line having an x-intercept of 4 and a y-intercept of -4

50. The line having an x-intercept of $\frac{1}{A}$ and a y-intercept of $\frac{1}{B}$

51. The line having an x-intercept of $\frac{C}{A}$ and a y-intercept of $\frac{C}{B}$

52. If $P(a, 0)$ and $Q(0, b)$ are the x- and y-intercepts, respectively, of $\overrightarrow{PQ}$ $(a \neq 0, b \neq 0)$, prove that $\frac{x}{a} + \frac{y}{b} = 1$ is an equation of $\overrightarrow{PQ}$.

SLOPE-INTERCEPT METHOD OF DRAWING GRAPHS

Objectives **To draw the graph of a linear equation in two variables, using the slope-intercept method**
To determine whether a point is on a line, given the coordinates of the point and an equation of the line

To draw the graph of $y = \frac{3}{5}x + 1$ using its slope and y-intercept, proceed as follows:

First, determine the slope and the y-intercept of the line.

$$y = \frac{3}{5}x + 1$$
$$\uparrow \quad \uparrow$$
$$m \quad b$$

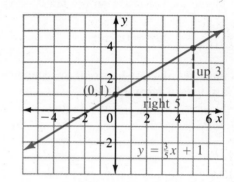

Then, plot the point whose y-intercept is 1. From this point, move to the right 5 units and up 3 units to locate a second point.

Draw the line that contains the two points. The line is the graph of $y = \frac{3}{5}x + 1$.

Example 1 **Draw the graph of $2x + 3y = -9$.**

Solve for y in terms of x.
$$2x + 3y = -9$$
$$3y = -2x - 9$$
$$y = -\frac{2}{3}x - 3$$

$m = \frac{-2}{3}$ or $\frac{2}{-3}$, and $b = -3$.

Plot the point whose y-intercept is -3. From this point, move to the right 3 units and down 2 units *or* move to the left 3 units and up 2 units.

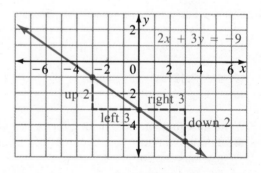

Recall that $m = \frac{y_2 - y_1}{x_2 - x_1}$. Thus, whenever the slope, m, is written as a fraction, the denominator indicates the change in the x-coordinates (distance and direction to move horizontally), and the numerator indicates the change in the y-coordinates (distance and direction to move vertically) as you move from one point to another point. Even though two points determine a line, when drawing graphs it is wise to locate a third point as a check.

Example 2   **Draw the graph of $y = 4$ and the graph of $x = -3$.**

$y = c$, where c is a constant, determines a horizontal line.

$x = c$, where c is a constant, determines a vertical line.

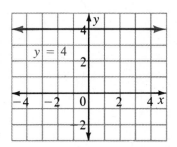

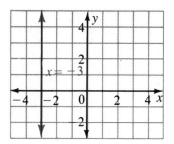

Given the coordinates of a point and an equation of a line, you can determine whether the point is on the line. The process involves substituting the x- and y- coordinate values of the point into the equation and checking to see if a true equation results.

Example 3   **Determine whether each point having the given coordinates is on the line described by the given equation.**

$P(-4, 7),\ 5x + 3y - 1 = 0$ | $Q(8, -2),\ 4y - x = 34$

Substitute −4 for x and 7 for y. ▶

$$\begin{array}{c|c} 5x + 3y - 1 & 0 \\ 5(-4) + 3(7) - 1 & 0 \\ -20 + 21 - 1 & \\ 0 & \end{array}$$

$$\begin{array}{c|c} 4y - x & 34 \\ 4(-2) - (8) & 34 \\ -8 - 8 & \\ -16 & \end{array}$$

◀ *Substitute 8 for x and −2 for y.*

Thus, $(-4, 7)$ satisfies the equation and so P is a point on the line.

Thus, $(8, -2)$ does not satisfy the equation and so Q is not a point on the line.

Point on a line	Each point $P(x, y)$ whose coordinates satisfy an equation of a given line is a point on the line.
	Each point $P(x, y)$ on a given line has coordinates that satisfy an equation of the line.

Oral Exercises

Match each equation to exactly one group of ordered pairs.

1. $3x - 8y = 24$
2. $6x + 4y = -24$
3. $x - 24y = 24$
4. $12x + 2y = -24$

A $(0, -1)$
$(24, 0)$
$(-24, -2)$

B $(8, 0)$
$(-8, -6)$
$(0, -3)$

C $(-2, -3)$
$(0, -6)$
$(-4, 0)$

D $(-1, -6)$
$(-2, 0)$
$(0, -12)$

Tell whether each equation describes a horizontal line, a vertical line, or neither.

5. $y = 5$ **6.** $x = -4$ **7.** $2y = 7$ **8.** $y = 0$ **9.** $-x = -6$ **10.** $y = 7 + x$

Written Exercises

Draw the graph of each equation.

Ⓐ **1.** $y = \frac{2}{3}x + 3$ **2.** $y = -\frac{2}{3}x - 2$ **3.** $y = \frac{1}{3}x + 1$ **4.** $y = -\frac{1}{2}x + 4$

5. $y = -2x - 1$ **6.** $y = -2x + 1$ **7.** $y = -3x - 2$ **8.** $y = 5x + 3$

9. $y = -2x + 4$ **10.** $y = -3x - 1$ **11.** $y = 4x - 3$ **12.** $y = -x + 5$

13. $3x + 2y = 4$ **14.** $x - 3y - 9 = 0$ **15.** $3y = 15$ **16.** $3x - 4y = 12$

17. $x + y - 4 = 0$ **18.** $x = 4$ **19.** $5x + 20 = 0$ **20.** $2x + 5y + 15 = 0$

Determine whether a point having the given coordinates is on the line described by the given equation.

21. $(1, 2)$, $3x + 2y = 7$ **22.** $(2, 3)$, $x - 3y = -7$ **23.** $(-3, -2)$, $3x - 2y = -6$

24. $(-2, 1)$, $2x - y = -5$ **25.** $(-1, -2)$, $-2x + y + 4 = 0$ **26.** $(-2, 1)$, $-2x + y - 5 = 0$

Ⓑ **27.** $(2, 3)$, $\frac{2}{3}y - \frac{1}{2}x = 1$ **28.** $(-6, -4)$, $-\frac{1}{3}x + \frac{3}{4}y = -1$ **29.** $(-2, 2)$, $\frac{3}{4}x - \frac{3}{4}y = -3$

30. $(-5, -6)$, $\frac{3}{5}x + \frac{1}{3}y = 5$ **31.** $(8, 4)$, $\frac{3}{4}y - \frac{7}{8}x = -4$ **32.** $(10, -6)$, $\frac{2}{5}x + \frac{5}{3}y = 6$

Draw the graph of each equation.

33. $2(x - 3) - 4(3 - 2y) = 8$ **34.** $6x - 2(5 - 3y) = 12$ **35.** $-4x - 3(2y - 1) = 6$

36. $2y - (5 - 2x) = -2x - 3y$ **37.** $-4(3 - 2x) - (8 - 2y) = x - y$ **38.** $-(x - 3y) - 2(x + y) = -10$

Find the coordinates of a point on the line that satisfy the given conditions. Do not draw a graph.

Ⓒ **39.** An equation of the line is $x + 4y = 18$ and the ordinate of the point is twice the abscissa of the point.

40. An equation of the line is $3x - 2y = 12$ and the ordinate of the point is one-half the abscissa of the point.

Find each equation that satisfies the given description. Draw the graph of each equation.

41. A set of points equidistant from both axes

42. The set of all points whose ordinate is two more than three times its abscissa

NON-ROUTINE PROBLEMS

What number does the sequence of numbers below seem to approach?

$$\left(1 - \frac{1}{2^2}\right), \left(1 - \frac{1}{3^2}\right), \left(1 - \frac{1}{4^2}\right), \left(1 - \frac{1}{5^2}\right), \ldots$$

THE DISTANCE BETWEEN TWO POINTS   9.5

Objectives   **To find the length of each side of a right triangle in a coordinate plane**
To find the distance between any two points in a coordinate plane
To determine if a triangle with given lengths for its sides is a right triangle

Recall that the square of the length of the hypotenuse of a right triangle is equal to the sum of the squares of the lengths of the two legs: That is, for right triangle ABC with right angle C, $c^2 = a^2 + b^2$, where c is the length of the hypotenuse and a and b are the lengths of the legs. This relationship, known as the *Pythagorean relation*, is illustrated below.

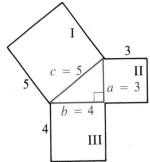

$$\underbrace{\frac{\text{Area of}}{\text{Square I}}}_{25} = \underbrace{\frac{\text{Area of}}{\text{Square II}}}_{9} + \underbrace{\frac{\text{Area of}}{\text{Square III}}}_{16}$$

$$c^2 \;=\; a^2 \;+\; b^2$$

The Pythagorean relation can be used to find the length of the hypotenuse of a right triangle where the coordinates of the vertices of the triangle are given.

Example 1   **The vertices of right triangle *PRS* are $P(-1, 2)$, $R(6, 5)$, and $S(6, 2)$. Find the length of $\overline{PR}$.**

Let $c = PR$, $a = RS$, and $b = PS$.

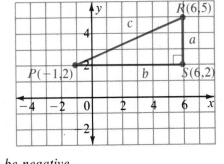

$$c^2 = a^2 + b^2$$
$$(PR)^2 = (RS)^2 + (PS)^2$$

$\overline{RS}$ *is vertical.*
$\overline{PS}$ *is horizontal.*   ▶

$$= |5 - 2|^2 + |6 - (-1)|^2$$
$$= 3^2 + 7^2$$
$$= 9 + 49$$
$$(PR)^2 = 58$$
$$PR = \pm\sqrt{58} \quad ◀ \; A \; length \; cannot \; be \; negative.$$

Thus, the length of $\overline{PR}$ is $\sqrt{58}$.

In Example 1, notice that for $P(x_1, y_1)$, $R(x_2, y_2)$, and $S(x_2, y_1)$, $(PS)^2 = (x_2 - x_1)^2$ and $(RS)^2 = (y_2 - y_1)^2$. Therefore, $(PR)^2 = (x_2 - x_1)^2 + (y_2 - y_1)^2$, or $PR = \sqrt{(x_2 - x_1)^2 + (y_2 - y_1)^2}$. On page 224, it is shown that this relationship remains true for any two points in a coordinate plane.

In the figure at the right, points $A(x_1, y_1)$ and $B(x_2, y_2)$ determine $\overline{AB}$. A horizontal ray from A and a vertical ray from B intersect at a point C. Then, C has the same x-coordinate as B and the same y-coordinate as A. Since ABC is a right triangle,

$$(AB)^2 = (AC)^2 + (BC)^2 \text{ by the Pythagorean relation.}$$
$$= |x_2 - x_1|^2 + |y_2 - y_1|^2 \text{ by the definition of length of a horizontal and a vertical segment.}$$
$$= (x_2 - x_1)^2 + (y_2 - y_1)^2 \text{ since } |a - b|^2 = (a - b)^2.$$

So, $AB = \sqrt{(x_2 - x_1)^2 + (y_2 - y_1)^2}$ by the definition of square root.

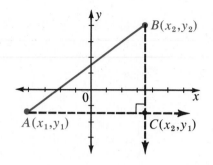

Distance formula	The distance d between any two points $A(x_1, y_1)$ and $B(x_2, y_2)$ in a coordinate plane is given by the formula $$d = \sqrt{(x_2 - x_1)^2 + (y_2 - y_1)^2}.$$

Notice that the distance between any two points A and B is the length of the segment connecting the two points. The symbol d or AB may be used to indicate this distance.

Example 2

Find the distance between $A(2, -4)$ and $B(4, 6)$.

Let $A(2, -4) = A(x_1, y_1)$ and $B(4, 6) = B(x_2, y_2)$.
$$d = \sqrt{(x_2 - x_1)^2 + (y_2 - y_1)^2}$$
$$= \sqrt{(4 - 2)^2 + [6 - (-4)]^2}$$
$$= \sqrt{4 + 100} = \sqrt{104}, \text{ or } 2\sqrt{26}$$

Thus, the distance between A and B is $2\sqrt{26}$.

The converse of the Pythagorean relation is also true. If $c^2 = a^2 + b^2$ where a, b, and c are the lengths of the sides of a triangle, then the triangle is a right triangle.

Example 3

A triangle has sides of lengths 5, $\sqrt{7}$, and $4\sqrt{2}$. Is the triangle a right triangle?

The hypotenuse is the longest side, so $c = 4\sqrt{2}$. ▶

c^2	$a^2 + b^2$
$(4\sqrt{2})^2$	$(5)^2 + (\sqrt{7})^2$
$16 \cdot 2$	$25 + 7$
32	32

Thus, since $c^2 = a^2 + b^2$, the triangle is a right triangle.

Written Exercises

Find the length of each side of right triangle ABC with the given coordinates for its vertices. C is the vertex of the right angle.

(A) 1. $A(-3, 4)$, $B(6, 5)$, $C(6, 4)$

2. $A(2, -4)$, $B(4, 6)$, $C(4, -4)$

3. $A(4, 1)$, $B(10, 8)$, $C(10, 1)$

4. $A(-5, -1)$, $B(-2, 5)$, $C(-2, -1)$

5. $A(-6, -4)$, $B(-2, -1)$, $C(-2, -4)$

6. $A(8, 10)$, $B(10, 15)$, $C(10, 10)$

Find the distance between the given points. Give each answer in simplest radical form.

7. $A(2, 4)$, $B(6, 6)$

8. $R(-2, -3)$, $S(-7, 5)$

9. $P(-2, 4)$, $Q(7, 8)$

10. $M(7, 8)$, $N(-2, -4)$

11. $A(6, -2)$, $B(-3, 4)$

12. $A(5, 1)$, $B(-2, 4)$

13. $M(8, 4)$, $N(-2, -3)$

14. $A(0, 3)$, $B(4, 0)$

15. $M(0, -4)$, $N(5, 0)$

16. $M(9, 1)$, $N(-2, -1)$

17. $A(5, 1)$, $B(7, -3)$

18. $P(10, 2)$, $Q(-1, -3)$

19. $R(-5, -4)$, $S(-2, -3)$

20. $P(-1, 4)$, $Q(5, -3)$

21. $R(-2, -2)$, $S(-4, -4)$

22. $R(-10, 12)$, $S(2, -5)$

(B) 23. $P(1, -3)$, $Q(\frac{2}{3}, -\frac{1}{2})$

24. $A(\frac{1}{2}, \frac{1}{4})$, $B(-3, 2)$

25. $A(1, 2)$, $B(\frac{1}{6}, -\frac{1}{3})$

26. $A(\frac{1}{3}, \frac{1}{4})$, $B(-\frac{2}{3}, \frac{3}{4})$

27. $P(\frac{1}{5}, -\frac{1}{4})$, $Q(\frac{3}{5}, \frac{1}{2})$

28. $A(\frac{2}{3}, \frac{1}{3})$, $B(\frac{1}{3}, \frac{2}{3})$

29. $M(a, b)$, $N(3a, 2b)$

30. $M(2x, 3y)$, $N(-3, y)$

31. $M(s, p)$, $N(-3s, -2p)$

32. $P(0, 0)$, $Q(a, 0)$

33. $A(b, c)$, $B(a + b, c)$

34. $P(\frac{b}{2}, \frac{c}{2})$, $Q(\frac{2+a}{2}, \frac{c}{2})$

35. $A(\frac{a}{2}, \frac{b}{3})$, $B(\frac{a}{6}, \frac{b}{4})$

36. $M(\frac{c}{5}, \frac{d}{3})$, $N(\frac{c}{3}, -\frac{d}{2})$

37. $M(\frac{a}{3}, -\frac{b}{2})$, $N(-\frac{a}{6}, \frac{b}{8})$

The lengths of the sides of a triangle are given. Is the triangle a right triangle?

38. 6, 8, 11

39. $\sqrt{5}$, 1, 2

40. 18, $3\sqrt{5}$, 19

41. 15, 12, 9

42. $2\sqrt{3}$, $4\sqrt{3}$, 6

43. 4, 6, 9

44. 6, $2\sqrt{3}$, $2\sqrt{6}$

45. 12, 20, 16

46. 10, 24, 26

47. $\sqrt{19}$, $3\sqrt{5}$, 8

For each of the following, find the values of x, if any, that make the distance from A to B equal to the given distance.

(C) 48. $A(x, 4)$, $B(-2, 3)$; $d = 4$

49. $A(3, x)$, $B(6, 2)$; $d = 5$

50. $A(3, -1)$, $B(-x, 4)$; $d = 3$

51. $A(-2, 1)$, $B(-4, x)$; $d = 6$

Find the perimeter of each polygon with the given coordinates for its vertices. Give each answer in simplest radical form.

52. $P(3, -1)$, $Q(6, -5)$, $R(-4, -5)$

53. $P(-b, 0)$, $Q(c, 0)$, $R(c - d, e)$

54. $P(-a, 0)$, $Q(a, 0)$, $R(0, a)$

55. $P(-\frac{2}{3}, -\frac{1}{2})$, $Q(2, -\frac{4}{5})$, $R(\frac{5}{6}, -5)$

56. $P\left(\frac{a + c}{2}, \frac{b + d}{2}\right)$, $Q\left(\frac{c + e}{2}, \frac{f + d}{2}\right)$, $R\left(\frac{g + e}{2}, \frac{h + f}{2}\right)$, $S\left(\frac{a + g}{2}, \frac{h + b}{2}\right)$

57. Prove that the midpoint of the hypotenuse of a right triangle is equidistant from the vertices of the triangle.

CUMULATIVE REVIEW

1. One number is 8 more than another number. Their product is 20. Find the numbers.

2. Find three consecutive integers so that 5 times the second, decreased by the third, is 15.

THE MIDPOINT OF A SEGMENT

Objectives
To determine the coordinates of the midpoint of a segment
To determine the coordinates of an endpoint of a segment, given the coordinates of the other endpoint and the midpoint

In the figure at the right, M is the **midpoint** of $\overline{AB}$ since $AM = MB = \frac{1}{2}(AB)$.

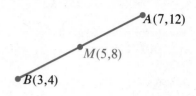

$AM = \sqrt{(7-5)^2 + (12-8)^2} = \sqrt{4+16} = \sqrt{20}$, or $2\sqrt{5}$
$MB = \sqrt{(5-3)^2 + (8-4)^2} = \sqrt{4+16} = \sqrt{20}$, or $2\sqrt{5}$
$AB = \sqrt{(7-3)^2 + (12-4)^2} = \sqrt{16+64} = \sqrt{80}$, or $4\sqrt{5}$

Notice that $M(5, 8) = M\left(\dfrac{7+3}{2}, \dfrac{12+4}{2}\right)$. In other words, the x-coordinate of M is the average (arithmetic mean) of the x-coordinates of A and B, and the y-coordinate of M is the average of the y-coordinates of A and B.

Midpoint formula	Given $P(x_1, y_1)$ and $Q(x_2, y_2)$, the midpoint of $\overline{PQ}$ is $$M\left(\frac{x_1 + x_2}{2}, \frac{y_1 + y_2}{2}\right).$$

Example 1
Determine the coordinates of M, the midpoint of $\overline{PQ}$, given $P(6, -3)$ and $Q(-14, 9)$.

$$M\left(\frac{x_1 + x_2}{2}, \frac{y_1 + y_2}{2}\right) = M\left(\frac{6 + (-14)}{2}, \frac{-3 + 9}{2}\right) = M(-4, 3)$$

Thus, $(-4, 3)$ are the coordinates of M.

If the coordinates of one endpoint of a segment and the coordinates of its midpoint are known, you can find the coordinates of the other endpoint.

Example 2
Determine the coordinates of Q, the other endpoint of $\overline{PQ}$, given $P(2, 3)$ and its midpoint $M(4, -7)$.

$P(2, 3)$, $M(4, -7)$, $Q(x_2, y_2)$

x-coordinate of the midpoint. ▶ $x_m = \dfrac{x_1 + x_2}{2}$ $\qquad y_m = \dfrac{y_1 + y_2}{2}$ ◀ *y-coordinate of the midpoint.*

$4 = \dfrac{2 + x_2}{2} \qquad\qquad -7 = \dfrac{3 + y_2}{2}$

$8 = 2 + x_2 \qquad\qquad -14 = 3 + y_2$

$x_2 = 6 \qquad\qquad\quad y_2 = -17$

Thus, the coordinates of the other endpoint are $(6, -17)$.

Oral Exercises

Determine the coordinates (x, y) of M, the midpoint of $\overline{AB}$.

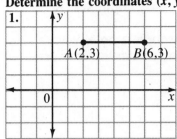
1. $A(2,3)$ $B(6,3)$

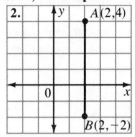
2. $A(2,4)$ $B(2,-2)$

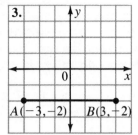
3. $A(-3,-2)$ $B(3,-2)$

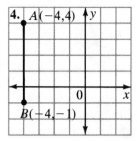
4. $A(-4,4)$ $B(-4,-1)$

Written Exercises

Determine the coordinates (x, y) of M, the midpoint of the segment joining P and Q.

(A)
1. $P(7, 4)$, $Q(3, 2)$
2. $P(-2, 1)$, $Q(8, 7)$
3. $P(-6, 1)$, $Q(-2, -2)$
4. $P(-6, 12)$, $Q(-2, -2)$
5. $P(0, 4)$, $Q(6, 0)$
6. $P(9, -2)$, $Q(1, 6)$
7. $P(-3, 5)$, $Q(7, -3)$
8. $P(19, 0)$, $Q(-4, 9)$
9. $P(5, 4)$, $Q(-1, -2)$
10. $P(8, 2)$, $Q(4, 6)$
11. $P(0, -6)$, $Q(-8, 0)$
12. $P(-5, 6)$, $Q(-1, -8)$
13. $P(10, 9)$, $Q(-3, 7)$
14. $P(-9, -3)$, $Q(-1, -7)$
15. $P(-11, -5)$, $Q(-2, -8)$
16. $P(3, -4)$, $Q(5, 7)$

(B)
17. $P(\frac{2}{3}, -\frac{1}{2})$, $Q(\frac{1}{6}, \frac{1}{5})$
18. $P(-\frac{1}{3}, \frac{1}{3})$, $Q(-\frac{1}{2}, -\frac{1}{6})$
19. $P(\frac{2}{5}, \frac{3}{4})$, $Q(-\frac{2}{3}, \frac{1}{3})$
20. $P(-\frac{1}{5}, -\frac{3}{4})$, $Q(-\frac{1}{2}, \frac{1}{8})$
21. $P((\frac{2}{5}, \frac{1}{5})$, $Q(-\frac{2}{3}, -\frac{1}{3})$
22. $P(-\frac{1}{9}, -\frac{1}{6})$, $Q(-\frac{1}{2}, -\frac{2}{3})$
23. $P(3a, 2b)$, $Q(5a, 4b)$
24. $P(m, n)$, $Q(n, m)$
25. $P(m, n)$, $Q(m + 4, n - 6)$

Determine the coordinates of Q, the other endpoint of $\overline{PQ}$, given P and the midpoint, M.

26. $P(4, 3)$, $M(6, 5)$
27. $P(-1, -2)$, $M(4, 3)$
28. $P(8, -1)$, $M(4, 6)$
29. $P(9, 2)$, $M(4, 0)$
30. $P(-3, -2)$, $M(1, 2)$
31. $P(-8, -6)$, $M(-2, -4)$
32. $P(8, -2)$, $M(-2, -1)$
33. $P(4, 3)$, $M(6, 2)$
34. $P(3, 4)$, $M(4, 8)$
35. $P(\frac{1}{3}, \frac{2}{3})$, $M(-\frac{2}{3}, -\frac{5}{4})$
36. $P(-\frac{7}{5}, \frac{1}{4})$, $M(3, \frac{5}{6})$
37. $P(-5, -\frac{9}{2})$, $M(\frac{4}{5}, -\frac{1}{2})$

(C)
38. Given triangle ABC with $A(-3, -1)$, $B(3, 5)$, and $C(-5, 13)$, find the length of the median from B. [A median of a triangle is a segment drawn from a vertex to the midpoint of the opposite side.]

39. Given triangle ABC with $A(5, -3)$, $B(-1, -1)$, and $C(-2, 6)$, find the length of the median from C.

40. In Exercise 38, if K, L, and M are the midpoints of $\overline{AB}$, $\overline{BC}$, and $\overline{AC}$, respectively, find the perimeter of triangle KLM.

41. Given $A(-5, 5)$, $B(5, 10)$, $C(3, -1)$, and $D(-7, -6)$, prove that the diagonals of $ABCD$ bisect each other.

42. Given quadrilateral $PQRS$ with $P(-1, 2)$, $Q(1, -1)$, $R(5, 1)$, and $S(1, 3)$, and points M, N, T, V, the midpoints of $\overline{PQ}$, $\overline{QR}$, $\overline{RS}$, and $\overline{SP}$, respectively, draw the quadrilateral on a coordinate plane and find the coordinates of M, N, T, and V. Prove that $MN = VT$ and $MV = NT$.

The Midpoint of a Segment

PARALLEL AND PERPENDICULAR LINES 9.7

Objectives
To determine whether two lines are parallel, perpendicular, or neither, using their slopes
To write an equation of a line that passes through a point whose coordinates are given and is parallel, or perpendicular, to a line whose equation is given

Recall that **parallel lines** are lines that lie in the same plane but do not intersect. In the figure at the right, $\overleftrightarrow{MN} \parallel \overleftrightarrow{PQ} \parallel \overleftrightarrow{RS}$.

$$m(\overleftrightarrow{MN}) = \frac{-5 - 0}{3 - (-1)} = -\frac{5}{4}$$

$$m(\overleftrightarrow{PQ}) = \frac{0 - 5}{4 - 0} = -\frac{5}{4}$$

$$m(\overleftrightarrow{RS}) = \frac{7 - 2}{2 - 6} = -\frac{5}{4}$$

Notice that each line has the same slope, $-\frac{5}{4}$.

Slope of parallel lines	In a plane, if two or more different nonvertical lines are parallel, then the lines have the same slope. If two or more different nonvertical lines have the same slope, then the lines are parallel.

Notice that the statements above about the slope of parallel lines refer only to *nonvertical* lines. The slope of any vertical line is undefined, but all vertical lines in the same plane are parallel.

You can use slope to find an equation of a line that is parallel to a second line.

Example 1

Write an equation, in standard form, of the line that passes through the point $P(4, 2)$ and is parallel to a line whose equation is $2y - 3x = 6$.

First, find the slope.

$$2y - 3x = 6$$
$$2y = 3x + 6$$
$$m = \frac{3}{2} \blacktriangleright \qquad y = \frac{3}{2}x + 3$$

The slope of each line is $\frac{3}{2}$ since the lines are parallel.

Then, use the point-slope form.

$$y - y_1 = m(x - x_1)$$
$$y - 2 = \frac{3}{2}(x - 4) \quad \blacktriangleleft (x_1, y_1) = (4, 2)$$
$$2y - 4 = 3x - 12$$
$$3x - 2y = 8 \qquad \blacktriangleleft \textit{Standard form}$$

Thus, an equation of the line is $3x - 2y = 8$.

Recall that two lines are **perpendicular** if they intersect to form a right angle. In the figure, $\overleftrightarrow{PQ} \perp \overleftrightarrow{RS}$.

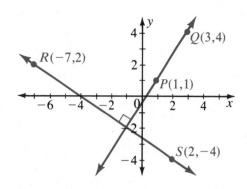

$$m(\overleftrightarrow{PQ}) = \frac{4 - 1}{3 - 1} = \frac{3}{2}$$

$$m(\overleftrightarrow{RS}) = \frac{-4 - 2}{2 - (-7)} = \frac{-6}{9}, \text{ or } -\frac{2}{3}$$

Notice that the product of their slopes is $\frac{3}{2} \cdot -\frac{2}{3}$, or -1. Hence, their slopes are *negative reciprocals* of each other.

Slope of perpendicular lines	If two lines are perpendicular, then the slopes of the lines are negative reciprocals.
	If the slopes of two lines are negative reciprocals, then the lines are perpendicular.

The statements above about the slope of perpendicular lines do not refer to horizontal or vertical lines. A vertical line and a horizontal line in the same plane are always perpendicular to each other.

Example 2

Write an equation, in standard form, of the line that passes through the point $Q(4, 2)$ and is perpendicular to a line whose equation is $2y - 3x = 6$.

First, find the slope.
$$2y - 3x = 6$$
$$m = \frac{3}{2} \blacktriangleright \qquad y = \frac{3}{2}x + 3$$
The slope of the given line is $\frac{3}{2}$ and so the slope of the line perpendicular to it is $-\frac{2}{3}$.

Then, use the point-slope form.
$$y - y_1 = m(x - x_1)$$
$$y - 2 = -\frac{2}{3}(x - 4)$$
$$3y - 6 = -2x + 8$$
$$2x + 3y = 14 \quad \blacktriangleleft \text{ Standard form}$$

Thus, $2x + 3y = 14$ is an equation of the line.

Example 3

Determine whether the lines with equations of $3x = -4y + 7$ and $6x + 8y = 11$ are parallel, perpendicular, or neither.

$$3x = -4y + 7 \qquad\qquad 6x + 8y = 11$$
$$4y = -3x + 7 \qquad\qquad 8y = -6x + 11$$

Rewrite in slope-intercept form. $\blacktriangleright$
$$y = -\frac{3}{4}x + \frac{7}{4} \qquad\qquad y = -\frac{6}{8}x + \frac{11}{8}$$
$$m = -\frac{3}{4} \qquad\qquad m = -\frac{6}{8}, \text{ or } -\frac{3}{4}$$

Thus, the lines are parallel since their slopes are the same.

Example 4   **Write an equation, in standard form, of the line that passes through the point $B(-3, 4)$ and is perpendicular to a line whose equation is $y = 2$.**

$y = 2$ is the equation of a horizontal line.
A line perpendicular to this line is vertical. A vertical line has an equation of the form $x = c$, where c is a constant.

Thus, $x = -3$ is an equation of the line.

Written Exercises

Write an equation, in standard form, of the line that passes through the given point and is parallel to the line described by the given equation.

(A)   **1.** $A(5, 7)$, $x + 4y = 5$
4. $D(3, -4)$, $y = 6$
2. $B(6, 3)$, $5x + 4y = 16$
5. $E(-3, -8)$, $x = -5$
3. $C(-3, -4)$, $2x - 5y = 15$
6. $F(6, -10)$, $y = x$

Determine whether the lines whose equations are given are parallel, perpendicular, or neither.

7. $y = 2x + 3$; $2y = 4x + 7$
10. $4x - 5y + 3 = 0$; $x = 2$
8. $6x - 4y = 3$; $8y - 12x = -7$
11. $7x + 3y = 7$; $8x + 4y = 6$
9. $x = 5$; $y = 4$
12. $y = 2$; $x = 0$

Write an equation, in standard form, of the line that passes through the given point and is perpendicular to the line described by the given equation.

13. $G(6, -2)$, $x = 5$
16. $J(-3, 2)$, $2x - 5y = 4$
14. $H(3, -4)$, $y = -4$
17. $K(-5, -8)$, $3x + 2y + 7 = 0$
15. $I(6, -10)$, $y = -x$
18. $L(8, 0)$, $3y + 5 = 0$

Write an equation, in standard form, of the line that is perpendicular to $\overline{AB}$ at point B.

(B)   **19.** $A(-3, 5)$, $B(6, -2)$
22. $A(-6, -4)$, $B(10, -3)$
20. $A(-2, -3)$, $B(-1, 4)$
23. $A(8, -7)$, $B(6, -4)$
21. $A(6, -1)$, $B(8, 2)$
24. $A(-3, -5)$, $B(0, 0)$

Write an equation, in standard form, of the line that is the perpendicular bisector of the segment joining P and Q.

25. $P(-3, -4)$, $Q(6, 5)$
28. $P(-2, 3)$, $Q(4, 7)$
26. $P(7, -3)$, $Q(-8, 2)$
29. $P(1, 5)$, $Q(7, -3)$
27. $P(-7, -3)$, $Q(4, 8)$
30. $P(6, -3)$, $Q(4, 7)$

(C)   **31.** Given $P(a, 2b)$, $Q(2a, b)$ and $R(a, b)$, write an equation, in standard form, of the line through R that is parallel to $\overrightarrow{PQ}$ if $a \neq 0$.

32. Given $P(a, b)$, $Q(3a, 3b)$, $a \neq 0$ and $b \neq 0$, write an equation, in standard form, of the line through P that is perpendicular to $\overrightarrow{PQ}$.

33. Given $P(2a, 2b)$, $Q(2c, 2d)$, $a \neq c$, and $b \neq d$, write an equation, in standard form, of the line that is the perpendicular bisector of $\overline{PQ}$.

34. Given triangle PQR with $P(2a, 2b)$, $Q(2a + 2c, 2b)$, and $R(2a + c, 2d)$, write an equation of the line that contains the median from R.

CUMULATIVE REVIEW

Solve each equation for x.

1. $ax + b = c$

2. $ax - bx = c$

3. $\dfrac{1}{a} - \dfrac{1}{b} = \dfrac{1}{x}$

USING COORDINATE GEOMETRY 9.8

Objective   **To prove that a given property of a geometric figure is true, by using the slope, distance, or midpoint formula**

Coordinate geometry can often be used to prove that a given property of a geometric figure is true. The geometric figure is drawn on a coordinate plane, usually with coordinates assigned to each of the vertices of the figure. The slope, distance, or midpoint formula can then be applied.

Example 1   **Prove that the segments joining the points $A(-3, -4)$, $B(2, 5)$ and $C(-5, 3)$ form a right triangle.**

Draw a diagram and find the slopes.

$$m(\overline{AB}) = \frac{-4 - 5}{-3 - 2} = \frac{9}{5}$$

$$m(\overline{AC}) = \frac{-4 - 3}{-3 - (-5)} = -\frac{7}{2}$$

$$m(\overline{BC}) = \frac{5 - 3}{2 - (-5)} = \frac{2}{7}$$

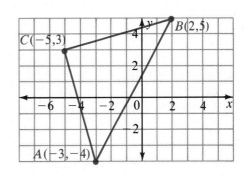

Since the slopes of $\overline{AC}$ and $\overline{BC}$ are negative reciprocals, $\overline{AC} \perp \overline{BC}$ and angle C is a right angle.

Thus, $\triangle ABC$ is a right triangle.

Points that lie on the same line are said to be *collinear*. You can use the slope formula to prove that certain points are collinear. Recall that the coordinates of any two points of a line can be used to find its slope, and that the slope should be the same regardless of which points are used.

Example 2   **Prove that the points $A(-3, -2)$, $B(0, 0)$, and $C(6, 4)$ are collinear.**

$$m(\overrightarrow{AB}) = \frac{0 - (-2)}{0 - (-3)} = \frac{2}{3}$$

$$m(\overrightarrow{BC}) = \frac{4 - 0}{6 - 0} = \frac{4}{6}, \text{ or } \frac{2}{3}$$

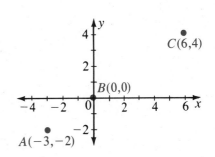

Thus, $m(\overrightarrow{AB}) = m(\overrightarrow{BC})$ and so the points A, B, and C are collinear.

Using Coordinate Geometry **231**

If you studied geometry, you may have proved that the line connecting the midpoints of two sides of a triangle is parallel to the third side of the triangle. This can be proved algebraically using coordinate geometry.

Example 3   **In triangle ABC, D is the midpoint of $\overline{AB}$ and E is the midpoint of $\overline{BC}$. Prove that $\overleftrightarrow{DE} \parallel \overline{AC}$.**

First, find the coordinates of the midpoints.

$$D(x, y) = D\left(\frac{a + f}{2}, \frac{b + g}{2}\right) \qquad E(x, y) = E\left(\frac{f + d}{2}, \frac{g + e}{2}\right)$$

Next, find the slope of $\overleftrightarrow{DE}$ and $\overline{AC}$.

$$m = \frac{y_1 - y_2}{x_1 - y_2}$$

$$m(\overleftrightarrow{DE}) = \frac{\dfrac{g + e}{2} - \dfrac{b + g}{2}}{\dfrac{f + d}{2} - \dfrac{a + f}{2}} = \frac{\dfrac{g + e - b - g}{2}}{\dfrac{f + d - a - f}{2}} = \frac{e - b}{d - a}$$

$$m(\overline{AC}) = \frac{e - b}{d - a}$$

Thus, $m(\overleftrightarrow{DE}) = m(\overline{AC})$, and so $\overleftrightarrow{DE} \parallel \overline{AC}$.

In the next example, you will use the slope formula to prove that a given quadrilateral is a parallelogram. Recall that if both pairs of opposite sides of a quadrilateral are parallel, then the quadrilateral is a parallelogram.

Example 4   **The vertices of a quadrilateral are $A(-1, 3)$, $B(0, 0)$, $C(6, 1)$, and $D(5, 4)$. Prove that $ABCD$ is a parallelogram.**

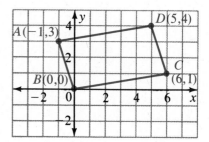

$$m(\overline{AB}) = \frac{0 - 3}{0 - (-1)} = -3; \; m(\overline{DC}) = \frac{1 - 4}{6 - 5} = -3$$

$$m(\overline{BC}) = \frac{1 - 0}{6 - 0} = \frac{1}{6}; \; m(\overline{AD}) = \frac{4 - 3}{5 - (-1)} = \frac{1}{6}$$

So, $\overline{AB} \parallel \overline{DC}$ and $\overline{BC} \parallel \overline{AD}$.

Thus, $ABCD$ is a parallelogram.

Written Exercises

For each of the following, prove that the segments joining the points A, B, and C form a right triangle.

(A) 1. $A(1, 2)$, $B(2, 5)$, $C(-1, 6)$

2. $A(-1, -2)$, $B(5, 1)$, $C(3, 5)$

3. $A(5, -5)$, $B(7, 3)$, $C(-1, 5)$

4. $A(2, 3)$, $B(4, -1)$, $C(8, 1)$

5. Prove that the points $A(1, 2)$, $B(-1, -8)$, and $C(0, -3)$ are collinear.

6. The vertices of a quadrilateral are $A(0, 0)$, $B(1, 2)$, $C(6, 5)$, and $D(5, 3)$. Prove that $ABCD$ is a parallelogram.

7. The vertices of triangle ABC are $A(2, 1)$, $B(8, 3)$, and $C(4, 5)$. Prove that the segments joining the midpoints of $\overline{AB}$ and $\overline{BC}$ is one-half the length of $\overline{AC}$.

8. The vertices of a triangle are $A(2, 1)$, $B(7, 4)$, and $C(2, 7)$. Prove that triangle ABC is isosceles.

(B) 9. The vertices of a quadrilateral are $A(0, 0)$, $B(a, 0)$, $C(a, b)$, and $D(0, b)$. Prove that $ABCD$ is a rectangle by proving that $\overline{AB} \perp \overline{BC}$, $\overline{AB} \perp \overline{AD}$, $\overline{CD} \perp \overline{AD}$, and $\overline{CD} \perp \overline{BC}$.

10. The vertices of a rectangle are $A(0, 0)$, $B(a, 0)$, $C(a, b)$, and $D(0, b)$. Prove that the segments joining the midpoints of $\overline{AB}$, $\overline{BC}$, $\overline{CD}$, and $\overline{DA}$ form a rhombus.

11. Prove that the midpoints of the sides of trapezoid $ABCD$ determine a parallelogram. The coordinates of the vertices are $A(0, 0)$, $B(a, 0)$, $C(b + d, c)$, and $D(b, c)$.

12. Prove that the diagonals of rhombus $STUV$ are perpendicular to each other. The coordinates of the vertices are $S(0, 0)$, $T(a, 0)$, $U(a + \sqrt{a^2 - b^2}, b)$, and $V(\sqrt{a^2 - b^2}, b)$.

13. Prove that the diagonals of a rectangle bisect each other.

(C) 14. Given $A(0, 0)$, $B(2a, 0)$, and $C(2b, 2c)$, prove that the three medians of triangle ABC intersect at $T\left(\dfrac{2a + 2b}{3}, \dfrac{2c}{3}\right)$. [*Hint:* Use the point-slope form to write an equation of each median and then solve each equation for y.]

15. Prove that the diagonals of a square are perpendicular.

16. Prove that the diagonals of a rectangle are equal in length.

17. Prove that the segments successively joining the midpoints of a quadrilateral form a parallelogram.

18. Prove that the medians of an equilateral triangle are equal in length.

19. Prove that if two lines in the same plane are perpendicular to a given line, then the two lines are parallel.

APPLICATIONS

Read → Plan → Solve → Interpret

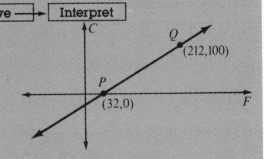

1. Use the slope-intercept form $y = mx + b$ and the two points given in the figure at the right to show that $C = \frac{5}{9}(F - 32)$, the relationship between Fahrenheit and Celsius.

2. Show $F = \frac{9}{5}C + 32$.

3. At what temperature do the Celsius and Fahrenheit scales register the same? Show your derivation.

4. Use $y = mx + b$ and the two points given in the figure at the right to show that $C = k - 273.15°$, the relationship between Kelvin and Celsius.

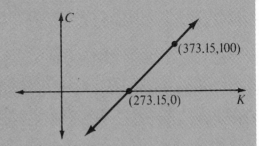

5. Thermometers sell for $3.50 each. Let n represent the number of thermometers and C the cost of n thermometers. Write an equation representing C in terms of n. Draw the graph showing this relationship.

6. A car traveled at an average speed of 70 km/h. How far did it travel in 3 h? in 5 h? Draw the graph showing this relationship.

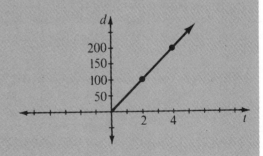

7. Use the graph at the right and write an equation describing the relationship among distance, rate, and time. How many units would you travel in 3 h? in 10 h?

COMPUTER ACTIVITIES

Linear Equations

OBJECTIVE: **To determine the equation of a line in which the only INPUTs are two points on the line**

To write an equation of a line through two points is not very difficult. Using the coordinates of the two points, you determine the slope and write the equation. The following three-line program INPUTs two points and outputs the resulting equation.

```
100    INPUT A,B
102    INPUT C,D
103    PRINT B - D"X + "C - A"Y = "B * (C - A) - A *
       (D - B)
```

The above program appears to work well enough. However, what if the line is either vertical or horizontal? Then the equation is not totally valid. In fact, when the line is vertical, the computer will output an error message. Therefore, the program must include two additional tests. One tests for lines parallel to the x-axis, and the other tests for lines parallel to the y-axis.

In addition to the two tests, it is a good idea to include a PRINT statement that identifies the output as an equation through two particular points. Add the following lines to the program above, and the new program will do it all.

```
110    PRINT "THE EQUATION THROUGH ("A","B") AND ("; 
120    PRINT C","D") IS "
130    IF A = C THEN 160
140    IF B = D THEN 180
150    GOTO 100
160    PRINT "X = "A
170    GOTO 100
180    PRINT "Y = "B
190    INPUT "MORE POINTS? (YES OR NO);K$
200    IF K$ = "YES" THEN 100
210    END
```

See the Computer Section beginning on page 574 for more information.

EXERCISES

1. Write a program that calculates the slope of a line through two given points.

2. Write a program that calculates the length and determines the midpoint of a segment, given the coordinates of its endpoints.

3. Write a program that finds the missing endpoints of a segment, from the INPUT of the midpoint and the known endpoint.

CHAPTER NINE REVIEW

Vocabulary
abscissa [9.1]
coordinate plane [9.1]
coordinates [9.1]
linear equation [9.1]
  point-slope form of [9.3]
  slope-intercept form of [9.3]
ordered pair [9.1]
ordinate [9.1]
origin [9.1]
quadrant [9.1]
slope [9.2]
y-intercept [9.3]

On the same coordinate plane, graph each ordered pair. [9.1]
1. $(-2, -3)$ 2. $(4, -6)$ 3. $(-7, 0)$
4. $(0, -3)$ 5. $(5, -5)$ 6. $(0, 2)$

Give the coordinates of each point. [9.1]
7. A
8. B
9. C
10. D
11. E

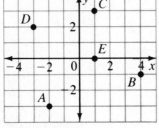

Draw the graph of each equation.
12. $2x + 4y = 12$ [9.1]
13. $-(x - y) - 3(x - y) = 6$ [9.4]

Find the length of $\overline{AB}$. [9.2]
14. $A(2, -3), B(-6, -3)$
15. $A(-4, 1), B(-4, 5)$

Find the slope of $\overleftrightarrow{PQ}$. [9.2]
16. $P(6, 4), Q(-2, 3)$
17. $P(-2, -2), Q(-3, -4)$
18. $P(\frac{1}{2}, \frac{1}{4}), Q(-\frac{3}{4}, \frac{2}{3})$
★ 19. $P(2x, -3y), Q(-3x, 2y)$

Find the missing coordinate for the given data. [9.2]
★ 20. $A(-3, y), B(6, -8), m(\overline{AB}) = -2$

A line has the given slope, y-intercept, or contains the indicated point(s). Write an equation, in standard form, of each line. [9.3]
21. $P(2, 5), m = \frac{2}{3}$ 22. $m = 3, b = 7$
23. $A(-3, 4), B(-1, 5)$

Find the slope and the y-intercept of the line described by each given equation. [9.3]
24. $y = -\frac{2}{3}x + 4$ 25. $y - 2(3 - 2x) = 7$
★ 26. $6 - (3 - 2y) - 2[-3(2x - 1)] = 5$

Is the given point on the line described by the given equation? [9.4]
27. $P(-1, -3), 2x - 3y = 7$
28. $Q(-4, 2), \frac{3}{5}x + \frac{2}{3}y = 1$

Find the distance between the given points. Answer in simplest radical form. [9.5]
29. $P(-1, 2), Q(7, 4)$
30. $M(3, 0), N(10, 8)$
31. $P(\frac{1}{2}, -\frac{1}{3}), Q(\frac{2}{3}, \frac{5}{6})$

The lengths of the sides of a triangle are given. Is the triangle a right triangle?
32. $8, 12, 4\sqrt{13}$ [9.5]

Determine the coordinates (x, y) of M, the midpoint of $\overline{PQ}$. [9.6]
33. $P(4, -3), Q(7, 4)$
34. $P(-\frac{1}{4}, \frac{1}{2}), Q(\frac{2}{5}, -\frac{1}{3})$

Determine the coordinates of Q, the other endpoint of $\overline{PQ}$, given P and the midpoint, M. [9.6]
35. $P(-3, 6), M(1, 2)$

Write an equation, in standard form, of the line that passes through the given point and is parallel to the given line. [9.7]
36. $C(1, 2), 3x - 5y = 15$
37. $D(4, -1), -2x - 3y = 10$

Write an equation, in standard form, of the line that passes through the given point and is perpendicular to the given line. [9.7]
38. $A(-1, -3), 2x + 5y = 7$
39. $B(2, -7), 3y + 6x = 8$

40. The vertices of a quadrilateral are $A(-5, 3), B(4, 0), C(6, 6),$ and $D(-3, 9)$. Prove that $ABCD$ is a parallelogram. [9.8]

Give the coordinates of each point.

1. A
2. B
3. C
4. D

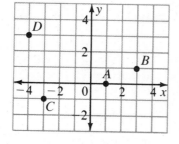

Draw the graph of each equation.
5. $y = -2x + 5$
6. $6x - 2(-3 + 2x) - (3 - y) = 6$

Find the length of $\overline{AB}$.
7. $A(3, 4)$, $B(6, -3)$

Find the slope of $\overrightarrow{PQ}$.
8. $P(3, -6)$, $Q(5, -2)$
9. $P(-\frac{1}{2}, \frac{3}{4})$, $Q(\frac{5}{6}, -\frac{1}{2})$

10. Write an equation, in standard form, of the line that passes through the point $P(-1, 2)$ and whose slope is $\frac{1}{2}$.

11. Write an equation, in standard form, of the line that passes through the points $A(-3, 2)$ and $B(-4, 4)$.

Find the slope and the y-intercept of the line described by each given equation.
12. $2x + 4y = 12$ 13. $y - (2x - 5) = -8$

Is the given point on the line described by the given equation?
14. $R(2, -3)$, $3x - 2y = -12$
15. $S(-1, 0)$, $\frac{4}{5}x - \frac{1}{3}y = -\frac{4}{5}$

16. Find the distance between points $P(-1, -7)$ and $Q(7, -3)$. Give the answer in simplest radical form.

17. The lengths of the sides of a triangle are $4\sqrt{3}$, $8\sqrt{3}$, and 12. Is the triangle a right triangle?

18. Determine the coordinates of M, the midpoint of $\overline{PQ}$, given $P(4, -3)$ and $Q(6, -7)$.

19. Determine the coordinates of Q, the other endpoint of $\overline{PQ}$, given $P(-2, -3)$ and its midpoint $M(3, -1)$.

20. Write an equation, in standard form, of the line the passes through the point $A(-1, -2)$ and is parallel to a line whose equation is $3x + 5y = 15$.

21. Write an equation, in standard form, of the line that passes through the point $N(-4, 1)$ and is perpendicular to a line whose equation is $5x - 2y = 10$.

22. The vertices of a quadrilateral are $A(3, 0)$, $B(9, 4)$, $C(6, 9)$, and $D(0, 5)$. Prove that $ABCD$ is a parallelogram.

★ 23. Find the missing coordinate for the given data: $P(x, -3)$, $Q(9, 6)$, $m(\overline{PQ}) = 3$.

★ 24. Prove that the midpoints of the sides of an isosceles triangle determine an isosceles triangle.

DIRECTIONS: Choose the *one* best answer to each question or problem.

1. In the figure, the coordinates of M are $(6, 0)$ and the area of triangle MON is 24. What are the coordinates of N?

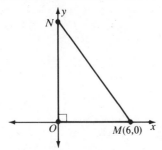

   (A) $(4, 0)$ (B) $(0, 4)$ (C) $(0, 8)$
   (D) $(8, 0)$ (E) None of these

2. In the figure, triangles PQR and PMN are isosceles right triangles. If the areas of the triangles are the same, what are the coordinates of point M?

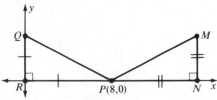

   (A) $(8, 16)$ (B) $(16, 8)$ (C) $(8, 8)$
   (D) $(16, 4)$ (E) $(8, 4)$

3. The distance between $R(-5, 0)$ and S is 4. The coordinates of point S could be any of the following except

   (A) $(-5, 4)$ (B) $(-5, -4)$ (C) $(-9, 0)$
   (D) $(0, -1)$ (E) $(-1, 0)$

4. The point $(a, -b)$ is on the line described by which linear equation in x and y?

   (A) $x + y = a + b$ (B) $x - y = a - b$

   (C) $2x - b = 2a - y$ (D) $x - \dfrac{y + b}{2} = a$

   (E) None of these

5. Which point is the farthest away from the origin?

   (A) $P(7, -5)$ (B) $Q(5, -7)$
   (C) $R(8, -3)$ (D) $S(2, 8)$
   (E) $T(0, -9)$

6. Find the area of right triangle PQR, with right angle R, if the equation of $\overrightarrow{PQ}$ is $y = 3x + 9$ and the coordinates of R are $(7, 0)$.

   (A) 70   (B) 100   (C) 150
   (D) 300   (E) None of these

7. Choose the table of ordered pairs (x, y) for which $5ay - 10bx = 15ab$.

   (A)

x	y
$3b$	0
b	$-a$
$-b$	$-2a$

   (B)

x	y
a	$5b$
$-3a$	$-3b$
$-a$	b

   (C)

x	y
0	$3b$
a	b
$-a$	$-b$

   (D)

x	y
$-2a$	$-b$
$-a$	b
a	$-2b$

   (E) None of these

8. Given $P(a, b)$, $Q(3a, 3b)$, $R(-c, -d)$, and $S(c, d)$, where a, b, c, and d are nonzero real numbers, which statement is true?

   (A) $PQ = RS$ if $a = d$ and $b = c$.
   (B) $\overline{PQ} \parallel \overline{RS}$ if $a = b = c$.
   (C) $\overline{PQ} \perp \overline{RS}$ if $ac = bd$.
   (D) $\overline{PQ}$ and $\overline{RS}$ have the same midpoint if $a = b$.
   (E) None of these

10 LINEAR SYSTEMS, MATRICES, AND DETERMINANTS

Non-Routine Problem Solving

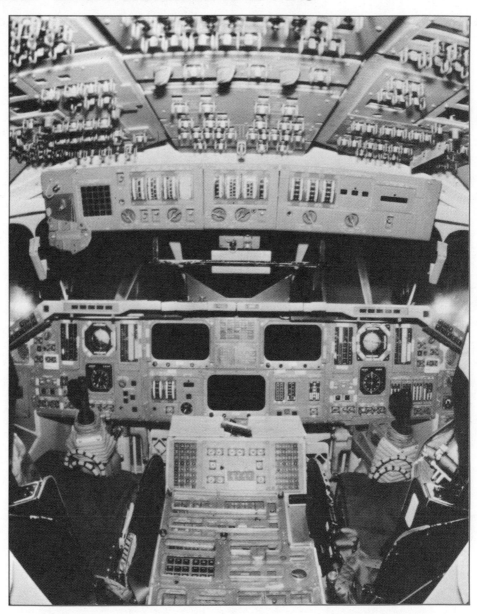

Problem 1. A two-digit number is three times the product of the two digits. Find all such numbers.

Problem 2. Find the number of integers between 1 and 10,000 that can be expressed in the form $13y^2$, where y is a positive integer.

Problem 3. There are only 4 four-digit numbers that are perfect squares with all even digits. Find as many of these 4 numbers as you can.

SOLVING LINEAR SYSTEMS GRAPHICALLY

Objectives

To solve a system of two linear equations in two variables graphically
To determine whether a system of two linear equations is consistent, inconsistent, dependent, or independent

A pair of linear equations in two variables is called a **system of equations.** A solution of a system of two linear equations is an ordered pair of real numbers that makes both equations true. You can solve a system of linear equations by graphing each equation. You should draw the graph of each equation in the same coordinate plane and then locate the point of intersection of the two graphs, if any. The ordered pair associated with this point is a solution of the system.

Example 1

Solve the system $2x - y = 5$ graphically.
$$-x + y = -3$$

Write each equation in slope-intercept form.

$$2x - y = 5 \qquad\qquad -x + y = -3$$
$$y = 2x - 5 \qquad\qquad y = x - 3$$

Graph each equation in the same coordinate plane.

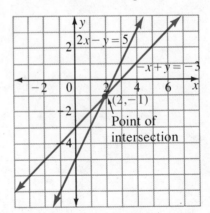

Check that the ordered pair $(2, -1)$ is a solution of both equations.

Thus, $(2, -1)$ is the solution of the system.

In Example 1, the two lines have exactly one point in common, and so the two equations have one common solution. When a system of linear equations has exactly one solution, the system is called an **independent system.** When a system has at least one solution, the system is called a **consistent system.**

Example 2

Solve the system $5x - 3(y + x) = -6$ graphically.
$$2x - 3y = 3$$

$$5x - 3(y + x) = -6$$
$$5x - 3y - 3x = -6$$
$$-3y = -2x - 6$$
$$y = \frac{2}{3}x + 2$$

$$2x - 3y = 3$$
$$-3y = -2x + 3$$
$$y = \frac{2}{3}x - 1$$

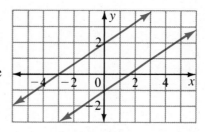

The slopes of the two lines are the same but their y-intercepts are different. The lines are parallel. There is no point of intersection of the graphs.

Thus, there is no solution of the system.

A system of linear equations which has no solution is called an **inconsistent system.**

Example 3

Solve the system $3x + y = 1$ graphically.
$$2y = 2 - 6x$$

$$3x + y = 1$$
$$y = -3x + 1$$

$$2y = 2 - 6x$$
$$y = -3x + 1$$

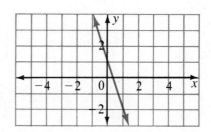

The slopes of the two lines and their y-intercepts are the same. The lines coincide. The two graphs have all of their points in common.

Thus, all (x, y) such that $y = -3x + 1$ are solutions of the system.

In Example 3, the two lines have all of their points in common and so each solution of one equation is also a solution of the other. A system of linear equations that has all solutions in common is called a **dependent system.** A dependent system has infinitely many solutions.

Reading in Algebra

True or false? Give a reason for your answer.
1. A system of two linear equations has at least one solution.
2. A system of two linear equations has at most one solution.
3. An inconsistent system of equations is one that has more than one solution.
4. A dependent system of equations is also a consistent system.

Written Exercises

Solve each system of equations graphically. Indicate whether the system is consistent, inconsistent, dependent, or independent.

(A)

1. $2x + y = 3$
$y - x = -3$

2. $3x + 2y = 8$
$3x = 10 - 2y$

3. $4x - y + 3 = 0$
$x - y = 0$

4. $y = 2 - 3x$
$x = \dfrac{2}{3} - \dfrac{y}{3}$

5. $\dfrac{x}{2} = \dfrac{y}{3}$
$y - x = 1$

6. $\dfrac{x + 2y}{2} = x$
$2y = x$

7. $3y + 4 = 5x$
$10x - 3 = 6y$

8. $-4x + 2y - 3 = 0$
$6y - 9 = 12x$

9. $2x + y = 3$
$2y - 3x = 6$

(B)

10. $2(x + y) = x + 2$
$2y + 3(x - 2y + 8) = 0$

11. $x + 2(y + 5) = 3(x + y + 1)$
$0 = 5y - 3(x - 10)$

12. $-3(x - y) = 2(y - x) + 8$
$-(2x - y) = 3(y + 3x - 1)$

13. $-x - 3y = 2(x - 3y) + 4$
$y - x - 2(x - y) = -3$

14. $0.2x - 0.3y = -0.8$
$0.3x + 0.5y = 0.7$

15. $0.3x - 0.2y = 0.7$
$0.4x - 0.1y = 0.6$

Write a system of two linear equations in two variables that has the given number of solutions.

(C)

16. exactly one solution

17. no solution

18. infinitely many solutions

Find the value(s) of k, if any, that satisfies the given condition for each system.

19. $\left.\begin{array}{l} y = kx + 8 \\ y = 3x - 1 \end{array}\right\}$ is inconsistent

20. $\left.\begin{array}{l} kx - 3y = 9 \\ 2x + 3y = 12 \end{array}\right\}$ is consistent

21. $\left.\begin{array}{l} 4x - 2y = 7 \\ 12x - ky = 21 \end{array}\right\}$ has infinitely many solutions

22. $\left.\begin{array}{l} 2x + ky = 6 \\ 3x - 4y = -8 \end{array}\right\}$ has exactly one solution

CALCULATOR ACTIVITIES

To check to see if $(1.3, 2.5)$ is a solution of the system
$12.6x + 15.7y = 55.63$,
$-5.3x + 11.8y = 22.61$
substitute 1.3 for x and 2.5 for y in each equation.

$$12.6 \otimes 1.3 \qquad 15.7 \otimes 2.5 \qquad\qquad -5.3 \otimes 1.3 \qquad 11.8 \otimes 2.5$$
$$\ominus \qquad\qquad \ominus \qquad\qquad\qquad \ominus \qquad\qquad \ominus$$
$$16.38 \quad \oplus \quad 39.25 \qquad\qquad -6.89 \quad \oplus \quad 29.50$$
$$55.63 \qquad\qquad\qquad\qquad 22.61$$

Check to determine if the given ordered pair is a solution of the system of equations.

1. $4.6x + 3.5y = 29.23$
$5.1x - 1.7y = 17.34$, $(4.3, 2.7)$

2. $15.9x - 13.3y = 21.44$
$-18.1x + 21.7y = 13.64$, $(6.2, 5.8)$

SOLVING LINEAR SYSTEMS ALGEBRAICALLY

Objectives **To solve a system of two linear equations by the substitution method**
To solve a system of two linear equations by the addition method

A system of two linear equations can be solved algebraically. In this lesson, you will use two algebraic methods, *substitution* and *addition*.

You can solve a system of two linear equations in two variables by substitution using the following procedure:
(1) Solve one of the equations for either variable.
(2) Substitute this value in the other equation.
(3) Solve for the other variable.
(4) Substitute in either original equation to find the value of the second variable.
(5) Check the ordered pair in both original equations.

Example 1 **Solve the system $2x - y = 5$ by substitution.**
$$-x + y = -3$$

Solve one equation for y. ▶
$$2x - y = 5$$
$$y = 2x - 5$$

Substitute in the other equation. ▶
$$-x + y = -3$$
$$-x + (2x - 5) = -3$$
$$-x + 2x - 5 = -3$$

Solve for x. ▶
$$x = 2$$

Substitute in one of the original equations. ▶
Solve for y. ▶
$$2x - y = 5$$
$$2(2) - y = 5$$
$$y = -1$$

Check in both original equations. ▶

$2x - y$	5
$2(2) - (-1)$	5
$4 + 1$	
5	

$-x + y$	-3
$-(2) + (-1)$	-3
$-2 - 1$	
-3	

Thus, $(2, -1)$ is the solution of the system.

Notice that Example 1 above is an algebraic solution to Example 1 of the preceding lesson.

A system of two linear equations can also be solved algebraically by addition, as shown on page 244.

Example 2 Solve the system $2x + 3y = 9$ by addition.
$$-2x - 5y = 1$$

Add the two equations and solve for y.
$$2x + 3y = 9$$
$$-2x - 5y = 1$$
$$\overline{\quad\quad -2y = 10}$$
$$y = -5$$

Substitute -5 for y in either equation. Solve for x.
$$2x + 3y = 9$$
$$2x + 3(-5) = 9$$
$$2x - 15 = 9$$
$$2x = 24$$
$$x = 12$$

Check in both original equations. Substitute 12 for x and -5 for y.

$2x + 3y$	9
$2(12) + 3(-5)$	9
$24 - 15$	
9	

$-2x - 5y$	1
$-2(12) - 5(-5)$	1
$-24 + 25$	
1	

Thus, $(12, -5)$ is the solution of the system.

In Example 2, notice that the coefficients of x in the original equations are additive inverses. Thus, when the equations are added the x-variable is eliminated.

Sometimes it is necessary to multiply one or both of the linear equations by some number or numbers so that the coefficients of one variable are additive inverses. This is shown in Examples 3 and 4 which follow.

Example 3 Solve the system $5x - 2y = 20$ by addition.
$$7x + 4y = 11$$

Multiply the first equation by 2. ▶
$$5x - 2y = 20$$
$$2(5x - 2y) = 2(20)$$

Add the equations. ▶
$$10x - 4y = 40$$
$$7x + 4y = 11$$
$$\overline{17x \quad\quad = 51}$$
$$x = 3$$

Substitute 3 for x in one of the original equations. ▶
$$5x - 2y = 20$$
$$5(3) - 2y = 20$$
$$-2y = 5$$

Solve for y. ▶
$$y = -\frac{5}{2}$$

Check $(3, -\frac{5}{2})$ in both original equations.

Thus, $(3, -\frac{5}{2})$ is the solution of the system.

Example 4   **Solve the system $3x + 2y = 5$ by addition.**
$$4x - 5y = 22$$

<u>First Way</u>

Eliminate x. Use $12x$ and $-12x$.

$4(3x + 2y) = 4(5)$		$12x + 8y = 20$
$-3(4x - 5y) = -3(22)$		$-12x + 15y = -66$
		$23y = -46$
		$y = -2$

Solve for x. $3x + 2y = 5$
$3x + 2(-2) = 5$
$3x = 9$
$x = 3$

<u>Second Way</u>

Eliminate y. Use $10y$ and $-10y$.

$5(3x + 2y) = 5(5)$		$15x + 10y = 25$
$2(4x - 5y) = 2(22)$		$8x - 10y = 44$
		$23x = 69$
		$x = 3$

Solve for y. $3x + 2y = 5$
$3(3) + 2y = 5$
$2y = -4$
$y = -2$

Check in both original equations. Substitute 3 for x and -2 for y.

$3x + 2y$	5
$3(3) + 2(-2)$	5
$9 - 4$	
5	

$4x - 5y$	22
$4(3) - 5(-2)$	22
$12 + 10$	
22	

Thus, $(3, -2)$ is the solution of the system.

When you solve a system of two linear equations either by the **method of** substitution or addition, sometimes both variables are eliminated. **If the** resulting equation is a false statement, the system is inconsistent **and has no** solution. If the resulting equation is a true statement, the system is **dependent** and has infinitely many solutions.

Example 5   **Solve each system by addition.**

$2x - 4y = 5$
$-x + 2y = 8$

$-3x + 2y = 4$
$9x - 6y = -12$

Multiply the second equation by 2. ▶
$-x + 2y = 8$
$2(-x + 2y) = 2(8)$
$-2x + 4y = 16$

$-3x + 2y = 4$
$3(-3x + 2y) = 3(4)$
$-9x + 6y = 12$
◀ *Multiply the first equation by 3.*

Add the equations. ▶
$2x - 4y =$	5
$-2x + 4y =$	16
$0 + 0 =$	21
	$0 = 21$ False

$-9x + 6y =$	12
$9x - 6y =$	-12
$0 + 0 = 0$	
$0 = 0$ True	
◀ *Add the equations.*

Thus, there is no solution.

Thus, all (x, y) such that $-3x + 2y = 4$
are the solutions.

Solving Linear Systems Algebraically

Example 6   **A riverboat travels downstream 16 km in 2 h and returns the same distance**
upstream in 3 h. Find the rate of the boat in still water and the rate of the
READ ▶   **current.**

PLAN ▶   Let x = rate of boat in still water and y = rate of current

	rate(r)	time(t)	distance(d)
Downstream	$x + y$	2	$2(x + y)$
Upstream	$x - y$	3	$3(x - y)$

◀ *Use $d = rt$.*

$$\begin{array}{l} 2(x + y) = 16 \\ 3(x - y) = 16 \end{array} \Rightarrow \begin{array}{l} 2x + 2y = 16 \\ 3x - 3y = 16 \end{array} \Rightarrow \begin{array}{l} x + y = 8 \\ 3x - 3y = 16 \end{array}$$

SOLVE ▶   Multiply the first equation by 3. Then add the equations.

$$\begin{array}{r} 3x + 3y = 24 \\ 3x - 3y = 16 \\ \hline 6x \quad\quad = 40 \\ x = 6\frac{2}{3} \end{array} \qquad \begin{array}{l} 2(x + y) = 16 \\ 2(6\frac{2}{3} + y) = 16 \\ 6\frac{2}{3} + y = 8 \\ y = 1\frac{1}{3} \end{array}$$

INTERPRET ▶   Check to see that distance downstream, $2(x + y)$, is equal to distance
upstream, $3(x - y)$.

Thus, the rate of the boat in still water is $6\frac{2}{3}$ km/h and the rate of the
current is $1\frac{1}{3}$ km/h.

Written Exercises

Solve each system of equations by substitution.

(A) **1.** $4x - 3y = 2$
$2x + y = -4$

2. $5x + 6y = 14$
$4x - y = 17$

3. $2a - b + 1 = 0$
$3b - 5a + 1 = 0$

4. $2r - 3s = 1.3$
$s - r = -0.5$

5. $2x - 4y = -6$
$-x + 2y = 3$

6. $2y - z = 8$
$-4y + 6z = -16$

7. $2 = x + 4y$
$8y = 7 + x$

8. $5x - 2y = 0$
$2x + y = 3$

Solve each system of equations by addition.

9. $2x + 2y = 3$
$-5x + 4y = 15$

10. $3x - 2y = 10$
$5x + 3y = -15$

11. $7c + 9d - 3 = 0$
$9c - 7d - 2 = 0$

12. $2y + 3x - 8.1 = 0$
$2x - 3y + 3.7 = 0$

13. $4t - 7u = -13$
$-3u - 5 = -7t$

14. $5x - 7y + 16 = 0$
$x + 4y - 13 = 0$

15. $9m = 21 - 7n$
$12n = 36 - m$

16. $3x - 5y + 10 = 0$
$-9x + 15y = -30$

Solve each problem by using a system of two linear equations in two variables.

(B) **17.** An airplane flew 112 km in 21 min with a
tailwind and returned the same distance in
24 min against the wind. What was the rate
of the tailwind?

18. A first number is 10 less than 4 times a
second number. The first number,
decreased by the second number, is -19.
Find the numbers.

19. Alicia is 4 years older than Bill. Two years
ago, she was $1\frac{1}{2}$ times as old as he was.
Find their present ages.

20. How many milliliters of a 25% iodine
solution must be added to 90 mL of a 65%
iodine solution to obtain a 55% iodine
solution?

AIR-SPACE EXPLORATION

Air and Space explorations, which extend our view from the earth's crust outward, require complex and accurate equipment to correlate and classify all the information gathered.

Deep Space Network Control Center. This is the control and data center for all deep-space probes, including Voyager and Pioneer. On the large screen is a computer-enhanced picture of Saturn.

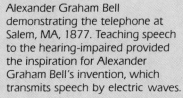

Alexander Graham Bell demonstrating the telephone at Salem, MA, 1877. Teaching speech to the hearing-impaired provided the inspiration for Alexander Graham Bell's invention, which transmits speech by electric waves.

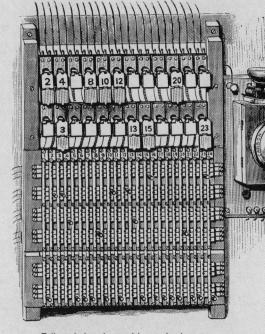

Edison's loud-speaking telephone. Thomas Edison improved Bell's telephone by adding the carbon transmitter, which markedly increased its audibility.

Canada's domestic communications satellite. Fixed-orbit satellites offer the possibility of reliable communications without the need for cumbersome equipment.

Parabolic dish antenna in stand. A parabolic antenna, by virtue of its shape, concentrates minute amounts of energy to provide data from farther out in space than do other types of telescopes.

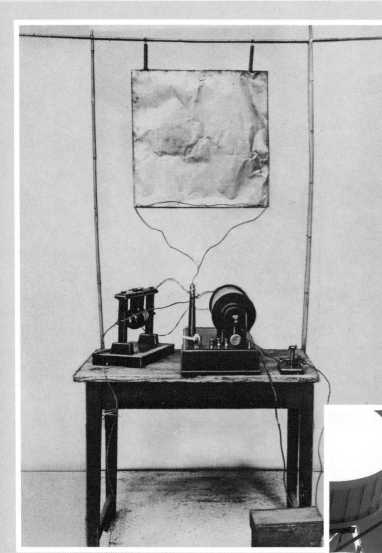

Replica of Marconi's first transmitter, Italy, 1895. Guglielmo Marconi established the first wireless communication, and by 1901 trans-Atlantic radio communications were in operation.

Microwave tower. Microwave transmissions, on a very tight beam capable of accurate focusing on a distant receiver, are simplifying earthbound communications systems.

Replica of Isaac Newton's reflecting telescope, 1671. Using mirrors rather than prisms, Newton's reflecting telescope eliminated the spectrum of colors that obscured images.

Astronomers in the Istanbul Observatory, late sixteenth century. Adherence to Ptolemy's geocentric view of the universe impeded much advancement in the science of astronomy, even when observed phenomena did not fit in with what were believed to be facts.

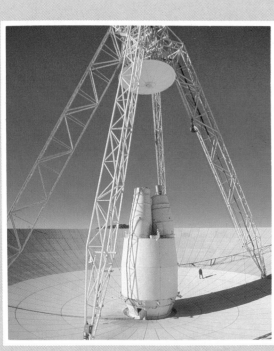

Radio telescope. Radio telescopes are used to pick up signals from and measure distances between other bodies within and beyond our solar system.

Ancient Chinese Observatory. Ancient Chinese astronomers were hindered in their exploration of the mechanics of solar, lunar, and planetary movement because they had not yet acquired a system of geometry.

Eighteenth-century celestial observatory, Jaipur, India. After Galileo's defense of the Copernican theory was verified, astronomical research proceeded with increasingly accurate instruments to plumb the depths of space.

Kitt's Peak, AZ. The National Observatory at Kitt's Peak houses the McMath Solar Tower telescope, the largest of its type; a 158-inch reflecting telescope; and various other telescopes.

Airplane Propeller. With its central hub and radiating blades forming a helical surface, the propeller produces thrust and forward motion when it is rotated.

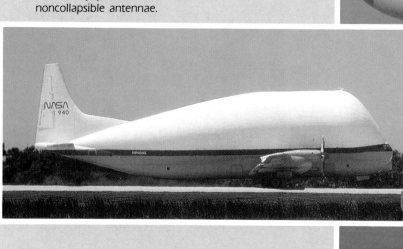

Guppy. An airplane designed to haul large, bulky objects such as satellite equipment and noncollapsible antennae.

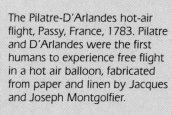

The Pilatre-D'Arlandes hot-air flight, Passy, France, 1783. Pilatre and D'Arlandes were the first humans to experience free flight in a hot air balloon, fabricated from paper and linen by Jacques and Joseph Montgolfier.

Hang glider. A source of inspiration to the Wright brothers, Otto Lilienthal was one of the earliest inventors of the glider; he made over 2,000 flights.

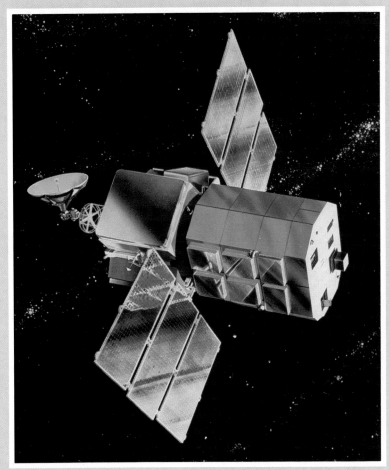

Blimp. By 1906, Fernand von Zeppelin had built a rigid airship capable of a speed of 30 miles per hour.

Solar Maximum Mission spacecraft, solar powered. Research into the collection of solar energy by satellite may soon make available tight beam transmissions to earth that can be converted to electrical energy.

Technicolor tornado. The atmospheric variations inherent in a tornado suggest the vast areas yet to be researched within our atmosphere as well as the artistic colorations that are possible in computer-enhanced photography.

SOLVING SYSTEMS OF THREE LINEAR EQUATIONS

Objective | **To solve a system of three linear equations**

You have learned that a system of two linear equations may have no solution, infinitely many solutions, or exactly one solution. A solution of an independent system of two linear equations is always an ordered pair of real numbers. In this lesson, you will learn to solve a system containing three linear equations. The equations in the system contain three variables, and so a solution of an independent system is an **ordered triple** of real numbers.

Example 1

Solve the system:
$$x + 2y - z = 1$$
$$2x + y + 3z = 5$$
$$3x + y + 2z = 8.$$

Choose *any two* equations and eliminate one of the variables.

$$\begin{array}{l} x + 2y - z = 1 \\ 2x + y + 3z = 5 \end{array} \Rightarrow \begin{array}{l} 3x + 6y - 3z = 3 \\ 2x + y + 3z = 5 \\ \hline 5x + 7y \quad = 8 \end{array}$$

Next, choose a *different* pair of equations and eliminate the same variable.

$$\begin{array}{l} x + 2y - z = 1 \\ 3x + y + 2z = 8 \end{array} \Rightarrow \begin{array}{l} 2x + 4y - 2z = 2 \\ 3x + y + 2z = 8 \\ \hline 5x + 5y \quad = 10 \end{array}$$

A system of 2 linear equations in x and y remains. Solve the system.

$$\begin{array}{l} 5x + 7y = 8 \\ 5x + 5y = 10 \end{array} \Rightarrow \begin{array}{l} 5x + 7y = 8 \\ -5x - 5y = -10 \\ \hline 2y = -2 \\ \boxed{y = -1} \end{array}$$

Substitute for y in either equation to find x.

$$5x + 5y = 10$$
$$5x + 5(-1) = 10$$
$$5x = 15$$
$$\boxed{x = 3}$$

Substitute for x and y in one of the original equations to find z.

$$x + 2y - z = 1$$
$$3 + 2(-1) - z = 1$$
$$1 - z = 1$$
$$-z = 0$$
$$\boxed{z = 0}$$

Check by substituting 3 for x, -1 for y, and 0 for z in the original equations.

$x + 2y - z$	1	$2x + y + 3z$	5	$3x + y + 2z$	8
$3 + 2(-1) - 0$	1	$2(3) + (-1) + 3(0)$	5	$3(3) + (-1) + 2(0)$	8
$3 - 2$		$6 - 1$		$9 - 1$	
1		5		8	

Thus, $(3, -1, 0)$ is the solution of the system.

Example 2   **Solve the system:** $x + 2y = -6$
$y + 2z = 11$
$2x + z = 16.$

Use the addition method with the second and third original equations. Eliminate z. ▶

$$\begin{array}{l} y + 2z = 11 \\ 2x + z = 16 \end{array} \Rightarrow \begin{array}{r} y + 2z = 11 \\ -4x \quad -2z = -32 \\ \hline -4x + y \quad\quad = -21 \end{array}$$

The equation $-4x + y = -21$ and the first original equation, $x + 2y = -6$, form a linear system in two variables. Eliminate x and solve for y. ▶

$$\begin{array}{l} x + 2y = -6 \\ -4x + y = -21 \end{array} \Rightarrow \begin{array}{r} 4x + 8y = -24 \\ -4x + y = -21 \\ \hline 9y = -45 \\ \boxed{y = -5} \end{array}$$

Substitute -5 for y in the first original equation. ▶

$$\begin{array}{r} x + 2y = -6 \\ x + 2(-5) = -6 \\ \hline \boxed{x = 4} \end{array}$$

Substitute 4 for x in the third original equation. ▶

$$\begin{array}{r} 2x + z = 16 \\ 2(4) + z = 16 \\ \hline \boxed{z = 8} \end{array}$$

Check in all three original equations.

Thus, $(4, -5, 8)$ is the solution of the system.

A system of linear equations in three variables may be inconsistent.

Example 3   **Solve the system:** $2x + 3y - 2z = 4$
$3x - 3y + 2z = 16$
$6x - 12y + 8z = 5.$

Use the addition method with the first and third original equations. Two variables, y and z, are eliminated. ▶

$$\begin{array}{l} 2x + 3y - 2z = 4 \\ 6x - 12y + 8z = 5 \end{array} \Rightarrow \begin{array}{r} 8x + 12y - 8z = 16 \\ 6x - 12y + 8z = 5 \\ \hline 14x \quad\quad = 21 \\ x = \frac{3}{2} \end{array}$$

Use the addition method with the first and second original equations. Two variables, y and z, are eliminated. ▶

$$\begin{array}{r} 2x + 3y - 2z = 4 \\ 3x - 3y + 2z = 16 \\ \hline 5x \quad\quad = 20 \\ x = 4 \end{array}$$

Two different values are obtained for x, which is a contradiction. The system is inconsistent.

Thus, there is no solution of the system.

Reading in Algebra

True or false? Give a reason for your answer.

1. A system of three linear equations in three variables has at least one solution.
2. A solution of a system of three linear equations in three variables is an ordered triple of real numbers.
3. A system of three linear equations in three variables may be independent, inconsistent, or dependent.
4. If an ordered triple of real numbers is a solution of one of the linear equations in a system, it is a solution of all equations in the system.

Written Exercises

Solve each system of equations.

(A)

1. $-x + 3y + z = -10$
   $3x + 2y - 2z = 3$
   $2x - y - 4z = -7$

2. $a + b - 3c = 8$
   $3a + 4b - 2c = 20$
   $2a - 3b + c = -6$

3. $m + n + p = 1$
   $m + 3n + 7p = 13$
   $m + 2n + 3p = 4$

4. $x + 2y + 3z = 9$
   $-3x + 5y - 4z = -7$
   $3x - y + 2z = -1$

5. $a + b + c = 2$
   $2a + b + 2c = 3$
   $3a - b + c = 4$

6. $2m + 2n + 6p = 9$
   $m - 3n + 2p = 5$
   $-m - 5n - 4p = 4$

(B)

7. $x - 2y = 14$
   $y + 2z = 11$
   $2x + z = 16$

8. $y + z = -3$
   $-x - 2z = 5$
   $3x + 2y = -5$

9. $2x + 3y = -5$
   $4y - 5z = -32$
   $3x + 2z = 14$

Solve each problem by using a system of three linear equations in three variables.

10. Find three numbers in decreasing order such that their sum is 3, the difference of the first two numbers is 4, and the sum of the smallest number and the greatest number is 2.

11. Find three positive numbers in increasing order such that the difference of the first two numbers is 2, the difference of the first and the last numbers is 4, and the quotient of the last two numbers is $1\frac{1}{5}$.

12. A coin box contains pennies, nickels, and dimes. The pennies and nickels are worth 35¢, the nickels and dimes are worth 80¢, and the value of the dimes and the pennies is 75¢. Find the number of each kind of coin in the box.

13. Find the number of nickels, dimes, and quarters in a collection of 80 such coins if the nickels and the quarters are worth $4.50 and the value of the quarters and the dimes is $5.50.

Solve each system of equations.

(C)

14. $\dfrac{m}{4} - \dfrac{3n}{2} + \dfrac{p}{2} = -6$
    $\dfrac{m}{6} - \dfrac{n}{4} - \dfrac{p}{3} = 1$
    $\dfrac{m}{3} + \dfrac{n}{2} - p = 7$

15. $\dfrac{1}{a} + \dfrac{1}{b} - \dfrac{2}{c} = 9$
    $\dfrac{3}{a} - \dfrac{2}{b} + \dfrac{1}{c} = -1$
    $\dfrac{2}{a} - \dfrac{1}{b} + \dfrac{3}{c} = 7$

16. $\dfrac{3}{x} - \dfrac{1}{y} + \dfrac{4}{z} = -3$
    $\dfrac{2}{x} + \dfrac{3}{y} - \dfrac{1}{z} = 6$
    $-\dfrac{1}{x} + \dfrac{2}{y} - \dfrac{3}{z} = 2$

Solving Systems of Three Linear Equations **249**

TWO BY TWO DETERMINANTS   10.4

Objectives   **To find the value of a 2 by 2 determinant**
To solve a system of two linear equations in two variables using determinants

A **matrix** is an array of numbers or other elements arranged in rows and columns. A matrix that has the same number of rows as columns is called a **square matrix.** Each square matrix has a corresponding real number assigned to it called a **determinant.**

The square matrix $\begin{bmatrix} 3 & 7 \\ 2 & 8 \end{bmatrix}$ contains four elements, which are arranged in 2 rows and 2 columns. It is called a 2×2 (read: two by two) matrix. The determinant of the matrix is written in the same form as the matrix but with vertical bars instead of brackets. The value of the 2 by 2 determinant $\begin{vmatrix} 3 & 7 \\ 2 & 8 \end{vmatrix}$ is found in the following way.

$$\begin{vmatrix} 3 & 7 \\ 2 & 8 \end{vmatrix} = 3 \cdot 8 - 2 \cdot 7 = 10$$

Definition: 2 by 2 determinant	The determinant of $\begin{bmatrix} a & b \\ c & d \end{bmatrix}$ is $\begin{vmatrix} a & b \\ c & d \end{vmatrix}$. Its value is $ad - cb$.

Example 1   **Find the value of each determinant.**

$$\begin{vmatrix} 2 & 1 \\ -3 & -5 \end{vmatrix} \qquad\qquad \begin{vmatrix} m+n & m-n \\ -3 & 5 \end{vmatrix}$$

$$\begin{vmatrix} 2 & 1 \\ -3 & -5 \end{vmatrix} = 2(-5) - (-3)(1)$$
$$= -10 + 3$$
$$= -7$$

Thus, the value is -7.

$$\begin{vmatrix} m+n & m-n \\ -3 & 5 \end{vmatrix} = (m+n)(5) - (-3)(m-n)$$
$$= 5m + 5n + 3m - 3n$$
$$= 8m + 2n$$

Thus, the value is $8m + 2n$.

Using determinants to solve a system of linear equations is known as **Cramer's rule,** named for the Swiss mathematician Gabriel Cramer (1704–1752). Cramer's rule is used to solve a system such as $5x + 2y = 4$
$$2x - 3y = 13.$$

$$\boxed{5}x + \bigcirc{2}y = \bigcirc{4}$$
$$\boxed{2}x + \bigcirc{-3}y = \bigcirc{13}$$

$$x = \frac{\begin{vmatrix} 4 & 2 \\ 13 & -3 \end{vmatrix}}{\begin{vmatrix} 5 & 2 \\ 2 & -3 \end{vmatrix}} = \frac{\begin{vmatrix} 4 & 2 \\ 13 & -3 \end{vmatrix}}{\begin{vmatrix} 5 & 2 \\ 2 & -3 \end{vmatrix}} = \frac{4(-3) - 13(2)}{5(-3) - 2(2)} = \frac{-12 - 26}{-15 - 4} = \frac{-38}{-19} = 2$$

$$y = \frac{\begin{vmatrix} 5 & 4 \\ 2 & 13 \end{vmatrix}}{\begin{vmatrix} 5 & 2 \\ 2 & -3 \end{vmatrix}} = \frac{\begin{vmatrix} 5 & 4 \\ 2 & 13 \end{vmatrix}}{\begin{vmatrix} 5 & 2 \\ 2 & -3 \end{vmatrix}} = \frac{5(13) - 2(4)}{5(-3) - 2(2)} = \frac{65 - 8}{-15 - 4} = \frac{57}{-19} = -3$$

So, $x = 2$ and $y = -3$. The solution of the system is $(2, -3)$.

The solutions of a system of linear equations of the form $a_1x + b_1y = c_1$
$$a_2x + b_2y = c_2,$$
where all coefficients are real numbers, are found as follows:

$$x = \frac{\begin{vmatrix} c_1 & b_1 \\ c_2 & b_2 \end{vmatrix}}{\begin{vmatrix} a_1 & b_1 \\ a_2 & b_2 \end{vmatrix}} = \frac{c_1b_2 - c_2b_1}{a_1b_2 - a_2b_1} \qquad y = \frac{\begin{vmatrix} a_1 & c_1 \\ a_2 & c_2 \end{vmatrix}}{\begin{vmatrix} a_1 & b_1 \\ a_2 & b_2 \end{vmatrix}} = \frac{a_1c_2 - a_2c_1}{a_1b_2 - a_2b_1} \qquad [a_1b_2 - a_2b_1 \neq 0]$$

Each solution is an ordered pair (x, y) of real numbers.

Example 2

Solve the system $2x + 3y = 3$ using determinants.
$$6x = y - 11$$

Write each equation in the general form.

$$\underset{\underset{a_1}{\uparrow} \quad \underset{b_1}{\uparrow} \quad \underset{c_1}{\uparrow}}{2x + 3y = 3} \qquad\qquad \underset{\underset{a_2}{\uparrow} \quad \underset{b_2}{\uparrow} \quad \underset{c_2}{\uparrow}}{6x + (-1)y = -11}$$

$$x = \frac{\begin{vmatrix} c_1 & b_1 \\ c_2 & b_2 \end{vmatrix}}{\begin{vmatrix} a_1 & b_1 \\ a_2 & b_2 \end{vmatrix}} \qquad\qquad y = \frac{\begin{vmatrix} a_1 & c_1 \\ a_2 & c_2 \end{vmatrix}}{\begin{vmatrix} a_1 & b_1 \\ a_2 & b_2 \end{vmatrix}}$$

$$= \frac{\begin{vmatrix} 3 & 3 \\ -11 & -1 \end{vmatrix}}{\begin{vmatrix} 2 & 3 \\ 6 & -1 \end{vmatrix}} \qquad\qquad = \frac{\begin{vmatrix} 2 & 3 \\ 6 & -11 \end{vmatrix}}{\begin{vmatrix} 2 & 3 \\ 6 & -1 \end{vmatrix}}$$

$$= \frac{3(-1) - (-11)(3)}{2(-1) - 6(3)} \qquad\qquad = \frac{2(-11) - 6(3)}{2(-1) - 6(3)}$$

$$= \frac{-3 + 33}{-2 - 18} \qquad\qquad = \frac{-22 - 18}{-2 - 18}$$

$$= \frac{30}{-20}, \text{ or } -\frac{3}{2} \qquad\qquad = \frac{-40}{-20}, \text{ or } 2$$

Check in the original equations.

Thus, $\left(-\dfrac{3}{2}, 2\right)$ is the solution of the system.

Two by Two Determinants

Oral Exercises

To solve the system $2x - 3y = -4$ using determinants, state the determinant that
$$-4x + 5y = 7$$
you would use for each of the following. Do not find the value.

1. the denominator for x **2.** the numerator for x

3. the numerator for y **4.** the denominator for y

Written Exercises

Find the value of each determinant.

(A)

1. $\begin{vmatrix} 2 & 3 \\ -2 & 5 \end{vmatrix}$ **2.** $\begin{vmatrix} 6 & -1 \\ 4 & 8 \end{vmatrix}$ **3.** $\begin{vmatrix} -3 & -2 \\ 5 & 3 \end{vmatrix}$ **4.** $\begin{vmatrix} -8 & 6 \\ 4 & -2 \end{vmatrix}$

5. $\begin{vmatrix} -10 & 3 \\ -4 & 5 \end{vmatrix}$ **6.** $\begin{vmatrix} 1 & 0 \\ 0 & 1 \end{vmatrix}$ **7.** $\begin{vmatrix} -3 & 5 \\ -9 & -2 \end{vmatrix}$ **8.** $\begin{vmatrix} -2 & 0 \\ 4 & 2 \end{vmatrix}$

9. $\begin{vmatrix} 3\sqrt{2} & 5 \\ -2 & 2\sqrt{2} \end{vmatrix}$ **10.** $\begin{vmatrix} 0 & 7 \\ 0 & 9 \end{vmatrix}$ **11.** $\begin{vmatrix} \frac{1}{4} & \frac{5}{6} \\ \frac{1}{2} & \frac{2}{3} \end{vmatrix}$ **12.** $\begin{vmatrix} \frac{1}{3} & \frac{1}{6} \\ -\frac{1}{5} & \frac{3}{10} \end{vmatrix}$

13. $\begin{vmatrix} m+n & 3 \\ m-n & 5 \end{vmatrix}$ **14.** $\begin{vmatrix} c-d & c+d \\ 3 & 4 \end{vmatrix}$ **15.** $\begin{vmatrix} -r & p+r \\ 6 & 10 \end{vmatrix}$

Solve each system of linear equations using determinants.

16. $2x + 3y = 7$
    $x + 2y = 3$

17. $5x - 12y = 4$
    $4x - 7y = -2$

18. $3x - 4y = 7$
    $4x + 6y = 15$

19. $2x + 3y = 4$
    $4x + y = -2$

20. $3x + 2y = 1$
    $3x - 2y = -5$

21. $-x + y = 1$
    $3x - 4y = -3$

22. $4x + y = 0$
    $6x - y = 5$

23. $2x + 3y = -2$
    $x + 5y = 3$

(B)

24. $2x - 3y = 2.5$
    $-x - 2y = 0.5$

25. $5x - y = 0.4$
    $x - y = 0.2$

26. $0.6x - 0.1y = -1$
    $-4x - 0.5y = 2$

Find the value of each determinant.

27. $\begin{vmatrix} p+r & -r \\ r & p-r \end{vmatrix}$ **28.** $\begin{vmatrix} m+n & m+n \\ m+n & m-n \end{vmatrix}$ **29.** $\begin{vmatrix} m^2 - n^2 & 1 \\ m+n & \frac{1}{m+n} \end{vmatrix}$

Solve each system of linear equations for (x, y) using determinants. Assume that all
variables represent real numbers.

(C)

30. $mx - 2y = 1$
    $nx + 3y = 4$

31. $x - my = r$
    $2x + ny = s$

32. $a_1 x + b_1 y = 1$
    $a_2 x + b_2 y = 1$

33. $a_1 x + b_1 y = c_1$
    $a_2 x + b_2 y = c_2$

THREE BY THREE DETERMINANTS 10.5

Objectives

To find the value of a 3 by 3 determinant
To solve a system of three linear equations in three variables using determinants and minors

The 3 by 3 square matrix $\begin{bmatrix} a_1 & b_1 & c_1 \\ a_2 & b_2 & c_2 \\ a_3 & b_3 & c_3 \end{bmatrix}$ contains 3 rows and 3 columns.

You can find the value of its corresponding determinant as shown below.

(1) Repeat columns 1 and 2 as columns 4 and 5.

(2) Multiply on the *down* diagonals.

(3) Multiply on the *up* diagonals.

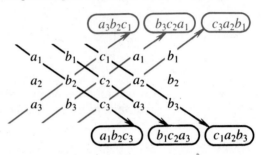

(4) Subtract the sum of the up diagonal products from the sum of the down diagonal products.

$$[a_1b_2c_3 + b_1c_2a_3 + c_1a_2b_3] -$$
$$[a_3b_2c_1 + b_3c_2a_1 + c_3a_2b_1]$$

Definition:
3 by 3
determinant

The determinant of $\begin{bmatrix} a_1 & b_1 & c_1 \\ a_2 & b_2 & c_2 \\ a_3 & b_3 & c_3 \end{bmatrix}$ is $\begin{vmatrix} a_1 & b_1 & c_1 \\ a_2 & b_2 & c_2 \\ a_3 & b_3 & c_3 \end{vmatrix}$. Its value is

$$[a_1b_2c_3 + b_1c_2a_3 + c_1a_2b_3] - [a_3b_2c_1 + b_3c_2a_1 + c_3a_2b_1].$$

Example 1

Find the value of $\begin{vmatrix} 2 & 3 & 5 \\ 4 & 2 & 1 \\ -1 & -3 & 2 \end{vmatrix}$.

Repeat columns 1 and 2 and follow the steps above. ▶

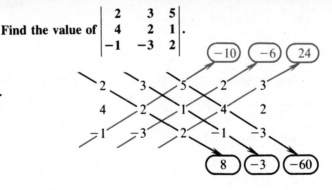

$$[a_1b_2c_3 + b_1c_2a_3 + c_1a_2b_3] - [a_3b_2c_1 + b_3c_2a_1 + c_3a_2b_1]$$
$$[2(2)(2) + 3(1)(-1) + 5(4)(-3)] - [-1(2)(5) + (-3)(1)(2) + 2(4)(3)]$$

$$[8 + (-3) + (-60)] - [-10 + (-6) + 24] = (-55) - (8) = -63$$

Thus, the value of the determinant is -63.

You can also use Cramer's rule to solve a system of three linear equations of the form

$$a_1x + b_1y + c_1z = d_1$$
$$a_2x + b_2y + c_2z = d_2$$
$$a_3x + b_3y + c_3z = d_3$$

All coefficients are real numbers, as shown below.

$$x = \frac{\begin{vmatrix} d_1 & b_1 & c_1 \\ d_2 & b_2 & c_2 \\ d_3 & b_3 & c_3 \end{vmatrix}}{D} \qquad y = \frac{\begin{vmatrix} a_1 & d_1 & c_1 \\ a_2 & d_2 & c_2 \\ a_3 & d_3 & c_3 \end{vmatrix}}{D} \qquad z = \frac{\begin{vmatrix} a_1 & b_1 & d_1 \\ a_2 & b_2 & d_2 \\ a_3 & b_3 & d_3 \end{vmatrix}}{D},$$

$$\text{where } D = \begin{vmatrix} a_1 & b_1 & c_1 \\ a_2 & b_2 & c_2 \\ a_3 & b_3 & c_3 \end{vmatrix}$$

Each solution is an ordered triple (x, y, z) of real numbers. The value of the determinant in the denominator must not be equal to 0.

Example 2

Solve the system $2x + 3y + 4z = 4$ **using determinants.**
$$5x + 7y + 8z = 9$$
$$3x - 2y - 6z = 7$$

$$x = \frac{\begin{vmatrix} 4 & 3 & 4 \\ 9 & 7 & 8 \\ 7 & -2 & -6 \end{vmatrix}}{D} \qquad y = \frac{\begin{vmatrix} 2 & 4 & 4 \\ 5 & 9 & 8 \\ 3 & 7 & -6 \end{vmatrix}}{D} \qquad z = \frac{\begin{vmatrix} 2 & 3 & 4 \\ 5 & 7 & 9 \\ 3 & -2 & 7 \end{vmatrix}}{D}$$

$$D = \begin{vmatrix} 2 & 3 & 4 \\ 5 & 7 & 8 \\ 3 & -2 & -6 \end{vmatrix} \begin{matrix} 2 & 3 \\ 5 & 7 \\ 3 & -2 \end{matrix} = [-84 + 72 + (-40)] - [84 + (-32) + (-90)]$$
$$= -14$$

$$x = \frac{[-168 + 168 + (-72)] - [196 + (-64) + (-162)]}{-14} = \frac{-42}{-14}, \text{ or } 3$$

$$y = \frac{[-108 + 96 + 140] - [108 + 112 + (-120)]}{-14} = \frac{28}{-14}, \text{ or } -2$$

$$z = \frac{[98 + 81 + (-40)] - [84 + (-36) + 105]}{-14} = \frac{-14}{-14}, \text{ or } 1$$

Check in the original equations.

Thus, $(3, -2, 1)$ **is the solution of the system.**

Minors can also be used to find the determinant of a matrix and for solving systems of equations. The **minor** of an element of a determinant is the determinant found by deleting the column and row in which the element lies.

The value of the determinant $\begin{vmatrix} a_1 & b_1 & c_1 \\ a_2 & b_2 & c_2 \\ a_3 & b_3 & c_3 \end{vmatrix}$ can be found by finding

the minors of all the elements of any row or column. Using Column 1, the expansion by minors is shown below.

$$D = +a_1 \begin{vmatrix} b_2 & c_2 \\ b_3 & c_3 \end{vmatrix} - a_2 \begin{vmatrix} b_1 & c_1 \\ b_3 & c_3 \end{vmatrix} + a_3 \begin{vmatrix} b_1 & c_1 \\ b_2 & c_2 \end{vmatrix}$$
$$= a_1[b_2c_3 - b_3c_2] - a_2[b_1c_3 - b_3c_1] + a_3[b_1c_2 - b_2c_1]$$
$$= a_1b_2c_3 - a_1b_3c_2 - a_2b_1c_3 + a_2b_3c_1 + a_3b_1c_2 - a_3b_2c_1$$

The product of an element and its minor is multiplied by -1 only if the sum of the number of the row and the number of the column is an odd integer. The value of the determinant is the same value shown in the definition of a 3 by 3 determinant.

Example 3 **Using minors, show that the value of** $\begin{vmatrix} 2 & 3 & 5 \\ 4 & 2 & 1 \\ -1 & -3 & 2 \end{vmatrix}$ **is -63.**

See Example 1. ▶ $+2 \begin{vmatrix} 2 & 1 \\ -3 & 2 \end{vmatrix} - 4 \begin{vmatrix} 3 & 5 \\ -3 & 2 \end{vmatrix} + (-1) \begin{vmatrix} 3 & 5 \\ 2 & 1 \end{vmatrix}$ ◀ *Use Column 1.*

$= 2[4 + 3] - 4[6 + 15] + (-1)[3 - 10] = 14 - 84 + 7 = -63$

Thus, the value of the determinant is -63.

Example 4 **Using minors, show that $(3, -2, 1)$ is the solution of the system** $2x + 3y + 4z = 4$
$5x + 7y + 8z = 9$
$3x - 2y - 6z = 7.$

See Example 2. ▶

$$x = \frac{4 \begin{vmatrix} 7 & 8 \\ -2 & -6 \end{vmatrix} - 9 \begin{vmatrix} 3 & 4 \\ -2 & -6 \end{vmatrix} + 7 \begin{vmatrix} 3 & 4 \\ 7 & 8 \end{vmatrix}}{D}$$

$$= \frac{4[-42 + 16] - 9[-18 + 8] + 7[24 - 28]}{D} = \frac{-42}{D}$$

$$y = \frac{+2 \begin{vmatrix} 9 & 8 \\ 7 & -6 \end{vmatrix} - 5 \begin{vmatrix} 4 & 4 \\ 7 & -6 \end{vmatrix} + 3 \begin{vmatrix} 4 & 4 \\ 9 & 8 \end{vmatrix}}{D}$$

$$= \frac{2[-54 - 56] - 5[-24 - 28] + 3[32 - 36]}{D} = \frac{28}{D}$$

$$z = \frac{2\begin{vmatrix} 7 & 9 \\ 2 & 7 \end{vmatrix} - 5\begin{vmatrix} 3 & 4 \\ -2 & 7 \end{vmatrix} + 3\begin{vmatrix} 3 & 4 \\ 7 & 9 \end{vmatrix}}{D}$$

$$= \frac{2[49 + 18] - 5[21 + 8] + 3[27 - 28]}{D} = \frac{-14}{D}$$

$$D = 2\begin{vmatrix} 7 & 8 \\ -2 & -6 \end{vmatrix} - 5\begin{vmatrix} 3 & 4 \\ -2 & -6 \end{vmatrix} + 3\begin{vmatrix} 3 & 4 \\ 7 & 8 \end{vmatrix}$$

$$= 2[-42 + 16] - 5[-18 + 8] + 3[24 - 28] = -14$$

$$x = \frac{-42}{-14}, \text{ or } 3; \quad y = \frac{28}{-14}, \text{ or } -2; \quad z = \frac{-14}{-14}, \text{ or } 1$$

Thus, $(3, -2, 1)$ is the solution.

Written Exercises

Find the value of each determinant.

(A) **1.** $\begin{vmatrix} 2 & -4 & -3 \\ 4 & -1 & -2 \\ -3 & 4 & -2 \end{vmatrix}$ **2.** $\begin{vmatrix} -3 & 2 & -5 \\ 4 & -1 & 2 \\ -1 & -2 & -3 \end{vmatrix}$ **3.** $\begin{vmatrix} 4 & -1 & -2 \\ -3 & 2 & -1 \\ 2 & -1 & 3 \end{vmatrix}$ **4.** $\begin{vmatrix} -2 & -3 & 2 \\ -1 & 2 & -2 \\ 3 & -2 & 1 \end{vmatrix}$

Using minors, find the value of each determinant.

5. $\begin{vmatrix} 2 & 3 & 5 \\ 4 & 2 & 1 \\ -1 & -3 & 2 \end{vmatrix}$ **6.** $\begin{vmatrix} -3 & 1 & 5 \\ -2 & 0 & 2 \\ 6 & 3 & 4 \end{vmatrix}$ **7.** $\begin{vmatrix} 1 & 3 & -2 \\ 1 & -4 & 5 \\ 1 & 2 & 3 \end{vmatrix}$ **8.** $\begin{vmatrix} 1 & -1 & 4 \\ 0 & 1 & -7 \\ 0 & 0 & 1 \end{vmatrix}$

Using determinants, solve each system of linear equations.

(B) **9.** $\begin{aligned} 2x - 3y + z &= 7 \\ 3x - 2y + 2z &= 9 \\ -2x - y + 2z &= 3 \end{aligned}$ **10.** $\begin{aligned} -x + 2y - 3z &= 11 \\ 2x + y + 2z &= -4 \\ 3x + 3y - z &= 5 \end{aligned}$ **11.** $\begin{aligned} 3x - 2y + 4z &= 11 \\ -2x + 4y + z &= -10 \\ 2x - y + 3z &= 7 \end{aligned}$

12. $\begin{aligned} -3x + 2y - 2z &= -10 \\ 2x + 3y + z &= -1 \\ -x - y + 2z &= -8 \end{aligned}$ **13.** $\begin{aligned} -3x + 4y + 5z &= 3 \\ x - y + 3z &= 8 \\ -2x + 3y - z &= -7 \end{aligned}$ **14.** $\begin{aligned} 4x - 3y + 2z &= 0 \\ -2x + y - 3z &= 4 \\ 5x - 6y + 2z &= 1 \end{aligned}$

Using minors, solve each system of linear equations.

15. $\begin{aligned} a + b + c &= -1 \\ 2a - b + c &= 19 \\ 3a - 2b - 4c &= 16 \end{aligned}$ **16.** $\begin{aligned} 3x - 4z &= 7 \\ 2y + 5z &= 2 \\ 6x + 5y &= 10 \end{aligned}$ **17.** $\begin{aligned} 5n + 10d &= 70 \\ 5n + 25q &= 270 \\ 10d + 25q &= 300 \end{aligned}$

(C) **18.** $\begin{aligned} 0.6x + 0.4y - 0.6z &= 5.8 \\ 0.8x - 1y - 0.4z &= 16 \\ 1.5x - 1.5y &= 19.5 \end{aligned}$ **19.** $\begin{aligned} \frac{2x}{5} + \frac{3y}{4} - \frac{2z}{3} &= -\frac{9}{10} \\ -\frac{x}{3} - \frac{2y}{5} + \frac{3z}{5} &= \frac{4}{3} \\ \frac{3x}{2} - \frac{3y}{5} - \frac{z}{5} &= -\frac{33}{10} \end{aligned}$ **20.** $\begin{aligned} a_1x + b_1y &= d_1 \\ b_2y + c_2z &= d_2 \\ a_3x + c_3z &= d_3 \end{aligned}$

MATRIX ADDITION

Objectives   **To find the sum of two matrices**
To determine the properties of matrix addition

Recall that a matrix (pl. matrices) is an array of real numbers or other elements enclosed by brackets.

$$\begin{bmatrix} a_{11} & a_{12} & a_{13} \\ a_{21} & a_{22} & a_{23} \\ a_{31} & a_{32} & a_{33} \end{bmatrix} \quad \begin{bmatrix} a_{11} & a_{12} \\ a_{21} & a_{22} \\ a_{31} & a_{32} \end{bmatrix} \quad \begin{bmatrix} a_{11} \\ a_{21} \\ a_{31} \end{bmatrix} \quad [a_{11} \quad a_{12} \quad a_{13}]$$

3 by 3 square matrix	3 by 2 matrix	3 by 1 matrix	1 by 3 matrix
3 rows, 3 columns	3 rows, 2 columns	3 rows, 1 column	1 row, 3 columns

In the matrices above, the first number of the subscript indicates the row and the second number of the subscript indicates the column; a_{32} means the number or element in the third row and second column position.

The number of rows and columns determines the **dimensions** of a matrix. The dimensions of the matrices above, in order, are 3 by 3, 3 by 2, 3 by 1, and 1 by 3.

It is possible for two matrices to be equal. The matrix $\begin{bmatrix} 1 & -2 \\ -1 & 4 \end{bmatrix}$ is equal to

the matrix $\begin{bmatrix} \frac{2}{2} & -\frac{4}{2} \\ -\frac{6}{6} & \frac{12}{3} \end{bmatrix}$ since they have the same dimensions, 2 by 2, and their

corresponding elements are equal.

Equality of matrices	Two matrices are equal if they have the same dimensions and the elements in corresponding positions are equal.

Example 1   **Determine which pairs of matrices are equal, if any.**

$$\begin{bmatrix} -2 & 1 & 6 \\ 3 & -1 & 2 \end{bmatrix} \quad \begin{bmatrix} -\frac{6}{3} & \frac{7}{7} & \frac{-12}{-2} \\ \frac{12}{4} & -\frac{8}{8} & \frac{4}{2} \end{bmatrix} \qquad\qquad [1 \quad -3 \quad 2] \quad \begin{bmatrix} 1 \\ -3 \\ 2 \end{bmatrix}$$

Dimensions of each matrix are 2 by 3.
Elements in corresponding positions are equal:

$-2 = -\frac{6}{3} \quad 1 = \frac{7}{7} \quad 6 = \frac{-12}{-2}$

$3 = \frac{12}{4} \quad -1 = -\frac{8}{8} \quad 2 = \frac{4}{2}$

Dimensions of one matrix are 1 by 3.
Dimensions of other matrix are 3 by 1.

Thus, the first pair are equal; the second pair are not.

You can find the sum of two matrices with the same dimensions by using the following property.

Sum of two matrices	The *sum* of two matrices, A and B, having the same dimensions is a matrix whose elements are the sums of the corresponding elements of A and B. If $A = \begin{bmatrix} a_{11} & a_{12} \\ a_{21} & a_{22} \end{bmatrix}$ and $B = \begin{bmatrix} b_{11} & b_{12} \\ b_{21} & b_{22} \end{bmatrix}$, then $$A + B = \begin{bmatrix} a_{11} + b_{11} & a_{12} + b_{12} \\ a_{21} + b_{21} & a_{22} + b_{22} \end{bmatrix}.$$

Example 2

If $A = \begin{bmatrix} -3 & -2 \\ 5 & 6 \end{bmatrix}$, $B = \begin{bmatrix} 4 & 8 \\ -7 & -3 \end{bmatrix}$, and $C = \begin{bmatrix} 4 \\ -7 \\ 1 \end{bmatrix}$, find $A + B$ and $A + C$, if possible.

A and B have the same dimensions. ▶ $A + B = \begin{bmatrix} -3 + 4 & -2 + 8 \\ 5 + (-7) & 6 + (-3) \end{bmatrix} = \begin{bmatrix} 1 & 6 \\ -2 & 3 \end{bmatrix}$

Since A and C do not have the same dimensions, they cannot be added.

Thus, $A + B = \begin{bmatrix} 1 & 6 \\ -2 & 3 \end{bmatrix}$, but $A + C$ is not defined.

Recall that the *additive identity* is 0 since $a + 0 = a$ for any real number a. Also, the *additive inverse* of any real number a is $-a$ since $a + (-a) = 0$. Similarly, identities and inverses exist for matrix addition.

Additive identity matrix	I is called the *additive identity* (or *zero*) *matrix* for any matrix A if A and I have the same dimensions and $A + I = A$.
Additive inverse matrix	$-A$ is called the *additive inverse matrix* of any matrix A if A and $-A$ have the same dimensions and $A + (-A) = I$.

Example 3

If $A = \begin{bmatrix} 2 & -3 \\ 5 & -7 \end{bmatrix}$, $I = \begin{bmatrix} 0 & 0 \\ 0 & 0 \end{bmatrix}$, and $-A = \begin{bmatrix} -2 & 3 \\ -5 & 7 \end{bmatrix}$, show that $A + I = A$ and $A + (-A) = I$.

A and I have the same dimensions. ▶ $A + I = \begin{bmatrix} 2 + 0 & -3 + 0 \\ 5 + 0 & -7 + 0 \end{bmatrix} = \begin{bmatrix} 2 & -3 \\ 5 & -7 \end{bmatrix} = A$

A and $-A$ have the same dimensions. ▶ $A + (-A) = \begin{bmatrix} 2 + -2 & -3 + 3 \\ 5 + -5 & -7 + 7 \end{bmatrix} = \begin{bmatrix} 0 & 0 \\ 0 & 0 \end{bmatrix} = I$

Thus, $A + I = A$ and $A + (-A) = I$.

Matrix addition is *associative* and *commutative*.

Associative and commutative property of addition

For any matrices A, B, and C having the same dimensions,
$(A + B) + C = A + (B + C)$ and $A + B = B + A$.

Example 4   If $A = \begin{bmatrix} -3 & 2 \\ -1 & 4 \end{bmatrix}$, $B = \begin{bmatrix} 5 & -3 \\ 3 & -6 \end{bmatrix}$, and $C = \begin{bmatrix} -1 & 8 \\ 4 & -2 \end{bmatrix}$, show that
$(A + B) + C = A + (B + C)$ and that $A + B = B + A$.

$$(A + B) + C = \left(\begin{bmatrix} -3 + 5 & 2 + -3 \\ -1 + 3 & 4 + -6 \end{bmatrix} \right) + \begin{bmatrix} -1 & 8 \\ 4 & -2 \end{bmatrix} = \begin{bmatrix} 2 + -1 & -1 + 8 \\ 2 + 4 & -2 + -2 \end{bmatrix} = \begin{bmatrix} 1 & 7 \\ 6 & -4 \end{bmatrix}$$

$$A + (B + C) = \begin{bmatrix} -3 & 2 \\ -1 & 4 \end{bmatrix} + \begin{bmatrix} 5 + -1 & -3 + 8 \\ 3 + 4 & -6 + -2 \end{bmatrix} = \begin{bmatrix} -3 + 4 & 2 + 5 \\ -1 + 7 & 4 + -8 \end{bmatrix} = \begin{bmatrix} 1 & 7 \\ 6 & -4 \end{bmatrix}$$

$$A + B = \begin{bmatrix} -3 + 5 & 2 + -3 \\ -1 + 3 & 4 + -6 \end{bmatrix} = \begin{bmatrix} 2 & -1 \\ 2 & -2 \end{bmatrix}$$

$$B + A = \begin{bmatrix} 5 + -3 & -3 + 2 \\ 3 + -1 & -6 + 4 \end{bmatrix} = \begin{bmatrix} 2 & -1 \\ 2 & -2 \end{bmatrix}$$

Written Exercises

Determine if the pairs of matrices are equal.

(A) **1.** $\begin{bmatrix} -1 & 0 \\ 3 & -2 \end{bmatrix}$, $\begin{bmatrix} -\frac{2}{2} & 5 - 5 \\ \frac{-6}{-2} & 2 - 4 \end{bmatrix}$ **2.** $\begin{bmatrix} -2 & 3 \\ 1 & -2 \\ -3 & 4 \end{bmatrix}$ $\begin{bmatrix} -\frac{4}{2} & \frac{6}{2} \\ \frac{3}{3} & -\frac{8}{4} \\ \frac{9}{-3} & 2^2 \end{bmatrix}$ **3.** $\begin{bmatrix} \frac{1}{2} \\ \frac{1}{3} \\ \frac{1}{4} \end{bmatrix}$ $\begin{bmatrix} -\frac{4}{8} & -\frac{3}{9} & -\frac{5}{20} \end{bmatrix}$,

Find each sum, if possible.

4. $\begin{bmatrix} -2 & -3 \\ 4 & -6 \\ -5 & 1 \end{bmatrix} + \begin{bmatrix} 4 & 6 \\ -1 & 7 \\ 8 & -2 \end{bmatrix}$

5. $\begin{bmatrix} -1 & 2 & 3 \\ -2 & 3 & -4 \\ 7 & -6 & 5 \end{bmatrix} + \begin{bmatrix} 8 & -2 & 7 \\ 6 & -5 & 3 \\ -1 & 0 & -8 \end{bmatrix}$

6. $\begin{bmatrix} 4 & -2 & 1 \\ 3 & -6 & -4 \end{bmatrix} + \begin{bmatrix} 3 & 8 \\ 1 & 7 \\ 5 & -2 \end{bmatrix}$

7. $\begin{bmatrix} 7 & -3 & 6 \\ -1 & 4 & -8 \end{bmatrix} + \begin{bmatrix} -7 & 3 & -6 \\ 1 & -4 & 8 \end{bmatrix}$

8. $\begin{bmatrix} 9 & 8 & -7 \end{bmatrix} + \begin{bmatrix} 6 \\ -3 \\ -2 \end{bmatrix}$

9. $\begin{bmatrix} 0 & -8 \\ 9 & -1 \end{bmatrix} + \begin{bmatrix} -9 & -7 \\ -6 & -5 \end{bmatrix}$

Let $A = \begin{bmatrix} 3 & 5 \\ -2 & -6 \end{bmatrix}$, $B = \begin{bmatrix} 8 & -3 \\ -7 & -6 \end{bmatrix}$, and $C = \begin{bmatrix} 1 & -2 \\ 7 & -9 \end{bmatrix}$.

10. Find $-A$, $-B$, and $-C$.

11. Show that $B + I = B$.

(B) **12.** Show that $B + C = C + B$.

13. Show that $(B + C) + A = B + (C + A)$.

14. Show that $C + A = A + C$.

15. Show that $(C + A) + B = C + (A + B)$.

16. Show that $(A + C) + B = A + (C + B)$.

Matrix Addition

MATRIX MULTIPLICATION

Objectives
To find the product of a scalar and a matrix
To find the product of two matrices
To determine the properties of matrix multiplication

When working with matrices, any real number is called a **scalar.** You will now learn how to find the product of a scalar and a matrix.

Product of scalar and matrix	The product of a scalar k and a matrix A is the matrix kA obtained by multiplying each element of A by k.

$$\text{If } A = \begin{bmatrix} a_{11} & a_{12} & a_{13} \\ a_{21} & a_{22} & a_{23} \\ a_{31} & a_{32} & a_{33} \end{bmatrix}, \text{ then } kA = \begin{bmatrix} ka_{11} & ka_{12} & ka_{13} \\ ka_{21} & ka_{22} & ka_{23} \\ ka_{31} & ka_{32} & ka_{33} \end{bmatrix}.$$

Example 1

Find kA if $k = -3$ and $A = \begin{bmatrix} -1 & 2 & -3 \\ 13 & -5 & 4 \end{bmatrix}$.

$$kA = \begin{bmatrix} -3(-1) & -3(2) & -3(-3) \\ -3(13) & -3(-5) & -3(4) \end{bmatrix} = \begin{bmatrix} 3 & -6 & 9 \\ -39 & 15 & -12 \end{bmatrix}$$

Thus, $kA = \begin{bmatrix} 3 & -6 & 9 \\ -39 & 15 & -12 \end{bmatrix}$.

Two matrices can be multiplied if the number of columns of one matrix equals the number of rows of the other.

Product of two matrices	The *product* of two matrices, A and B, where the number of columns of A equals the number of rows of B, is a matrix whose elements are obtained by multiplying the row elements of A by the column elements of B in the following manner.

$$\text{If } A = \begin{bmatrix} a_{11} & a_{12} & a_{13} \\ a_{21} & a_{22} & a_{23} \end{bmatrix} \text{ and } B = \begin{bmatrix} b_{11} & b_{12} \\ b_{21} & b_{22} \\ b_{31} & b_{32} \end{bmatrix}, \text{ then}$$

$$A \cdot B = \begin{bmatrix} a_{11}b_{11} + a_{12}b_{21} + a_{13}b_{31} & a_{11}b_{12} + a_{12}b_{22} + a_{13}b_{32} \\ a_{21}b_{11} + a_{22}b_{21} + a_{23}b_{31} & a_{21}b_{12} + a_{22}b_{22} + a_{23}b_{32} \end{bmatrix}.$$

In general, $A_{m \times n} \cdot B_{n \times r} = AB_{m \times r}$ where $m \times n$, $n \times r$, and $m \times r$ are the dimensions of the matrices.

Chapter Ten

Example 2   If $A = \begin{bmatrix} -1 & 2 & -6 \\ 3 & -1 & 4 \end{bmatrix}$ and $B = \begin{bmatrix} -6 & 4 \\ -2 & 3 \\ 1 & -4 \end{bmatrix}$, find AB, if possible.

$$AB = \begin{bmatrix} (-1)(-6) + 2(-2) + (-6)(1) & (-1)(4) + 2(3) + (-6)(-4) \\ 3(-6) + (-1)(-2) + 4(1) & 3(4) + (-1)(3) + 4(-4) \end{bmatrix} = \begin{bmatrix} -4 & 26 \\ -12 & -7 \end{bmatrix}$$

Thus, $AB = \begin{bmatrix} -4 & 26 \\ -12 & -7 \end{bmatrix}$.

Recall that the *multiplicative identity* is 1 since $a \cdot 1 = a$ for any real number a. Also, the *multiplicative inverse* of any nonzero real number a is $\frac{1}{a}$ since $a \cdot \frac{1}{a} = 1$. Similarly, identities and inverses exist for matrix multiplication.

Multiplicative identity matrix	I is called the *multiplicative identity matrix* for any square matrix A if $A \cdot I = A$.
Multiplicative inverse matrix	A^{-1} is called the *multiplicative inverse matrix* of any square matrix A if $A \cdot A^{-1} = I$.

For a 3 by 3 square matrix, the multiplicative identity matrix is $\begin{bmatrix} 1 & 0 & 0 \\ 0 & 1 & 0 \\ 0 & 0 & 1 \end{bmatrix}$.

Example 3   If $A = \begin{bmatrix} -1 & 2 & 3 \\ -2 & 4 & -1 \\ 5 & -3 & 2 \end{bmatrix}$ and $I = \begin{bmatrix} 1 & 0 & 0 \\ 0 & 1 & 0 \\ 0 & 0 & 1 \end{bmatrix}$, show that $AI = A$.

$$AI = \begin{bmatrix} -1(1) + 2(0) + 3(0) & -1(0) + 2(1) + 3(0) & -1(0) + 2(0) + 3(1) \\ -2(1) + 4(0) + (-1)(0) & -2(0) + 4(1) + (-1)(0) & -2(0) + 4(0) + (-1)(1) \\ 5(1) + (-3)(0) + 2(0) & 5(0) + (-3)(1) + 2(0) & 5(0) + (-3)(0) + 2(1) \end{bmatrix}$$

$$= \begin{bmatrix} -1 & 2 & 3 \\ -2 & 4 & -1 \\ 5 & -3 & 2 \end{bmatrix} = A$$

Thus, $AI = A$.

Example 4   If $A = \begin{bmatrix} 1 & 4 \\ 2 & 9 \end{bmatrix}$ and $B = \begin{bmatrix} 9 & -4 \\ -2 & 1 \end{bmatrix}$, show that B is the multiplicative inverse of A by showing that $A \cdot B = I$.

$$A \cdot B = \begin{bmatrix} 1(9) + 4(-2) & 1(-4) + 4(1) \\ 2(9) + 9(-2) & 2(-4) + 9(1) \end{bmatrix} = \begin{bmatrix} 1 & 0 \\ 0 & 1 \end{bmatrix} = I.$$

Thus, B is the multiplicative inverse of A, or $B = A^{-1}$.

Matrix Multiplication

The next example shows you how to find the multiplicative inverse matrix for a given square matrix.

Example 5   If $A = \begin{bmatrix} 3 & -2 \\ -2 & 1 \end{bmatrix}$, find A^{-1}.

Let $A^{-1} = \begin{bmatrix} a_1 & b_1 \\ a_2 & b_2 \end{bmatrix}$.

$\begin{bmatrix} 3 & -2 \\ -2 & 1 \end{bmatrix} \cdot \begin{bmatrix} a_1 & b_1 \\ a_2 & b_2 \end{bmatrix} = \begin{bmatrix} 1 & 0 \\ 0 & 1 \end{bmatrix}$ ◀ $A \cdot A^{-1} = I$

$\begin{bmatrix} 3a_1 - 2a_2 & 3b_1 - 2b_2 \\ -2a_1 + a_2 & -2b_1 + b_2 \end{bmatrix} = \begin{bmatrix} 1 & 0 \\ 0 & 1 \end{bmatrix}$

Since the two matrices are equal, the corresponding elements are equal.

$\begin{aligned} 3a_1 - 2a_2 &= 1 \\ -2a_1 + a_2 &= 0 \end{aligned}$ $\begin{aligned} 3b_1 - 2b_2 &= 0 \\ -2b_1 + b_2 &= 1 \end{aligned}$

Solve each system of equations.

$\begin{aligned} 3a_1 - 2a_2 &= 1 \\ -4a_1 + 2a_2 &= 0 \\ \hline -a_1 &= 1 \end{aligned}$ $\begin{aligned} 3b_1 - 2b_2 &= 0 \\ -4b_1 + 2b_2 &= 2 \\ \hline -b_1 &= 2 \end{aligned}$

$\boxed{a_1 = -1}$ $\boxed{b_1 = -2}$

$\begin{aligned} -2a_1 + a_2 &= 0 \\ -2(-1) + a_2 &= 0 \end{aligned}$ $\begin{aligned} 3b_1 - 2b_2 &= 0 \\ 3(-2) - 2b_2 &= 0 \end{aligned}$

$\boxed{a_2 = -2}$ $\boxed{b_2 = -3}$

Thus, $A^{-1} = \begin{bmatrix} -1 & -2 \\ -2 & -3 \end{bmatrix}$.

Matrix multiplication is *associative*.

Associative property of multiplication	For any matrices A, B, and C, if the products exist, $(A \cdot B) \cdot C = A \cdot (B \cdot C)$.

As shown in Example 6, matrix multiplication is *not commutative*.

Example 6   If $A = \begin{bmatrix} 2 & 1 \\ -1 & 3 \end{bmatrix}$ and $B = \begin{bmatrix} -1 & 5 \\ 4 & -2 \end{bmatrix}$, show that $A \cdot B \neq B \cdot A$.

$A \cdot B = \begin{bmatrix} 2(-1) + 1(4) & 2(5) + 1(-2) \\ -1(-1) + 3(4) & -1(5) + 3(-2) \end{bmatrix} = \begin{bmatrix} 2 & 8 \\ 13 & -11 \end{bmatrix}$

$B \cdot A = \begin{bmatrix} -1(2) + 5(-1) & -1(1) + 5(3) \\ 4(2) + -2(-1) & 4(1) + -2(3) \end{bmatrix} = \begin{bmatrix} -7 & 14 \\ 10 & -2 \end{bmatrix}$

Thus, $A \cdot B \neq B \cdot A$.

Written Exercises

Which of the following pairs of matrices can be multiplied? Why or why not?

(A) **1.** $[3 \quad 2 \quad -1], \begin{bmatrix} 2 \\ -1 \\ 5 \end{bmatrix}$

2. $\begin{bmatrix} 2 & -1 & 3 \\ 4 & 6 & -1 \end{bmatrix}, \begin{bmatrix} 1 & -3 \\ -5 & 4 \\ 6 & -2 \end{bmatrix}$

3. $\begin{bmatrix} 4 & 5 & -6 \\ 3 & -1 & 7 \end{bmatrix}, \begin{bmatrix} -1 & 4 & -6 \\ 5 & 9 & -2 \end{bmatrix}$

4. $\begin{bmatrix} -1 & -5 \\ 6 & -3 \\ -4 & 7 \end{bmatrix}, \begin{bmatrix} 3 & -1 & 5 \\ -7 & 6 & -7 \end{bmatrix}$

Find each product, if possible.

5. $-3 \begin{bmatrix} -2 & 3 & 5 \\ -6 & 10 & 6 \\ 8 & 12 & -4 \end{bmatrix}$

6. $9 \begin{bmatrix} -8 & -2 \\ 5 & -1 \\ 9 & 3 \end{bmatrix}$

7. $-1 \begin{bmatrix} 7 & 6 & 9 \\ 8 & -1 & -5 \end{bmatrix}$

8. $\begin{bmatrix} -2 & 1 \\ 4 & -3 \end{bmatrix} \cdot \begin{bmatrix} 4 & -5 \\ -1 & 2 \end{bmatrix}$

9. $\begin{bmatrix} -2 & 3 & 8 \\ 4 & -1 & -3 \end{bmatrix} \cdot \begin{bmatrix} 4 & 6 & 1 \\ -1 & 7 & -2 \\ 2 & -1 & 2 \end{bmatrix}$

10. $\begin{bmatrix} 1 & -3 & 2 \\ -1 & 2 & -4 \\ -2 & 5 & -3 \end{bmatrix} \cdot \begin{bmatrix} 1 & 4 & -1 \\ -2 & 5 & -4 \end{bmatrix}$

11. $\begin{bmatrix} 1 & -8 & -9 \\ 7 & 6 & -3 \end{bmatrix} \cdot \begin{bmatrix} -1 & 0 & 1 \\ 0 & 1 & -1 \\ -1 & 0 & 1 \end{bmatrix}$

Find the multiplicative identity and the multiplicative inverse for each matrix, if possible.

(B) **12.** $\begin{bmatrix} 6 & -7 \\ -4 & 5 \end{bmatrix}$

13. $\begin{bmatrix} -8 & 2 \\ -6 & -1 \end{bmatrix}$

14. $\begin{bmatrix} -1 & 2 & -3 \\ -2 & 4 & 2 \end{bmatrix}$

15. $\begin{bmatrix} 7 & -3 & 2 \\ -1 & -2 & 5 \\ 6 & -4 & 8 \end{bmatrix}$

$A = \begin{bmatrix} -3 & 1 \\ -2 & -4 \end{bmatrix}, B = \begin{bmatrix} 1 & -6 \\ 3 & -2 \end{bmatrix}$ and $C = \begin{bmatrix} 7 & -3 \\ 2 & 5 \end{bmatrix}$. **Prove that each of the following statements is true.**

16. $(A \cdot B) \cdot C = A \cdot (B \cdot C)$

17. $B \cdot C \neq C \cdot B$

18. $A \cdot B \neq B \cdot A$

19. $A \cdot (B + C) = A \cdot B + A \cdot C$

20. $A \cdot I = I \cdot A$

21. $B \cdot B^{-1} = I$

$A = \begin{bmatrix} a_{11} & a_{12} \\ a_{21} & a_{22} \end{bmatrix}, B = \begin{bmatrix} b_{11} & b_{12} \\ b_{21} & b_{22} \end{bmatrix}$, and $C = \begin{bmatrix} c_{11} & c_{12} \\ c_{21} & c_{22} \end{bmatrix}$. **All elements represent real numbers.**

Determine which are true. Justify your answer by performing the computations.

(C) **22.** $A \cdot B = B \cdot A$

23. $I \cdot A = A \cdot I = A$

24. $(A \cdot B)^{-1} = B^{-1} \cdot A^{-1}$

25. $A \cdot (B + C) = A \cdot B + A \cdot C$

26. $(A \cdot B)^2 = A^2 \cdot B^2$

27. $A \cdot A^{-1} = A^{-1} \cdot A = I$

CUMULATIVE REVIEW

1. Solve $\sqrt{x - 8} + \sqrt{x} = 4$.

2. Find the distance from $A(3, 5)$ to $B(-6, 4)$.

SOLVING LINEAR SYSTEMS USING MATRICES

Objectives

To solve a system of two linear equations in two variables using the inverse of a matrix

To solve a system of three linear equations in three variables using matrix transformations

The general system of two linear equations $\begin{array}{l} a_1x + b_1y = c_1 \\ a_2x + b_2y = c_2 \end{array}$ can be written in matrix form as

$$\begin{bmatrix} a_1 & b_1 \\ a_2 & b_2 \end{bmatrix} \begin{bmatrix} x \\ y \end{bmatrix} = \begin{bmatrix} c_1 \\ c_2 \end{bmatrix}.$$

This can be verified by multiplying the two matrices on the left side. Using the definition of matrix equality, you get

$$\begin{bmatrix} a_1x + b_1y \\ a_2x + b_2y \end{bmatrix} = \begin{bmatrix} c_1 \\ c_2 \end{bmatrix}, \text{ or } \begin{array}{l} a_1x + b_1y = c_1 \\ a_2x + b_2y = c_2. \end{array}$$

Let $A = \begin{bmatrix} a_1 & b_1 \\ a_2 & b_2 \end{bmatrix}$, $X = \begin{bmatrix} x \\ y \end{bmatrix}$, and $B = \begin{bmatrix} c_1 \\ c_2 \end{bmatrix}$, then $AX = B$.

$A^{-1}(AX) = A^{-1}B$ ◀ *Multiply both sides by A^{-1}.*

$(A^{-1}A)X = A^{-1}B$ ◀ $A^{-1}A = I$

$IX = A^{-1}B$ ◀ $IX = X$

$X = A^{-1}B$

Substituting, $\begin{bmatrix} x \\ y \end{bmatrix} = \begin{bmatrix} a_1 & b_1 \\ a_2 & b_2 \end{bmatrix}^{-1} \begin{bmatrix} c_1 \\ c_2 \end{bmatrix}$.

So, the solution can be found by multiplying the inverse of A by B.

Example 1

Using the inverse of a matrix, solve the system $\begin{array}{l} -2x + 3y = -8 \\ x + 2y = -3. \end{array}$

Write in matrix form.

$$\begin{bmatrix} -2 & 3 \\ 1 & 2 \end{bmatrix} \begin{bmatrix} x \\ y \end{bmatrix} = \begin{bmatrix} -8 \\ -3 \end{bmatrix}$$

Use $X = A^{-1}B$. ▶ $\begin{bmatrix} x \\ y \end{bmatrix} = \begin{bmatrix} -2 & 3 \\ 1 & 2 \end{bmatrix}^{-1} \begin{bmatrix} -8 \\ -3 \end{bmatrix}$

Let $A^{-1} = \begin{bmatrix} a_1 & b_1 \\ a_2 & b_2 \end{bmatrix}$. ▶ Find A^{-1}.

$$\begin{bmatrix} -2 & 3 \\ 1 & 2 \end{bmatrix} \begin{bmatrix} a_1 & b_1 \\ a_2 & b_2 \end{bmatrix} = \begin{bmatrix} 1 & 0 \\ 0 & 1 \end{bmatrix}$$

Multiply. ▶ $\begin{bmatrix} -2a_1 + 3a_2 & -2b_1 + 3b_2 \\ a_1 + 2a_2 & b_1 + 2b_2 \end{bmatrix} = \begin{bmatrix} 1 & 0 \\ 0 & 1 \end{bmatrix}$

$$\begin{array}{ll} -2a_1 + 3a_2 = 1 & -2b_1 + 3b_2 = 0 \\ a_1 + 2a_2 = 0 & b_1 + 2b_2 = 1 \end{array}$$

Solve each system of equations.

$$a_1 = \frac{\begin{vmatrix} 1 & 3 \\ 0 & 2 \end{vmatrix}}{D}, \; a_2 = \frac{\begin{vmatrix} -2 & 1 \\ 1 & 0 \end{vmatrix}}{D}, \; b_1 = \frac{\begin{vmatrix} 0 & 3 \\ 1 & 2 \end{vmatrix}}{D}, \; b_2 = \frac{\begin{vmatrix} -2 & 0 \\ 1 & 1 \end{vmatrix}}{D}$$

$$D = \begin{bmatrix} -2 & 3 \\ 1 & 2 \end{bmatrix} = [-4 - 3], \text{ or } -7$$

$$a_1 = \frac{2 - 0}{-7}, \text{ or } -\frac{2}{7} \qquad a_2 = \frac{0 - 1}{-7}, \text{ or } \frac{1}{7}$$

$$b_1 = \frac{0 - 3}{-7}, \text{ or } \frac{3}{7} \qquad b_2 = \frac{-2 - 0}{-7}, \text{ or } \frac{2}{7}$$

So, $A^{-1} = \begin{bmatrix} -\dfrac{2}{7} & \dfrac{3}{7} \\ \dfrac{1}{7} & \dfrac{2}{7} \end{bmatrix}.$

$X = A^{-1}B$ ▶ $\begin{bmatrix} x \\ y \end{bmatrix} = \begin{bmatrix} -\dfrac{2}{7} & \dfrac{3}{7} \\ \dfrac{1}{7} & \dfrac{2}{7} \end{bmatrix} \begin{bmatrix} -8 \\ -3 \end{bmatrix}$ or $\begin{bmatrix} \dfrac{16}{7} & -\dfrac{9}{7} \\ -\dfrac{8}{7} & -\dfrac{6}{7} \end{bmatrix}$ or $\begin{bmatrix} 1 \\ -2 \end{bmatrix}$

Thus, $(1, -2)$ is the solution of the system.
By applying a series of steps, called **transformations,** the coefficients and constants of a system of three linear equations can be changed until the system is in triangular form like the system of linear equations shown on the right.

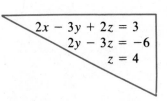

$$2x - 3y + 2z = 3$$
$$2y - 3z = -6$$
$$z = 4$$

The solution of a system is easily obtained from triangular form. $z = 4$; so, $2y - 3(4) = -6$, or $y = 3$; and $2x - 3 \cdot 3 + 2 \cdot 4 = 3$, or $x = 2$; and the solution is $(2, 3, 4)$.

The matrix form of the system of equations shown on the right is

$$2x - 3y - 3z = -1$$
$$-x + 2y + 4z = 5$$
$$3x - 4y - 5z = -3$$

$$\begin{bmatrix} 2 & -3 & -3 \\ -1 & 2 & 4 \\ 3 & -4 & -5 \end{bmatrix} \begin{bmatrix} x \\ y \\ z \end{bmatrix} = \begin{bmatrix} -1 \\ 5 \\ -3 \end{bmatrix}.$$

The matrix of coefficients is sometimes called the **matrix of detached coefficients.**

In working with matrix transformations, it is convenient to write the matrix of detached coefficients and the matrix of constants in one combined matrix, called an **augmented matrix.** The augmented matrix for

$$\begin{bmatrix} 2 & -3 & -3 \\ -1 & 2 & 4 \\ 3 & -4 & -5 \end{bmatrix} \begin{bmatrix} x \\ y \\ z \end{bmatrix} = \begin{bmatrix} -1 \\ 5 \\ -3 \end{bmatrix} \text{ is } \begin{bmatrix} 2 & -3 & -3 & -1 \\ -1 & 2 & 4 & 5 \\ 3 & -4 & -5 & -3 \end{bmatrix}.$$

There are three elementary transformations that can be used to change an augmented matrix into an equivalent augmented matrix.

You can solve a system of three linear equations by using elementary matrix transformations. Begin by writing an augmented matrix. Then, transform it to an equivalent augmented matrix in triangular form with ones in the main diagonal.

$$\begin{bmatrix} 1 & b_1 & c_1 & d_1 \\ 0 & 1 & c_2 & d_2 \\ 0 & 0 & 1 & d_3 \end{bmatrix}$$

Augmented matrix in triangular form

Example 2

Solve the system by matrix transformations:
$$5x - 2y + 3z = -7$$
$$-2x + 4y + z = 4$$
$$3x - 3y + 4z = 2.$$

Augmented matrix ▶ $\begin{bmatrix} 5 & -2 & 3 & -7 \\ -2 & 4 & 1 & 4 \\ 3 & -3 & 4 & 2 \end{bmatrix}$ $\begin{bmatrix} 1 & 6 & 5 & 1 \\ 0 & -5 & 0 & 5 \\ 0 & 21 & 11 & 1 \end{bmatrix}$ ◀ *Add row 3 to row 2.*

◀ *Multiply row 3 by -1.*

Add 2 times row 2 to row 1. ▶ $\begin{bmatrix} 1 & 6 & 5 & 1 \\ -2 & 4 & 1 & 4 \\ 3 & -3 & 4 & 2 \end{bmatrix}$ $\begin{bmatrix} 1 & 6 & 5 & 1 \\ 0 & 1 & 0 & -1 \\ 0 & 21 & 11 & 1 \end{bmatrix}$ ◀ *Multiply row 2 by $-\frac{1}{5}$.*

Add 2 times row 1 to row 2. ▶ $\begin{bmatrix} 1 & 6 & 5 & 1 \\ 0 & 16 & 11 & 6 \\ 3 & -3 & 4 & 2 \end{bmatrix}$ $\begin{bmatrix} 1 & 6 & 5 & 1 \\ 0 & 1 & 0 & -1 \\ 0 & 0 & 11 & 22 \end{bmatrix}$

Add -21 times row 2 to row 3. ◀

Add -3 times row 1 to row 3. ▶ $\begin{bmatrix} 1 & 6 & 5 & 1 \\ 0 & 16 & 11 & 6 \\ 0 & -21 & -11 & -1 \end{bmatrix}$ $\begin{bmatrix} 1 & 6 & 5 & 1 \\ 0 & 1 & 0 & -1 \\ 0 & 0 & 1 & 2 \end{bmatrix}$ ◀ *Multiply row 3 by $\frac{1}{11}$.*

Solve the system $1x + 6y + 5z = 1$ for x. $\qquad 1x + 6(-1) + 5(2) = 1$
$$\qquad\qquad\qquad 1y = -1 \qquad\qquad\qquad\qquad\qquad\qquad x = -3$$
$$\qquad\qquad\qquad 1z = 2$$
Thus, $(-3, -1, 2)$ is the solution of the system.

Whenever a matrix can be transformed into an augmented matrix of the form $\begin{bmatrix} 1 & 0 & 0 & d_1 \\ 0 & 1 & 0 & d_2 \\ 0 & 0 & 1 & d_3 \end{bmatrix}$, it is even easier to solve.

If you continue to transform the augmented matrix in Example 2 above, you get $\begin{bmatrix} 1 & 0 & 0 & -3 \\ 0 & 1 & 0 & -1 \\ 0 & 0 & 1 & 2 \end{bmatrix}$, which is equivalent to $\begin{bmatrix} 1 & 0 & 0 \\ 0 & 1 & 0 \\ 0 & 0 & 1 \end{bmatrix} \begin{bmatrix} x \\ y \\ z \end{bmatrix} = \begin{bmatrix} -3 \\ -1 \\ 2 \end{bmatrix}$.

So, $x = -3$, $y = -1$, and $z = 2$.

Written Exercises

Solve by using the inverse of a matrix.

Ⓐ
1. $2x - 3y = 5$
   $-x + 2y = -3$

2. $-2x + y = -6$
   $3x - 3y = 12$

3. $x - 4y = 3$
   $-3x + 5y = -2$

4. $4x - y = -6$
   $-2x - 3y = 10$

5. $-x + 3y = -6$
   $-2x + y = -7$

6. $3x - 2y = -1$
   $2x + 3y = 8$

Solve by using matrix transformations.

Ⓑ
7. $-x + 2y - 3z = -8$
   $2x + 3y + z = -3$
   $-2x - y + 2z = 2$

8. $3x - 2y + 2z = -1$
   $-2x + 3y - 3z = -1$
   $-x - y + 4z = 8$

9. $-2x + 3y - 2z = -1$
   $-3x + 2y - z = -5$
   $4x - 3y + 2z = 5$

10. $4x - 2y + z = 2$
    $-x - 3y + 2z = -5$
    $2x + y + 3z = -4$

11. $-3x + 4y + z = 5$
    $2x + 5y + 3z = 0$
    $-4x - 2y + 2z = 16$

12. $2x - 3y + z = 10$
    $-x + 2y + 3z = 1$
    $-3x + y - 2z = -9$

13. $-3x + 2y + 3z = 7$
    $2x + 3y - z = -7$
    $-x - y + z = 4$

14. $2x - 3y + 3z = -12$
    $3x - 2y - 2z = 4$
    $-3x + 2y - z = 5$

15. $-4x + 5y + 2z = 2$
    $3x + 2y + 3z = -4$
    $-3x - 5y + 2z = 20$

16. $6x - 3y + 4z = -4$
    $-2x + 5y + 3z = 2$
    $3x - 2y - 2z = -9$

17. $5x + 2y + 3z = 12$
    $3x - 5y + 2z = 13$
    $-2x + 3y + 4z = -19$

18. $-3x - 4y + 5z = 4$
    $2x + 5y + 3z = -1$
    $-4x - y + 2z = -8$

Ⓒ
19. $-2a + 3b - c + 2d = -11$
    $3a - 2b + 3c + d = 9$
    $-4a + 5b - 3c - 3d = -9$
    $4a - 3b + 4c + 4d = 7$

20. $5a + 2b + 3c - 2d = 4$
    $-3a + 3b - 2c + 4d = 9$
    $-2a - 3b + 4c - 2d = -14$
    $3a + 4b - 5c + 3d = 19$

21. $3r + 2s - 2t + 4v = 4$
    $-2r - 3s + 4t - 3v = 4$
    $4r + 2s - 3t + v = 1$
    $-3r - 4s + 3t + 5v = 9$

22. $-2r + 6s - 3t + 3v = -5$
    $4r - 3s + 2t - v = -3$
    $-3r - 4s - 3t + 4v = -15$
    $5r + 2s - 2t + 3v = -20$

23. If $A = \begin{bmatrix} a_1 & b_1 \\ a_2 & b_2 \end{bmatrix}$, then prove that $A^{-1} = \dfrac{1}{\text{Determinant } A} \begin{bmatrix} b_2 & -b_1 \\ -a_2 & a_1 \end{bmatrix}$.

Find A^{-1} by using the formula in Exercise 23.

24. $A = \begin{bmatrix} 3 & 5 \\ -2 & 4 \end{bmatrix}$

25. $A = \begin{bmatrix} -1 & -2 \\ 3 & -4 \end{bmatrix}$

26. $A = \begin{bmatrix} -1 & 4 \\ -2 & 3 \end{bmatrix}$

27. $A = \begin{bmatrix} c & -d \\ e & f \end{bmatrix}$

Solving Linear Systems Using Matrices

PROBLEM SOLVING USING LINEAR SYSTEMS

Objectives   **To solve a digit problem using a system of linear equations**
To solve various types of word problems using systems of linear equations

A two-digit number like 76 can be written as follows:
$$76 = 70 + 6$$
$$= 10(7) + 1(6),$$
where 7 is the *tens digit* and 6 is the *units digit*.

To *reverse the digits* of a two-digit number, interchange the positions of the tens and units digits as follows: $76 = 10(7) + 6$; its reverse, $67 = 10(6) + 7$.

The *sum of the digits* of a two-digit number like 76 is $7 + 6$, or 13.

Two-digit number	Tens digit	Units digit	Original number	Sum of the digits	Number with its digits reversed
	t	u	$10t + u$	$t + u$	$10u + t$

Systems of linear equations can often be used to solve problems involving two-digit numbers.

Example 1

READ ▶

In a certain two-digit number, the units digit is 24 less than 3 times the sum of the digits. If the digits are reversed, the new number is 18 more than the original number. Find the two-digit number.

PLAN ▶

Let t = tens digit and u = units digit.
The units digit is 24 less than 3 times the sum of the digits.

$$u = 3(t + u) - 24 \quad \text{or} \quad 2u + 3t = 24$$

When the digits are reversed, the new number is 18 more than the original.

$$10u + t = 10t + u + 18 \quad \text{or} \quad u - t = 2$$

SOLVE ▶

Solve the system $2u + 3t = 24$.
$\qquad\qquad\qquad\qquad\quad u - t = 2$

$$\begin{array}{r} 2u + 3t = 24 \\ \underline{3u - 3t = 6} \\ 5u = 30 \\ \boxed{u = 6} \end{array}$$

$$\begin{array}{r} u - t = 2 \\ 6 - t = 2 \\ \boxed{t = 4} \end{array}$$

INTERPRET ▶

The two-digit number is $10t + u = 10 \cdot 4 + 6$, or 46. (Check the answer in the problem.)

Thus, the two-digit number is 46.

A three-digit number like 658 can be written as follows:

$$658 = 600 + 50 + 8$$
$$= 100(6) + 10(5) + 1(8), \text{ where 6 is the hundreds digit,}$$

5 is the tens digit, and 8 is the units digit.

When the digits are reversed, you get

$$856 = 100(8) + 10(5) + 6.$$

The sum of the digits of the number is $6 + 5 + 8$, or 19.

Three-digit number	Hundreds digit	Tens digit	Units digit	Original number	Sum of the digits	Number with its digits reversed
	h	t	u	$100h + 10t + u$	$h + t + u$	$100u + 10t + h$

Example 2

In a certain three-digit number, if the sum of the digits is doubled, the result is 16 more than 3 times the tens digit. One more than twice the hundreds digit is the units digit. The hundreds digit is 11 decreased by twice the tens digit. Find the three-digit number.

Represent ▶
the data.

Let h = hundreds digit, t = tens digit, and u = units digit.

Write a ▶
system of
equations.

(1) $2(h + t + u) = 3t + 16$ (2) $2h + 1 = u$ (3) $h = 11 - 2t$
   $\quad 2h + 2t + 2u = 3t + 16$
   $\quad 2h - t + 2u = 16$ $2h - u = -1$ $h + 2t = 11$

$$2h - t + 2u = 16$$
$$2h - u = -1$$
$$h + 2t = 11$$

Solve the ▶
system.

Solve the system for h using determinants.

$$h = \frac{\begin{vmatrix} 16 & -1 & 2 \\ -1 & 0 & -1 \\ 11 & 2 & 0 \end{vmatrix} \begin{matrix} 16 & -1 \\ -1 & 0 \\ 11 & 2 \end{matrix}}{\begin{vmatrix} 2 & -1 & 2 \\ 2 & 0 & -1 \\ 1 & 2 & 0 \end{vmatrix} \begin{matrix} 2 & -1 \\ 2 & 0 \\ 1 & 2 \end{matrix}} = \frac{[0 + 11 + (-4)] - [0 + (-32) + 0]}{[0 + 1 + 8] - [0 + (-4) + 0]} = \frac{39}{13}, \text{ or } 3$$

$$2h + 1 = u \qquad\qquad h + 2t = 11$$
$$2 \cdot 3 + 1 = u \qquad\qquad 3 + 2t = 11$$
$$\boxed{h = 3} \qquad \boxed{7 = u} \qquad\qquad \boxed{t = 4}$$

Check the ▶
answer.

The three-digit number is $100h + 10t + u = 100 \cdot 3 + 10 \cdot 4 + 7$, or 347. (Check the answer in the problem.)

Thus, the three-digit number is 347.

In Chapter 2, you learned how to solve various types of mixture problems by writing and then solving one equation in one variable. Many of these same problems can be solved by using a system of two equations in two variables.

Example 3   **Some dried fruit worth $3.45/kg is to be mixed with some raisins worth $2.10/kg to make a mixture worth $2.64/kg. How many kilograms of dried fruit and how many kilograms of raisins should be used to make 10.5 kg of the mixture?**

Represent ▶
the data.

Represent the data in a table. Use two variables.

	No. of kg	Value/kg	Value in ¢
Dried fruit	x	345	$345x$
Raisins	y	210	$210y$
Mixture	10.5	264	$264(10.5)$

Write a ▶
system of
equations.

$$\text{(kg of fruit)} + \text{(kg of raisins)} = \text{(kg of mixture)}$$
$$x \qquad + \qquad y \qquad = \qquad 10.5$$

$$\text{(Value of fruit)} + \text{(Value of raisins)} = \text{(Value of mixture)}$$
$$345x \qquad + \qquad 210y \qquad = \qquad 264(10.5)$$

Solve the ▶
system.

Solve the system $x + y = 10.5$ by substitution.
$$345x + 210y = 2772$$

$x + y = 10.5$ $345x + 210y = 2772$ $y = 10.5 - x$
$y = 10.5 - x$ $345x + 210(10.5 - x) = 2772$ $y = 10.5 - 4.2$
                     $345x + 2205 - 210x = 2772$ $\boxed{y = 6.3}$
                          $135x = 567$
                        $\boxed{x = 4.2}$

Check the ▶
answer.

(4.2 kg at $3.45/kg) + (6.3 kg at $2.10/kg)	(10.5 kg at $2.64/kg)
$14.49 + $13.23	$27.72
$27.72	

Thus, 4.2 kg of dried fruit and 6.3 kg of raisins should be used.

In the exercises on page 271, you will be asked to solve a variety of word problems using a system of two (three) linear equations in two (three) variables.

Written Exercises

Use a system of two (three) linear equations in two (three) variables to solve each problem.

(A) 1. The tens digit of a two-digit number is 3 more than twice the units digit. If the digits are reversed, the resulting number is 45 less than the original number. Find the original number.

2. A two-digit number is 9 less than 4 times the sum of its digits. If the digits are reversed, the new number is 54 more than the original number. Find the original number.

3. A two-digit number is 27 more than the number obtained by reversing the digits. The number is also 38 more than twice the sum of the digits. Find the two-digit number.

4. The sum of the digits of a two-digit number is 8. If the digits are reversed, the new number is 18 more than the original number. Find the two-digit number.

5. Brand A coffee worth $5.60/kg is to be mixed with Brand B coffee worth $6.10/kg to make 15 kg of a mixture worth $5.90/kg. How many kilograms of each brand should be mixed?

6. Powdered milk worth $3.60/kg and cocoa worth $5.20/kg are to be mixed so that their combination is worth $82.40. How many kilograms of milk should be used if the weight of the milk is 10 times that of the cocoa?

(B) 7. The tens digit of a three-digit number is 7 less than 3 times the units digit. Twice the hundreds digit is 10 less than 3 times the tens digit. Three times the sum of the three digits is 4 more than 8 times the hundreds digit. Find the three-digit number.

8. A three-digit number is 198 less than the number obtained by reversing the digits. Twice the sum of the digits is 5 more than 7 times the tens digit. The tens digit is 5 less than twice the hundreds digit. Find the original number.

9. Ralph is 8 years younger than Lisa but 6 years ago, she was three times as old as he was. How old is each now?

10. A certain boat can travel 24 km downstream in 48 minutes and 21 km upstream in 72 minutes. Find the rate of the boat in still water and the rate of the current in kilometers per hour.

11. Mrs. Brown invested a total of $6,400, one part at 9% per year and the rest at 8% per year. At the end of 2 years 6 months, she had earned a total of $1,380 in simple interest. How much did she invest at each rate?

12. Sunflower seeds worth $2.90/kg, peanuts worth $2.50/kg, and almonds worth $3.30/kg are to be mixed to make 40 kg worth $3.00/kg. If the combined weight of the peanuts and the almonds is triple that of the seeds, how many kilograms of each should be used?

(C) 13. A chemist mixed iodine solutions of 25%, 40%, and $37\frac{1}{2}$% to obtain 15 L of a $33\frac{1}{3}$% iodine solution. Find the volume of each solution if the volume of the 25% solution was $1\frac{1}{2}$ times the volume of the $37\frac{1}{2}$% solution.

CHAPTER TEN REVIEW

Vocabulary
determinant [10.4]
  2 by 2 [10.4]   3 by 3 [10.5]
matrix [10.4]
  additive identity [10.6]
  additive inverse [10.6]
  augmented [10.8]
  dimensions of [10.6]
  multiplicative identity [10.7]
  multiplicative inverse [10.7]
  square [10.4]
ordered triple [10.3]   scalar [10.7]
system of equations [10.1]
  consistent [10.1]   dependent [10.1]
  inconsistent [10.1] independent [10.1]

Solve each system of equations graphically. Classify each system as consistent, inconsistent, dependent, or independent. [10.1]

1. $3x + 2y - 14 = 0$
    $2y - 3x + 10 = 0$

2. $5y - 2x = 15$
    $2x - 5y = 15$

3. $3(x + 2y) = 2y + 8$
    $4y = 8 - 3x$

4. $-2(y - 3x) = (y + 5x) + 9$
    $2(3x - 2y) = -(5x - 7y) + 11$

Solve each system of equations by substitution. [10.2]

5. $5x + 2y = 7$
   $4x - y = 3$

6. $4x + 5y = 12$
   $6x - y = 18$

Solve each system of equations by addition. [10.2]

7. $4x - 3y - 10 = 0$
   $5x + 6y + 7 = 0$

8. $5x + 4y + 8 = 0$
   $7y = 3x + 33$

9. $4x - 6y = 11$
   $-2x + 3y = 9$

★ **10.** $c_1x + d_1y = e_1$
    $c_2x + d_2y = e_2$

Solve the system of equations. [10.3]

11. $x + 3y + 2z = 11$
    $4x + 2y + 5z = 15$
    $-2x + 3y - z = 5$

Solve by using Cramer's rule. [10.4]

12. $-3x + 4y = -6$
    $5x - 3z = -22$
    $3y + 2z = -1$

Find the value of each determinant. [10.4]

13. $\begin{vmatrix} 3 & -2 \\ 4 & 8 \end{vmatrix}$

14. $\begin{vmatrix} a - b & -b \\ & -3 & 2 \end{vmatrix}$

15. $\begin{vmatrix} m + n & m \\ m & m - n \end{vmatrix}$

16. $\begin{vmatrix} a - b & a + b \\ a - b & a - b \end{vmatrix}$

Using minors, find the value of each determinant. [10.5]

17. $\begin{vmatrix} -2 & 1 & 2 \\ 3 & -1 & 4 \\ 5 & 2 & 3 \end{vmatrix}$

18. $\begin{vmatrix} a & 0 & b \\ 0 & 1 & 0 \\ c & 0 & d \end{vmatrix}$

Solve each system of equations by using determinants. [10.4], [10.5]

19. $-2x + 5y = 19$
    $4x - 7y = -29$

20. $3x + 2y - 5z = 5$
    $2x - y + z = 6$
    $x + 4y - 6z = -4$

Solve by using the inverse of a matrix. [10.8]

21. $2x - y = 0.7$
    $3x - 4y = -0.2$

22. Find the sum, if possible: [10.6]
$$\begin{bmatrix} -1 & 2 & 3 \\ 2 & 3 & -1 \\ 4 & -2 & 5 \end{bmatrix} + \begin{bmatrix} 6 & -1 & 2 \\ -3 & 4 & 5 \\ 2 & -1 & -3 \end{bmatrix}$$

23. Find the product, if possible: [10.7]
$$\begin{bmatrix} -2 & 3 & 1 \\ 4 & -1 & 2 \end{bmatrix} \cdot \begin{bmatrix} 1 & -1 & 2 \\ -3 & 4 & 1 \\ 2 & -2 & 3 \end{bmatrix}$$

Find the multiplicative identity, additive inverse, and the multiplicative inverse for each, if possible. [10.6], [10.7]

24. $\begin{bmatrix} 7 & -3 \\ 2 & 1 \end{bmatrix}$

25. $\begin{bmatrix} 3 & -1 & 2 \\ 2 & 1 & 2 \\ -3 & 2 & -1 \end{bmatrix}$

Solve by using matrix transformations. [10.8]

26. $3x - 2y + 2z = 10$
    $-2x + y - 3z = -8$
    $x + 3y - 2z = -3$

Use a system of two equations in two variables to solve the problem. [10.9]

27. A two-digit number is 45 less than the number obtained by reversing the digits. Find the original number if twice the sum of the digits is 1 more than 7 times the tens digit.

Solve each system graphically. Classify each system as consistent, inconsistent, dependent, or independent.

1. $4y - 3x - 8 = 0$
   $x + 2y + 6 = 0$

2. $4y - 5x = 8$
   $5x - 4y = 12$

3. Solve the system by substitution.
   $4x - 3y = 3$
   $x - 2y = -8$

4. Solve the system by addition.
   $2x + 3y = 10$
   $3x + 5y = 19$

Solve each system of equations.

5. $2x - 3y + 5z = 13$
   $-3x + 3y - 2z = -8$
   $5x - 2y + 4z = 2$

6. $2x + 5y = -2$
   $5x - 2z = 14$
   $3y + 4z = 6$

Find the value of each determinant.

7. $\begin{vmatrix} -3 & 4 \\ 6 & -2 \end{vmatrix}$

8. $\begin{vmatrix} a + b & 2a + b \\ b & a + b \end{vmatrix}$

Using minors, find the value of the determinant.

9. $\begin{vmatrix} 3 & -1 & 2 \\ -2 & 3 & 4 \\ 2 & 1 & -2 \end{vmatrix}$

Solve the system of equations by using determinants.

10. $3x + 5y = 1$
    $-2x + 3y = 12$

Solve by using Cramer's rule.

11. $3x + 2y + 2z = -3$
    $2x + 3y + 3z = -2$
    $-3x - 5y + z = -9$

12. Find the sum, if possible:
    $$\begin{bmatrix} 3 & -1 & 2 \\ 4 & 5 & 3 \\ -2 & -6 & -1 \end{bmatrix} + \begin{bmatrix} -2 & 3 & 4 \\ -3 & -2 & 1 \\ 4 & 6 & -5 \end{bmatrix}$$

13. Find the product, if possible:
    $$\begin{bmatrix} 2 & 1 & 3 \\ -3 & 2 & -4 \end{bmatrix} \cdot \begin{bmatrix} -1 & 4 & 6 \\ 2 & -1 & -3 \\ 3 & -2 & 1 \end{bmatrix}$$

Find the additive inverse and the multiplicative identity for each, if possible.

14. $\begin{bmatrix} 3 & 2 \\ -1 & 2 \end{bmatrix}$

15. $\begin{bmatrix} 4 & -1 & 6 \\ -3 & 2 & 1 \\ 2 & -4 & -3 \end{bmatrix}$

Solve by using the inverse of a matrix.

16. $2x - 3y = 5$
    $-3x + y = -4$

17. Solve by using matrix transformations.
    $4x - 3y + z = -9$
    $-3x + 2y - 5z = -10$
    $2x - 4y + 3z = 8$

Use a system of two linear equations in two variables to solve each problem.

18. If 3 more than a first number is divided by a second number and if 5 less than the second number is divided by 1 more than the first number, both quotients are $\frac{3}{4}$. Find the numbers.

19. A certain two-digit number is 27 less than the number obtained by reversing the digits. Three times the sum of the digits is 3 more than 7 times the tens digit. Find the two-digit number.

★20. Solve this system using matrix transformations:
   $3a + 4b - 2c + 5d = -5$
   $-2a - 3b + c + 2d = 14$
   $4a + 2b + 3c - 3d = -8$
   $-5a - 2b + 4c + 3d = 23$

COLLEGE PREP TEST

DIRECTIONS: Choose the *one* best answer to each question or problem.

1. Which ordered pair (x, y) is a solution of
   $$ax + by = 7ab?$$
   $$3ax - 2by = 6ab$$

   (A) $(3b, 4a)$ **(B)** $(4b, 3a)$
   (C) $(2b, 0)$ **(D)** all of these
   (E) None of these

2. The equation of
   the graph shown is

   (A) $y = -\frac{1}{2}x + \frac{1}{2}$
   (B) $y = -\frac{2}{3}x + \frac{1}{2}$
   (C) $y = \frac{1}{2}x - 1$
   (D) $y = -\frac{3}{2}x + \frac{1}{2}$
   (E) $y = \frac{1}{2}x - \frac{3}{2}$

   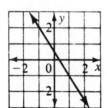

3. $\dfrac{\begin{vmatrix} a & b \\ c & d \end{vmatrix}}{\begin{vmatrix} 2b & -a \\ 2d & -c \end{vmatrix}} =$

   (A) $-\frac{1}{2}$ **(B)** $-\frac{1}{4}$ **(C)** $\frac{1}{2}$
   (D) $\frac{1}{4}$ **(E)** None of these

4. If the value of $\begin{vmatrix} 2x & -4 \\ x & x \end{vmatrix}$ is 6, then x
   equals

   (A) 1   **(B)** -1 **(C)** -1 or 3
   (D) -3 **(E)** 1 or -3

5. The solution of the system
   $$\tfrac{x}{2} + y + z = 6$$
   $$x + \tfrac{y}{3} - z = 6$$
   $$x + y + 2z = 5 \text{ is}$$

   (A) $(4, 1, 1)$ **(B)** $(6, 3, 1)$ **(C)** $(4, 3, -1)$
   (D) $(1, 1, 2)$ **(E)** None of these

6. In a collection of pennies, nickels, and
   dimes, the pennies and nickels are worth
   35¢, the nickels and dimes are worth 80¢,
   and the pennies and dimes are worth 75¢.
   What is the value of the collection?

   (A) $1.90   **(B)** $1.55   **(C)** $1.15
   (D) 95¢   **(E)** None of these

7. Which of the following matrices are
   equivalent?

   $$R = \begin{bmatrix} 1 & 5 & 6 \\ 2 & 3 & -1 \end{bmatrix} \quad S = \begin{bmatrix} 2 & 3 & -1 \\ 1 & 5 & 6 \end{bmatrix}$$

   $$T = \begin{bmatrix} 3 & 8 & -5 \\ 2 & 3 & -1 \end{bmatrix} \quad U = \begin{bmatrix} 2 & 6 & 7 \\ 2 & 3 & -1 \end{bmatrix}$$

   (A) R and S **(B)** R and T
   (C) T and U **(D)** R, S, and T
   (E) R, S, T, and U

8. Consider square matrices A, B, C, and I,
   each having the same dimensions and with
   I the multiplicative identity element. Which
   of the following statements is false?

   (A) $A + B = B + A$.
   (B) $AB = BA$
   (C) $AI = IA = A$
   (D) $A + (B + C) = (A + B) + C$
   (E) $(A \cdot B) \cdot C = A \cdot (B \cdot C)$

11 FUNCTIONS

The primary responsibility of a pharmacist is to dispense prescribed medicines. Although the pharmacist may be called on to mix ingredients to obtain a certain form of a given medicine, most medicines prescribed today are produced in consumable form by the drug manufacturers. When situations require the pharmacist to mix ingredients, he or she must possess a clear understanding of the mathematics that is needed to obtain the precise results.

All states require pharmacists to obtain a license to practice. Five years of study and a degree from an accredited college or university are usually the minimum requirements. Most pharmacists study sciences, such as chemistry, biology, and social sciences, as well as mathematics.

Project

A pharmacist has to measure exactly 22 oz of water. How can she do it if only 5-oz beakers and 8-oz beakers are available for measuring liquids?

5 OZ

8 OZ

RELATIONS AND FUNCTIONS

Objectives   **To determine the domain and the range of a relation**
To determine whether a relation is a function

The points shown on the coordinate plane at the right are associated with the following set of ordered pairs of real numbers:

$$A = \{(-1, -2), (-1, 1), (2, -1), (3, 2)\}$$

A set of ordered pairs is called a **relation**. The set of all first coordinates of the ordered pairs is called the **domain** of the relation, and the set of all second coordinates is called the **range** of the relation.

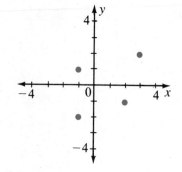

The domain of relation A above is $\{-1, 2, 3\}$ and the range is $\{-2, 1, -1, 2\}$. Notice that even though -1 is the first coordinate of two different points, it is listed only once in the domain.

Definition: Relation	A *relation* is a set of ordered pairs of real numbers.
Domain	The *domain* of a relation is the set of all first coordinates of the ordered pairs.
Range	The *range* of a relation is the set of all second coordinates of the ordered pairs.

Example 1   **Write the relation G whose graph is given. Determine the domain and the range of G.**

$$G = \{(-3, -2), (-1, 3), (3, 2), (4, 1)\}$$

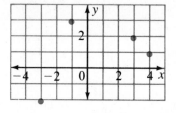

The domain of G is $\{-3, -1, 3, 4\}$.
The range of G is $\{-2, 3, 2, 1\}$.

For relation G in Example 1, notice that each ordered pair has a different first coordinate. However, this is not true for all relations. When it is true, the relation is called a **function**.

Definition: Function	A *function* is a relation in which different ordered pairs have different first coordinates

Example 2

Determine which of the following relations is a function.

$A = \{(-1, 2), (2, 2), (3, -1)\}$ $B = \{(2, 3), (3, 4), (2, 6)\}$

All ordered pairs have
different first coordinates.

Two ordered pairs have a first
coordinate of 2.

Thus, relation A is a function.

Thus, relation B is not a function.

A given relation may contain infinitely many ordered pairs (x, y) of real
numbers. A graph of one such relation is shown in Example 3.

Example 3

**Determine the domain and the range of the
relation C whose graph is given. Is the
relation a function?**

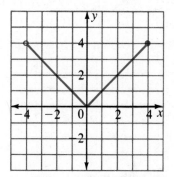

The domain is $\{x | -4 < x \leq 4\}$.
The range is $\{y | 0 \leq y \leq 4\}$.

For each value of x, there is exactly one
value for y. All ordered pairs (x, y) have
different first coordinates.

Thus, relation C is a function.

The open circle on the segment in the figure above indicates that this point is
not a point of the graph. Notice that -4 is not included in the domain of the
relation.

You can tell whether a given relation is a function by examining its graph. If
any vertical line intersects the graph of a relation in more than one point, the
relation is not a function.

Example 4

Determine which graphs are the graphs of functions.
**Draw vertical lines. Check to see if any of the lines intersect the graph in more
than one point.**

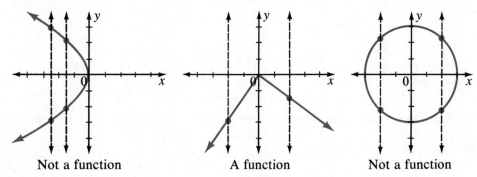

Not a function   A function   Not a function

Relations and Functions   **277**

Oral Exercises

State the domain and the range of each relation.
1. {(2, −1), (3, 2), (2, −5)}   **2.** {(−2, −2), (6, −4), (5, 2)}   **3.** {(−2, 3), (2, −3), (3, −2), (−3, 2)}

Written Exercises

Determine the domain and the range of each relation. Is the relation a function?

(A)
1. {(3, 1), (2, −3), (−1, 5), (−2, −2), (0, 2)}   **2.** {(8, 7), (7, 8), (−7, 8), (−8, 7), (5, −2)}
3. {(0, 0), (8, 7), (−3, 6), (4, −1), (5, −1)}   **4.** {(3, −3), (4, −4), (5, −5), (−3, 3), (−4, 4)}
5. {(3, 2), (4, 2), (−4, −2), (−3, 2), (2, −3)}   **6.** {(0, 1), (3, 1), (0, 4), (1, −3), (2, −3)}
7. {(1, −2), (−1, 0), (2, −4), (−2, 3), (3, 0)}   **8.** {(9, 8), (7, 9), (8, 9), (−9, 8), (−8, 7)}
9. {(4, −1), (−1, 4), (−1, 3), (4, −3), (−3, 0)}   **10.** {(6, −3), (−3, 3), (−6, −3), (6, 5), (−3, 6)}

Write the relation G whose graph is given. Determine the domain and the range of G.

11. 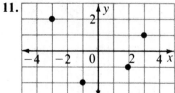   **12.** 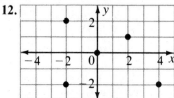   **13.**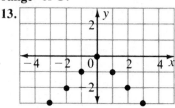

Determine the domain and the range of each relation whose graph is given. Is the relation a function?

(B) **14.** 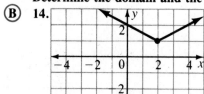   **15.** 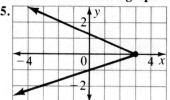   **16.**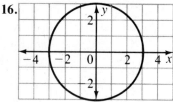

17. 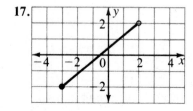   **18.** 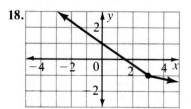   **19.**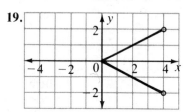

(C) **20.** 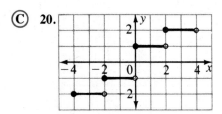   **21.** 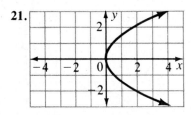   **22.**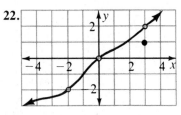

Determine the domain and the range of each relation. Is the relation a function?
23. {. . . , (−4, −2), (−2, −1), (0, 0), (2, 1), . . .}   **24.** {(2, 4), (3, 9), (4, 16), (5, 25), . . .}
25. {(4, 2), (9, 3), (16, 4), (25, 5), (36, 6), . . .}   **26.** {. . . , (−2, −1), (−1, 1), (0, 3), (1, 5), (2, 7), . . .}

VALUES AND COMPOSITION OF FUNCTIONS

Objectives

To find the value of a function, given an element in the domain of the function
To find the range and domain of a function
To find a value of a function that is composed of two other functions
To find the zeros of a rational function

The graph of the function, $f = \{(-3, -2),$ $(-2, 3), (3, 2), (4, 1)\}$, is shown at the right. A lowercase letter is usually used to name a function. Notice that each element of the domain is paired with exactly one element of the range.

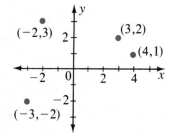

Domain of f: $\{-3, -2, 3, 4\}$

Range of f: $\{-2, \quad 3, 2, 1\}$

Each element of the range is called a **value of the function.** The expression $f(-3) = -2$ is read "the value of f at negative 3 equals negative two," or "f at negative three equals negative two," or "f of negative three equals negative two." In general, the notation $f(x)$ means the value of f at x.

Definition:
Value of a
function

For any ordered pair, (x, y) of a function, f,
$f(x) = y$. The *value of f at x equals y.*

Example 1   **Let $g = \{(-1, -2), (0, 3), (2, 3), (3, -1)\}$. Find $g(-1)$, $g(0)$, $g(2)$ and $g(3)$.**

$(-1, -2) \qquad (0, 3) \qquad (2, 3) \qquad (3, -1)$

$\qquad g(-1) \qquad\quad g(0) \qquad\quad g(2) \qquad\quad g(3)$

Thus, $g(-1) = -2$, $g(0) = 3$, $g(2) = 3$, and $g(3) = -1$.

A function f is often defined by giving an equation or formula for its range.

Example 2   **If $f(x) = 2x - 3$, find $f(1)$, $f(-2)$, and $f(-a)$.**

Substitute for x in $f(x) = 2x - 3$.

$f(x) = 2x - 3$	$f(x) = 2x - 3$	$f(x) = 2x - 3$
$f(1) = 2(1) - 3$	$f(-2) = 2(-2) - 3$	$f(-a) = 2(-a) - 3$
$= -1$	$= -7$	$= -2a - 3$

Example 3   **If $f(x) = 3x + 2$, find $f(2a - 5)$, $f(a + b)$, and $f(a + h) - f(a)$.**

$$f(x) = 3x + 2$$
$$f(2a - 5) = 3(2a - 5) + 2$$
$$= 6a - 15 + 2$$
$$= 6a - 13$$

$$f(x) = 3x + 2$$
$$f(a + b) = 3(a + b) + 2$$
$$= 3a + 3b + 2$$

$$f(x) = 3x + 2$$
$$f(a + h) - f(a) = 3(a + h) + 2 - [3(a) + 2]$$
$$= 3a + 3h + 2 - 3a - 2$$
$$= 3h$$

Example 4   **If $h(x) = -x^2 + 3$, find the range of h for D (domain) = $\{-2, 0, 1\}$.**

$$h(x) = -x^2 + 3$$
$$h(-2) = -(-2)^2 + 3$$
$$= -4 + 3$$
$$= -1$$

$$h(x) = -x^2 + 3$$
$$h(0) = -(0)^2 + 3$$
$$= 0 + 3$$
$$= 3$$

$$h(x) = -x^2 + 3$$
$$h(1) = -(1)^2 + 3$$
$$= -1 + 3$$
$$= 2$$

Thus, R (range) = $\{-1, 3, 2\}$.

Definition: Rational function	A *rational function* is a quotient of two polynomial functions of the form $f(x) = \dfrac{p(x)}{q(x)}$, where $q(x) \neq 0$.

The **domain** of a rational function is the set of all real numbers except those numbers that make the denominator equal to 0.

Example 5   **If $g(x) = \dfrac{x^2 - 5x + 6}{x^2 - 2x}$, determine the domain of g.**

Factor: $g(x) = \dfrac{(x - 3)(x - 2)}{x(x - 2)}$.

If $x(x - 2) = 0$, then $x = 0$ or $x = 2$. So, the domain must exclude 0 and 2.
Thus, the domain is {all real numbers except 0 and 2}.

Definition: Zeros of a rational function	The number r is called a *zero* of a rational function if and only if for $f(x) = \dfrac{p(x)}{q(x)}$, $p(r) = 0$ and $q(r) \neq 0$.

Example 6   **If $f(x) = \dfrac{x^2 + x - 2}{x(x - 3)}$, determine the zeros of f.**

$f(x) = \dfrac{p(x)}{q(x)}$ ▶   Factor: $f(x) = \dfrac{(x - 1)(x + 2)}{x(x - 3)}$, where $p(x) = (x - 1)(x + 2)$ and

$q(x) = x(x - 3)$.

If $p(x) = (x - 1)(x + 2) = 0$, then $x = 1$ or $x = -2$. So, $p(1) = 0$, $p(-2) = 0$, $q(1) \neq 0$, and $q(-2) \neq 0$.
Thus, the zeros are -2 and 1.

Example 7

If $f(x) = x + 2$ and $g(x) = 2x^2 - 3$, find $f(g(-4))$ and $g(f(-4))$.

$$g(x) = 2x^2 - 3 \qquad f(g(-4)) = f(29) \qquad \vert \qquad f(x) = x + 2 \qquad g(f(-4)) = g(-2)$$
$$g(-4) = 2(-4)^2 - 3 \qquad f(29) = 29 + 2 \qquad \vert \qquad f(-4) = -4 + 2 \qquad g(-2) = 2(-2)^2 - 3$$
$$= 2 \cdot 16 - 3 \qquad \qquad = 31 \qquad \vert \qquad \qquad = -2 \qquad \qquad = 5$$
$$= 29$$

Thus, $f(g(-4)) = 31$ and $g(f(-4)) = 5$.

Written Exercises

Let $g = \{(2, -3), (-1, 5), (-4, -2), (0, 5), (8, -1), (-7, 9)\}$. Find each of the following function values.

Ⓐ **1.** $g(2)$　　　**2.** $g(-1)$　　　**3.** $g(-4)$　　　**4.** $g(0)$　　　**5.** $g(8)$　　　**6.** $g(-7)$

Let $f(x) = -2x + 5$. Find each of the following function values.

　7. $f(2)$　　　**8.** $f(-1)$　　**9.** $f(0)$　　　**10.** $f(-2)$　　**11.** $f(5)$　　　**12.** $f(-a)$　　**13.** $f(b)$

Let $t(x) = x^2 + 3x - 1$. Find each of the following function values.

14. $t(0)$　　　　　**15.** $t(-1)$　　　　　**16.** $t(-5)$　　　　　**17.** $t(a + b)$　　　　　**18.** $t(2a + 3)$

Find the range of each function for the given domain.

19. $h(x) = x + 5; D = \{-1, 0, 2\}$ 　　　　**20.** $f(x) = -3x + 2; D = \{-2, 2, 3\}$
21. $t(x) = 2x - 3; D = \{0, 1, 3\}$ 　　　　**22.** $g(x) = 4x + 3; D = \{-3, 2, 0\}$
23. $g(x) = x^2 - 3; D = \{-2, 0, 2\}$ 　　　　**24.** $t(x) = (x - 2)^2; D = \{-1, 0, 2\}$
25. $f(x) = |x|; D = \{-3, -2, 0, 1, 2\}$ 　　　　**26.** $h(x) = \sqrt{x}; D = \{0, 1, 4, 9\}$

Let $f(x) = x - 1$ and $g(x) = 2x$. Find each of the following.

27. $f(g(2))$　　　　　**28.** $g(f(2))$　　　　　**29.** $g(f(-3))$　　　　　**30.** $f(g(-3))$

Find the domain and zeros of each function.

31. $h(x) = \dfrac{1}{x}$ 　　　　**32.** $t(x) = \dfrac{x - 1}{x(x + 2)}$ 　　　　**33.** $f(x) = \dfrac{3x - 6}{(x - 3)(x - 4)}$

Ⓑ **34.** $g(x) = -\dfrac{4x^2 - 3x}{(x + 1)(x + 3)}$ 　　**35.** $f(x) = \dfrac{2x - 4}{x^2 + 4x - 12}$ 　　**36.** $t(x) = \dfrac{-3x^2}{x^2 - 5x - 14}$

Let $f(x) = 2x - 3$ and $g(x) = x^2 - 2$. Find each of the following.
Assume that no denominator has the value of 0. (Ex. 37–52)

37. $f(2) - g(3)$ 　　　**38.** $g(-1) - f(-2)$ 　　　**39.** $f(-3) - g(0)$ 　　　**40.** $f(a) + g(b)$
41. $f(a + h) - f(a)$ 　　**42.** $g(a + h) - g(a)$ 　　**43.** $f(a) - f(t)$ 　　　**44.** $g(a) - f(b)$
45. $f(g(a))$ 　　　　**46.** $g(f(a))$ 　　　　**47.** $g(f(-3a))$ 　　　**48.** $f(g(-3a))$

Ⓒ **49.** $\dfrac{f(a + h) - f(a)}{h}$ 　　**50.** $\dfrac{g(a) - g(b)}{a - b}$ 　　**51.** $\dfrac{f(b) - f(a)}{b - a}$ 　　**52.** $\dfrac{g(c + h) - g(c)}{h}$

53. If $s(x) = -x + 2$ and $t(x) = -x - 2$, find $t(s(x))$ and $s(t(x))$. Does $t(s(x)) = s(t(x))$?

54. If $f(x) = 3x - 4$ and $g(x) = \dfrac{x + 4}{3}$, find $f(g(x))$ and $g(f(x))$. Does $f(g(x)) = g(f(x))$?

55. If $f(x) = x^2 - 3$, $g(x) = 5x + 2$, and $h(x) = 4 - x$, find $f(g(h(3)))$ and $h(g(f(-2)))$.

56. If $t(x) = x^2 - 3x + 2$ and $s(x) = -x - 3$, and $v(x) = 5 - x$, find $t(s(v(-a)))$ and $s(v(t(a + 3)))$.

Values and Composition of Functions

LINEAR AND CONSTANT FUNCTIONS 11.3

Objectives   **To determine whether a given function is a linear function or a constant function, given either an equation of the function or a graph of the function**
To draw the graphs of some special functions

As shown in the figures below, the graph of a **linear function** may be a line, a ray, or a segment, depending upon the domain of the function.

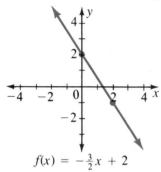

$$f(x) = -\frac{3}{2}x + 2$$

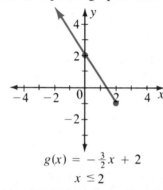

$$g(x) = -\frac{3}{2}x + 2$$
$$x \leq 2$$

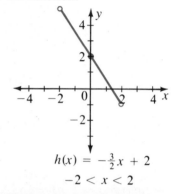

$$h(x) = -\frac{3}{2}x + 2$$
$$-2 < x < 2$$

Definition: Linear function	A *linear function* is a function whose graph is a nonvertical line, ray, or segment. The equation $y = mx + b$, where m and b are real numbers, describes a linear function.

Example 1   **Determine which graphs are the graphs of linear functions.**

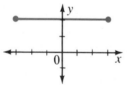

Horizontal
segment
A linear function

Ray
A linear function

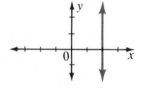

Vertical line
Not a function
Not a linear function

The graph of a linear function described by the equation $y = mx + b$, where $m = 0$, is a horizontal line, ray, or segment, depending upon the domain of the function. A function of this type is called a **constant function.**

Definition: Constant function	A constant function is a linear function whose graph is a horizontal line, ray, or segment. The equation $y = mx + b$, where $m = 0$ and b is a real number, or $y = b$, describes a constant function.

Example 2

Determine which graphs are the graphs of constant functions.

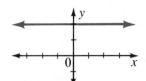

Horizontal line
A linear function
A constant function

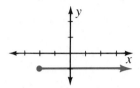

Line
A linear function
Not a constant function

Horizontal ray
A linear function
A constant function

Example 3

Determine whether each equation describes a linear function, a constant function, or neither of these.

$$3x + 5y - 15 = 0 \qquad 4(x^2 - x + 2) + y = (2x - 1)^2 \qquad xy = 12$$

Solve for y. ▶

$5y = -3x + 15$ | $4x^2 - 4x + 8 + y = 4x^2 - 4x + 1$ |
$y = -\dfrac{3}{5}x + 3$ | $y = -7$ | $y = \dfrac{12}{x}$

A linear function | Both linear and constant | Neither of these

The graphs of two special types of functions are shown in the examples that follow. The function in Example 5 is the absolute value function.

Example 4

Graph the function $f(x) = \begin{cases} 1, & \text{if } x \geq 0 \\ -1, & \text{if } x < 0. \end{cases}$

x	$f(x)$
0	1
1	1
2	1
−1	−1
−2	−1
−3	−1

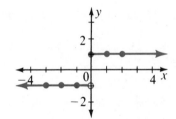

Example 5

Graph the function $f(x) = \begin{cases} x, & \text{if } x > 0 \\ 0, & \text{if } x = 0 \\ -x, & \text{if } x < 0. \end{cases}$

x	$f(x)$
0	0
1	1
2	2
3	3
−1	−(−1), or 1
−2	−(−2), or 2
−3	−(−3), or 3

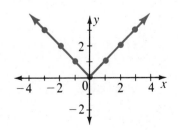

Reading in Algebra

True or false?
1. A constant function is a linear function.
2. A vertical line is the graph of a function that is not a linear function.
3. The equation $f(x) = b$, where b is a real number, describes a constant function.
4. The equation $x = 0$ describes a linear function.

Written Exercises

Determine whether each of the following describes a linear function, a constant function, or neither of these.

(A) 1. $y = 2x - 5$ 2. $y = -2x$ 3. $y = -5$ 4. $f(x) = 3x + 1$
     5. $h(x) = x^2$ 6. $t(x) = \frac{x}{2}$ 7. $x = -7$ 8. $g(x) = 3$
     9. $y = x$ 10. $x = y + 2$ 11. $xy = 9$ 12. $h(x) = 6, -2 \leq x \leq 2$
     13. $5(7x - 2y) - 15x = 20x - 4$ 14. $(2x + 3)^2 + 5 = (2x - 3)^2 + 4$ 15. $h(x) = 3x - 2, x \geq 1$

(B) 16. $(x + 5)^2 - y = 10x$ 17. $(x - 3)^2 + y = (x + 1)^2$ 18. $2x - 3y - 6 = 3(-2y + x)$

19. $\{(5, 3), (7, 3), (9, 3), \ldots\}$ 20. $\{(1, 7), (2, 9), (3, 11), (4, 13), \ldots\}$ 21. $\{(3, 9), (4, 16), (5, 25)\}$

22. 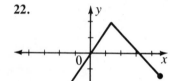   23. 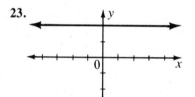   24.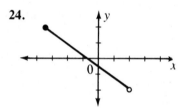

Graph each of the following functions.

25. $f(x) = \begin{cases} 0, & \text{if } x \leq 0 \\ -1, & \text{if } x > 0 \end{cases}$

26. $f(x) = \begin{cases} -x, & \text{if } x > 0 \\ 0, & \text{if } x = 0 \\ x, & \text{if } x < 0 \end{cases}$

27. $f(x) = \begin{cases} 2, & \text{if } x \geq 2 \\ 1, & \text{if } 0 \leq x < 2 \\ -1, & \text{if } x < 0 \end{cases}$

(C) 28. The **greatest integer function** is the function that is described by the equation $y = [x]$ or $f(x) = [x]$, where $[x]$ is the greatest integer less than or equal to any real number x. Draw the graph of $y = [x]$ for $-3 \leq x \leq 3$.
29. Graph $y = -[x]$ for $-3 \leq x \leq 3$.   30. Graph $y = 2 + [x]$ for $-4 \leq x \leq 4$.
31. Graph $y = [x] - 2$ for $-4 \leq x \leq 4$.   32. If $f(x) = [x]$, find $f(0)$, $f(4.5)$, $f(-\frac{5}{2})$, and $f(\sqrt{6})$.

33. Graph $h(x) = \begin{cases} 0, & \text{if } x \text{ is rational} \\ 1, & \text{if } x \text{ is irrational.} \end{cases}$

CUMULATIVE REVIEW

Solve each proportion.

1. $\dfrac{5}{n + 4} = \dfrac{3}{n - 2}$ 2. $\dfrac{4a + 3}{3} = \dfrac{2a + 5}{4}$ 3. $\dfrac{5}{n - 7} = \dfrac{5}{7 - n}$

INVERSE RELATIONS AND FUNCTIONS 11.4

Objectives
To find the inverse of a relation
To draw the graph of a function and its inverse
To determine whether the inverse of a function is a function

If you reverse the order of the elements of each ordered pair in a relation A, you will obtain the **inverse** of relation A, or A^{-1}.
For example, if $A = \{(1, 2), (2, -3), (5, 2)\}$,

then $A^{-1} = \{(2, 1), (-3, 2), (2, 5)\}$.

Notice that the domain of A is the range of A^{-1}, and the range of A is the domain of A^{-1}. Also, A is a function, but A^{-1} is not a function.

Definition: Inverse relations	The inverse of a relation A is the relation A^{-1} obtained by reversing the order of the elements of each ordered pair in A. A and A^{-1} are called *inverse relations*.

When both a relation and its inverse are functions, they are called **inverse functions.**

Example 1   **Let $A = \{(-3, -2), (-1, 2), (3, -5)\}$. Find A^{-1}. Is A^{-1} a function?**

Interchange the elements of each ordered pair in A.

$$A^{-1} = \{(-2, -3), (2, -1), (-5, 3)\}$$

All ordered pairs in A^{-1} have different first coordinates.

Thus, A^{-1} is a function.

When the inverse of a given function f is a function, the symbol f^{-1} is used to denote it. If a function is defined by an equation, the equation of the inverse is obtained by interchanging x and y in the original equation.

Example 2   **A function f is defined by $y = -\frac{2}{3}x + 4$. Find an equation of f^{-1}.**

$$y = -\tfrac{2}{3}x + 4$$
$$x = -\tfrac{2}{3}y + 4 \quad \blacktriangleleft \text{ Interchange } x \text{ and } y.$$
$$3x = -2y + 12$$
$$2y = -3x + 12$$
$$y = -\tfrac{3}{2}x + 6$$

Thus, an equation of f^{-1} is $y = -\frac{3}{2}x + 6$.

In the next two examples, you will learn how to graph the inverse of a function, and to determine whether the inverse is a function by using the vertical line test.

Example 3   **Graph the function defined by $y = |x|$ and its inverse. Determine if the inverse is a function.**

Make a table of ordered pairs of the function and its inverse.

x	$|x|$	function	inverse
0	0	$(0, 0)$	$(0, 0)$
1	1	$(1, 1)$	$(1, 1)$
2	2	$(2, 2)$	$(2, 2)$
-1	1	$(-1, 1)$	$(1, -1)$
-2	2	$(-2, 2)$	$(2, -2)$

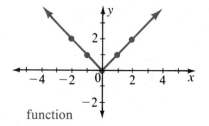

function

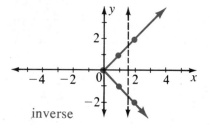

inverse

Since a vertical line intersects the graph of the inverse in two points, the inverse is not a function.

Example 4   **Graph the function defined by $4x - 2y = 8$ and its inverse in the same coordinate plane. Determine if the inverse is a function.**

$$4x - 2y = 8$$
$$-2y = -4x + 8$$
$$y = 2x - 4$$

x	$2x - 4$	function	inverse
0	-4	$(0, -4)$	$(-4, 0)$
1	-2	$(1, -2)$	$(-2, 1)$
2	0	$(2, 0)$	$(0, 2)$

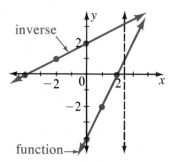

inverse

function→

Since no vertical line intersects the graph of the inverse in more than one point, the inverse is a function.

An equation of the inverse function in Example 4 is $x = 2y - 4$, or $y = \frac{1}{2}x + 2$. If a function is defined by an equation of the form $y = mx + b$, $m \neq 0$, then its inverse can be written in the same form, and the inverse is a function.

Reading in Algebra

True or false? Give a reason for your answer.
 1. Each relation A has an inverse relation, A^{-1}.
 2. Each function f has an inverse function, f^{-1}.
 3. If $G = \{(3, 7), (5, 5), (7, 3)\}$, then $G = G^{-1}$.
 4. For any relations A and B, if $B = A^{-1}$, then $A = B^{-1}$.
 5. For any relation A, $(A^{-1})^{-1} = A$.

Written Exercises

Find the inverse of each relation. Determine if the inverse is a function.

(A)   1. $\{(-2, 3), (-1, 3), (3, -1)\}$ 2. $\{(0, 6), (1, -3), (1, 4)\}$ 3. $\{(4, -1), (5, -2), (6, -3)\}$
     4. $\{(4, 9), (5, -1), (6, 9)\}$ 5. $\{(7, 11), (8, 10), (9, 9)\}$ 6. $\{(5, -2), (-6, -6), (-2, 5)\}$

Graph each function and its inverse in the same coordinate plane. Determine if the inverse is a function.

  7. $x + y = 6$ 8. $x - y = 6$ 9. $5x - 2y = 10$
10. $2y - 3x = 4$ 11. $3x + 4y = -8$ 12. $4y - x - 10 = 0$
13. $y = 6$ 14. $x = y$ 15. $y = 0.5x - 10$
16. $y = |x| - 2$ 17. $y = 2|x| - 1$ 18. $y = |x| + 3$

A relation is defined by each given equation. Find an equation of the inverse of each relation. Determine if the inverse is a function.

(B)   19. $\frac{2}{3}x - \frac{3}{5}y = 10$ 20. $\frac{5}{4}y + \frac{3}{4}x = 12$ 21. $7x - \frac{2}{3}y - 8 = 0$
     22. $x^2 + y^2 = 1$ 23. $x^2 - y^2 = 4$ 24. $y = x^2$
     25. $y = x^2 + 5$ 26. $x = y^2$ 27. $xy = 9$

(C)   28. Show that the graph of $y = mx + b$ and the graph of its inverse are symmetric with respect to the line determined by the equation $y = x$. [Hint: Select points $P(x, y)$ and $Q(y, x)$ on the graphs. Show that the line determined by $y = x$ is the perpendicular bisector of $\overline{PQ}$.]

If $f(x) = 3x - 5$, find each of the following function values.
29. $f^{-1}(1)$ 30. $f^{-1}(4)$ 31. $f^{-1}(-5)$ 32. $f^{-1}(16)$
33. $f^{-1}(2.5)$ 34. $f^{-1}(3a - 5)$ 35. $f(f^{-1}(10))$ 36. $f^{-1}(f(30))$
37. $f(f^{-1}(a))$ 38. $f^{-1}(f(a))$

NON-ROUTINE PROBLEMS

If f is any linear function, is the following statement true or false? Prove your answer.

For each real number x, $f(x + 2) - f(x + 1) = f(x + 1) - f(x)$.

DIRECT VARIATION

Objectives
To determine whether a relation expresses direct variation
To solve a word problem using direct variation

Frequently in mathematics, one variable is related to another variable in such a way that a change in the value of one produces a corresponding change in the other. For example, if you drive an automobile at a constant speed of 60 km/h, the distance traveled is directly related to the time traveled. When the time increases, the distance traveled increases. In the table, notice that the ratio $\frac{d}{t}$ is 60 in each case.

Time in hours(t)	1	2	3	4	5
Distance in kilometers(d)	60	120	180	240	300

The table determines a set of ordered pairs with coordinates (t, d) that expresses **direct variation** between the variables t and d.

Definition: Direct variation	A linear function defined by an equation of the form $y = kx$, or $\frac{y}{x} = k$, where k is a nonzero constant, is called a *direct variation*. The constant k is called the *constant of variation*.

The equation $\frac{y}{x} = k$, or $y = kx$, is read, "y varies directly as x," or "y varies directly with x."

Example 1

Determine if the table expresses a direct variation between the variables y and x. If so, find the constant of variation and an equation that defines the function.

x	y	ordered pair
1	-6	$(1, -6)$
-2	12	$(-2, 12)$
3	-18	$(3, -18)$

$$\frac{y}{x} = \frac{-6}{1} = \frac{12}{-2} = \frac{-18}{3} = -6$$

$$\frac{y}{x} = -6, \text{ or } y = -6x$$

The ratio $\frac{y}{x} = -6$ for each ordered pair in the function.

Thus, y varies directly as x, the constant of variation is -6, and an equation of the function is $y = -6x$.

Example 2

Determine if the equation expresses a direct variation between the variables y and x, in each case.

$y = 2x$ $y = 3x + 4$

The equation is of the form
$y = kx$.

The equation is not of the form
$y = kx$.

Thus, $y = 2x$ expresses a direct variation; $y = 3x + 4$ does not.

Notice that an equation of a function that expresses direct variation is an equation of the form $y = mx + b$, where m is a constant and $b = 0$.

Example 3

The voltage V of a given electrical circuit varies directly as the current I. If V is 120 volts when I is 9 amperes, find the voltage V when the current I is 15 amperes.

(1) Write a formula and find the value of k.

$V = kI$
$120 = k \cdot 9$
$k = \dfrac{120}{9} = \dfrac{40}{3}$

(2) Substitute for k and I. Find the value of V.

$V = kI$
$V = \dfrac{40}{3} \cdot 15 = 200$

Thus, the voltage is 200 volts when the current is 15 amperes.

A direct variation problem may be solved by using a proportion. If (x_1, y_1) and (x_2, y_2) are any two ordered pairs of a direct variation and neither is $(0, 0)$, then $\dfrac{y_1}{x_1} = k$ and $\dfrac{y_2}{x_2} = k$ for some constant $k \neq 0$. Thus, $\dfrac{y_1}{x_1} = \dfrac{y_2}{x_2}$.

The constant k is called the **constant of proportionality,** and y is said to be "directly proportional to" x.

Example 4

If y varies directly as the cube of x, and $y = 54$ when $x = 3$, find y when $x = 5$.

y varies directly as the cube of x means that y is directly proportional to the cube of x. Write a proportion.

$$\frac{54}{3^3} = \frac{y}{5^3}$$

$$\frac{54}{27} = \frac{y}{125}$$

$$\frac{2}{1} = \frac{y}{125}$$ ◀ *In a proportion, product of means equals product of extremes.*

$$1 \cdot y = 2 \cdot 125$$
$$y = 250$$

Thus, $y = 250$ when $x = 5$.

Oral Exercises

If c is a constant, determine whether each equation expresses a direct variation.

1. $\dfrac{a}{b} = c$ **2.** $m = nc$ **3.** $mn = c$ **4.** $x = cy$ **5.** $xy = c$

6. $p = cq$ **7.** $\dfrac{1}{p} = \dfrac{c}{q}$ **8.** $\dfrac{c}{1} = \dfrac{q}{p}$ **9.** $x = \dfrac{c}{y}$ **10.** $\dfrac{y}{x} - c = 0$

Written Exercises

Determine if the table expresses a direct variation between the variables. If so, find the constant of variation and an equation that defines the function.

Ⓐ

1.

x	y
-3	1
-9	3
3	-1
6	-2

2.

c	d
7	3
14	6
-21	-9
28	-12

3.

A	r
-15	-3
10	2
-20	-4
25	5

4.

R	S
2	1
3	2
4	3
5	4

5.

M	N
1	1
2	2
3	3
-4	-4

Determine if the equation expresses a direct variation between the variables. If so, find the constant of variation.

6. $y = -x$ **7.** $y = 3x$ **8.** $y = -2x + 5$ **9.** $c = 2d$ **10.** $c = 2\pi r$

11. $A = 3r + 6$ **12.** $A = \dfrac{4}{l}$ **13.** $\dfrac{y}{x} - 6 = 0$ **14.** $y = \dfrac{4}{x}$ **15.** $3y = 2x$

In Exercises 16–23, y varies directly as x. Find the value as indicated.

16. If $y = 12$ when $x = 4$, find y when $x = 12$.
17. If $y = -81$ when $x = 9$, find y when $x = 7$.
18. If $y = 16$ when $x = -4$, find y when $x = 8$.
19. If $y = -3$ when $x = -15$, find x when $y = 2$.
20. If $y = -12$ when $x = -3$, find y when $x = -8$.
21. If $y = 5$ when $x = 25$, find y when $x = 30$.
22. If $y = 3$ when $x = 10$, find x when $y = 1.2$.
23. If $y = 3$ when $x = 10$, find x when $y = \frac{1}{2}$.

Ⓑ **24.** If y varies directly as the square of x, and $y = 25$ when $x = 3$, find y when $x = 6$.

25. If y is directly proportional to $\sqrt{x}$, and $y = 3.5$ when $x = 4$, find x when $y = 8.75$.

26. If y is directly proportional to $\sqrt[3]{x}$, and $x = 27$ when $y = 7.2$, find x when $y = 12$.

27. If y varies directly as the cube of x, and $x = 0.2$ when $y = 16$, find y when $x = 0.5$.

28. If you earn \$22.75 for 7 hours work, how long must you work to earn \$39.65?

29. If a batter averages 81 hits for each 216 times at bat, approximately how many hits will he get in 400 times at bat?

30. The scale on a certain blueprint is $3'' = 8'$. Find the dimensions of a floor that is $5\frac{1}{2}''$ by $7\frac{3}{4}''$ on the blueprint.

31. The stretch S in a spring balance varies directly as the applied weight w. If $S = 5$ in. when $w = 15$ lb, find the weight needed for a stretch of $7\frac{1}{2}$ in.

Ⓒ **32.** In a statement of direct variation in which y varies directly as x, if the value of y remains constant, $(y > 0)$, what happens to the value of k when x increases? when x decreases?

33. If A varies directly as B, and if A is multiplied by c, how is the corresponding value of B affected?

INVERSE VARIATION

Objectives   **To determine whether a relation expresses inverse variation**
To solve a word problem using inverse variation

In some mathematical situations, when the value of one variable increases, the value of a second variable decreases. For example, if you drive an automobile a distance of 180 km, the time required for the trip decreases as the speed increases. In the table, notice that the product rt is 180 in each case.

Speed in kilometers per hour (r)	90	72	60	45	40	36	30
Time in hours (t)	2	2.5	3	4	4.5	5	6

The table determines a set of ordered pairs with coordinates (r, t) that expresses **inverse variation** between the variables r and t.

Definition: Inverse variation	A function defined by an equation of the form $xy = k$, or $y = \dfrac{k}{x}$, where k is a nonzero constant, is called an *inverse variation*. The constant k is called the *constant of variation*.

The equation $xy = k$, or $y = \dfrac{k}{x}$, is read, "y varies inversely as x," or "y varies inversely with x."

Example 1   **Determine if the table expresses an inverse variation between the variables y and x. If so, find the constant of variation and an equation that defines the function.**

x	4	8	1	-1	-2	$-\frac{1}{2}$
y	-2	-1	-8	8	4	16

$$
\begin{array}{cccccc}
4 \cdot (-2) & 8 \cdot (-1) & 1 \cdot (-8) & -1 \cdot 8 & -2 \cdot 4 & -\frac{1}{2} \cdot 16 \\
-8 & -8 & -8 & -8 & -8 & -8
\end{array}
$$

$x \cdot y = -8$ for each ordered pair (x, y) in the function.

Thus, y varies inversely as x, the constant of variation is -8, and an equation of the function is $xy = -8$.

Example 2   **Determine if the equation expresses an inverse variation between the variables y and x, in each case.**

$$y = 5x - 3 \qquad\qquad xy = -12$$

The equation is not of the form $xy = k$.

The equation is of the form $xy = k$.

Thus, $y = 5x - 3$ does not express an inverse variation; $xy = -12$ does.

Example 3   **The height h of a triangle varies inversely as the length of the base b. If $h = 10$ cm when $b = 4.8$ cm, find h when $b = 9$ cm.**

(1) Write a formula and find the value of k.

$$bh = k$$
$$4.8 \cdot 10 = k$$
$$48 = k$$

(2) Substitute for k and b. Find the value of h.

$$bh = k$$
$$9 \cdot h = 48$$
$$h = 5\tfrac{1}{3}, \text{ or } 5.3$$

Thus, $h = 5.3$ cm when $b = 9$ cm.

If (x_1, y_1) and (x_2, y_2) are any two ordered pairs of an inverse variation and neither is $(0, 0)$, then $x_1 y_1 = k$ and $x_2 y_2 = k$. Thus, $x_1 y_1 = x_2 y_2$.
This equation is often written as a proportion of the form $\frac{x_1}{x_2} = \frac{y_2}{y_1}$; y is said to be "inversely proportional to" x.

Example 4   **If y varies inversely as the square of x, and $y = 10$ when $x = 4$, find x when $y = 5$.**

y varies inversely as the square of x means that y is inversely proportional to the square of x. Write a proportion.

$$\frac{4^2}{x^2} = \frac{5}{10}$$
$$\frac{16}{x^2} = \frac{1}{2}$$
$$x^2 \cdot 1 = 16 \cdot 2 \qquad \blacktriangleleft \quad \textit{In a proportion, product of means}$$
$$x^2 = 32 \qquad\qquad\qquad \textit{equals product of extremes.}$$
$$x = \pm\sqrt{32}, \text{ or } \pm 4\sqrt{2}$$

Thus, $x = \pm 4\sqrt{2}$ when $y = 5$.

Example 5   **If y is inversely proportional to $\sqrt[3]{x}$ and $y = 2.7$ when $x = 8$, find y when $x = 27$.**

$$\sqrt[3]{8} \cdot 2.7 = \sqrt[3]{27} \cdot y \qquad \blacktriangleleft \quad \textit{Write an equation in the}$$
$$2 \cdot 2.7 = 3 \cdot y \qquad\qquad \textit{product form.}$$
$$5.4 = 3y$$
$$1.8 = y$$

Thus, $y = 1.8$ when $x = 27$.

Oral Exercises

If c is a constant, determine whether each equation expresses an inverse variation.

1. $\dfrac{x}{y} = c$

2. $xc = y$

3. $x = \dfrac{c}{y}$

4. $x = \dfrac{y}{c}$

5. $y = cx + x$

6. $\dfrac{1}{c} = \dfrac{x}{y}$

7. $\dfrac{c}{x} = \dfrac{1}{y}$

8. $1 = \dfrac{c}{xy}$

9. $xc = \dfrac{c}{y}$

10. $\dfrac{c}{x} = \dfrac{y}{c}$

Written Exercises

Determine if the table expresses an inverse variation between the variables. If so, find the constant of variation and an equation that defines the function.

Ⓐ

1.

s	t
12	3
16	4
20	5
24	6

2.

x	y
-3	-4
-2	-6
4	3
12	1

3.

c	d
-4	6
8	-3
-12	-2
-24	-1

4.

x	y
1.0	0.32
0.4	0.80
0.2	1.60
0.1	3.20

5.

a	b
$\frac{1}{2}$	$\frac{1}{6}$
$\frac{1}{3}$	$\frac{1}{4}$
$-\frac{1}{6}$	$-\frac{1}{2}$
$-\frac{1}{12}$	-1

Determine if the equation expresses an inverse variation between the variables. If so, find the constant of variation.

6. $x = 10y$

7. $xy = 12$

8. $5st = 12$

9. $c = 5r$

10. $a \cdot b = \pi$

11. $A = 6.28r$

12. $\dfrac{r}{4} = t$

13. $6m = \dfrac{4}{n}$

14. $x = \dfrac{10}{y}$

15. $y = \dfrac{10}{x}$

In Exercises 16–19, y varies inversely as x. Find the value as indicated.

16. If $y = 27$ when $x = 3$, find y when $x = 9$.

17. If $y = 15$ when $x = -2$, find y when $x = -5$.

18. If $y = 9$ when $x = 4$, find y when $x = 6$.

19. If $y = 81$ when $x = -9$, find x when $y = 27$.

20. The illumination i from a light varies inversely as the square of its distance d from an object. If $i = 8$ foot-candles when $d = 3$ ft, find i when $d = 4$ ft.

21. The pressure P of a gas at a constant temperature varies inversely as the volume V. If $V = 450$ in.^{3} when $P = 30$ lb/in.^{2}, find P when $V = 750$ in.^{3}.

Ⓑ

22. If y varies inversely as the square of x, and $y = 3$ when $x = 4$, find y when $x = 2$.

23. If y is inversely proportional to $\sqrt{x}$, and $y = 6$ when $x = 9$, find y when $x = 16$.

24. If x is inversely proportional to $\sqrt[3]{y}$, and $x = 1.8$ when $y = 64$, find x when $y = 27$.

25. If x varies inversely as the cube of y, and $x = 3$ when $y = 2$, find x when $y = 5$.

26. The frequency of a radio wave is inversely proportional to its wave length. If a radio wave, 30 m long, has a frequency of 1200 kilocycles per second, what is the length of a wave with a frequency of 900 kilocycles per second?

Inverse Variation

JOINT AND COMBINED VARIATION 11.7

Objectives
To determine whether a relation expresses joint variation or combined variation
To solve a word problem using joint variation or combined variation

In some mathematical situations, one variable is a constant multiple of the product of two or more variables. For example, simple interest, i, received on an investment of a specific amount of money is related to both the rate of interest, r, and the time, t, that the money was invested. Specifically, as the value of $r \cdot t$ increases, the value of i increases. The variation of the variables is called a **joint variation**.

Definition:
Joint variation

A function defined by an equation of the form $y = kxz$, or $\frac{y}{xz} = k$, where k is a nonzero constant, is called a *joint variation*.
The constant k is called the *constant of variation*.

The equation $\frac{y}{xz} = k$, or $y = kxz$, is read, "y varies jointly as x and z," or "y varies jointly with x and z." The variation in this case is a direct variation between y and the product xz.

Example 1

Determine if the equation expresses a joint variation between the variables, in each case.
$A = \frac{1}{2}bh$ $m = \frac{3}{4}n + \frac{1}{3}$

The equation is of the form $y = kxz$. | The equation is not of the form
                                          | $y = kxz$.

Thus, $A = \frac{1}{2}bh$ expresses a joint variation; $m = \frac{3}{4}n + \frac{1}{3}$ does not.

Example 2

If y varies jointly as x and z and if $y = -24$ when $x = 4$ and $z = 3$, find y when $x = -6$ and $z = 2$.

| First Method: | Second Method: |

Use the equation $y = kxz$. ▶

$$-24 = k \cdot 4 \cdot 3$$
$$-24 = 12k$$
$$-2 = k$$

$$\frac{-24}{4 \cdot 3} = \frac{y}{-6 \cdot 2}$$
$$\frac{-2}{1} = \frac{y}{-12}$$

◀ *Use a proportion of the form $\frac{y_1}{x_1 z_1} = \frac{y_2}{x_2 z_2}$.*

$$y = -2(-6)(2)$$
$$y = 24$$

$$1 \cdot y = -2(-12)$$
$$y = 24$$

Thus, $y = 24$ when $x = -6$ and $z = 2$.

Example 3 The area A of a triangle varies jointly as the length of a base b and the length of a corresponding altitude h. If $A = 15$ cm when $b = 10$ cm and $h = 3$ cm, find A when $b = 25$ cm and $h = 6$ cm.

$$A = kbh$$
$$15 = k \cdot 10 \cdot 3$$
◄ *Write a formula and solve for k.*
$$15 = 30k$$
$$\tfrac{1}{2} = k$$

$$A = kbh$$
$$A = \tfrac{1}{2} \cdot 25 \cdot 6$$
◄ *Substitute for k, b, and h. Find the value of A.*
$$A = 75$$

Thus, A is 75 cm^2 when $b = 25$ cm and $h = 6$ cm.

In some mathematical applications, both direct variation and inverse variation occur at the same time. This type of variation is called **combined variation**.

Definition: Combined variation	A function defined by an equation of the form $y = \dfrac{kxz}{w}$, or $\dfrac{yw}{xz} = k$, where k is a nonzero constant, is called a *combined variation*. The constant k is called the *constant of variation*.

In the equation $y = \dfrac{kxz}{w}$, or $\dfrac{yw}{xz} = k$, the variable y varies jointly with the variables x and z and inversely with the variable w.

Example 4 If y varies jointly as x and z and inversely as $\sqrt{w}$, and $y = 12$ when $x = 2$, $z = 6$, and $w = 9$, find y when $x = 5$, $z = 7$, and $w = 25$.

Use an equation of the form $y = \dfrac{kxz}{\sqrt{w}}$.
►

First Method:

$$12 = \frac{k \cdot 2 \cdot 6}{\sqrt{9}} \qquad y = \frac{3 \cdot 5 \cdot 7}{\sqrt{25}}$$

$$12 = \frac{12k}{3} \qquad y = \frac{3 \cdot 5 \cdot 7}{5}$$

$$12 = 4k \qquad y = 21$$

$$3 = k$$

Second Method:

$$\frac{12 \cdot \sqrt{9}}{2 \cdot 6} = \frac{y \cdot \sqrt{25}}{5 \cdot 7}$$

$$\frac{12 \cdot 3}{2 \cdot 6} = \frac{y \cdot 5}{5 \cdot 7}$$

$$\frac{3}{1} = \frac{y}{7}$$

$$1 \cdot y = 3 \cdot 7$$

$$y = 21$$

◄ *Use a proportion of the form $\dfrac{y_1 \sqrt{w_1}}{x_1 z_1} = \dfrac{y_2 \sqrt{w_2}}{x_2 z_2}$.*

Thus, $y = 21$ when $x = 5$, $z = 7$, and $w = 25$.

Joint and Combined Variation

Written Exercises

Determine whether each equation expresses a joint variation, a combined variation, or neither. If either joint or combined variation exists, find the constant of variation.

Ⓐ 1. $x = \dfrac{-2}{y}$ 2. $mn = 5$ 3. $\dfrac{r}{s} = \dfrac{6t}{v}$ 4. $-6x = \dfrac{y}{z}$

5. $w = 4u$ 6. $a \cdot b \cdot c = 15$ 7. $m = \dfrac{5n^2}{2p}$ 8. $y = \dfrac{8x}{\sqrt{t}}$

In Exercises 9–15, y varies jointly as x and z. Find the value as indicated.

9. If $y = 12$ when $x = 4$ and $z = 3$, find y when $x = 9$ and $z = 8$.
10. If $y = 72$ when $x = 3$ and $z = 8$, find y when $x = -2$ and $z = -3$.
11. If $y = 24$ when $x = 2$ and $z = 3$, find y when $x = 4$ and $z = 7$.
12. If $y = 18$ when $x = 1$ and $z = 6$, find y when $x = 4$ and $z = 10$.
13. If $y = 30$ when $x = 3$ and $z = 15$, find y when $x = -5$ and $z = -12$.

14. If y varies directly as the square of x and inversely as z, and if $y = 12$ when $x = 2$ and $z = 7$, find y when $x = 3$ and $z = 9$.

15. If y varies directly as $\sqrt[3]{x}$ and inversely as the square of z, and if $y = 3$ when $x = 8$ and $z = 4$, find y when $x = 27$ and $z = 6$.

16. The area A, of a parallelogram varies jointly as the length of a base, b, and the length of a corresponding altitude, h. If $A = 16$ when $b = 2$ and $h = 8$, find A when $b = 8$ and $h = 16$.

17. The distance, D, traveled at a uniform rate varies jointly as the rate, r, and the time, t. If $D = 120$ when $r = 60$ and $t = 2$, find D when $r = 80$ and $t = 3$.

Ⓑ 18. If y varies jointly as x and $\sqrt{z}$ and inversely as the cube of w, and if $y = 6$ when $x = 3$, $z = 9$, and $w = 3$, find y when $x = 4$, $z = 36$, and $w = 4$.

19. If y varies jointly as x and z and inversely as the square of w, and if $y = 6$ when $x = 3$, $z = 4$, and $w = 49$, find y when $x = 6$, $z = 8$, and $w = 4$.

20. Interest i varies jointly as the principal p, the rate r, and the time t. If the value of r is halved, what happens to the value of i?

21. The volume V of a cube varies directly as the cube of its edge e. If the value of e is doubled, what happens to the value of V?

Ⓒ 22. If r varies directly as the square of s and inversely as t, what happens to the value of r if the value of s is doubled and the value of t is tripled?

23. If t varies directly as s and inversely as the square of r, how is the value of t affected if the value of r is increased by 25%?

24. If 6 people take 14 days to assemble 18 computers, how many days will it take 10 people to assemble 30 computers?

25. If y varies directly as x and inversely as z, how is the value of y affected if the value of z is decreased by 25%?

CUMULATIVE REVIEW

Solve each equation.

1. $\sqrt{x - 3} + \sqrt{x} = 3$ 2. $\sqrt{x} + \sqrt{x - 5} = 5$ 3. $\sqrt{2x} - \sqrt{\tfrac{x}{2}} = 2$

APPLICATIONS

Read → Plan → Solve → Interpret

1. The frequency of a string under constant tension is inversely proportional to its length. If a string, 40 cm long, vibrates 680 times per second, what length must the string be to vibrate 850 times per second under the same tension?

2. The frequency of a radio wave is inversely proportional to its wavelength. If a radio wave, 30 m long, has a frequency of 1,200 kilocycles per second, what is the length of a wave with a frequency of 900 kilocycles per second?

3. Boyle's Law states that the volume of a given mass of gas varies inversely as the absolute pressure, provided the temperature remains constant. If the volume of a gas is 40 in.^{3} when the absolute pressure is 50 lb/in.^{2}, what will be the volume when the pressure is 70 lb/in.^{2}?

4. The kinetic energy, E, of a body varies jointly as its weight, w, and the square of its velocity, V. If a 10-lb body moving at 6 ft/s has 3 ft-lb of kinetic energy, what will the kinetic energy be of a 4-T truck traveling at 45 mi/h?

5. The load, l, that a beam of fixed length can support varies jointly as its width, w, and the square of its depth, d. A beam with $w = 4$ and $d = 2$ can support a load of 1,760 kg. Find l when $w = 4$ and $d = 8$.

6. The pressure, P, of a gas varies directly as the temperature, T, and inversely as the volume, V. If $P = 50$ when $T = 25$ and $V = 2$, find P when $T = 40$ and $V = 4$.

7. The speed, V, at which a motorcycle must travel on the inside cylinder in order not to fall varies directly as the square root of the radius of the cylinder. If the motorcycle must travel at least 45 km/h to stay on the side of the cylinder with a radius of 12 m, how fast must it travel on a cylinder with radius of 30 m? Answer to the nearest km/h.

8. A conservationist can approximate the number of deer in a forest. Suppose 650 deer are caught, tagged and released. Suppose later 216 deer are caught and 54 of them were tagged. How many deer are in the forest? Assume that the number of tagged deer caught varies directly as the number of deer caught later.

Applications

CHAPTER ELEVEN REVIEW

Vocabulary
combined variation [11.7]
constant function   [11.3]
direct variation   [11.5]
domain [11.1]   function [11.1]
inverse function [11.4]   inverse relation [11.4]
inverse variation [11.6]
joint variation [11.7]   range [11.1]
rational function [11.2]   relation [11.1]
zeros of a rational function [11.2]

Determine the domain and the range of each relation. Is the relation a function? [11.1]
1. $\{(-2, 3), (0, -7), (8, -3), (4, 3)\}$
2. $\{(0, 6), (1, -3), (4, -6), (0, 0)\}$
★ 3. $\{\ldots (-1, 1), (0, 0), (1, 1), (2, 4), (3, 9), \ldots\}$

4.
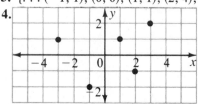

5.
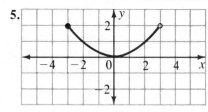

Let $f = \{(2, -3), (1, 0), (-3, 5), (-1, -2)\}$. **Find each of these function values.** [11.2]
6. $f(1)$ 7. $f(-1)$ 8. $f(-3)$ 9. $f(2)$

Let $h(x) = x^2 - 2$. **Find each of the following function values.** [11.2]
10. $h(0)$ 11. $h(-4)$ 12. $h(a - 1)$

Find the range of each function for the given domain. [11.2]
13. $h(x) = 2x + 3$, $D = \{0, 1, -2\}$
14. $g(x) = x^2 + 1$, $D = \{-1, 0, 4\}$

Find the domain and zeros of each function. [11.2]
15. $h(x) = \dfrac{x - 2}{x - 1}$ 16. $t(x) = \dfrac{-4x - 12}{x^2 - 3x - 10}$

Let $f(x) = 3x - 5$ **and** $g(x) = x^2 + 6$. **Find each of the following.** [11.2]
17. $f(-7)$ 18. $g(2\sqrt{3})$ 19. $g(f(\tfrac{2}{3}))$
★ 20. $\dfrac{f(a) - f(b)}{a - b}; a \neq b$ ★ 21. $\dfrac{g(c + h) - g(c)}{h}; h \neq 0$

Determine whether each equation describes a linear function, a constant function, or neither of these. [11.3]
22. $2x - 3y = 6$ 23. $y = 2$ 24. $x = -3$
25. $-3(2x - 3y) - 8x = 3(2x - 1)$

Find the inverse of each relation. Determine if the inverse is a function. [11.4]
26. $\{(-1, -2), (6, -2), (1, 2)\}$
27. $\{(3, 1), (3, 6), (3, 9)\}$

Graph each function and its inverse in the same coordinate plane. Determine if the inverse is a function. [11.4]
28. $3x - 2y = 8$ 29. $y = |x| - 1$
30. $xy = 4$ 31. $x^2 = y - 1$

Determine if the table expresses a direct variation or an inverse variation. Find the constant of variation. [11.5–11.6]

32.

x	y
8	3
12	2
-6	-4

33.

s	t
-6	2
-3	1
6	-2

34.

m	n
12	1
6	2
-4	-3

In Exercises 35–36, y varies directly as x. [11.5]
35. If $y = 18$ when $x = 3$, find y when $x = 5$.
36. If $y = -8$ when $x = 2$, find y when $x = -4$.

In Exercises 37–38, y varies inversely as x. [11.6]
37. If $y = 12$ when $x = 2$, find y when $x = 12$.
38. If $y = -3$ when $x = 5$, find y when $x = -3$.
39. If y varies jointly as x and z, and $y = -12$ when $x = 3$ and $z = -1$, find y when $x = 5$ and $z = 6$. [11.7]
40. If y varies directly as the square of x and inversely as the positive square root of z, and if $y = 9$ when $x = 3$ and $z = 16$, find y when $x = -1$ and $z = 4$. [11.7]

CHAPTER ELEVEN TEST

Determine the domain and the range of each relation. Is the relation a function?

1. $\{(-2, -1), (3, -2), (4, -3), (2, 1)\}$

2. $\{(2, 5), (3, 0), (1, 3), (-2, 0)\}$

Let $h = \{(-1, 4), (0, -3), (1, 4), (2, -6)\}$. Find each of these function values.

3. $h(-1)$ **4.** $h(0)$ **5.** $h(2)$ **6.** $h(1)$

Let $f(x) = x^2 + 3$. Find each of the following function values.

7. $f(-1)$ **8.** $f(0)$ **9.** $f(a + 2)$

Find the range of each function for the given domain.

10. $h(x) = x^2 + 2$, $D = \{-1, 0, 4\}$

11. $g(x) = 3x - 5$, $D = \{0, -1, 2\}$

Find the domain and zeros of each function.

12. $g(x) = \dfrac{x - 3}{x - 2}$ **13.** $t(x) = \dfrac{4x - 4}{x^2 - 4x - 5}$

Let $f(x) = 2x + 4$ and $g(x) = x^2 - 1$. Find each of the following.

14. $f(-1)$ **15.** $g(3\sqrt{2})$ **16.** $f(g(\tfrac{1}{3}))$

★17. $\dfrac{f(a) - f(b)}{a - b}$; $a \neq b$

Determine whether each equation describes a linear function, a constant function, or neither of these.

18. $-2x + 3y = 12$ **19.** $y = -6$ **20.** $x = 0$

21. $-2(-2y - x) - 3x = 2(3x - 1)$

Graph each function and its inverse in the same coordinate plane. Determine if the inverse is a function.

22. $y = x^2 + 2$ **23.** $xy = -2$

Determine if the table expresses a direct variation or an inverse variation. Find the constant of variation.

24.

x	y
-4	-5
-5	-4
1	20

25.

r	s
-2	8
16	-1
-32	$\frac{1}{2}$

26.

m	n
2	24
-3	-36
$\frac{1}{2}$	6

In Exercises 27–28, y varies directly as x.

27. If $y = 10$ when $x = 2$, find y when $x = 6$.

28. If $y = -10$ when $x = 5$, find y when $x = -6$.

In Exercises 29–30, y varies inversely as x.

29. If $y = 5$ when $x = 6$, find y when $x = 15$.

30. If $y = 8$ when $x = 3$, find y when $x = 12$.

31. If r varies directly as the square of s and if $r = 32$ when $s = 4$, find r when $s = 6$.

32. If x varies jointly as s and t, and $x = 24$ when $s = 4$ and $t = 2$, find x when $s = -2$ and $t = -3$.

33. If x varies directly as the positive square root of y and inversely as the square of z, and if $x = 2$ when $y = 4$ and $z = 5$, find x when $y = 9$ and $z = 6$.

COMPUTER ACTIVITIES

User Defined Functions

OBJECTIVE: **To evaluate composite functions, using computer programs that contain user defined functions**

As you have progressed through the computer pages in this text, you have learned about many different functions that are part of the BASIC programming language. These functions are immediately available for you to use. However, you are not limited to only those functions that are part of BASIC. You can use the DEF to DEFine your own functions for use in programs.

To define your own functions using the DEF command, you write DEF, the function name FN followed by a letter of your choice, the variable of the function within parentheses, an equal sign, and the function itself. For example, to DEFine a function $A(X) = 2X + 3$, you would write the statement DEF FN $A(X) = 2 * X + 3$. Once you have DEFined a function in a program, you can use that defined function in the program. The following program DEFines the function $F(X) = X^2 + 2X - 5$ and allows you to insert values into it.

```
10   DEF   FN A(X) = X ^ 2 + 2 * X - 5
20   INPUT B
30   PRINT "A("B") = " FN A(B)
40   END
```

If you have DEFined two or more functions in a program, then in addition to the functions themselves, you can use composites of those functions in the program. The following program illustrates how composite functions can be used.

```
10   DEF   FN F(X) = X + 3
20   DEF   FN G(X) = 2 * X - 5
30   INPUT "FOR WHAT VALUE OF X DO YOU WANT
         COMPOSITES? ";X
40   PRINT "F(G("X")) = " FN F( FN G(X))
50   PRINT "G(F("X")) = " FN G( FN F(X))
60   PRINT "F(F("X")) = " FN F( FN F(X))
70   PRINT "G(G("X")) = " FN G( FN G(X)): PRINT
80   PRINT : GOTO 10
```

See the Computer Section beginning on page 574 for more information.

1. Use the new DEF function to evaluate the polynomials in Exercises 30–33 on page 40. Use one program that INPUTs values one at a time, to evaluate all four Exercises.

COLLEGE PREP TEST

DIRECTIONS: Choose the *one* best answer to each question or problem.

1. A bag of rabbit food will feed 24 rabbits for 72 days. How many days will it feed 16 rabbits?

   (A) 108 (B) 48 (C) 71
   (D) 49 (E) 96

2. Which of the following equations gives the relationship between x and y in the table?

x	1	2	3	4	5	6
y	3	7	11	15	19	23

   (A) $y = 3x$ (B) $y = x^2 + 2$
   (C) $y = x^2 + 3$ (D) $y = 3x + 5$
   (E) $y = 4x - 1$

3. In making egg salad, a recipe calls for 4 cups of diced eggs to $\frac{1}{4}$ cup of chopped onions. How much onion should be used for 16 cups of egg salad?

   (A) $\frac{1}{4}$ (B) $\frac{1}{8}$ (C) $\frac{3}{4}$
   (D) 4 (E) 1

4. A person sews m boxes of shirts in n days. If there are p shirts in a box, how many shirts does the person sew in a day?

   (A) $\frac{m}{pn}$ (B) $\frac{n}{pm}$ (C) $\frac{pn}{m}$
   (D) $\frac{pm}{n}$ (E) $\frac{mn}{p}$

5. A bell chimes every c minutes and a horn blasts every b minutes. Assume c and b are prime numbers and the bell and horn start simultaneously. In how many minutes will they sound simultaneously?

   (A) $\frac{b}{c}$ (B) $\frac{c}{b}$ (C) bc
   (D) $\frac{cb}{2}$ (E) $2bc$

6. If $2y - 6$ varies directly as the square of $x + 2$ and inversely as $z - 3$, and if $y = 1$ when $x = 2$ and $z = 11$, find y when $x = 4$ and $z = 9$.

   (A) -213 (B) $\frac{17}{6}$ (C) -3
   (D) $\frac{18 - \sqrt{6}}{6}$ (E) None of these

7. A machine can produce 140 valves in 20 seconds. At this rate, how many valves will it produce in 30 minutes?

   (A) 210 (B) 12,600 (C) 114
   (D) 93 (E) None of these

8. If the tax on a $125 bike is $10, at the same rate, what is the tax on a $300 bike?

   (A) $30 (B) $21 (C) $27
   (D) $24

12 CONIC SECTIONS

**Non-Routine
Problem
Solving**

Problem 1.   How many positive integers less than or equal to 3,300 are divisible by neither 3 nor 5 nor 11?

Problem 2.   For the real numbers A, B, and C, in that order, subtraction is associative. What are the possible values for A, B, and C?

Problem 3.   Let $s = ax + by$, $t = bx - ay$, and $a^2 + b^2 = 1$. Express $x^2 + y^2$ in terms of s and t. (*Hint:* Square both equations involving s and t.)

SYMMETRIC POINTS IN A COORDINATE PLANE

Objective

To determine whether two given points are symmetric with respect to the x-axis, the y-axis, or the origin

Two points in a coordinate plane are **symmetric** with respect to a given line, *l*, in the plane if *l* is the perpendicular bisector of the segment connecting the two points. Line *l* is called the **line of symmetry** or the **axis of symmetry** for the two points.

In the figure at the right, the two pairs of points $P(-5, 4)$ and $Q(-5, -4)$, $R(4, 3)$ and $S(4, -3)$ are symmetric with respect to the *x*-axis. The *x*-axis, whose equation is $y = 0$, is the axis of symmetry. Notice that the points $T(1, 2)$ and $U(2, -1)$ are not symmetric with respect to the *x*-axis.

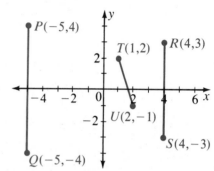

Symmetric points with respect to the x-axis	Two points, $P(x, y)$ and $Q(x, -y)$, in a coordinate plane are *symmetric with repect to the x-axis*. Their x-coordinates are the same and their y-coordinates are additive inverses.

Example 1

Determine whether the two points are symmetric with respect to the x-axis.
$C(2, -3)$ $D(2, 3)$ $E(5, -4)$ $F(-5, -4)$

Points *C* and *D* have the same *x*-coordinates. Their *y*-coordinates are additive inverses.
Thus, *C* and *D* are symmetric with respect to the *x*-axis.

Points *E* and *F* have different *x*-coordinates.

Thus, *E* and *F* are not symmetric with respect to the *x*-axis.

In the figure at the right, the two pairs of points $G(-4, 3)$ and $H(4, 3)$, $I(-5, -2)$ and $J(5, -2)$ are symmetric with respect to the *y*-axis. The *y*-axis, whose equation is $x = 0$, is the axis of symmetry. Notice that the points $K(-1, 2)$ and $L(2, 1)$ are not symmetric with respect to the *y*-axis.

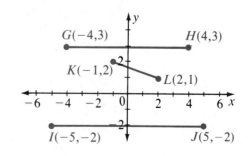

Symmetric Points in a Coordinate Plane

| Symmetric points with respect to the *y*-axis | Two points, $P(x, y)$ and $Q(-x, y)$, in a coordinate plane are *symmetric with respect to the y-axis*. Their *x*-coordinates are additive inverses and their *y*-coordinates are the same. |

Example 2

Determine whether the two points are symmetric with respect to the *y*-axis.
A(5, −6)   B(−5, −6)   M(−4, 6)   N(−4, −6)

The *x*-coordinates of points *A* and *B* are additive inverses. Their *y*-coordinates are the same.

Thus, *A* and *B* are symmetric with respect to the *y*-axis.

The *x*-coordinates of points *M* and *N* are not additive inverses.

Thus, *M* and *N* are not symmetric with respect to the *y*-axis.

It is possible for two given points in a coordinate plane to be symmetric with respect to the origin.
In the figure at the right, the two pairs of points, the endpoints of the diameters of the circle, $P(-3, 2)$ and $Q(3, -2)$, $R(3, 2)$ and $S(-3, -2)$ are symmetric with respect to the origin, whose coordinates are $(0, 0)$. The origin is a **point of symmetry.**

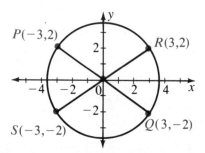

| Symmetric points with respect to the origin | Two points, $P(x, y)$ and $Q(-x, -y)$, in a coordinate plane are *symmetric with respect to the origin*. Their *x*-coordinates are additive inverses and their *y*-coordinates are additive inverses. |

Example 3

Determine whether the two points are symmetric with respect to the origin.
W(5, −9)   X(−5, 9)   Y(2, √5̄)   Z(−2, √5̄)

The *x*-coordinates of points *W* and *X* are additive inverse. Their *y*-coordinates are additive inverses.
Thus, *W* and *X* are symmetric with respect to the origin.

The *y*-coordinates of points *Y* and *Z* are not additive inverses.

Thus, *Y* and *Z* are not symmetric with respect to the origin.

Example 4

For P(3, −5), find the coordinates of the corresponding point symmetric with respect to the *x*-axis, the *y*-axis, and the origin.

x-axis	*y*-axis	origin
(3, 5)	(−3, −5)	(−3, 5)

Oral Exercises

Determine which point, A, B, C, or D, is symmetric to each given point with respect to the x-axis, y-axis, and the origin.

1. $P(7, -4)$: $A(-7, -4)$ $B(0, 0)$ $C(-7, 4)$ $D(7, 4)$
2. $Q(-2, 3)$: $A(-2, 0)$ $B(-2, -3)$ $C(2, -3)$ $D(2, 3)$
3. $R(6, 1)$: $A(-6, -1)$ $B(0, 1)$ $C(-6, 1)$ $D(6, -1)$
4. $S(-1, -5)$: $A(1, -5)$ $B(1, 5)$ $C(-1, 5)$ $D(5, 0)$

Written Exercises

Determine whether P and Q are symmetric with respect to the x-axis, the y-axis, the origin, or none of these.

Ⓐ
1. $P(-3, 2)$; $Q(3, -2)$
2. $P(-5, -4)$; $Q(5, 4)$
3. $P(-4, -3)$; $Q(4, -3)$
4. $P(7, 3)$; $Q(7, -3)$
5. $P(6, 8)$; $Q(-6, 8)$
6. $P(-4, -5)$; $Q(4, 5)$
7. $P(0, 5)$; $Q(0, -5)$
8. $P(8, 0)$; $Q(-8, 0)$
9. $P(-7, 6)$; $Q(6, -7)$

For each point, find the coordinates of the corresponding point symmetric with respect to the x-axis, the y-axis, and the origin.

10. $A(6, -4)$ 11. $B(-4, -5)$ 12. $C(6, -6)$ 13. $D(-3, 5)$ 14. $E(-1, -3)$ 15. $F(-2, 5)$
16. $G(-7, -2)$ 17. $H(5, 4)$ 18. $I(7, -2)$ 19. $J(-3, -4)$ 20. $K(2, -2)$ 21. $L(-8, -8)$

Ⓑ
22. $M(\frac{1}{3}, -\frac{1}{4})$ 23. $N(-\frac{2}{3}, \frac{4}{5})$ 24. $O(-\frac{1}{2}, -\frac{1}{3})$ 25. $P(\frac{5}{6}, -\frac{1}{3})$ 26. $Q(\frac{2}{3}, \frac{5}{7})$ 27. $R(-1, -\frac{2}{3})$
28. $S(a, -b)$ 29. $T(-a, -b)$ 30. $U(-a, b)$ 31. $V(x, -y)$ 32. $W(-x, y)$ 33. $Z(-x, -y)$

For the given point on each circle, find the coordinates of the corresponding point symmetric with respect to the x-axis, the y-axis, and the origin.

34.
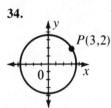

35.
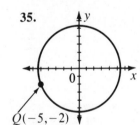

36.
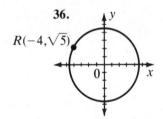

37.
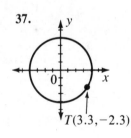

Determine if the graph of each equation is symmetric with respect to the x-axis, y-axis, or the origin. [Hint: For all real numbers a and b, if (a, b) and $(-a, -b)$ satisfy the equation, the graph is symmetric with respect to the origin.]

Ⓒ
38. $x^2 + y^2 = 7$ 39. $y = x^2$ 40. $x^2 + (-y)^2 = 7$ 41. $-y = x^2$ 42. $-y = (-x)^2$

A function is said to be *even* if its graph has y-axis symmetry and *odd* if it has origin symmetry. Which of the following are odd? even?

43. $y = x^2$ 44. $y = \frac{1}{x}$ 45. $y = |x|$ 46. $xy = 4$ 47. $y = x^3$

Symmetric Points in a Coordinate Plane

THE CIRCLE

Objectives
To write an equation of a circle given its center and radius or given its center and a point on the circle
To find the center and radius of a circle given an equation of a circle
To draw a graph of the equation of a circle

Definition: Circle	A *circle* is the set of all points in a coordinate plane that are a given distance from a given point in the plane. The given distance is the *radius* of the circle and the given point is the *center* of the circle.

For simplicity, the term "radius" is used to mean "length of the radius." That is, radius 6 means a radius has a length of 6 units.

You can write an equation of a circle with radius r and center at the origin. Let $P(x, y)$ be any point on a circle.

Use the distance formula to find the distance from P to the center of the circle.

$$d = \sqrt{(x_2 - x_1)^2 + (y_2 - y_1)^2}$$
$$r = \sqrt{(x - 0)^2 + (y - 0)^2}$$

Squaring both sides and simplifying, you get $x^2 + y^2 = r^2$.

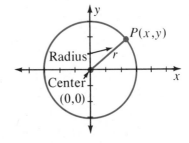

Standard form of equation of circle	The standard form of the equation of a circle with center at the origin $(0, 0)$ and radius of length r is $x^2 + y^2 = r^2$.

Example 1   **Write an equation of a circle whose center is at the origin and whose radius is 4.**

Use the standard form of the equation of a circle.

$$x^2 + y^2 = r^2$$
Substitute 4 for r. ▶ $\quad x^2 + y^2 = 4^2$
$$x^2 + y^2 = 16$$

Thus, $x^2 + y^2 = 16$ is an equation of the circle.

You can draw a graph of an equation of a circle after finding the center and the radius of the circle.

Example 2

Use $x^2 + y^2 = r^2$. ▶

Find the radius of each circle. Draw the graph of each equation.

$x^2 + y^2 = 9$

$$x^2 + y^2 = 9$$
$$r^2 = 9$$
$$r = 3$$

Plot some points 3 units from the origin.

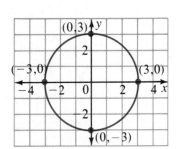

$x^2 + y^2 = 21$

$$x^2 + y^2 = 21$$
$$r^2 = 21$$
$$r = \sqrt{21} \doteq 4.6$$

Plot some points 4.6 units from the origin.

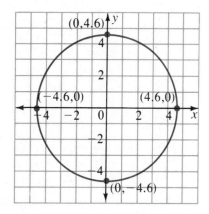

You can also write an equation of a circle with center at $C(h, k)$ and radius of length r, $r > 0$.

Let $P(x, y)$ be any point on the circle. Use the distance formula to find the distance from P to the center of the circle.

$$d = \sqrt{(x_2 - x_1)^2 + (y_2 - y_1)^2}$$
$$r = \sqrt{(x - h)^2 + (y - k)^2}$$

Squaring both sides and simplifying, you get

$$(x - h)^2 + (y - k)^2 = r^2.$$

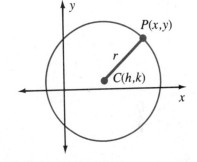

Standard form of equation of circle with center (h, k)	The standard form of the equation of a circle with center at $C(h, k)$ and radius of length r, $r > 0$, is $$(x - h)^2 + (y - k)^2 = r^2.$$

Example 3

Write an equation of a circle whose center is at $C(7, -6)$ and whose radius is 8.

Use $(x - h)^2 + (y - k)^2 = r^2.$
$$(x - 7)^2 + [y - (-6)]^2 = 8^2$$ ◀ *Substitute 7 for h, −6 for k, and 8 for r.*

Thus, $(x - 7)^2 + (y + 6)^2 = 64$ is an equation of the circle.

The Circle

Example 4

Write an equation of a circle whose center is at $C(-3, 5)$ and which passes through the point $P(7, -3)$.

Let $P(x, y) = P(7, -3)$ and $C(h, k) = C(-3, 5)$. Find r^2.

$$(x - h)^2 + (y - k)^2 = r^2$$
$$[7 - (-3)]^2 + (-3 - 5)^2 = r^2$$
$$\text{So, } r^2 = 100 + 64, \text{ or } 164$$

Thus, $(x + 3)^2 + (y - 5)^2 = 164$ is an equation of the circle.

You can rewrite equations like $x^2 + y^2 - 6x + 12y - 5 = 0$ in standard form for a circle by completing the square. Recall that to solve a quadratic equation of the form $x^2 + bx = c$ by completing the square, divide b by 2, square the result, then add $\left(\dfrac{b}{2}\right)^2$ to each side.

Example 5

Rewrite $x^2 + y^2 - 6x + 12y - 5 = 0$ in the standard form $(x - h)^2 + (y - k)^2 = r^2$. Find the center $C(h, k)$ and the radius r of the circle. Complete the square.

Group the terms with the same variable. ▶ $(x^2 - 6x + \underline{?}) + (y^2 + 12y + \underline{?}) = 5$

$\left(-\dfrac{6}{2}\right)^2 = 9; \left(\dfrac{12}{2}\right)^2 = 36$ ▶ $x^2 - 6x + 9 + y^2 + 12y + 36 = 5 + 9 + 36$

$(x - 3)^2 + (y + 6)^2 = 50$

$(x - h)^2 + (y - k)^2 = r^2$

So, $h = 3$, $k = -6$, and $r = \sqrt{50}$, or $5\sqrt{2}$

Thus, $(x - 3)^2 + (y + 6)^2 = 50$ is the equation in standard form for the circle with center $C(3, -6)$, and radius $r = 5\sqrt{2}$.

Example 6

Draw the graph of the equation $x^2 + y^2 + 2x - 4y + 1 = 0$.

Rewrite in the standard form $(x - h)^2 + (y - k)^2 = r^2$.

Complete the ▶ $x^2 + 2x + \underline{?} + y^2 - 4y + \underline{?} = -1$
square.

$x^2 + 2x + 1 + y^2 - 4y + 4 = -1 + 1 + 4$

$(x + 1)^2 + (y - 2)^2 = 4$

Thus, $C(h, k) = C(-1, 2)$ and $r = \sqrt{4}$, or 2.

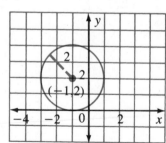

Oral Exercises

Find the center and the radius for each circle.

1. $(x - 2)^2 + (y - 8)^2 = 4$ **2.** $(x + 3)^2 + (y + 2)^2 = 21$ **3.** $(x - 1)^2 + (y + 2)^2 = 20$
4. $(x + 5)^2 + (y - 3)^2 = 18$ **5.** $x^2 + (y - 3)^2 = 16$ **6.** $(x + 6)^2 + y^2 = 15$

Written Exercises

Write an equation, in standard form, for each circle with its center at the origin and with the given radius.

Ⓐ   **1.** 1   **2.** 12   **3.** 3   **4.** 15   **5.** 2
     6. 13   **7.** $\sqrt{2}$ **8.** $\sqrt{5}$ **9.** $\sqrt{3}$ **10.** $\sqrt{7}$

Find each radius. Draw the graph of each equation.

11. $x^2 + y^2 = 25$ **12.** $x^2 + y^2 = 16$ **13.** $x^2 + y^2 = 81$ **14.** $x^2 + y^2 = 100$
15. $x^2 + y^2 = 36$ **16.** $x^2 + y^2 = 3$ **17.** $x^2 + y^2 = 12$ **18.** $x^2 + y^2 = 6$

Write an equation, in standard form, for each circle given its center C and radius r.

Ⓑ   **19.** $C(5, 6)$; $r = 7$ **20.** $C(-2, 8)$; $r = 9$ **21.** $C(11, -7)$; $r = 3\frac{1}{2}$
     22. $C(-3, -6)$; $r = 2$ **23.** $C(0, 0)$; $r = \sqrt{5}$ **24.** $C(0, 0)$; $r = \sqrt{14}$

Write an equation, in standard form, for each circle given its center C and a point P on the circle.

25. $C(2, -3)$; $P(4, 6)$ **26.** $C(7, 0)$; $P(5, 2)$ **27.** $C(-6, -3)$; $P(4, -1)$
28. $C(-4, -3)$; $P(-3, -4)$ **29.** $C(0, 0)$; $P(8, 0)$ **30.** $C(0, 0)$; $P(0, 4)$

Write each equation of a circle in standard form. Find the center $C(h, k)$ and the radius r of the circle.

31. $x^2 + y^2 - 6y - 7 = 0$ **32.** $x^2 + y^2 + 8x - 20 = 0$
33. $x^2 + y^2 + 8x + 4y - 5 = 0$ **34.** $x^2 + y^2 - 6x - 10y - 2 = 0$

Draw the graph of each equation.

35. $x^2 + y^2 + 4y - 12 = 0$ **36.** $x^2 + y^2 - 6x - 3 = 0$
37. $x^2 + y^2 + 10x - 6y + 25 = 0$ **38.** $x^2 + y^2 - 4x + 8y + 6 = 0$

Write an equation for each circle described. (Ex. 39–43)

Ⓒ   **39.** $P(5, -3)$ and $Q(-3, 11)$ are the endpoints of a diameter of the circle.
     40. The circle is tangent to the x-axis at $A(6, 0)$ and tangent to the y-axis at $B(0, 6)$.
     41. The circle intersects the axes at $Q(0, 3)$, $R(-1, 0)$, $S(0, -3)$, and $T(9, 0)$.
     42. $\overline{AB}$ is a diameter of the circle through $A(-2, 6)$ and $B(8, -4)$.
     43. Write an equation of the circle determined by $T(6, -2)$, $V(5, -9)$, and $W(-1, -1)$.

Determine the center $C(h, k)$ and the radius r for each circle.

44. $x^2 + y^2 + 2ax + 2by + c = 0$ **45.** $x^2 + y^2 + cx + dy + e = 0$

The Circle

THE ELLIPSE

Objectives

To write an equation of an ellipse given its center and its x- and y-intercepts
To find the center and x- and y-intercepts of an ellipse, given an equation of an ellipse
To draw a graph of the equation of an ellipse

The orbit of the earth around the sun is in the shape of an **ellipse.** An ellipse can be defined as a set of points as follows.

Definition: Ellipse	An *ellipse* is the set of all points in a coordinate plane such that for each point of the set, the sum of its distances from two fixed points is constant. Each of the fixed points is called a *focus* (plural: foci).

You can write an equation of an ellipse whose center is at the origin, foci are $F_1(-3, 0)$ and $F_2(3, 0)$, and where the sum of the distances from any point on the ellipse to the foci is 10.

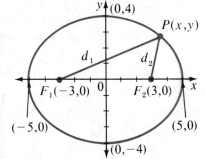

Let $P(x, y)$ be any point on the ellipse. Then find the distance d_1 from $P(x, y)$ to $F_1(-3, 0)$ and d_2 from $P(x, y)$ to $F_2(3, 0)$. Use the distance formula.

$$d_1 = \sqrt{[x - (-3)]^2 + (y - 0)^2}$$
$$d_2 = \sqrt{(x - 3)^2 + (y - 0)^2}$$

The sum of the distances is 10. Use this fact to write one radical equation. Then simplify the radical equation by squaring each side.

$$\sqrt{(x + 3)^2 + y^2} + \sqrt{(x - 3)^2 + y^2} = 10$$
$$\sqrt{x^2 + 6x + 9 + y^2} = 10 - \sqrt{x^2 - 6x + 9 + y^2}$$
$$(\sqrt{x^2 + 6x + 9 + y^2})^2 = (10 - \sqrt{x^2 - 6x + 9 + y^2})^2$$
$$x^2 + 6x + 9 + y^2 = 100 - 20\sqrt{x^2 - 6x + 9 + y^2} + x^2 - 6x + 9 + y^2$$
$$12x - 100 = -20\sqrt{x^2 - 6x + 9 + y^2}$$
$$3x - 25 = -5\sqrt{x^2 - 6x + 9 + y^2}$$
$$(3x - 25)^2 = (-5\sqrt{x^2 - 6x + 9 + y^2})^2$$
$$9x^2 - 150x + 625 = 25x^2 - 150x + 225 + 25y^2$$
$$400 = 16x^2 + 25y^2$$

Divide each side by 400 to get the following equation of the ellipse.

$$\frac{x^2}{25} + \frac{y^2}{16} = 1$$

Notice that the x-intercepts of the ellipse in the figure above are $\pm\sqrt{25}$, or ±5, and the y-intercepts are $\pm\sqrt{16}$, or ±4.

<table>
<tr>
<td>

Standard form
of equation
of ellipse with
center (0, 0)

</td>
<td>

The standard form of the equation of an ellipse with center at the origin
(0, 0) and whose x- and y-intercepts are $\pm a$ and $\pm b$, respectively, is
$$\frac{x^2}{a^2} + \frac{y^2}{b^2} = 1; \; a \neq 0, \, b \neq 0.$$

</td>
</tr>
</table>

Example 1

Write an equation, in standard form, of an ellipse whose center is at the origin and whose x-intercepts are ± 9 and whose y-intercepts are ± 7.

Use standard form of the equation of an ellipse.
$$\frac{x^2}{a^2} + \frac{y^2}{b^2} = 1$$

Substitute ± 9 for a and ± 7 for b. ▶ $\dfrac{x^2}{(\pm 9)^2} + \dfrac{y^2}{(\pm 7)^2} = 1$

Thus, $\dfrac{x^2}{81} + \dfrac{y^2}{49} = 1$ is an equation of the ellipse.

You can also write an equation of an ellipse whose center is at $C(h, k)$ and whose x'- and y'-intercepts are $\pm a$ and $\pm b$. The x', y' axes are obtained by *translating* the x, y axes. Then center (h, k) is moved h units horizontally and k units vertically.

<table>
<tr>
<td>

Standard form
of equation
of ellipse with
center (h, k)

</td>
<td>

The standard form of the equation of an ellipse with center at $C(h, k)$
and whose x'- and y'-intercepts are $\pm a$ and $\pm b$, respectively, is
$$\frac{(x - h)^2}{a^2} + \frac{(y - k)^2}{b^2} = 1.$$

</td>
</tr>
</table>

Example 2

Sketch the graph of an ellipse whose center is $C(2, -1)$, whose x'-intercepts are ± 4, and whose y'-intercepts are ± 6. Then, write an equation for the ellipse.

Use the standard form
$$\frac{(x - h)^2}{a^2} + \frac{(y - k)^2}{b^2} = 1.$$

$C(h, k) = C(2, -1)$
Substitute ± 4 for a and ± 6 for b. ▶ $\dfrac{(x - 2)^2}{(\pm 4)^2} + \dfrac{(y + 1)^2}{(\pm 6)^2} = 1$

Thus, $\dfrac{(x - 2)^2}{16} + \dfrac{(y + 1)^2}{36} = 1$
is an equation of the ellipse.

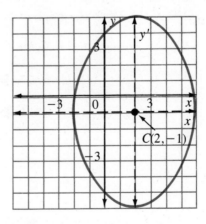

You can also rewrite equations like $4x^2 + 9y^2 + 32x - 90y + 253 = 0$ in standard form for an ellipse by completing the square.

The Ellipse

Example 3   **Rewrite $4x^2 + 9y^2 + 32x - 90y + 253 = 0$ in standard form for an ellipse. Find the center $C(h, k)$ and x'- and y'-intercepts. Draw the graph.**

Group the terms with the same variable. ▶ $(4x^2 + 32x + \underline{\,?\,}) + (9y^2 - 90y + \underline{\,?\,}) = -253$

Factor out the coefficients of $4x^2$ and $9y^2$. ▶ $4(x^2 + 8x + \underline{\,?\,}) + 9(y^2 - 10y + \underline{\,?\,}) = -253$

Complete the squares in x and y. ▶ $4(x^2 + 8x + 16) + 9(y^2 - 10y + 25) = -253 + 4(16) + 9(25)$
$4(x + 4)^2 + 9(y - 5)^2 = 36$

Divide each term by 36. ▶ $\dfrac{4(x + 4)^2}{36} + \dfrac{9(y - 5)^2}{36} = \dfrac{36}{36}$

$$\dfrac{(x - h)^2}{a^2} + \dfrac{(y - k)^2}{b^2} = 1$$

Thus, (1) $\dfrac{(x + 4)^2}{9} + \dfrac{(y - 5)^2}{4} = 1$ is the equation in standard form.

(2) $h = -4$ and $k = 5$, so the center is at $C(-4, 5)$.

(3) $a = \pm3$ and $b = \pm2$, so the x'-intercepts are ±3 and the y'-intercepts are ±2. The horizontal and vertical diameters are $2 \cdot 3$ and $2 \cdot 2$ units long, respectively.

(4) Draw the ellipse through the endpoints of the diameters.

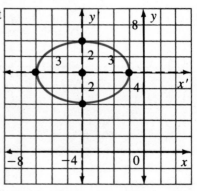

Example 4   **Draw the graph of $6x^2 + 9y^2 = 27$. Find the x- and y-intercepts.**

Write $6x^2 + 9y^2 = 27$ in standard form.

Divide by 27. ▶ $\dfrac{6x^2}{27} + \dfrac{9y^2}{27} = \dfrac{27}{27}$

Use $\dfrac{x^2}{a^2} + \dfrac{y^2}{b^2} = 1$. ▶ $\dfrac{x^2}{\frac{27}{6}} + \dfrac{y^2}{3} = 1$

$\dfrac{x^2}{\frac{9}{2}} + \dfrac{y^2}{3} = 1$

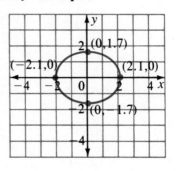

Find a and b. ▶ $a = \pm\dfrac{3\sqrt{2}}{2} \doteq \pm2.1 \qquad b = \pm\sqrt{3} \doteq \pm1.7$

The x-intercepts are ±2.1 and the y-intercepts are ±1.7.

Written Exercises

Write an equation, in standard form, of an ellipse whose center is at the orgin and whose intercepts are given.

(A)

1. x-intercepts ± 3; y-intercepts ± 2

2. x-intercepts $\pm \sqrt{8}$; y-intercepts ± 3

3. x-intercepts ± 2; y-intercepts $\pm \sqrt{3}$

4. x-intercepts $\pm \sqrt{3}$; y-intercepts $\pm \sqrt{5}$

5. x-intercepts ± 8; y-intercepts ± 6

6. x-intercepts ± 7; y-intercepts $\pm \sqrt{3}$

Find the x- and y-intercepts of the ellipse whose equation is given. Then draw the graph.

7. $\dfrac{x^2}{16} + \dfrac{y^2}{9} = 1$

8. $\dfrac{x^2}{36} + \dfrac{y^2}{25} = 1$

9. $\dfrac{x^2}{25} + \dfrac{y^2}{4} = 1$

10. $\dfrac{x^2}{81} + \dfrac{y^2}{9} = 1$

11. $\dfrac{x^2}{4} + \dfrac{y^2}{16} = 1$

12. $\dfrac{x^2}{100} + \dfrac{y^2}{25} = 1$

13. $\dfrac{x^2}{9} + \dfrac{y^2}{100} = 1$

14. $\dfrac{x^2}{4} + \dfrac{y^2}{25} = 1$

15. $4x^2 + y^2 = 100$

16. $3x^2 + y^2 = 75$

17. $x^2 + 9y^2 = 36$

18. $4x^2 + 25y^2 = 100$

Write an equation, in standard form, for each ellipse given its center C and its x'- and y'-intercepts.

19. $C(-1,3)$, $a = \pm 4$, $b = \pm 5$

20. $C(2,-1)$, $a = \pm 3$, $b = \pm 6$

21. $C(-2,-3)$, $a = \pm 2$, $b = \pm 4$

22. $C(4,3)$, $a = \pm 4$, $b = \pm 7$

23. $C(-5,2)$, $a = \pm \sqrt{7}$, $b = \pm 6$

24. $C(-1,-3)$, $a = \pm \sqrt{9}$, $b = \pm \sqrt{5}$

Determine the center of the ellipse whose equation is given. Find the lengths of the horizontal and vertical diameters. Draw the ellipse.

(B)

25. $4x^2 + 25y^2 + 16x - 150y + 141 = 0$

26. $16x^2 + 9y^2 - 96x + 72y + 144 = 0$

27. $4x^2 + y^2 - 32x - 4y + 52 = 0$

28. $9x^2 + 25y^2 + 54x + 200y + 256 = 0$

Write an equation, in standard form, of each ellipse with center at the origin, F_1 and F_2 the foci, and where c is the *sum of the distances* from any point on the ellipse to the foci.

(C)

29. $F_1(-3,0)$, $F_2(3,0)$; $c = 8$

30. $F_1(-2,0)$, $F_2(2,0)$; $c = 12$

31. $F_1(-4,0)$, $F_2(4,0)$; $c = 10$

32. $F_1(-5,0)$, $F_2(5,0)$; $c = 14$

Find the domain and range of the function defined by the given equation by solving the equation for each variable in terms of the other variable. [The *extent* of the graph of the equation is the set of all possible values of x and the set of all possible values of y. This is the region of the coordinate plane that the graph occupies.]

33. $\dfrac{x^2}{9} + \dfrac{y^2}{16} = 1$

34. $\dfrac{x^2}{36} + \dfrac{y^2}{25} = 1$

35. $\dfrac{x^2}{4} + \dfrac{y^2}{81} = 1$

Draw the graph of each equation. Determine the extent of the graph.

36. $y = \pm \dfrac{1}{3}\sqrt{81 - x^2}$

37. $y = \pm \dfrac{1}{2}\sqrt{16 - x^2}$

38. $y = \pm \dfrac{3}{2}\sqrt{4 - x^2}$

$(x - h)^2 + (y - k)^2 = r^2$ defines a circle whose area is πr^2.

$\dfrac{(x - h)^2}{a^2} + \dfrac{(y - k)^2}{b^2} = 1$ defines an ellipse whose area is πab.

Find the area of each figure defined.

39. $x^2 + y^2 + 8x + 4y - 5 = 0$

40. $x^2 + y^2 - 6x - 10y - 2 = 0$

41. $4x^2 + 25y^2 + 16x - 150y + 141 = 0$

42. $16x^2 + 9y^2 - 96x + 72y + 144 = 0$

The Ellipse

THE PARABOLA

Objectives **To draw the graph of the equation of a parabola whose vertex is not at the origin**
To find the coordinates of the vertex and write the equation of the axis of symmetry of a parabola
To estimate graphically the roots of a quadratic equation

The graph of a quadratic function is called a **parabola**. In general, any equation of the form $y = ax^2 + bx + c$, where a, b, and c, are real numbers with $a \neq 0$, describes a parabola.

You can graph a parabola determined by the equation $y = x^2$ or $y = -x^2$ as follows. First, make a table by choosing values for x and finding corresponding values for y. Then, plot as many points as needed to determine the parabola.

x	x^2	y
-2	$(-2)^2$	4
-1	$(-1)^2$	1
0	0^2	0
1	1^2	1
2	2^2	4

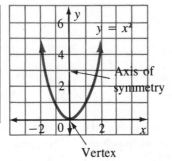

x	$-x^2$	y
-2	$-(-2)^2$	-4
-1	$-(-1)^2$	-1
0	0^2	0
1	$-(1)^2$	-1
2	$-(2)^2$	-4

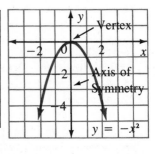

The **vertex**, or **turning point**, of each parabola above is the origin $(0,0)$. The **axis of symmetry** passes through the vertex and is parallel to—or in this case, is—the y-axis. You can draw the graph of an equation of the form $y = (x - h)^2$ by using the basic graph of $y = x^2$ for reference.

Example 1   **Graph $y = (x - 2)^2$ and $y = (x + 3)^2$. Compare with the graph of $y = x^2$.**

x	$(x - 2)^2$	y
0	$(0 - 2)^2$	4
1	$(1 - 2)^2$	1
2	$(2 - 2)^2$	0
3	$(3 - 2)^2$	1
4	$(4 - 2)^2$	4

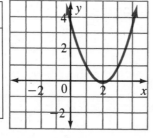

x	$(x + 3)^2$	y
-1	$(-1 + 3)^2$	4
-2	$(-2 + 3)^2$	1
-3	$(-3 + 3)^2$	0
-4	$(-4 + 3)^2$	1
-5	$(-5 + 3)^2$	4

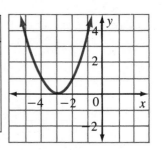

Move the graph of $y = x^2$ two units to the right.

Move the graph of $y = x^2$ three units to the left.

Chapter Twelve

Using the basic graph of $y = x^2$ for reference, you can draw the graph of an equation of the form $y = x^2 + k$.

Example 2   **Graph $y = x^2 + 2$ and $y = x^2 - 1$. Compare with the graph of $y = x^2$.**

x	$x^2 + 2$	y
-1	$1 + 2$	3
0	$0 + 2$	2
1	$1 + 2$	3

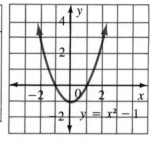
$y = x^2 + 2$

x	$x^2 - 1$	y
-2	$4 - 1$	3
-1	$1 - 1$	0
0	$0 - 1$	-1
1	$1 - 1$	0
2	$4 - 1$	3

$y = x^2 - 1$

Move the graph of $y = x^2$ two units up.   Move the graph of $y = x^2$ one unit down.

Standard form of equation of parabola symmetric to y-axis

The standard form of the equation of a parabola with vertex at (h, k) and axis of symmetry parallel to the y-axis is
$$y = (x - h)^2 + k, \text{ or } y - k = (x - h)^2.$$

Example 3   **Graph $y = x^2 + 8x + 6$. Estimate graphically, to the nearest tenth, the roots of the quadratic equation.**

Complete the square. ▶

Rewrite in the standard form $y = (x - h)^2 + k$.
$y = x^2 + 8x + \underline{\ ?\ } + 6$
$y = x^2 + 8x + 16 - 16 + 6$
$y = (x + 4)^2 - 10$
$y = [x - (-4)]^2 - 10$

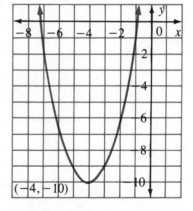
$(-4, -10)$

The vertex has coordinates $(-4, -10)$. Move the graph of $y = x^2$ four units to the left and ten units down.

Thus, the roots (x-intercepts) of the quadratic equation are approximately -7.2 and -0.8.

In Example 3, the graph intersects the x-axis in two points with coordinates of $(-7.2, 0)$ and $(-0.8, 0)$. Notice that the *equation of the axis of symmetry,* $x = -4$, is the *average of the roots.* That is, $x = \dfrac{-7.2 + (-0.8)}{2} = -4$. In general, if r and s are the solutions of the quadratic equation, then an equation of the axis of symmetry is $x = \dfrac{r + s}{2}$. Since the sum $(r + s)$ of the roots of a

quadratic equation is $-\dfrac{b}{a}$ and since $x = \dfrac{r + s}{2}$, then $x = \dfrac{-\dfrac{b}{a}}{2}$, or $x = -\dfrac{b}{2a}$.

The Parabola

Equation of axis of symmetry.	For a parabola with equation $y = ax^2 + bx + c$, $a \neq 0$, an equation of the axis of symmetry is $x = -\dfrac{b}{2a}$.

Example 4

Write an equation of the axis of symmetry of the parabola with equation $y = 4x + x^2 - 3$.

Write the equation in the form $y = ax^2 + bx + c$.

$y = 1x^2 + 4x - 3$

    ↑   ↑

    a b ◀ *Determine a and b.*

The equation for the axis of symmetry is $x = -\dfrac{b}{2a}$.

    Substitute 1 for a ▶ $x = -\dfrac{4}{2}$, or -2

    and 4 for b.

Thus, an equation of the axis of symmetry is $x = -2$.

Example 5

Graph $y = x^2 - 7x + 5$. Estimate graphically, to the nearest tenth, the roots of the quadratic equation. Determine the coordinates of the vertex, write the equation of the axis of symmetry, and find the x-intercepts.

Rewrite in the standard form $y = (x - h)^2 + k$.

Complete the ▶
square.

$y = x^2 - 7x + \underline{\ ?\ } + 5$

$y = x^2 - 7x + \dfrac{49}{4} - \dfrac{49}{4} + 5$

$y = \left(x - \dfrac{7}{2}\right)^2 - \dfrac{29}{4}$

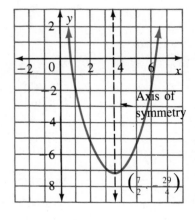

The vertex has coordinates $\left(\dfrac{7}{2}, -\dfrac{29}{4}\right)$. Move the graph of $y = x^2$ to the right 3.5 units and down 7.25 units.

The equation of the axis of symmetry is
$x = -\dfrac{b}{2a}$. $x = -\dfrac{-7}{2}$, or $x = \dfrac{7}{2}$, or 3.5.

The x-intercepts are the points of intersection of the graph and the x-axis. They are approximately 0.8 and 6.2.

Thus, the roots of the quadratic equation are approximately 0.8 and 6.2, the coordinates of the vertex are $\left(\dfrac{7}{2}, -\dfrac{29}{4}\right)$, and the equation of the axis of symmetry is $x = \dfrac{7}{2}$.

Example 6

Draw the graphs of $y = \frac{1}{2}x^2$, $y = x^2$, and $y = 2x^2$ on the same coordinate axes. Write an equation of the axis of symmetry and the coordinates of the vertex of each parabola.

x	x^2	$\frac{1}{2}x^2$	$2x^2$
-2	4	2	8
-1	1	.5	2
0	0	0	0
1	1	.5	2
2	4	2	8

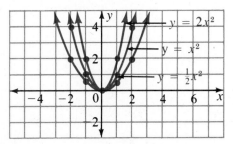

Thus, for each parabola, $x = 0$ is an equation of the axis of symmetry and $(0, 0)$ are the coordinates of the vertex.

In Example 6, notice that each parabola intersects the x-axis at the point whose coordinates are $(0, 0)$. Since there is only one point of intersection, the single x-intercept of each graph is 0. This means also that the corresponding quadratic equations each have exactly one real number solution.

Sometimes a parabola of the form $y = ax^2 + bx + c$ does not intersect the x-axis. In this case, there is no x-intercept and the corresponding quadratic equation has no real number solution. The solutions of such quadratic equations are imaginary numbers.

You can also graph a parabola determined by the equation $x = y^2$ or $x = -y^2$ by making a table of values. To make a table of values for these equations, choose values for y and find corresponding values for x.

y	y^2	x
-2	$(-2)^2$	4
-1	$(-1)^2$	1
0	0^2	0
1	1^2	1
2	2^2	4

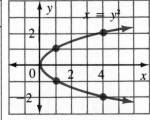

y	$-y^2$	x
-2	$-(-2)^2$	-4
-1	$-(-1)^2$	-1
0	-0^2	0
1	$-(1)^2$	-1
2	$-(2)^2$	-4

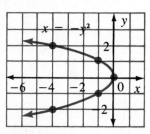

Notice that the graphs of $x = y^2$ and $x = -y^2$ are symmetric with respect to the x-axis. Parabolas of this form are relations but not functions. The standard form of equation for such parabolas is stated below.

Standard form of equation of parabola symmetric to x-axis	The standard form of the equation of a parabola with vertex at (h, k) and axis of symmetry parallel to the x-axis is $$x = (y - k)^2 + h, \text{ or } x - h = (y - k)^2.$$

The Parabola

Example 7 **Graph $x = 2y^2 + 8y + 9$. Determine the coordinates of the vertex and write the equation of the axis of symmetry.**

Rewrite $x = 2y^2 + 8y + 9$ in the standard form $x = (y - k)^2 + h$.

Factor. ▶ $x = 2(y^2 + 4y) + 9$

Complete the ▶ $x = 2(y^2 + 4y + 4) - 8 + 9$

square. $x = 2(y + 2)^2 + 1$

So, the vertex has coordinates $(1, -2)$. Move

$y = -\dfrac{b}{2a} = -\dfrac{8}{4}$,   the graph of $x = y^2$ one unit to the right and two units down.

or $y = -2$.   ▶   The axis of symmetry is $y = -2$.

Thus, the vertex has coordinates $(1, -2)$ and the equation of the axis of symmetry is $y = -2$.

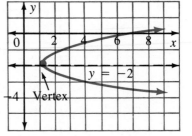

Written Exercises

Draw a graph of each of the following equations. Use a basic graph for reference.

Ⓐ **1.** $y = (x + 1)^2$ **2.** $y = (x - 3)^2$ **3.** $y = (x - 1)^2$ **4.** $y = (x + 3)^2$

5. $y = x^2 + 1$ **6.** $y = x^2 - 2$ **7.** $y = x^2 - 4$ **8.** $y = x^2 + 3$

9. $y = (x - 2)^2 + 1$ **10.** $y = (x + 2)^2 - 1$ **11.** $y = (x - 1)^2 - 2$ **12.** $y = (x - 1)^2 + 2$

Write an equation of the axis of symmetry of each parabola with the given equation. Determine the coordinates of the vertex. Do not draw the graphs.

13. $y = x^2 - 2x + 3$ **14.** $y = -x^2 + 3x - 1$ **15.** $y = 6x + 2x^2 - 8$

16. $y = -5x + 6x^2 - 4$ **17.** $y = -4x^2 - x + 7$ **18.** $y = -10x^2 + 4x - 6$

Graph each quadratic equation. Estimate, to the nearest tenth, the x-intercept(s) of each parabola.

Ⓑ **19.** $y = x^2 - 10x + 25$ **20.** $y = x^2 + 6x + 9$ **21.** $x = y^2 - 4y + 4$

22. $y = x^2 + 4x + 2$ **23.** $y = -x^2 + 4x - 1$ **24.** $x = y^2 + 6y + 7$

25. $y = 2x^2 - 12x + 18$ **26.** $y = -2x^2 - 8x - 8$ **27.** $x = 2y^2 - 16y + 32$

28. $y = 2x^2 + 12x + 13$ **29.** $x = 2y^2 - 8y + 3$ **30.** $y = 3x^2 - 18x + 19$

31. $x = -y^2 - 10y - 22$ **32.** $y = -3x^2 + 12x - 7$ **33.** $x = 3y^2 + 18y + 20$

Graph each set of functions on the same coordinate axes. Find an equation of the axis of symmetry and the coordinates of the vertex.

34. $y = -\frac{1}{4}x^2, \ y = -x^2, \ y = -4x^2$ **35.** $y = \frac{1}{4}x^2 - 2, \ y = x^2 - 2, \ y = 4x^2 - 2$

36. $y = x^2, \ y = 2x^2, \ y = 4x^2$ **37.** $y = x^2 + 1, \ y = 2x^2 + 1, \ y = 4x^2 + 1$

Find the inverse of the function determined by each given equation. Draw the graph of each inverse. Is the inverse an inverse function? Find an equation of the axis of symmetry for the inverse relation.

Ⓒ **38.** $y = x^2 - 3x - 4$ **39.** $y = x^2 + 2x - 6$ **40.** $y = -x^2 - 4x + 3$

41. $y = -2x^2 - 4x + 5$ **42.** $y = -3x^2 + 6x + 1$ **43.** $y = 4x^2 - 5x - 2$

44. Show that the coordinates of the vertex of the parabola with equation

$y = ax^2 + bx + c$ are $\left(-\dfrac{b}{2a}, \dfrac{4ac - b^2}{4a}\right)$.

THE HYPERBOLA

Objectives

To write an equation of a hyperbola
To determine the intercepts of a hyperbola
To write the equations of the asymptotes of a hyperbola
To draw the graph of an equation of a hyperbola

Another curve that can be defined as a set of points is a **hyperbola.**

Definition: Hyperbola	A *hyperbola* is the set of all points in a coordinate plane such that for each point of the set, the difference of its distances from two fixed points is constant. The fixed points are called the *foci*.

You can write an equation of a hyperbola whose center is at the origin, foci are $F_1(-5, 0)$ and $F_2(5, 0)$, and where the difference of the distances from any point on the hyperbola to the foci is 8.

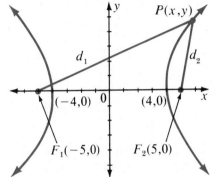

Let $P(x, y)$ be any point on the hyperbola. Then find the distance d_1 from $P(x, y)$ to $F_1(-5, 0)$ and d_2 from $P(x, y)$ to $F_2(5, 0)$. Use the distance formula.

$$d_1 = \sqrt{[x - (-5)]^2 + (y - 0)^2}$$
$$d_2 = \sqrt{(x - 5)^2 + (y - 0)^2}$$

The difference of the distance is 8. Use this fact to write an equation for the hyperbola.

$$d_1 - d_2 = 8 \text{ or } d_1 = 8 + d_2$$
$$\sqrt{(x + 5)^2 + y^2} = 8 + \sqrt{(x - 5)^2 + y^2}$$
$$\sqrt{x^2 + 10x + 25 + y^2} = 8 + \sqrt{x^2 - 10x + 25 + y^2}$$
$$(\sqrt{x^2 + 10x + 25 + y^2})^2 = (8 + \sqrt{x^2 - 10x + 25 + y^2})^2$$
$$x^2 + 10x + 25 + y^2 = 64 + 16\sqrt{x^2 - 10x + 25 + y^2} + x^2 - 10x + 25 + y^2$$
$$20x - 64 = 16\sqrt{x^2 - 10x + 25 + y^2}$$
$$5x - 16 = 4\sqrt{x^2 - 10x + 25 + y^2}$$
$$(5x - 16)^2 = (4\sqrt{x^2 - 10x + 25 + y^2})^2$$
$$25x^2 - 160x + 256 = 16x^2 - 160x + 400 + 16y^2$$
$$9x^2 - 16y^2 = 144$$

An equation of the hyperbola is $\dfrac{x^2}{16} - \dfrac{y^2}{9} = 1$.

Standard form of equation of hyperbola with x-intercepts	The standard form of the equation of a hyperbola with center at the origin $(0, 0)$ and whose x-intercepts are $\pm a$ is $\dfrac{x^2}{a^2} - \dfrac{y^2}{b^2} = 1$; $a \neq 0$, $b \neq 0$.

The Hyperbola

Example 1

Determine the intercepts of the hyperbola whose equation is $\dfrac{x^2}{16} - \dfrac{y^2}{9} = 1$.

If $y = 0$, then $\dfrac{x^2}{16} = 1$ and $x = \pm 4$.

If $x = 0$, then $-\dfrac{y^2}{9} = 1$ and $y^2 = -9$ or no real number.

Thus, the x-intercepts are ± 4 and the y-intercepts do not exist.

Example 2

Draw the graph of $\dfrac{x^2}{25} - \dfrac{y^2}{9} = 1$.

Use $\dfrac{x^2}{a^2} - \dfrac{y^2}{b^2} = 1$. ▶ $\dfrac{x^2}{25} - \dfrac{y^2}{9} = 1$

$\uparrow \qquad \uparrow$

$a^2 \qquad b^2$

Find a and b. ▶ $a = \pm 5 \quad b = \pm 3$

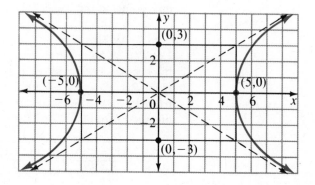

Draw a rectangle using $(5, 0)$, $(-5, 0)$, $(0, 3)$, and $(0, -3)$ as the coordinates of the midpoints of the sides.

Use the extended diagonals of the rectangle and the x-intercepts to draw a smooth curve.

The extended diagonal lines in Example 2 are called the **asymptotes** of the hyperbola. A hyperbola approaches but never intersects its asymptotes.

The graph of $\dfrac{x^2}{a^2} - \dfrac{y^2}{b^2} = 1$ is shown at the right. The y-intercept of each of the asymptotes is 0 and the slopes are $\dfrac{b}{a}$ and $-\dfrac{b}{a}$, respectively. Therefore, $y = \dfrac{b}{a}x$ and $y = -\dfrac{b}{a}x$ are the equations of the asymptotes, where a and b are positive real numbers.

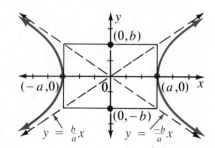

Example 3

Write the equations of the asymptotes of the hyperbola whose equation is $\dfrac{x^2}{25} - \dfrac{y^2}{9} = 1$.

$a^2 = 25 \qquad b^2 = 9$

$a = \pm 5 \qquad b = \pm 3$

Use $y = \dfrac{b}{a}x$ and $y = -\dfrac{b}{a}x$. ▶ **Thus,** $y = \dfrac{3}{5}x$ and $y = -\dfrac{3}{5}x$ are the equations of the asymptotes.

You can also graph a hyperbola that intersects the y-axis. Its equation is the inverse of the standard form $\frac{x^2}{a^2} - \frac{y^2}{b^2} = 1$, which is $\frac{y^2}{a^2} - \frac{x^2}{b^2} = 1$. The graph of the inverse is a hyperbola whose center is at the origin and whose y-intercepts are $\pm a$.

Standard form of equation of hyperbola with y-intercepts	The standard form of the equation of a hyperbola with center at the origin $(0, 0)$ and whose y-intercepts are $\pm a$ is $\frac{y^2}{a^2} - \frac{x^2}{b^2} = 1$; $a \neq 0, b \neq 0$.

The equations of the asymptotes for a hyperbola of this type are $y = \frac{a}{b}x$ and $y = -\frac{a}{b}x$.

Example 4

Draw the graph of $4y^2 - x^2 = 16$. Write the equations of the asymptotes.

Write $4y^2 - x^2 = 16$ in standard form.

$$\frac{4y^2}{16} - \frac{x^2}{16} = \frac{16}{16}, \text{ or}$$

$$\frac{y^2}{4} - \frac{x^2}{16} = 1$$

$Use \ \frac{y^2}{a^2} - \frac{x^2}{b^2} = 1.$ ▶
$$a^2 = 4 \quad b^2 = 16$$
$$a = \pm 2 \quad b = \pm 4$$

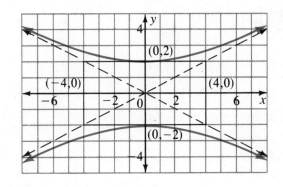

Draw a rectangle using $(0, 2)$, $(0, -2)$, $(4, 0)$, and $(-4, 0)$ as the coordinates of the midpoints of the sides.

$Use \ y = \frac{a}{b}x \ and \ y = -\frac{a}{b}x.$ ▶ $\quad y = \frac{2}{4}x \quad y = -\frac{2}{4}x$

Thus, the equations of the asymptotes are $y = \frac{1}{2}x$ and $y = -\frac{1}{2}x$.

The graph of an equation of the form $xy = k$ $(k \neq 0)$ is a hyperbola that does not intersect the x- or y-axis. A hyperbola of this type is called a **rectangular hyperbola.** The asymptotes of the hyperbola are the x- and y-axes. If $k > 0$, the branches of the hyperbola lie in Quadrants I and III. If $k < 0$, the branches lie in Quadrants II and IV.

Standard form of equation of rectangular hyperbola	The standard form of the equation of a hyperbola with center at the origin $(0, 0)$ and which does not intersect the x-axis or y-axis is $xy = k$, where k is a nonzero real number. The hyperbola is called a *rectangular hyperbola.*

Example 5   **Draw the graph of $xy = 4$. Write the equations of the asymptotes.**

Solve for y in terms of x and make a table.

$xy = 4$, or $y = \dfrac{4}{x}$ ▶

x	$\frac{4}{x}$	y
1	$\frac{4}{1}$	4
2	$\frac{4}{2}$	2
4	$\frac{4}{4}$	1
-1	$\frac{4}{-1}$	-4
-2	$\frac{4}{-2}$	-2
-4	$\frac{4}{-4}$	-1

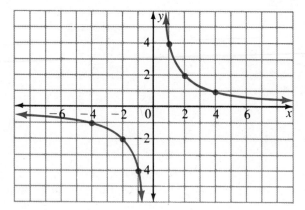

Thus, the equations of the asymptotes are $x = 0$ and $y = 0$.
You can also graph an equation of a hyperbola whose center is not at the origin.

Example 6   **Graph $\dfrac{(x + 1)^2}{9} - \dfrac{(y + 2)^2}{4} = 1$. Use the basic graph of $\dfrac{x^2}{9} - \dfrac{y^2}{4} = 1$.**

The hyperbola defined by $\dfrac{x^2}{9} - \dfrac{y^2}{4} = 1$ has

center at the origin and x-intercepts of ± 3.

The hyperbola defined by $\dfrac{(x + 1)^2}{9} - \dfrac{(y + 2)^2}{4} = 1$

has center at $C(-1, -2)$.
Move the basic graph one unit to the
left and two units down.

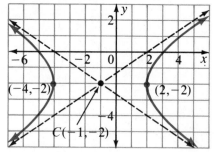

You can also rewrite equations like $25x^2 - 4y^2 - 150x - 16y + 109 = 0$ in
the standard form for a hyperbola by completing the square.

Example 7   **Draw the graph of the equation $25x^2 - 4y^2 - 150x - 16y + 109 = 0$.**

Rewrite in the standard form.

Complete the
square:
$25(9) = 225$ ▶
$-4(4) = -16$

$25x^2 - 150x + \underline{\ ?\ } - 4y^2 - 16y + \underline{\ ?\ } = -109$
$25(x^2 - 6x + \underline{\ ?\ }) - 4(y^2 + 4y + \underline{\ ?\ }) = -109$
$25(x^2 - 6x + 9) - 4(y^2 + 4y + 4) =$
$\qquad -109 + 225 - 16$
$25(x - 3)^2 - 4(y + 2)^2 = 100$

Divide by 100. ▶ $\dfrac{(x - 3)^2}{4} - \dfrac{(y + 2)^2}{25} = 1$

$\qquad\qquad \overset{\uparrow}{a^2} \qquad \overset{\uparrow}{b^2}$

$\qquad\quad a = \pm 2 \quad b = \pm 5$

Move the basic
graph three units
right and two
units down. ▶ The center is at $(3, -2)$.

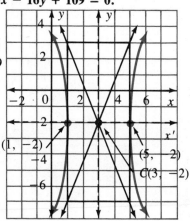

Written Exercises

Draw a graph of each equation. Determine the intercepts.

(A) 1. $\dfrac{x^2}{16} - \dfrac{y^2}{4} = 1$ 2. $\dfrac{x^2}{25} - \dfrac{y^2}{9} = 1$ 3. $\dfrac{x^2}{4} - \dfrac{y^2}{16} = 1$ 4. $\dfrac{x^2}{36} - \dfrac{y^2}{49} = 1$

5. $\dfrac{x^2}{9} - \dfrac{y^2}{16} = 1$ 6. $\dfrac{x^2}{49} - \dfrac{y^2}{81} = 1$ 7. $\dfrac{x^2}{49} - \dfrac{y^2}{4} = 1$ 8. $\dfrac{x^2}{100} - \dfrac{y^2}{64} = 1$

9. $\dfrac{x^2}{4} - \dfrac{y^2}{9} = 1$ 10. $\dfrac{x^2}{81} - \dfrac{y^2}{64} = 1$ 11. $\dfrac{x^2}{100} - \dfrac{y^2}{81} = 1$ 12. $\dfrac{x^2}{121} - \dfrac{y^2}{36} = 1$

13. $xy = 10$ 14. $xy = -6$ 15. $xy = -8$ 16. $4xy = 48$

17. $\dfrac{y^2}{16} - \dfrac{x^2}{9} = 1$ 18. $\dfrac{y^2}{25} - \dfrac{x^2}{16} = 1$ 19. $\dfrac{y^2}{81} - \dfrac{x^2}{25} = 1$ 20. $\dfrac{y^2}{36} - \dfrac{x^2}{4} = 1$

21. $\dfrac{y^2}{9} - \dfrac{x^2}{25} = 1$ 22. $\dfrac{y^2}{36} - \dfrac{x^2}{16} = 1$ 23. $\dfrac{y^2}{16} - \dfrac{x^2}{4} = 1$ 24. $\dfrac{y^2}{81} - \dfrac{x^2}{36} = 1$

Write the equations of the asymptotes.

25. $\dfrac{x^2}{25} - \dfrac{y^2}{9} = 1$ 26. $\dfrac{x^2}{36} - \dfrac{y^2}{16} = 1$ 27. $\dfrac{x^2}{4} - \dfrac{y^2}{25} = 1$ 28. $\dfrac{x^2}{16} - \dfrac{y^2}{81} = 1$

29. $\dfrac{y^2}{36} - \dfrac{x^2}{25} = 1$ 30. $\dfrac{y^2}{81} - \dfrac{x^2}{25} = 1$ 31. $\dfrac{y^2}{4} - \dfrac{x^2}{9} = 1$ 32. $\dfrac{y^2}{25} - \dfrac{x^2}{4} = 1$

33. $xy = 8$ 34. $xy = 4$ 35. $xy = -16$ 36. $xy = 20$

(B) 37. $xy = -6$ 38. $xy = -9$ 39. $-xy = 12$ 40. $-3xy = 12$

41. $4x^2 - y^2 = 36$ 42. $4x^2 - 9y^2 = 36$ 43. $16x^2 - 9y^2 = 144$ 44. $12x^2 - 3y^2 = 48$

45. $4y^2 - 16x^2 = 64$ 46. $5y^2 - 9x^2 = 45$ 47. $16y^2 - 4x^2 = 64$ 48. $4y^2 - 5x^2 = 100$

Draw the graph of each equation.

49. $\dfrac{(x-1)^2}{16} - \dfrac{(y+2)^2}{9} = 1$ 50. $\dfrac{(x+2)^2}{25} - \dfrac{(y+3)^2}{4} = 1$ 51. $\dfrac{(x-2)^2}{9} - \dfrac{(y-3)^2}{4} = 1$

Write each equation in the standard form for a hyperbola. Draw the graph of each equation.

52. $16x^2 - 9y^2 - 72y - 288 = 0$ 53. $25x^2 - 9y^2 + 200x + 175 = 0$

54. $y^2 - 4x^2 - 6y + 32x - 71 = 0$ 55. $y^2 - x^2 + 10y + 6x + 12 = 0$

Write an equation, in standard form, of each hyperbola with center at the origin, F_1 and F_2 the foci, and where c is the *difference of the distances* from any point on the hyperbola to the foci.

(C) 56. $F_1(-13, 0)$, $F_2(13, 0)$; $c = 24$ 57. $F_1(-10, 0)$, $F_2(10, 0)$; $c = 16$

58. $F_1(-6, 0)$, $F_2(6, 0)$; $c = 10$ 59. $F_1(-10, 0)$, $F_2(10, 0)$; $c = 18$

Write an equation, in standard form, of each hyperbola whose center C is given, and with the given values for a and b. Draw a graph of each equation.

60. $C(-3, 5)$; $a = \pm3$, $b = \pm2$ 61. $C(4, -1)$; $a = \pm5$, $b = \pm3$

62. $C(1, 7)$; $a = \pm8$, $b = \pm5$ 63. $C(-8, -2)$; $a = \pm10$, $b = \pm8$

NON-ROUTINE PROBLEMS

Find the average of the abscissas of all the points of $\dfrac{x^2}{9} + \dfrac{y^2}{4} = 1$ that lie in the first and second quadrants.

IDENTIFYING CONIC SECTIONS

Objective **To identify an equation of a circle, parabola, ellipse, and hyperbola**

As shown below, a circle, parabola, ellipse, or hyperbola can be formed by passing a plane through a hollow double cone. Therefore, each of these curves is called a **conic section.**

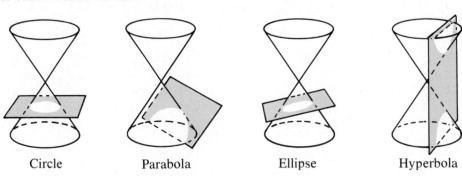

| Circle | Parabola | Ellipse | Hyperbola |

An equation of any conic section can be written in the form

$$Ax^2 + Bxy + Cy^2 + Dx + Ey + F = 0,$$
where A, B, and C are not all zero.

To identify the conic section described by a given *quadratic* equation of the form above, you should rewrite the equation in one of the standard forms below.

Conic Section	Standard Forms	
Circle	$x^2 + y^2 = r^2$	$(x - h)^2 + (y - k)^2 = r^2$
Parabola	$y = ax^2 + c$ $x = ay^2 + c$	$y = (x - h)^2 + k$ $x = (y - k)^2 + h$
Ellipse	$\dfrac{x^2}{a^2} + \dfrac{y^2}{b^2} = 1$	$\dfrac{(x - h)^2}{a^2} + \dfrac{(y - k)^2}{b^2} = 1$
Hyperbola	$\dfrac{x^2}{a^2} - \dfrac{y^2}{b^2} = 1$ $\dfrac{y^2}{a^2} - \dfrac{x^2}{b^2} = 1$	$\dfrac{(x - h)^2}{a^2} - \dfrac{(y - k)^2}{b^2} = 1$ $\dfrac{(y - k)^2}{a^2} - \dfrac{(x - h)^2}{b^2} = 1$
Rectangular Hyperbola	$xy = k; \ k \neq 0$	

Example 1

Determine whether $9x^2 - 16y^2 = 144$ is an equation of a parabola, a circle, an ellipse, or a hyperbola.

Write the equation in standard form.

$$9x^2 - 16y^2 = 144$$

Divide each term by 144. ▶ $\dfrac{9x^2}{144} - \dfrac{16y^2}{144} = \dfrac{144}{144}$

Standard form: $\dfrac{x^2}{a^2} - \dfrac{y^2}{b^2} = 1.$ ▶ $\dfrac{x^2}{16} - \dfrac{y^2}{9} = 1$

Thus, $9x^2 - 16y^2 = 144$ is an equation of a hyperbola.

Example 2

Determine whether $17 + 3x^2 = y + 5x$ is an equation of a parabola, a circle, an ellipse, or a hyperbola.

There is no y^2 term. ▶ $17 + 3x^2 = y + 5x$

Solve for y. ▶ $3x^2 - 5x + 17 = y$ ◀ *Standard form:* $y = ax^2 + bx + c.$

Thus, $17 + 3x^2 = y + 5x$ is an equation of a parabola.

When the x-intercepts and y-intercepts of an ellipse are equal, the standard form of the equation of the ellipse becomes $\dfrac{x^2}{a^2} + \dfrac{y^2}{a^2} = 1$. Multiplying by a^2, you get $x^2 + y^2 = a^2$. This is an equation of a circle with radius a. Thus, a circle can be considered to be a special type of ellipse.

Example 3

Identify each conic section whose equation is given.

$$9x^2 = 36 - 4y^2 \qquad 8x^2 + 8y^2 = 56$$

Standard form: $\dfrac{x^2}{a^2} + \dfrac{y^2}{b^2} = 1.$ ▶

$9x^2 = 36 - 4y^2$
$9x^2 + 4y^2 = 36$
$\dfrac{x^2}{4} + \dfrac{y^2}{9} = 1$

$8x^2 + 8y^2 = 56$
$\dfrac{x^2}{7} + \dfrac{y^2}{7} = 1$
$x^2 + y^2 = 7$

◀ *Standard form:* $x^2 + y^2 = r^2.$

Thus, $9x^2 = 36 - 4y^2$ is an equation of an ellipse and $8x^2 + 8y^2 = 56$ is an equation of a circle.

You may have to complete the square in x or y or *both* to rewrite an equation in the standard form for a conic section.

Example 4

Identify the conic section described by the equation $x^2 - 4x + y^2 + 6y = 1$.

Complete the square. ▶ $x^2 - 4x + 4 + y^2 + 6y + 9 = 1 + 4 + 9$

Standard form: $(x - h)^2 + (y - k)^2 = r^2$ ▶ $(x - 2)^2 + (y + 3)^2 = 14$

Thus, $x^2 - 4x + y^2 + 6y = 1$ is an equation of a circle.

Example 5

Identify the conic section described by each equation.

$$16x^2 - 9y^2 - 72y - 288 = 0 \qquad\qquad -5xy = 60$$

$$16x^2 - 9(y^2 + 8y) = 288 \qquad\qquad \frac{-5xy}{-5} = \frac{60}{-5}$$
$$16x^2 - 9(y^2 + 8y + 16) = 288 - 9(16) \qquad\qquad xy = -12$$
$$16x^2 - 9(y + 4)^2 = 144$$

Thus, $16x^2 - 9y^2 - 72y - 288 = 0$ and $-5xy = 60$ are equations of hyperbolas.

Written Exercises

Identify each conic section whose equation is given.

Ⓐ
1. $\dfrac{x^2}{16} + \dfrac{y^2}{4} = 1$
2. $\dfrac{x^2}{14} + \dfrac{y^2}{14} = 1$
3. $\dfrac{x^2}{16} - \dfrac{y^2}{16} = 1$
4. $\dfrac{x^2}{12} + \dfrac{y^2}{3} = 1$

5. $\dfrac{y^2}{4} - \dfrac{x^2}{16} = 1$
6. $x = \dfrac{-3}{y}$
7. $\dfrac{x^2}{50} + \dfrac{y^2}{50} = 1$
8. $y = \dfrac{8}{x}$

9. $6x^2 + 4y^2 = 24$
10. $x^2 = 4 + y$
11. $7x^2 - 3x = -3 + y$
12. $5x^2 - 6y^2 = 30$

13. $15x^2 + 15y^2 = 255$
14. $5x^2 - 3y^2 = 15$
15. $y - 3x^2 = 6 + x$
16. $30x^2 = -3y^2 + 15$

17. $y + x^2 = 3x + 2$
18. $x^2 + 2y^2 = 8$
19. $9x^2 - 4y^2 = 36$
20. $-y - 3x = x^2 + 5$

21. $3x^2 + 4y^2 = 48$
22. $7x^2 - 8y^2 = 35$
23. $4x^2 + 4y^2 = 15$
24. $2x^2 + 3y^2 = 45$

25. $16x^2 - 16y^2 = 5$
26. $12x^2 + 16y^2 = 17$
27. $19y^2 + 9x^2 = 5$
28. $6x^2 - 6y^2 = 8$

Ⓑ
29. $(x + 3y)(x - 3y) = 4$
30. $y - 3 = (x - 2)^2$
31. $(3x - 2y)(3x + 2y) = 1$
32. $\dfrac{x}{9} = \dfrac{1}{y}$

33. $x^2 - 6x + y^2 + 8y = 24$
34. $x^2 - 2x + y^2 - 6y = 26$
35. $x^2 + 10x + y^2 + 12y = 60$

36. $(2x - 3)^2 = y + 4$
37. $(3y - 2x)(3y + 2x) = 1$
38. $x = \dfrac{-4}{y}$

39. $2y^2 - x + 20y + 50 = 0$
40. $6x^2 + 6y^2 + 36y + 18 = 24x$

41. $3x^2 - 12x + y + 7 = 0$
42. $25x^2 + 200x + 175 = 9y^2$

43. $x^2 - 24 + 8y = 6x - y^2$
44. $5y^2 + 50y + 275 = 100x - 2x^2$

Ⓒ
45. $Ax^2 + Bxy + Cy^2 + Dx + Ey + F = 0$ if $A > 0$, $B = 0$, $C < 0$, and D, E, and F are real numbers, $F \neq 0$

46. $Ax^2 + Bxy + Cy^2 + Dx + Ey + F = 0$ if $A = C = D = E = 0$ and B and F are nonzero real numbers

Identify each conic section whose equation is given. Determine whether the graph is symmetric with respect to the x-axis, the y-axis, the origin, or none of these.

47. $5x^2 - 6y^2 = 30$
48. $x^2 + 3y^2 = 12$
49. $4x^2 + 4y^2 = 16$
50. $y - x^2 - 2x = 8$

51. $\dfrac{x^2}{16} + \dfrac{y^2}{16} = 1$
52. $\dfrac{x^2}{4} - \dfrac{y^2}{16} = 1$
53. $\dfrac{y^2}{25} - \dfrac{x^2}{4} = 1$
54. $\dfrac{x^2}{16} + \dfrac{y^2}{4} = 1$

55. $7y = \dfrac{-16}{x}$
56. $x^2 - 16 = y$
57. $\dfrac{16}{y} = \dfrac{-x}{7}$
58. $\dfrac{x^2}{18} + \dfrac{y^2}{18} = 23$

CUMULATIVE REVIEW

Find the slope of a line perpendicular to each line whose equation is given.

1. $-3x + 5y = 8$
2. $7x + 2y = 10$
3. $x = -4$
4. $y = 6$

EQUATIONS AND PROPERTIES OF CONIC SECTIONS

Objectives
To write an equation of a parabola, using the coordinates of the focus and the equation of the directrix
To draw the graph of an equation of a parabola
To write an equation of an ellipse, using the coordinates of the foci, length of major axis, or length of minor axis

In the figure at the right, the distance from the point P to the fixed point F is the same as the distance from P to the fixed line l. Notice that the curve traced by all such points P is a parabola.

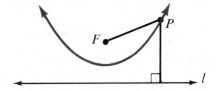

Definition: Parabola	A *parabola* is the set of all points in a coordinate plane that are equidistant from a fixed point and a fixed line in the plane. The fixed point is called the *focus* and the fixed line is called the *directrix*.

Let $P(x, y)$ be any point on a parabola with $F(0, p)$ the focus, and $y = -p$ the equation of the directrix. You can use the distance formula to find an equation of the parabola. Let d_1 = distance from P to F and d_2 = distance from P to A.

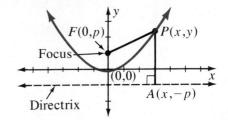

$$d_1 = d_2$$
$$\sqrt{(x - 0)^2 + (y - p)^2} = \sqrt{(x - x)^2 + [y - (-p)]^2}$$
$$\sqrt{x^2 + y^2 - 2py + p^2} = \sqrt{y^2 + 2py + p^2}$$
$$x^2 + y^2 - 2py + p^2 = y^2 + 2py + p^2$$
$$x^2 = 4py$$

There are three other types of parabolas with vertex at the origin. Information on all four types is presented in the table.

Coordinates of Focus	$(0, p)$	$(0, -p)$	$(p, 0)$	$(-p, 0)$
Equation of directrix	$y = -p$	$y = p$	$x = -p$	$x = p$
Opening	Up	Down	Right	Left
Equation of the Parabola	$x^2 = 4py$ $p > 0$	$x^2 = -4py$ $p > 0$	$y^2 = 4px$ $p > 0$	$y^2 = -4px$ $p > 0$

Example 1 **Write an equation of a parabola with $F(-3, 0)$ the focus and $x = 3$ the equation of the directrix.**

The coordinates of the vertex are of the form $(-p, 0)$.
Use $y^2 = -4px$ from the table on page 327.
$$y^2 = -4px$$
Substitute 3 for p. ▶ $\quad y^2 = -4 \cdot 3x$

Thus, $y^2 = -12x$ is an equation of the parabola.

Example 2 **Draw a graph of the equation $y^2 = -6x$.**

The equation $y^2 = -6x$ is of the form $y^2 = -4px$.

$$y^2 = -6x$$
$$-4p = -6$$
$$p = \tfrac{3}{2}$$

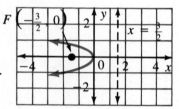

The coordinates of the focus, $(-p, 0)$, are $(-\tfrac{3}{2}, 0)$.
The equation of the directrix, $x = p$, is $x = \tfrac{3}{2}$.
The graph opens to the left.

Recall that an ellipse is the set of all points in a plane the sum of whose distances from two fixed points called the foci is a constant. $\dfrac{x^2}{a^2} + \dfrac{y^2}{b^2} = 1$ is the standard form of the equation of an ellipse with center at the origin and foci on the x-axis. This equation is derived as follows.

Let $P(x, y)$ be any point on an ellipse with $F_1(-c, 0)$ and $F_2(c, 0)$ the foci. Let $d_1 =$ distance from P to F_1 and $d_2 =$ distance from P to F_2.
Then, $d_1 + d_2 = 2a$, a constant.

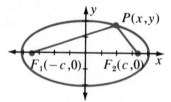

$$\sqrt{[x - (-c)]^2 + (y - 0)^2} + \sqrt{(x - c)^2 + (y - 0)^2} = 2a$$
$$\sqrt{(x + c)^2 + y^2} = 2a - \sqrt{(x - c)^2 + y^2}$$
$$x^2 + 2cx + c^2 + y^2 = 4a^2 - 4a\sqrt{(x - c)^2 + y^2} + x^2 - 2cx + c^2 + y^2$$
$$cx - a^2 = -a\sqrt{(x - c)^2 + y^2}$$
$$c^2x^2 - 2a^2cx + a^4 = a^2x^2 - 2a^2cx + a^2c^2 + a^2y^2$$
$$a^2(a^2 - c^2) = x^2(a^2 - c^2) + a^2y^2$$

$$\frac{x^2\cancel{(a^2 - c^2)}}{a^2\cancel{(a^2 - c^2)}} + \frac{\cancel{a^2}y^2}{\cancel{a^2}(a^2 - c^2)} = \frac{\cancel{a^2}\cancel{(a^2 - c^2)}}{\cancel{a^2}\cancel{(a^2 - c^2)}}$$

Let $a^2 - c^2 = b^2$.
Therefore, $\dfrac{x^2}{a^2} + \dfrac{y^2}{b^2} = 1$ is an equation of the ellipse.

<table>
<tr><td>

Properties of
an ellipse of
the form
$\dfrac{x^2}{a^2} + \dfrac{y^2}{b^2} = 1$

</td><td>

Vertices: $V_1(-a, 0)$ and $V_2(a, 0)$
Major axis: $\overline{V_1 V_2}$ of length $2a$
Minor axis: $\overline{B_1 B_2}$ of length $2b$
x-intercepts: a and $-a$
y-intercepts: b and $-b$
Foci: $F_1(-c, 0)$ and $F_2(c, 0)$
Relationship between intercepts
and foci: $b^2 = a^2 - c^2$

</td><td>

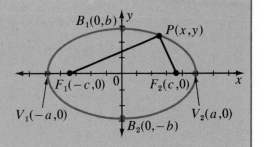

</td></tr>
</table>

Example 3

Write an equation, in standard form, of an ellipse whose center is at the origin, coordinates of vertices are $(\pm 4, 0)$, and coordinates of foci are $(\pm 3, 0)$.

The major axis of the ellipse is on the x-axis. Use $\dfrac{x^2}{a^2} + \dfrac{y^2}{b^2} = 1$.

$$b^2 = a^2 - c^2$$
$$b^2 = 4^2 - 3^2 \quad \blacktriangleleft \; \text{Substitute 4 for}$$
$$b^2 = 16 - 9 \qquad \qquad a \text{ and 3 for } c.$$
$$b^2 = 7$$

Thus, $\dfrac{x^2}{16} + \dfrac{y^2}{7} = 1$ is an equation of the ellipse.

Example 4

Write an equation, in standard form, of an ellipse whose center is at the origin, coordinates of foci are $(\pm 4, 0)$ and the length of the minor axis is 6.

Use $\dfrac{x^2}{a^2} + \dfrac{y^2}{b^2} = 1$. $\quad \blacktriangleleft$ *Major axis is on the x-axis.*

Length of minor $\qquad b^2 = a^2 - c^2$
axis = 2b = 6. $\quad \blacktriangleright \quad 3^2 = a^2 - 4^2 \quad \blacktriangleleft$ *The foci has coordinates $(\pm c, 0)$.*
Substitute 3 for b. $\qquad \; 9 = a^2 - 16 \qquad \quad$ *Substitute 4 for c.*
$\qquad \qquad \qquad \qquad \quad 25 = a^2$

Thus, $\dfrac{x^2}{25} + \dfrac{y^2}{9} = 1$ is an equation of the ellipse.

When the major axis of an ellipse is on the y-axis, the general form of the equation of the ellipse becomes $\dfrac{x^2}{b^2} + \dfrac{y^2}{a^2} = 1$.

<table>
<tr><td>

Properties of
an ellipse of
the form
$\dfrac{x^2}{b^2} + \dfrac{y^2}{a^2} = 1$

</td><td>

Vertices: $V_1(0, a)$ and $V_2(0, -a)$
Major Axis: $\overline{V_1 V_2}$ of length $2a$
Minor Axis: $\overline{B_1 B_2}$ of length $2b$
x-intercepts: b and $-b$
y-intercepts: a and $-a$
Foci: $F_1(0, c)$ and $F_2(0, -c)$
Relationship between intercepts
and foci: $b^2 = a^2 - c^2$.

</td><td>

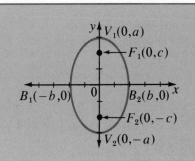

</td></tr>
</table>

Equations and Properties of Conic Sections

Example 5   **Write an equation, in standard form, of an ellipse whose center is at the origin and whose x-intercepts are ±3 and y-intercepts are ±4. Find the coordinates of the foci.**

The major axis of the ellipse is on the y-axis.

$$\text{Use } \frac{x^2}{b^2} + \frac{y^2}{a^2} = 1.$$

$$b = 3; b^2 = 9 \qquad\qquad b^2 = a^2 - c^2$$
$$a = 4; a^2 = 16 \qquad\qquad 9 = 16 - c^2$$
$$\qquad\qquad\qquad\qquad c^2 = 7, \text{ or } c = \sqrt{7}$$

Thus, $\dfrac{x^2}{9} + \dfrac{y^2}{16} = 1$ is an equation of the ellipse and $(0, \sqrt{7})$ and $(0, -\sqrt{7})$ are the coordinates of the foci.

Written Exercises

Find an equation of each parabola.

Ⓐ   **1.** Focus $(2, 0)$, directrix $x = -2$
  3. Focus $(-2, 0)$, directrix $x = 2$
  5. Focus $(0, -1)$, directrix $y = 1$
  7. Focus $(0, 2)$, directrix $y = -2$

  2. Focus $(0, -2)$, directrix $y = 2$
  4. Focus $(0, 3)$, directrix $y = -3$
  6. Focus $(4, 0)$, directrix $x = -4$
  8. Focus $(-4, 0)$, directrix $x = 4$

Draw the graph of each equation.

  9. $x^2 = -4y$
  13. $y^2 = 20x$

  10. $y^2 = -3x$
  14. $x^2 = 6y$

  11. $y^2 = 4x$
  15. $-x^2 = 3y$

  12. $x^2 = 10y$
  16. $-y^2 = 2x$

Write an equation, in standard form, of an ellipse with center at the origin and which satisfies the following conditions.

  17. Vertices $(\pm 5, 0)$, Foci $(\pm 4, 0)$
  19. Vertices $(\pm 4, 0)$, Foci $(\pm 2, 0)$
  21. Vertices $(0, \pm 6)$, Foci $(0, \pm 5)$
  23. Foci $(\pm 4, 0)$, length of minor axis 6
  25. x-intercepts ± 7, y-intercepts ± 9

  18. Vertices $(\pm 7, 0)$, Foci $(\pm 5, 0)$
  20. Vertices $(\pm \sqrt{7}, 0)$, Foci $(\pm 2, 0)$
  22. Vertices $(0, \pm 8)$, Foci $(0, \pm 4)$
  24. Foci $(0, \pm 6)$, length of minor axis 8
  26. x-intercepts ± 11, y-intercepts ± 10

Ⓑ   **27.** length of major axis 10
    length of minor axis 6

  28. length of major axis 8
    length of minor axis 4

  29. length of major axis 6
    length of minor axis 4

  30. length of major axis 12
    length of minor axis 10

Ⓒ   **31.** A *latus rectum* of an ellipse is a chord that is perpendicular to the major axis through one of the foci. Show that the length of a latus rectum for the ellipse defined by $\dfrac{x^2}{a^2} + \dfrac{y^2}{b^2} = 1$ is $\dfrac{2b^2}{a}$.

  32. The standard form of the equation of a hyperbola with x-intercepts is $\dfrac{x^2}{a^2} - \dfrac{y^2}{b^2} = 1$.

  Let $P(x, y)$ be any point on a hyperbola with $F_1(-c, 0)$ and $F_2(c, 0)$ the foci. Show how the standard form of the equation is derived.

COMPUTER ACTIVITIES

Graphs of Parabolas

OBJECTIVE: To draw graphs of vertical parabolas

The program below draws graphs of parabolas of the form $y = (x + a)^2 + b$, where a and b are restricted to values between -10 and 10.

Ordinarily, you would insert the values for x into the equation $y = (x + a)^2 + b$ to find corresponding values for y. However, since the computer must graph from top to bottom, the equation is converted into a form that finds values for x when values for y are plugged into it. The program below uses the equation $x = \pm\sqrt{y - b} - a$, and inserts values for y from 12 to -10 into it.

```
110   PRINT "VALUE OF A";: INPUT A
120   IF  ABS (A) > 10 THEN 110
130   PRINT "VALUE OF B";: INPUT B
140   IF  ABS (B) > 10 THEN 110
150   FOR Y = 12 TO  - 10 STEP  - 1
160   IF Y <  > 0 THEN 180
170   FOR I = 1 TO 39: PRINT "*";: NEXT I: PRINT :
        GOTO 340
180  Y1 = Y - B: IF Y1 < 0 THEN 320
190  A1 =  - A -  INT ( SQR (Y1)):A2 =  - A +  INT
        ( SQR (Y1))
200   IF A1 < 0 THEN 220
210  C = 0:D = A1:E = A2: GOTO 250
220   IF A2 < 0 THEN 240
230  C = A1:D = 0:E = A2: GOTO 250
240  C = A1:D = A2:E = 0
250   IF C <  - 19 THEN 330
260   PRINT  TAB( 20 + C);"*";
270   IF D = C THEN 290
280   PRINT  TAB( 20 + D);"*";
290   IF E > 19 THEN 330
300   IF D = E THEN 330
310   PRINT  TAB( 20 + 3)"*";: GOTO 330
320   PRINT  TAB( 20)"*";
330   PRINT : NEXT Y: GOTO 110
```

See the Computer Section beginning on page 574 for more information.

EXERCISES

1. Transform equations 19, 20, and 22 on page 318 to the form $Y = (X + A)^2 + B$ and then run the values through the program above.

2. Make all necessary changes in the program above so that it will graph $Y = -(X + A)^2 + B$. Be sure to specify appropriate parameters.

Graphs of Parabolas

331

CHAPTER TWELVE REVIEW

Vocabulary

asymptotes [12.5]   axis of symmetry [12.1]
circle   [12.2]   conic section   [12.6]
ellipse   [12.3]   hyperbola   [12.5]
parabola   [12.4]   rectangular hyperbola
symmetric points [12.1]   [12.5]

Determine whether P and Q are symmetric with respect to the x-axis, the y-axis, the origin, or none of these. [12.1]

1. $P(-2, 6)$; $Q(2, -6)$ **2.** $P(-3, 5)$; $Q(3, 5)$
3. $P(8, -1)$; $Q(8, 1)$ **4.** $P(-5, 3)$; $Q(3, -5)$

For each point find the coordinates of the corresponding point symmetric with respect to the x-axis, the y-axis, and the origin. [12.1]

5. $A(-5, 1)$ **6.** $B(7, -2)$ **7.** $C(-3, -8)$
8. $D\left(-\dfrac{1}{7}, -\dfrac{1}{8}\right)$ **9.** $E(m, -n)$ **10.** $F(-r, -s)$

Determine whether the following functions are odd, even, or neither. [12.1]
★**11.** $y = x^2$ ★**12.** $x - 4y^2 = 1$

Write an equation of each circle whose center is at the origin and whose radius is given. [12.2]
13. 13   **14.** 17   **15.** $\sqrt{8}$

Write an equation of each circle whose center C and radius r are given. [12.2]
16. $C(-2, 3)$, $r = 5$ **17.** $C(3, -2)$, $r = 4$

Write an equation of each circle whose center C is given and which passes through point P. [12.2]
18. $C(-6, -5)$, $P(5, -1)$
19. $C(1, -8)$, $P(-1, -6)$

Draw the graph of each equation. [12.2]
20. $x^2 + y^2 = 49$
21. $x^2 + y^2 - 8x + 6y + 11 = 0$
22. $(x - 1)^2 + (y + 2)^2 = 16$

Find the x- and y-intercepts of the ellipse whose equation is given. Then draw the graph. [12.3]
23. $\dfrac{x^2}{36} + \dfrac{y^2}{4} = 1$ **24.** $\dfrac{(x - 3)^2}{16} + \dfrac{(y + 5)^2}{4} = 1$
25. $9x^2 + y^2 = 36$
26. Draw the graph of $3x^2 + 4y^2 + 12x - 8y + 4 = 0$. [12.3]

27. Write an equation, in standard form, of an ellipse whose center is at the origin, whose x-intercepts are ± 5, and whose y-intercepts are ± 3. [12.3]

Graph each quadratic function. Draw the axis of symmetry and label the coordinates of the vertex. Write an equation of the axis of symmetry. [12.4]
28. $y = x^2 - 6x + 12$ **29.** $y = -\dfrac{1}{3}x^2$

Graph each equation. [12.4]
30. $y = (x - 2)^2 - 3$ **31.** $y = (x + 1)^2 + 4$

Graph each function. Estimate, to the nearest tenth, the x-intercepts of each parabola. [12.4]
32. $y = x^2 + 4x - 3$ **33.** $y = -2x^2 + 7x + 1$

Draw the graph of each equation. Find the equations of the asymptotes. [12.5]
34. $\dfrac{x^2}{9} - \dfrac{y^2}{4} = 1$ **35.** $xy = -6$

Draw the graph of each equation. [12.5]
36. $\dfrac{(x - 2)^2}{9} - \dfrac{(y + 1)^2}{4} = 1$
37. $9x^2 - 4y^2 - 18x - 16y - 43 = 0$

Identify each conic section. [12.6]
38. $\dfrac{x^2}{16} - \dfrac{y^2}{4} = 1$ **39.** $\dfrac{x^2}{8} + \dfrac{y^2}{8} = 1$
40. $4x^2 + 5y^2 = 20$ **41.** $10x^2 + 20y^2 = 20$
42. $x^2 - y^2 - 2x - 4y - 4 = 0$
43. $x^2 + y^2 + 4x - 4y + 7 = 0$
44. $2x^2 + 3y^2 + 6x - 9y - 15 = 0$
45. Write an equation of a parabola with $F(-4, 0)$ the focus and $x = 4$ the equation of the directrix. Draw a graph of the equation. [12.7]

Write an equation, in standard form, of an ellipse with center at the origin and which satisfies the following conditions. [12.7]
46. Vertices $(\pm 6, 0)$, foci $(\pm 5, 0)$
47. Foci $(0, \pm 7)$, length of minor axis 12

CHAPTER TWELVE TEST

Determine whether P and Q are symmetric with respect to the x-axis, the y-axis, the origin, or none of these.

1. $P(3, -4); Q(3, 4)$ **2.** $P(-2, -1); Q(2, 1)$

For each point, find the coordinates of the corresponding point symmetric with respect to the x-axis, the y-axis, and the origin.

3. $A(-8, 1)$ **4.** $B(-2, -3)$

Write an equation of each circle whose center is at the origin and whose radius is given.

5. 6   **6.** $\sqrt{20}$

7. Write an equation of a circle whose center is at $C(1, -4)$ and whose radius r is 3.

8. Write an equation of a circle whose center is at $C(-8, -5)$ and which passes through the point $P(3, -2)$.

Draw the graph of each equation.

9. $x^2 + y^2 = 4$ **10.** $(x - 3)^2 + (y + 1)^2 = 16$

Find the x- and y-intercepts of the ellipse whose equation is given. Then draw the graph.

11. $\dfrac{x^2}{16} + \dfrac{y^2}{4} = 1$ **12.** $9x^2 + 4y^2 = 36$

13. Draw the graph of $x^2 + 4y^2 + 4x - 8y - 8 = 0$.

14. Write an equation, in standard form, of an ellipse whose center is at the origin, whose x-intercepts are ± 3, and whose y-intercepts are ± 2.

Graph each quadratic function. Draw the axis of symmetry and label the coordinates of the vertex. Write an equation of the axis of symmetry.

15. $y = x^2 + 8x + 3$ **16.** $y = -\frac{2}{3}x^2$

17. $y = (x + 3)^2 - 2$

18. Graph the quadratic function $y = x^2 + 6x + 2$. Estimate, to the nearest tenth, its x-intercepts.

Draw the graph of each equation. Write the equations of the asymptotes.

19. $\dfrac{x^2}{16} - \dfrac{y^2}{4} = 1$ **20.** $xy = -2$

21. Draw the graph of $\dfrac{(x - 1)^2}{16} - \dfrac{(y + 2)^2}{25} = 1$.

Identify each conic section whose equation is given.

22. $\dfrac{x^2}{49} + \dfrac{y^2}{81} = 1$ **23.** $xy = -4$

24. $\dfrac{x^2}{9} - \dfrac{y^2}{25} = 1$

25. $3x^2 + 5y^2 + 6x - 15y + 12 = 0$

26. $x^2 - 3y^2 - 4x + 9y - 1 = 0$

27. Write an equation of a parabola with $F(0, 3)$ the focus and $y = -3$ the equation of the directrix.

28. Write an equation, in standard form, of the ellipse with center at the origin, foci $(0, \pm 10)$, and length of minor axis 8.

Determine whether the following functions are odd, even, or both.

★ **29.** $y = 7x^2$ ★ **30.** $x^2 + 16y^2 = 1$

★ **31.** Find the coordinates of the center of a circle and the radius of the circle whose equation is $x^2 - 10x + y^2 + 6y = -33$.

DIRECTIONS: Choose the *one* best answer to each question or problem.

1.

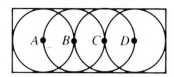

Each circle in the above figure has a diameter of 4. What is the area of the rectangle?

(A) 64   (B) 32   (C) 80
(D) 16   (E) 40

2. The circumference of a circle in a coordinate plane is 18π units. The center of the circle is at the origin. Which point is on the circle?

(A) $P(9, 0)$ (B) $P(-3, -3)$ (C) $P(0, -3)$
(D) $P(3, 0)$ (E) $P(9, 9)$

3. The area of a circle is 36π square units. The center of the circle is at $C(-2, 3)$. Which point does the circle not pass through?

(A) $P(-2, -3)$ (B) $P(4, 3)$ (C) $P(-2, 9)$
(D) $P(-8, 3)$ (E) $P(-2, 3)$

4. $\overline{AB}$ is a diameter of a circle whose center is point O. The diameter is 6, C is a point on the circle, and the measure of $\angle BOC$ is 60°. What is the length of chord $\overline{AC}$?

(A) 3   (B) $3\sqrt{3}$ (C) 6
(D) $3\sqrt{2}$ (E) None of these

5. A radius of a given circle measures 10 cm. If the length of the radius is decreased by 2 cm, what is the percent of area decrease?

(A) 81   (B) 19   (C) 64
(D) 36   (E) None of these

6. The set of all points in a coordinate plane 10 units from the origin is represented by which of the following equations?

(A) $x = 10$ (B) $x^2 + y^2 = 10$
(C) $x^2 + y^2 = \sqrt{10}$ (D) $x^2 + y^2 = 100$
(E) $y = 10$

7. An equation of a circle with center at $C(-3, 4)$, and radius 5 is

(A) $(x - 3)^2 + (y + 4)^2 = 25$
(B) $(x + 3)^2 + (y - 4)^2 = 25$
(C) $x^2 + y^2 = 25$
(D) $x^2 + y^2 = 1$ (E) None of these

8. The x- and y-intercepts of the ellipse defined by $\dfrac{x^2}{9} + \dfrac{y^2}{16} = 1$ are

(A) 3 and 4   (B) 4 and 3
(C) ± 3 and ± 4 (D) ± 4 and ± 3
(E) None of these

9. The equations of the asymptotes of a hyperbola defined by $\dfrac{x^2}{9} - \dfrac{y^2}{4} = 1$ are

(A) $y = \pm\dfrac{2}{3}x$ (B) $y = \pm 2x$
(C) $y = \pm 3x$ (D) $y = \pm\dfrac{3}{2}x$
(E) None of these

10. The one equation whose graph is not a conic section is

(A) $(x + 2)^2 + (y - 6)^2 = 4$
(B) $\dfrac{x}{4} + \dfrac{y}{9} = 16$ (C) $\dfrac{x^2}{4} + \dfrac{y^2}{16} = 1$
(D) $\dfrac{x^2}{9} - \dfrac{y^2}{25} = 1$ (E) $xy = 4$

CUMULATIVE REVIEW

DIRECTIONS: Choose the *one* best answer to each question or problem.

1. Factor $2x^3 - 5x^2 - 3x$ completely.

    (A) $(x^2 - 3x)(2x + 1)$
    (B) $(2x^2 + x)(x - 3)$
    (C) $x(x - 3)(2x + 1)$
    (D) $x(x + 3)(2x - 1)$

2. Solve $16^{x-2} = 64$.

    (A) $\frac{7}{2}$ **(B)** 6   **(C)** 3   **(D)** $\frac{9}{4}$

3. Solve $-3(x - 2c) = 2 + a(bx + c)$ for x.

    (A) $\dfrac{-(ac + 6c + 2)}{ab + 3}$ **(B)** $\dfrac{6c - ac - 2}{ab + 3}$

    (C) $\dfrac{6c - ac - 2}{3ab}$ **(D)** $\dfrac{5ac - 2}{ab + 3}$

4. Find the value of $2 \cdot 27^{-\frac{4}{3}}$.

    (A) $\dfrac{2}{81}$ **(B)** 162   **(C)** $\dfrac{1}{6}$ **(D)** $-\dfrac{2}{3}$

5. Solve $n + 2 = \sqrt{2n + 12}$

    (A) -4 **(B)** 2
    (C) -4 and 2   **(D)** 10

6. Factor $9x^2 + 6x + 1 - 16y^2$ as a difference of two squares.

    (A) $(3x - 1 + 4y)(3x - 1 - 4y)$
    (B) $(3x - 2x + 1 - 4y)^2$
    (C) $(3x + 1 + 4y)(3x + 1 - 4y)$
    (D) $(3x - 4y - 1)(3x - 4y + 1)$

7. Find the slope of $\overrightarrow{PQ}$ for $P(-3, 5)$ and $Q(6, -2)$.

    (A) $-\dfrac{9}{7}$ **(B)** $-\dfrac{1}{3}$
    (C) $-\dfrac{7}{9}$ **(D)** $-\dfrac{7}{3}$

8. Write an equation, in standard form, of the line that passes through the point $P(-2, 1)$ and is parallel to the line described by the equation $3x - 2y = 4$.

    (A) $2x + 3y = -1$ **(B)** $2x - 3y = 1$
    (C) $3x - 2y = -8$ **(D)** $3x - 2y = -2$

9. Find the product, if possible.

$$\begin{bmatrix} -3 & 1 & -2 \\ 2 & 5 & -1 \end{bmatrix} \cdot \begin{bmatrix} 1 & -2 & 2 \\ 2 & 3 & 4 \\ -1 & 2 & -2 \end{bmatrix}$$

    (A) $\begin{bmatrix} -2 & -1 & 0 \\ 4 & 8 & 0 \\ -1 & 2 & -2 \end{bmatrix}$ **(B)** $\begin{bmatrix} 1 & 13 \\ 5 & 9 \\ -1 & 11 \end{bmatrix}$

    (C) $\begin{bmatrix} 1 & 5 & -1 \\ 13 & 9 & 11 \end{bmatrix}$ **(D)** None of these

10. Find the domain of the function defined by

$$t(x) = \frac{-2x}{x^2 + 2x - 3}.$$

    (A) {all real numbers}
    (B) {all real numbers except 0}
    (C) {all real numbers except -3, 0, and 1}
    (D) {all real numbers except -3 and 1}

11. Let $f(x) = 3x - 1$ and $g(x) = x^2 + 1$. Find $f(g(-2))$.

    (A) 14   **(B)** -10 **(C)** 50   **(D)** 5

12. Write an equation of the inverse of the function defined by $y = 3x - 2$.

    (A) $y = -3x + 2$ **(B)** $x - 3y = -2$
    (C) $3y = x - 2$ **(D)** $-3x + y = -2$

13. Which of the following is an equation of a parabola?

    (A) $y = 2x + 3$ **(B)** $y = x^2 + 2$
    (C) $y^2 - x^2 = 3$ **(D)** $y^2 + x^2 = 2$

14. Which of the following is an equation of a hyperbola?

    (A) $5x^2 + 4y^2 = 20$ **(B)** $4x^2 + 5y^2 = 20$
    (C) $16x^2 - 16y^2 = 1$ **(D)** $8x^2 + 8y^2 = 8$

15. Write an equation of a parabola with $F(-2, 0)$ the focus and $x = 2$ the equation of the directrix.

    (A) $x^2 = -8y$ **(B)** $x^2 = 8y$
    (C) $y^2 = 8x$ **(D)** $y^2 = -8x$

Solve each equation.

16. $|3x - 2| = 13$

17. $\dfrac{10}{n - 1} - \dfrac{3}{n} = 4$

18. $3^{x-5} = \dfrac{1}{9}$

19. $n^2 - 9 = -5n$

20. Solve $\dfrac{-4}{a} = \dfrac{3}{b} + \dfrac{1}{x}$ for x.

Factor each polynomial completely.

21. $36a^2 - 49b^2$ 22. $2x^3 - 16$

Simplify and write each expression with positive exponents.

23. $(-2a^{-3}b^4)^2$ 24. $\dfrac{-25x^{-3}y^4}{15x^{-2}y^6}$

25. Divide: $\dfrac{8 - 2n}{4n^2 + 8n} \div \dfrac{n^2 - 16}{n^3 - n^2 - 6n}$.

26. Simplify: $ab\sqrt{18ab^3} - 2b\sqrt{8a^3b^3} + a\sqrt{32ab^5}$; $a \geq 0$, $b \geq 0$.

27. Simplify: $-2 - \sqrt{-8} - (5 - 3\sqrt{-2})$.

28. Some sunflower seeds worth \$3.80/kg are mixed with some raisins worth \$3.20/kg to make 12 kg of a mixture worth \$3.60/kg. How many kg of each kind are there in the mixture?

29. A second number is 6 more than 3 times a first number. Find the numbers if their difference is 26.

Draw the graph of each equation.

30. $3x - 2y = 8$ 31. $\frac{2}{3}y = \frac{1}{2}x + 4$

A line has the given slope, y-intercept, or contains the indicated point(s). Write an equation, in standard form, of each line.

32. $A(-1, 2)$, $m = -\frac{3}{5}$
33. $m = \frac{2}{3}$, $b = -1$
34. $P(4, -1)$, $Q(-5, 2)$

35. The vertices of a quadrilateral are $A(2, -3)$, $B(5, 6)$, $C(3, 2)$ and $D(-2, -1)$. Prove that the segments connecting the midpoints of $ABCD$ form a parallelogram.

Solve each system of equations.

36. $3(x - 2) = 2y - 7$
     $-2(y + 4) = -(x + 5)$

37. $2x - 3y + z = 7$
    $-x + 2y - 3z = -9$
    $3x - y + 2z = 8$

38. A two-digit number is 54 less than the number obtained by reversing the digits. Find the original number if 4 times the sum of the digits is 3 more than 5 times the units digit.

Let $g(x) = x^2 - 2x + 3$. Find each of the following function values.

39. $g(-2)$ 40. $g(2n)$

41. If y varies inversely as x, and if $y = 6$ when $x = 5$, find y when $x = 10$.

42. If y varies directly as the positive square root of x and inversely as the cube of z, and if $y = 6$ when $x = 9$ and $z = 2$, find y when $x = 4$ and $z = 3$.

43. For the point whose coordinates are $(-6, 1)$, find the corresponding point symmetric with respect to the x-axis, the y-axis, and the origin.

44. Draw the graph of $y = 2x^2 + 6x - 1$.

45. Find an equation of a circle whose center is at $C(3, -1)$ and which passes through $P(5, 4)$.

Graph each of the following equations.

46. $x^2 - y^2 = 9$ 47. $y = (x + 2)^2 - 3$

48. $\dfrac{(x - 1)^2}{9} + \dfrac{(y + 2)^2}{4} = 1$

49. $(x + 2)^2 + (y - 2)^2 = 9$

13 LINEAR-QUADRATIC SYSTEMS

Non-Routine Problem Solving

Problem 1.   For 3 h a fishing boat traveled at the same rate. For the next hour, the fishing boat traveled at one-half that rate. If the boat traveled a total distance of 105 mi, what was the initial rate?

Problem 2.   A swimmer can swim at a rate of 80 yd/min in still water. In a river with a current of 30 ft/min, she can swim round-trip, upstream and downstream, between points A and B in 4 min. Find the distance between points A and B.

SOLVING LINEAR-QUADRATIC SYSTEMS OF EQUATIONS

Objective

To find the real number solutions of a system of one quadratic and one linear equation graphically and algebraically

The general form of the quadratic equations of some conic sections you have studied are given below.

Parabola: $y = ax^2 + bx + c$ Circle: $x^2 + y^2 = r^2$

Hyperbola: $\dfrac{x^2}{a^2} - \dfrac{y^2}{b^2} = 1$ and $xy = k$ Ellipse: $\dfrac{x^2}{a^2} + \dfrac{y^2}{b^2} = 1$

You have also learned that the general form of a linear equation in two variables is $ax + by = c$, and that the graph of each such equation is a line.

A system of equations may contain one quadratic equation and one linear equation. Such a system is called a **linear-quadratic system.** A common solution of a linear-quadratic system of equations is an ordered pair of real numbers corresponding to a point of intersection of their graphs. As illustrated by the graphs below, a system of one quadratic and one linear equation may have 0, 1, or 2 solutions.

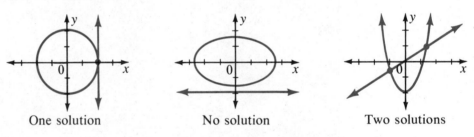

One solution   No solution   Two solutions

The solutions of a linear-quadratic system may be found graphically.

Example 1   **Solve the system** $\dfrac{x^2}{16} + \dfrac{y^2}{9} = 1$ **graphically.**

$$2x - 3y = 6$$

Estimate the solutions to the nearest tenth.

Ellipse ▶ $\dfrac{x^2}{16} + \dfrac{y^2}{9} = 1$
$a^2 = 16 \qquad b^2 = 9$
$a = \pm 4, \ b = \pm 3$

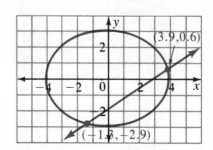

$2x - 3y = 6$ ◀ *Line*
$y = \dfrac{2}{3}x - 2$
$m = \dfrac{2}{3}, \ b = -2$

Thus, $(-1.3, -2.9)$ **and** $(3.9, 0.6)$ **are the solutions of the system.**

                                                       Chapter Thirteen

Example 2   **Solve the system** $x^2 + y^2 = 13$ **algebraically.**
$$-2x + y = -1$$

Step 1   Solve the linear equation for $-2x + y = -1$
one of its variables. $y = 2x - 1$

Step 2   Substitute this result in the $x^2 + y^2 = 13$
quadratic equation. $x^2 + (2x - 1)^2 = 13$
                                 $x^2 + 4x^2 - 4x + 1 = 13$

Step 3   Solve the resulting quadratic $5x^2 - 4x - 12 = 0$ ◀ *Solve by factoring.*
equation by factoring, $(5x + 6)(x - 2) = 0$
completing the square, or
using the quadratic formula. $\boxed{x = -\frac{6}{5}}$ or $\boxed{x = 2}$

Step 4   Find the value of the other $-2x + y = -1$ $-2x + y = -1$
variable by substituting in
the linear equation. $-2\left(-\frac{6}{5}\right) + y = -1$ $-2(2) + y = -1$

                                 $\frac{12}{5} + y = -1$ $-4 + y = -1$

Step 5   Check by substituting each
ordered pair in both $\boxed{y = -\frac{17}{5}}$ $\boxed{y = 3}$
equations.

$x^2 + y^2 = 13$		$-2x + y = -1$		$x^2 + y^2 = 13$		$-2x + y = -1$	
$\left(-\frac{6}{5}\right)^2 + \left(-\frac{17}{5}\right)^2$	13	$-2\left(-\frac{6}{5}\right) + \left(-\frac{17}{5}\right)$	-1	$2^2 + 3^2$	13	$-2(2) + 3$	-1
$\frac{36}{25} + \frac{289}{25}$		$+\frac{12}{5} - \frac{17}{5}$		$4 + 9$		$-4 + 3$	
$\frac{325}{25}$		$-\frac{5}{5}$		13		-1	
13		-1					

Thus, $\left(-\frac{6}{5}, -\frac{17}{5}\right)$ and $(2, 3)$ are the solutions of the system.

Oral Exercises

State the number of solutions of each system of equations whose graphs are given.
Give the solution(s) of each system, if any.

1.
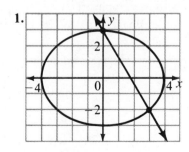

2.
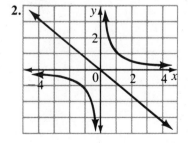

3.
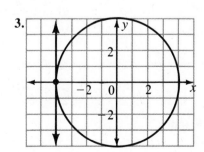

Written Exercises

Solve each system of equations graphically. Estimate the decimal solutions to the nearest tenth.

(A) **1.** $\dfrac{x^2}{16} + \dfrac{y^2}{9} = 1$
$2x - 3y = 6$

2. $\dfrac{x^2}{25} - \dfrac{y^2}{4} = 1$
$x = 5$

3. $\dfrac{x^2}{16} + \dfrac{y^2}{16} = 1$
$x + y = 2$

4. $xy = 4$
$2x + 3y = 9$

5. $y = x^2 - 6x + 8$
$x - 2y = 2$

6. $x^2 + y^2 = 16$
$3x - 2y = 4$

7. $\dfrac{x^2}{4} + \dfrac{y^2}{4} = 1$
$2x - 6y = 12$

8. $y = x^2 - 9$
$x + y = 3$

Solve each system of equations algebraically. Give the irrational solutions in simplest radical form.

9. $xy = 8$
$y = x + 2$

10. $y = x^2 - 7x - 3$
$2x - y = 3$

11. $y = 5(x - 3)$
$2y = x^2 - 6$

12. $\dfrac{x^2}{10} + \dfrac{y^2}{40} = 1$
$y = \dfrac{2}{3}x$

13. $y = 3(x - 3)$
$2y = x^2 - 10$

14. $y = 2x - 3$
$x = -\dfrac{y^2}{4} + \dfrac{21}{4}$

15. $y - 2x = 5$
$x^2 + y^2 = 25$

16. $\dfrac{x^2}{1} - \dfrac{y^2}{2} = 1$
$y = 2x - 2$

(B) **17.** $y = -\dfrac{x}{2} + \dfrac{1}{2}$
$x = -3y^2 - 12y + 8$

18. $y = x^2 - 8x + 2$
$x - 3y = 5$

19. $3x - 2y = 8$
$y = 2x^2 - 5x - 3$

20. $3y - 2x = -5$
$y = -3x^2 + 7x + 4$

21. $4x + y = 1$
$3x^2 + y^2 = 12$

22. $y = 2x + 3$
$x^2 + y^2 = 25$

23. $x^2 + (y - 5)^2 = 49$
$y - x = -2$

24. $\dfrac{x^2}{16} + \dfrac{y^2}{3} = 1$
$y = x - 1$

(C) **25.** $(x - 2)^2 + (y + 3)^2 = 96$
$y = 2x$

26. $x^2 + 4x + y^2 - 6y = 12$
$y = -x + 3$

27. $2x^2 + y^2 = 8x + 4y + 3$
$2x - 3y = 7$

28. $3x^2 - 2y^2 - 6x + 5y = -12$
$3x - 2y = -8$

NON-ROUTINE PROBLEMS

The number of distinct points in the coordinate plane common to the graphs of
$(2x - y + 1)(2x + 3y - 11) = 0$ and $(x + 2y - 7)(3x - 2y + 3) = 0$ is
(A) 0   **(B)** 1   **(C)** 2   **(D)** 3   **(E)** 4   **(F)** infinitely many

---

CUMULATIVE REVIEW

Simplify.

1. $3a^3b(-2a^4b^5)$

2. $\dfrac{16x^6yz^2}{-4xy^5z^2}$

3. $(-3a^2b^3c)^3$

SOLVING QUADRATIC SYSTEMS OF EQUATIONS

Objective

To find the real number solutions of a system of two quadratic equations graphically and algebraically

A system of equations may contain two quadratic equations. A system of this type is called a **quadratic system.** As illustrated by the graphs below, a system of two quadratic equations may have 4, 3, 2, 1, or 0 solutions.

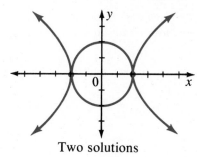

Two solutions

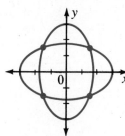

Four solutions

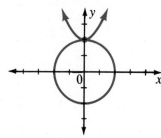

One solution

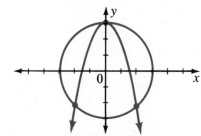

Three solutions

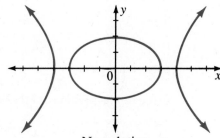

No solution

Example 1

Solve the system $\dfrac{x^2}{34} + \dfrac{y^2}{15} = 1$ graphically.

$$x^2 + y^2 = 34$$

Estimate the solutions to the nearest tenth.

Ellipse ▶ $\dfrac{x^2}{34} + \dfrac{y^2}{15} = 1$ $x^2 + y^2 = 34$ ◀ *Circle*

$a^2 = 34$ $b^2 = 15$ $r^2 = 34$
$a \doteq \pm 5.8$ $b \doteq \pm 3.9$ $r \doteq 5.8$

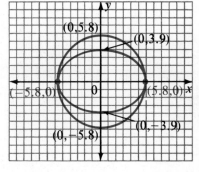

The graphs intersect at two points with coordinates of $(5.8, 0)$ and $(-5.8, 0)$. Check by substituting $(5.8, 0)$ and $(-5.8, 0)$ in the two equations.

Thus, $(5.8, 0)$ and $(-5.8, 0)$ are the solutions of the system.

The next examples show you how to find the solutions of a system of two quadratic equations algebraically. Notice that either the addition method or substitution method may be used.

Example 2   **Solve the system $\dfrac{x^2}{36} + \dfrac{y^2}{25} = 1$ algebraically.**

$$x^2 + y^2 = 25$$

$\dfrac{x^2}{36} + \dfrac{y^2}{25} = 1$ ◀ *Multiply the first equation by 900 to eliminate fractions.*

$$25x^2 + 36y^2 = 900$$
$$\underline{-25x^2 - 25y^2 = -625}$$ ◀ *Multiply the second equation by −25.*
$$11y^2 = 275$$ ◀ *Add the equations to eliminate the x^2*
$$y^2 = 25$$ *term.*

$$\boxed{y = 5} \qquad \boxed{y = -5}$$

Solve for x. Substitute 5 and −5 for y. ▶

$x^2 + y^2 = 25$	$x^2 + y^2 = 25$
$x^2 + (5)^2 = 25$	$x^2 + (-5)^2 = 25$
$x^2 = 0$	$x^2 = 0$
$\boxed{x = 0}$	$\boxed{x = 0}$

Check by substituting $(0, 5)$ and $(0, -5)$ in the original equations.

Thus, $(0, 5)$ and $(0, -5)$ are the solutions of the system.

Example 3   **Solve the system $x^2 + y^2 = 49$ algebraically.**

$$-2y^2 = 30 - 2x^2$$

Estimate the solutions to the nearest tenth.

$$-2y^2 = 30 - 2x^2$$ ◀ *Solve the second equation for y^2.*
$$y^2 = x^2 - 15$$

$$x^2 + y^2 = 49$$ ◀ *Substitute in the first equation.*
$$x^2 + (x^2 - 15) = 49$$
$$2x^2 - 15 = 49$$
$$2x^2 = 64$$
$$x^2 = 32$$

$$\boxed{x = 4\sqrt{2} \doteq 5.7} \qquad \boxed{x = -4\sqrt{2} \doteq -5.7}$$

Solve for y. Substitute $4\sqrt{2}$ and $-4\sqrt{2}$ for x. ▶

$x^2 + y^2 = 49$	$x^2 + y^2 = 49$
$(4\sqrt{2})^2 + y^2 = 49$	$(-4\sqrt{2})^2 + y^2 = 49$
$32 + y^2 = 49$	$32 + y^2 = 49$
$y^2 = 17$	$y^2 = 17$
$\boxed{y \doteq \pm 4.1}$	$\boxed{y \doteq \pm 4.1}$

Check by substituting the four ordered pairs in the original equations.

Thus, $(5.7, 4.1)$, $(5.7, -4.1)$, $(-5.7, 4.1)$, and $(-5.7, -4.1)$ are the solutions of the system.

Oral Exercises

State the number of solutions of each system of equations whose graphs are given.
Give the solution(s) of each system, if any.

1.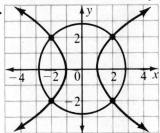

2.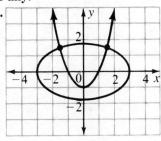

3.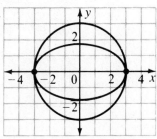

Written Exercises

Solve each system of equations graphically. Estimate the decimal solutions to the nearest tenth.

(A)

1. $\dfrac{x^2}{16} + \dfrac{y^2}{4} = 1$
$x^2 + y^2 = 16$

2. $\dfrac{x^2}{25} + \dfrac{y^2}{9} = 1$
$x^2 - y^2 = 4$

3. $\dfrac{x^2}{9} - \dfrac{y^2}{4} = 1$
$xy = 5$

4. $\dfrac{x^2}{16} + \dfrac{y^2}{25} = 1$
$x^2 + y^2 = 4$

Solve each system of equations algebraically. Give the irrational solutions as decimals to the nearest tenth.

5. $x^2 + y^2 = 16$
$x^2 - y^2 = 16$

6. $x^2 + y^2 = 1$
$y = x^2 + 5$

7. $x^2 + y^2 = 38$
$x^2 - y^2 = 12$

8. $x^2 + y^2 = 13$
$xy = -6$

Solve each system of equations algebraically. Give the irrational solutions in simplest radical form.

(B)

9. $\dfrac{x^2}{36} + \dfrac{y^2}{9} = 1$
$x^2 - y^2 = 16$

10. $\dfrac{x^2}{2} + \dfrac{y^2}{4} = 1$
$8x^2 + 3y^2 = 12$

11. $3x^2 + 2y^2 = 33$
$3x^2 + y^2 = 17$

12. $\dfrac{x^2}{9} + \dfrac{y^2}{4} = 1$
$3x^2 - 4y^2 = 24$

13. $\dfrac{x^2}{16} + \dfrac{y^2}{4} = 1$
$\dfrac{x^2}{4} - \dfrac{y^2}{3} = 1$

14. $\dfrac{x^2}{10} + \dfrac{y^2}{4} = 1$
$2y^2 = x^2 - 8$

15. $x^2 + y^2 = 13$
$\dfrac{x^2}{5} - \dfrac{y^2}{5} = 1$

16. $\dfrac{x^2}{5} + \dfrac{y^2}{25} = 1$
$x^2 + y^2 = 9$

Solve each system of equations graphically. Give the solutions as decimals to the nearest tenth.

(C)

17. $\dfrac{(x-3)^2}{9} + \dfrac{(y+2)^2}{4} = 1$
$(x+1)^2 + (y-3)^2 = 16$

18. $\dfrac{(x+2)^2}{9} - \dfrac{(y-3)^2}{4} = 1$
$\dfrac{(x-1)^2}{25} + \dfrac{(y+2)^2}{16} = 1$

19. $(x-3)^2 + (y+4)^2 = 25$
$\dfrac{(x+2)^2}{9} + \dfrac{(y-1)^2}{16} = 1$

APPLICATIONS

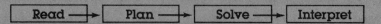

Read ⟶ Plan ⟶ Solve ⟶ Interpret

There are many applications that use a system of equations to determine a solution. The equations can be both linear, both quadratic, or one of each. The following example illustrates such an application using one linear and one quadratic equation.

Example

Find the length of each side of both squares if the sum of the perimeters is 56 cm and the sum of the areas is 106 cm^2.

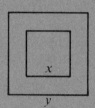

Let x = length of the side of the small square
  y = length of the side of the large square.

The sum of the perimeters is 56 cm, so
  $4x + 4y = 56$.
The sum of the areas is 106 cm^2, so
  $x^2 + y^2 = 106$.

Solve the system $4x + 4y = 56$.
  $x^2 + y^2 = 106$

Solve $4x + 4y = 56$ for y. $y = \dfrac{56 - 4x}{4}$, or $y = 14 - x$

Substitute in $x^2 + y^2 = 106$. $x^2 + (14 - x)^2 = 106$
Square. ▶ $x^2 + 196 - 28x + x^2 = 106$
                                    $2x^2 - 28x + 90 = 0$
                                    $x^2 - 14x + 45 = 0$
Factor. ▶ $(x - 9)(x - 5) = 0$
                                    $x = 9$ or $x = 5$

Find y. ▶ $4x + 4y = 56$ $4x + 4y = 56$
              $4(9) + 4y = 56$ $4(5) + 4y = 56$
                   $4y = 20$ $4y = 36$
                    $y = 5$ $y = 9$

Reject $y = 5$ since y represents the length of the side of the large square. Thus, it appears that the length of each side of both squares is 5 and 9, respectively.
The sum of the perimeters: $(4 \times 5) + (4 \times 9)$ is 56.
The sum of the areas is $5^2 + 9^2$ or $25 + 81$ is 106.
Thus, both measures check in the words of the problem.

Solve.

1. The perimeter of a rectangular lot is 26 cm. The area is 36 cm². Find the lengths of the sides.

2. Find the length of each side of two squares if the sum of their perimeters is 56 cm and the sum of their areas is 100 cm².

3. The area of a right triangle is 25 cm². The length of its hypotenuse is $5\sqrt{5}$ cm. Find the lengths of the legs.

4. The difference of the squares of two numbers is 84. The first number is 2 more than twice the second. What are the numbers?

5. Senior Girl Scout Troop 195 hiked 8 km to the top of High Peak and hiked the same distance down in a total time of 6 hours and 40 minutes. Their rate of ascent was 10 km/h slower than their rate of descent. What was their rate of ascent?

6. Ms. Stanton received $480 interest on an investment for one year. If the interest rate had been $1\frac{1}{2}\%$ higher and the principal $1,000 less, she would have received the same amount of interest. Find the principal and the rate of interest.

7. A rectangular sheet of metal has 4 squares with sides 8 cm long cut from the corners of the sheet. The edges are bent up to form an open box with a volume of 12,672 cm³. The area of the sheet is 3,120 cm². Find the dimensions of the box.

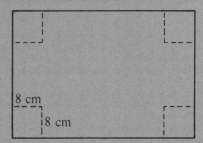

8. Find two numbers whose product is 18 and the sum of their reciprocals is $\frac{1}{2}$.

9. The sum of the squares of two numbers is 80 and their product is 32. What are the numbers?

10. If the numerator of a fraction is increased by 3 and the denominator is decreased by 3, the resulting fraction is the reciprocal of the original fraction. The numerator of the original fraction is 1 more than one half its denominator. What was the original fraction?

11. The product of a two-digit number and the number obtained by reversing its digits is 2,268. If the difference of the numbers is 27, find the numbers.

SOLVING LINEAR INEQUALITY SYSTEMS

Objectives **To draw the graph of a linear inequality in two variables**
To find the real number solutions of a system of two linear inequalities graphically

The general form of a **linear inequality** in two variables, x and y, is $ax + by > c$ or $ax + by < c$, where a and b are not both equal to 0.

The graph of the linear inequality $3x - 2y < 2$ is a region of the coordinate plane above the line determined by the equation $3x - 2y = 2$. You should use the following steps to draw the graph of $3x - 2y < 2$:

(1) Solve for y. Division by a negative number reverses the order of the inequality.

$$3x - 2y < 2$$
$$-2y < -3x + 2$$
$$y > \tfrac{3}{2}x - 1$$

(2) Draw the graph of the equation $y = \tfrac{3}{2}x - 1$. Use a dashed line with $m = \tfrac{3}{2}$ and $b = -1$.

(3) Shade the region above the line.

(4) Check. Pick a point in the shaded region and one below the line. Test $(-2, 1)$ and $(4, -2)$.

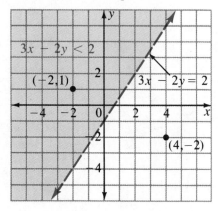

$3x - 2y < 2$		$3x - 2y < 2$	
$3(-2) - 2(1)$	2	$3(4) - 2(-2)$	2
$-6 - 2$		$12 + 4$	
-8		16	
$-8 < 2$		$16 < 2$	
True		False	

Notice that since $y > \tfrac{3}{2}x - 1$, the region *above* the line is shaded. Also notice that a dashed line is used to indicate that the line is not part of the graph. The graph would include the line if the inequality contained the symbol $\geq$. When the line is part of the graph, a solid line should be used.

Example 1 **Draw the graph of $-x - 2y \geq 4$.**

Solve for y. ▶
$$-x - 2y \geq 4$$
$$-2y \geq x + 4$$
$$y \leq -\tfrac{1}{2}x - 2$$

Graph the equation. Use a solid line. ▶
$$y = -\tfrac{1}{2}x - 2$$
$$m = -\tfrac{1}{2}, b = -2$$

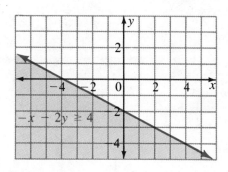

Shade the region below the line.

Chapter Thirteen

Example 2

Draw the graph of $y \leq -2$.

Graph the equation $y = -2$.
Use a solid line.

Shade the region below the line.

Check. Pick a point in the shaded region and one above the line.

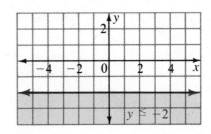

The next two examples show you how to solve a system of two linear inequalities graphically.

Example 3

Solve the system $x + y > -1$ graphically.
$$3x - 2y > 4$$

Graph $x + y > -1$ or $y > -x - 1$:
(a) Graph $y = -x - 1$. Use a dashed line.
(b) Shade above the line.

Graph $3x - 2y > 4$ or $y < \frac{3}{2}x - 2$:
(a) Graph $y = \frac{3}{2}x - 2$. Use a dashed line.
(b) Shade below the line with a different color.

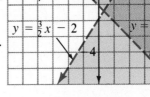

Check. Pick a point in the double-shaded region and one not in the region.

Thus, the solution set is the set of all ordered pairs for the points in the double-shaded region.

Example 4

Solve the system $2x - y \leq -3$ graphically.
$$x \leq -2$$

Graph $2x - y \leq -3$ or $y \geq 2x + 3$:
(a) Graph $y = 2x + 3$. Use a solid line.
(b) Shade above the line.

Graph $x \leq -2$:
(a) Graph $x = -2$. Use a solid line.
(b) Shade to the left of the line with a different color.

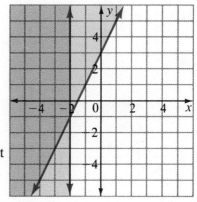

Check. Pick a point in the double-shaded region and one not in the region.

Thus, the solution set is the set of all ordered pairs for the points in the double-shaded region.

Solving Linear Inequality Systems

Reading in Algebra

Examine the inequalities on the left and the expressions describing the graphs of the inequalities on the right. Match each inequality with one or more of the descriptions.

1. $x \leq 3$
2. $y > -4$
3. $y \leq 2x + 3$
4. $y > 3x + 5$
5. $y \leq -7$
6. $x \geq -8$
7. $y < x + 8$
8. $y \geq -2x + 1$

A includes the line
B excludes the line
C above the line
D below the line
E right of the line
F left of the line

Written Exercises

Draw the graph of each of the following inequalities. Check.

(A)
1. $3x - 2y \leq 6$
2. $2y - 3x + 4 \geq 0$
3. $2x + 5y < 10$
4. $5x - 3y > 15$
5. $7x - 3y \geq 21$
6. $2x - 7y > 35$
7. $-y \leq 2x - 8$
8. $-5x - 2y \leq 12$
9. $3x < -y + 8$
10. $-5x > 4y + 8$
11. $-8 + x \geq 2y$
12. $y - 2x < -7$
13. $x \geq 8$
14. $y \geq 4$
15. $y < -4$
16. $x < 5$

Solve each system of inequalities graphically. Check.

(B)
17. $y \leq -2x + 3$
    $y > 4x - 1$
18. $y \leq 3x - 1$
    $y > -2x + 4$
19. $2x + 3y \geq 6$
    $2y - x < -4$
20. $5x + 2y \geq 12$
    $3x + 4y \leq 8$

21. $2x + 3y \geq 9$
    $-3x + 4y \geq -4$
22. $5x + 6y < 30$
    $4x + 3y \geq 9$
23. $-4x + 5y \leq 15$
    $4y \geq 5x + 12$
24. $-9 \leq 2x + 3y$
    $3y + 4x \leq 6$

25. $-2x - 3y \geq 6$
    $-2y + x > -4$
26. $x - 2y > 6$
    $x + 2y \leq 4$
27. $2x - 3y \geq 9$
    $-2y + x > 6$
28. $-3x - y \geq 6$
    $y - 2x > 1$

29. $2x + y > 3$
    $y \geq x$
30. $y \leq 2$
    $x \leq 4$
31. $y \geq -3$
    $x \leq -2$
32. $y \geq 5$
    $x \geq -3$

Solve each system of three inequalities graphically. Check.

(C)
33. $x + y \leq 3$
    $4x - 5y \geq 10$
    $-2x + y > 2$
34. $2x - y \geq -4$
    $x + 3y < 6$
    $x + 2y \geq 4$
35. $-x + 3y < 5$
    $3x - 2y \geq 6$
    $-4x + y < 3$
36. $4x - 2y \leq 2$
    $-4x + y \geq 1$
    $2x - y \leq -3$

SOLVING LINEAR-QUADRATIC INEQUALITY SYSTEMS

Objectives

To draw the graph of a quadratic inequality
To find the real number solutions of a system of one linear and one quadratic inequality graphically
To find the real number solutions of a system of two quadratic inequalities graphically

The graph of the **quadratic inequality** $x^2 + y^2 < 16$ is a region of the coordinate plane in the interior of the circle determined by the equation $x^2 + y^2 = 16$. You should use the following steps to draw the graph of $x^2 + y^2 < 16$:

(1) Draw the graph of the equation $x^2 + y^2 = 16$. The graph is a circle. Use a dashed curve.

(2) Shade the region in the interior of the circle.

(3) Check. Pick a point in the shaded region, one not in the region, and one on the circle.
Test $(2, 3)$, $(4, -3)$, and $(4, 0)$.

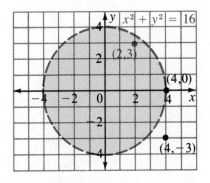

Notice that since $x^2 + y^2 < 16$, the region *in the interior of* the circle is shaded. The dashed curve is used to show that the circle is not part of the graph of the inequality. The graph would include the circle if the inequality contained the symbol $\leq$. The graph of $x^2 + y^2 > 16$ is the region *in the exterior of* the circle.

Example 1

Draw the graph of $xy \geq 4$.

Make a table of x and y values for $xy = 4$.

Graph the equation $xy = 4$. Use a solid curve.

Shade the region in the interior of each branch of the hyperbola.

Check. Pick a point in the shaded region, one not in the region, and one on the hyperbola.

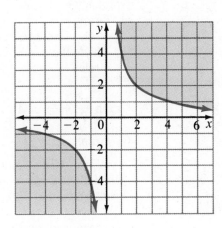

Systems containing linear or quadratic inequalities can be solved graphically.

Example 2

Solve the system $x^2 + y^2 \leq 9$ graphically.
$$y \geq 3x - 2$$

Graph $x^2 + y^2 \leq 9$:
(a) Graph $x^2 + y^2 = 9$. The graph is a circle. Use a solid curve.
(b) Shade the region in the interior of the circle.

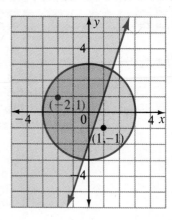

Graph $y \geq 3x - 2$:
(a) Graph $y = 3x - 2$. Use a solid line.
(b) Shade the region above the line with a different color.

Check. Pick a point in the double-shaded region and one not in the region. Test $(-2, 1)$ and $(1, -1)$ in both inequalities.

$x^2 + y^2 \leq 9$		$y \geq 3x - 2$		$x^2 + y^2 \leq 9$		$y \geq 3x - 2$	
$(-2)^2 + 1^2$	9	1	$3(-2) - 2$	$1^2 + (-1)^2$	9	-1	$3 \cdot 1 - 2$
5			-8	2			1
$5 \leq 9$		$1 \geq -8$		$2 \leq 9$		$-1 \geq 1$	
True		True		True		False	

Thus, the solution set is the set of all ordered pairs for the points in the double-shaded region, including the points of the boundary of the region.

Example 3

Solve the system $x^2 + y^2 \geq 4$ graphically.
$$\frac{x^2}{9} + \frac{y^2}{4} \leq 1$$

Graph $x^2 + y^2 \geq 4$:
(a) Graph $x^2 + y^2 = 4$. The graph is a circle. Use a solid curve.
(b) Shade the region in the exterior of the circle.

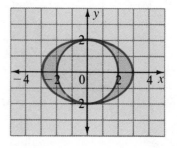

Graph $\dfrac{x^2}{9} + \dfrac{y^2}{4} \leq 1$:

(a) Graph $\dfrac{x^2}{9} + \dfrac{y^2}{4} = 1$. The graph is an ellipse. Use a solid curve.
(b) Shade the region in the interior of the ellipse with a different color.

Check. Pick a point in the double-shaded region and one not in the region.

Thus, the solution set is the set of all ordered pairs for the points in the double-shaded region, including the points of the boundary of the region.

Oral Exercises

State the inequality that describes the shaded region.

1.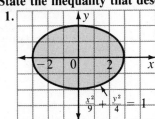

2.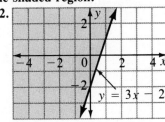

3.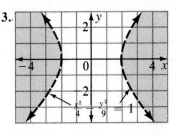

Written Exercises

Draw the graph of each of the following inequalities. Check.

(A)
1. $x^2 + y^2 \leq 9$ **2.** $x^2 + y^2 \geq 4$ **3.** $x^2 + y^2 < 25$ **4.** $x^2 + y^2 > 16$

5. $xy \geq 6$ **6.** $xy < 8$ **7.** $xy > 3$ **8.** $xy \leq -2$

9 $\dfrac{x^2}{25} + \dfrac{y^2}{4} \leq 1$ **10.** $\dfrac{x^2}{36} + \dfrac{y^2}{16} \geq 1$ **11.** $\dfrac{x^2}{100} + \dfrac{y^2}{9} < 1$ **12.** $\dfrac{x^2}{16} + \dfrac{y^2}{1} > 1$

13. $\dfrac{x^2}{9} - \dfrac{y^2}{4} \leq 1$ **14.** $\dfrac{x^2}{25} - \dfrac{y^2}{8} \geq 1$ **15.** $\dfrac{x^2}{4} - \dfrac{y^2}{9} < 1$ **16.** $\dfrac{x^2}{25} - \dfrac{y^2}{81} > 1$

(B)
17. $2x^2 + 3y^2 \geq 18$ **18.** $3x^2 + 9y^2 \leq 27$ **19.** $9x^2 - 4y^2 < 36$ **20.** $16x^2 - 8y^2 \geq 64$

21. $27x^2 + 9y^2 \geq 81$ **22.** $25x^2 - 4y^2 > 100$ **23.** $36x^2 - 18y^2 \leq 72$ **24.** $25x^2 + 4y^2 < 100$

Solve each system of inequalities graphically. Check.

25. $x^2 + y^2 \leq 16$
$2x - 3y \leq 6$

26. $x^2 + y^2 > 25$
$-3x + 4y > 12$

27. $xy \geq 1$
$4x + 5y \geq 15$

28. $xy < -5$
$-2x - 3y \leq 6$

29. $\dfrac{x^2}{16} + \dfrac{y^2}{4} \geq 1$
$-2x - 3y \leq 2$

30. $\dfrac{x^2}{4} - \dfrac{y^2}{2} \leq 2$
$3x - y > 2$

31. $\dfrac{x^2}{25} + \dfrac{y^2}{81} \leq 1$
$2x + 4y \leq -3$

32. $\dfrac{x^2}{9} - \dfrac{y^2}{16} \geq 3$
$-x - 4y \geq -5$

(C)
33. $x^2 + y^2 \geq 25$
$xy < -5$

34. $x^2 - y^2 \leq 12$
$x^2 + y^2 \leq 36$

35. $25x^2 + 4y^2 \geq 100$
$x^2 + y^2 \leq 36$

36. $3x^2 - 27y^2 \leq 81$
$4x^2 + 25y^2 \geq 100$

37. $4x^2 + 16y^2 \geq 64$
$3x^2 - 9y^2 \leq 27$

38. $-3x^2 + 4y^2 \leq 36$
$x^2 + 5y^2 \geq 25$

39. $4x^2 + 6y^2 \geq 24$
$-x^2 - y^2 \geq -9$

40. $2x^2 + 4y^2 \leq 12$
$3x^2 - 5y^2 \geq 15$

CUMULATIVE REVIEW

1. Denine has nickels, dimes, and quarters in her coin collection. The number of quarters is 4 less than twice the number of dimes. The number of dimes is 3 times the number of nickels. If the value of her collection is $8.25, how many nickels, dimes, and quarters does she have?

2. Some dry peanuts worth $2.20/kg are mixed with sunflower seeds worth $3.80/kg to make 30 kg of a mixture worth $3.00/kg. How many kilograms of peanuts and how many kilograms of sunflower seeds are there in the mixture?

CHAPTER THIRTEEN REVIEW

Vocabulary
linear inequality   [13.3]
quadratic inequality   [13.4]
system of equations
  linear-quadratic   [13.1]
  quadratic   [13.2]
system of inequalities
  linear-quadratic   [13.4]
  quadratic   [13.4]

Solve each system of equations graphically. Estimate the decimal solutions to the nearest tenth.

1. $\dfrac{x^2}{25} + \dfrac{y^2}{4} = 1$
$y = 2x + 1$

2. $x^2 - y^2 = 9$ [13.1]
$x - 3y = 6$

3. $4x^2 + 9y^2 = 100$
$2x - 9y = -10$
[13.1]

4. $\dfrac{x^2}{9} + \dfrac{y^2}{4} = 1$ [13.2]
$xy = 2$

5. $\dfrac{x^2}{8} - \dfrac{y^2}{4} = 1$ [13.2]
$x^2 + 4y^2 = 10$

★ 6. $(x - 1)^2 + (y - 3)^2 = 9$ [13.2]
$\dfrac{(x - 2)^2}{4} + \dfrac{(y + 2)^2}{9} = 1$

Solve each system of equations algebraically. Give the irrational solutions in simplest radical form.

7. $y = x^2 + 7x - 3$
$x + 2y = -3$

8. $x^2 - y^2 = -5$ [13.1]
$2x + y = -1$

★ 9. $x^2 + 2x + y^2 - 4y = 8$ [13.1]
$x - y = 2$

10. $y = 2x^2 - x - 5$ [13.2]
$y = x^2 + 1$

11. $x^2 + y^2 = 10$
$xy = 3$

12. $\dfrac{x^2}{16} + \dfrac{y^2}{16} = 1$ [13.2]
$\dfrac{x^2}{16} - \dfrac{y^2}{9} = 1$

Draw the graph of each of the following inequalites. Check.   [13.3]

13. $y \geq 2x - 3$

14. $y < -3x + 5$

15. $3x - 2y \leq 8$

16. $y > -3$

17. $x^2 + y^2 \leq 4$

18. $xy > 4$ [13.4]

19. $xy \leq -5$

20. $3x^2 + 6y^2 \geq 18$

21. $\dfrac{x^2}{36} - \dfrac{y^2}{45} \leq 1$

22. $9x^2 + 4y^2 > 36$

Solve each system of inequalities graphically. Check.

23. $y \geq -x + 3$
$y < 3x - 4$

24. $5x - 2y < 4$ [13.3]
$x \geq y$

25. $y - 3x < 5$
$2x + 4y \geq 15$

26. $3x + 3y > 7$
$5x - 2y \leq 10$

★27. $2x - y \leq 2$
$x + y \geq 4$
$-2x + 3y \leq 6$ [13.3]

28. $x^2 + y^2 \leq 25$
$-x + 2y > 6$ [13.4]

29. $xy < 6$
$y \geq x$

30. $\dfrac{x^2}{36} + \dfrac{y^2}{4} \geq 1$ [13.4]
$-3x + 2y \leq 8$

31. $4x^2 - 25y^2 < 100$
$x + 2y \leq 4$

★32. $4x^2 + 5y^2 \geq 60$
$2x^2 - 3y^2 \leq 18$

Solve each system of equations graphically. Estimate the decimal solutions to the nearest tenth.

1. $\dfrac{x^2}{4} - \dfrac{y^2}{9} = 1$

   $-2x + 3y = 6$

2. $x^2 + y^2 = 25$

   $-2y = 8 - 4x$

3. $x^2 + y^2 = 16$

   $xy = 6$

4. $\dfrac{x^2}{25} + \dfrac{y^2}{36} = 1$

   $x^2 - y^2 = 9$

Solve each system of equations algebraically. Give the irrational solutions in simplest radical form.

5. $x^2 - y^2 = 36$

   $y = 6 - x$

6. $y = x^2 - 2x + 5$

   $4x - y = 4$

7. $x^2 + y^2 = 16$

   $\dfrac{x^2}{16} + \dfrac{y^2}{4} = 1$

8. $x^2 + 2y^2 = 24$

   $x^2 - y^2 = 3$

Draw the graph of each of the following inequalities. Check.

9. $y \leq 3x + 5$

10. $2x - 3y \geq 9$

11. $x > -1$

12. $x^2 - y^2 \geq 6$

13. $xy > 2$

14. $\dfrac{x^2}{36} + \dfrac{y^2}{4} \leq 1$

Solve each system of inequalities graphically. Check.

15. $y < 2x - 1$

    $y \geq -3x + 2$

16. $3x - 4y \leq -8$

    $y - x \leq 2$

17. $2x - 5y \geq 10$

    $3x + 7y < 14$

18. $x^2 - y^2 \geq 4$

    $2x - 3y \leq 6$

19. $\dfrac{x^2}{16} + \dfrac{y^2}{4} \leq 1$

    $4x + 5y \geq 10$

20. $xy \leq 4$

    $-y \leq x$

★ 21. Solve the system $(x - 2)^2 + (y + 1)^2 = 16$

   $y = x + 1$

   algebraically.

Solve each system of inequalities graphically. Check.

★ 22. $3x + y \geq 4$

   $x - 2y \leq 6$

   $-2x + 3y < 9$

★ 23. $5x^2 + 25y^2 \leq 125$

   $3x^2 - 9y^2 \geq 27$

COMPUTER ACTIVITIES

Equation of a Circle

OBJECTIVE: **To determine whether a given point is on, inside, or outside a circle whose equation is given**

The program below generates different equations for circles by randomly choosing a value for r^2 (lines 110–130). The program then asks you to enter any point in the coordinate plane (lines 150–180). Then the computer uses the distance formula to calculate the distance from the center, whereupon it describes the location of the point with respect to the circle (lines 190–220).

The empty loop in line 225 works as a delaying device that allows the program output to remain displayed on the screen for a few moments before the HOME statement in line 230 erases the screen. The loop makes the computer count to a certain number before continuing with the rest of the program. To see the difference line 225 makes, run the program without it before doing so with it.

```
110 R =  INT ( RND (1) * 50) + 1
120  PRINT "THE EQUATION OF A CIRCLE IS"
130  PRINT " X^2 + Y^2 = ";R
140  PRINT
150  PRINT "INPUT THE COORDINATES OF A POINT AND I"
160  PRINT "WILL TELL YOU ITS LOCATION WITH"
170  PRINT "RESPECT TO THE CIRCLE."
180  INPUT A,B
190  PRINT "THE POINT (";A;",";B") IS ";
200  IF A ^ 2 + B ^ 2 = R THEN  PRINT "ON THE CIRCLE."
210  IF A ^ 2 + B ^ 2 > R THEN  PRINT "OUTSIDE THE
       CIRCLE."
220  IF A ^ 2 + B ^ 2 < R THEN  PRINT "INSIDE THE
       CIRCLE."
225  FOR P = 1 TO 2500: NEXT P
230  HOME : INPUT "MORE COORDINATES? (ENTER 1 FOR
       YES, 2 FOR NO.)";K
240  IF K = 1 THEN 110
250  END
```

See the Computer Section beginning on page 574 for more information.

EXERCISES

1. Alter the program above so that if the point is outside the circle, the computer will tell how far outside.

2. Rewrite the program above so that it generates two concentric circles and describes whether a point lies on either circle, within the smaller, outside the larger, or outside the smaller but inside the larger.

COLLEGE PREP TEST

DIRECTIONS: Choose the *one* best answer to each question or problem.

1. Find the area of the region described by $\frac{x^2}{2} + \frac{y^2}{2} \leq 8$.

   (A) 4π (B) 8π (C) 16π
   (D) 32π (E) None of these

2. The graph of $2(x + y) < y - 3$ does not pass through

   (A) quadrant I   (B) quadrant II
   (C) quadrant III   (D) quadrant IV
   (E) None of these

3. Find the area of the region determined by the graph of the solutions of the system
   $x^2 + y^2 \geq 4$
   $x^2 + y^2 \leq 25$.

   (A) 4π (B) 21π (C) 25π
   (D) 29π (E) None of these

4. Find a solution of the system
   $3x + 2y < 12$
   $4x + \ y > 11$.

   (A) $(2, 3)$ (B) $(3, 2)$ (C) $(2, 2)$
   (D) $(3, 3)$ (E) None of these

5. Find the area of the region determined by the graph of the solutions of the system
   $y \leq x$
   $x \leq 6$
   $y \geq 0$.

   (A) 12   (B) 18   (C) 24
   (D) 30   (E) 36

6. Two distinct parallel lines may intersect another two distinct parallel lines in the same plane in exactly

   (A) one point   (B) two points
   (C) three points   (D) four points
   (E) None of these

7. The system $x^2 + y^2 \geq a^2$
   $x^2 + y^2 \leq b^2$
   has no solution if

   (A) $a^2 > b^2$ (B) $a^2 < b^2$ (C) $a^2 = b^2$
   (D) All of these   (E) None of these

8. Which ordered pair (x, y) is a solution of the system
   $ax + \ by = 7ab$
   $3ax - 2by = 6ab?$

   (A) $(3b, 4a)$ (B) $(4b, 3a)$
   (C) $(2b, 0)$ (D) All of these
   (E) None of these

9. Find the area of the triangle determined by the lines described by
   $x = -4$
   $y = 0$
   $x + 3y = 8.$

   (A) 12   (B) 24   (C) 32
   (D) 48   (E) None of these

10. Circle I has a radius of 4 with center at $P(-3, 4)$. Circle II has a radius of 2 with center at $Q(2, 4)$. Circles I and II intersect in exactly

    (A) one point   (B) two points
    (C) infinitely many points
    (D) no points   (E) None of these

14 PROGRESSIONS, SERIES, AND BINOMIAL EXPANSIONS

HISTORY

Blaise Pascal 1623–1662

Blaise Pascal was born at Clermont-Ferrand in Auvergne on June 19, 1623. Blaise showed phenomenal ability in mathematics at an early age.

Even though his father, Etienne, was an able and noted mathematician, he wanted his son to study languages rather than mathematics. Therefore, he took away from Blaise all books on mathematics. Blaise secretly studied geometry on his own, and he succeeded considerably before his endeavors were realized by his father.

```
                    1
                  1   1
                1   2   1
              1   3   3   1
            1   4   6   4   1
          1   5  10  10   5   1
        1   6  15  20  15   6   1
            PASCAL'S TRIANGLE
```

Blaise Pascal is credited with many mathematical discoveries. At the age of 19, he invented a computing machine that served as a starting point in the development of the modern-day calculator. Blaise wrote extensively on the triangular arrangement of the coefficients of the powers of a binomial, $(a + b)^n$. This arrangement is known today as Pascal's Triangle.

Project   Write a paper on one of Pascal's major works.

ARITHMETIC PROGRESSIONS 14.1

Objectives
**To write several consecutive terms of an arithmetic progression
To find a specified term of a given arithmetic progression**

$$12, 17, 22, 27, \ldots \qquad 2, -4, -10, -16, \ldots \qquad -2, -0.5, 1, 2.5$$

An ordered list of numbers formed according to some pattern is called a **progression** or a **sequence.** Three progressions are shown above. The numbers in each progression are called the **terms** of the progression.

In the first progression above, notice that the **common difference** between consecutive terms is 5.

$$17 - 12 = 5 \qquad 22 - 17 = 5 \qquad 27 - 22 = 5$$

A progression formed in this way is called an **arithmetic progression** (abbreviated as A.P.). Note: When "arithmetic" is used as an adjective as in "arithmetic progression," it is pronounced as air-ith-MEH-tik.

Definition:
Arithmetic
progression

An *arithmetic progression* is a progression in which the difference found by subtracting any term from the next term is constant.

In general, if $x_1, x_2, x_3, x_4, \ldots$ is an arithmetic progression (A.P.) with common difference d, then $d = x_2 - x_1 = x_3 - x_2 = x_4 - x_3 = \ldots$.

Example 1
For each progression that is an A.P., find the common difference d. Give a reason for each answer.

Progression	Answer	Reason
5, 7, 10, 14, 19, . . .	Not an A.P.	$7 - 5 \neq 10 - 7$
$-6, -1, 4, 9, \ldots$	$d = 5$	$-1 - (-6) = 5; 4 - (-1) = 5$; and so on
$2, -4, -10, -16, \ldots$	$d = -6$	$-4 - 2 = -6; -10 - (-4) = -6$; and so on
6, 12, 24, 48, . . .	Not an A.P.	$12 - 6 \neq 24 - 12$
$-2, -0.5, 1, 2.5, \ldots$	$d = 1.5$	$-0.5 - (-2) = 1.5; 1 - (-0.5) = 1.5$; and so on
$3, 3 + 2\sqrt{5}, 3 + 4\sqrt{5}, \ldots,$	$d = 2\sqrt{5}$	$(3 + 2\sqrt{5}) - 3 = 2\sqrt{5}; (3 + 4\sqrt{5}) - (3 + 2\sqrt{5}) = 2\sqrt{5}$; and so on

You can write several consecutive terms of an arithmetic progression by using the following procedure. First find the common difference d. Then add d to any term to get the next term.

Example 2
Write the next three terms of the A.P.: 16, 4, -8, -20, ...

$d = 4 - 16$, or -12. Add -12 to obtain consecutive terms.

Thus, the next three terms are -32, -44, and -56.

Example 3

Write the first four terms of the A.P. whose first term a is 5 and common difference d is -3.

$$a = 5 \qquad 5 + (-3) = 2 \qquad 2 + (-3) = -1 \qquad -1 + (-3) = -4$$

Thus, the first four terms are 5, 2, -1, and -4.

The arithmetic progression 12, 17, 22, 27, . . . with common difference 5 can be written in a way that will help you find a specified term, such as the 81st term, of the progression.

12	17	22	27	. . .	412
↓	↓	↓	↓		↑
$12 + 0 \cdot 5$	$12 + 1 \cdot 5$	$12 + 2 \cdot 5$	$12 + 3 \cdot 5$	. . .	$12 + 80 \cdot 5$
1st	2nd	3rd	4th	. . .	81st

Notice that the 3rd term is $12 + (3 - 1) \cdot 5$ and that the 4th term is $12 + (4 - 1) \cdot 5$. Thus, the 81st term is $12 + (81 - 1) \cdot 5$, or $12 + 80 \cdot 5$. This leads to the following formula for the nth term of an arithmetic progression.

nth term of an arithmetic progression	The nth term of an arithmetic progression is given by the formula $$l = a + (n - 1)d, \qquad n \geq 1,$$ where l is the nth term, a is the first term, and d is the common difference.

You can use this formula to find a specified term of an arithmetic progression (A.P.).

Example 4

Find the specified term of each A.P.

26th term of 8, 5.4, 2.8, 0.2, . . .

$d = 5.4 - 8 = -2.6$
$a = 8$
$n = 26$
$l = a + (n - 1)d$
$\quad = 8 + (26 - 1)(-2.6)$
$\quad = 8 + 25(-2.6)$
$\quad = 8 - 65$
$\quad = -57$

Thus, the 26th term is -57.

31st term of $3 - \sqrt{2}, 1, -1 + \sqrt{2}, . . .$

$d = 1 - (3 - \sqrt{2}) = -2 + \sqrt{2}$
$a = 3 - \sqrt{2}$
$n = 31$
$l = a + (n - 1)d$
$\quad = (3 - \sqrt{2}) + (31 - 1)(-2 + \sqrt{2})$
$\quad = (3 - \sqrt{2}) + 30(-2 + \sqrt{2})$
$\quad = 3 - \sqrt{2} - 60 + 30\sqrt{2}$
$\quad = -57 + 29\sqrt{2}$

Thus, the 31st term is $-57 + 29\sqrt{2}$.

Reading in Algebra

Match each arithmetic progression at the left with exactly one statement at the right.

1. 7, 5, 3, 1, . . .
2. 4, 11, 18, 25, . . .
3. 5, 6, 7, 8, . . .
4. −1, 1, 3, 5, . . .
5. 2, 4, 6, 8, . . .

  A 7 is the common difference of the A.P.
  B 7 is the first term of the A.P.
  C 7 is the next term of the A.P.
  D 7 is the counting number corresponding to the term 14 of the A.P.
  E 7 is the third term of the A.P.

Oral Exercises

For each progression that is an A.P., find the common difference d.

1. 23, 35, 47, 59, . . .
2. 36, 23, 10, −3, . . .
3. −10, −8, −6, −4, . . .
4. 2, 4, 8, 16, . . .
5. 4, 6.6, 9.2, 11.8, . . .
6. 9.7, 7.4, 6.7, 5, . . .
7. 16, $15\frac{1}{3}$, $14\frac{2}{3}$, 14, . . .
8. $7\frac{1}{8}$, $8\frac{3}{4}$, $10\frac{3}{8}$, 12, . . .
9. −2, 4, −6, 8, . . .

Written Exercises

Write the next three terms of each arithmetic progression.

Ⓐ   1. 22, 23.2, 24.4, 25.6, . . .
    2. 7, $6\frac{3}{4}$, $6\frac{1}{2}$, $6\frac{1}{4}$, . . .
    3. $\frac{1}{4}$, $\frac{1}{3}$, $\frac{5}{12}$, $\frac{1}{2}$, . . .

Write the first four terms of each A.P. whose first term and common difference are given.

4. $a = 18$ and $d = 7$
5. $a = 9$ and $d = -6$
6. $a = -7$ and $d = -3$
7. $a = \$5.00$ and $d = \$2.40$
8. $a = -6°C$ and $d = 4°C$
9. $a = 15$ kg and $d = -3.4$ kg

Find the specified term of each arithmetic progression.

10. 21st term of 11, 15, 19, 23, . . .
11. 36th term of 10, 4, −2, −8, . . .
12. 31st term of 8.6, 8.3, 8, 7.7, . . .
13. 46th term of −9, −7.6, −6.2, −4.8, . . .
14. 81st term of 7, $7\frac{1}{2}$, 8, $8\frac{1}{2}$, . . .
15. 26th term of $12\frac{4}{5}$, $11\frac{2}{5}$, 10, $8\frac{3}{5}$, . . .

For each progression that is an A.P., write the next three terms.

Ⓑ   16. 3, $3 + \sqrt{2}$, $3 + 2\sqrt{2}$, . . .
    17. 5, $3 + \sqrt{3}$, $1 + 2\sqrt{3}$ . . .
    18. $1 + i$, 3, $5 - i$ . . .
    19. −2, $4 - \sqrt{2}$, $8 - \sqrt{2}$, . . .
    20. $\sqrt{15}$, $\sqrt{17}$, $\sqrt{19}$, $\sqrt{21}$, . . .
    21. $\sqrt{2}$, $\sqrt{8}$, $\sqrt{18}$, $\sqrt{32}$, . . .
    22. $5x + 2$, $7x$, $9x - 2$, . . .
    23. $8x - y$, $5x + 2y$, $2x + 5y$, . . .
    24. $x^2 + 1$, $x^3 + 1$, $x^4 + 1$, . . .

Write the first four terms of each A.P. whose first term and common difference are given.

25. $a = -3$ and $d = 3 + \sqrt{6}$
26. $a = 4 + \sqrt{5}$ and $d = -2 + 3\sqrt{5}$
27. $a = x^2 - y$ and $d = x^2 + y$

Find the 21st term of each arithmetic progression in Exercises 28–31.

28. 4, $4 + 2\sqrt{3}$, $4 + 4\sqrt{3}$, . . .
29. $5 - \sqrt{2}$, $6 - 2\sqrt{2}$, $7 - 3\sqrt{2}$, . . .
30. $4 + 3i$, 5, $6 - 3i$, . . .
31. $3x + 2y + 1$, $2x + 3y$, $x + 4y - 1$, . . .

Arithmetic Progressions

32. Some boxes are stacked so that there are 3 in the top row, 5 in the second row, 7 in the third row, and so on. How many boxes are in the fifteenth row?

33. A mechanic's starting salary is $12,500. He is guaranteed a minimum annual increase of $600. What minimum salary may he expect in his twelfth year?

34. Mrs. Prince deposited $200 in a bank on January 15. Then, each month she deposited $35 more than she had deposited the preceding month. What was her deposit on December 15?

35. A parachutist in free fall travels 5 m in the first second, 15 m during the second, 25 m during the third second. How far would the chutist travel in free fall during the eighth second?

Example   **If 76 is the *n*th term of −4, −2, 0, 2, . . . , find the value of *n*.**

$$-4, -2, 0, 2, \ldots, 76, \ldots$$

$\leftarrow$ *n*th term

$$l = a + (n - 1)d$$
$$76 = -4 + (n - 1) \cdot 2 \quad \blacktriangleleft \quad d = -2 - (-4) = 2$$
$$80 = (n - 1) \cdot 2$$
$$40 = n - 1 \qquad n = 41$$

Thus, 76 is the 41st term of the A.P.

Find the value of *n* in each exercise.

36. 62 is the *n*th term of −4, −1, 2, 5, ...

37. 129 is the *n*th term of 5, 9, 13, 17, ...

38. −54 is the *n*th term of 11, 6, 1, −4, ...

39. −106 is the *n*th term of −26, −30, −34, ...

40. 7 is the *n*th term of 1, 1.2, 1.4, 1.6, ...

41. 8 is the *n*th term of −2, −1.6, −1.2, −0.8, ...

Ⓒ **42.** Find the value of y so that $y + 3$, $4y + 1$, $8y - 3$ is an A.P. [HINT: $d = x_2 - x_1 = x_3 - x_2$ if x_1, x_2, x_3 is an A.P.]

43. Find two values of c such that $\dfrac{1}{c}$, 1, $\dfrac{10}{c + 3}$ is an A.P.

44. Find the first term of the A.P. whose 16th term is 110 and common difference is 7.

45. Find the 1st term of the A.P. whose 41st term is −124 and common difference is −3.

46. Find the first three terms of the A.P. whose 9th term is 26 and 20th term is 59.

47. Find the first three terms of the A.P. whose 48th term is −189 and 26th term is −79.

48. Prove that the progression formed by every odd-numbered term of an A.P. is also an A.P.

49. Prove: The progression formed by reversing the order in a finite A.P. is also an A.P.

CALCULATOR ACTIVITIES

Find the 93rd term of the A.P. where *a* = 62.34 and *d* = −1.25.

$$l = a + (n - 1)d = 62.34 + (93 - 1)(-1.25) = \ominus 1.25 \otimes 92 \oplus 62.34 \ominus -52.66$$

Find the *n*th term of each A.P. for the given values of *n*, *a*, and *d*.

1. $n = 119$, $a = -73$, $d = 6.75$

2. $n = 43$, $a = 72.58$, $d = -2.5$

3. $n = 2,000$, $a = -8.6$, $d = -5.4$

THE ARITHMETIC MEAN
14.2

Objectives **To find arithmetic means between two given terms of an arithmetic progression**
To find the arithmetic mean of two numbers

The terms between two given terms of an arithmetic progression are called **arithmetic means** between the given terms. For example, three arithmetic means between 2 and 18 in the progression below are 6, 10, and 14 since 2, 6, 10, 14, 18, . . . is an arithmetic progression.

$$2, \underline{6}, \underline{10}, \underline{14}, 18, . . .$$

As shown in the examples below, you can find any specified number of arithmetic means between two given numbers.

Example 1 **Find the four arithmetic means between 24 and 20.**

$$24, \underline{\hspace{1cm}}, \underline{\hspace{1cm}}, \underline{\hspace{1cm}}, \underline{\hspace{1cm}}, 20$$

When the four arithmetic means are found, an A.P. with six terms will be formed.

$n = 6 \qquad a = 24 \qquad l = 20$

Use $l = a + (n - 1)d$ to find d.

$20 = 24 + (6 - 1)d$ ◀ *Substitute for l, a, and n.*
$-4 = 5d$
$-0.8 = d$

Add -0.8 to each term.

$$24, \underline{23.2}, \underline{22.4}, \underline{21.6}, \underline{20.8}, 20$$

Thus, the four arithmetic means are 23.2, 22.4, 21.6, and 20.8.

Example 2 **Find the one arithmetic mean between 5 and 17.**

$$5, \underline{\hspace{1cm}}, 17$$

$l = a + (n - 1)d$
$17 = 5 + (3 - 1)d$ ◀ *Substitute 17 for l, 5 for a, and 3 for n.*
$12 = 2d$
$6 = d$

$$5, \underline{11}, 17$$

Thus, the one arithmetic mean is 11.

In Example 2, notice that the one arithmetic mean is the ''average'' of 5 and 17, since $(5 + 17) \div 2 = 11$. In this case, 11 is called **the arithmetic mean** of 5 and 17.

| Arithmetic mean | The *arithmetic mean* of the real numbers x and y is $\dfrac{x+y}{2}$. |

In Exercise 32, you will prove that the statement above is true.

Example 3

The arithmetic mean of two numbers is 12. If the greater number is decreased by 5 times the smaller number, the result is 6. Find the two numbers.

Represent
the data.

Let x = the smaller number and y = the greater number.

Write a system
of equations.

(1) $\dfrac{x+y}{2} = 12 \rightarrow x + y = 24$ (2) $y - 5x = 6 \rightarrow y = 5x + 6$

Solve the
system.

$$x + (5x + 6) = 24 \qquad\qquad y - 5x = 6$$
$$6x = 18 \qquad\qquad\qquad y - 5 \cdot 3 = 6$$
$$x = 3 \qquad\qquad\qquad\qquad y = 21$$

Check.

Arithmetic mean	12
$\dfrac{3 + 21}{2}$	12
$\dfrac{24}{2}$, or 12	

$y - 5x$	6
$21 - 5 \cdot 3$	6
$21 - 15$	
6	

Answer the
question.

Thus, the two numbers are 3 and 21.

Oral Exercises

For each progression that is an A.P., state the arithmetic means between the first term
and the last term.

1. 10, 7, 4   **2.** 1, 2, 4, 8   **3.** 4, 1, −2, −5
4. −4, −2, 2, 4   **5.** −7, −4, −1, 2, 5   **6.** 3, 5, 7, 9, 11, 13

Written Exercises

Find the indicated number of arithmetic means between each pair of numbers.

(A)
1. Two, between 12 and 33   **2.** Three, between 30 and 2
3. Four, between −2 and 18   **4.** Five, between 22 and −8
5. Three, between 6 and 8   **6.** Four, between 15 and 7
7. Two, between 3.4 and 10.3   **8.** Five, between 10.6 and −8.6

Find the arithmetic mean of each pair of numbers.

9. 26 and 72   **10.** 40 and −12   **11.** 2.1 and 3.5   **12.** 12.8 and 4.2

Solve each problem.

13. The arithmetic mean of two numbers is 15. Two less than the larger number is 3 times the smaller number. Find the two numbers.

14. The arithmetic mean of two numbers is 25. One of the numbers increased by twice the other number is 79. Find the two numbers.

Ⓑ 15. A teacher's annual salary for ten consecutive years was in arithmetic progression from $12,200 to $19,400. Find the salary for each of the ten years.

16. Eight different weights are used in a science laboratory to balance small masses. The weights are in arithmetic progression. Find each weight if the heaviest is 22 g and the lightest is 1 g.

17. If the arithmetic mean of two numbers is increased by one-fourth of the greater number, the sum is four times the smaller number. Find the two numbers if five times the smaller number is three more than the greater number.

18. The reciprocal of the arithmetic mean of two numbers is $\frac{4}{21}$ and the difference of the two numbers is the reciprocal of $\frac{2}{5}$. Find the two numbers.

Find the indicated number of arithmetic means between each pair of terms.

19. Two, between $3\sqrt{7}$ and $15\sqrt{7}$
20. Three, between 4 and $16 + 4\sqrt{5}$
21. Four, between $-5i$ and $-10 + 10i$
22. Two, between $1 + \sqrt{3}$ and $7 - 5\sqrt{3}$
23. Three, between $-8a + 9b$ and $4a - 3b$
24. Four, between $10x$ and $15y$.

For Exercises 25–30, find the arithmetic mean of each pair of terms.

25. $-6x$ and $14x$
26. $15x - 3y + 2z$ and $9x + 15y - 18z$
27. $3 + \sqrt{2}$ and $7 - 7\sqrt{2}$
28. $9\sqrt{21} - 5\sqrt{10}$ and $9\sqrt{21} + 5\sqrt{10}$

Ⓒ 29. The complex number $a + bi$ and its conjugate
30. $(x + y)^2$ and $(x - y)^2$

31. Find the two arithmetic means between x and y. Write each mean as a rational expression.

32. Prove: The arithmetic mean of x and y is $\dfrac{x + y}{2}$. $\begin{bmatrix} \text{Hint: Use the formula} \\ l = a + (n - 1)d. \end{bmatrix}$

33. Prove: The arithmetic mean of any two consecutive odd integers is an even integer. [Hint: If x is an integer, then $2x - 1$ is an odd integer and $2x$ is an even integer.]

CUMULATIVE REVIEW

1. Solve the formula $T = \frac{n}{2}(x + y)$ for y. Then find the value of y if $T = 750$, $n = 25$, and $x = 20$.

2. Solve the formula $2T + nd = 2an + n^2d$ for d. Then find the value of d if $T = 210$, $n = 10$, and $a = 3$.

The Arithmetic Mean

ARITHMETIC SERIES AND Σ-NOTATION   14.3

Objectives   **To find the sum of the first *n* terms of an arithmetic series**
To write an arithmetic series using Σ-notation

If you take the arithmetic progression 10, 6, 2, −2, −6, . . . and write it as the indicated sum 10 + 6 + 2 − 2 − 6 − . . ., the result is an **arithmetic series.**

Definition: Arithmetic series	An *arithmetic series* is an indicated sum of the terms of an arithmetic progression.

Example 1   **Write each A.P. as an arithmetic series. Find the common difference *d* of the series.**

$$-7, 2, 11, 20, \ldots \qquad\qquad 8, 3, -2, -7, \ldots$$

The arithmetic series is   The arithmetic series is
$-7 + 2 + 11 + 20 + \ldots$ $8 + 3 - 2 - 7 - \ldots$
$d = 2 - (-7)$, or 9 $d = 3 - 8$, or -5

You can find the sum *S* of the first 20 terms of the arithmetic series
$3 + 6 + 9 + 12 + \ldots$ as follows: Write the series in its given order and also in reverse order. Then add the two series.

$$
\begin{array}{rcl}
S &=& 3 + 6 + 9 + \ldots + 54 + 57 + 60 \\
S &=& 60 + 57 + 54 + \ldots + 9 + 6 + 3 \\
\hline
2S &=& 63 + 63 + 63 + \ldots + 63 + 63 + 63
\end{array}
$$

$$2S = 20 \cdot 63$$
$$S = 630$$

The steps above suggest a method for obtaining a formula for the sum *S* of the first *n* terms of an arithmetic series. Let *l* represent the *n*th term.

$$
\begin{array}{rcl}
S &=& a + (a + d) + (a + 2d) + \ldots + (l - 2d) + (l - d) + l \\
S &=& l + (l - d) + (l - 2d) + \ldots + (a + 2d) + (a + d) + a \\
\hline
2S &=& (a + l) + (a + l) + (a + l) + \ldots + (a + l) + (a + l) + (a + l)
\end{array}
$$

$$n \text{ binomials}$$
$$2S = n(a + l)$$
$$S = \frac{n}{2}(a + l)$$

Sum of n terms of arithmetic series	The sum *S* of the first *n* terms of an arithmetic series is given by the formula $$S = \frac{n}{2}(a + l),$$ where *a* is the first term and *l* is the *n*th term.

Example 2

Find the sum of the arithmetic series in which the number of terms $n = 35$, the first term $a = 12$, and the last term $l = 114$.

$$S = \frac{n}{2}(a + l) = \frac{35}{2}(12 + 114)$$
$$= \frac{35}{2} \cdot 126$$
$$= 35 \cdot 63, \text{ or } 2{,}205$$

Thus, the sum of the arithmetic series is 2,205.

You can find the sum of the first 34 terms of the arithmetic series $24.5 + 21.5 + 18.5 + 15.5 + \ldots$ without finding the 34th term, or l. To do this, another sum formula is used. It is obtained by substitution, as shown below:

$$S = \frac{n}{2}[a + l] \text{ and } l = a + (n - 1)d$$

$$S = \frac{n}{2}[a + \underbrace{a + (n - 1)d}_{l}], \text{ or } S = \frac{n}{2}[2a + (n - 1)d]$$

Use this second sum formula when l, the last term, is not known but a, the first term, and d, the common difference, are easily determined for a given arithmetic series.

Example 3

Find the sum of the first 34 terms of the arithmetic series:
$24.5 + 21.5 + 18.5 + 15.5 + \ldots$.

$$n = 34 \qquad a = 24.5 \qquad d = -3$$
$$S = \frac{n}{2}[2a + (n - 1)d] = \frac{34}{2}[2(24.5) + 33(-3)]$$
$$= 17[49 - 99], \text{ or } -850$$

Thus, the sum of the first 34 terms is -850.

The indicated sum, $4 + 8 + 12 + 16 + 20$, can be written in a more compact form as $\sum_{k=1}^{5} 4k$. Recall that the Greek letter Σ (read: sigma) is used as a summation sign and k is called the **index of summation.**

$$\underbrace{\sum_{k=1}^{5} 4k = 4 \cdot 1 + 4 \cdot 2 + 4 \cdot 3 + 4 \cdot 4 + 4 \cdot 5}_{\text{Expanded Form}} = \underbrace{4 + 8 + 12 + 16 + 20}_{\text{Series Form}} = \underbrace{60}_{\text{Sum}}$$

$\sum_{k=1}^{5} 4k$ is read "the summation from 1 to 5 of $4k$." Notice that the five terms of the series are found by substituting 1, 2, 3, 4, and 5 for k in the general term, $4k$.

Example 4

Write the expanded form and the series form of $\sum_{j=3}^{6} (5j - 22)$. Find the sum of the series.

Substitute 3, 4, 5, and 6 for j in $(5j - 22)$.

$\sum_{j=3}^{6} (5j - 22) = (5 \cdot 3 - 22) + (5 \cdot 4 - 22) + (5 \cdot 5 - 22) + (5 \cdot 6 - 22)$ ◀ *Expanded form*

$= -7 - 2 + 3 + 8$ ◀ *Series form*

$= 2$ ◀ *Sum of the series*

Example 5

Write the expanded form and the series form of $\sum_{c=1}^{35} (90 - 2c)$. Find the sum of the series.

$\sum_{c=1}^{35} (90 - 2c) = (90 - 2 \cdot 1) + (90 - 2 \cdot 2) + (90 - 2 \cdot 3) + \ldots + (90 - 2 \cdot 35)$ ◀ *Expanded form*

$= 88 + 86 + 84 + \ldots + 20$ ◀ *Series form*

$= \frac{35}{2} (88 + 20)$ ◀ *Use $S = \frac{n}{2} (a + l)$ to find the sum.*

$= 1,890$ ◀ *Sum of the series*

You can reverse the procedure above and write certain arithmetic series using Σ-notation. For example,

$$-2 - 4 - 6 - 8 - \ldots - 40 = -2 \cdot 1 - 2 \cdot 2 - 2 \cdot 3 - 2 \cdot 4 - \ldots - 2 \cdot 20 = \sum_{n=1}^{20} - 2n$$

Notice that the common difference d is -2 and that n goes from 1 to 20.

Example 6

Write $10 + 13 + 16 + 19 + \ldots + 52$ using Σ-notation.

The series is arithmetic and the common difference is 3.
Write each term in the form $3k + t$ where

$$k = 1, 2, 3, 4, \ldots \qquad \text{and } t \text{ is a real number.}$$

Notice that the first term is 10 and that $10 = 3 \cdot 1 + 7$. So, $t = 7$.

$$10 \quad + \quad 13 \quad + \quad 16 \quad + \quad 19 \quad + \ldots + \quad 52$$
$$\downarrow \qquad\qquad \downarrow \qquad\qquad \downarrow \qquad\qquad \downarrow \qquad\qquad\qquad \downarrow$$
$$(3 \cdot 1 + 7) + (3 \cdot 2 + 7) + (3 \cdot 3 + 7) + (3 \cdot 4 + 7) + \ldots + (3 \cdot 15 + 7)$$
$$\sum_{k=1}^{15} (3k + 7)$$

Oral Exercises

State each A.P. as an arithmetic series.

1. 5, 9, 13, 17, . . . **2.** 5, 1, -3, -7, . . . **3.** -2, -3, -4, -5, . . .

Read each summation aloud. State the series form.

4. $\displaystyle\sum_{k=1}^{10} 5k$ 5. $\displaystyle\sum_{j=8}^{12} -j$ 6. $\displaystyle\sum_{c=1}^{6} (c+2)$ 7. $\displaystyle\sum_{n=5}^{7} (17-n)$ 8. $\displaystyle\sum_{p=1}^{4} \frac{p}{5}$

Find the common difference d of each arithmetic series.

9. $12 + 15 + 18 + 21 + \dots$ 10. $3 + 1 - 1 - 3 - \dots$ 11. $1 + 1.4 + 1.8 + 2.2 + \dots$

12. $\displaystyle\sum_{j=1}^{4} 5j$ 13. $\displaystyle\sum_{k=1}^{3} (k+2)$ 14. $\displaystyle\sum_{c=1}^{3} -2c$

Written Exercises

Find the sum of each arithmetic series for the given data.

(A) 1. $n = 20, a = 1, l = 154$ 2. $n = 40, a = 2, l = -115$
3. $n = 25, a = 16, l = 184$ 4. $n = 45, a = 3, l = -50$
5. $n = 35, a = -5.2, l = 60$ 6. $n = 30, a = -7, l = -60.2$
7. $n = 30; 5 + 15 + 25 + 35 + \dots$ 8. $n = 40; 6 + 2 - 2 - 6 - \dots$
9. $n = 65; 7 + 7.2 + 7.4 + 7.6 + \dots$ 10. $n = 45; -2 - 0.9 + 0.2 + 1.3 + \dots$

11. $\displaystyle\sum_{k=1}^{6} 3k$ 12. $\displaystyle\sum_{j=1}^{5} -j$ 13. $\displaystyle\sum_{c=1}^{50} (c+1)$

14. $\displaystyle\sum_{j=5}^{8} (3j-4)$ 15. $\displaystyle\sum_{m=40}^{43} (8-m)$ 16. $\displaystyle\sum_{k=1}^{25} (15-3k)$

Write each arithmetic series using Σ-notation.

17. $6 + 12 + 18 + 24$ 18. $-8 - 16 - 24 - 32 - \dots - 80$
19. $7 + 8 + 9 + 10 + \dots + 26$ 20. $12 + 11 + 10 + 9 + \dots - 5$

(B) 21. $8 + 10 + 12 + 14 + 16$ 22. $15 + 19 + 23 + 27 + \dots + 47$
23. $9 + 4 - 1 - 6 - \dots - 46$ 24. $-7 - 10 - 13 - 16 - \dots - 43$

Find the sum of each arithmetic series for the given data.

25. $n = 41, a = 2\sqrt{3}, l = 198\sqrt{3}$ 26. $n = 16, a = 3x - 10y, l = -47x + 30y$
27. $n = 26; 0.3 + 0.8 + 1.3 + 1.8 + \dots$ 28. $n = 37; -1.3 - 0.1 + 1.1 + 2.3 + \dots$

29. $\displaystyle\sum_{n=1}^{100} \frac{n}{4}$ 30. $\displaystyle\sum_{n=1}^{30} \frac{n+4}{3}$

31. Cartons are stacked in 20 rows with 2 in the top row, 5 in the 2nd row, 8 in the 3rd row, and so on. How many cartons are stacked?

32. Gerry saved 1 dime the 1st day, 2 dimes the 2nd day, 3 on the 3rd day, and so on. How much money did she save in 30 days?

(C) 33. Prove: $\displaystyle\sum_{k=1}^{50} 10kx = 10x \cdot \sum_{k=1}^{50} k$. 34. Prove: $\displaystyle\sum_{j=1}^{20} (jx + jy) = \sum_{j=1}^{20} jx + \sum_{j=1}^{20} jy$.

35. Prove: $\displaystyle\sum_{c=1}^{30} (c+t) = 30t + \sum_{c=1}^{30} c$. 36. Solve $\displaystyle\sum_{k=4}^{7} (kx - 3) = 32$ for x.

37. Prove: $\displaystyle\sum_{n=1}^{k} (2n-1) = k^2$. What does this say about adding odd integers?

38. Prove: $\displaystyle\sum_{n=1}^{k} 2n = k(k+1)$. What does this say about adding even integers?

GEOMETRIC PROGRESSIONS 14.4

Objectives   **To write several consecutive terms of a geometric progression**
To find a specified term of a given geometric progression

Another type of progression (or sequence) is one such as

$$5, 10, 20, 40, \ldots$$

in which each consecutive term is found by multiplying the preceding term by a given number. Notice that in the progression 5, 10, 20, 40, . . . , the **common ratio** of consecutive terms is 2.

$$\frac{10}{5} = 2 \qquad \frac{20}{10} = 2 \qquad \frac{40}{20} = 2$$

A progression formed in this way is called a **geometric progression** (abbreviated as G. P.). Note: Pronounce "geometric" as gee-oh-MEH-trik.

Definition: Geometric progression	A *geometric progression* is a progression in which the ratio found by dividing any term by the preceding term is constant.

In general, if $x_1, x_2, x_3, x_4, \ldots$ is a geometric progression (G. P.) with common ratio $r \neq 0$, then $r = \dfrac{x_2}{x_1} = \dfrac{x_3}{x_2} = \dfrac{x_4}{x_3} = \ldots$.

Example 1   **For each progression that is a G. P., find the common ratio r. Give a reason for each answer.**

Progression	Answer	Reason
3, 6, 9, 12, . . .	Not a G. P.	$6 \div 3 \neq 9 \div 6$
$-2, -8, -32, -128, \ldots$	$r = 4$	$\dfrac{-8}{-2} = 4; \dfrac{-32}{-8} = 4$; and so on
$64, -16, 4, -1, \ldots$	$r = -\dfrac{1}{4}$	$\dfrac{-16}{64} = -\dfrac{1}{4}; \dfrac{4}{-16} = -\dfrac{1}{4}$; and so on
6, 0, 0, 0, . . .	Not a G. P.	$0 \div 6 \neq 0 \div 0$ ($0 \div 0$ is undefined)
7, 0.7, 0.07, 0.007, . . .	$r = 0.1$	$0.7 \div 7 = 0.1; 0.07 \div 0.7 = 0.1$; and so on
$5, 5\sqrt{2}, 10, 10\sqrt{2}, \ldots$	$r = \sqrt{2}$	$\dfrac{5\sqrt{2}}{5} = \sqrt{2}; \dfrac{10}{5\sqrt{2}} = \sqrt{2}$; and so on

You can write several consecutive terms of a G.P. as shown below.

Example 2   **Write the next three terms of the G. P.: 128, -64, 32, -16, ...**

$r = \dfrac{-64}{128}$, or $-\dfrac{1}{2}$. Multiply by $-\dfrac{1}{2}$ to obtain consecutive terms.

Thus, the next three terms are 8, -4, and 2.

Example 3   Write the first four terms of the G.P. whose first term a is -2 and common ratio r is -5.

$$a = -2 \qquad -2(-5) = 10 \qquad 10(-5) = -50 \qquad -50(-5) = 250$$

Thus, the first four terms are -2, 10, -50, and 250.

The geometric progression 5, 10, 20, 40, 80, ... with common ratio 2 can be written in a way that will help you find a specified term, such as the 11th term, of the progression.

5	10	20	40	80	... 5,120
↓	↓	↓	↓	↓	↑
5	$5 \cdot 2^1$	$5 \cdot 2^2$	$5 \cdot 2^3$	$5 \cdot 2^4$ ... $5 \cdot 2^{10}$	
1st	2nd	3rd	4th	5th ...	11th

Notice that the 4th term is $5 \cdot 2^{4-1}$, or $5 \cdot 2^3$, and that the 5th term is $5 \cdot 2^{5-1}$, or $5 \cdot 2^4$. Thus, the 11th term is $5 \cdot 2^{11-1}$, or $5 \cdot 2^{10}$. This leads to the following formula for the nth term of a geometric progression.

nth term of a geometric progression	The nth term of a geometric progression is given by the formula $$l = a \cdot r^{n-1}, n \geq 1,$$ where l is the nth term, a is the first term, and r is the common ratio.

Example 4   Find the specified term of each geometric progression, given the first term a and the common ratio r.

5th term: $a = 12$ and $r = 0.1$

$$\begin{aligned} l = a \cdot r^{n-1} &= 12(0.1)^{5-1} \\ &= 12(0.1)^4 \\ &= 12(0.0001) \\ &= 0.0012 \end{aligned}$$

Thus, the 5th term is 0.0012.

7th term: $a = -8$ and $r = -\dfrac{1}{2}$

$$\begin{aligned} l = a \cdot r^{n-1} &= -8\left(-\frac{1}{2}\right)^{7-1} \\ &= -8 \cdot \frac{1}{64} = -\frac{1}{8} \end{aligned}$$

Thus, the 7th term is $-\dfrac{1}{8}$.

Example 5   Find the 10th term of the G.P.: $\dfrac{1}{2}$, -1, 2, -4, ...

$$r = -1 \div \frac{1}{2}, \text{ or } -2 \qquad a = \frac{1}{2} \qquad n = 10 \qquad l \text{ is the 10th term.}$$

$$l = a \cdot r^{n-1} = \frac{1}{2}(-2)^9 = \frac{1}{2}(-512) = -256$$

Thus, the 10th term is -256.

Reading in Algebra

Match each geometric progression at the left with exactly one statement at the right.

1. 0.003, 0.03, 0.3, ...

2. 5, 15, 45, ...

3. $3, 1, \frac{1}{3}, \ldots$

4. 48, 24, 12, ...

5. $x, x^3, x^5, \ldots$

A 3 is the common ratio of the G.P.

B 3 is the first term of the G.P.

C 3 is the next term of the G.P.

D 3 is the counting number for the term, x^5.

E 3 is the 5th term of the G.P.

Oral Exercises

For each progression that is a G.P., find the common ratio r.

1. 2, 8, 32, 128, ...

2. 2, −6, 18, −54, ...

3. 16, 8, 4, 2, ...

4. 5, 10, 15, 20, ...

5. −40, −20, −10, −5, ...

6. 3, −3, 3, −3, ...

7. $-9, 3, -1, \frac{1}{3}, \ldots$

8. 4, 0.4, 0.04, 0.004, ...

9. $2\sqrt{3}, 6, 6\sqrt{3}, 18, \ldots$

10. $\frac{1}{8}, \frac{1}{4}, \frac{1}{2}, 1, \ldots$

11. $3, 2, \frac{4}{3}, \frac{8}{9}, \ldots$

12. $5x, 10x^2, 20x^3, 40x^4, \ldots$

Written Exercises

Write the next three terms of each geometric progression.

Ⓐ **1.** 30, 60, 120, ...

2. −0.1, −0.2, −0.4, ...

3. $-\frac{1}{3}, 1, -3, \ldots$

Write the first four terms of each G.P. whose first term and common ratio are given.

4. $a = 20$ and $r = 2$

5. $a = -\frac{1}{4}$ and $r = 4$

6. $a = 0.25$ and $r = 4$

7. $a = 32$ and $r = \frac{1}{2}$

8. $a = -8$ and $r = \frac{1}{4}$

9. $a = 6$ and $r = 0.1$

10. $a = \frac{1}{8}$ and $r = -2$

11. $a = 64$ and $r = -\frac{3}{8}$

12. $a = -2$ and $r = -3$

Find the specified term of each geometric progression.

13. 5th term: $a = -10, r = 3$

14. 6th term: $a = 20, r = -5$

15. 7th term: $a = 34, r = 10$

16. 7th term: $a = 900, r = 0.1$

17. 6th term: $a = -30, r = 0.2$

18. 5th term: $a = 72, r = -0.1$

19. 10th term: 4, 8, 16, ...

20. 9th term: 10, −20, 40, ...

21. 10th term: $-\frac{1}{4}, -\frac{1}{2}, -1, \ldots$

22. 11th term: −8, −4, −2, ...

23. 12th term: −32, 16, −8, ...

24. 11th term: 96, 48, 24, ...

25. 8th term: 370, 37, 3.7, ...

26. 9th term: 0.24, 2.4, 24, ...

27. 7th term: 1, 0.2, 0.04, ...

Ⓑ **28.** 6th term: $a = -9, r = \frac{1}{3}$

29. 4th term: $a = 80, r = -\frac{1}{4}$

30. 5th term: $a = 25, r = \frac{2}{5}$

31. 8th term: $a = 6x, r = 2x$

32. 7th term: $a = 4y, r = 2y^2$

33. 8th term: $a = 8xy, r = x^2y^3$

34. 7th term: $a = 10, r = \sqrt{3}$

35. 6th term: $a = -4, r = \sqrt{5}$

36. 5th term: $a = \sqrt{2}, r = 3\sqrt{2}$

Write the next three terms of each geometric progression in Exercises 37–42.

37. $2x^2, 8x^5, 32x^8, \ldots$

38. $1, i, -1, -i, \ldots$

39. $3 + 2\sqrt{5}, 6 + 4\sqrt{5}, 12 + 8\sqrt{5}, \ldots$

40. $2 + \sqrt{6}, 6 + 2\sqrt{6}, 12 + 6\sqrt{6}, \ldots$ **41.** $-3x^3, 6x^5, -12x^7, \ldots$ **42.** $3i, -6, -12i, 24, \ldots$

43. A tennis ball, dropped from a height of 128 dm, rebounds on each bounce one-half the distance from which it fell. How high does it go on its 9th rebound?

44. A tank contains 4,000 L of water. Each day, one-half of the water will be removed. How much water will be in the tank at the end of the eighth day?

45. A golf ball, dropped from a height of 81 m, rebounds on each bounce two-thirds of the distance from which it fell. How far does it fall on its 6th descent?

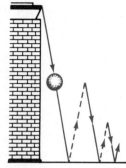

46. A tank contains 243 L of gasoline. Each time that a valve is opened, one-third of the gasoline is released. How much gas will be in the tank after the valve is operated 6 times?

47. Find the first term of the geometric progression whose 7th term is 512 and 8th term is 1,024. [Hint: Use the formula $l = ar^{n-1}$.]

$a, \ldots, 256, 512, 1024, \ldots$
$1st \qquad 7th \; 8th$

48. Find the first term of the G.P. whose 9th term is $\frac{1}{64}$ and 10th term is $\frac{1}{128}$.

49. What is the first term of a G.P., if its 8th and 9th terms are 6,250 and $-31,250$, respectively?

Ⓒ 50. Find two values of t so that $\frac{1}{3}, t - 1, 4t$ is a geometric sequence.

$$\left[\text{Hint:} \, r = \frac{x_2}{x_1} = \frac{x_3}{x_2} \text{ if } x_1, x_2, x_3 \text{ is a geometric sequence.} \right]$$

51. Find two values of t so that $6t, t + 1, \frac{3}{4}$ is a G.P.

52. Which term of $3, -6, 12, \ldots$ is -384?

53. Which term of $\frac{1}{8}, \frac{1}{4}, \frac{1}{2}, \ldots$ is 128?

CALCULATOR ACTIVITIES

Find the 7th term of the G.P. where $a = 12.5$ and $r = 3$.

$$l = a \cdot r^{n-1} = 12.5 \times 3^6$$
$l = 3 \otimes \ominus \ominus \ominus \ominus \ominus \ominus \otimes 12.5 \ominus 9112.5$ ◀ *The 7th term*

Find the nth term of each G.P. for the given values of n, a, and r.
  1. $n = 9, a = -7.2, r = 2$ **2.** $n = 6, a = 640, r = 4$ **3.** $n = 8, a = 0.2, r = 0.5$

CUMULATIVE REVIEW

Solve each equation for n.

  1. $2^n = 64$ **2.** $3^n = \frac{1}{81}$ **3.** $5^{3n+1} = 625$ **4.** $8^n = 128$ **5.** $9^n = \frac{1}{27}$

THE GEOMETRIC MEAN 14.5

Objectives   **To find geometric means between two given terms of a geometric progression**
To find the geometric mean (mean proportional) of two numbers

The terms between two given terms of a geometric progression are called **geometric means** between the given terms. For example, four geometric means between 3 and 96 in the progression below are 6, 12, 24, and 48, since 3, 6, 12, 24, 48, 96, . . . is a geometric progression.

$$3, \underline{6}, \underline{12}, \underline{24}, \underline{48}, 96, \ldots$$

Example 1   **Find three geometric means between -2 and -162.**

$$-2, \underline{\hspace{1cm}}, \underline{\hspace{1cm}}, \underline{\hspace{1cm}}, -162$$

When the three geometric means are found, a G.P. with five terms will be formed.

$n = 5 \qquad a = -2 \qquad l = -162$

Use $\quad l = a \cdot r^{n-1}$ to find r.

$-162 = -2 \cdot r^{5-1}$ ◀ *Substitute for l, a, and n.*

$81 = r^4$

$r = \sqrt[4]{81} \quad$ or $\quad r = -\sqrt[4]{81}$

$r = 3 \qquad\qquad r = -3$ ◀ *Two real numbers for r*

Multiply each term by either 3 or -3.

$-2, \underline{-6}, \underline{-18}, \underline{-54}, -162 \quad$ or $\quad -2, \underline{6}, \underline{-18}, \underline{54}, -162$

Thus, the means are $-6, -18, -54 \quad$ or $\quad 6, -18, 54$.

Example 2   **Find two real geometric means between 9 and $-\dfrac{8}{3}$.**

$$9, \underline{\hspace{1cm}}, \underline{\hspace{1cm}}, -\frac{8}{3}$$

$l = a \cdot r^{n-1}$

$-\dfrac{8}{3} = 9 \cdot r^3$ ◀ *Substitute $-\dfrac{8}{3}$ for l, 9 for a, and 4 for n.*

$-\dfrac{8}{27} = r^3$

$r = \sqrt[3]{-\dfrac{8}{27}} = -\dfrac{2}{3}$

Multiply each term by $-\dfrac{2}{3}$: $9, \underline{-6}, \underline{4}, -\dfrac{8}{3}$.

Thus, the two real geometric means are -6 and 4.

A single geometric mean between two numbers is called **the geometric mean** or **mean proportional** of the two numbers.

Example 3   **Find the geometric mean (mean proportional) of 5 and 10.**

$$5, \underline{\hspace{1cm}}, 10$$

$l = ar^{n-1}$

$10 = 5 \cdot r^2$ ◀ *Substitute 10 for l, 5 for a, and 3 for n.*

$2 = r^2$

$r = \sqrt{2}$ or $r = -\sqrt{2}$

Multiply by either $\sqrt{2}$ or $-\sqrt{2}$.

$$5, \underline{5\sqrt{2}}, 10 \quad \text{or} \quad 5, \underline{-5\sqrt{2}}, 10$$

Thus, the geometric mean (mean proportional) is $5\sqrt{2}$ or $-5\sqrt{2}$.

In Example 3, the *positive* geometric mean of 5 and 10 is $\sqrt{5 \cdot 10}$, or $5\sqrt{2}$, and the *negative* geometric mean of 5 and 10 is $-\sqrt{5 \cdot 10}$, or $-5\sqrt{2}$. This suggests the following statement about the geometric mean of two nonzero real numbers x and y. You will prove this statement in Exercise 45.

Geometric mean	The geometric mean (mean proportional) of the real numbers x and y ($xy > 0$) is $\sqrt{xy}$ or $-\sqrt{xy}$.

Some word problems that involve the geometric mean of two numbers can be solved by using a system of two equations as shown below.

Example 4   **One number is 30 more than another number and the positive geometric mean of the two numbers is 8. Find the two numbers.**

Let x and y represent the two numbers. Write a system of equations.

(1) $y = x + 30$ (2) $\sqrt{xy} = 8$

Solve the system by substitution.

$\sqrt{x(x + 30)} = 8$ ◀ *Substitute for y in equation (2).*

$(\sqrt{x^2 + 30x})^2 = 8^2$

$x^2 + 30x = 64$

$x^2 + 30x - 64 = 0$

$(x - 2)(x + 32) = 0$

$x = 2$ or $x = -32$

If $x = 2$, then $y = x + 30 = 32$. If $x = -32$, then $y = x + 30 = -2$. Check each pair of numbers in both equations, (1) and (2).

Thus, the numbers are 2 and 32 or -32 and -2.

When you are asked to find "n real geometric means" between a pair of given numbers, there are two sets of means if n is odd, as in Examples 1 and 3. If n is even, there is only one set of means, as in Example 2.

The Geometric Mean   **373**

Written Exercises

Find the indicated number of real geometric means between each pair of numbers.

(A)
1. Three, between 1 and 81
2. One positive, between 3 and 12
3. One negative, between 2 and 50
4. Two, between 5 and 320
5. Two, between -3 and 24
6. Two, between -2 and -128
7. One positive, between -4 and -36
8. One negative, between -5 and -80
9. Two, between 16 and -2
10. Three, between -64 and -4
11. Three, between -5 and -405
12. Three, between 2 and 1250
13. One positive, between 8 and $\frac{1}{2}$
14. One negative, between $\frac{2}{5}$ and 10
15. Two, between 9 and $\frac{1}{3}$
16. Two, between $\frac{3}{5}$ and 75
17. Four, between 11 and 352
18. Four, between 2 and 200,000
19. One positive, between 3 and 6
20. One negative, between 4 and 12
21. One negative, between $5\sqrt{2}$ and $10\sqrt{2}$
22. One positive, between $3\sqrt{5}$ and $15\sqrt{5}$

23. One number is 4 times another number and the positive geometric mean of the two numbers is 6. Find the two numbers.
24. Eight more than the positive geometric mean of two numbers is 20. Find the two numbers if one number is 9 times the other number.
25. If one number is 6 more than another number and the positive geometric mean of the two numbers is 4, what are the two numbers?
26. Find two numbers whose difference is 5 and whose positive geometric mean is 6.

(B)
27. Seven different weights are used to balance small masses in a laboratory. These weights are in geometric progression. Find each weight if the lightest is 0.5 g and the median (middle) weight is 4 g.
28. Crude oil is pumped from a tanker so that the amounts remaining at the end of successive hours are in G.P. At the end of 3 hours, 6,400 kiloliters (kL) remain in the tanker, and at the end of 6 hours, there are 2,700 kL. How much oil remains in the tanker at the end of 4 hours?

Find the indicated number of real geometric means between each pair of terms.

29. One positive, between 0.1 and 0.016
30. Two, between 0.2 and 0.0128
31. Two, between 3 and $21\sqrt{7}$
32. Two, between $3\sqrt{5}$ and 75
33. Three, between $5x^3$ and $80x^{11}$
34. Five, between y^2 and $64y^{20}$
35. One positive, between $2i$ and $-18i$
36. One negative, between $-3i$ and $15i$

Find the geometric mean of each two numbers. (Each exercise has two answers.)

(C)
37. i and $4i$
38. 3 and -27
39. -4 and 12
40. $5i$ and $8i$
41. $1 + i$ and $1 - i$
42. $3 - 4i$ and $3 + 4i$
43. x^3 and y^3
44. $\sqrt{x}$ and $\sqrt{y}$

Use the formula $l = a \cdot r^{n-1}$ to prove each statement.

45. The geometric mean of the real numbers x and $y(xy > 0)$ is $\sqrt{xy}$ or $-\sqrt{xy}$.
46. The geometric mean of the complex number $a + bi$ and its conjugate is a real number.

GEOMETRIC SERIES   14.6

Objective   **To find the sum of the first *n* terms of a geometric series**

If you take the geometric progression 2, 10, 50, 250, . . .
and write it as the indicated sum 2 + 10 + 50 + 250 + . . . ,
the result is a **geometric series**.

Definition: Geometric series	A *geometric series* is an indicated sum of the terms of a geometric progression.

Example 1   **Write each G.P. as a geometric series. Find the common ratio *r* of the series.**

$-2, 8, -32, 128, \ldots$ $-8, -2, -\dfrac{1}{2}, -\dfrac{1}{8}, \ldots$

The geometric series is   The geometric series is

$-2 + 8 - 32 + 128 - \ldots.$ $-8 - 2 - \dfrac{1}{2} - \dfrac{1}{8} - \ldots.$

$r = \dfrac{8}{-2}$, or -4 $r = \dfrac{-2}{-8}$, or $\dfrac{1}{4}$

You can find the sum S of the terms of the geometric series $5 + 15 + 45 + \ldots + 3{,}645$ with common ratio 3 in the following way:

(1) Write the series in the form $a + ar + ar^2 + ar^3 + \ldots + l$.
(2) Multiply each term by -3 to get a second series.
(3) Then add the two series.

$$S = 5 + 5 \cdot 3 + 5 \cdot 3^2 + 5 \cdot 3^3 + \ldots + 3{,}645$$

$$-3S = \quad - 5 \cdot 3 - 5 \cdot 3^2 - 5 \cdot 3^3 - \ldots - 3{,}645 - 3(3{,}645)$$

$$-2S = 5 + \quad 0 \quad + 0 \quad + 0 \quad + \ldots + 0 \quad - 3(3{,}645)$$

$$S = \frac{5 - 3(3{,}645)}{-2}, \text{ or } 5{,}465$$

The steps above suggest a method for obtaining a formula for the sum S of the terms of a finite geometric series with first term a, common ratio r, and last term l.

$$S = a + ar + ar^2 + ar^3 + \ldots + l$$

$$-r \cdot S = \quad - ar - ar^2 - ar^3 - \ldots - l - rl$$

$$S - rS = a + 0 \quad + 0 \quad + 0 \quad + \ldots + 0 - rl$$

$$S - rS = a - rl$$
$$(1 - r)S = a - rl$$
$$S = \frac{a - rl}{1 - r}$$

In this book, "the sum of a geometric series" will be used to mean "the sum of the terms of a geometric series."

Example 2 Find the sum of the geometric series in which the first term $a = 3$, the common ratio $r = -5$, and the last term $l = -375$.

$$S = \frac{a - rl}{1 - r} = \frac{3 - (-5)(-375)}{1 - (-5)} = \frac{3 - 1875}{6} = -312$$

Thus, the sum of the geometric series is -312.

Example 3 Find the sum of the geometric series: $8 - 4 + 2 - \ldots - \frac{1}{16}$.

$$a = 8 \qquad r = \frac{-4}{8}, \text{ or } -\frac{1}{2} \qquad l = -\frac{1}{16}$$

$$S = \frac{8 - \left(-\frac{1}{2}\right)\left(-\frac{1}{16}\right)}{1 - \left(-\frac{1}{2}\right)} = \left(8 - \frac{1}{32}\right) \div \frac{3}{2} \qquad \blacktriangleleft \; \begin{array}{l} \textit{Use the sum formula.} \\ \textit{Simplify.} \end{array}$$

$$= \frac{255}{32} \cdot \frac{2}{3}$$

$$= \frac{85}{16}, \text{ or } 5\frac{5}{16} \qquad \blacktriangleleft \; \textit{Sum of the series}$$

You can find the sum of the first 9 terms of the geometric series $-\frac{1}{4} + \frac{1}{2} - 1 + 2 - \ldots$ without finding the 9th term, or l. To do this, another sum formula is used. This formula is obtained by substitution.

$$S = \frac{a - r \cdot l}{1 - r} \text{ and } l = a \cdot r^{n-1}$$

$$S = \frac{a - r(a \cdot r^{n-1})}{1 - r} = \frac{a - a \cdot r^1 \cdot r^{n-1}}{1 - r} = \frac{a - a \cdot r^n}{1 - r}, \text{ or } \frac{a(1 - r^n)}{1 - r}$$

Example 4 Find the sum of the first 9 terms of the geometric series: $-\frac{1}{4} + \frac{1}{2} - 1 + 2 - \ldots$.

$$S = \frac{a(1 - r^n)}{1 - r} = \frac{-\frac{1}{4}[1 - (-2)^9]}{1 - (-2)} = \frac{-\frac{1}{4}(1 + 512)}{3} = -\frac{1}{4} \cdot \frac{513}{3} = -\frac{1}{4} \cdot 171 = -42\frac{3}{4}$$

$\sum\limits_{k=3}^{6} 2k$ is the arithmetic series $2 \cdot 3 + 2 \cdot 4 + 2 \cdot 5 + 2 \cdot 6$, or

$6 + 8 + 10 + 12$, whose common difference is 2. In a similar way, $\sum\limits_{k=3}^{6} 2^k$ is

the geometric series $2^3 + 2^4 + 2^5 + 2^6$, or $8 + 16 + 32 + 64$, whose common ratio is 2.

Example 5   **Write the expanded form and the series form of $\sum\limits_{j=1}^{5} 5(-3)^j$. Find the sum of the series.**

Substitute 1, 2, 3, 4, and 5 for j in $5(-3)^j$.

$\sum\limits_{j=1}^{5} = 5(-3)^1 + 5(-3)^2 + 5(-3)^3 + 5(-3)^4 + 5(-3)^5$ ◀ *Expanded form*

$= -15 + 45 - 135 + 405 - 1215$ ◀ *Series form*

$= \dfrac{-15 - (-3)(-1215)}{1 - (-3)}$ ◀ *Use* $S = \dfrac{a - rl}{1 - r}$ *to find the sum.*

$= -915$ ◀ *Sum of the series*

Reading in Algebra

Determine whether each expression is (A) an arithmetic progression, (B) an arithmetic series, (C) a geometric progression, or (D) a geometric series.

1. $5 + 10 + 15 + 20 + \ldots$ **2.** $5, 15, 45, 135, \ldots$ **3.** $4 - 12 + 36 - 108 + \ldots$

4. $4, -1, -6, -11, \ldots$ **5.** $\sum\limits_{k=1}^{6} 3k$ **6.** $\sum\limits_{k=1}^{6} 3^k$

Oral Exercises

State each G.P. as a geometric series.

1. $4, 12, 36, 108, \ldots$ **2.** $3, -1, \dfrac{1}{3}, -\dfrac{1}{9}, \ldots$ **3.** $-\dfrac{2}{5}, -2, -10, -50, \ldots$

Find the common ratio r of each geometric series.

4. $-\dfrac{1}{9} + \dfrac{1}{3} - 1 + 3 - \ldots$ **5.** $6 + 0.6 + 0.06 + 0.006 + \ldots$ **6.** $14 + 0.14 + 0.0014 + \ldots$

7. $25 - 5 + 1 - \dfrac{1}{5} + \ldots$ **8.** $\sum\limits_{j=3}^{6} (-2)^j$ **9.** $\sum\limits_{c=1}^{5} 4(5^c)$

Written Exercises

Find the sum of each geometric series for the given data.

Ⓐ **1.** $a = 6, r = 10, l = 60{,}000$ **2.** $a = -2, r = 3, l = -1{,}458$

3. $a = -4, r = -2, l = -1{,}024$ **4.** $a = 256, r = -\dfrac{1}{2}, l = 4$

5. $16 + 8 + 4 + \ldots + \dfrac{1}{8}$ **6.** $\dfrac{1}{9} + \dfrac{1}{3} + 1 + \ldots + 243$

7. $0.3 + 3 + 30 + \ldots + 300{,}000$

8. $-1{,}024 - 512 - 256 - \ldots - \dfrac{1}{4}$

9. $0.2 + 0.4 + 0.8 + \ldots + 51.2$

10. $900 - 90 + 9 - \ldots + 0.0009$

11. $n = 10; \; 5 + 10 + 20 + 40 + \ldots$

12. $n = 9; \; -3 - 6 - 12 - 24 - \ldots$

13. $n = 8; \; 4 - 8 + 16 - 32 + \ldots$

14. $n = 10; \; -\dfrac{1}{8} + \dfrac{1}{4} - \dfrac{1}{2} + 1 - \ldots$

15. $n = 7; \; 0.06 + 0.6 + 6 + 60 + \ldots$

16. $n = 8; \; -1 + 10 - 100 + 1{,}000 - \ldots$

17. $a = 12, \; r = 3, \; n = 6$

18. $a = -4, \; r = -3, \; n = 5$

19. $a = 20, \; r = 0.2, \; n = 6$

20. $a = -300, \; r = 0.3, \; n = 5$

21. $a = \dfrac{1}{5}, \; r = 10, \; n = 7$

22. $a = \dfrac{3}{5}, \; r = -10, \; n = 8$

23. $\displaystyle\sum_{k=1}^{6} 2^k$ **24.** $\displaystyle\sum_{j=3}^{6} (-3)^j$ **25.** $\displaystyle\sum_{c=1}^{5} \left(\dfrac{1}{2}\right)^c$ **26.** $\displaystyle\sum_{k=4}^{7} (0.2)^{k-3}$

Ⓑ **27.** $\displaystyle\sum_{j=1}^{4} \dfrac{1}{2}(4^j)$ **28.** $\displaystyle\sum_{k=1}^{4} 6(-2)^{k-1}$ **29.** $\displaystyle\sum_{c=1}^{5} -64\left(\dfrac{1}{4}\right)^{c-1}$ **30.** $\displaystyle\sum_{j=1}^{6} -27\left(-\dfrac{1}{3}\right)^{j-1}$

31. $n = 9; \; -32 + 16 - 8 + 4 - \ldots$

32. $n = 7; \; 125 + 25 + 5 + 1 + \ldots$

33. $n = 7; \; \dfrac{2}{3} + 2 + 6 + 18 + \ldots$

34. $n = 6; \; \dfrac{2}{5} + 2 + 10 + 50 + \ldots$

35. $a = -243, \; r = \dfrac{1}{3}, \; n = 6$

36. $a = 64, \; r = -\dfrac{1}{4}, \; n = 4$

37. $a = 2{,}500, \; r = -\dfrac{1}{5}, \; n = 5$

38. $a = -128, \; r = -\dfrac{1}{4}, \; n = 6$

39. $a = 5, \; r = 4, \; n = 6$

40. $a = -3, \; r = 5, \; n = 6$

41. A golf ball, dropped from a height of 128 m, rebounds on each bounce one-half the distance from which it fell. When the ball hits the ground the 8th time, how far has it traveled since it was first dropped?

42. A steel ball rebounds on each bounce two-thirds the distance from which it fell. If the ball is dropped from a height of 81 m, how far will it travel before it hits the ground the 6th time?

43. Smaller and smaller squares are formed consecutively as shown at the right. Find the sum of the perimeters of the first 8 squares so formed, if the width of the first square is 20 cm.

44. Find the sum of the areas of the first 5 squares formed, as shown at the right, if the first square is 20 cm wide.

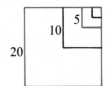

Ⓒ **45.** Prove: $\displaystyle\sum_{k=1}^{4} ax^{k-1} = a + \left(a \cdot \sum_{k=1}^{3} x^k\right)$.   **46.** Solve $\displaystyle\sum_{k=3}^{5} (x^k + 3) = 9$ for x.

CALCULATOR ACTIVITIES

Find the sum of each geometric series for the given data.

(1) $3 + 15 + 75 + \ldots + 234{,}375$ **(2)** $n = 7: \; 2 - 6 + 18 - 54 + \ldots$

(1) $S = \dfrac{a - r \cdot l}{1 - r} = \dfrac{3 - 5 \times 234{,}375}{-4}$: $\ominus 5 \otimes 234375 \oplus 3 \oslash 4$ Answer: 292968

(2) $S = \dfrac{a(1 - r^n)}{1 - r} = \dfrac{2[1 - (-3)^7]}{1 - (-3)} = \dfrac{2(1 + 3^7)}{4} = 3 \otimes \ominus \ominus \ominus \ominus \ominus \ominus \ominus$
$\oplus 1 \otimes 2 \oslash 4 \ominus 1094$

INFINITE GEOMETRIC SERIES 14.7

Objectives **To find the sum, if it exists, of an infinite geometric series**
To write a repeating decimal as an infinite geometric series or an infinite indicated sum
To write a repeating decimal in the form $\frac{x}{y}$ where x and y are integers

The geometric series, $2 + 1 + \frac{1}{2} + \frac{1}{4} + \frac{1}{8} + \ldots$, is an **infinite geometric series.**

The sum of the first 10 terms of the series is $3\frac{255}{256}$. This sum is found by

using the formula $S = \frac{a(1 - r^n)}{1 - r}$, where $S = \dfrac{2\left[1 - \left(\frac{1}{2}\right)^{10}\right]}{1 - \frac{1}{2}}$. The value of r^n

in this sum is $\left(\frac{1}{2}\right)^{10}$, or $\frac{1}{1024}$. Observe how rapidly $\left(\frac{1}{2}\right)^n$ approaches 0 as n increases by 5.

$$\left(\frac{1}{2}\right)^0 = 1 \qquad \left(\frac{1}{2}\right)^5 = \frac{1}{32} \qquad \left(\frac{1}{2}\right)^{10} = \frac{1}{1024} \qquad \left(\frac{1}{2}\right)^{15} = \frac{1}{32,768}$$

The sum formula $S = \frac{a(1 - r^n)}{1 - r}$ can be written as $S = \frac{a}{1 - r} \cdot (1 - r^n)$.
If $-1 < r < 1$ and n increases without limit, then
r^n approaches 0 as a limit and $\frac{a}{1 - r} \cdot (1 - r^n)$ approaches $\frac{a}{1 - r} \cdot (1 - 0)$, or

$\frac{a}{1 - r}$, as a limit. This limit, $\frac{a}{1 - r}$, is defined to be the sum of the infinite geometric series with first term a and common ratio r.

Sum of infinite geometric series	The sum S of the terms of an infinite geometric series is given by the formula $$S = \frac{a}{1 - r}, \qquad [-1 < r < 1, r \neq 0]$$ where a is the first term and r is the common ratio.

You can see that the infinite geometric series, $2 + 1 + \frac{1}{2} + \frac{1}{4} + \ldots$,

appears to **converge** to 4 by finding the sums of the series for $n = 1, 2, 3, \ldots$.

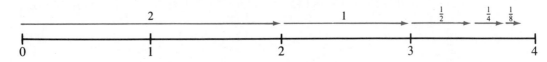

Example 1 **Find the sum, if it exists, of each infinite geometric series.**

$$2 + 1 + \frac{1}{2} + \frac{1}{4} + \ldots \qquad 8 - 2 + \frac{1}{2} - \frac{1}{8} + \ldots \qquad \frac{1}{4} + \frac{1}{2} + 1 + 2 + \ldots$$

$r = \frac{1}{2}$

$S = \dfrac{a}{1 - r} = \dfrac{2}{1 - \dfrac{1}{2}}$

$= 2 \div \dfrac{1}{2}$

$= 4$

$r = -\dfrac{1}{4}$

$S = \dfrac{a}{1 - r} = \dfrac{8}{1 - \left(-\dfrac{1}{4}\right)}$

$= 8 \div \dfrac{5}{4}$

$= \dfrac{32}{5}$, or $6\dfrac{2}{5}$

$r = 2$

Since r is not between -1 and 1, the sum of the series does not exist.

The decimal 0.56 is a terminating decimal while 0.434343 . . . and 0.27$\overline{7}$ are repeating decimals. Any repeating decimal can be written as an infinite geometric series or as the indicated sum of a terminating decimal and an infinite geometric series.

$\quad 0.434343 \ldots = 0.43 + 0.0043 + 0.000043 + \ldots$, an infinite geometric series
$\qquad\qquad\qquad\qquad\qquad\qquad\qquad\qquad\qquad$ with $a = 0.43$ and $r = 0.01$

$\quad 0.27\overline{7} = 0.2 + (0.07 + 0.007 + 0.0007 + \ldots)$, an infinite indicated sum
$\qquad\quad \uparrow \qquad\qquad\qquad \uparrow$
$\qquad$ terminating $\qquad$ infinite geometric series
$\qquad$ decimal $\qquad\qquad$ with $a = 0.07$ and $r = 0.1$

After a repeating decimal is written in series form, the formula $S = \dfrac{a}{1 - r}$ can be used to write the decimal in the form $\dfrac{x}{y}$ where x and y are integers.

Example 2 **Write 0.434343 . . . in the form $\dfrac{x}{y}$ where x and y are integers.**

$0.434343 \ldots = 0.43 + 0.0043 + 0.000043 + \ldots$ ◀ *Infinite geometric series*
$\qquad\qquad\qquad a = 0.43 \qquad\quad r = 0.01$

$$S = \frac{a}{1 - r} = \frac{0.43}{1 - 0.01} = \frac{0.43}{0.99} = \frac{43}{99}$$

Thus, $0.434343 \ldots = \dfrac{43}{99}$.

Example 3 **Write 6.012$\overline{12}$ in the form $\dfrac{x}{y}$ where x and y are integers.**

$6.012\overline{12} = 6.0 + (0.012 + 0.00012 + 0.0000012 + \ldots)$ ◀ *Infinite indicated sum*
$\qquad\qquad\qquad\quad a = 0.012 \qquad\quad r = 0.01$

$$S = \frac{a}{1 - r} = \frac{0.012}{1 - 0.01} = \frac{0.012}{0.99} = \frac{2}{165} \qquad 6 + \frac{2}{165} = \frac{992}{165}$$

Thus, $6.012\overline{12} = \dfrac{992}{165}$.

Written Exercises

Find the sum, if it exists, of each infinite geometric series.

(A)
1. $\frac{1}{2} + 1 + 2 + 4 + \ldots$
2. $8 + 4 + 2 + 1 + \ldots$
3. $9 - 3 + 1 - \frac{1}{3} + \ldots$
4. $9 + 6 + 4 + \frac{8}{3} + \ldots$
5. $\frac{1}{2} + \frac{1}{4} + \frac{1}{8} + \frac{1}{16} + \ldots$
6. $\frac{1}{4} - \frac{1}{8} + \frac{1}{16} - \frac{1}{32} + \ldots$
7. $1 - \frac{1}{3} + \frac{1}{9} - \frac{1}{27} + \ldots$
8. $2 + 0.2 + 0.02 + 0.002 + \ldots$
9. $70 + 7 + 0.7 + 0.07 + \ldots$

Write each repeating decimal as an infinite geometric series or as an infinite indicated sum and then in the form $\frac{x}{y}$ where x and y are integers.

10. $0.555\ldots$
11. $0.\overline{77}$
12. $3.333\ldots$
13. $22.2\overline{2}$
14. $0.04\overline{4}$
15. $4.666\ldots$
16. $7.54\overline{4}$
17. $6.2555\ldots$

Solve each problem.

(B)
18. The midpoints of the sides of an equilateral triangle are connected to form a second triangle. This procedure is repeated on each resulting triangle. Find the sum of the perimeters of all triangles so formed, if one side of the first triangle measures 60 cm.

19. The midpoints of the sides of a square are joined to form a second square. This process is repeated on each resulting square. What is the sum of the areas of all squares so formed, if one side of the first square measures 8 m?

20. A golf ball, dropped from a height of 128 m, rebounds on each bounce one-half the distance from which it falls. How far will it travel before coming to rest?

21. A steel ball was dropped from a height of 81 m and rebounded, on each bounce, two-thirds of the distance from which it fell. How far did it travel before coming to rest?

Find the sum, if it exists, of each infinite geometric series.

22. $\frac{1}{5} + 1 + 5 + 25 + \ldots$
23. $-16 + 12 - 9 + \frac{27}{4} - \ldots$
24. $25 - 10 + 4 - \frac{8}{5} + \ldots$

Find the sum of the series for $n = 1, 2, 3, \ldots, 8$. To what value does the series converge?

25. $27 + 9 + 3 + \ldots$
26. $-4 - 2 - 1 - \ldots$
27. $0.9 + 0.09 + 0.009 + \ldots$

Use the formula $S = \dfrac{a}{1 - r}$ to write each repeating decimal in the form $\dfrac{x}{y}$ where x and y are integers.

28. $53.5\overline{353}$
29. $67.47\overline{474}$
30. $3.02\overline{929}$
31. $17.17\overline{17}$
32. $8.03\overline{232}$
33. $63.53\overline{535}$

For what values of x does each infinite geometric series have a sum? [Hint: $-1 < r < 1,\ r \neq 0$]

(C)
34. $(x - 1)^0 + (x - 1)^1 + (x - 1)^2 + \ldots$
35. $\dfrac{1}{x} + \dfrac{1}{x^3} + \dfrac{1}{x^5} + \ldots$
36. $\dfrac{1}{x^2} + \dfrac{1}{x^3} + \dfrac{1}{x^4} + \ldots$

BINOMIAL EXPANSIONS

Objectives

To expand a positive integral power of a binomial
To find a specified term of a binomial expansion

The first few positive integral powers of $(a + b)$ reveal patterns that can be used to **expand** any positive integral power of a binomial.

$$(a + b)^0 = 1$$
$$(a + b)^1 = a + b$$
$$(a + b)^2 = a^2 + 2ab + b^2$$
$$(a + b)^3 = a^3 + 3a^2b + 3ab^2 + b^3$$
$$(a + b)^4 = a^4 + 4a^3b + 6a^2b^2 + 4ab^3 + b^4$$
$$(a + b)^5 = a^5 + 5a^4b + 10a^3b^2 + 10a^2b^3 + 5ab^4 + b^5$$

Notice in the expansion of $(a + b)^n$ that:

(1) The number of terms is always $n + 1$.
(2) The first term is a^n and the last term is b^n.
(3) The exponent of a decreases by 1, from one term to the next.
(4) The exponent of b increases by 1, from one term to the next.
(5) For each term, the sum of the exponents of a and b is n.
(6) The coefficients are symmetrical. (They read the same from left to right as right to left.)
(7) There is a pattern for finding the coefficients in an expansion.

Notice that $(a + b)^4 = a^4 + \dfrac{4}{1}a^3b + \dfrac{4 \cdot 3}{1 \cdot 2}a^2b^2 + \dfrac{4 \cdot 3 \cdot 2}{1 \cdot 2 \cdot 3}ab^3 + b^4$

$$= a^4 + 4a^3b + 6a^2b^2 + 4ab^3 + b^4.$$

All of the patterns above can be seen in the statement of the **binomial theorem.**

Binomial theorem	If n is a positive integer, then $(a + b)^n =$
	$a^n + \dfrac{n}{1}a^{n-1}b + \dfrac{n(n-1)}{1 \cdot 2}a^{n-2}b^2 + \dfrac{n(n-1)(n-2)}{1 \cdot 2 \cdot 3}a^{n-3}b^3 + \ldots + b^n.$

Example 1

Expand $(3x - y^2)^5$. Simplify each term.

Use the binomial theorem with $n = 5$, $a = 3x$, and $b = -y^2$: $[3x + (-y^2)]^5$.

$(3x)^5 + \dfrac{5}{1}(3x)^4(-y^2) + \dfrac{5 \cdot 4}{1 \cdot 2}(3x)^3(-y^2)^2 + \dfrac{5 \cdot 4 \cdot 3}{1 \cdot 2 \cdot 3}(3x)^2(-y^2)^3 + \dfrac{5 \cdot 4 \cdot 3 \cdot 2}{1 \cdot 2 \cdot 3 \cdot 4}$
$(3x)(-y^2)^4 + (-y^2)^5$

Thus, $(3x - y^2)^5 = 243x^5 - 405x^4y^2 + 270x^3y^4 - 90x^2y^6 + 15xy^8 - y^{10}.$

Example 1 suggests the following properties for the signs in a binomial expansion.
(1) In the expansion of $(a + b)^n$, the signs will all be $+$.
(2) In the expansion of $(a - b)^n$, the signs will alternate $+$ and $-$.

The notation $k!$ is read as "k factorial" and is defined below as is $0!$.

$$k! = 1 \cdot 2 \cdot 3 \cdot \ldots \cdot k, \text{ where } k \text{ is a positive integer, and } 0! = 1$$

Factorial notation can be used with the binomial theorem to rewrite the denominators of the coefficients, since $1 = 1!$, $1 \cdot 2 = 2!$, $1 \cdot 2 \cdot 3 = 3!$, and so on.

Thus, the binomial theorem can be expressed as $(a + b)^n =$

$$a^n + \frac{n}{1!}a^{n-1}b + \frac{n(n-1)}{2!}a^{n-2}b^2 + \frac{n(n-1)(n-2)}{3!}a^{n-3}b^3 + \ldots + b^n.$$

The 7th term is $\dfrac{n(n-1)(n-2) \ldots (n - [7-2])}{(7-1)!}a^{n-7+1}b^{7-1}$, or

$$\frac{n(n-1)(n-2) \ldots (n-5)}{6!}a^{n-6}b^6.$$

rth term in a binomial expansion	The rth term in the expansion of $(a + b)^n$ is $$\frac{n(n-1)(n-2) \ldots (n-r+2)}{(r-1)!}a^{n-r+1}b^{r-1}.$$

The formula for the rth term permits you to find a specified term of an expansion without writing the preceding terms.

Example 2   **Find the 4th term in the expansion of $(2x - y^3)^7$. Simplify the term.**

Use the 4th term of $(a + b)^7$, or $\dfrac{7 \cdot 6 \cdot 5}{(4-1)!}a^{7-4+1}b^{4-1}$, to find the 4th term of $(2x - y^3)^7$ where $a = 2x$, $b = -y^3$, $n = 7$, and $r = 4$.

$$\frac{7 \cdot 6 \cdot 5}{1 \cdot 2 \cdot 3}(2x)^{7-3}(-y^3)^3 = 35 \cdot 2^4x^4(-y^9)$$

Thus, the 4th term is $-560x^4y^9$.

Written Exercises

Expand each power of a binomial. Simplify.

(A)
1. $(a + b)^8$
2. $(a - b)^9$
3. $(x + 2)^6$
4. $(2c - 3)^5$
5. $(x^2 + y)^8$
6. $(m^2 - 2n)^5$
7. $(2r + 5t)^4$
8. $(y - 0.2)^4$
9. $(2x + 0.1)^6$

Find the specified term of each expansion. Simplify.

10. 4th term of $(c + 3d)^5$
11. 4th term of $(x^2 - y)^6$
12. middle term of $(2x + y^2)^4$

(B)
13. 3rd term of $(2x^3 + 3y)^5$
14. 5th term of $(3c^2 - \sqrt{d})^6$
15. middle term of $(4x^2 + \frac{x}{2})^6$

Expand each power of a binomial. Simplify.

16. $(x^3 - 2y)^4$
17. $(2c^3 + 3d^2)^5$
18. $(x + \sqrt{2})^6$
19. $(2x - \sqrt{y})^4$
20. $(2r^2 + \frac{t}{2})^4$
21. $(\frac{1}{x} + \frac{1}{y})^5$

Find the first four terms of each expansion. Simplify.

22. $(x + y)^{12}$
23. $(x - 2y)^{11}$
24. $(x^2 + 3y^3)^{10}$
25. $(1 + 0.01)^7$
26. $(1 + 0.02)^8$
27. $(1 + 0.03)^6$

The Binomial Theorem is true for $(a + b)^q$, where q is a rational number. For each expansion, write the first four terms in simplest radical form.

(C)
28. $(x + 4)^{\frac{1}{2}}$
29. $(x + 9)^{\frac{2}{3}}$
30. $(x - 9)^{-\frac{1}{3}}$

Binomial Expansions

SPECIAL INFINITE SERIES 14.9

Objective   **To apply an example of a special infinite series**

Some special infinite series involve a product of consecutive counting numbers beginning with 1, such as

$$1 \times 2, \quad 1 \times 2 \times 3, \quad 1 \times 2 \times 3 \times 4, \quad \text{and so on.}$$

These products are abbreviated as follows.

$$1 \times 2 = 2! \qquad 1 \times 2 \times 3 \times 4 = 4!$$
$$1 \times 2 \times 3 = 3! \qquad 1 \times 2 \times 3 \times 4 \times 5 = 5!$$

Thus, $1 \times 2 \times 3 \times \ldots \times (n-1) \times n = n!$

Reciprocals of factorials, $\frac{1}{n!}$, occur in some special series. Shown below is a table of factorials ($n!$) and reciprocals of factorials $\left(\frac{1}{n!}\right)$ for some values of n.

n	$n!$	$\frac{1}{n!}$	n	$n!$	$\frac{1}{n!}$
2	2	0.5	6	720	0.0013888 ...
3	6	0.1666666 ...	7	5,040	0.0001984 ...
4	24	0.0416666 ...	8	40,320	0.0000248 ...
5	120	0.0083333 ...	9	362,880	0.0000027 ...

The trigonometric functions called *sine (sin)* and *cosine (cos)* are defined by the following infinite series, where x is a real number.

$$\sin x = x - \frac{x^3}{3!} + \frac{x^5}{5!} - \frac{x^7}{7!} + \frac{x^9}{9!} - \cdots$$
$$\cos x = 1 - \frac{x^2}{2!} + \frac{x^4}{4!} - \frac{x^6}{6!} + \frac{x^8}{8!} - \cdots$$

Example 1   **Use the infinite series for sin x to compute sin 0.2 to four decimal places.**

Replace x by 0.2 in the series until you find a term whose first five decimal places are each 0.

$$\sin x = x - \frac{x^3}{3!} + \frac{x^5}{5!} - \cdots$$

$$\sin 0.2 = 0.2 - \frac{(0.2)^3}{3!} + \frac{(0.2)^5}{5!} - \cdots$$
$$= 0.2 - \frac{0.008}{6} + \frac{0.00032}{120} - \cdots$$
$$= 0.2 - 0.0013333 \ldots + 0.0000026 \ldots - \cdots$$
$$= 0.19866 \ldots$$

Thus, $\sin 0.2 = 0.1987$, correct to four decimal places.

Another number related to an infinite series is the number **e,** which is used as the base of **natural** logarithms. The number e is defined by the infinite series:

$$e = 2 + \frac{1}{2!} + \frac{1}{3!} + \frac{1}{4!} + \ldots + \frac{1}{n!} + \ldots$$

Example 2   **Use the infinite series for e to compute e to three decimal places.**

Use the table for $\frac{1}{n!}$ until you find a term whose first four decimal places are each 0.

$$e = 2 + .5 + .1666 \ldots + .041666 \ldots + .008333 \ldots$$
$$+ .0013888 \ldots + .0001984 \ldots + .0000248 \ldots + \ldots$$

Thus, $e = 2.718$, correct to three decimal places.

Written Exercises

Ⓐ 1. Use the infinite series for sin x to compute sin 1 to three decimal places.
2. Use the infinite series for cos x to compute cos 1 to three decimal places.
3. Use the infinite series for e to compute e to four decimal places.

4. Compute $\sum\limits_{n=2}^{6} n!$.

5. Compute $\sum\limits_{n=2}^{6} \frac{1}{n!}$.

Use the infinite series for sin x or for cos x to compute each of the following to four decimal places.

Ⓑ **6.** sin 0.1   **7.** sin 0.3   **8.** sin 0   **9.** cos 0.1   **10.** cos 0.2   **11.** cos 0

Use the infinite series $e^x = 1 + x + \dfrac{x^2}{2!} + \dfrac{x^3}{3!} + \ldots + \dfrac{x^{n-1}}{(n-1)!} + \ldots$ to compute each of the following. (Ex. 12–13)
12. e^2, to two decimal places
13. $\sqrt{e}$, to four decimal places [*Hint:* Recall that $\sqrt{x} = x^{\frac{1}{2}}$, or $x^{0.5}$.]

Ⓒ **14.** Divide 1 by $1 - x$ [$1 - x \overline{)1\quad}$] to show that
$$\frac{1}{1 - x} = 1 + x + x^2 + x^3 + \ldots + x^{n-1} + \ldots, \text{ where } |x| < 1.$$

A series of the form $1 + x + x^2 + x^3 + \ldots$, where $|x| < 1$, is called an infinite power series. This series is an infinite geometric series with a common ratio $r = x$ and $|r| < 1$, $r \neq 0$.

15. Use the formula $S = \dfrac{a}{1 - r}$ for an infinite geometric series, $|r| < 1$, $r \neq 0$,

to show that $1 + x + x^2 + x^3 + \ldots = \dfrac{1}{1 - x}$.

16. Simplify the infinite power series $1 + (y + 1) + (y + 1)^2 + (y + 1)^3 + \ldots$.
17. Simplify the infinite power series $1 + (z - 2) + (z - 2)^2 + (z - 3)^3 + \ldots$.

CHAPTER FOURTEEN REVIEW

Vocabulary
arithmetic means [14.2]
arithmetic progression [14.1]
arithmetic series [14.3]
binomial expansion [14.8]
geometric means [14.5]
geometric progression [14.4]
geometric series [14.6]
mean proportional [14.5]
progression [14.1]   sequence [14.1]

Write the first four terms of each A.P. [14.1]
1. $a = 14$ and $d = -8$
2. $a = 3x - 4$ and $d = -x + 2$
3. Find the 36th term of the A.P.:
   $-7.2, -3.4, 0.4, 4.2, \ldots$ [14.1]
4. Find the three arithmetic means between 12 and 22.8. [14.2]
5. Find the arithmetic mean of -11.8 and 19.6. [14.2]

Find the sum of each arithmetic series [14.3]
6. $n = 35$, $a = -15.4$, $l = 43.8$
7. $n = 26$; $2.5 + 2.9 + 3.3 + 3.7 + \ldots$
8. $\sum\limits_{k=1}^{14} -5k$ 9. $\sum\limits_{j=3}^{6} (2j - 5)$

Write each arithmetic series using Σ-notation.
10. $-10 - 20 - 30 - 40 - 50$ [14.3]
11. $18 + 22 + 26 + 30 + \cdots + 58$

Write the next three terms of each G.P. [14-4]
12. $-\frac{1}{3}, 1, -3, \ldots$ [14.4]
13. $-4x, -8x^2, -16x^3, \ldots$

Find the specified term of each G.P. [14.4]
14. 10th term: $6, -12, 24, \ldots$
15. 5th term: $a = -4$, $r = 3$

Find the indicated number of real geometric means between each pair of numbers. [14.5]
16. One positive, between $\frac{1}{5}$ and 20
17. Three, between -3 and -48
18. Two, between $2\sqrt{6}$ and 72

Find the sum of each geometric series for the given data. [14.6]
19. $64 + 32 + 16 + \ldots + \frac{1}{4}$
20. $a = 3$, $r = 5$, $l = 1{,}875$
21. $n = 8$; $32 - 16 + 8 - 4 + \ldots$
22. $a = -2$, $r = -4$, $n = 5$
23. $\sum\limits_{k=2}^{6} (-5)^{k-1}$ 24. $\sum\limits_{j=1}^{4} 25\left(\frac{1}{5}\right)^j$

Find the sum, if it exists, of each infinite geometric series. [14.7]
25. $\frac{1}{3} + 1 + 3 + 9 + \ldots$
26. $64 - 16 + 4 - 1 + \ldots$
27. $16\sqrt{3} + 8\sqrt{3} + 4\sqrt{3} + 2\sqrt{3} + \ldots$

28. Use the formula $S = \dfrac{a}{1 - r}$ to write $83.73\overline{737}$ in the form $\dfrac{x}{y}$, where x and y are integers. [14.7]

Expand each power of a binomial. Simplify. [14.8]
29. $(x + 3y)^4$ 30. $(2x^3 - y)^5$ 31. $\left(\dfrac{1}{m} + \dfrac{1}{n}\right)^4$

32. Find the 5th term of $(3x^2 - \sqrt{5})^6$. Simplify.
33. Find the first four terms of $(x - 2y^2)^{10}$. Simplify. [14.8]

Solve each problem.
34. Mr. Edwards began a 30-day training program by jogging 440 m. Each day, he ran 50 m more than he did the preceding day. How far did Mr. Edwards jog on the 30th day? [14.1]
35. The arithmetic mean of two numbers is 24. Six less than the larger number is 5 times the smaller number. Find the numbers. [14.2]
36. If one number is 9 less than another number and their positive geometric mean is 6, what are the two numbers? [14.5]

1. Write the first four terms of the A.P. whose first term is -2 and common difference is 1.5.
2. Find the 31st term of the A.P.: 6.0, 6.4, 6.8, 7.2 . . . .
3. Find the three arithmetic means between 18 and 29.
4. Find the arithmetic mean of -5.4 and 8.2.

Find the sum of each arithmetic series.

5. $n = 26$; $7.6 + 7.2 + 6.8 + 6.4 + \ldots$
6. $\displaystyle\sum_{k=1}^{20} (2k + 3)$

7. Cartons are stacked in 16 rows with 4 cartons in the top row, 7 in the second row, 10 in the next row, and so on. How many cartons are in the stack?
8. Write the arithmetic series, $10 + 13 + 16 + 19 + 22$, using Σ-notation.
9. Write the next three terms of the G.P.: $\dfrac{3}{4}, \dfrac{3}{2}, 3, \ldots$
10. Find the 10th term of the G.P.: $\dfrac{1}{4}, \dfrac{1}{2}, 1, 2, \ldots$

Find the indicated number of real geometric means between each pair of numbers.

11. Two between -4 and 108
12. One positive, between $\dfrac{1}{6}$ and 24

Find the sum of each geometric series.

13. $a = 5$, $r = 4$, $l = 5{,}120$
14. $n = 6$; $100 - 50 + 25 - 12\frac{1}{2} + \ldots$
15. $a = 5$, $r = -10$, $n = 4$
16. $\displaystyle\sum_{j=1}^{4} 81 \left(\dfrac{1}{3}\right)^{j}$

17. Find the sum of the infinite geometric series: $16 - 4 + 1 - \dfrac{1}{4} + \ldots$

18. Write 7.2888 . . . as an infinite indicated sum and then in the form $\dfrac{x}{y}$, where x and y are integers.
19. Expand $(3x - y^2)^4$. Simplify.
20. Find the 5th term of $(4x - \sqrt{2})^6$. Simplify.
21. Find the first four terms of $(x^2 + \frac{1}{2}y)^{10}$. Simplify.

Solve each problem.

22. An object, shot upward at 200 m/s, travels 195 m during the 1st second, 185 m during the 2nd second, 175 m during the 3rd second, and so on. How far does it travel during the 14th second?
23. The arithmetic mean of two numbers is 20. Four less than one number is twice the other number. Find the two numbers.
24. Two more than the geometric mean of two numbers is 10. Find the two numbers if one number is 12 more than the other number.

Smaller and smaller squares are formed, one after the other, as shown at the right. The width of the first square is 32 m. Use this information in Ex. 25–26.

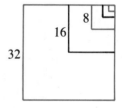

25. Find the sum of the perimeters of the first eight squares.
26. Find the sum of the areas of all such squares.

★ 27. Find two values of c such that $\dfrac{2}{c}, 4, \dfrac{15}{c-2}$ is an A.P.
★ 28. Find two values of x such that $x - 9$, 12, $4x$ is a G.P.
★ 29. Find the geometric mean of $-2i$ and $10i$.
★ 30. For what values of x does the infinite geometric series $\dfrac{x}{2} - 2x^3 + 8x^5 - 32x^7 + \ldots$ have a sum?

COLLEGE PREP TEST

DIRECTIONS: In each item, you are to compare a quantity in Column 1 with a quantity in Column 2. Write the letter of the correct answer from these choices:

A The quantity in Column 1 is greater than the quantity in Column 2.
B The quantity in Column 2 is greater than the quantity in Column 1.
C The quantity in Column 1 is equal to the quantity in Column 2.
D The relationship cannot be determined from the given information.

Notes: Information centered over both columns refers to one or both of the quantities to be compared.
A symbol that appears in both columns has the same meaning in each column.
All variables represent real numbers.

SAMPLE ITEM AND ANSWER

Column 1	Column 2
$x = 2$ and $y = -2$	
The value of the next term of the A.P.: x, $2x$, $3x$, $4x$	The value of the next term of the G.P.: y, y^2, y^3, y^4

The answer is A because $5x = 5 \cdot 2 = 10$, $y^5 = (-2)^5 = -32$, and $10 > -32$.

Column 1	Column 2
1. Arithmetic mean of $\frac{1}{2}$ and $\frac{1}{8}$	Arithmetic mean of $-\frac{1}{8}$ and $\frac{3}{4}$
2. The next term in the G.P.: 9, 3, 1	The next term in the G.P.: $\frac{1}{16}$, $\frac{1}{8}$, $\frac{1}{4}$
3. Next term in the sequence: 0, 1, 1, 2, 3, 5, 8, 13	Next term in the sequence: 75, 64, 53, 42, 31
4. Sum of the series: $-20 - 15 - 10 - \ldots + 30$	Sum of the series: $-25 - 20 - 15 - \ldots + 35$

Column 1	Column 2
5. Arithmetic mean of 2 and 8	Geometric mean of 2 and 8
6. $(x + 5)^3$	$(x + 5)^4$
7. $\sum\limits_{k=1}^{5} 2^{k-1}$	$\sum\limits_{j=3}^{7} 2^{j-3}$
8. The value of x if $x + 3$, $2x$, $x - 7$ is an arithmetic progression	The value of n if $2n$, $n - 1$, $n + 1$ is an arithmetic progression
	$x > 0$ and $n > 0$
9. Value of x if 3, $3x$, $12x$ is a G.P.	Value of n if -2, $-2n$, $-10n$ is a G.P.
10. Coefficient of the 3rd term in the expansion of $(x + 2)^{12}$	Coefficient of the 3rd term in the expansion of $(y - 2)^{12}$

15 EXPONENTIAL, LOGARITHMIC, AND POLYNOMIAL FUNCTIONS

Non-Routine Problem Solving

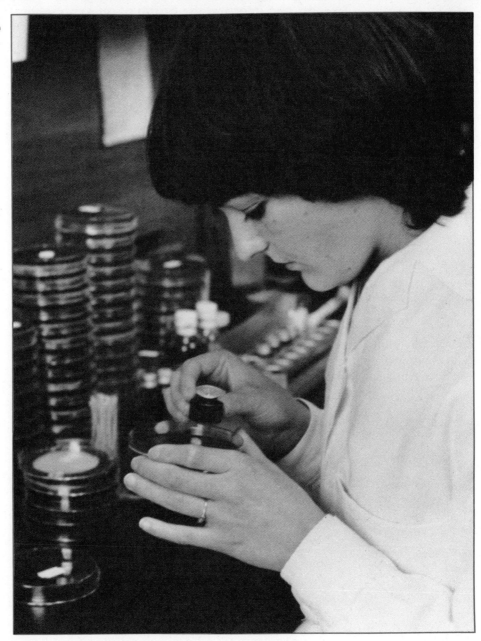

Problem 1. At the end of x hours, the number y of bacteria in a culture is $(0, 100)$, $(1, 145)$, $(2, 210)$, $(3, 305)$. Write an exponential equation of the form $y = a \cdot b^x$.

Problem 2. Find four integers x, y, z, and w such that $2x + 3y - z - w = 84$ and $x:y = 3:4$, $y:z = 2:3$, and $x:w = 5:6$.

GRAPHING EXPONENTIAL FUNCTIONS   15.1

Objectives   **To draw the graph of an exponential function**
To read and interpret the graph of an exponential function

The equation $y = 3 \cdot 2^x$ describes a base-2 **exponential function**. The graph of this function is drawn using a table of ordered pairs (x, y).

x	2^x	$3 \cdot 2^x = y$
-3	$\dfrac{1}{8}$	$\dfrac{3}{8}$
-2	$\dfrac{1}{4}$	$\dfrac{3}{4}$
-1	$\dfrac{1}{2}$	$\dfrac{3}{2}$
0	1	3
1	2	6
2	4	12
3	8	24

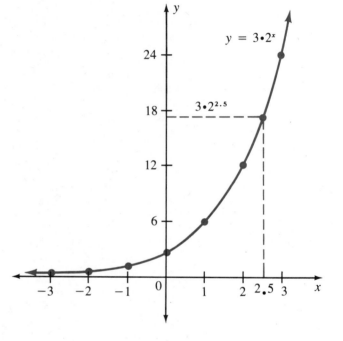

The ordered pairs (x, y) in the table are plotted, and a smooth curve is drawn. All rational numbers and all irrational numbers are in the domain of the function above.

To draw the graph for $y = 3 \cdot 2^x$, integers were used as the values of x. You can use a calculator with the y^x function to verify the following:

If $x = 2.5$, then $y = 3 \cdot 2^x = 3 \cdot 2^{2.5} = 16.970562 \ldots$, and
If $x = \sqrt{3}$, then $y = 3 \cdot 2^x = 3 \cdot 2^{\sqrt{3}} = 9.9659913$

From the graph above, you can list four properties of the exponential function described by $y = 3 \cdot 2^x$.

(1) The domain is {all real numbers}.

(2) The range is $\{y \mid y > 0\}$.

(3) The function is *constantly increasing*. That is, if $x_2 > x_1$, then $y_2 > y_1$.

(4) The ordered pair $(0, 3)$ belongs to the function. That is, $3 \cdot 2^0 = 3$.

Example

Draw the graphs of the exponential functions $A: y = 4^x$ **and** $B: y = \left(\frac{1}{4}\right)^x$ **in the same coordinate plane with convenient scales for the axes. List four properties of each function.**

1. Rewrite $y = \left(\frac{1}{4}\right)^x$ as $y = 4^{-x}$. ◀ $y = \left(\frac{1}{4}\right)^x = \frac{1}{4^x} = 4^{-x}$

2. Construct a table of ordered pairs (x, y) for A and for B.

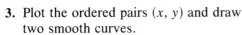

x	$4^x = y$
-2	$\frac{1}{16}$
-1	$\frac{1}{4}$
0	1
0.5	2
1	4
$\sqrt{2}$	7.10

x	$\left(\frac{1}{4}\right)^x = 4^{-x} = y$
-2	16
-1	4
0	1
0.5	0.5
1	$\frac{1}{4}$
$\sqrt{2}$	0.14

3. Plot the ordered pairs (x, y) and draw two smooth curves.

The graphs show the following four

properties for $A: y = 4^x$ and $B: y = \left(\frac{1}{4}\right)^x$.

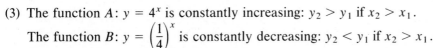

(1) The domain for A and B is {all real numbers}.

(2) The range for A and B is $\{y \mid y > 0\}$.

(3) The function $A: y = 4^x$ is constantly increasing: $y_2 > y_1$ if $x_2 > x_1$.

The function $B: y = \left(\frac{1}{4}\right)^x$ is constantly decreasing: $y_2 < y_1$ if $x_2 > x_1$.

(4) The ordered pair $(0, 1)$ belongs to both functions A and B.

Notice that the graphs of A and B are *symmetrical* with respect to the y-axis.

Written Exercises

Graph the exponential function described by each equation. Choose convenient scales for the axes. List four properties of the function.

Ⓐ
1. $y = 2^x$ **2.** $y = 3^x$ **3.** $y = 5^x$
4. $y = \left(\frac{1}{2}\right)^x$ **5.** $y = \left(\frac{1}{3}\right)^x$ **6.** $y = \left(\frac{1}{5}\right)^x$

Ⓑ
7. $y = 5 \cdot 2^x$ **8.** $y = -1 \cdot 3^x$ **9.** $y = 3 \cdot 4^x$
10. $y = 3\left(\frac{1}{2}\right)^x$ **11.** $y = 2\left(\frac{1}{3}\right)^x$ **12.** $y = -\left(\frac{1}{4}\right)^x$
13. $y = 2^x + 4$ **14.** $y = 4^x - 8$ **15.** $y = 3\left(\frac{1}{2}\right)^x + 6$

Ⓒ
16. $y = 2^{x-3}$ **17.** $y = \left(\frac{1}{2}\right)^{x+2}$ **18.** $y = 9^{\sqrt{x}}$
for $1 \le x \le 7$ for $-4 \le x \le 2$ for $0 \le x \le 4$
Use multiples of 0.5 for x.

LOGARITHMIC FUNCTIONS

Objectives
To find the base-*b* logarithm of a positive number
To solve a base-*b* logarithmic equation

The base-2 exponential function $A: y = 2^x$ can be viewed as a *mapping* from the exponent (x) to the power $(y$, or $2^x)$.

The *inverse* of the base-2 exponential function is a mapping from the power to the exponent.

Exponent $\longrightarrow$	Power
5 $\longrightarrow$	2^5, or 32
3 $\longrightarrow$	2^3, or 8
0 $\longrightarrow$	2^0, or 1
-1 $\longrightarrow$	2^{-1}, or $\frac{1}{2}$

Power $\longrightarrow$	Exponent
32, or 2^5 $\longrightarrow$	5
8, or 2^3 $\longrightarrow$	3
1, or 2^0 $\longrightarrow$	0
$\frac{1}{2}$, or 2^{-1} $\longrightarrow$	-1

The *inverse* of the base-2 *exponential function* is the base-2 **logarithmic function**. The base-2 logarithm of 8 ($\log_2 8$) is found as follows.

$$8 = \quad 2^3$$
$$\downarrow \qquad \downarrow$$
$$\log_2 8 = \log_2 2^3 = 3$$

The four sentences below show how some base-5 logarithms ($\log_5$) are found.

$$\log_5 125 = \log_5 5^3 = 3 \quad \blacktriangleleft \quad 125 = 5 \cdot 5 \cdot 5$$
$$\log_5 1 = \log_5 5^0 = 0 \quad \blacktriangleleft \quad 1 = 5^0$$
$$\log_5 \frac{1}{25} = \log_5 5^{-2} = -2 \quad \blacktriangleleft \quad \frac{1}{25} = \frac{1}{5^2} = 5^{-2}$$
$$\log_5 \sqrt{5} = \log_5 5^{\frac{1}{2}} = \frac{1}{2} \quad \blacktriangleleft \quad \sqrt[n]{x} = x^{\frac{1}{n}}$$

Definition: **Base-*b* logarithm**	For each positive number y and each positive number $b \neq 1$, if $y = b^x$, then $\log_b y = \log_b b^x = x$.

Example 1 **Find $\log_{10} 100$ and $\log_2 \frac{1}{16}$.**

$100 = 10^2$; so, $\log_{10} 100 = \log_{10} 10^2 = 2$.

$\frac{1}{16} = \frac{1}{2^4} = 2^{-4}$; so, $\log_2 \frac{1}{16} = \log_2 2^{-4} = -4$.

Example 2 **Find $\log_3 \sqrt[4]{3}$ and $\log_2 \sqrt[5]{8}$.**

$\sqrt[4]{3} = 3^{\frac{1}{4}}$; so, $\log_3 \sqrt[4]{3} = \log_3 3^{\frac{1}{4}} = \frac{1}{4}$.

$\sqrt[5]{8} = \sqrt[5]{2^3} = 2^{\frac{3}{5}}$; so, $\log_2 \sqrt[5]{8} = \log_2 2^{\frac{3}{5}} = \frac{3}{5}$.

You can solve some logarithmic equations of the form $\log_b y = x$ by rewriting them in the exponential form, $y = b^x$.

Example 3   **Solve each logarithmic equation for the variable.**

Logarithmic equation	Exponential equation	Solution	
1. $\log_b 81 = 4$	$81 = b^4$	3	◀ $3^4 = 81$
2. $\log_5 625 = x$	$625 = 5^x$	4	◀ $5^4 = 625$
3. $\log_4 y = 3$	$y = 4^3$	64	
4. $\log_b \frac{1}{8} = -3$	$\frac{1}{8} = b^{-3}$	2	◀ $2^{-3} = \frac{1}{2^3} = \frac{1}{8}$

Written Exercises

Find each logarithm.

(A) **1.** $\log_7 49$ **2.** $\log_4 64$ **3.** $\log_3 81$ **4.** $\log_6 1$

5. $\log_3 3$ **6.** $\log_{12} 1$ **7.** $\log_2 \frac{1}{2}$ **8.** $\log_4 \frac{1}{16}$

9. $\log_3 \frac{1}{81}$ **10.** $\log_2 \sqrt[3]{2}$ **11.** $\log_3 \sqrt[5]{3}$ **12.** $\log_6 \sqrt{6}$

Solve each equation for the variable.

13. $\log_b 1000 = 3$ **14.** $\log_2 y = 8$ **15.** $\log_b 32 = 5$ **16.** $\log_3 243 = x$

(B) **17.** $\log_{\frac{1}{2}} y = 3$ **18.** $\log_b \frac{1}{9} = -2$ **19.** $\log_{\frac{1}{3}} 27 = x$ **20.** $\log_b \frac{1}{125} = -3$

Find each logarithm.

21. $\log_2 \sqrt[5]{16}$ **22.** $\log_3 \sqrt[4]{27}$ **23.** $\log_5 \sqrt{125}$ **24.** $\log_{10} \sqrt[3]{100}$

25. $\log_{\frac{1}{2}} \frac{1}{16}$ **26.** $\log_{\frac{3}{5}} \frac{27}{125}$ **27.** $\log_{\frac{1}{5}} 25$ **28.** $\log_{\frac{3}{4}} \frac{16}{9}$

(C) **29.** $\log_c c^d$ **30.** $\log_b (b^3)^a$ **31.** $\log_{5n} 25n^2$ **32.** $\log_{3n} \frac{1}{9n^2}$

Find the sum of each series for the given data.

33. $\displaystyle\sum_{k=1}^{5} \log_2 2^{k-2}$ **34.** $\displaystyle\sum_{n=2}^{5} \log_{10} \sqrt[n]{10}$ **35.** $\displaystyle\sum_{j=3}^{6} \log_2 4^{5-j}$

CUMULATIVE REVIEW

1. Box type A holds 18 fewer vitamin capsules than box type B, and box type C contains three times as many capsules as box type A. If 10 type A boxes, 8 type B boxes, and 4 type C boxes hold 27 dozen capsules, find the number of capsules that each box type contains.

2. A rectangular garden is 9 m longer than it is wide. A second rectangular garden is planned so that it will be 6 m wider and twice as long as the first garden. Find the area of the first garden if the sum of the areas of both gardens will be 528 m².

GRAPHING LOGARITHMIC FUNCTIONS 15.3

Objectives **To write an equation for the inverse of an exponential function or a logarithmic function**
To draw the graph of a logarithmic function
To read and interpret the graph of a logarithmic function

Consider the exponential function A: $y = 2^x$ and its *inverse* function, A^{-1}. You can write the y-form of the equation for the inverse as follows.

$$A: \qquad y = 2^x$$

To obtain the inverse of A, replace x by y and y by x.

$$A^{-1}, \text{ inverse of } A: \qquad x = 2^y$$

$$\log_2 x = \log_2 2^y = y$$

Change the exponential equation to a logarithmic equation.

$$y = \log_2 x$$

Thus, the inverse of the base-2 exponential function ($y = 2^x$) is the base-2 logarithmic function ($y = \log_2 x$).

Inverse of an exponential function

The *inverse* of the exponential function
$$y = b^x \quad [y > 0, b > 0, b \neq 1, x \text{ is a real number}]$$
is the logarithmic function described by
$$y = \log_b x \quad [x > 0, b > 0, b \neq 1, y \text{ is a real number}].$$

Example 1 **Write the y-form of the equation for the inverse of A: $y = \left(\dfrac{1}{4}\right)^x$ and the inverse of B: $y = \log_3 x$.**

$y = b^x$ and $y = \log_b x$ describe inverse functions.

Thus, the inverse of A: $y = \left(\dfrac{1}{4}\right)^x$ is described by $y = \log_{\frac{1}{4}} x$.

The inverse of B: $y = \log_3 x$ is described by $y = 3^x$.

You can draw the graph for B: $y = \log_3 x$ by constructing a table of ordered pairs (x, y). First, rewrite $y = \log_3 x$ as $x = 3^y$. Second, choose values of y and use $x = 3^y$ to find the corresponding values of x.

x	$\frac{1}{9}$	$\frac{1}{3}$	1	3	9
$y = \log_3 x$	-2	-1	0	1	2

◀ *Second, use $x = 3^y$ to find x.*

◀ *First, choose values of y.*

The inverse of B: $y = \log_3 x$ is the exponential function B^{-1}: $y = 3^x$.
The graphs of these inverse functions are shown on page 395.

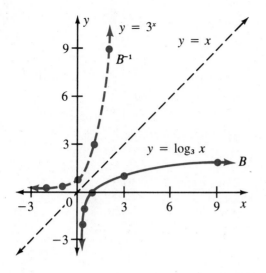

Recall that the graphs of inverse functions are symmetric with respect to the line $y = x$. This can be seen for the graphs of the inverse functions B and B^{-1}.

From the graph of B you can list four properties of the logarithmic function described by $y = \log_3 x$.
(1) The domain is $\{x \mid x > 0\}$.
(2) The range is {all real numbers}.
(3) The function is *constantly increasing*. That is, if $x_2 > x_1$, then $y_2 > y_1$.
(4) The ordered pair (1, 0) belongs to the function. That is, $\log_3 1 = 0$.

Example 2

Draw the graphs of the logarithmic functions $A: y = \log_4 x$ and $B: y = \log_{\frac{1}{4}} x$ in the same coordinate plane with convenient scales for the axes. List four properties of each function.

1. Rewrite $y = \log_4 x$ as $x = 4^y$.

A:

x	$\frac{1}{16}$	$\frac{1}{4}$	1	4	16
y	-2	-1	0	1	2

◀ $x = 4^y$

2. Rewrite $y = \log_{\frac{1}{4}} x$ as $x = \left(\frac{1}{4}\right)^y = 4^{-y}$.

B:

x	16	4	1	$\frac{1}{4}$	$\frac{1}{16}$
y	-2	-1	0	1	2

◀ $x = 4^{-y}$

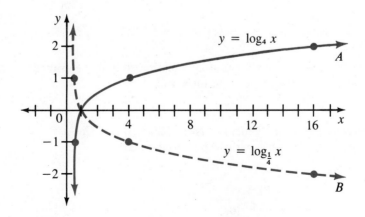

The graphs show the following four properties for
A: $y = \log_4 x$ and B: $y = \log_{\frac{1}{4}} x$.

(1) The domain for A and B is $\{x \mid x > 0\}$.
(2) The range for A and B is {all real numbers}.
(3) The function A: $y = \log_4 x$ is constantly increasing.
   The function B: $y = \log_{\frac{1}{4}} x$ is constantly decreasing.

(4) The ordered pair $(1, 0)$ belongs to both functions A and B.

Notice that the graphs of A: $y = \log_4 x$ and B: $y = \log_{\frac{1}{4}} x$
are symmetric with respect to the x-axis.

Reading in Algebra

Determine whether the graphs of the two functions described are symmetric
with respect to the x-axis, the y-axis, or the line described by $y = x$. Are the two
functions inverses of each other?

1. $y = \log_6 x$, $y = \log_{\frac{1}{6}} x$

2. $y = \log_6 x$, $y = 6^x$

3. $y = 6^x$, $y = \left(\dfrac{1}{6}\right)^x$

4. $y = \log_{\frac{1}{6}} x$, $y = \left(\dfrac{1}{6}\right)^x$

Written Exercises

For each function described, write the y-form of the equation for the inverse
function.

(A) **1.** $y = 5^x$ **2.** $y = \log_{\frac{1}{3}} x$ **3.** $y = \log_8 x$ **4.** $y = \left(\dfrac{1}{5}\right)^x$

Graph each function described. Choose convenient scales for the axes. List four
properties of the function.

5. $y = \log_2 x$ **6.** $y = \log_{\frac{1}{2}} x$ **7.** $y = \log_5 x$

8. $y = \log_{\frac{1}{3}} x$ **9.** $y = \log_6 x$ **10.** $y = \log_{\frac{1}{5}} x$

(B) **11.** $y = 4 \cdot \log_2 x$ **12.** $y = 2 \cdot \log_3 x$ **13.** $y = 3 \cdot \log_{\frac{1}{2}} x$

14. $y = 4 + \log_2 x$ **15.** $y = 2 + \log_3 x$ **16.** $y = 3 + \log_{\frac{1}{2}} x$

For each function described, write the y-form of an equation so that the graphs
of both equations are symmetric with respect to the x-axis.

17. $y = \log_5 x$ **18.** $y = \log_{\frac{1}{5}} x$

For each function described, write the y-form of an equation so that the
graphs of both equations are symmetric with respect to the y-axis.

19. $y = 5^x$ **20.** $y = \left(\dfrac{1}{5}\right)^x$

(C) **21.** Graph the function described by $y = \log_2 (x - 3)$.
22. Graph the function described by $y = \log_3 (x + 2)$.

LOGS OF PRODUCTS AND QUOTIENTS 15.4

Objectives
To expand the logarithm of a product or quotient
To simplify a sum or difference of logarithms
To solve a logarithmic equation

One of the laws of logarithms concerns the logarithm of the **product** of two factors.

First law of logarithms: $\log_b (x \cdot y) = \log_b x + \log_b y$

As an example, you can show that $\log_2 (8 \cdot 32) = \log_2 8 + \log_2 32$.

$\log_2 (8 \cdot 32)$	$\log_2 8 + \log_2 32$
$\log_2 (2^3 \cdot 2^5)$	$\log_2 2^3 + \log_2 2^5$
$\log_2 2^8$	$3 + 5$
8	8

Proof Let: $\log_b x = p$ and $\log_b y = q$
Then: $x = b^p$ and $y = b^q$
$x \cdot y = b^p \cdot b^q$
$x \cdot y = b^{p+q}$
$\log_b (x \cdot y) = \log_b b^{p+q}$
$\log_b (x \cdot y) = p + q$
$\log_b (x \cdot y) = \log_b x + \log_b y$

In a similar way, you can prove another law of logarithms. It concerns the logarithm of the **quotient** of two terms.

Second law of logarithms: $\log_b \dfrac{x}{y} = \log_b x - \log_b y$

As an example, you can show that $\log_2 \dfrac{128}{8} = \log_2 128 - \log_2 8$, since

(1) $\log_2 \dfrac{128}{8} = \log_2 \dfrac{2^7}{2^3} = \log_2 2^4 = 4$ and

(2) $\log_2 128 - \log_2 8 = \log_2 2^7 - \log_2 2^3 = 7 - 3 = 4.$

Log$_b$ of a product	$\log_b (x \cdot y) = \log_b x + \log_b y \; [x > 0, y > 0, b > 0, b \neq 1]$
Log$_b$ of a quotient	$\log_b \dfrac{x}{y} = \log_b x - \log_b y \; [x > 0, y > 0, b > 0, b \neq 1]$

You can use the first two laws of logarithms to *expand* an expression like $\log_3 \dfrac{5c}{\pi d}$.

Example 1   **Write $\log_3 \dfrac{5c}{\pi d}$ in expanded form.**

$\log_3 \dfrac{5 \cdot c}{\pi \cdot d} = \quad \log_3 5 \cdot c \quad - \quad \log_3 \pi \cdot d \quad \blacktriangleleft \; log_b \dfrac{x}{y} = log_b x - log_b y$

$= (\log_3 5 + \log_3 c) - (\log_3 \pi + \log_3 d) \quad \blacktriangleleft \; log_b x \cdot y = log_b x + log_b y$

$= \log_3 5 + \log_3 c - \log_3 \pi - \log_3 d$

Thus, $\log_3 \dfrac{5c}{\pi d} = \log_3 5 + \log_3 c - \log_3 \pi - \log_3 d.$

Example 2

Given the formula $L = 2\pi rh$, write $\log_b r$ in expanded form.

$$r = \frac{L}{2\pi h} \quad \blacktriangleleft \text{ Solve } L = 2\pi rh \text{ for } r.$$

$$\log_b r = \log_b \frac{L}{2\pi h} \quad \blacktriangleleft \text{ If } x = y, \text{ then } \log_b x = \log_b y.$$

$$= \log_b L - \log_b 2\pi h$$

$$= \log_b L - (\log_b 2 + \log_b \pi + \log_b h)$$

Thus, $\log_b r = \log_b L - \log_b 2 - \log_b \pi - \log_b h$, if $L = 2\pi rh$.

Reversing the steps in Examples 1 and 2 yields the logarithm of the original expression since the first two laws of logarithms are reversible. That is,

(1) $\log_b x + \log_b y = \log_b (x \cdot y)$ and (2) $\log_b x - \log_b y = \log_b \dfrac{x}{y}$.

Example 3

Write $\log_3 7 + \log_3 t - \log_3 4 - \log_3 v$ as the $\log_3$ of *one* expression.

$$\log_3 7 + \log_3 t - \log_3 4 - \log_3 v = \underbrace{\log_3 7 + \log_3 t}_{} - \underbrace{(\log_3 4 + \log_3 v)}_{} \quad \blacktriangleleft \text{ Rewrite.}$$

$$= \log_3 (7 \cdot t) \quad - \quad \log_3 (4 \cdot v) \quad \blacktriangleleft \text{ First law}$$

$$= \log_3 \frac{7t}{4v} \quad \blacktriangleleft \text{ Second law}$$

Thus, $\log_3 7 + \log_3 t - \log_3 4 - \log_3 v = \log_3 \dfrac{7t}{4v}$.

You can solve a logarithmic equation by using the **property of equality for logarithms:** If $\log_b x = \log_b y$, then $x = y$.

Each *apparent* solution of a logarithmic equation must be checked since $\log_b x$ is defined only for $x > 0$.

Example 4

Solve $\log_5 2t + \log_5 (t - 4) = \log_5 48 - \log_5 2$. Check.

$$\underbrace{\log_5 2t + \log_5 (t - 4)}_{} = \underbrace{\log_5 48 - \log_5 2}_{}$$

$$\textit{First law} \blacktriangleright \quad \log_5 2t(t - 4) \quad = \quad \log_5 \frac{48}{2} \quad \blacktriangleleft \textit{ Second law}$$

$$2t(t - 4) = \frac{48}{2} \quad \blacktriangleleft \begin{array}{l} \textit{If } \log_b x = \log_b y, \\ \textit{then } x = y. \end{array}$$

$$2t^2 - 8t = 24$$

$$t^2 - 4t - 12 = 0$$

$$(t - 6)(t + 2) = 0$$

$$t = 6 \text{ or } t = -2$$

Check: $t = 6$. $\log_5 2t + \log_5 (t - 4) = \log_5 48 - \log_5 2$
$\log_5 12 + \log_5 2 \qquad = \log_5 48 - \log_5 2$
$\log_5 24 \qquad\qquad\quad = \log_5 24$
Thus, 6 is the only solution.

Check: $t = -2$.
$\log_5 2t = \log_5 (-4)$
and $\log_5 (-4)$ is undefined
since $-4 < 0$.

Oral Exercises

True or false?

1. $\log_2 (3 \cdot 7) = \log_2 3 + \log_2 7$

2. $\log_5 12 = \log_5 4 + \log_5 3$

3. $\log_b 8 = \log_b 3 + \log_b 5$

4. $\log_5 2 + \log_5 3 = \log_5 6$

5. $\log_b 30 = \log_b 2 + \log_b 3 + \log_b 5$

6. $\log_b 5 + \log_b 5 + \log_b 5 = \log_b 15$

7. $\log_3 \frac{17}{4} = \log_3 17 - \log_3 4$

8. $\log_b \frac{18}{6} = \log_b 18 \div \log_b 6$

9. $\log_2 \frac{3 \cdot 5}{7} = \log_2 3 - \log_2 7 - \log_2 5$

10. $\log_2 \frac{6 \cdot 9}{5} = \log_2 6 + \log_2 9 - \log_2 5$

11. $\log_b \frac{35}{3 \cdot 8} = \log_b 35 - \log_b 3 - \log_b 8$

12. $\log_3 \frac{7}{5 \cdot 4} = \log_3 7 - \log_3 5 + \log_3 4$

13. If $\log_3 x + \log_3 y = \log_3 z$, then $x \cdot y = z$.

14. If $\log_2 m - \log_2 n = \log_2 p$, then $m \div n = p$.

Written Exercises

Write each logarithmic expression in expanded form.

Ⓐ **1.** $\log_2 5a$

2. $\log_3 7tw$

3. $\log_b 2mnp$

4. $\log_4 \frac{a}{3}$

5. $\log_3 \frac{4}{a}$

6. $\log_b \frac{8f}{g}$

For each formula, write the indicated logarithm in expanded form.

7. $d = 2r$; $\log_b r$

8. $T = \frac{1}{2}A$; $\log_5 A$

9. $3B = 2V$; $\log_5 B$

Write each expression as the logarithm of one expression.

10. $\log_2 9 + \log_2 c$

11. $\log_3 7 - \log_3 t$

12. $\log_b 8 + \log_b m - \log_b n$

13. $\log_5 6c^2 - \log_5 2c$

14. $\log_2 3t + \log_2 5t$

15. $\log_b 9c + \log_b 8c - \log_b 6c$

Solve each equation. Check.

16. $\log_2 6 + \log_2 k = \log_2 18$

17. $\log_b 12k - \log_b 4 = \log_b 15$

18. $\log_b 2t + \log_b 3t = \log_b 24$

19. $\log_5 t + \log_5 (t - 4) = \log_5 12$

Ⓑ **20.** $\log_3 y^2 + \log_3 y^3 = \log_3 16y$

21. $\log_b (c^2 - 4c) - \log_b c = \log_b 2$

22. $\log_5 3 + \log_5 t + \log_5 (t + 4) = \log_5 63$

23. $\log_b (y - 3) - \log_b (y - 4) = \log_b (y + 3) - \log_b y$

Write each logarithmic expression in expanded form.

24. $\log_2 \frac{3}{mn}$

25. $\log_b \frac{5c}{3d}$

26. $\log_3 \frac{1}{cd}$

Write each expression as the logarithm of one expression.

27. $\log_4 7 - \log_4 m + \log_4 3 - \log_4 n$

28. $\log_b 2t - \log_b 3v - \log_b 6t^2 + \log_b 9v^3$

For each formula, write the indicated logarithm in expanded form.

29. $L = 2\pi rh$; $\log_b h$

30. $V = \frac{1}{3}Bh$; $\log_2 B$

31. $A = \frac{3}{4}rt$; $\log_5 r$

Ⓒ **32.** Prove the second law of logarithms: $\log_b \frac{x}{y} = \log_b x - \log_b y$.

Logs of Products and Quotients

LOGS OF POWERS AND RADICALS

Objectives   **To expand a logarithm of a power or a radical**
To simplify a multiple of a logarithm
To solve a logarithmic equation

The third law of logarithms involves the logarithm of a **power.**

Third law of logarithms: $\log_b x^r = r \cdot \log_b x$

As an example, you can show that $\log_2 5^3 = 3 \cdot \log_2 5$, where $r = 3$, since

$$\log_2 5^3 = \log_2 (5 \cdot 5 \cdot 5) = \log_2 5 + \log_2 5 + \log_2 5 = 3 \cdot \log_2 5.$$

Proof Let $\log_b x = p$. Then, $x = b^p$ and $x^r = (b^p)^r = b^{r \cdot p}$.
If $x^r = b^{r \cdot p}$, then $\log_b x^r = \log_b b^{r \cdot p} = r \cdot p = r \cdot \log_b x$.

Recall that $\sqrt[n]{x} = x^{\frac{1}{n}}$. Thus, the third law can be used with the logarithm of a **radical.** For example,

$$\log_2 \sqrt[3]{5} = \log_2 5^{\frac{1}{3}} = \frac{1}{3} \cdot \log_2 5, \text{ since } \log_2 5^r = r \cdot \log_2 5, \text{ where } r = \frac{1}{3}.$$

Log$_b$ of a power	$\log_b x^r = r \cdot \log_b x$ $[x > 0, r$ is a real number$]$ $\log_b \sqrt[n]{x} = \dfrac{1}{n} \cdot \log_b x$ $[x > 0, n$ is a positive integer$]$.

This third law is applied in Examples 1 and 2.

Example 1   **Write $\log_5 (mn^3)^2$ and $\log_b \sqrt[4]{\dfrac{m^3}{n}}$ in expanded form.**

$\log_5 (mn^3)^2 = 2 \cdot \log_5 mn^3 = 2(\log_5 m + \log_5 n^3) = 2(\log_5 m + 3 \cdot \log_5 n)$

$\log_b \sqrt[4]{\dfrac{m^3}{n}} = \log_b \left(\dfrac{m^3}{n}\right)^{\frac{1}{4}} = \dfrac{1}{4} \cdot \log_b \dfrac{m^3}{n} = \dfrac{1}{4}(\log_b m^3 - \log_b n) = \dfrac{1}{4}(3 \cdot \log_b m - \log_b n)$

Example 2   **Given $d = \dfrac{1}{2}gt^2$, write $\log_b t$ in expanded form.**

Solve $d = \dfrac{1}{2}gt^2$ for t, $t > 0$.

$$\dfrac{1}{2}gt^2 = d$$

$$gt^2 = 2d$$

$$t^2 = \dfrac{2d}{g}$$

$$t = \sqrt{\dfrac{2d}{g}}$$

$\log_b t = \log_b \sqrt{\dfrac{2d}{g}} = \dfrac{1}{2} \cdot \log_b \dfrac{2d}{g}$

$$= \dfrac{1}{2}(\log_b 2d - \log_b g)$$

$$= \dfrac{1}{2}(\log_b 2 + \log_b d - \log_b g)$$

Thus, $\log_b t = \dfrac{1}{2}(\log_b 2 + \log_b d - \log_b g)$ if $d = \dfrac{1}{2}gt^2$.

You can simplify some logarithmic expressions by "reversing" the third law of logarithms.

$$r \cdot \log_b x = \log_b x^r \text{ and } \frac{1}{n} \cdot \log_b x = \log_b x^{\frac{1}{n}} = \log_b \sqrt[n]{x}$$

Example 3

Write $5 \log_2 c + 3 \log_2 d$ as the $\log_2$ of one expression.

$$5 \cdot \log_2 c + 3 \cdot \log_2 d = \log_2 c^5 + \log_2 d^3 = \log_2 c^5 d^3$$

Example 4

Write $\frac{1}{3}(\log_5 t + 4 \log_5 v - \log_5 w)$ as the $\log_5$ of one expression.

$$\frac{1}{3}(\log_5 t + 4 \cdot \log_5 v - \log_5 w) = \frac{1}{3}(\log_5 t + \log_5 v^4 - \log_5 w)$$

$$= \frac{1}{3} \cdot \log_5 \frac{tv^4}{w}$$

$$= \log_5 \sqrt[3]{\frac{tv^4}{w}}$$

Logarithmic equations with terms of the form $r \cdot \log_b x$ can be solved by using the third law of logarithms as shown below.

Example 5

Solve $2 \log_3 t = \log_3 2 + \log_3(t + 12)$. Check.

$$2 \cdot \log_3 t = \log_3 2 + \log_3(t + 12)$$
$$\log_3 t^2 = \log_3 2(t + 12) \quad \blacktriangleleft \text{ If } \log_b x = \log_b y, \qquad \qquad t^2 - 2t - 24 = 0$$
$$t^2 = 2t + 24 \qquad \blacktriangleleft \text{ then } x = y. \qquad \qquad \qquad (t - 6)(t + 4) = 0$$
$$t = 6 \text{ or } t = -4$$

Check: $t = 6$

$2 \log_3 t$	$\log_3 2 + \log_3 (t + 12)$
$2 \log_3 6$	$\log_3 2 + \log_3 18$
$\log_3 6^2$	$\log_3 2 \cdot 18$
$\log_3 36$	$\log_3 36$

Check: $t = -4$

$2 \log_3 t$	$\log_3 2 + \log_3 (t + 12)$
$2 \log_3 (-4)$	$\log_3 2 + \log_3 8$

$\text{Log}_3 (-4)$ is undefined.
Thus, 6 is the solution.

Example 6

Solve $\frac{1}{2} \log_b a + \frac{1}{2} \log_b (a - 6) = \log_b 4$. Check.

$\frac{1}{2} log_b x = log_b \sqrt{x}$ ▶

$$\log_b \sqrt{a} + \log_b \sqrt{a - 6} = \log_b 4$$
$$\log_b (\sqrt{a} \cdot \sqrt{a - 6}) = \log_b 4$$
$$\sqrt{a^2 - 6a} = 4$$
$$a^2 - 6a = 16$$
$$a^2 - 6a - 16 = 0$$
$$(a - 8)(a + 2) = 0$$
$$a = 8 \text{ or } a = -2$$
Thus, 8 is the solution.

Check: $a = 8$
$$\frac{1}{2} \log_b a + \frac{1}{2} \log_b (a - 6)$$
$$\frac{1}{2} \log_b 8 + \frac{1}{2} \log_b 2$$
$$\log_b \sqrt{8} + \log_b \sqrt{2}$$
$$\log_b \sqrt{8} \ \sqrt{2}$$
$$\log_b \sqrt{16}, \text{ or } \log_b 4$$
Check: $a = -2$
$\text{Log}_b (-2)$ is undefined.

Written Exercises

Write each logarithmic expression in expanded form.

(A)

1. $\log_2 a^5$

2. $\log_3 \sqrt[5]{a}$

3. $\log_5 (mn)^2$

4. $\log_b \sqrt{2g}$

5. $\log_4 \left(\dfrac{v}{h}\right)^3$

6. $\log_b \sqrt[3]{\dfrac{d}{g}}$

For each formula, write the indicated logarithm in expanded form.

7. $v^2 = 2gh$; $\log_3 v$

8. $d = \frac{1}{2}gt^2$; $\log_b g$

9. $A = \pi r^2$; $\log_4 r$

Write each expression as the logarithm of one expression.

10. $2\log_3 x + 4\log_3 y$

11. $\frac{1}{3}(\log_2 6 + \log_2 c)$

12. $3(\log_b 2 + \log_b t)$

13. $\frac{1}{2}\log_5 2c + \frac{1}{2}\log_5 d$

14. $4\log_b t - 3\log_b v$

15. $\frac{1}{4}\log_3 6y - \frac{1}{4}\log_3 5z$

Solve each equation. Check.

16. $2\log_3 y = \log_3 4 + \log_3 (y + 8)$

17. $2\log_b t - \log_b 2 = \log_b (2t + 6)$

18. $\frac{1}{2}\log_b a + \frac{1}{2}\log_b (a + 5) = \log_b 6$

19. $\frac{1}{3}\log_5 y + \frac{1}{3}\log_5 (y + 2) = \log_5 2$

(B)

20. $2\log_b (2y + 2) = \log_b 16 + 2\log_b (y - 2)$

21. $2\log_5 (3a + 1) = \log_5 4 - 2\log_5 (2a - 1)$

22. $\frac{1}{2}\log_3 (c + 1) + \frac{1}{2}\log_3 (c - 4) = \log_3 6$

23. $\frac{1}{2}\log_b (y - 1) - \frac{1}{2}\log_b (2y - 1) = \log_b 2 - \log_b 3$

24. $2\log_3 (y + 5) - \log_3 (y + 2) = \log_3 (y + 10)$

25. $\log_5 12 - \frac{1}{2}\log_5 3 = \frac{1}{2}\log_5 2y$

Write each logarithmic expression in expanded form.

26. $\log_b (t^3 v^2 w)^2$

27. $\log_5 \left(\dfrac{mn^2}{p^3}\right)^3$

28. $\log_b \sqrt[4]{\dfrac{5d}{cg^3}}$

For each formula, write the indicated logarithm in expanded form.

29. $A = P(1.09)^{12}$; $\log_{10} A$

30. $V = \frac{4}{3}\pi r^3$; $\log_b r$

31. $V = \frac{1}{3}\pi r^2 h$; $\log_2 r$

Write each expression as the logarithm of one expression.

32. $\log_5 7 + \frac{1}{2}\log_5 x - 2\log_5 y$

33. $\frac{1}{3}\log_4 2a + \frac{1}{3}\log_4 5b - 2\log_4 a - \log_4 b$

34. $3(\log_b m + 2\log_b n - \log_b p)$

35. $\frac{1}{4}(\log_b 3 + 3\log_b x - \log_b 2 - \log_b 5y)$

Solve each equation. Check (Ex. 36–39) [*Hint*: If $\log_b x = y$, then $b^y = x$.]

(C)

36. $\log_2 (y + 4) + \log_2 y = 5$

37. $\log_4 (3y + 1) + \log_4 (y - 1) = 3$

38. $\log_5 (y^2 + 2y + 5) - \log_5 (y - 5) = 2$

39. $2\log_3 (y - 2) - \log_3 (y - 4) = 2$

40. Prove: $\log_b \sqrt[3]{y^2} - \log_b \sqrt[4]{y^3} + 4\log_b \sqrt[3]{y} - 5\log_b \sqrt[4]{y} = 0$.

41. Prove: $\frac{2}{5}\log_b 3 - \frac{1}{5}\log_b 2 - \frac{2}{5}\log_b 4 + \frac{1}{5}\log_b 27 = \log_b 3 - \log_b 2$.

COMMON LOGARITHMS AND ANTILOGARITHMS

Objectives

To find the base-10 logarithm of a positive rational number
To find the base-10 antilogarithm of a rational number

The table of **common (base-10)** logarithms on page 593 can be used to find that $\log_{10} 3.26 \doteq 0.5132$, correct to four decimal places.

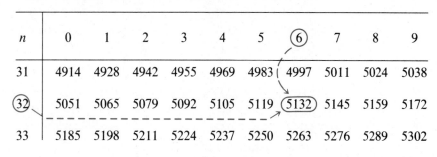

n	0	1	2	3	4	5	6	7	8	9
31	4914	4928	4942	4955	4969	4983	4997	5011	5024	5038
32	5051	5065	5079	5092	5105	5119	5132	5145	5159	5172
33	5185	5198	5211	5224	5237	5250	5263	5276	5289	5302

$$\log_{10} \underline{3.2\ 6} \dashrightarrow \underline{0.5132}$$

row 32 column 6   Found in row 32 and column 6

Decimal points must be placed in "326" and "5132" so that $1 \le 3.26 < 10$ and $0 \le 0.5132 < 1$. If $1 \le n < 10$, then $0 \le \log_{10} n < 1$.

You can approximate $\log_{10} 3260$ and $\log_{10} 0.0326$ to four decimal places using $\log_{10} 3.26 = 0.5132$ and *scientific notation*. Recall that

A number is in **scientific notation** when it is written in the form $a \times 10^c$ where $1 \le a < 10$ and c is an integer.

You can proceed as shown below.

(1) **$\log_{10} 3260$**

$$3260 = 3.26 \times 10^3$$
$$\log_{10} 3260 = \log_{10} (3.26 \times 10^3)$$
$$= \log_{10} 3.26 + \log_{10} 10^3$$
$$= 0.5132 + 3$$
$$= 3 + 0.5132, \text{ or } 3.5132$$
$$\log_{10} 3260 = 3 + 0.5132, \text{ or } 3.5132$$

The integer 3 is called the **characteristic** of the log. The decimal .5132 is called the **mantissa** of the log.

(2) **$\log_{10} 0.0326$**

$$0.0326 = 3.26 \times 10^{-2}$$
$$\log_{10} 0.0326 = \log_{10} (3.26 \times 10^{-2})$$
$$= \log_{10} 3.26 + \log_{10} 10^{-2}$$
$$= 0.5132 + (-2)$$
$$= -2 + 0.5132, \text{ or } 8.5132 - 10$$
$$\log_{10} 0.0326 = -2 + 0.5132, \text{ or } 8.5132 - 10,$$

where -2 is rewritten as $8 - 10$.
The characteristic is -2, or $8 - 10$.
The mantissa is .5132.

Definition:
Characteristic
Mantissa

If $y = a \times 10^c$ where $1 \le a < 10$ and c is an integer, then $\log_{10} y = c + \log_{10} a$. The integer c is the *characteristic* and $\log_{10} a$ is the *mantissa* of $\log_{10} y$.

It is customary to use the *equal* sign (=), rather than the *approximately equal* sign (≐), when approximating common logarithms.

You can omit the *subscript 10* from expressions like $\log_{10} 3260$ when working with common logarithms. That is,

log 3260 (without a written base) means $\log_{10} 3260$.

Example 1   **Find log 294 and log 0.00294.**

$$294 = 2.94 \times 10^2$$
$$\log 2.94 = 0.4683$$
Thus, log 294 = 2 + 0.4683, or 2.4683.

$$0.00294 = 2.94 \times 10^{-3}$$
$$\log 2.94 = 0.4683$$
Thus, log 0.00294 = −3 + 0.4683, or
7.4683 − 10.

The base-10 logarithmic function, where $\log 1000 = \log 10^3 = 3$, can be viewed as a mapping.

$$1000, \text{ or } 10^3 \longrightarrow 3$$
$$10, \text{ or } 10^1 \longrightarrow 1$$
$$0.01, \text{ or } 10^{-2} \longrightarrow -2$$
$$10^x \longrightarrow x$$

The **inverse** of the base-10 logarithmic function is the base-10 **antilogarithmic** function.

$$3 \longrightarrow 10^3, \text{ or } 1000$$
$$1 \longrightarrow 10^1, \text{ or } 10$$
$$-2 \longrightarrow 10^{-2}, \text{ or } 0.01$$
$$x \longrightarrow 10^x$$

So, $\log 10^x = x$ and antilog $x = 10^x$ for each real number x.

<table>
<tr><td>**Definition:**
Base-10
antilogarithm</td><td>antilog $x = 10^x$, for each real number x.</td></tr>
</table>

The table on the preceding page shows how to find that
$$\log 3.26 = \log 10^{0.5132} = 0.5132.$$

You can reverse this process to find that
$$\text{antilog } 0.5132 = 10^{0.5132} = 3.26.$$

Example 2   **Find antilog 4.9309 and antilog 7.9309 − 10.**

The mantissa is .9309. Use the table on page 593 to find that antilog $0.9309 = 10^{0.9309} = 8.53$.

	antilog 4.9309	antilog 7.9309 − 10	
	antilog 4 + 0.9309	antilog −3 + 0.9309	◀ *Rewrite*
antilog $x = 10^x$ ▶	$10^{4+0.9309}$	$10^{-3+0.9309}$	*7 − 10 as −3.*
$10^{m+n} = 10^m \cdot 10^n$ ▶	$10^4 \times 10^{0.9309}$	$10^{-3} \times 10^{0.9309}$	
	$10^4 \times 8.53$	$10^{-3} \times 8.53$	◀ *Scientific*
Ordinary notation ▶	85,300	0.00853	*notation*

Thus, antilog 4.9309 = 85,300 and antilog 7.9309 − 10 = 0.00853.

Reading in Algebra

Using the equation log 346 = 2.5391, find each of the following.
1. log 346
2. mantissa of log 346
3. characteristic of log 346
4. log 3.46
5. antilog 2.5391
6. $10^{2.5391}$
7. log 0.0346
8. antilog 0.5391
9. x, if $10^x = 34.6$

Write the characteristic of each logarithm in two ways.
10. log 0.0023 = 7.3617 − 10
11. log 0.23 = −1 + 0.3617

Express each number in scientific notation.
12. 654,000
13. 4.32
14. 0.000789

Express each number in ordinary notation.
15. 3.21×10^{-1}
16. 5.43×10^0
17. 8.76×10^3

Find each logarithm or antilogarithm.
18. log 100,000
19. log 1
20. log 0.01
21. antilog 2
22. antilog −3
23. antilog 0

Find the characteristic of each logarithm.
24. log 7520
25. log 7.52
26. log 0.000752

Written Exercises

Find each logarithm or antilogarithm.

(A)
1. log 19,800
2. antilog 2.5132
3. log 0.207
4. antilog 7.8129 − 10
5. log 0.00613
6. antilog −2 + 0.2227

(B)
7. log 32,600,000
8. antilog 6.9345 − 10
9. antilog 4.9206
10. log $\sqrt{10}$
11. antilog $\frac{1}{3}$
12. log $\sqrt[3]{1000}$

Solve each equation for x.

(C)
13. $x^2 = \log 10^{5x-6}$
14. $\dfrac{x + 5}{\log 100} = \dfrac{\log 1,000}{x}$
15. $\dfrac{\log 10^{\frac{1}{x}}}{x} = \dfrac{1}{9}$

NON-ROUTINE PROBLEMS

1. Find log (antilog (log (antilog 2.0899))).
2. Find $\displaystyle\sum_{k=1}^{4}$ antilog $(k - 2)$.

CUMULATIVE REVIEW

1. Solve $\dfrac{y}{10} = \dfrac{7}{11}$.
2. Solve $\dfrac{9}{10} = \dfrac{x}{12}$.
3. Solve $\dfrac{y}{10} = \dfrac{11}{25}$.

Common Logarithms and Antilogarithms

USING EXPONENTIAL EQUATIONS 15.7

Objectives **To solve an exponential equation using base-10 logarithms**
To find the base-*b* log of a positive number, using a table of base-10 logarithms
To solve a word problem leading to an exponential equation

An equation like $4^{2x-1} = 360$ is called an **exponential equation** since the variable appears in the exponent. You can estimate a range for x by knowing some powers of 4.

$4^4 = 256$ and $4^5 = 1{,}024$ $\quad\quad$ So, $\quad 4 < 2x - 1 < 5$
$\quad 256 < 360 < 1{,}024$ $\quad\quad\quad\quad\quad 5 < 2x < 6$
$\quad 4^4 < 4^{2x-1} < 4^5$ $\quad\quad\quad$ **Thus,** $2.5 < x < 3.0$.

You can find x to three significant digits using base-10 logarithms.

Example 1 **Solve $4^{2x-1} = 360$ for x to three significant digits.**

$$4^{2x-1} = 360$$
$$\log 4^{2x-1} = \log 360 \quad \blacktriangleleft \textit{ If } p = q, \textit{ then } \log p = \log q.$$
$$(2x - 1) \log 4 = \log 360 \quad \blacktriangleleft \textit{ } \log a^r = r \cdot \log a$$
$$2x - 1 = \frac{\log 360}{\log 4} \quad \blacktriangleleft \textit{ Divide each side by } \log 4.$$
$$2x - 1 = \frac{2.5563}{0.6021}$$
$$2x - 1 = 4.246 \quad \blacktriangleleft \textit{ Divide to 4 significant digits.}$$
$$2x = 5.246$$
$$x = 2.623$$
Thus, $x = 2.62$ to three significant digits.

In Example 1, the value of x is 2.62. Notice that this agrees with the estimate (above) that $2.5 < x < 3.0$.

A base-10 log such as $\log 4$ is found in the table on page 593. A base-*b* log such as $\log_3 4$ can be found to three significant digits by solving an exponential equation using base-10 logs as shown in Example 2.

Example 2 **Find $\log_3 4$ to three significant digits.**

Let $\quad \log_3 4 = x$. $\quad \blacktriangleleft \textit{ Base-3 log}$
$$3^x = 4 \quad \blacktriangleleft \textit{ Change the equation to exponential form.}$$
$$\log 3^x = \log 4 \quad \blacktriangleleft \textit{ Base-10 logs}$$
$$x \cdot \log 3 = \log 4$$
$$x = \frac{\log 4}{\log 3} = \frac{0.6021}{0.4771} = 1.262$$
Thus, $\log_3 4 = 1.26$ to three significant digits.

When an investment is earning *simple* interest, the principal remains constant. If it is earning *compound* interest that is compounded *annually,* the principal is increased each year by the amount of interest earned in the previous year. If $500 is invested at a rate of 10%, compounded annually, the total amounts at the end of years 0 through 3 are as follows.

$$\$500, \$500 + \$50, \$550 + \$55, \$605 + \$60.50, \ldots$$

If the interest on an investment is compounded *semiannually,* then every 6 months the interest earned for a half year is added to the previous principal.

Recall that the formula for the total amount A of an investment earning simple interest is $A = p + prt = p(1 + rt)$. The formula for compound interest is given below.

Compound interest formula	$A = p\left(1 + \dfrac{r}{n}\right)^{nt}$

A is the total amount at the end of t years when p dollars are invested at $r\%$ per year, compounded n times each year.

Example 3

A bond that pays 12% per year, compounded quarterly (4 times each year, or every 3 mo), is bought for $6,000. How much will the bond be worth 10 yr from now? How much will the bond have earned in compound interest? Use logarithms.

$$p = 6,000 \qquad r = 0.12 \qquad t = 10 \qquad n = 4$$

$$A = p\left(1 + \frac{r}{n}\right)^{nt} = 6,000\left(1 + \frac{0.12}{4}\right)^{40} = 6,000(1.03)^{40}$$

$$\begin{aligned}
\log A = \log 6,000(1.03)^{40} &= \log 6,000 + 40 \cdot \log 1.03 \\
&= 3.7782 + 40(0.0128) \\
&= 4.2902 \\
A &= \text{antilog } 4.2902 = 10^4 \times 1.95 = 19,500 \\
A - p &= 19,500 - 6,000 = 13,500
\end{aligned}$$

Thus, the $6,000 bond will be worth $19,500 and will have earned $13,500 in compound interest.

Some bacteria reproduce by splitting themselves into two new bacteria at the end of a growth period. This process of splitting is called *simple fission.* Suppose there are 400 bacteria at the start of an experiment and simple fission takes place every 20 min, or 3 times in each hour. The number of bacteria increases rapidly.

Growth period	0	1	2	3	4	...
Number of bacteria	400	800	1600	3200	6400	...
	400	$400 \cdot 2$	$400 \cdot 2^2$	$400 \cdot 2^3$	$400 \cdot 2^4$	...

In 6 h, there are $3 \cdot 6$, or 18, twenty-minute periods. **Thus,** at the end of 6 h the number of bacteria is $400 \cdot 2^{18}$, or more than 104 million.

Example 4

To three significant digits, find the number of bacteria produced from 500 bacteria at the end of 4 h if simple fission occurs every 12 min.

$$N = N_0 \cdot 2^k = 500 \cdot 2^{20} \quad \blacktriangleleft \quad \textit{20 twelve-minute periods = 4 hours.}$$
$$\log N = \log (500 \cdot 2^{20}) = \log 500 + 20 \cdot \log 2 = 2.6990 + 20(0.3010) = 8.7190$$
$$N = \text{antilog } 8.7190 = 10^8 \times 5.24$$

Thus, there will be 524 million bacteria.

Reading in Algebra

For each equation, estimate a range for x.

(A) **1.** $2^{x+1} = 70$ **2.** $3^{2x} = 90$ **3.** $5^{2x-1} = 200$ **4.** $2^{-x} = 18$ **5.** $4^{2x+1} = 200$

Written Exercises

Solve each equation for x to three significant digits.

1. $4^x = 25$ **2.** $4^{2x} = 25$ **3.** $4^{x-1} = 25$ **4.** $7^{x+1} = 56$ **5.** $8^{3x} = 72$

Find each logarithm to three significant digits.

6. $\log_2 6$ **7.** $\log_6 2$ **8.** $\log_5 14$

(B) **9.** $\log_3 864$ **10.** $\log_4 7250$ **11.** $\log_2 \sqrt[3]{35}$

Find the value of each bond and the amount of interest earned at the end of the time period for the given data.

12. $5,000 bond paying 8%, compounded semiannually, for 6 yr.
13. $8,000 bond paying 9%, compounded annually, for 7 yr.
14. $7,500 bond paying 12%, compounded quarterly, for 15 yr.

Find N, the number of bacteria present, for the given data. Use the formula $N = N_0 \cdot 2^k$.

15. $N_0 = 750$ and simple fission occurs every 15 min for 3 h.
16. $N_0 = 3,000$ and simple fission occurs every 90 min for 18 h.
17. $N_0 = 25,000$ and simple fission occurs every 30 min for 24 h.

Solve each equation for x to three significant digits.

18. $9^{2x+1} = 624$ **19.** $7^{3x-2} = 834$ **20.** $2^{\sqrt{5}} = x$

(C) **21.** $3^x = 4 \cdot 2^x$ **22.** $2 \cdot 5^{x+1} = 5 \cdot 2^{x+2}$ **23.** $4^{\sqrt{6}} = 2x - 1$

Prove that each statement is true. Each variable represents a positive number $\neq 1$.
[*Hint*: Begin each proof with "Let $\log_a y = x$."]

24. $\log_a y = \dfrac{\log y}{\log a}$ **25.** $\log_a y = \dfrac{\log_b y}{\log_b a}$ **26.** $\log_a y = \dfrac{1}{\log_y a}$

HIGHER-DEGREE POLYNOMIAL EQUATIONS

Objectives

To find the zeros of an integral polynomial
To factor an integral polynomial in one variable into first-degree factors
To solve an integral polynomial equation of degree greater than two

The polynomial $2x^4 + 5x^3 - 11x^2 - 20x + 12$ is an **integral polynomial** since all the coefficients are *integers*. The notation $P(x)$ is used to represent such a polynomial. The value of the polynomial $P(x)$ is represented by $P(r)$, when x is replaced by any complex number r.

If $P(x) = 2x^4 + 5x^3 - 11x^2 - 20x + 12$, then
$$P(1) = 2 \cdot 1^4 + 5 \cdot 1^3 - 11 \cdot 1^2 - 20 \cdot 1 + 12, \text{ or } P(1) = -12.$$

You can factor $P(x) = 2x^4 + 5x^3 - 11x^2 - 20x + 12$ into *first-degree factors*, given that $x - 2$ and $x + 3$ are first-degree factors of $P(x)$, using *synthetic substitutions* as shown below.

$$
\begin{array}{r|rrrrr}
 & 2 & 5 & -11 & -20 & 12 \\
\hline
2 & 2 & 9 & 7 & -6 & \boxed{0} \\
\hline
-3 & 2 & 3 & -2 & \boxed{0} &
\end{array}
$$

◀ $P(x) = 2x^4 + 5x^3 - 11x^2 - 20x + 12$
◀ $P(x) = (x - 2)(2x^3 + 9x^2 + 7x - 6)$
◀ $P(x) = (x - 2)(x + 3)(2x^2 + 3x - 2)$

Factor $2x^2 + 3x - 2$ into $(x + 2)(2x - 1)$.

Thus, $P(x) = (x - 2)(x + 3)(x + 2)(2x - 1)$.

Notice the following for $P(x) = (x - 2)(x + 3)(x + 2)(2x - 1)$ above.
(1) $P(2) = (2 - 2)(2 + 3)(2 + 2)(2 \cdot 2 - 1) = 0$. The number 2 is called a **zero** of the polynomial $P(x)$ because $P(2) = 0$.
(2) $P(-3) = 0$, $P(-2) = 0$, and $P(\frac{1}{2}) = 0$. So, -3, -2, and $\frac{1}{2}$ are zeros of $P(x)$.

Definition: Zero of a polynomial	The number r is called a *zero* of a polynomial $P(x)$ if and only if $P(r) = 0$.

The number of "potential" integral zeros of an integral polynomial is *finite*. This can be illustrated as follows.

Let $P(x) = 2x^4 + 5x^3 - 11x^2 - 20x + 12$.
$$2r^4 + 5r^3 - 11r^2 - 20r + 12 = 0$$
$$2r^4 + 5r^3 - 11r^2 - 20r = -12$$
$$r(2r^3 + 5r^2 - 11r - 20) = -12$$

If r is an integral zero of $P(x)$, then $2r^3 + 5r^2 - 11r - 20$ is an integer because r is an integer. So, r is an integral factor of -12.

The "potential" integral zeros of $P(x) = 2x^4 + 5x^3 - 11x^2 - 20x + 12$ are the integral factors of the *constant term* 12: $1, -1, 2, -2, 3, -3, 4, -4, 6, -6, 12, -12$.

Integral zero theorem	If an integer r is a zero of an integral polynomial $P(x)$, then r is a factor of the constant term in $P(x)$. That is, if r is an integral zero of $a_0x^n + a_1x^{n-1} + a_2x^{n-2} + \ldots + a_{n-1}x^1 + a_n$, then r is a factor of a_n.

Some of the zeros of an integral polynomial can be irrational numbers.

Example 1

Find the zeros of $P(x) = x^4 - x^3 - 8x^2 + 2x + 12$.
Factor $P(x)$ into first-degree factors.

1. List the potential integral zeros: $\pm 1, \pm 2, \pm 3, \pm 4, \pm 6, \pm 12$.
2. Test each potential zero until a *second-degree* factor is found.

$$
\begin{array}{r|rrrrr}
 & 1 & -1 & -8 & 2 & 12 \\
\hline
-2 & 1 & -3 & -2 & 6 & \boxed{0} \\
\hline
3 & 1 & 0 & -2 & \boxed{0} &
\end{array}
$$

◀ $P(x) = x^4 - x^3 - 8x^2 + 2x + 12$
◀ $P(x) = (x + 2)(x^3 - 3x^2 - 2x + 6)$
◀ $P(x) = (x + 2)(x - 3)(x^2 - 2)$

3. Solve $x^2 - 2 = 0$.
$x^2 = 2 \qquad x = \pm\sqrt{2}$
$\qquad\qquad P(x) = (x + 2)(x - 3)(x - \sqrt{2})(x + \sqrt{2})$

Thus, the zeros of $P(x)$ are $-2, 3, \sqrt{2}$, and $-\sqrt{2}$, and $P(x) = (x + 2)(x - 3)(x - \sqrt{2})(x + \sqrt{2})$.

In Example 1, the four zeros of the *polynomial $P(x)$* are also the solutions of the *equation $x^4 - x^3 - 8x^2 + 2x + 12 = 0$*. To solve a polynomial equation $P(x) = 0$, factor $P(x)$ into first-degree factors and find the zero of each factor.

Example 2

Solve $x^4 - 5x^2 - 36 = 0$.

Let $P(x) = x^4 - 5x^2 - 36$. Factor $P(x)$ into first-degree factors.
1. The potential integral zeros of $P(x)$ are the factors of 36: $\pm 1, \pm 2, \ldots, \pm 36$.
2. Test each potential zero, in turn, until a *second-degree* factor is found.

$$
\begin{array}{r|rrrrr}
 & 1 & 0 & -5 & 0 & -36 \\
\hline
3 & 1 & 3 & 4 & 12 & \boxed{0} \\
\hline
-3 & 1 & 0 & 4 & \boxed{0} &
\end{array}
$$

◀ $P(x) = x^4 + 0x^3 - 5x^2 + 0x - 36 = 0$
◀ $P(x) = (x - 3)(x^3 + 3x^2 + 4x + 12)$
◀ $P(x) = (x - 3)(x + 3)(x^2 + 4)$

3. Solve $x^2 + 4 = 0$.
$x^2 = -4 \qquad x = \pm 2i$
$\qquad\qquad P(x) = (x - 3)(x + 3)(x - 2i)(x + 2i) = 0$

Thus, the solutions are $3, -3, 2i$, and $-2i$.

The **Fundamental Theorem of Algebra** assures you that each polynomial of degree $n > 0$ has *at least one* complex zero.

Fundamental theorem of algebra	If $P(x)$ is a polynomial of degree greater than 0, then there is a complex number r for which $P(r) = 0$.

Example 3

Find the zeros of $P(x) = 2x^3 - 17x^2 + 40x - 16$. Solve the equation $2x^3 - 17x^2 + 40x - 16 = 0$.

$$
\begin{array}{r|rrrr}
 & 2 & -17 & 40 & -16 \\
\hline
4 & 2 & -9 & 4 & \boxed{0}
\end{array}
$$

◀ $P(x) = 2x^3 - 17x^2 + 40x - 16 = 0$
◀ $P(x) = (x - 4)(2x^2 - 9x + 4)$

$\underbrace{\qquad\qquad}$
$2x^2 - 9x + 4$
$\qquad P(x) = (x - 4)(x - 4)(2x - 1) = 0$

Thus, 4 and $\frac{1}{2}$ are the zeros of the polynomial $P(x)$ and the solutions of the equation $P(x) = 0$.

In Example 3, notice the following three facts.
(1) The degree of $P(x)$ is 3 and $P(x)$ has 3 first-degree factors.
(2) $P(x)$ has *only* 2 distinct zeros: 4 and $\frac{1}{2}$. The zero 4 has a *multiplicity of* 2 since the factor $x - 4$ appeared *twice*. The zero $\frac{1}{2}$ has a multiplicity of 1. The sum of these two multiplicities is 3, which is the degree of $P(x)$.
(3) $P(x)$ can be factored *uniquely* into three first-degree factors and the constant factor 2, which is the coefficient of $2x^3$, as shown below.

$$P(x) = 2x^3 - 17x^2 + 40x - 16 = (x - 4)(x - 4)(2x - 1) = (x - 4)(x - 4) \cdot 2(x - \tfrac{1}{2})$$

The theorem below guarantees that a polynomial $P(x)$ of degree $n > 0$ has exactly n first-degree factors, and thus, the *sum of the multiplicities* of the zeros of $P(x)$ is equal to n.

Unique factorization theorem	Each polynomial $P(x)$ of degree $n > 0$ can be factored uniquely into n first-degree factors (not all necessarily distinct) and a constant factor which is the coefficient of the highest-degree term in $P(x)$.

Reading in Algebra

1. List the potential integral zeros of $x^4 + 23x^2 - 50$.
2. If -4 is a zero of a polynomial $P(x)$, what is one binomial factor of $P(x)$? What is one solution of the equation $P(x) = 0$?
3. The zeros of $x^4 - 16$ are the solutions of what equation?
4. Rewrite $(x + 5)(3x + 2)(2x - 5)$ to show three first-degree factors and a constant factor 6.

Find the complex zeros and give the multiplicity of each zero.

5. $5x - 3$ 6. $(x + 3)(2x + 6)$ 7. $x^2 - 8$
8. $x^2 - 10x + 25$ 9. $x^2 + 8$ 10. $(x - 6)^3(x^2 - 4)$

Written Exercises

Find the zeros of each polynomial. Factor the polynomial into first-degree factors.

(A)   1. $3x^3 - 10x^2 - 9x + 4$ 2. $6x^3 + 11x^2 - 4x - 4$
     3. $x^3 - 2x^2 - 7x + 14$ 4. $x^3 - 3x^2 + 4x - 12$
     5. $2x^3 + 3x^2 - 32x + 15$ 6. $x^3 + x^2 + 8x + 8$

Solve each equation.

     7. $2x^3 - 3x^2 - 17x + 30 = 0$ 8. $x^3 + 3x^2 - 5x - 15 = 0$
     9. $x^3 + 5x^2 - 16x - 80 = 0$ 10. $x^3 - 7x^2 - 5x + 75 = 0$
(B) 11. $2x^4 + x^3 - 8x^2 - x + 6 = 0$ 12. $3x^4 - x^3 - 37x^2 + 9x + 90 = 0$
   13. $x^4 - 6x^2 - 27 = 0$ 14. $2x^4 + 12x + 8 = 3x^3 + 10x^2$

Find the zeros of each polynomial.

   15. $x^4 + 3x^3 - 30x^2 - 6x + 56$ 16. $x^4 - 3x^3 - 43x^2 + 9x + 120$
(C) 17. $x^5 - 4x^3 - x^2 + 4$ 18. $x^5 - 4x^3 - 8x^2 + 32$
   19. $x^5 - 9x^3 - x^2 + 9$ 20. $2x^4 + 6x^3 - 15x^2 + 15x - 50$

RATIONAL ZERO THEOREM

Objectives
To find an upper bound and a lower bound for the real zeros of an integral polynomial
To find the zeros of an integral polynomial using the rational zero theorem
To solve an integral polynomial equation using the rational zero theorem

Let $P(x) = (2x + 1)(3x - 2)(x - \sqrt{3})(x + \sqrt{3}) = 6x^4 - x^3 - 20x^2 + 3x + 6$.
The zeros of $P(x)$ are $-\frac{1}{2}, \frac{2}{3}, \sqrt{3}$, and $-\sqrt{3}$. Each of these zeros is in the
interval from -2 to 2.

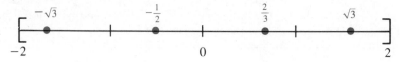

2 is called an **upper bound** and -2 is called a **lower bound** for the four zeros.

If you did not know the zeros of $P(x)$ above, you could find an upper bound and a lower bound for the real zeros as follows.

Upper Bounds
Find $P(x)$ for $x = 0, 1, 2, 3, \ldots$
by synthetic substitution.

x	6	-1	-20	3	6	$P(x)$
0	6	-1	-20	3	6	6
1	6	5	-15	-12	-6	-6
2	6	11	2	7	20	20

All nonnegative

3	6	17	31	96	294	294

All nonnegative

These numbers will continue to be all nonnegative, and $P(x)$ will continue to increase. So, $P(x)$ cannot be zero for $x \geq 2$.

Thus, 2 and 3 are upper bounds for the real zeros of $P(x)$.

Lower Bounds
Find $P(x)$ for $x = 0, -1, -2, -3, \ldots$
by synthetic substitution.

x	6	-1	-20	3	6	$P(x)$
0	6	-1	-20	3	6	6
-1	6	-7	-13	16	-10	-10
-2	6	-13	6	-9	24	24

Signs alternate

-3	6	-19	37	-108	330	330

Signs alternate

These signs will continue to alternate, and $P(x)$ will continue to be positive. So, $P(x)$ cannot be zero for $x \leq -2$.

Thus, -2 and -3 are lower bounds for the real zeros of $P(x)$.

Upper and lower bound theorem

Let $P(x)$ be an integral polynomial, U and L real numbers, $U \geq 0$ and $L \leq 0$.
(1) Find $P(U)$ by synthetic substitution. If the last row of numbers is *all nonnegative* or *all nonpositive*, then U is an **upper bound** for the real zeros of $P(x)$. (2) Find $P(L)$ by synthetic substitution. If the last row of numbers *alternates in sign*, then L is a **lower bound** for the real zeros of $P(x)$.

Example 1   **Find an upper bound U and a lower bound L for the real zeros of $P(x) = 2x^3 - 5x^2 - 4x + 10$.**

1. For U, try $x = 0, 1, 2, \ldots$

x	2	-5	-4	10
0	2	-5	-4	10
1	2	-3	-7	3
2	2	-1	-6	-2
3	2	1	-1	7
4	2	3	8	42

All nonnegative ▶ $U = 4$

2. For L, try $x = 0, -1, -2, \ldots$

x	2	-5	-4	10
0	2	-5	-4	10
-1	2	-7	3	7
-2	2	-9	14	-18

Signs alternate ▶ $L = -2$

Thus, 4 is an upper bound and -2 is a lower bound.

An integral polynomial might not have enough integral zeros to yield its first-degree factors by using the integral zero theorem. In such a case, you can use the **rational zero theorem** to find some of the rational zeros of the polynomial.

Rational zero theorem

If $\frac{c}{d}$ (c and d relatively prime integers) is a zero of an integral polynomial $P(x)$, then c is a factor of the constant term in $P(x)$ and d is a factor of the coefficient of the highest-degree term. That is, if $\frac{c}{d}$ is a zero of
$$a_0x^n + a_1x^{n-1} + a_2x^{n-2} + \ldots + a_{n-1}x^1 + a_n,$$ then c is a factor of a_n and d is a factor of a_0.

If $P(x) = 6x^4 - x^3 - 14x^2 + 2x + 4$, you can list its "potential" rational zeros $\frac{c}{d}$ that are not integers, as shown below.

c is a factor of 4: 1, 2, 4 $\left.\begin{array}{l}\\ \\ \end{array}\right\}$ $\frac{c}{d}$ can be $\pm\frac{1}{2}, \pm\frac{1}{3}, \pm\frac{1}{6}, \pm\frac{2}{3}, \pm\frac{4}{3}$.
d is a factor of 6: 1, 2, 3, 6

Example 2   **Find the zeros of $P(x) = 6x^4 - x^3 - 14x^2 + 2x + 4$.**

1. Test the potential *integral* zeros: $\pm 1, \pm 2, \pm 4$.

	6	-1	-14	2	4
1	6	5	-9	-7	-3
2	6	11	8	18	40

All nonnegative

So, 2 is an upper bound.
Do not test 4.

	6	-1	-14	2	4
-1	6	-7	-7	9	-5
-2	6	-13	12	-22	48

Signs alternate

So, -2 is a lower bound.
Do not test -4.

So, $P(x)$ has no integral zeros.

2. Test the potential *rational* zeros: $\pm\frac{1}{2}, \pm\frac{1}{3}, \pm\frac{1}{6}, \pm\frac{2}{3}, \pm\frac{4}{3}$.

$$
\begin{array}{rrrrr}
6 & -1 & -14 & 2 & 4 \\
\end{array} \blacktriangleleft P(x)
$$

$$
-\frac{1}{2} \Big| \begin{array}{rrrr} 6 & -4 & -12 & 8 \end{array} \boxed{0} \blacktriangleleft P(x) = (x + \tfrac{1}{2})(6x^3 - 4x^2 - 12x + 8)
$$

$$
2(3x^3 - 2x^2 - 6x + 4) \qquad \blacktriangleleft P(x) = (x + \tfrac{1}{2}) \cdot 2(3x^3 - 2x^2 - 6x + 4)
$$

$$
\begin{array}{rrrr}
3 & -2 & -6 & 4 \\
\end{array}
$$

$$
\frac{2}{3} \Big| \begin{array}{rrr} 3 & 0 & -6 \end{array} \boxed{0} \qquad \blacktriangleleft P(x) = (x + \tfrac{1}{2}) \cdot 2(x - \tfrac{2}{3})(3x^2 - 6)
$$

$$
3(x^2 + 0x - 2) \qquad \blacktriangleleft P(x) = (x + \tfrac{1}{2}) \cdot 2(x - \tfrac{2}{3}) \cdot 3(x^2 - 2)
$$

3. Solve $x^2 - 2 = 0$. $\qquad \blacktriangleleft P(x) = 6(x + \tfrac{1}{2})(x - \tfrac{2}{3})(x - \sqrt{2})(x + \sqrt{2})$

$\qquad x^2 = 2 \qquad x = \pm\sqrt{2}$

Thus, the zeros of $P(x)$ are $-\frac{1}{2}, \frac{2}{3}, \sqrt{2}$, and $-\sqrt{2}$.

You can use the rational zero theorem to solve an integral polynomial equation, $P(x) = 0$, as shown in Example 3.

Example 3

Solve $3x^4 - 5x^3 + 10x^2 - 20x - 8 = 0$.

Let $P(x) = 3x^4 - 5x^3 + 10x^2 - 20x - 8$. Find the zeros of $P(x)$.

1. Test the potential integral zeros: $\pm 1, \pm 2, \pm 4, \pm 8$.

$$
\begin{array}{rrrrr}
3 & -5 & 10 & -20 & -8 \\
\end{array}
$$

$$
1 \Big| \begin{array}{rrrrr} 3 & -2 & 8 & -12 & -20 \end{array}
$$

$$
2 \Big| \begin{array}{rrrrr} 3 & 1 & 12 & 4 & \boxed{0} \end{array} \blacktriangleleft \text{2 is a zero and an upper bound.}
$$

$\quad$ (All nonnegative) $\qquad\qquad P(x) = (x - 2)(3x^3 + x^2 + 12x + 4)$

$$
\begin{array}{rrrr}
3 & 1 & 12 & 4 \\
\end{array} \quad \blacktriangleleft \text{−1 is a lower bound, since the}
$$

$$
-1 \Big| \begin{array}{rrrr} 3 & -2 & 14 & -10 \end{array} \qquad \text{signs alternate.}
$$

2. List the potential rational zeros for $Q(x) = 3x^3 + x^2 + 12x + 4$.

c is a factor of 4: 1, 2, 4 $\qquad \blacktriangleleft \dfrac{c}{d}$ can be $\pm\frac{1}{3}, \pm\frac{2}{3}, \frac{4}{3} \left(not -\frac{4}{3} \right)$.
d is a factor of 3: 1, 3

3. Test the potential rational zeros for $Q(x) = 3x^3 + x^2 + 12x + 4$.

$$
\begin{array}{rrrr}
3 & 1 & 12 & 4 \\
\end{array} \qquad P(x) = (x - 2)(3x^3 + x^2 + 12x + 4)
$$

$$
-\frac{1}{3} \Big| \begin{array}{rrr} 3 & 0 & 12 \end{array} \boxed{0} \blacktriangleleft P(x) = (x - 2)(x + \tfrac{1}{3})(3x^2 + 12)
$$

$$
3(x^2 + 0x + 4) \qquad \blacktriangleleft P(x) = (x - 2)(x + \tfrac{1}{3}) \cdot 3(x^2 + 4)
$$

4. Solve $x^2 + 4 = 0$. $\quad \blacktriangleleft P(x) = 3(x - 2)(x + \tfrac{1}{3})(x - 2i)(x + 2i)$

$\quad x^2 = -4 \qquad x = \pm 2i$

Thus, the solutions are $2, -\frac{1}{3}, 2i$, and $-2i$.

Written Exercises

Find an upper bound and a lower bound for the real zeros of each polynomial.

(A) 1. $2x^3 - 5x^2 - 8x + 18$

2. $2x^3 - 7x^2 - 12x + 42$

3. $6x^4 + x^3 - 8x^2 - x + 2$

4. $-x^3 - 2x^2 + 10x + 20$

Find the zeros of each polynomial.

5. $3x^3 - 2x^2 - 9x + 6$

6. $2x^3 + x^2 + 8x + 4$

7. $3x^4 - 5x^3 - 13x^2 + 25x - 10$

8. $4x^4 + x^3 - 27x^2 - 6x + 18$

Solve each equation.

9. $2x^4 - 7x^3 + 8x^2 - 7x + 6 = 0$

10. $4x^4 + x^3 + 5x^2 + 2x - 6 = 0$

11. $6x^4 - 5x^3 - 11x^2 + 10x - 2 = 0$

12. $6x^4 + x^3 - 31x^2 - 5x + 5 = 0$

(B) 13. $12x^4 - 20x^3 - 11x^2 + 5x + 2 = 0$

14. $8x^4 + 20x^3 - 14x^2 - 5x + 3 = 0$

15. $8x^4 + 2x^3 + 3 = 25x^2 + 6x$

16. $6x^4 - 29x^2 = 5x^3 - 25x + 5$

Find an upper bound and a lower bound for the real zeros of each polynomial.

17. $x^4 - 7x^2 + 10$

18. $-x^4 + 2x^3 + 4x^2 - 10x + 5$

19. $x^5 - 3x^4 - 9x^3 + 2$

20. $x^6 - 13x^4 - 28x^2 - 20$

Find the zeros of each polynomial.

21. $4x^4 - 4x^3 + 17x^2 - 16x + 4$

22. $12x^4 - 5x^3 + 106x^2 - 45x - 18$

23. $36x^4 - 12x^3 - 11x^2 + 2x + 1$

24. $36x^4 - 60x^3 - 5x^2 + 25x - 6$

(C) 25. $6x^5 + x^4 + 20x^3 + 5x^2 - 16x + 4$

26. $6x^5 - 7x^4 - 19x^3 + 23x^2 + 3x - 6$

CUMULATIVE REVIEW

Solve each problem.

1. In a bag of coins, there are 12 more quarters than dimes and 4 times as many nickels as quarters. The nickels and dimes together have the same value as the quarters. Find the number of each kind of coin.

2. Ben is twice as old as Al, and Carol is two years older than Ben. Five years ago, the sum of the two younger ages was the same as the age of the oldest. Find the present age of all three people.

3. Some 50¢ red pens and some 30¢ blue pens are mixed to make a package of 20 pens. If the package is worth $8.40, how many red pens and how many blue pens are in the package?

4. Machine A can do a job in 15 h. If machines A and B work together, the job can be done in 10 h. How many hours would it take machine B to do the job if it works alone?

5. Find four consecutive multiples of 4 such that the product of the second and the third numbers is 192.

6. How much water must be added to 4 L of a 30% iodine solution to dilute it to a 10% iodine solution?

Objective **To sketch the graph of an integral polynomial function of degree *n* with *n* distinct real zeros**

To "sketch" the graph for $y = x^3 - x^2 - 12x + 12$, you can use synthetic substitution to find ordered pairs (x, y), then plot the points and draw a "smooth" curve.

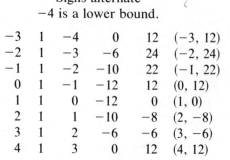

x	1	−1	−12	12	(x, y)
−4	1	−5	8	−20	(−4, −20)

Signs alternate
−4 is a lower bound.

x	1				(x, y)
−3	1	−4	0	12	(−3, 12)
−2	1	−3	−6	24	(−2, 24)
−1	1	−2	−10	22	(−1, 22)
0	1	−1	−12	12	(0, 12)
1	1	0	−12	0	(1, 0)
2	1	1	−10	−8	(2, −8)
3	1	2	−6	−6	(3, −6)
4	1	3	0	12	(4, 12)

All nonnegative
4 is an upper bound.

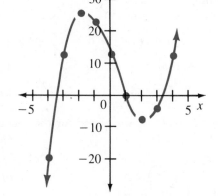

Notice that:
1. The graph of the third-degree function has a characteristic shape.
2. The "turning" points in quadrants II and IV cannot be determined at this time.
3. The polynomial has three real zeros. One is 1, one is between −4 and −3, and one is between 3 and 4.

Example 1   **Sketch the graph for $y = 8x^4 - 6x^3 - 77x^2 + 54x + 45$.**

x	8	−6	−77	54	45	(x, y)
−4	8	−38	75	−246	1029	(−4, 1029)

−4 is a lower bound.

x	8					(x, y)
−3	8	−30	13	15	0	(−3, 0)
−2	8	−22	−33	120	−195	(−2, −195)
−1	8	−14	−63	117	−72	(−1, −72)
0	8	−6	−77	54	45	(0, 45)
1	8	2	−75	−21	24	(1, 24)
2	8	10	−57	−60	−75	(2, −75)
3	8	18	−23	−15	0	(3, 0)
4	8	26	27	162	693	(4, 693)

4 is an upper bound.

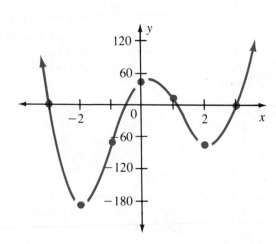

In Example 1, notice that (1) the graph of the fourth-degree function with four real zeros has a characteristic 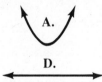 shape and (2) the turning points cannot be determined at this time.

The graph of a polynomial function of degree 2 is a *parabola*. Its turning point can be determined as follows.

If $y = ax^2 + bx + c$, $a \neq 0$, the minimum or maximum value of the function will occur when $x = \dfrac{-b}{2a}$.

Example 2

Sketch the graph for $y = 4x^2 - 12x - 27$.

$a = 4$, $b = -12$, $c = -27$, and $x = \dfrac{-b}{2a} = \dfrac{12}{8} = \dfrac{3}{2}$

If $x = \dfrac{3}{2}$, then $y = 4 \cdot \dfrac{9}{4} - 12 \cdot \dfrac{3}{2} - 27 = -36$, so the turning point is $(1.5, -36)$.

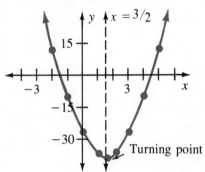

x	4	-12	-27	(x, y)
-2	4	-20	13	$(-2, 13)$
-1	4	-16	-11	$(-1, -11)$
0	4	-12	-27	$(0, -27)$
1	4	-8	-35	$(1, -35)$
2	4	-4	-35	$(2, -35)$
3	4	0	-27	$(3, -27)$
4	4	4	-11	$(4, -11)$
5	4	8	13	$(5, 13)$

Reading in Algebra

Match the degree of a polynomial function at the left with a characteristic graph at the right.

1. degree 4
2. degree 3
3. degree 2
4. degree 1
5. degree 0

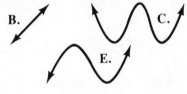

Written Exercises

Sketch the graph of each polynomial function. For each parabola, give the coordinates of the turning point.

1. $y = 3$
2. $y = 2x - 5$
3. $y = x^2 + 4x - 5$
4. $y = x^2 - x - 6$
5. $y = x^3 - 7x^2 + 4x + 12$
6. $y = x^4 - x^3 - 14x^2 + 2x + 24$
7. $y = 4x^2 - 16x + 15$
8. $y = x^3 - x^2 - 10x + 10$
9. $y = 2x^4 - x^3 - 35x^2 + 16x + 48$
10. $y = -x^2 + 3x + 10$
11. $y = -x^3 + x^2 + 9x - 9$
12. $y = -x^4 + 10x^2 - 9$

Graphing Polynomial Functions

APPLICATIONS: DESCARTES' RULE OF SIGNS

There are three *changes in sign* in $P(x)$, where

$$P(x) = +2x^5 \; -9x^4 \; -5x^3 \; +51x^2 \; -9x \; -54$$

There are two sign changes in $P(-x)$ as shown below.

$$P(-x) = 2(-x)^5 - 9(-x)^4 - 5(-x)^3 + 51(-x)^2 - 9(-x) - 54$$
$$= -2x^5 \quad - 9x^4 \quad + 5x^3 + 51x^2 \quad + 9x \quad - 54$$

Recall that the *sum* of the multiplicities of the zeros is 5 for a polynomial of degree 5. The *positive* real zeros of $P(x)$ are 3, with a multiplicity of *two*, and 1.5, with a multiplicity of *one*. The *sum* of these multiplicities is *three*, the same as the number of sign changes in $P(x)$. The *negative* real zeros of $P(x)$ are -1 and -2, *each* with a multiplicity of *one*. The *sum* of these multiplicities is *two*, which is the number of sign changes in $P(-x)$.

Consider $Q(x) = x^3 - 6x^2 + 10x - 8$ and $Q(-x) = (-x)^3 - 6(-x)^2 + 10(-x) - 8$
$$= -x^3 - 6x^2 - 10x - 8.$$

The zeros of $Q(x)$ are 4, $1 - i$, and $1 + i$, each of multiplicity one. Notice:
1. $Q(x)$ has *three* sign changes and *one* positive real zero.
2. $Q(-x)$ has *no* sign changes and $Q(x)$ has *no* negative real zeros.
3. $Q(x)$ has *two imaginary* zeros and they are a pair of *conjugates*.

Descartes' rule of signs	The sum of the multiplicities of the positive real zeros of an integral polynomial $P(x)$ is either equal to the number of sign changes in $P(x)$ or less than that number by a multiple of two.
	The sum of the multiplicities of the negative real zeros of $P(x)$ is either equal to the number of sign changes in $P(-x)$ or less than that number by a multiple of two.

Example 1

Find the possible combinations of the sums of the multiplicities of the positive real zeros, negative real zeros, and imaginary zeros of $P(x)$, where
$$P(x) = x^5 - 4x^3 - x^2 + 4.$$

$$P(x) = +x^5 \quad -4x^3 \quad -x^2 \quad + 4 \qquad \blacktriangleleft \; 2 \text{ sign changes in } P(x)$$
$$P(-x) = (-x)^5 \quad -4(-x)^3 \quad -(-x)^2 \quad + 4$$
$$= -x^5 \quad +4x^3 \quad -x^2 \quad + 4 \qquad \blacktriangleleft \; 3 \text{ sign changes in } P(-x)$$

The sum of the multiplicities of the zeros must be 5, the degree of $P(x)$. Imaginary zeros, if any, must appear in pairs.

Thus, there are four possible combinations (I–IV).

	I	II	III	IV
Positive real	2	2	0	0
Negative real	3	1	3	1
Imaginary	0	2	2	4
All zeros	5	5	5	5

Sums of the multiplicities of each type of zero. ▶

According to the **conjugate zero theorem,** stated below, $2 + 3i$ is a zero of $P(x) = x^4 - 5x^3 + 15x^2 - 5x - 26$ if $2 - 3i$ is a zero of $P(x)$. With this information, you can find all the zeros of $P(x)$.

Conjugate zero theorem	If $a + bi$ is a zero of an integral polynomial $P(x)$, then $a - bi$ is also a zero of $P(x)$.

Example 2

Given $P(x) = x^4 - 5x^3 + 15x^2 - 5x - 26$ and $2 - 3i$ is a zero of $P(x)$, find all the zeros of $P(x)$.

$$
\begin{array}{r|rrrrr}
 & 1 & -5 & 15 & -5 & -26 \\
\hline
2 - 3i & 1 & -3 - 3i & 3i & 4 + 6i & \boxed{0} \\
\hline
2 + 3i & 1 & -1 & -2 & \boxed{0} & \\
\end{array}
$$

$$x^2 - x - 2 = (x + 1)(x - 2) = 0$$
$$x = -1 \text{ or } x = 2$$

Thus, the zeros of $P(x)$ are $2 - 3i$, $2 + 3i$, -1, and 2.

Written Exercises

Find the possible combinations of the sums of the multiplicities of the positive real, negative real, and imaginary zeros of each polynomial.

1. $x^3 - x^2 + 6$
2. $2x^3 + 3x^2 + 4x + 6$
3. $x^4 - 2x^3 + 2x^2 - 10$
4. $x^4 + 16$
5. $x^4 - x^3 + x^2 + x + 5$
6. $x^5 - x^3 + x^2 - x - 6$
7. $x^5 + 32$
8. $x^7 - x^6 + x^4 + x^2 + 10$
9. $2x^6 + x^5 - 3x^4 + x^3 + 8$

Use the conjugate zero theorem to find all the zeros of $P(x)$ if x_1 is a zero of $P(x)$.
10. $P(x) = x^3 - 5x^2 + 9x - 45, \ x_1 = 3i$
11. $P(x) = 2x^3 - x^2 + 50x - 25, \ x_1 = -5i$
12. $P(x) = x^4 - 6x^3 + 9x^2 + 24x - 52, \ x_1 = 3 - 2i$
13. $P(x) = x^4 - 6x^3 + 29x^2 - 24x + 100, \ x_1 = 3 + 4i$

CUMULATIVE REVIEW

1. Write $\sqrt[3]{x}$ in exponential form.
2. Write $a^{\frac{1}{4}}$ in radical form.
3. Write $\sqrt[4]{y^3}$ in exponential form.
4. Write $c^{\frac{2}{3}}$ in radical form.

Applications: Descartes' Rule of Signs

CHAPTER FIFTEEN REVIEW

Vocabulary
antilogarithm [15.6]
common logarithms [15.6]
compound interest [15.7]
exponential equation [15.7]
exponential function [15.1]
integral polynomial [15.8]
logarithmic equation [15.2]
logarithmic function [15.2]
lower bound [15.9]
multiplicity of a zero [15.8]
upper bound [15.9]
zero of a polynomial [15.8]

Draw the graph of each function and list four of its properties. Choose convenient scales for the axes.

1. $y = \left(\frac{1}{6}\right)^x$ [15.1]   **2.** $y = \log_6 x$ [15.3]

Find each logarithm. [15.2]

3. $\log_5 625$ **4.** $\log_3 \frac{1}{9}$ **5.** $\log_3 \sqrt[4]{27}$

Write the y-form of the equation for the inverse of each function. [15.3]

6. $y = 6^x$ **7.** $y = \log_{\frac{1}{4}} x$

Write each expression in expanded form.

8. $\log_2 \frac{7c}{bd}$ [15.4]   **9.** $\log_b \sqrt[3]{\frac{3v^2}{k}}$ [15.5]

Write each expression as the log of one expression. [15.4]
10. $\log_b 5 + \log_b a - \log_b 3 - \log_b t$

11. $\frac{1}{3} \log_4 a + 2 \log_4 c$ [15.5]

Solve each equation. Check.
12. $\log_5 (4y + 3) - \log_5 9 = \log_5 (y - 3)$ [15.4]
13. $2 \log_4 x = \log_4 3 + \log_4 (x + 6)$ [15.5]
14. $\log_b 125 = 3$ **15.** $\log_2 y = 5$ [15.2]

Find the base-10 log or antilog. (Ex. 16–17)
[15.6]
16. log 3,040   **17.** antilog $(8.2227 - 10)$

18. Solve $5^{2x-3} = 17$ for x to three significant digits. [15.7]

19. Find $\log_3 542$ to three significant digits. [15.7]

20. Find the value of a $6,000 bond paying 8%, compounded quarterly, for 12 yr. Use the formula $A = p\left(1 + \frac{r}{n}\right)^{nt}$. [15.7]

For each polynomial $P(x)$, (1) find an upper bound and a lower bound for the real zeros of $P(x)$, (2) factor $P(x)$ into first-degree factors, and (3) solve the equation $P(x) = 0$. (Ex. 21–23)
21. $x^4 + x^3 - 5x^2 - 3x + 6$ [15.8]
22. $4x^4 - 9x^3 + 18x^2 - 36x + 8$ [15.9]
23. $12x^3 + 8x^2 - x - 1$

24. Choose convenient scales for the axes and sketch the graph of $y = 2x^3 - x^2 - 10x + 5$. [15.10]

★**25.** Find $\log_t (t^3)^a$. [15.2]

★**26.** Find the sum of the series:
$3 \cdot \sum_{k=2}^{6} \log_{10} (0.1)^{k-5}$ [15.2]

★**27.** Solve $\log_2 (y - 3) + \log_2 (2y + 2) = 6$. [15.5]

Prove that each statement is true.
★**28.** $\frac{2}{5} \log_b 4 - \frac{1}{5} \log_b 2 = \frac{1}{5} \log_b 8$ [15.5]

★**29.** $\log_b 69 = \frac{\log_{10} 69}{\log_{10} b}$ for each positive number $b \neq 1$. [15.7]

★**30.** The functions defined by $y = \log_4 x$ and by $y = 4^x$ are inverse functions. [15.3]

CHAPTER FIFTEEN TEST

Draw the graph of each function and list four of its properties. Choose convenient scales for the axes.

1. $y = 2^x$

2. $y = \log_{\frac{1}{2}} x$

Solve each equation. Check.

3. $\log_5 (y + 1) - \log_5 8 = \log_5 (y - 3) - \log_5 6$

4. $2 \log_5 x = \log_5 4 + \log_5 (x + 3)$

5. $\log_b 64 = 3$

Find each logarithm. (Ex. 6–8)

6. $\log_5 125$

7. $\log_4 \dfrac{1}{16}$

8. $\log_2 \sqrt[3]{4}$

9. Write the y-form of the equation for the inverse of $y = \log_3 x$.

10. Write $\log_2 \dfrac{a^3 \sqrt{y}}{5x}$ in expanded form.

11. Write $\dfrac{1}{2}(\log_b 7 + \log_b c) - 2 \log_b d$ as the $\log_b$ of one expression.

Find the base-10 log or antilog.

12. log 0.0374

13. antilog 1.7818

14. Solve $3^{x-1} = 42$ for x to three significant digits.

15. Find $\log_6 724$ to three significant digits.

16. Find the value of a $10,000 bond paying 8%, compounded semiannually, for 10 yr. Use the formula $A = p\left(1 + \dfrac{r}{n}\right)^{nt}$.

Let $P(x) = 2x^4 - 7x^3 + 5x^2 - 7x + 3$ for Exercises 17–19.

17. Find an upper bound and a lower bound for the real zeros of $P(x)$.

18. Factor $P(x)$ into first-degree factors.

19. Solve the equation where $P(x) = 0$.

20. Choose convenient scales for the axes and sketch the graph of $y = 2x^3 - x^2 - 10x + 5$.

★ **21.** Find $\log_{2a} (4a^2)^3$.

★ **22.** Solve $\log_2 (y + 1) + \log_2 (3y - 1) = 5$.

★ **23.** Find the sum of the series $\displaystyle\sum_{j=1}^{4} \text{antilog}_{10} (3 - j)$.

Prove that each statement is true.

★ **24.** $2 \log_b xy - \log_b xy^4 = \log_b x - 2 \log_b y$

★ **25.** $\dfrac{3}{4} \log_b 9 - \dfrac{1}{2} \log_b 3 = \dfrac{1}{3} \log_b 27$

★ **26.** $\log_8 59 = \dfrac{\log_{10} 59}{\log_{10} 8}$

★ **27.** The functions defined by $y = \log_6 x$ and by $y = 6^x$ are a pair of inverse functions.

COLLEGE PREP TEST

DIRECTIONS: Choose the one best answer to each question or problem.

1. If $x^2 = \log_2 64$ and $x > 0$, then $x =$

(A) 2   (B) $\sqrt{6}$ (C) 6   (D) 8   (E) 16

2. $\log_b \dfrac{1}{xy} =$

(A) $\log_b x + \log_b y$ (B) $\log_b x - \log_b y$
(C) $1 - \log_b x - \log_b y$ (D) $-\log_b x - \log_b y$
(E) None of these

3. If $\log_5 3x + \log_5 4x = \log_5 24$, then

(A) $x = \dfrac{24}{7}$ (B) $x = 2$

(C) $x^2 = 2$ (D) $x^2 = 12$

(E) None of these

4. If $P(x) = 8x^3 - 4x^2 + 6x + 1$, then $P(x)$ has

(A) no integral zeros.
(B) 1 integral zero.
(C) 2 integral zeros.
(D) 3 integral zeros.
(E) None of these

5. $4\left(\log_3 a + 2 \log_3 b - \dfrac{1}{2} \log_3 c\right) =$

(A) $\log_3 \left(\dfrac{ab}{c}\right)^4$ (B) $\log_3 \dfrac{4ab^2}{\sqrt{c}}$

(C) $\log_3 \dfrac{ab^8}{\sqrt{c}}$ (D) $\log_3 \dfrac{a^4 b^8}{c^2}$

(E) None of these

6. If $2^{x+1} = 50$, then

(A) $x = 24$ (B) $4 < x < 5$
(C) $5 < x < 6$ (D) $6 < x < 7$
(E) None of these

7. If $a = \log_b c$, then

(A) $b^a = c$ (B) $b^c = a$
(C) $a^b = c$ (D) $a^c = b$
(E) None of these

8. Find the number of distinct (different) real zeros of $P(x)$ if $P(x) = (x^2 + 1)(x - 2)^2$.

(A) 4   (B) 3
(C) 2   (D) 1
(E) None of these

9. Find the number of *potential* rational nonintegral zeros of $x^4 + x^3 - x^2 - 1$.

(A) 2   (B) 4
(C) 6   (D) 8
(E) None of these

10. If $\log_2 y = x$, then

(A) $y < x$ (B) $y > x$
(C) $y = x$ (D) $y = 2x$
(E) None of these

11. Two is a zero of the polynomial $x^5 - 4x^4 + x^3 + 10x^2 - 4x - 8$. Find the multiplicity of the zero 2.

(A) 1   (B) 2
(C) 3   (D) 4
(E) None of these

16 PERMUTATIONS, COMBINATIONS, AND PROBABILITY

**Non-Routine
Problem
Solving**

Problem 1. In how many different ways can a sports fan enter the stadium by one entrance and leave by another if the stadium has ten entrances numbered 1–10?

Problem 2. In how many ways can 12 people be seated, 3 at a time, on a bench with room for only 3 people?

FUNDAMENTAL COUNTING PRINCIPLE 16.1

Objective

To find the number of possible arrangements of objects by using the Fundamental Counting Principle

A truck driver must drive from Miami to Orlando and then continue on to Lake City. There are 4 different routes that he can take from Miami to Orlando and 3 different routes from Orlando to Lake City.

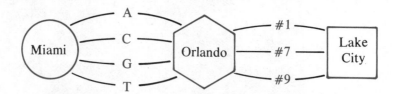

One Miami-Orlando-Lake City route is *first* to take A and *then* take #1, or simply $A1$. All of the possible routes are listed below.

$A1$ $A7$ $A9$ $C1$ $C7$ $C9$ $G1$ $G7$ $G9$ $T1$ $T7$ $T9$

Notice that each route is identified by a letter and then a digit, in that order. By counting, you can find that there are 12 possible Miami-Orlando-Lake City routes. The number of possible routes can also be found without listing any of the routes.

The driver had to make 2 <u>decisions</u>. Mark 2 places.   <u>   </u> <u>   </u>
First decision: Choose a Miami-Orlando route.
                 There are 4 <u>choices</u>: A, C, G, or T.   <u>4</u> <u>   </u>
Second decision: Choose an Orlando-Lake City route.
                 There are 3 <u>choices</u>: 1, 7, or 9.   <u>   </u> <u>3</u>

$$\underset{\begin{pmatrix}\text{number of}\\\text{letter-choices}\end{pmatrix}}{\underline{4}} \times \underset{\begin{pmatrix}\text{number of}\\\text{digit-choices}\end{pmatrix}}{\underline{3}} = 12$$

$$\begin{pmatrix}\text{number of}\\\text{letter-choices}\end{pmatrix} \times \begin{pmatrix}\text{number of}\\\text{digit-choices}\end{pmatrix} = \text{number of possible routes}$$

This suggests a principle for finding the number of possible arrangements of objects without listing all of the arrangements.

Fundamental counting principle	If one choice can be made in a ways and a second choice can be made in b ways, then the choices *in order* can be made in $a \times b$ different ways.

This Fundamental Counting Principle can be extended to three or more choices made *in order*. If the truck driver above had to make a round trip, Miami-Orlando-Lake City-Orlando-Miami, there would be 144 possible routes. One such route is $C79T$. The driver would have to make 4 decisions. The number of choices for each decision is shown below:

$$\underline{4} \times \underline{3} \times \underline{3} \times \underline{4} = 144$$

The Fundamental Counting Principle is used to find the number of possible arrangements of n objects taken r at a time. In Example 1, 6 digits are available and they are taken 3 at a time to form three-digit numbers.

Example 1   **How many three-digit numbers can be formed from the 6 digits, 1, 2, 6, 7, 8, 9, if no digit may be repeated in a number?**

There are three decisions to make. Choose (1) a 100s digit, then (2) a 10s digit, and then (3) a 1s digit.

$$\underline{\hspace{2cm}} \quad \underline{\hspace{2cm}} \quad \underline{\hspace{2cm}}$$
$$(100s) \quad (10s) \quad (1s)$$

1 2 6 ⑦ 8 9 ▶ There are 6 choices for a 100s digit. $\underline{\;6\;}$ $\underline{\hspace{1cm}}$ $\underline{\hspace{1cm}}$

1 ② 6   8 9 ▶ This leaves 5 choices for a 10s digit. $\underline{\hspace{1cm}}$ $\underline{\;5\;}$ $\underline{\hspace{1cm}}$

1   6   ⑧ 9 ▶ Four choices remain for a 1s digit. $\underline{\hspace{1cm}}$ $\underline{\hspace{1cm}}$ $\underline{\;4\;}$

Use the Fundamental Counting Principle. $\underline{\;6\;} \times \underline{\;5\;} \times \underline{\;4\;}$

Thus, 120 three-digit numbers can be formed.

Six of the 120 three-digit numbers from Example 1 are listed below.

267   276   627   672   726   762

Each of these numbers is a **permutation** of the digits 2, 6, and 7.

Definition:
Permutation

A *permutation* is an arrangement of objects in a definite order.

In Example 1, it was specified that a digit should not be repeated in a number. If such repetition is allowed, more than 120 three-digit numbers are possible. Four of these numbers, with a repeated digit, are shown below.

222   227   272   722

Example 2   **How many three-digit numbers can be formed from the digits 2, 4, 6, 8, 9, if a digit may be repeated in a number?**

$$\underline{\;5\;} \times \underline{\;5\;} \times \underline{\;5\;} = 125$$

Thus, 125 three-digit numbers can be formed.

The Fundamental Counting Principle can be used to find the number of permutations of objects other than digits and letters.

Example 3   **A manufacturer makes sweaters in 6 different colors. Each sweater is available with choices of 3 fabrics, 4 kinds of collars, and with or without buttons. How many different types of sweaters does the manufacturer make?**

There are 4 decisions to be made.

$$\underset{\text{color}}{6} \times \underset{\text{fabric}}{3} \times \underset{\text{collar}}{4} \times \underset{\text{buttons}}{2} = 144$$

Thus, 144 types of sweaters are made by the manufacturer.

The number of permutations of n distinct objects taken n at a time forms a pattern. This is shown in the next example.

Example 4   **Find the number of possible batting orders (permutations) for the nine starting players on a baseball team.**

The coach has to make 9 decisions.

$$\underset{\substack{\text{Leadoff} \\ \text{batter}}}{9} \times \underset{\substack{\text{Second} \\ \text{batter}}}{8} \times \underset{\substack{\text{Third} \\ \text{batter}}}{7} \times \underset{\substack{\text{Cleanup} \\ \text{batter}}}{6} \times \underset{\substack{\text{No.} \\ 5}}{5} \times \underset{\substack{\text{No.} \\ 6}}{4} \times \underset{\substack{\text{No.} \\ 7}}{3} \times \underset{\substack{\text{No.} \\ 8}}{2} \times \underset{\substack{\text{No.} \\ 9}}{1} = 362{,}880$$

Thus, a baseball coach can select his batting order from among 362,880 possible orders.

In Example 4, the number of permutations, $9 \times 8 \times 7 \times \ldots \times 3 \times 2 \times 1$, can be written as 9! Recall that the expression 9! is read "9 factorial." The value of $n!$ increases rapidly as n increases.

$4! = 4 \times 3 \times 2 \times 1 = 24$
$8! = 8 \times 7 \times 6 \times 5 \times 4 \times 3 \times 2 \times 1 = 40{,}320$
$10! = 10 \times 9 \times 8 \times \ldots \times 3 \times 2 \times 1 = 3{,}628{,}800$

You should notice that the number of arrangements (permutations) of n objects taken n at a time is always $n!$ for each positive integer n.

Oral Exercises

State each of the following expressions in factored form.

1. 6! **2.** 2! **3.** 9! **4.** 21!

Written Exercises

Ⓐ **1.** How many four-digit numbers can be formed from the digits 1, 3, 5, 7, and 9, if no digit may be repeated in a number?

2. If a digit may be repeated in a number, how many 4-digit numbers can be formed from the digits 2, 4, 6, 8, and 9?

3. In how many ways can 6 different books be placed side by side on a shelf?

4. Find the number of different ways that 8 waiters can be assigned to 8 tables if each waiter services one table.

5. Find the number of 3-letter permutations of the letters in the word CAR if no letter is repeated in an arrangement.

6. How many 4-letter "words" can be made from the letters in WEST if no letter is repeated in a "word"?

7. How many different signals can be shown by arranging 3 flags in a row if 7 different flags are available?

8. If a signal consists of 5 flags arranged in a row, how many signals can be made from 9 different flags?

9. Find the number of different automobile license-plate numbers that can be formed using a letter followed by five digits.

10. Find the number of 7-digit phone numbers that can be formed if the first digit of a telephone number cannot be zero.

A sports stadium has eleven gates with six on the north side and five on the south side. Solve Exercises 11–14 using this information.

Ⓑ **11.** In how many ways can a person enter the stadium through a north gate and later leave the stadium through a south gate?

12. Find the number of ways that you can enter the stadium through a south gate and then exit through a north gate.

13. Find the number of ways that you can enter from the north side and then exit from the north side.

14. In how many different ways can a person enter and then leave the stadium?

15. A company makes trucks in 6 different sizes. Each size comes with a choice of 5 exterior colors, a choice of 3 interior colors, with or without a CB radio, and with or without an extra horn. How many different types of trucks does the company make?

16. A school cafeteria offers each student 2 choices of meat, 4 choices of vegetable, 3 choices of drink, and 6 choices of fruit. How many different four-item lunch trays are available?

17. Find the number of 3-letter permutations of the letters in STUDY if a letter may be repeated in a permutation.

18. How many 5-letter code words can be formed from the letters in MONETARY if no letter is repeated in a code word?

19. How many different auto license-plate numbers can be formed by 3 different letters followed by 3 digits?

20. In how many different ways can 6 multiple-choice questions be answered if each question has 5 choices for the answer?

CONDITIONAL PERMUTATIONS 16.2

Objective   **To find the number of permutations of objects when conditions are attached to the arrangement**

Problems involving permutations may have specific conditions attached to the arrangement of the objects. Some examples follow:
(1) The numbers are <u>odd</u> numbers.
(2) The numbers contain <u>one</u> <u>or</u> <u>more</u> digits.
(3) The words <u>end</u> with the letter *t*.
(4) The numbers are <u>less than</u> 600.

Example 1   **How many permutations of all the letters in the word *MONEY* end with either the letter *E* or the letter *Y*?**

The first decision is to choose the 5th letter, which must be *E* or *Y*.

$$\underline{\quad 4 \quad} \times \underline{\quad 3 \quad} \times \underline{\quad 2 \quad} \times \underline{\quad 1 \quad} \times \underline{\quad 2 \quad} = 48$$

Two choices: *E* or *Y*

Thus, 48 permutations of *M, O, N, E, Y* end with *E* or *Y*.

From the digits 7, 8, 9, you can form 10 *odd* numbers containing *one or more* digits if no digit may be repeated in a number. Since the numbers are odd, there are two choices for the units digit, 7 or 9. In this case, the numbers may contain one, two, or three digits.

One-digit numbers: 7	9		
Two-digit numbers: 79	87	89	97
Three-digit numbers: 789	879	897	987

Notice that there are 2 one-digit numbers, 4 two-digit numbers, 4 three-digit numbers and that $2 + 4 + 4 = 10$. This suggests that an "or" decision, like <u>one</u> <u>or</u> <u>more</u> digits, involves <u>addition</u>.

Example 2   **How many even numbers containing one or more digits can be formed from 2, 3, 4, 5, 6 if no digit may be repeated in a number?**

There are 3 choices for a units digit: 2, 4, or 6. ▶

$$3 = 3$$
$$4 \times 3 = 12 \qquad 3 + 12 + 36 + 72 + 72 = 195$$
$$4 \times 3 \times 3 = 36$$
$$4 \times 3 \times 2 \times 3 = 72$$
$$4 \times 3 \times 2 \times 1 \times 3 = 72$$

Thus, there are 195 such even numbers.

In some situations, the total number of permutations is the *product* of two or more numbers of permutations. For example, there are 12 permutations of A, B, X, Y, Z with A, B to the left "and" X, Y, Z to the right.

ABXYZ	ABXZY	ABYXŻ	ABYZX	ABZXY	ABZYX
BAXYZ	BAXZY	BAYXZ	BAYZX	BAZXY	BAZYX

Notice that (1) A, B can be arranged in 2!, or 2 ways;
(2) X, Y, Z can be arranged in 3!, or 6 ways; and
(3) A, B, X, Y, Z can be arranged in $2! \times 3!$, or 12 ways.

An "and" decision involves multiplication.

Example 3   **Four different algebra books and three different geometry books are to be displayed on a shelf with the algebra books together and to the left of the geometry books. How many such arrangements are possible?**

Algebra books (left)   Geometry books (right)

$$\underbrace{\frac{4}{\underset{I}{ALG}} \times \frac{3}{\underset{II}{ALG}} \times \frac{2}{\underset{III}{ALG}} \times \frac{1}{\underset{IV}{ALG}}} \times \underbrace{\frac{3}{\underset{I}{GEOM}} \times \frac{2}{\underset{II}{GEOM}} \times \frac{1}{\underset{III}{GEOM}}} = 144$$

Thus, there are $4! \times 3!$, or 144, possible arrangements.

Written Exercises

Solve each problem.

(A) **1.** How many even five-digit numbers can be formed from the digits 2, 3, 4, 5, 6, if no digit may be repeated in a number?

2. How many odd three-digit numbers can be formed from the digits 1, 2, 3, 4, 5, if a digit may be repeated in a number?

3. Find the number of four-digit numbers that can be formed from the digits 0, 2, 4, 5, 6, 9, if a digit may be repeated in a number. (Consider 0495 as a three-digit number.)

4. Find the number of three-digit numbers less than 500 that can be formed from the digits 0, 2, 4, 6, 8, if no digit may be repeated in a number.

5. How many permutations of all the letters in the word JUNIOR begin with N?

6. How many permutations of all the letters in the word NUMBERS do not end with N?

7. Six different biology books and four different chemistry books are to be placed on a shelf with the chemistry books together and to the left of the biology books. How many such arrangements are possible?

8. A grocery store displays five brands of ground coffee and four brands of instant coffee on a shelf in a row. How many different displays of the nine brands are possible if all of the ground coffees must be shown to the right of the instant coffees?

9. How many three-digit or five-digit numbers can be formed from 1, 3, 5, 7, 9, if a digit may be repeated in a number?

10. Find the number of two-digit, three-digit, or four-digit numbers that may be formed from 0, 2, 4, 6, if repetition of digits is not allowed. (Consider 024 and 24 as the same number.)

(B) 11. How many numbers of one or more digits can be formed from the digits 6, 7, 8, 9, if no digit is repeated in a number?

12. How many numbers of one or more digits can be formed from the digits 1, 2, 3, if the numbers are less than 400 and repetition of digits is allowed?

13. Find the number of permutations of *a*, *e*, *i*, *o*, *u*, *y* that end with *a*, or *i*, or *y*.

14. How many permutations of the letters in *UNTIL* begin with the prefix *UN-*?

15. Five novels and six short stories are to be displayed. In how many ways can this be done if the novels are kept together and the short stories are kept together on a shelf? (Note: The novels can be at the left *or* the right.)

16. A 3-volume dictionary, a 4-volume atlas, and a 6-volume collection of plays are placed on a shelf. In how many ways can this be done if volumes of the same type are kept together?

17. How many even numbers of one or more digits can be formed from 2, 4, 6, 7, 8, 9, if repetition of digits is not allowed in a number?

18. Find the number of odd numbers less than 1,000 that can be formed from the digits 1, 2, 3, 4, 5, 6, 7, if repetition of digits is allowed.

(C) 19. In how many ways can a family of five stand together in line for tickets if the twins, Mary and Mark, are not to be separated?

20. In how many ways can 6 students be seated in a row of 6 chairs if two of them are a brother and sister who do not want to sit together?

21. A secretary typed 3 letters and then addressed 3 envelopes for the letters. The letters were placed in the envelopes at random with one letter per envelope. In how many different ways could this be done?

22. How many even numbers of one or more digits can be formed from 0, 1, 2, 3, 4, if no digit may be repeated in a number? Note: Do not count a number like 034. [Hint: First, count the numbers ending with 0.]

NON-ROUTINE PROBLEMS

In how many ways can a tennis game of mixed doubles be arranged from a group of 6 males and 4 females?

DISTINGUISHABLE PERMUTATIONS 16.3

To find the quotient of numbers given in factorial notation
To find the number of distinguishable permutations when some of the objects in an arrangement are alike

Some permutation problems lead to expressions like $\frac{8!}{4! \times 3!}$. To find the value of such an expression, begin by dividing out common factors of the numerator and denominator.

Example 1

Find the value of $\frac{8!}{4! \times 3!}$.

One Method	Short Method
$\frac{8 \times 7 \times 6 \times 5 \times (4 \times 3 \times 2 \times 1)}{(4 \times 3 \times 2 \times 1) \times 3 \times 2 \times 1}$	$\frac{8 \times 7 \times \cancel{6} \times 5 \times \cancel{(4!)}}{\cancel{(4!)} \times \cancel{3 \times 2 \times 1}}$
$\frac{8 \times 7 \times \cancel{6} \times 5}{\cancel{3 \times 2 \times 1}}$	280
280	

The letters in the word *Pop* are distinguishable since one of the two *p*'s is a capital letter. There are 3!, or 6, **distinguishable permutations** of *P, o, p*.

$$Pop \qquad Ppo \qquad oPp \qquad opP \qquad poP \qquad pPo$$

In the word *pop*, the two *p*'s are alike and can be permuted in 2! ways. The number of distinguishable permutations of *p, o, p* is $\frac{3!}{2!}$, or 3.

$$pop \qquad ppo \qquad opp$$

The number of distinguishable permutations of the 5 letters in *daddy* is $\frac{5!}{3!}$ since the three *d*'s are alike and can be permuted in 3! ways. This suggests the following rule.

Number of distinguishable permutations	Given *n* objects in which *a* of them are alike, the number of distinguishable permutations of the *n* objects is $\frac{n!}{a!}$.

You can extend this rule to find the number of distinguishable permutations of *n* objects in which more than one group of the objects are alike. For example, the word *pepper* contains three *p*'s which are alike and two *e*'s which are alike. The three *p*'s can be permuted in 3! ways, the two *e*'s can be permuted in 2! ways, and the number of distinguishable permutations of the 6 letters *p, e, p, p, e, r* is $\frac{6!}{3! \times 2!}$, or 60.

Example 2   **How many distinguishable six-digit numbers can be formed from the digits of 747457?**

The 7's can be permuted in 3! ways and the 4's can be permuted in 2! ways.

$$\frac{6!}{3! \times 2!} = \frac{6 \times 5 \times 4 \times (3!)}{(3!) \times 2 \times 1} = 6 \times 5 \times 2 = 60$$

Thus, 60 such numbers can be formed.

Example 3   **How many distinguishable signals can be formed by displaying eleven flags if 3 of the flags are red, 5 are green, 2 are yellow, and 1 is white?**

$$\frac{11!}{3! \times 5! \times 2!} = \frac{11 \times 10 \times 9 \times \overset{4}{\cancel{8}} \times 7 \times \overset{1}{\cancel{6}} \times \overset{1}{\cancel{5!}}}{(3 \times 2 \times 1) \times (2 \times 1) \times \cancel{5!}} = 27{,}720$$

Thus, 27,720 signals can be formed.

Written Exercises

Find the value of each expression.

(A) **1.** $\dfrac{7!}{3!}$ **2.** $\dfrac{10!}{7!}$ **3.** $\dfrac{12!}{8! \times 4!}$ **4.** $\dfrac{10!}{2! \times 6! \times 3!}$ **5.** $\dfrac{12!}{2! \times 3! \times 8!}$

How many distinguishable eight-digit numbers can be formed from the digits of each number?

  6. 33553533   **7.** 24227242   **8.** 19116136   **9.** 88775599

Find the number of distinguishable permutations of all the letters in each word.

  10. *root*   **11.** *tepee*   **12.** *puppet*   **13.** *scissors*

(B) **14.** *divided*   **15.** *murmur*   **16.** *nonsense*   **17.** *Tennessee*

18. How many distinguishable signals can be formed by displaying 9 flags if 3 of the flags are blue, 2 are orange, and 4 are black?

19. In how many distinguishable ways can 4 nickels, 3 dimes, 2 quarters, and 1 penny be distributed to 10 children if each child is to receive one coin?

NON-ROUTINE PROBLEMS

1. In how many distinguishable ways can $1.14 in pennies, nickels, dimes, quarters, and half-dollars be distributed among 12 children if each child is to receive one coin?

Simplify.

2. $\dfrac{n!}{(n-3)!}$ **3.** $\dfrac{(n+2)!}{n!}$ **4.** $\dfrac{(n-4)!}{(n-6)!}$ **5.** $\dfrac{(n+2)!}{(n-1)!}$

CIRCULAR PERMUTATIONS 16.4

Objective **To find the number of possible permutations of objects in a circle**

Three objects may be arranged in a line in 3!, or 6, ways. Any one of the objects may be placed in the first position.

ABC   ACB   BAC   BCA   CAB   CBA

In a **circular permutation** of objects, there is no first position. Only the positions of the objects relative to one another are considered. In the figures below, Al, Betty, and Carl are seated in a circular arrangement with each person facing the center of the circle.

In each of the first three figures, Al has Betty to his left and Carl to his right. This is *one circular permutation* of Al, Betty, and Carl.

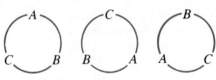

The remaining three figures each show Al with Betty to his right and Carl to his left. Again, these count as only one circular permutation of the three people.

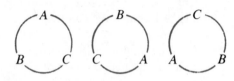

Thus, there are 2!, or 2, circular permutations of the 3 people.

To find the circular permutations of n distinct objects, begin by placing one object on the circle. Then arrange the remaining $(n - 1)$ objects in $(n - 1)!$ ways relative to the first object.

Number of circular permutations	The number of circular permutations of n distinct objects is $(n - 1)!$

Example **A married couple invites 3 other couples to an anniversary dinner. In how many different ways can all of the 8 people be seated around a circular table?**

$$(n - 1)! = (8 - 1)! = 7! = 5,040$$

Thus, there are 5,040 different ways to seat the 8 people.

Reading in Algebra

Tell which number of permutations is the greater for each pair below.
1. (a) 5 people in a line
   (b) 5 people in a circle
2. (a) all the letters in *HOBO*
   (b) all the letters in *HOPE*
3. (a) 6 people in a circle
   (b) all the letters in *GEESE*
4. (a) all the digits in 52,466
   (b) all the digits in 63,777

Oral Exercises

Tell whether each arrangement is <u>linear</u>, <u>circular</u>, or <u>either</u> of these.
1. Digits in a 5-digit number
2. Keys on a key ring
3. Football players in a huddle
4. Letters in your last name
5. Books on a table
6. Charms on a bracelet
7. Flags in a signal
8. Trees in a park

Written Exercises

Solve each problem.

Ⓐ 1. In how many different ways can seven children be seated around a circular table at a birthday party?

2. The president of a company and his seven vice-presidents are to be seated around a circular conference table. In how many ways can this be done?

3. A football team of eleven players forms a circular huddle before each play. In how many different ways can the players be arranged?

4. Find the number of different ways that a gardener may plant a dozen different bushes around a circular flower bed.

Ⓑ 5. A husband and wife invite four other couples to dinner. In how many ways can they all be seated around the circular dining table?

6. A woman and her daughter each invite three friends to lunch. Find the number of ways that all of them can be seated around one circular table.

7. Four teachers and four students are seated in a circular discussion group. Find the number of ways this can be done if teachers and students must be seated alternately.

8. In how many different ways can five girls and five boys be seated around a circular table if no two girls may be seated next to each other?

Example **In how many different ways can 3 keys be arranged on a ring?**

A key ring has a front view and a rear view, but the two views are considered to be one arrangement.

The 3 keys can be arranged in $\dfrac{(3 - 1)!}{2}$ ways.

Thus, the 3 keys can be arranged in only 1 way.

Front View Rear View

Ⓒ 9. In how many ways can 4 keys be arranged on a ring?

10. Find the number of ways that 7 charms can be arranged on a bracelet.

NON-ROUTINE PROBLEMS

Simplify.
1. $(n + 1) \cdot n \cdot (n - 1)!$

2. $(n - 3)(n - 4)(n - 5)!$

COMBINATIONS

Objectives

To find the number of possible selections of *n* objects taken *r* at a time without regard to order

To find the value of $\binom{n}{r}$ for nonnegative integers *n* and *r* where $n \geq r$

In a permutation, the order of the objects is important. Another type of problem involves selecting objects where the order in which they are selected or arranged does not matter. A selection of this type is called a **combination** of the objects. Some examples are listed below.

Committee: The Al-Betty-Carl committee is the same as the Carl-Al-Betty committee.

Vertices of a triangle: $\triangle DEF$ is the same as $\triangle EFD$, which is the same as $\triangle FDE$.

Value of 3 coins: A dime, a nickel, and a penny have the same total value as a nickel, a penny, and a dime.

Definition: Combination	A *combination* is a selection of objects without regard to order.

You can list all of the three-letter *combinations* that can be formed from the 5 letters *a*, *b*, *c*, *d*, and *e* and count the number of combinations.

a-b-c	a-c-d	a-d-e	b-c-d	b-d-e	c-d-e
a-b-d	a-c-e		b-c-e		
a-b-e					

Thus, there are *ten* combinations of 5 things taken 3 at a time.

$\binom{5}{3}$ is the symbol for "the number of combinations of 5 things taken 3 at a time." The value of $\binom{5}{3}$ is found in the following way.

$$\binom{5}{3} = \frac{5!}{3!(5-3)!} = \frac{5!}{3! \times 2!} = \frac{5 \times 4 \times 3!}{3! \times 2 \times 1} = 10$$

Number of combinations	$\binom{n}{r}$ is the number of combinations of *n* things taken *r* at a time. $$\binom{n}{r} = \frac{n!}{r!(n-r)!}$$

Example 1

Find the value of $\binom{7}{4}$.

$$\binom{7}{4} = \frac{7!}{4!(7-4)!} = \frac{7!}{4! \times 3!} = \frac{7 \times 6 \times 5 \times 4!}{4! \times 3 \times 2 \times 1} = 35$$

Example 2

How many different 4-member committees can be formed if 10 people are available for appointment to a committee?

$$\binom{10}{4} = \frac{10!}{4! \times 6!} = \frac{10 \times 9 \times 8 \times 7 \times 6!}{4 \times 3 \times 2 \times 1 \times 6!} = 210$$

Thus, 210 committees can be formed.

Example 3

Seven points, A through G, are located on a circle. How many triangles are determined by the 7 points?

A triangle is determined by any 3 points that are not on the same line.

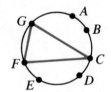

$$\binom{7}{3} = \frac{7!}{3! \times 4!} = \frac{7 \times 6 \times 5 \times 4!}{3 \times 2 \times 1 \times 4!} = 35$$

Thus, 35 triangles are determined.

The number of combinations of 5 things taken 5 at a time is 1. Notice the consequence when the formula for $\binom{n}{r}$ is used to find $\binom{5}{5}$.

$$\binom{5}{5} = \frac{5!}{5! \times 0!} = \frac{1}{0!}, \text{ which must be equal to 1.}$$

Recall the definition of 0!

$$0! = 1$$

With this definition of 0!, the formula for $\binom{n}{r}$ can be used with any nonnegative integers n and r where $n \geq r$.

Reading in Algebra

Tell whether each collection of objects is a *permutation* or a *combination* of the objects in the collection. (Ex. 1–6)

1. A four-digit number

2. A two-letter name of a line

3. A five-person committee

4. A six-flag signal

5. A five-letter code word

6. A six-card hand in a card game

7. List all of the different amounts of money that can be determined by selecting three coins from 1 penny, 1 nickel, 1 dime, and 1 quarter.

8. List a two-letter name for each line determined by the points A, B, C, and D which are located on a circle.

Written Exercises

Find the value of each expression. (Ex. 1–6)

(A)   **1.** $\binom{7}{2}$ **2.** $\binom{10}{6}$ **3.** $\binom{18}{3}$ **4.** $\binom{6}{6}$ **5.** $\binom{63}{1}$ **6.** $\binom{8}{0}$

7. How many 4-letter combinations can be formed from the first half of our alphabet?

8. How many different 5-member committees can be formed if 20 people are available for membership?

9. Find the number of lines determined by 10 points, A through J, on a circle.

10. Find the number of triangles determined by 15 points, A through O, on a circle.

(B) **11.** An algebra student must solve 7 out of 10 word problems. In how many different ways can the 7 problems be selected?

12. In how many different ways can a team of 33 football players choose three captains for its first game of the season?

13. During the baseball season, an 8-school athletic conference requires that each school play 3 games with each of the other schools. Find the total number of baseball games played.

One bag contains 5 red marbles numbered 1 through 5 and another bag contains 7 blue marbles numbered 1 through 7. (Ex. 14–16)

14. In how many different ways can you choose 3 red marbles *and* 3 blue marbles?

15. In how many different ways can you choose 3 red marbles *or* 3 blue marbles?

16. In how many different ways can you choose *one or more* red marbles?

17. There are 5 pellets, lettered A, B, C, D, and E, that weigh 1 g, 2 g, 4 g, 8 g, and 16 g, respectively. How many different masses can be measured by using one or more of the 5 weights on a balance scale?

18. How many different amounts of money can be determined by selecting one or more coins from 1 penny, 1 nickel, 1 dime, 1 quarter, 1 half-dollar, and 1 silver dollar?

You can verify that $\binom{7}{2} = \binom{7}{5}$ and that $\binom{4}{2} + \binom{4}{3} = \binom{5}{3}$.
This suggests the theorems in Exercises 19 and 20.

(C) **19.** Prove that $\binom{n}{r} = \binom{n}{n-r}$ for all integers r and n, $0 \le r \le n$.

20. Prove that $\binom{n}{r} + \binom{n}{r+1} = \binom{n+1}{r+1}$ for all integers r and n where $0 \le r < n$.

21. Find three pairs of integers (n, r) for which $\binom{n}{4} = \binom{12}{r}$.

NON-ROUTINE PROBLEMS

1. Rewrite the binomial theorem (page 382) using $\binom{n}{r}$ notation for the coefficients of the terms.

2. Find the number of different 5-card hands that may be drawn from a deck of 52 cards.

Combinations

PROBABILITY: SIMPLE EVENT

Objective   **To determine the probability of a simple event**

When a "fair" coin is tossed it seems reasonable to expect that the coin is just as likely to come up heads (H) as tails (T). These two outcomes, heads and tails, have equal chances of occurring. In everyday language, we say the coin has 1 chance in 2 of coming up heads and 1 chance in 2 of coming up tails. In mathematical language, we say the **probability** of obtaining a head in one flip of a coin is $\frac{1}{2}$. In symbols, $P(\text{head}) = \frac{1}{2}$.

One way of obtaining the probability of a simple event occurring is to determine the number of ways the event can occur, called *favorable outcomes,* and then divide by the *total number of all possible outcomes.*

Probability of simple event	The probability of a simple event is given by the formula: $$P(\text{simple event}) = \frac{\text{Number of favorable outcomes}}{\text{Total number of all possible outcomes}}.$$

Example 1   **What is the probability that in a throw of a die (plural: dice) a face with 4 dots will appear? What is the probability that a face with an even number of dots will appear?**

$$\text{Use } P(\text{simple event}) = \frac{\text{Number of favorable outcomes}}{\text{Total number of all possible outcomes}}.$$

One face of a die contains 4 dots.
There is only 1 favorable outcome.
There are 6 faces on a die.
There are 6 possible outcomes.

Thus, $P(\text{a face with 4 dots}) = \frac{1}{6}$.

Three faces of a die contain even numbers: 2, 4, 6.
There are 6 faces on a die.
The number of favorable outcomes is 3.
The number of possible outcomes is 6.

Thus, $P(\text{even number of dots}) = \frac{3}{6}$, or $\frac{1}{2}$.

Example 2   **A bag contains 5 red balls and 4 black balls. One ball is drawn from the bag. What is the probability that the ball is red? What is the probability that it is black? What is the probability that it is white?**

There are 5 + 4, or 9 possible outcomes.
When looking for a red ball, there are 5 favorable outcomes.
When looking for a black ball, there are 4 favorable outcomes.
When looking for a white ball, there are 0 favorable outcomes.

Thus, $P(\text{red ball}) = \frac{5}{9}$, $P(\text{black ball}) = \frac{4}{9}$, and $P(\text{white ball}) = \frac{0}{9}$, or 0.

There are times when an event is certain to happen and times when it is impossible for an event to happen.

As indicated, the probability of a simple event E occurring is greater than or equal to 0 and less than or equal to 1:

$$0 \le P(E) \le 1.$$

Example 3

A deck of cards contains 52 cards, 13 cards of each suit. One card is drawn. What is the probability that the card is an ace? that it is the ace of hearts? that it is not an ace?

There are 52 possible outcomes.
When looking for an ace, there are 4 favorable outcomes.
When looking for the ace of hearts, there is 1 favorable outcome.
When looking for a card that is not an ace, there are 48 favorable outcomes.

Thus, $P(\text{ace}) = \dfrac{4}{52}$, or $\dfrac{1}{13}$; $P(\text{ace of hearts}) = \dfrac{1}{52}$; and $P(\text{not an ace}) = \dfrac{48}{52}$, or $\dfrac{12}{13}$.

In Example 3, observe that the probability of getting an ace is $\dfrac{1}{13}$ and the probability of not getting an ace is $\dfrac{48}{52}$, or $\dfrac{12}{13}$.

The sum of the two probabilities is $\dfrac{1}{13} + \dfrac{12}{13} = 1$. If the probability of an event occurring is P, then the probability of the event not occurring is $1 - P$.

| Definition: Odds | The *odds* that a simple event E will occur are $\dfrac{P(E)}{1 - P(E)}$, or $\dfrac{P(E)}{P(\overline{E})}$, where $P(\overline{E})$ is the probability of E not occurring. |

Example 4

A bag contains 30 balls; 18 of them are red. What are the odds that the ball drawn will be red?

Use $\dfrac{P(E)}{P(\overline{E})}$. ▶

$$P(E) = P\,(\text{red ball}) = \frac{18}{30}, \text{ or } \frac{3}{5}$$

$$P(\overline{E}) = 1 - P(E), \text{ or } 1 - \frac{3}{5} = \frac{2}{5}$$

Thus, the odds of getting a red ball are $\dfrac{\frac{3}{5}}{\frac{2}{5}}$, or $\dfrac{3}{2}$, or 3 to 2.

Written Exercises

(A) 1. What is the probability that in a throw of a die, the 6 will appear? the 3 will appear? an odd number will appear?

2. What is the probability that in a throw of a die, the 1 will appear? a prime number will appear? a number divisible by 2 will appear?

3. A bag contains 8 white balls and 6 blue balls. A ball is drawn from the bag. What is the probability that the ball is white? that it is blue?

4. A bag contains 12 red balls and 18 black balls. A ball is drawn from the bag. What is the probability that the ball is red? that it is black?

5. A deck of cards contains 52 cards, 13 cards of each suit. One card is drawn. What is the probability that the card is a jack? that it is the jack of clubs? that it is not a jack?

6. A box contains 1,000 light bulbs, 900 good ones and 100 defective ones. A bulb is drawn. What is the probability that the bulb is defective?

7. A bag contains 4 red, 3 white, and 5 blue balls. A ball is drawn from the bag. What is the probability that the ball is red? that it is white? that it is blue? that it is not red? that it is not blue?

8. A deck of cards contains 52 cards, 13 cards of each suit. There are 2 black suits (spades and clubs) and 2 red suits (hearts and diamonds). One card is drawn. What is the probability that the card is red? that it is black? that it is a red ace? that it is a black 7?

(B) 9. A die is rolled 100 times. About how many times should it stop on 4? About how many times should an even number appear? About how many times should a number less than 3 appear? [Hint: First determine the probability of each event.]

10. A die is rolled 6,000 times. About how many times should it not stop on 6? About how many times should it stop on 5? About how many times should a number greater than 2 appear?

11. The probability that it will rain on any given day in December in the town of Los Altos is $\frac{8}{10}$. About how many days will it not rain in December?

12. The probability of producing a defective part by a given machine is $\frac{1}{900}$. If 36,000 parts are produced in one day, about how many of the parts will be defective?

13. A bag contains 16 red balls and 32 white balls. A ball is drawn from the bag. What are the odds that the ball is red?

14. If the probability that it will snow on a given day is $\frac{2}{3}$, what is the probability that it will not snow? What are the odds that it will snow? that it will not snow?

The probability of an event E occurring if the odds are m to n that it will occur is $P(E) = \dfrac{m}{m + n}$.

(C) 15. A high school senior feels her odds of getting into a college of her choice are 5 to 1. Find the probability that she will be accepted.

16. The odds that the team will win are 3 to 2. Find the probability of winning.

PROBABILITY: COMPOUND EVENTS 16.7

Objective   **To determine the probability of an event using a sample space**

In the previous lesson, you learned how to find the probability of a simple event by determining the number of favorable outcomes and the number of all possible outcomes. In some situations, it is not easy to identify the number of outcomes. Making a list of all possible outcomes may be helpful. Such a listing is called a **sample space.**

A sample space showing all possible outcomes of tossing three coins is shown below.

1st coin ▶   Head (*H*)   or   Tail (*T*)

2nd coin ▶   *H*   or   *T*   *H*   or   *T*

3rd coin ▶   *H* or *T*   *H* or *T*   *H* or *T*   *H* or *T*

Sample ▶   *HHH*   *HTH*   *THH*   *TTH*
space   *HHT*   *HTT*   *THT*   *TTT*

Example 1   **In a toss of three coins, what is the probability of obtaining two heads and a tail? What is the probability of obtaining a head, a tail, and a head, in that order?**

As shown in the sample space above, two heads and a tail can occur three times:
HHT, HTH, THH.

A head, a tail, and a head can occur in that order only once: *HTH.*

There are 8 possible outcomes. ▶   **Thus,** P(2 heads and 1 tail) $= \dfrac{3}{8}$, and $P(HTH) = \dfrac{1}{8}$.

Example 2   **In a toss of three coins, what is the probability of obtaining exactly two tails? What is the probability of obtaining at least one head?**

Exactly two tails can occur three times: *HTT, THT, TTH.*

At least one head can occur seven times: *HHH, HHT, HTH, HTT, THH, THT, TTH.*

Thus, P(exactly 2 tails) $= \dfrac{3}{8}$, and P(at least 1 head) $= \dfrac{7}{8}$.

Example 3   **In a throw of two dice, a red one and a white one, find $P(r \leq 2 \text{ or } w \leq 3)$.**

There are 36 possible outcomes.
There are 12 ways to get $r \leq 2$ and
18 ways to get $w \leq 3$, but 6 ordered
pairs are common to both sets and
cannot be counted twice.

For $r \leq 2$ or $w \leq 3$, there are
$12 + 18 - 6$, or 24, favorable outcomes.

	White					
	1	2	3	4	5	6
1	(1, 1)	(1, 2)	(1, 3)	(1, 4)	(1, 5)	(1, 6)
2	(2, 1)	(2, 2)	(2, 3)	(2, 4)	(2, 5)	(2, 6)
Red 3	(3, 1)	(3, 2)	(3, 3)			
4	(4, 1)	(4, 2)	(4, 3)			
5	(5, 1)	(5, 2)	(5, 3)			
6	(6, 1)	(6, 2)	(6, 3)			

Thus, $P(r \leq 2 \text{ or } w \leq 3) = \dfrac{24}{36}$, or $\dfrac{2}{3}$.

In Example 3, notice that
$$P(r \leq 2) = \frac{12}{36}, \text{ or } \frac{1}{3}, \qquad P(w \leq 3) = \frac{18}{36}, \text{ or } \frac{1}{2}, \qquad P(r \leq 2 \text{ and } w \leq 3) = \frac{6}{36},$$
or $\dfrac{1}{6}$ and that $\dfrac{1}{3} + \dfrac{1}{2} - \dfrac{1}{6} = \dfrac{2}{3}$, which is equal to $P(r \leq 2 \text{ or } w \leq 3)$.

This leads to the following statement about the probability of two events A
or B occurring.

If two events can both occur at the same time, they are called **inclusive events.**
If they cannot both occur at the same time, they are called **mutually exclusive
events.**

Probability of A or B	If A and B are inclusive events, then $$P(A \text{ or } B) = P(A) + P(B) - P(A \text{ and } B).$$ If A and B are mutually exclusive events, then $$P(A \text{ or } B) = P(A) + P(B).$$

Example 4   **In a throw of 2 dice, what is the probability of obtaining a sum of 7? a sum of 11?
a sum of either 7 or 11?**

The sum 7 occurs 6 ways: (6, 1), (5, 2), (4, 3), (3, 4), (2, 5), (1, 6).
The sum 11 occurs 2 ways: (5, 6), (6, 5).
But both sums cannot occur at the same time.

The events are mutually exclusive. Use $P(A \text{ or } B) = P(A) + P(B)$.

$P(\text{sum of 7 or sum of 11}) = \dfrac{6}{36} + \dfrac{2}{36} = \dfrac{8}{36}$, or $\dfrac{2}{9}$.

Thus, $P(\text{sum of 7}) = \dfrac{1}{6}$, $P(\text{sum of 11}) = \dfrac{1}{18}$, and $P(\text{sum of 7 or sum of 11}) = \dfrac{2}{9}$.

Recall that when the word "and" is used to connect two statements, the conditions of both statements must be met simultaneously if the compound statement is to be true. If you wish to find $P(A \text{ and } B)$, then you are interested in those ways in which event A and event B can both occur at the same time.

Example 5

In a throw of two dice, a red one and a white one, find $P(r \leq 3 \text{ and } w \leq 2)$.

There are 36 possible outcomes.
There are 18 ways to get $r \leq 3$ and 12 ways to get $w \leq 2$, but only 6 ways to get both.

Six ordered pairs occur in both sets of ordered pairs.

Thus, $P(r \leq 3 \text{ and } w \leq 2) = \dfrac{6}{36}$, or $\dfrac{1}{6}$.

	w					
	1	2	3	4	5	6
1	(1, 1)	(1, 2)	(1, 3)	(1, 4)	(1, 5)	(1, 6)
2	(2, 1)	(2, 2)	(2, 3)	(2, 4)	(2, 5)	(2, 6)
r 3	(3, 1)	(3, 2)	(3, 3)	(3, 4)	(3, 5)	(3, 6)
4	(4, 1)	(4, 2)				
5	(5, 1)	(5, 2)				
6	(6, 1)	(6, 2)				

In Example 5, the outcome of throwing the red die does not affect the outcome of throwing the white die. Events of this type in which neither event depends upon the other are called **independent events**.

Notice that $P(r \leq 3) = \dfrac{18}{36}$, or $\dfrac{1}{2}$, $P(w \leq 2) = \dfrac{12}{36}$, or $\dfrac{1}{3}$, and that $\dfrac{1}{2} \cdot \dfrac{1}{3} = \dfrac{1}{6}$, which is equal to $P(r \leq 3 \text{ and } w \leq 2)$. This leads to the following statement about the probability of two events A and B occurring.

Probability of A and B	If A and B are independent events, then $$P(A \text{ and } B) = P(A) \cdot P(B).$$

In the next example, the outcome of throwing each die is affected by the outcome of throwing the other. Events of this type in which one event depends upon the other are called **dependent events**. The sample space for dependent events is usually a reduced sample space of the original space.

Example 6

Find $P(r = 2, \text{ given that } r + w \leq 5)$.

There are 10 ways to get $r + w \leq 5$ and so there are 10 possible outcomes. The 10 ordered pairs make up the reduced sample space.

There are 3 ways to get $r = 2$, so there are 3 favorable outcomes.

Thus, $P(r = 2, \text{ given that } r + w \leq 5) = \dfrac{3}{10}$.

	w					
	1	2	3	4	5	6
1	(1, 1)	(1, 2)	(1, 3)	(1, 4)	(1, 5)	(1, 6)
2	(2, 1)	(2, 2)	(2, 3)	(2, 4)	(2, 5)	(2, 6)
r 3	(3, 1)	(3, 2)	(3, 3)	(3, 4)	(3, 5)	(3, 6)
4	(4, 1)	(4, 2)	(4, 3)	(4, 4)	(4, 5)	(4, 6)
5	(5, 1)	(5, 2)	(5, 3)	(5, 4)	(5, 5)	(5, 6)
6	(6, 1)	(6, 2)	(6, 3)	(6, 4)	(6, 5)	(6, 6)

Probability: Compound Events

Probability of A, given B	If A and B are dependent events, $P(A \text{ given } B) = \dfrac{P(A \text{ and } B)}{P(B)}$.

Written Exercises

For the following exercises, use the sample space showing all possible outcomes of tossing 3 coins. Find the indicated probability.

(A)
1. $P(3 \text{ heads})$
2. $P(2 \text{ tails and a head})$
3. $P(1 \text{ head and 2 tails})$
4. $P(3 \text{ tails})$
5. $P(\text{TTH})$
6. $P(\text{THT})$
7. $P(\text{HHT})$
8. $P(\text{HTT})$
9. $P(\text{at least 1 tail})$
10. $P(\text{at least 2 heads})$
11. $P(\text{at least 2 tails})$
12. $P(\text{at least 1 head and 1 tail})$
13. $P(\text{exactly 2 tails})$
14. $P(\text{exactly 1 head})$

For the following exercises, use the sample space showing all possible outcomes of rolling 2 dice, a red one, r, and a white one, w. Find each indicated probability.

15. $P(r = 4)$
16. $P(r \le 3)$
17. $P(w \ge 5)$
18. $P(r + w = 8)$
19. $P(r + w = 11)$
20. $P(r + w = 2)$
21. $P(r + w \ge 10)$
22. $P(r + w \le 7)$
23. $P(r + w < 3)$
24. $(P(r + w \le 1)$
25. $P(r \le 3 \text{ or } w = 2)$
26. $P(r \ge 5 \text{ or } w \ge 5)$
27. $P(r \le 2 \text{ or } w \le 5)$
28. $P(r \le 4 \text{ or } w \ge 4)$
29. $P(\text{sum of 6 or sum of 10})$
30. $P(\text{sum of 2 or sum of 12})$
31. $P(\text{sum of 7 or sum of 11})$
32. $P(\text{sum of 11 or sum of 12})$
33. $P(r = 2 \text{ and } w \ge 5)$
34. $P(r \ge 4 \text{ and } w = 4)$
35. $P(r \ge 5 \text{ and } w \le 2)$
36. $P(r \le 2 \text{ and } w \le 4)$
37. $P(r + w = 7, \text{ given that } r = 4)$
38. $P(r = 5, \text{ given that } r + w = 9)$
39. $P(w = 4, \text{ given that } r + w = 6)$
40. $P(r + w = 9, \text{ given that } w = 6)$

A bag contains 6 red, 8 white, and 4 blue balls. A ball is drawn from the bag. Find each indicated probability for drawing a ball.

(B)
41. $P(\text{red or white})$
42. $P(\text{red or blue})$
43. $P(\text{white or blue})$
44. $P(\text{red, white, or blue})$
45. $P(\text{blue or green})$
46. $P(\text{green or yellow})$

One card is drawn from an ordinary deck. Find the indicated probability for drawing a card.

47. $P(\text{red or a queen})$
48. $P(\text{black king or a club})$
49. $P(\text{an even number or black})$
50. $P(\text{red face or a jack})$.

(C)
51. Prepare a sample space showing all possible outcomes for boys and girls in a family with three children.

Using the sample space in Exercise 51, find each of the following probabilities.

52. $P(3 \text{ boys})$
53. $P(2 \text{ girls and a boy})$
54. $P(2 \text{ boys and a girl})$

55. $P(\text{1st is a girl, 2nd is a boy, 3rd is a boy})$
56. $P(\text{1st is a boy, 2nd is a girl, 3rd is a boy})$

A family has 4 children. Find each indicated probability.

57. Find $P(4 \text{ boys})$
58. $P(3 \text{ girls and a boy})$
59. $P(2 \text{ girls and 2 boys})$

In a given town, there are 1,000 families, and each family has 4 children. Answer each question.

60. Approximately how many families have exactly 3 girls? exactly 3 boys?
61. Approximately how many families had 2 boys first and then 2 girls?

PROBABILITY AND ARRANGEMENTS 16.8

Objective

To determine the probability of an event using the Fundamental Counting Principle, permutations, and combinations

The Fundamental Counting Principle as well as permutations and combinations can often be used to determine the probability of a given event.

Example 1

A matching test is given. There are 5 statements in the left column and 6 possible answers in the right column. If an answer can be used only once, what is the probability that a student can guess all the answers correctly?

$$P(\text{simple event}) = \frac{\text{Number of favorable outcomes}}{\text{Total number of all possible outcomes}}$$

There is 1 way all answers can be guessed correctly, so there is 1 favorable outcome. The total number of all possible outcomes is the number of 5-answer permutations that can be formed from the 6 answers.

Thus, $P\left(\begin{array}{l}\text{guessing all answers using}\\\text{each answer only once}\end{array}\right) = \frac{1}{6 \times 5 \times 4 \times 3 \times 2} = \frac{1}{720}.$

In Example 1, to find the probability of guessing all the answers correctly if each answer can be used at least once, the Fundamental Counting Principle can be applied. The five answer choices in order can be made in $6 \times 6 \times 6 \times 6 \times 6$, or 7,776 different ways. So, $P\left(\begin{array}{l}\text{guessing all answers using}\\\text{each answer at least once}\end{array}\right) = \frac{1}{7,776}.$

The next two examples show you how to determine the probability of an event using combinations.

Example 2

A deck of cards contains 52 cards, 13 cards of each suit. Five cards are drawn. What is the probability that all 5 cards are spades?

There are $\binom{13}{5}$ ways to draw 5 spades from 13 cards.

$\binom{n}{r} = \frac{n!}{r!(n-r)!}$ ▶ $\binom{13}{5} = \frac{13!}{5!8!} = 1,287.$

There are $\binom{52}{5}$ ways to draw 5 spades from a deck of 52 cards.

$\binom{52}{5} = \frac{52!}{5!47!} = 2,598,960.$

Thus, $P\left(\begin{array}{l}\text{5-card hand}\\\text{of all spades}\end{array}\right) = \frac{1,287}{2,598,960}, \text{ or } \frac{33}{66,640}.$

Example 3

A mathematics class of 25 students consists of 15 girls and 10 boys. A committee of 6 to represent the class is chosen at random. What is the probability that all 6 committee members are girls? that all 6 committee members are boys? that exactly 2 committee members are boys?

There are $\binom{15}{6}$ ways to choose 6 girls from a group of 15 girls:

$$\binom{15}{6} = \frac{15!}{6!9!} = \frac{15 \cdot 14 \cdot 13 \cdot 12 \cdot 11 \cdot 10 \cdot \cancel{9!}}{6 \cdot 5 \cdot 4 \cdot 3 \cdot 2 \cdot 1 \cdot \cancel{9!}} = 5,005.$$

There are $\binom{10}{6}$ ways to choose 6 boys from a group of 10 boys:

$$\binom{10}{6} = \frac{10!}{6!4!}, \text{ or } 210.$$

To choose exactly 2 boys means that 4 committee members are girls. So, 2 boys and 4 girls are chosen. By the Fundamental Counting Principle, there are $\binom{10}{2}\binom{15}{4}$ ways to choose exactly 2 boys from a group of 10 boys and 4 girls from a group of 15 girls:

$$\binom{10}{2}\binom{15}{4} = \frac{10!}{2!8!} \times \frac{15!}{4!11!} = \frac{10 \cdot 9 \cdot \cancel{8!}}{2 \cdot 1 \cdot \cancel{8!}} \times \frac{15 \cdot 14 \cdot 13 \cdot 12 \cdot \cancel{11!}}{4 \cdot 3 \cdot 2 \cdot 1 \cdot \cancel{11!}}$$
$$= \quad 45 \quad \times 1,365, \text{ or } 61,425.$$

There are $\binom{25}{6}$ ways to choose a committee of 6 students from a group of 25 students: $\binom{25}{6} = \frac{25!}{6!19!} = \frac{25 \cdot 24 \cdot 23 \cdot 22 \cdot 21 \cdot 20 \cdot \cancel{19!}}{6 \cdot 5 \cdot 4 \cdot 3 \cdot 2 \cdot 1 \cdot \cancel{19!}} = 177,100.$

So, $P\left(\begin{array}{c}\text{committee of} \\ \text{6 girls}\end{array}\right) = \frac{5,005}{177,100}$, or $\frac{13}{460}$, $P\left(\begin{array}{c}\text{committee of} \\ \text{6 boys}\end{array}\right) = \frac{210}{177,100}$, or $\frac{3}{2,530}$,

and $P\left(\begin{array}{c}\text{committee with} \\ \text{exactly 2 boys}\end{array}\right) = \frac{61,425}{177,100}$, or $\frac{351}{1,012}$.

Example 4

A box contains 5 red marbles and 15 white marbles. What is the probability of drawing a red marble and a white marble with replacement? Without replacement? What is the probability of drawing a red marble, then a white marble with replacement? Without replacement?

$P(\text{red and white}) = P[\underbrace{\text{red 1st, then white}} \text{ or } \underbrace{\text{white first, then red}}]$
$$= P(\text{red})P(\text{white}) + P(\text{white})P(\text{red})$$
$$= 2P(\text{red})P(\text{white})$$

Without replacement	*With replacement*
$P(\text{red}) = \frac{5}{20}, P(\text{white}) = \frac{15}{19}$	$P(\text{red}) = \frac{5}{20}, P(\text{white}) = \frac{15}{20}$

$P(\text{red and white})$ ▶ $\quad 2\left(\frac{5}{20}\right)\left(\frac{15}{19}\right) = \frac{15}{38} \qquad\qquad 2\left(\frac{5}{20}\right)\left(\frac{15}{20}\right) = \frac{15}{40}$

$P(\text{red then white})$ ▶ $\quad \frac{5}{20}\left(\frac{15}{19}\right) = \frac{15}{76} \qquad\qquad\quad \frac{5}{20}\left(\frac{15}{20}\right) = \frac{15}{80}$

So, $P(\text{red and white})$ is $\frac{15}{40}$, or $\frac{3}{8}$, with replacement and $\frac{15}{38}$ without replacement.

$P(\text{red then white})$ is $\frac{15}{80}$, or $\frac{3}{16}$, with replacement and $\frac{15}{76}$ without replacement.

446

Chapter Sixteen

Written Exercises

(A) 1. An envelope contains three slips of paper with each of the numerals 1, 2, and 3 printed on exactly one slip. If one slip of paper is drawn at a time, without replacement, what is the probability that the three-digit number 231 is drawn in that order?

2. A bag contains one red, one white, and one blue cube. Each cube is drawn at random exactly once. What is the probability that the order of the cubes drawn is red, blue, and white?

3. A deck of cards contains 52 cards, 13 cards of each suit. Six cards are drawn. What is the probability that all 6 cards are hearts?

4. A deck of cards contains 52 cards, 13 cards of each suit. Five cards are to be drawn. What is the probability that exactly two of the cards are diamonds?

From a group of 5 boys and 3 girls, three violin students are to be selected at random to represent their school in a regional orchestra. Determine each probability.

(B) 5. The probability that all 3 students selected are girls

6. The probability that all 3 students selected are boys

7. The probability that exactly 1 student selected is a boy

8. The probability that exactly 1 student selected is a girl

9. The probability that 2 students selected are boys and 1 a girl

10. The probability that 1 student selected is a boy and 2 are girls

11. A box contains 6 red, 5 white, and 7 black marbles. What is the probability of drawing a black marble, then a red one, if the 1st marble is replaced before drawing the 2nd?

12. A box contains 4 red, 8 white, and 10 black marbles. What is the probability of drawing a red marble, then a white marble, then a black marble, if each marble is not replaced before the next draw?

Draw 2 cards *without* replacement. Find the probability that:

13. both are diamonds.

14. the first is a club and the second is a heart.

15. one is a club and one is a heart.

16. both are red.

17. the first is red and the second is black.

18. one is red and one is black.

Draw 2 cards, *replacing* the first before the second is drawn. Find the probability that:

19. one is a diamond and the other is a spade.

20. the first is a diamond and the second is a spade.

21. both are black.

22. both are clubs.

23. one is black and the other is red.

24. *exactly one* is a diamond.

(C) 25. A card is drawn from a deck of 52 cards. What is the probability that the card will be an ace, or red queen, or a ten of hearts?

26. A card is drawn from a deck of 52 cards. What is the probability that the card will be an ace, king, or queen?

27. A box contains 8 red marbles and 10 black marbles. Two marbles are drawn at random without replacement. What is the probability that both are the same color? That one will be red and one black? That both will be red? That both will be black?

28. A box contains 4 red, 6 white, and 8 blue marbles. What is the probability of drawing a white marble, and then a red one, if the 1st marble is replaced before drawing the 2nd? If the 1st marble is not replaced before drawing the 2nd?

Probability and Arrangements

447

CHAPTER SIXTEEN REVIEW

Vocabulary
circular permutation [16.4]
combination [16.5]
dependent events [16.7]
inclusive events [16.7]
independent events [16.7]
mutually exclusive events [16.7]
$n!$ [16.1]   permutation [16.1]
probability [16.6]   sample space [16.7]

Find the value of each expression.

1. $\dfrac{12!}{3! \times 8!}$ [16.3]   2. $\dbinom{100}{3}$ [16.5]

3. $\dbinom{9}{0}$ [16.5]

4. How many four-digit numbers can be formed from the digits 4, 5, 6, 7, 8, 9, if (a) no digit may be repeated in a number and (b) a digit may be repeated in a number? [16.1]

5. How many odd four-digit numbers can be formed from the digits 4, 5, 6, 7, 8, 9, if no digit may be repeated in a number? [16.2]

6. How many even numbers of one or more digits can be formed from the digits 4, 5, 6, 7, 8, if no digit may be repeated in a number? [16.2]

7. How many distinguishable six-digit numbers can be formed from the digits of 789778? [16.3]

8. How many different auto license-plate numbers can be formed by 3 different letters followed by 4 digits? [16.1]

9. Find the number of 7-letter permutations of all the letters in the word COMBINE. [16.1]

10. Tom and Mary invite three other couples to a cookout. In how many ways can they all be seated around the fire? [16.4]

★ 11. In how many ways can 6 keys be arranged on a ring? [16.4]

12. How many different 6-member committees can be formed if 9 people are available for appointment to a committee? [16.5]

13. A company makes sweaters in 7 colors. Each color is available with 5 choices of fabric and either with or without sleeves. How many different types of sweaters does the company make? [16.1]

14. How many tennis games are played by 8 people if each person plays each of the others one game? [16.5]

15. Three brands of canned soup and 4 brands of dried soup are to be displayed on a shelf with the dried soups together and to the right of the canned soups. How many displays are possible? [16.2]

★ 16. In how many different ways can a family of seven be photographed in a line if the father and mother are not to be separated? [16.2]

17. A deck of cards contains 52 cards, 13 cards of each suit. One card is drawn. What is the probability that the card is a king? that it is not a king? [16.6]

18. A die is rolled 600 times. About how many times should it stop on 2? About how many times should an even number appear? [16.6]

For Exercises 19–21, use the sample space showing all possible outcomes of rolling 2 dice, a red one, r, and a white one, w. Find each indicated probability. [16.7]

19. $P(w \le 4)$ 20. $P(r + w = 7)$
21. $P(r + w = 6$, given that $r = 3)$

★ 22. A family has 4 children. Find $P(3$ boys and a girl). [16.7]

23. A box contains 12 red, 8 white, and 10 blue cubes. What is the probability of drawing a blue cube, then a red one, if the first cube is replaced before drawing the second? If the first cube is not replaced before drawing the second? [16.8]

CHAPTER SIXTEEN TEST

Find the value of each expression.

1. $\dfrac{12!}{9! \times 2!}$ 2. $\dbinom{12}{8}$ 3. $\dbinom{20}{20}$

4. How many three-digit numbers can be formed from the digits 2, 3, 4, 5, 6, 7, if no digit may be repeated in a number?

5. A state uses 2 different letters followed by 5 digits on its auto license-plates. How many different license-plate numbers are possible in that state?

6. How many distinguishable six-digit numbers can be formed from 335533?

7. Find the number of 6-letter permutations of all the letters in EUCLID that end with either the letter E or the letter D.

8. In how many ways can 6 bushes, all of different types, be planted around a statue?

9. How many different 12-member juries can be selected from 15 qualified people?

10. Find the number of distinguishable permutations of all the letters in the word BANANA.

11. An amusement park has 9 gates with 5 on the north side and 4 on the south side. In how many different ways can you enter through a south gate and later exit through a north gate?

12. Aluminum chips A, B, C, and D weigh 1 g, 5 g, 10 g, and 20 g, respectively. How many different masses can be measured by using one or more of the 4 weights on a balance scale?

13. A box contains 15 green marbles, 12 orange marbles, and 23 purple marbles. A marble is drawn from the box. What is the probability that the marble is green? that it is not green?

14. A deck of cards contains 52 cards, 13 cards of each suit. One card is drawn. What is the probability that the card is an ace? that it is the ace of spades?

15. An 8-sided die is rolled 400 times. About how many times should it stop on 7? About how many times should an odd number appear?

For Exercises 16–18, use the sample space showing all possible outcomes of rolling 2 dice, a blue one, b, and a red one, r. Find each indicated probability.

16. $P(b = 2)$ 17. $P(b \geq 4 \text{ or } r \leq 2)$

18. $P(b \geq 2 \text{ and } r \leq 2)$

19. A bag contains 22 orange and 18 green balls. What is the probability of drawing a green ball, then an orange one, if the first ball is replaced before drawing the second? if the first ball is not replaced before drawing the second?

★ 20. In how many ways can 6 cheerleaders be arranged in a line if 3 of them do not want to be separated?

★ 21. In how many ways can 5 charms be arranged on a bracelet?

★ 22. Solve $\dbinom{n}{4} = \dbinom{n}{1}$ for n.

★ 23. A family has 4 children. Find $P(4 \text{ boys})$.

COMPUTER ACTIVITIES

Factorials and $\binom{n}{r}$

OBJECTIVES: **To evaluate factorials**
To evaluate combinations

From your study of factorials, you know that $n!$ grows very large as n increases. The computer displays numbers 10^9 and larger in BASIC scientific notation. For example, $1.234E+11$ is the computer display for the number 1.234×10^{11}. Values greater than approximately 10^{38} will cause the computer to display an overflow error. Run the following program, noting what value of $n!$ makes the computer switch to scientific notation, and what value of $n!$ causes an overflow error.

```
10   PRINT "WHAT IS THE NUMBER";: INPUT N
20 P = 1: FOR I = 1 TO N:P = P * I: NEXT I
30   PRINT N" FACTORIAL = "P
40   GOTO 10
```

See the Computer Section beginning on page 574 for more information.

The combination $\binom{n}{r}$ is defined to mean $\dfrac{n!}{r!(n-r)!}$.

Note that $\dfrac{n!}{(n-r)!} = n(n-1)(n-2)\ldots(n-r+1)$.

The program below uses these two facts to evaluate combinations.

```
10   PRINT "UPPER NUMBER";: INPUT U
20   PRINT "LOWER NUMBER";: INPUT L
30 P = 1: IF U = L THEN 80
40   IF L = 0 THEN 80
50   FOR I = U TO U - L + 1 STEP  - 1
60 P = P * I: NEXT I
70   FOR I = 1 TO L:P = P / I: NEXT I
80   PRINT "THE VALUE OF ";U;" OVER ";L;" IS ";P
90   GOTO 10
```

EXERCISES

1. Test the program above by using the values from the three examples on pages 435–436.

2. Write a program to evaluate $a!/b!$ (Be careful.)

COLLEGE PREP TEST

DIRECTIONS: In each item, you are to compare a quantity in Column 1 with a quantity in Column 2. Write the letter of the correct answer from these choices:

    A The quantity in Column 1 is greater than the quantity in Column 2
    B The quantity in Column 2 is greater than the quantity in Column 1
    C The quantity in Column 1 is equal to the quantity in Column 2
    D The relationship cannot be determined from the given information

Information centered over both columns refers to one or both of the quantities to be compared.

SAMPLE ITEMS AND ANSWERS

Column 1	Column 2
S1. $3! + 2!$	$3! \cdot 2!$

The answer is B: $3! + 2! = 6 + 2 = 8$, $3! \cdot 2! = 6 \cdot 2 = 12$, and $12 > 8$.

n and r are positive integers and $n > r$.

Column 1	Column 2
S2. $\binom{n}{r-1}$	$\binom{n-1}{r}$

The answer is D: If $n = 3$ and $r = 1$, then $\binom{n}{r-1} < \binom{n-1}{r}$, since $\binom{3}{0} < \binom{2}{1}$.
If $n = 3$ and $r = 2$, then $\binom{n}{r-1} > \binom{n-1}{r}$, since $\binom{3}{1} > \binom{2}{2}$, or $3 > 1$.

	Column 1	Column 2
1.	$(5 - 5)!$	$5! - 5!$
2.	$\binom{8}{3}$	$\binom{8}{5}$
3.	$\binom{48}{2}$	$\binom{49}{3}$

	Column 1	Column 2
4.	$\binom{20}{0}$	$\binom{40}{40}$
5.	$\dfrac{46! \times 29!}{15!}$	$\dfrac{47! \times 28!}{16!}$
6.	$\binom{6}{0} + \binom{6}{1} + \binom{6}{2}$	$\binom{6}{6} + \binom{6}{5} + \binom{6}{4}$

n and r are positive integers and $n > r$.

	Column 1	Column 2
7.	$\binom{n}{r}$	$\binom{n+1}{r+1}$
8.	$\binom{n}{r}$	$\binom{n}{r+1}$
9.	$\binom{n}{r-1} + \binom{n}{r}$	$\binom{n+1}{r+1}$

$B \bullet$ $\bullet C$
$A \bullet$ $\bullet D$
$F \bullet$ $\bullet E$

	Column 1	Column 2
10.	Number of triangles determined by the points	Number of lines determined by the points
11.	Number of quadrilaterals determined	Number of polygons determined

CUMULATIVE REVIEW

DIRECTIONS: Choose the *one* best answer to each question or problem. (Exercises 1–15)

1. Find the solution set of $|x - 3| < 2$.
   (A) $\{x|-1 < x < -5\}$ (B) $\{x|x < 3\}$
   (C) $\{x|x < 2\}$ (D) $\{x|1 < x < 5\}$

2. Solve $3^{3x-2} = 81$ for x.
   (A) $\frac{5}{3}$ (B) -2 (C) 9 (D) 2

3. Evaluate $-2xy + 3x^2y - y^2$ if $x = -2$ and $y = 3$.
   (A) -33 (B) 15 (C) 39 (D) 57

4. Factor $6x^3 + 4x^2 - 2x$ completely.
   (A) $2x(3x^2 + 2x - 1)$
   (B) $x(3x - 1)(2x + 2)$
   (C) $2x(3x - 1)(x + 1)$
   (D) $2x(3x + 1)(x - 1)$

5. Simplify $\dfrac{1}{n - 2} - \dfrac{3}{n^2 - 4}$.
   (A) $\dfrac{-2}{n + 2}$ (B) $\dfrac{n - 1}{(n - 2)(n + 2)}$
   (C) $\dfrac{n + 5}{n - 2}$ (D) $\dfrac{-2}{(n - 2)(n + 2)}$

6. Solve $\dfrac{2a}{x - b} = \dfrac{-c}{x + b}$ for x.
   (A) $\dfrac{c - 2ab}{2ac}$ (B) $\dfrac{bc - 2ab}{2a + c}$
   (C) $\dfrac{2ab - bc}{c + 2a}$ (D) $\dfrac{2ab - bc}{2ac}$

7. Simplify $\dfrac{x^{\frac{2}{3}}y^{-\frac{1}{4}}}{x^{-\frac{1}{3}}y^{-\frac{7}{4}}}$.
   (A) $x^{\frac{1}{3}}y^{\frac{3}{2}}$ (B) $x^{\frac{1}{3}}y^2$
   (C) $xy^{\frac{3}{2}}$ (D) xy^2

8. Compute the discriminant and determine the nature of the solutions of $3x^2 - 7x + 6 = 0$.
   (A) real and irrational (B) real and equal
   (C) real and rational (D) imaginary

9. Write an equation, in standard form, of the line that passes through $P(2, -1)$ and that is parallel to the line described by $2x - 7y = 12$.
   (A) $7x + 2y = 12$ (B) $2x - 7y = 11$
   (C) $2x - 7y = -11$ (D) $7x - 2y = 16$

10. Multiply, if possible.
    $$\begin{bmatrix} 3 & 1 & -2 \\ -1 & 2 & 1 \end{bmatrix} \cdot \begin{bmatrix} 1 & -1 & 2 \\ 3 & 1 & -2 \\ -4 & 2 & -3 \end{bmatrix}$$
    (A) $[15 \quad -1 \quad 1]$ (B) $\begin{bmatrix} 14 & -6 & 10 \\ 1 & 5 & -9 \end{bmatrix}$
    (C) $\begin{bmatrix} 3 & -1 & -4 \\ 9 & 1 & -4 \\ 4 & 4 & 3 \end{bmatrix}$ (D) Not possible

11. A function f is defined by $y = -\frac{1}{2}x - 4$. Find an equation of f^{-1}.
    (A) $y = -2x - 8$ (B) $y = 2x + 4$
    (C) $y = 2x + 8$ (D) $y = \frac{1}{2}x + 4$

12. Write an equation, in standard form, of an ellipse whose center is at the origin and whose x-intercepts are ± 6 and whose y-intercepts are ± 4.
    (A) $\dfrac{x^2}{36} + \dfrac{y^2}{16} = 1$ (B) $\dfrac{x^2}{16} + \dfrac{y^2}{36} = 1$
    (C) $\dfrac{x^2}{36} - \dfrac{y^2}{16} = 1$ (D) $\dfrac{x^2}{16} - \dfrac{y^2}{36} = 1$

13. Convert 18 lb/ft^2 to ounces per square inch (oz/in.^{2}).
    (A) 24 oz/in.^{2} (B) 0.5 oz/in.^{2}
    (C) 2 oz/in.^{2} (D) 1.5 oz/in.^{2}

14. Find the value of $\dbinom{14}{12}$.
    (A) 182 (B) $\dfrac{91}{6}$ (C) 91 (D) $\dfrac{91}{3}$

15. A deck of cards contains 52 cards, 13 cards of each suit. One card is drawn. What is the probability that the card is a red ace?
    (A) $\dfrac{1}{26}$ (B) $\dfrac{2}{13}$ (C) $\dfrac{1}{13}$ (D) $\dfrac{1}{52}$

16. Solve $3x - 3(2x - 1) = -(x + 15)$.

17. Simplify $\dfrac{12a^5b^{-6}}{-2a^{-3}b^{-7}}$ and write with positive exponents.

18. Factor $25a^2 - 100$ completely.

19. The height of a triangle is 12 cm more than 4 times the length of its base. Find the length of the base and the height of the triangle if its area is 20 cm^2.

20. Alberta can pick the apples in 8 days and Elias can pick them in 12 days. How long would it take them to pick the apples if they work together?

21. Simplify $\sqrt[3]{16a^5} + \sqrt[3]{54a^5}$.

22. Find the value of $81^{\frac{3}{4}}$.

23. Write $3x^{\frac{2}{3}}$ in radical form.

24. Solve $x^2 - 8x = 5$ by completing the square.

25. Simplify $-4 \div (3i)$ by rationalizing the denominator.

26. For what value(s) of k will $3x^2 + 6x - (3k - 2) = 0$ have exactly one solution?

27. Draw the graph of $-2x - 3y = 6$.

28. Find the slope and y-intercept of the line described by $y - 3(2 - x) = 6$.

29. Find the distance between $P(-3, -5)$ and $Q(5, -6)$.

30. Determine the coordinates of M, the midpoint of $\overline{AB}$, given $A(3, -1)$ and $B(7, 9)$.

31. Solve the system: $-3x + 2y + z = 8$
    $x - 3y + 2z = -5$
    $-x + y - 3z = 0$.

32. Let $f(x) = 2x^2 - 1$. Find $f(-3)$.

33. If $f(x) = \dfrac{2x}{x^2 + 2x - 15}$, determine the domain of f.

34. Draw the graph of $y = -x^2 - 6x + 2$.

35. Find an equation of a circle whose center is at $C(-3, 2)$ and which passes through the point $P(1, -7)$.

36. Solve the system $x^2 + y^2 \le 36$ graphically.
    $\dfrac{x^2}{9} + \dfrac{y^2}{4} \ge 1$

37. Find the sum, if it exists, of the infinite geometric series
    $$1 + \frac{1}{3} + \frac{1}{9} + \cdots.$$

38. Find the 5th term in the expansion of $(2x - 3y)^7$. Simplify.

39. Solve $\log_3 x + \log_3(x - 2) = \log_3 15$. Check.

40. How many even numbers containing one or more digits can be formed from the digits 1, 2, 3, 4 if no digit may be repeated in a number?

17 TRIGONOMETRIC FUNCTIONS

At age 19, he graduated from the Catholic Gymnasium (school) at Pearborn with honors in German, Latin, Greek, and mathematics.

At age 24, Karl enrolled at the Academy of Munster to prepare for the state teacher's examination to pursue a career as a secondary school teacher of mathematics. Two years later, at age 26, he passed the examination.

For the next 15 years, he taught in various small villages during the day and did mathematical research at night. This after-school work resulted in his being recognized as the leading mathematical analyst in Europe. Karl spent the years 1864–1897 as Professor of Mathematics at the University of Berlin.

Problem

Show that $\log_{10} 2$ is irrational.

ANGLES OF ROTATION   17.1

Objectives   **To sketch an angle of rotation associated with a directed degree measure**
To determine a reference angle and reference triangle for any angle

An **angle of rotation** formed by rotating a ray about its end point, or vertex. Each angle formed is composed of two rays, the initial ray or side and terminal ray or side. A ray can be rotated in either a counterclockwise or clockwise direction and the terminal ray can lie in any quadrant or on any axis. Angles formed by a *counterclockwise* rotation have a *positive measure*. The figure at the right illustrates a counterclockwise rotation of a ray forming an angle with a positive measure of 150°.

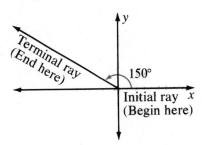

Example 1   **Sketch angles of rotation with measures of 90°, 225°, and 330°.**

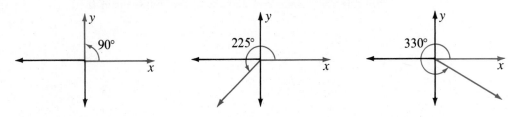

Angles formed by a *clockwise rotation* of a ray have *negative measures*. The figure at the right illustrates a clockwise rotation of a ray forming an angle with a measure of −45°.

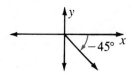

Example 2   **Sketch angles of rotation with measures of −120°, −210°, and −315°.**

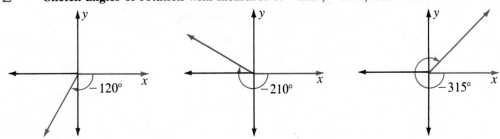

Angles of Rotation   **455**

Rays may be rotated through more than one revolution forming angles that measure more than 360°. The position of the terminal side of an angle determines the quadrant of the angle. Angles with terminal sides on an axis are called **quadrantal angles**. Angles with the same terminal sides are called **coterminal angles**.

Example 3   **Sketch angles of rotation with measures of 270°, −570°, and 420°. Determine the quadrant of each angle or identify as a quadrantal angle.**

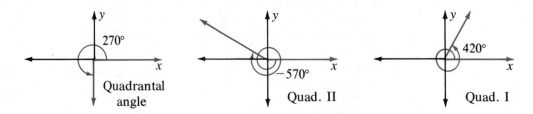

Each angle of rotation whose terminal side is not on an axis has a reference angle. A **reference angle** is the positive acute angle formed between the terminal side and the *x*-axis. The next example illustrates this idea.

Example 4   **Sketch angles of rotation with measures of 135°, −120°, and 330°. Identify the reference angle and determine the quadrant of the angle of rotation.**

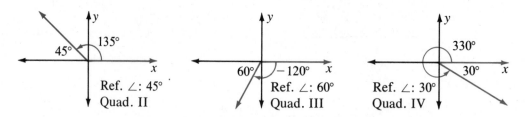

The triangle formed by drawing a perpendicular from a point on the terminal side to the *x*-axis is called a **reference triangle**. A *reference triangle* always includes the reference angle.

Example 5   **Sketch angles of rotation with measures of 60°, 200°, and −310°. Identify the reference triangle and reference angle.**

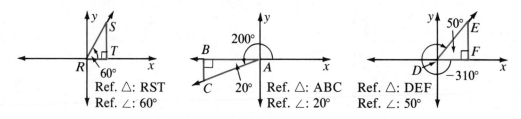

Reading in Algebra

Each item at the left can be described by one or more items at the right. Determine all items at the right that can be used to describe each item at the left. Justify.

1. 120°
2. −300°
3. 210°
4. −30°
5. 270°

A. Clockwise rotation
B. Counterclockwise rotation
C. Ref. ∠: 60°
D. Ref. ∠: 30°
E. Quad. I

F. Quad. II
G. Quad. III
H. Quad. IV
I. Quadrantal ∠

Oral Exercises

Determine the quadrant of each angle of rotation, and state the measure of the reference angle.

1. 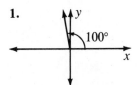 100°
2. 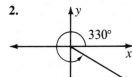 330°
3. 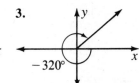 −320°
4. 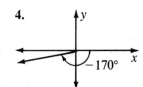 −170°

Written Exercises

Sketch angles of rotation with the given measures. Determine the quadrant of each angle of rotation or identify as a quadrantal angle (Quad. ∠).

Ⓐ
1. 30°
2. 300°
3. 210°
4. 135°
5. −30°
6. −150°
7. −360°
8. −225°
9. −90°
10. 225°
11. 330°
12. 10°
13. 515°
14. −190°
15. 400°
16. −650°
17. 100°
18. 720°

Sketch angles of rotation with the given measures. Determine the quadrant of each angle of rotation. Identify the reference triangle.

19. 75°
20. 160°
21. 225°
22. 330°
23. −40°
24. −195°
25. −245°
26. −300°
27. 495°
28. −780°
29. −510°
30. −360°

Sketch angles of rotation with the given measures. Identify the reference triangle and reference angle of each nonquadrantal angle.

31. −50°
32. 150°
33. −135°
34. −225°
35. 315°
36. 210°
37. −405°
38. −315°
39. 780°
40. 510°
41. −90°
42. 350°

State two more measures, one positive and one negative, for angles of rotation that have the same terminal side as the one shown.

Ⓑ
43. 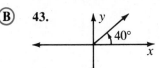 40°
44. 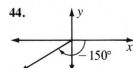 −150°
45. 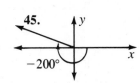 −200°
46. 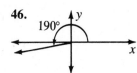 190°

Angles of Rotation

457

SPECIAL RIGHT TRIANGLES

Objectives   **To find lengths of sides in a 30°–60° or 45°–45° right triangle**
To find points on a circle using 30°–60° or 45°–45° right triangles

Recall that in an equilateral △, an altitude bisects both the base and the vertex angle from which it is drawn. You can see that in equilateral △ABC, altitude $\overline{AD}$ bisects ∠A and base $\overline{BC}$. You can also see that in rt. △ABD, the leg opposite the 30° angle, $\overline{BD}$, has $\frac{1}{2}$ the measure of hypotenuse $\overline{AB}$.

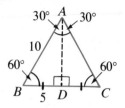

To find AD use the Pythagorean theorem.
$$(AB)^2 = (BD)^2 + (AD)^2$$
$$10^2 = 5^2 + (AD)^2$$
$$100 = 25 + (AD)^2$$
$$(AD)^2 = 75$$
$$(AD) = \sqrt{75}, \text{ or } 5\sqrt{3}$$

30°–60° right triangle rule

In a 30°–60° right △:
1. the leg opposite the 30° angle has $\frac{1}{2}$ the measure of the hypotenuse
2. the leg opposite the 60° angle has $\frac{1}{2}$ the measure of the hypotenuse times $\sqrt{3}$

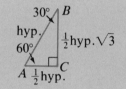

Example 1   **In right △RST, RS = 8. Find ST and RT.**

$\overline{ST}$ is opposite the 30° angle. $\overline{RT}$ is opposite the 60° angle.

$$\begin{array}{c|c}
ST = \frac{1}{2}(RS) & RT = \frac{1}{2}(RS)\sqrt{3} \\
\quad = \frac{1}{2}(8) & \quad = \frac{1}{2}(8)\sqrt{3} \\
\quad = 4 & \quad = 4\sqrt{3}
\end{array}$$

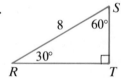

Thus, ST = 4 and RT = 4√3.

Example 2   **In right △XYZ, XZ = 5. Find XY and YZ.**

$\overline{XZ}$ is opposite the 30° angle. $\overline{YZ}$ is opposite the 60° angle.

$$\begin{array}{c|c}
XZ = \frac{1}{2}(XY) & YZ = \frac{1}{2}(XY)\sqrt{3} \\
5 = \frac{1}{2}XY & \quad = \frac{1}{2}(10)\sqrt{3} \\
10 = XY & \quad = 5\sqrt{3}
\end{array}$$

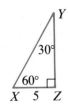

Thus, XY = 10 and YZ = 5√3.

Example 3

In right $\triangle ABC$, $BC = 6$.
Find AB and AC.

$\overline{AB}$ is opposite the 30° angle.
$\overline{BC}$ is opposite the 60° angle.

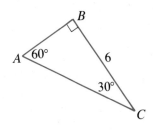

$$BC = \tfrac{1}{2}(AC)\sqrt{3} \qquad AB = \tfrac{1}{2}(AC)$$
$$6 = \tfrac{1}{2}(AC)\sqrt{3} \qquad\quad = \tfrac{1}{2}(4\sqrt{3})$$
$$12 = (AC)\sqrt{3} \qquad\qquad = 2\sqrt{3}$$
$$AC = \frac{12}{\sqrt{3}}, \text{ or } 4\sqrt{3}$$

Thus, $AC = 4\sqrt{3}$ and $AB = 2\sqrt{3}$.

Recall that if two angles of a triangle have the same measure, the triangle is **isosceles** and the sides opposite the two angles have the same measure. $\triangle ABC$ is an isosceles right triangle and $AC = BC$. To find AC or BC, use the Pythagorean theorem.

$$(AC)^2 + (BC)^2 = (AB)^2$$
$$(AC)^2 + (AC)^2 = (AB)^2$$
$$2(AC)^2 = 8^2$$
$$(AC)^2 = 32$$
$$AC = BC = \sqrt{32}, \text{ or } 4\sqrt{2}$$

45°–45° right triangle rule	In a 45°–45° right triangle, each leg has $\frac{1}{2}$ the measure of the hypotenuse times $\sqrt{2}$.	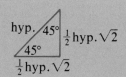

Example 4

In a 45°–45° right $\triangle$, one leg measures 12.
Find the length of the hypotenuse.

$$BC = \tfrac{1}{2}(AC)\sqrt{2}$$
$$12 = \tfrac{1}{2}(AC)\sqrt{2}$$
$$24 = (AC)\sqrt{2}$$
$$AC = \frac{24}{\sqrt{2}}, \text{ or } 12\sqrt{2}$$

Thus, the length of the hypotenuse is $12\sqrt{2}$.

You have learned that each nonquadrantal angle of rotation has a corresponding reference angle and reference triangle. There is also a circle that is associated with an angle of rotation. The examples that follow illustrate the use of a circle with reference triangles and special triangles.

Example 5

For a circle with radius of 8 units and a reference triangle of 30°–60°–90° as shown, find the point (x, y).

y is opposite the 60° angle.
x is opposite the 30° angle.

$$|y| = \tfrac{1}{2}(8)\sqrt{3} \qquad |x| = \tfrac{1}{2}(8)$$
$$|y| = 4\sqrt{3} \qquad\quad |x| = 4$$
$$y = 4\sqrt{3} \qquad\qquad x = 4$$

◄ *In Quad. I,* $x > 0, y > 0.$

Thus, the point (x, y) is $(4, 4\sqrt{3})$.

Example 6

For the following circle, determine the radius r and the point (x, y) if $y = 3$.

y is opposite a 45° angle.
x is opposite a 45° angle.

$$y = \tfrac{1}{2}r\sqrt{2} \qquad |x| = |y|$$
$$3 = \tfrac{1}{2}r\sqrt{2} \qquad |x| = 3$$
$$6 = r\sqrt{2} \qquad\quad x = -3$$

◄ *In Quad. II,* $x < 0, y > 0.$

$$r = \frac{6}{\sqrt{2}}, \text{ or } 3\sqrt{2}$$

Thus, $r = 3\sqrt{2}$ and the point (x, y) is $(-3, 3)$.

Oral Exercises

State the length of the indicated side.

1.

2.

3.

4.

Written Exercises

Determine the lengths of the indicated sides. Rationalize all denominators.

(A) **1.**

2.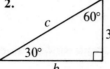

3.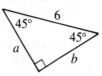

4.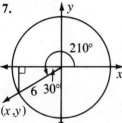

5.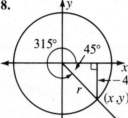

6.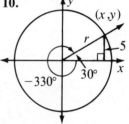

Determine the point (x, y) and the radius r.

7.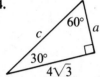

8.

9.

10.

Complete the chart.

	Length of hypotenuse	Length of leg opposite 30° ∠	Length of leg opposite 60° ∠
(B) **11.**	16	a	b
12.	c	9	b
13.	c	a	6

	Length of hypotenuse	Length of leg opposite 45° ∠	Length of leg opposite other 45° ∠
14.	c	3	b
15.	c	a	12
16.	14	a	b

Find the reference angle if its terminal side intersects a circle at the given point and the center of the circle is at the origin. Determine the radius of the circle.

(C) **17.** $(-3, 3)$ **18.** $(-\sqrt{3}, -1)$ **19.** $(4, -4\sqrt{3})$ **20.** $(5, -5)$ **21.** $(-2\sqrt{3}, -2)$
22. $(4, 4\sqrt{3})$ **23.** $(-5, 5\sqrt{3})$ **24.** $(-7, -7)$ **25.** $(4\sqrt{3}, 4)$ **26.** $(3, -3\sqrt{3})$
27. $(-a, a)$ **28.** $(m, m\sqrt{3})$ **29.** $(-r, r\sqrt{3})$ **30.** $(n\sqrt{3}, -n)$ **31.** $(-r, -r)$

32. Prove the 45°–45° right triangle rule.   **33.** Prove the 30°–60° right triangle rule.

SINE AND COSINE FUNCTIONS 17.3

Objectives
To find the sine and cosine of degree measures of 0°, 30°, 45°, 60°, 90° and their multiples
To determine if the sine or cosine is positive or negative for a given degree measure
To find the measure of an angle whose sine or cosine is given

Examine the four reference triangles (one in each quadrant) shown in the figures below.

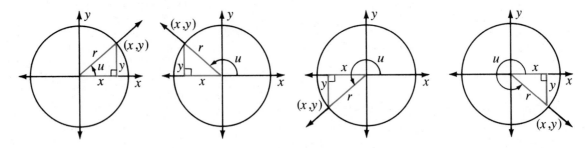

The ratios $\frac{y}{r}$ and $\frac{x}{r}$ in each of the figures determine trigonometric functions that are defined in terms of u as follows.

Definition:
Sine (sin) and
cosine (cos)
functions

For any angle of rotation u and any radius r, the sine of u is $\frac{y}{r}$ $\left(\sin u = \frac{y}{r}\right)$ and the cosine of u is $\frac{x}{r}$ $\left(\cos u = \frac{x}{r}\right)$. For a radius of 1, a unit circle, $\sin u = y$ and $\cos u = x$.

Example 1

Find sin 30°, cos 30°, sin 150°, and cos 150°.

For each angle, sketch a circle with any radius to show the angle of rotation and reference triangle.

Circle with radius 2

$$|y| = \tfrac{1}{2}r$$
$$= \tfrac{1}{2}(2) = 1$$

$$|x| = \tfrac{1}{2}(r)\sqrt{3}$$
$$= \tfrac{1}{2}(2)\sqrt{3} = \sqrt{3}$$

So, $y = 1$, $x = \sqrt{3}$,
$$\frac{y}{r} = \frac{1}{2} \text{ and } \frac{x}{r} = \frac{\sqrt{3}}{2}.$$

Circle with radius 1

$$|y| = \tfrac{1}{2}r$$
$$= \tfrac{1}{2}(1) = \tfrac{1}{2}$$

$$|x| = \tfrac{1}{2}(r)\sqrt{3}$$
$$= \frac{1}{2}(1)\sqrt{3} = \frac{\sqrt{3}}{2}$$

So, $y = \frac{1}{2}$, $x = -\frac{\sqrt{3}}{2}$

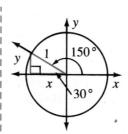

Thus, $\sin 30° = \dfrac{1}{2}$, $\cos 30° = \dfrac{\sqrt{3}}{2}$, $\sin 150° = \dfrac{1}{2}$, and $\cos 150° = -\dfrac{\sqrt{3}}{2}$.

Example 2   **Find sin 225°, cos 225°, sin 315°, and cos 315°.**

For each angle, sketch a circle with a radius of 2. Show the angle of rotation and reference triangle.

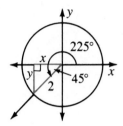

Ref. ∠: 45°

$$|y| = \tfrac{1}{2}(r)\sqrt{2}$$
$$= \tfrac{1}{2}(2)\sqrt{2} = \sqrt{2}$$

$$|x| = \tfrac{1}{2}(r)\sqrt{2}$$
$$= \tfrac{1}{2}(2)\sqrt{2} = \sqrt{2}$$

So, $y = -\sqrt{2}$, $x = -\sqrt{2}$,
$$\frac{y}{r} = \frac{-\sqrt{2}}{2} \text{ and } \frac{x}{r} = \frac{-\sqrt{2}}{2}.$$

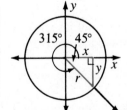

Ref. ∠: 45°

$$|y| = \tfrac{1}{2}(r)\sqrt{2}$$
$$= \tfrac{1}{2}(2)\sqrt{2} = \sqrt{2}$$

$$|x| = \tfrac{1}{2}(r)\sqrt{2}$$
$$= \tfrac{1}{2}(2)\sqrt{2} = \sqrt{2}$$

So, $y = -\sqrt{2}$, $x = \sqrt{2}$,
$$\frac{y}{r} = \frac{-\sqrt{2}}{2} \text{ and } \frac{x}{r} = \frac{\sqrt{2}}{2}.$$

Thus, $\sin 225° = \dfrac{-\sqrt{2}}{2}$, $\cos 225° = \dfrac{-\sqrt{2}}{2}$, $\sin 315° = -\dfrac{\sqrt{2}}{2}$ and $\cos 315° = \dfrac{\sqrt{2}}{2}$.

Examples 1 and 2 suggest that the sine is positive (+) in the first and second quadrants and negative (−) in the third and fourth quadrants. They also suggest that the cosine is positive in the first and fourth quadrants and negative in the second and third quadrants.

	1st Quad.	2nd Quad.	3rd Quad.	4th Quad.
Sine	+	+	−	−
Cosine	+	−	−	+

The next example shows how to express the sine and cosine of any angle as a function of an acute angle.

Example 3   **Write each in terms of the same trigonometric function of an acute angle: sin 240° and cos 315°.**

For each angle, sketch a circle. Show the reference triangle.

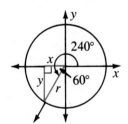

Ref. ∠: 60°

sin is negative
in quadrant III.

sin 240° = −sin 60°

Ref. ∠: 45°

cos is positive
in quadrant IV.

cos 315° = cos 45°

Thus, sin 240° = −sin 60° and cos 315° = cos 45°.

Recall that angles whose terminal sides are on an axis are called *quadrantal* angles. The quadrantal angles 0°, 90°, 180°, and 270° are shown below. Their sines and cosines are summarized in the table.

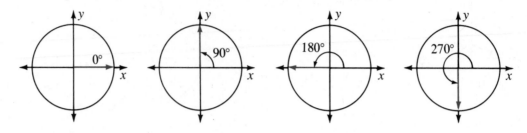

	0°	90°	180°	270°
sine: $\frac{y}{r}$	0	1	0	−1
cosine: $\frac{x}{r}$	1	0	−1	0

The following example illustrates how to solve an equation involving sin u or cos u.

Example 4

Find two values of u between 0° and 360° for which sin $u = -\dfrac{\sqrt{2}}{2}$.

First, determine the quadrant of the angle: sine is negative in the 3rd and 4th quadrants.

Second, determine the reference angle: $\sin 45° = \dfrac{\sqrt{2}}{2}$.

Finally, determine the angle of rotation.

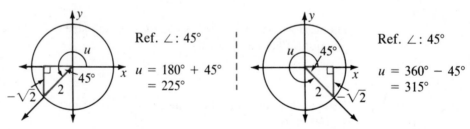

Ref. ∠: 45°

$u = 180° + 45°$
$\quad = 225°$

Ref. ∠: 45°

$u = 360° - 45°$
$\quad = 315°$

Thus, $\sin u = \dfrac{-\sqrt{2}}{2}$ for $u = 225°$ and $u = 315°$.

Summary

To find the measure of an angle whose sine or cosine is given:
1. Determine the quadrant from the sign (+ or −).

2. Determine the reference angle.

3. Use the reference angle to find the angle of rotation.

Reading in Algebra

Which value in each group is different from the others?

1. a) $\sin 45°$ **b)** $\sin 135°$ **c)** $\dfrac{\sqrt{2}}{2}$ **d)** $-\dfrac{\sqrt{2}}{2}$

2. a) $\cos 300°$ **b)** $\cos 60°$ **c)** $\cos 120°$ **d)** $\dfrac{\sqrt{3}}{2}$

3. a) $\sin 150°$ **b)** $\sin 210°$ **c)** $\sin 330°$ **d)** $-\dfrac{1}{2}$

4. a) $\dfrac{1}{2}$ **b)** $-\dfrac{1}{2}$ **c)** $\cos 120°$ **d)** $\cos 240°$

5. a) $\sin 0°$ **b)** 0 **c)** $\sin 180°$ **d)** 1

Written Exercises

(A) **Find each of the following. Rationalize any radical denominator.**

1. $\sin 60°$	**2.** $\cos 60°$	**3.** $\sin 45°$	**4.** $\cos 45°$	**5.** $\sin 30°$	**6.** $\cos 30°$
7. $\cos 270°$	**8.** $\cos 360°$	**9.** $\sin 180°$	**10.** $\cos 90°$	**11.** $\sin 0°$	**12.** $\sin 270°$
13. $\sin 135°$	**14.** $\cos 210°$	**15.** $\cos 150°$	**16.** $\sin 150°$	**17.** $\cos 180°$	**18.** $\cos 225°$
19. $\sin 315°$	**20.** $\sin 120°$	**21.** $\cos 330°$	**22.** $\sin 330°$	**23.** $\sin 390°$	**24.** $\cos 480°$

Write each in terms of the same trigonometric function of an acute angle.

25. $\cos 135°$	**26.** $\sin 330°$	**27.** $\cos 210°$	**28.** $\sin 225°$	**29.** $\sin 135°$	**30.** $\sin 300°$
31. $\sin 150°$	**32.** $\sin 210°$	**33.** $\sin 315°$	**34.** $\cos 300°$	**35.** $\cos 150°$	**36.** $\cos 330°$

(B) **Find two values of u between 0° and 360° for which each is true.**

37. $\sin u = \dfrac{\sqrt{2}}{2}$ **38.** $\cos u = \dfrac{1}{2}$ **39.** $\sin u = \dfrac{1}{2}$ **40.** $\cos u = 0$

41. $\cos u = -\dfrac{\sqrt{2}}{2}$ **42.** $\sin u = \dfrac{\sqrt{3}}{2}$ **43.** $\cos u = -\dfrac{\sqrt{3}}{2}$ **44.** $\sin u = -\dfrac{\sqrt{3}}{2}$

In which quadrants is each of the following true?

45. $\cos u < 0$ **46.** $\sin u > 0$ **47.** $\sin u < 0$ **48.** $\cos u > 0$

(C) **Find two values of u between 0° and 720° for which each is true.**

49. $\cos u = -\dfrac{1}{2}$ and $\sin u > 0$ **50.** $\sin u = \dfrac{1}{2}$ and $\cos u > 0$

51. $\cos u = \dfrac{\sqrt{3}}{2}$ and $\sin u < 0$ **52.** $\sin u = -\dfrac{\sqrt{2}}{2}$ and $\cos u < 0$

53. $\sin u = -\dfrac{\sqrt{3}}{2}$ and $\cos u > 0$ **54.** $\cos u = -\dfrac{\sqrt{2}}{2}$ and $\sin u < 0$

CUMULATIVE REVIEW

Complete the chart.

	0°	90°	180°	270°	360°
sin					
cos					

TANGENT AND COTANGENT FUNCTIONS

Objectives

To find the tangent and cotangent of degree measures 0°, 30°, 45°, 60°, 90°, and their multiples
To determine if the tangent or cotangent is positive or negative for a given degree measure
To find the measure of an angle whose tangent or cotangent is given
To find tangents and cotangents using trigonometric relationships

Recall that for any given angle of rotation, u, the two ratios $\frac{y}{r}$ and $\frac{x}{r}$ were defined as sin u and cos u, respectively. Using a reference triangle, the ratios $\frac{y}{x}$ and $\frac{x}{y}$ determine the trigonometric functions tangent and cotangent.

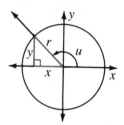

Definition: Tangent (tan) and cotangent (cot) functions	For any angle of rotation u and any radius r, the tangent of u is $\frac{y}{x}$ $\left(\tan u = \frac{y}{x}\right)$ and the cotangent of u is $\frac{x}{y}$ $\left(\cot u = \frac{x}{y}\right)$, $(x \neq 0, y \neq 0)$. For a unit circle, $\tan u = \frac{y}{x}$ and $\cot u = \frac{x}{y}$.

Example 1

Find tan 45°, cot 45°, tan 135°, and cot 135°.

For each angle, sketch a circle. Show the angle of rotation and reference triangle.

Circle with radius 2   *Unit circle*

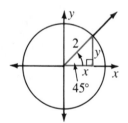

$|y| = \frac{1}{2}(r)\sqrt{2}$
$\quad = \frac{1}{2}(2)\sqrt{2} = \sqrt{2}$

$|x| = \frac{1}{2}(r)\sqrt{2}$
$\quad = \frac{1}{2}(2)\sqrt{2} = \sqrt{2}$

So, $y = \sqrt{2}$, $x = \sqrt{2}$,

$\frac{y}{x} = \frac{\sqrt{2}}{\sqrt{2}} = 1$ and

$\frac{x}{y} = \frac{\sqrt{2}}{\sqrt{2}} = 1$.

$|y| = \frac{1}{2}(r)\sqrt{2}$
$\quad = \frac{1}{2}(1)\sqrt{2} = \frac{\sqrt{2}}{2}$

$|x| = \frac{1}{2}(r)\sqrt{2}$
$\quad = \frac{1}{2}(1)\sqrt{2} = \frac{\sqrt{2}}{2}$

So, $y = \frac{\sqrt{2}}{2}$, $x = -\frac{\sqrt{2}}{2}$,

$\frac{y}{x} = \frac{\frac{\sqrt{2}}{2}}{\frac{-\sqrt{2}}{2}} = -1$, and $\frac{x}{y} = -1$.

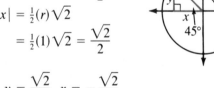

Thus, tan 45° = 1, cot 45° = 1, tan 135° = −1 and cot 135° = −1.

Example 2

Find tan 210°, cot 210°, tan 300°, and cot 300°.

For each angle, sketch a circle with a radius of 2. Show the angle of rotation and reference angle.

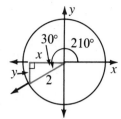

$$|y| = \tfrac{1}{2}(r)$$
$$= \tfrac{1}{2}(2) = 1$$

$$|x| = \tfrac{1}{2}(r)\sqrt{3}$$
$$= \tfrac{1}{2}(2)\sqrt{3} = \sqrt{3}$$

So, $y = -1$, $x = -\sqrt{3}$

$$\frac{y}{x} = \frac{-1}{-\sqrt{3}} = \frac{\sqrt{3}}{3} \text{ and}$$

$$\frac{x}{y} = \frac{-\sqrt{3}}{-1} = \sqrt{3}.$$

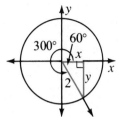

$$|y| = \tfrac{1}{2}(r)\sqrt{3}$$
$$= \tfrac{1}{2}(2)\sqrt{3} = \sqrt{3}$$

$$|x| = \tfrac{1}{2}(r)$$
$$= \tfrac{1}{2}(2) = 1$$

So, $y = -\sqrt{3}$, $x = 1$

$$\frac{y}{x} = \frac{-\sqrt{3}}{1} = -\sqrt{3} \text{ and}$$

$$\frac{x}{y} = \frac{1}{-\sqrt{3}} = -\frac{\sqrt{3}}{3}.$$

Thus, $\tan 210° = \dfrac{\sqrt{3}}{3}$, $\cot 210° = \sqrt{3}$, $\tan 300° = -\sqrt{3}$, and $\cot 300° = -\dfrac{\sqrt{3}}{3}$.

Examples 1 and 2 suggest that the tangent and cotangent are positive (+) in the first and third quadrants and negative (−) in the second and fourth quadrants.

	1st Quad.	2nd Quad.	3rd Quad.	4th Quad.
tangent	+	−	+	−
cotangent	+	−	+	−

The next example shows how to express the tan or cot of any angle as a function of an acute angle.

Example 3

Write each in terms of the same trigonometric function of an acute angle: tan 240° and cot 330°.

For each angle, sketch a circle. Show the reference triangle.

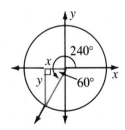

Ref. ∠: 60°

Tan is positive in quadrant III.

tan 240° = tan 60°

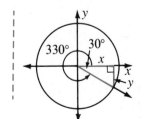

Ref. ∠: 30°

Cot is negative in quadrant IV.

cot 330° = −cot 30°.

Thus, tan 240° = tan 60° and cot 330° = −cot 30°.

Tangent and Cotangent Functions

Example 4

Find the tan and cot of the quadrantal angles 0°, 90°, 180°, and 270°.

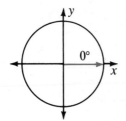 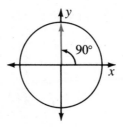 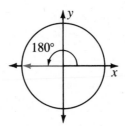 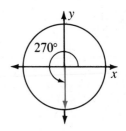

$x = r, y = 0$

$\tan 0° = \dfrac{y}{x} = \dfrac{0}{r}$

$\cot 0° = \dfrac{x}{y} = \dfrac{r}{0}$

$x = 0, y = r$

$\tan 90° = \dfrac{y}{x} = \dfrac{r}{0}$

$\cot 90° = \dfrac{x}{y} = \dfrac{0}{r}$

$x = -r, y = 0$

$\tan 180° = \dfrac{y}{x} = \dfrac{0}{-r}$

$\cot 180° = \dfrac{x}{y} = \dfrac{-r}{0}$

$x = 0, y = -r$

$\tan 270° = \dfrac{y}{x} = \dfrac{-r}{0}$

$\cot 270° = \dfrac{x}{y} = \dfrac{0}{-r}$

Thus, the tan and cot of 0°, 90°, 180°, and 270° are summarized in the table below.

	0°	90°	180°	270°
tangent	0	undefined	0	undefined
cotangent	undefined	0	undefined	0

Notice that $\tan u \cdot \cot u = \dfrac{y}{x} \cdot \dfrac{x}{y} = 1$. Since their product is 1, $\tan u$ and $\cot u$ are called *reciprocal* functions. That is, $\tan u = \dfrac{1}{\cot u}$ or $\cot u = \dfrac{1}{\tan u}$, for any *nonquadrantal* angle of rotation u.

Example 5

If $\tan x = -\dfrac{5}{3}$, find $\cot x$.

The reciprocal of $-\dfrac{5}{3}$ is $-\dfrac{3}{5}$.   **Thus,** $\cot x = -\dfrac{3}{5}$.

Example 6

Find two values of u between 0° and 360° for which $\tan u = \dfrac{-\sqrt{3}}{3}$.

Use the three steps in the summary on page 464.

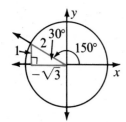

Tan is negative in quadrant II.

Ref. $\angle$: 30°

$u = 180° - 30°$
$\quad = 150°$

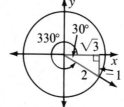

Tan is negative in quadrant IV.

Ref. $\angle$: 30°

$u = 360° - 30°$
$\quad = 330°$

Thus, $\tan u = -\dfrac{\sqrt{3}}{3}$, for $u = 150°$ and $u = 330°$.

Notice that $\dfrac{\sin u}{\cos u} = \dfrac{\frac{y}{r}}{\frac{x}{r}} = \dfrac{y}{x} \cdot \dfrac{r}{r} = \dfrac{y}{x} = \tan u$. So, $\tan u = \dfrac{\sin u}{\cos u}$.

Similarly, $\cot u = \dfrac{\cos u}{\sin u}$.

Quotient identities	$\tan u = \dfrac{\sin u}{\cos u}$	$\cot u = \dfrac{\cos u}{\sin u}$

Example 7   **If $\sin w = \dfrac{3}{5}$ and $\cos w = \dfrac{4}{5}$, find $\tan w$ and $\cot w$.**

$$\tan w = \frac{\sin w}{\cos w} = \frac{\frac{3}{5}}{\frac{4}{5}} = \frac{3}{4} \qquad \cot w = \frac{\cos w}{\sin w} = \frac{\frac{4}{5}}{\frac{3}{5}} = \frac{4}{3}$$

Written Exercises

(A) **Find each of the following. Rationalize all radical denominators.**

1. $\tan 45°$ **2.** $\cot 45°$ **3.** $\tan 60°$ **4.** $\cot 60°$ **5.** $\tan 30°$ **6.** $\cot 30°$

7. $\tan 90°$ **8.** $\cot 90°$ **9.** $\cot 315°$ **10.** $\tan 330°$ **11.** $\cot 180°$ **12.** $\tan 180°$

13. $\tan 135°$ **14.** $\cot 150°$ **15.** $\cot 300°$ **16.** $\cot 225°$ **17.** $\tan 210°$ **18.** $\cot 240°$

19. $\cot 450°$ **20.** $\tan 120°$ **21.** $\tan 390°$ **22.** $\tan 150°$ **23.** $\cot 360°$ **24.** $\tan 405°$

Write each of the following in terms of the same trigonometric function of an acute angle.

25. $\cot 110°$ **26.** $\tan 310°$ **27.** $\cot 230°$ **28.** $\tan 210°$ **29.** $\tan 340°$ **30.** $\cot 95°$

31. $\tan 250°$ **32.** $\cot 120°$ **33.** $\tan 150°$ **34.** $\cot 300°$ **35.** $\cot 250°$ **36.** $\tan 100°$

37. If $\cot w = -\dfrac{4}{3}$, find $\tan w$.   **38.** If $\tan u = \dfrac{5}{4}$, find $\cot u$.

39. If $\tan v = -\dfrac{17}{8}$, find $\cot v$.   **40.** If $\cot u = \dfrac{12}{5}$, find $\tan u$.

41. If $\sin m = \dfrac{4}{5}$ and $\cos m = -\dfrac{3}{5}$, find $\tan m$ and $\cot m$.

42. If $\cos w = -\dfrac{12}{13}$ and $\sin w = -\dfrac{5}{13}$, find $\tan w$ and $\cot w$.

43. If $\cos w = \dfrac{4\sqrt{41}}{41}$ and $\sin w = \dfrac{15\sqrt{41}}{41}$, find $\tan w$ and $\cot w$.

(B) **Find two values of u between 0° and 360° for which each is true.**

44. $\tan u = \sqrt{3}$ **45.** $\cot u = -\sqrt{3}$ **46.** $\tan u = -1$ **47.** $\cot u = 1$

48. $\cot u = -\dfrac{\sqrt{3}}{3}$ **49.** $\tan u = -\dfrac{\sqrt{3}}{3}$ **50.** $\cot u = -1$ **51.** $\tan u = 1$

(C) **Find two values of u between 0° and 720° for which each is true.**

52. $\cot u = -\sqrt{3}$ and $\cos u < 0$ **53.** $\tan u = -\sqrt{3}$ and $\sin u > 0$

54. $\tan u = \dfrac{\sqrt{3}}{3}$ and $\sin u < 0$ **55.** $\cot u = -\dfrac{\sqrt{3}}{3}$ and $\cos u > 0$

SECANT AND COSECANT FUNCTIONS   17.5

Objectives
To find the secant and cosecant of degree measures of 30°, 45°, 60°, 90°, and their multiples
To determine if the secant or cosecant is positive or negative for a given degree measure
To find the measure of an angle whose secant or cosecant is given
To find secants and cosecants using trigonometric relationships

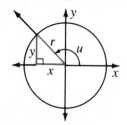

Thus far, four trigonometric functions have been defined for any angle of rotation u. Using a reference triangle, the ratios $\frac{r}{x}$ and $\frac{r}{y}$ determine the trigonometric functions **secant** and **cosecant**. Notice that these functions are the reciprocals of the cosine and sine functions, respectively.

Definition:
Secant (sec) and cosecant (csc) functions

For any angle of rotation u and any radius r, the secant of u is $\frac{r}{x}$ $\left(\sec u = \frac{r}{x}\right)$ and the cosecant of u is $\frac{r}{y}$ $\left(\csc u = \frac{r}{y}\right)$, $(x \neq 0, y \neq 0)$.

For a unit circle, $\sec u = \frac{1}{x}$ and $\csc u = \frac{1}{y}$. Also, $\sec u = \frac{1}{\cos u}$ or $\cos u = \frac{1}{\sec u}$, and $\csc u = \frac{1}{\sin u}$ or $\sin u = \frac{1}{\csc u}$.

Example 1

Find sec 120° and csc 120°.

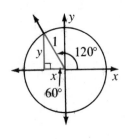

First Method: Unit circle
Use Ref. $\triangle$: 30–60 rt. $\triangle$.

$|y| = \frac{1}{2}(1)\sqrt{3} = \frac{\sqrt{3}}{2}$

$|x| = \frac{1}{2}(1) = \frac{1}{2}$

So, $y = \frac{\sqrt{3}}{2}$, $x = -\frac{1}{2}$

$\frac{r}{x} = \frac{1}{-\frac{1}{2}} = -2$, and

$\frac{r}{y} = \frac{1}{\frac{\sqrt{3}}{2}} = \frac{2\sqrt{3}}{3}$.

Second Method
Use reciprocal functions.

$\sec u = \frac{1}{\cos u}$ $\csc u = \frac{1}{\sin u}$

$\cos 120° = -\frac{1}{2}$ $\sin 120° = \frac{\sqrt{3}}{2}$

$\sec 120° = \frac{1}{-\frac{1}{2}}$ $\csc 120° = \frac{1}{\frac{\sqrt{3}}{2}}$

$\sec 120° = -2$ $\csc 120° = \frac{2}{\sqrt{3}}$ or $\frac{2\sqrt{3}}{3}$

Thus, $\sec 120° = -2$ and $\csc 120° = \frac{2\sqrt{3}}{3}$.

The sine, and therefore its reciprocal cosecant, are positive in the first and second quadrants. The cosine, and therefore its reciprocal secant, are positive in the first and fourth quadrants.

	1st Quad.	2nd Quad.	3rd Quad.	4th Quad.
secant	+	−	−	+
cosecant	+	+	−	−

The next example shows how to express the secant or cosecant of any angle as a function of an acute angle.

Example 2   **Write sec 135° and csc 240° as the same trigonometric function of an acute angle.**

For each angle, sketch a circle. Show the reference triangle.

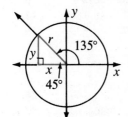

Ref. ∠: 45°
Sec is negative
in quadrant II.

sec 135° = −sec 45°

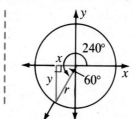

Ref. ∠: 60°
Csc is negative
in quadrant III.

csc 240° = −csc 60°

Thus, sec 135° = −sec 45° and csc 240° = −csc 60°.

Since the sec and csc are reciprocals of the cos and sin respectively, the following table can be constructed.

	0°	90°	180°	270°
sec	1	undefined	−1	undefined
csc	undefined	1	undefined	−1

Example 3   **If $\sin u = -\dfrac{3}{4}$, find csc u. If $\cos w = -\dfrac{5}{7}$, find sec w.**

$$\text{Csc } u = \frac{1}{\sin u} = \frac{1}{-\dfrac{3}{4}}, \text{ or } -\frac{4}{3}.$$

$$\text{Sec } w = \frac{1}{\cos w} = \frac{1}{-\dfrac{5}{7}}, \text{ or } -\frac{7}{5}.$$

Thus, csc $u = -\dfrac{4}{3}$.

Thus, sec $w = -\dfrac{7}{5}$.

Example 4   **Find two values of u between 0° and 360° for which sec $u = -\sqrt{2}$.**

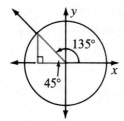

Sec is negative
in quadrant II.
Ref. ∠: 45°

$u = 180° − 45°$
$\quad = 135°$

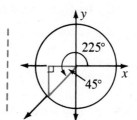

Sec is negative
in quadrant III.
Ref. ∠: 45°

$u = 180° + 45°$
$\quad = 225°$

Thus, sec $u = -\sqrt{2}$ for $u = 135°$ and $u = 225°$.

Oral Exercises

State whether each of the following is positive or negative.

1. sec 45°   **2.** csc 135°   **3.** sec 120°   **4.** csc 300°   **5.** sec 330°   **6.** sec 225°

7. csc 225°   **8.** sec 150°   **9.** csc 240°   **10.** sec 315°   **11.** csc 120°   **12.** csc 330°

Written Exercises

(A) **Find each of the following. Rationalize any radical denominator.**

1. csc 45°   **2.** sec 45°   **3.** sec 60°   **4.** csc 60°   **5.** sec 30°   **6.** csc 30°

7. csc 210°   **8.** sec 240°   **9.** csc 225°   **10.** sec 225°   **11.** csc 150°   **12.** sec 150°

13. sec 300°   **14.** csc 300°   **15.** sec 315°   **16.** csc 315°   **17.** sec 330°   **18.** csc 330°

Write each of the following in terms of the same trigonometric function of an acute angle.

19. sec 120°   **20.** csc 150°   **21.** csc 300°   **22.** sec 210°   **23.** sec 240°   **24.** csc 315°

25. csc 120°   **26.** csc 210°   **27.** sec 190°   **28.** sec 340°   **29.** csc 170°   **30.** csc 215°

31. If $\sin u = -\dfrac{1}{2}$, find csc u.   **32.** If $\cos u = -\dfrac{1}{3}$, find sec u.

33. If $\cos u = \dfrac{2}{3}$, find sec u.   **34.** If $\sin u = \dfrac{4}{5}$, find csc u.

35. If $\sin u = -\dfrac{5}{6}$, find csc u.   **36.** If $\cos u = -\dfrac{3}{7}$, find sec u.

(B) **Find all values of u between and including 0° and 360° for which each is true.**

37. sec $u = -1$ **38.** csc $u = -1$ **39.** csc $u = -\sqrt{2}$ **40.** sec $u = -2$

41. csc $u = \dfrac{2\sqrt{3}}{3}$ **42.** sec $u = \dfrac{-2\sqrt{3}}{3}$ **43.** sec $u = 2$ **44.** csc $u = 1$

(C) **Find all values of u between 0° and 720° for which each is true.**

45. sec $u = -1$ and $\sin u = 0$ **46.** sec $u = 2$ and $\cos u > 0$

47. csc $u = \dfrac{-2\sqrt{3}}{3}$ and $\tan u > 0$ **48.** csc $u = -2$ and $\cot u < 0$

NON-ROUTINE PROBLEMS

A triangle has two angles with measure 30° and 45°. The side opposite the 30° angle has length 12. Find the length of the side opposite the 45° angle.

CUMULATIVE REVIEW

Solve each of the following using the quadratic formula.

1. $x^2 - 9x + 3 = 0$ **2.** $3x^2 + 5x - 4 = 0$ **3.** $-2x^2 - 3x + 7 = 0$

FUNCTIONS OF NEGATIVE ANGLES 17.6

Objective

To find the sine, cosine, tangent, cotangent, secant, and cosecant of angles of negative measure

The figure below illustrates two angles of rotation, u and $(-u)$, where $P(-a, b)$ is a point on the terminal side of angle u and $Q(-a, -b)$ is a point on the terminal side of the angle $(-u)$. The two reference triangles formed are congruent. Using the definitions of sin, cos, and tan of an angle, it follows that

$$\sin u = \frac{y}{r} \text{ or } \frac{b}{r} \text{ and } \sin(-u) = \frac{y}{r} \text{ or } \frac{-b}{r}$$

$$\cos u = \frac{x}{r} \text{ or } \frac{-a}{r} \text{ and } \cos(-u) = \frac{x}{r} = \frac{-a}{r}$$

$$\tan u = \frac{y}{x} \text{ or } \frac{b}{-a} \text{ and } \tan(-u) = \frac{y}{x} \text{ or } \frac{-b}{-a} = \frac{b}{a}.$$

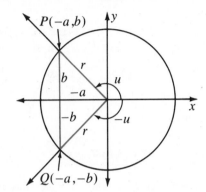

Since the cosecant, secant, and cotangent are reciprocal functions of the sine, cosine, and tangent, respectively, the following definitions are suggested.

Trigonometric functions of angles of negative measure	For any angles of rotation u and $(-u)$,	
	$\sin(-u) = -\sin u$	$\csc(-u) = -\csc u$
	$\cos(-u) = \cos u$	$\sec(-u) = \sec u$
	$\tan(-u) = -\tan u$	$\cot(-u) = -\cot u$.

Example 1 **Find sin (−45°), cos (−45°), and tan (−45°).**

Ref. ∠: 45°

$\sin 45° = \dfrac{\sqrt{2}}{2}$

$\cos 45° = \dfrac{\sqrt{2}}{2}$

$\tan 45° = 1$

$\sin(-45°) = -\sin 45°$

$\quad = -\dfrac{\sqrt{2}}{2}$

$\cos(-45°) = \cos 45°$

$\quad = \dfrac{\sqrt{2}}{2}$

$\tan(-45°) = -\tan 45°$

$\quad = -1$

Example 2 **Find cot (−330°), sec (−330°), and csc (−330°).**

Ref. ∠: 30°

$\cot 30° = \sqrt{3}$

$\sec 30° = \dfrac{2\sqrt{3}}{3}$

$\csc 30° = 2$

$\cot(-330°) = -\cot 330°$

$\quad = -(-\cot 30°)$

$\quad = \sqrt{3}$

$\sec(-330°) = \sec 330°$

$\quad = \sec 30°$

$\quad = \dfrac{2\sqrt{3}}{3}$

$\csc(-330°) = -\csc 330°$

$\quad = -(-\csc 30°)$

$\quad = 2$

Example 3   **Express sin (−315°) and cos (−315°) as functions of a positive acute angle.**

$$\sin(-315°) = -\sin 315°$$
$$= -(-\sin 45°)$$
$$= \sin 45°$$

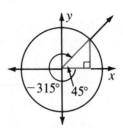

$$\cos(-315°) = \cos 315°$$
$$= \cos 45°$$

Example 4   **Express sec (−190°), tan (−300°), and csc (−210°) as functions of a positive acute angle.**

$$\sec(-190°) = \sec 190° \qquad \tan(-300°) = -\tan 300° \qquad \csc(-210°) = -\csc 210°$$
$$= -\sec 10° \qquad\qquad = \tan 60° \qquad\qquad = -(-\csc 30°)$$
$$= \csc 30°$$

Oral Exercises

Express each as a function of a positive acute angle.

1. $\sin(-10°)$	**2.** $\cos(-25°)$	**3.** $\tan(-50°)$	**4.** $\sec(-20°)$
5. $\csc(-70°)$	**6.** $\sin(-35°)$	**7.** $\cos(-40°)$	**8.** $\tan(-5°)$

Written Exercises

(A) **Find each of the following. Rationalize any radical denominator.**

1. $\sin(-30°)$	**2.** $\cos(-30°)$	**3.** $\tan(-30°)$	**4.** $\cot(-30°)$	**5.** $\sec(-30°)$
6. $\csc(-45°)$	**7.** $\sin(-60°)$	**8.** $\cos(-60°)$	**9.** $\sec(-60°)$	**10.** $\csc(-60°)$
11. $\cos(-150°)$	**12.** $\tan(-150°)$	**13.** $\sin(-150°)$	**14.** $\csc(-150°)$	**15.** $\cot(-150°)$
16. $\sec(-210°)$	**17.** $\csc(-210°)$	**18.** $\tan(-210°)$	**19.** $\sin(-210°)$	**20.** $\cos(-210°)$
21. $\cos(-315°)$	**22.** $\sin(-315°)$	**23.** $\cot(-315°)$	**24.** $\tan(-315°)$	**25.** $\sec(-315°)$

Express as a function of a positive acute angle.

26. $\cot(-15°)$	**27.** $\tan(-190°)$	**28.** $\cos(-280°)$	**29.** $\cot(-95°)$	**30.** $\sin(-260°)$
(B) **31.** $\sin(-100°)$	**32.** $\csc(-15°)$	**33.** $\sec(-18°)$	**34.** $\csc(-190°)$	**35.** $\cos(-35°)$
36. $\sec(-310°)$	**37.** $\cos(-350°)$	**38.** $\tan(-100°)$	**39.** $\sin(-350°)$	**40.** $\tan(-390°)$
41. $\csc(-195° \, 35')$		**42.** $\sin(-18° \, 20')$	**43.** $\cos(-300° \, 45')$	

(C) **For any angle of rotation u with negative measure show the following.**

44. $\tan(-u) = -\tan u$ **45.** $\cot(-u) = -\cot u$

46. $\sec(-u) = \sec u$ **47.** $\csc(-u) = -\csc u$

CUMULATIVE REVIEW

Solve each of the following by factoring.

1. $3x^2 - x = 0$ **2.** $9x^2 - 1 = 0$ **3.** $6x^2 + 7x + 2 = 0$

VALUES OF TRIGONOMETRIC FUNCTIONS

Objective

To find the other five trigonometric functions of an angle of rotation u, given one of the six functions and the quadrant of the terminal side of u

To find the value of five trigonometric functions of any angle of rotation u, given the value of one of them and the quadrant of the terminal side, do the following.
1. State two of the values of x, y, and r, using the definition of the given trigonometric function.
2. Find the remaining value of x, y, or r, using the Pythagorean theorem and the reference triangle.
3. Write the ratios for the other five trigonometric functions using x, y, and r.

Example 1

Find the value of each of the other five trigonometric functions of u if $\sin u = \frac{4}{5}$ and the terminal side is in the 2nd quadrant.

$$\sin u = \frac{4}{5} = \frac{y}{r} \text{ or } y = 4 \text{ and } r = 5$$

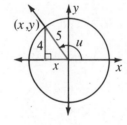

$\begin{aligned} x^2 + y^2 &= r^2 \\ x^2 + 16 &= 25 \\ x^2 &= 9 \\ |x| &= 3 \\ x &= -3 \end{aligned}$ ◀ *Use the Pythagorean theorem to find x.*

◀ *The terminal side is in the 2nd quadrant.*

Thus, $\cos u = -\frac{3}{5}$, $\tan u = -\frac{4}{3}$, $\cot u = -\frac{3}{4}$, $\sec u = -\frac{5}{3}$, and $\csc u = \frac{5}{4}$.

Example 2

Find the value of each of the other five trigonometric functions of u if $\tan u = -\frac{5}{12}$ and $\cos u < 0$.

Both the tan and cos are negative in the 2nd quadrant, so the terminal side lies in the 2nd quadrant.

$$\tan u = -\frac{5}{12} = \frac{y}{x} \text{ or } y = 5 \text{ and } x = -12$$

$\begin{aligned} x^2 + y^2 &= r^2 \\ 144 + 25 &= r^2 \\ r^2 &= 169 \text{ or } r = 13 \end{aligned}$ ◀ *Use the Pythagorean theorem.*

Thus, $\sin u = \frac{5}{13}$, $\cos u = -\frac{12}{13}$, $\cot u = -\frac{12}{5}$, $\sec u = -\frac{13}{12}$, and $\csc u = \frac{13}{5}$.

Oral Exercises

For each of the following, tell in which quadrant(s) the terminal side of the angle may lie.

1. $\sin u = \dfrac{4}{5}$ **2.** $\sin u = -\dfrac{3}{4}$ **3.** $\cos u = -\dfrac{5}{7}$ **4.** $\cos u = \dfrac{7}{8}$ **5.** $\tan u = -\dfrac{3}{4}$

6. $\tan u = \dfrac{\sqrt{5}}{3}$ **7.** $\cot u = -\dfrac{5}{12}$ **8.** $\cot u = \dfrac{12}{5}$ **9.** $\sec u = \dfrac{13}{5}$ **10.** $\sec u = -\dfrac{13}{12}$

11. $\sin u > 0$ and $\tan u < 0$ **12.** $\sin u < 0$ and $\cos u < 0$

Written Exercises

(A) **Find the value of each of the other five trigonometric functions of u.**

1. $\tan u = -\dfrac{12}{5}$ in 2nd quad.   **2.** $\sin u = \dfrac{12}{13}$ in 2nd quad.   **3.** $\cos u = -\dfrac{5}{13}$ in 3rd quad.

4. $\sin u = -\dfrac{3}{5}$ in 4th quad.   **5.** $\cot u = -\dfrac{3}{4}$ in 4th quad.   **6.** $\sec u = -\dfrac{13}{5}$ in 2nd quad.

7. $\cos u = -\dfrac{12}{13}$ in 2nd quad.   **8.** $\csc u = \dfrac{13}{5}$ in 1st quad.   **9.** $\tan u = \dfrac{4}{3}$ in 3rd quad.

10. $\sec u = \dfrac{13}{12}$ in 4th quad.   **11.** $\tan u = -1$ in 2nd quad.   **12.** $\cot u = -\sqrt{3}$ in 4th quad.

(B) **13.** $\sin u = \dfrac{5}{13}$ and $\sec u > 0$ **14.** $\cos u = -\dfrac{3}{5}$ and $\tan u < 0$

15. $\cot u = 1$ and $\cos u > 0$ **16.** $\tan u = -\dfrac{3}{4}$ and $\sin u > 0$

17. $\tan u = \dfrac{\sqrt{3}}{3}$ and $\cos u > 0$ **18.** $\cos u = \dfrac{\sqrt{2}}{2}$ and $\csc u < 0$

19. $\sec u = -\dfrac{2\sqrt{3}}{3}$ and $\sin u > 0$ **20.** $\csc u = -\sqrt{2}$ and $\tan u < 0$

(C) **Find the value(s) of each of the other trigonometric functions of u.**

21. $\sin u = \dfrac{\sqrt{3}}{2}$ and $\tan u = -\sqrt{3}$ **22.** $\cos u = -\dfrac{1}{2}$ and $\cot u = \dfrac{\sqrt{3}}{3}$

23. $\cos u = \dfrac{\sqrt{2}}{2}$ and $\sin u = -\dfrac{\sqrt{2}}{2}$ **24.** $\tan u = -1$ and $\sec u = -\sqrt{2}$

25. $\sec u = -\dfrac{13}{5}$ and $\tan u = \dfrac{12}{5}$ **26.** $\csc u = \dfrac{13}{12}$ and $\sin u = \dfrac{12}{13}$

27. $\tan u = -\dfrac{5}{4}$ and $\cos u = \dfrac{4\sqrt{41}}{41}$ **28.** $\cot u = -\sqrt{3}$ and $\cos u = -\dfrac{\sqrt{3}}{2}$

29. $\sin u = -\dfrac{7}{25}$ and $\csc u = -\dfrac{25}{7}$ **30.** $\sec u = -\dfrac{5}{4}$ and $\tan u = -\dfrac{3}{4}$

CUMULATIVE REVIEW

Graph each of the following. Find the domain and the range. Determine the coordinates of the vertex and an equation of the axis of symmetry.

1. $y = x^2$ **2.** $y = -x^2$ **3.** $y = x^2 - 2x - 8$

USING TRIGONOMETRIC TABLES 17.8

Objectives   **To determine the sin, cos, tan, and cot of any angle *u* reading a trigonometric table**
To rewrite trigonometric functions using cofunctions
To find angle *u* by reading a trigonometric table when the sin *u*, cos *u*, tan *u*, or cot *u* is given

You have been finding trigonometric functions of angles with special measures like 30°, 45°, 60° and their multiples. You will now learn how to find trigonometric functions of angles with any measure by using a table. In the table, angle measure is given in multiples of 10 minutes. One degree equals sixty minutes (1° = 60′).

Example 1   **Find sin 23° 30′ and cot 24° 50′ using the portion of a trigonometric table shown below.**

The functions at the *top* of the table must be used with degree measure in the left-hand column, (0° to 45°).

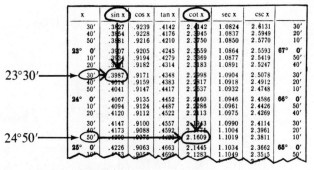

Thus, sin 23° 30′ = .3987 and cot 24° 50′ = 2.1609.

Example 2   **Find cos 45° 10′ and tan 46° 40′.**

The functions at the *bottom* of the table must be used with degree measure in the right-hand column, (45° to 90°).

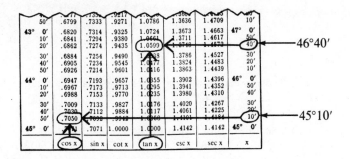

Thus, cos 45° 10′ = .7050 and tan 46° 40′ = 1.0599.

Since the trigonometric tables only show measures of acute angles, functions of larger angles must be written as functions of acute angles.

Example 3

Find sin 143° 20′ and cos 250° 40′.

Write each as the same trigonometric function of an acute angle. Sketch the angle of rotation and reference triangle.

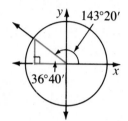

$$\begin{array}{r} 179°\ 60' \\ -\ 143°\ 20' \\ \hline 36°\ 40' \end{array}$$

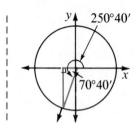

$$\begin{array}{r} 250°\ 40' \\ -\ 180° \\ \hline 70°\ 40' \end{array}$$

sin 143° 20′ = sin 36° 40′
            = .5972

cos 250° 40′ = −cos 70° 40′
              = −.3311

Thus, sin 143° 20′ = .5972, and cos 250° 40′ = −.3311.

Example 4

Find the value of each pair from the table.

sin 70° and cos 20°   tan 33° 40′ and cot 56° 20′

.9397   .6661

Thus, sin 70° = cos 20° = .9397, and tan 33° 40′ = cot 56° 20′ = .6661.

Recall that two angles are complementary if the sum of their measures is 90°. The sine and *co*sine functions are called *complementary functions, or cofunctions.* Other examples of cofunctions are tangent and *co*tangent as well as secant and *co*secant. Example 4 suggests the following relationships concerning cofunctions.

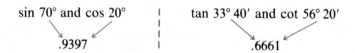

Cofunction relationships	$\sin u = \cos (90° - u)$ and $\cos u = \sin (90 - u)$
	$\tan u = \cot (90° - u)$ and $\cot u = \tan (90 - u)$
	$\sec u = \csc (90° - u)$ and $\csc u = \sec (90 - u)$

Example 5

Express each function in terms of its cofunction.
     cos 63°   **sec 13° 40′**   **cot 38° 10′**

Use the cofunction and the complement.
     cos 63°   sec 13° 40′   cot 38° 10′
       ↓   ↓   ↓
    sin 27°   csc 76° 20′   tan 51° 50′

Example 6

Express each as a function of an angle measure less than 45°. Use cofunctions.

$$\text{sin } 100° \qquad\qquad \text{tan } 290°$$

$sin\ (90° - u) = cos\ u$ ▶

$$\begin{aligned} \text{sin } 100° &= \text{sin } 80° & \text{tan } 290° &= -\text{tan } 70° \\ &= \text{cos } 10° & &= -\text{cot } 20° \end{aligned}$$ ◀ $tan\ (90° - u) = cot\ u$

Thus, sin 100° = cos 10° and tan 290° = −cot 20°.

You can find u if sin u, cos u, tan u, or cot u is given.

Example 7

sin u = .5200	**cos u = .8465**	**tan u = 1.7205**
Find u if u < 90°.	**Find u if u < 90°.**	**Find u if u < 90°.**

Look in the table to find the value, then find the corresponding angle u.

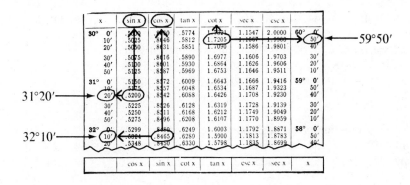

Thus, if sin u = .5200, u = 31° 20′; if cos u = .8465, u = 32° 10′; and if tan u = 1.7205, u = 59° 50′.

The next example illustrates how to solve a trigonometric equation using a table of values.

Example 8

Find all values of u, 0° < u < 360°, for which tan u = −.6168.

Read the table in Example 7. The measure of the ref. ∠ is 31° 40′.

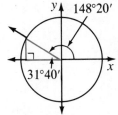

The tangent is negative in the 2nd and 4th quadrants.

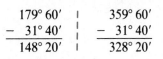

$$\begin{array}{r} 179°\ 60' \\ -\ 31°\ 40' \\ \hline 148°\ 20' \end{array} \qquad \begin{array}{r} 359°\ 60' \\ -\ 31°\ 40' \\ \hline 328°\ 20' \end{array}$$

Thus, u = 148° 20′ or u = 328° 20′.

Written Exercises

(A) Use the table to find each of the following.
1. sin 24°
2. tan 78°
3. cos 23°
4. cot 43°
5. tan 47° 40′
6. sin 70° 30′
7. cot 23° 20′
8. cos 43° 30′
9. sin 60° 40′
10. tan 82° 50′
11. tan 39° 50′
12. sin 42° 30′
13. tan 52° 50′
14. cos 73° 40′
15. cot 50° 10′
16. cos 152°
17. cos 138°
18. sin 340°
19. cot 155°
20. sin 260°
21. cot 190° 40′
22. tan 340° 50′
23. cot 140° 10′
24. tan 100° 40′
25. cos 350° 30′

Use the table to find u, $u < 90°$.
26. sin u = .9293
27. cos u = .9652
28. tan u = .5930
29. cot u = .3607
30. tan u = .7954
31. sin u = .7373
32. cos u = .7333
33. sin u = .3773
34. cot u = 6.3138
35. tan u = 3.8667
36. cot u = .3249
37. cos u = .2363

Express each in terms of its cofunction.
38. sin 35°
39. cos 25°
40. tan 68°
41. cot 10°
42. sin 48°
43. cos 50° 10′
44. tan 40° 40′
45. cot 55° 20′
46. sin 17° 40′
47. cos 34° 30′

(B) Find all values of u, $0° < u < 360°$, for which each is true.
48. cos u = .9636
49. tan u = −2.7725
50. cot u = 1.2723
51. cot u = −.2648
52. cos u = −.4874
53. sin u = −.9283
54. sin u = .8124
55. sin u = .1650
56. cos u = .7808

Express as a function of an angle measure less than 45°.
57. tan 120°
58. cot 200°
59. sin 316°
60. cos 110°
61. sin 224°
62. sin 324° 20′
63. cos 125° 10′
64. tan 212° 40′
65. cot 331° 50′
66. tan 134° 40′

(C) Express each as a reciprocal function of an acute angle.
67. cos 112°
68. sin 140° 10′
69. sin 176° 40′
70. tan 190° 40′
71. tan 304° 20′
72. cot 123° 40′
73. cos 340° 30′
74. cot 129° 20′

Find each of the following.
75. sec 35° 20′
76. csc 150° 20′
77. csc 50° 40′
78. sec 190° 40′

Find u if $u < 90°$.
79. sec u = 1.0003
80. csc u = 1.0712
81. csc u = 1.9801
82. sec u = 2.1657

Find all values of u, $0° < u < 360°$, for which each is true.
83. sec u = 1.1969
84. csc u = −6.8998
85. sec u = −1.2015
86. csc u = 1.6123
87. sec u = 2.7904
88. csc u = −2.5770

CUMULATIVE REVIEW

Write each expression in logarithmic form. Do not solve.
1. $\dfrac{45\sqrt{13.5}}{\sqrt[3]{38}}$

2. $\dfrac{98^3\sqrt{48.7}}{24(105)}$

3. $\dfrac{86.45(425.6)^3}{.00735}$

LINEAR TRIGONOMETRIC EQUATIONS   17.9

Objective   **To solve linear trigonometric equations**

You have learned how to determine all
values of u, $0° \leq u \leq 360°$, that make equations
like $\sin u = -\frac{1}{2}$ true. For $\sin u = -\frac{1}{2}$,
the reference angle is 30°. The sin is
negative in the 3rd and 4th quadrants. So,
$u = 180° + 30°$ or 210° and $360° - 30°$
or 330°.

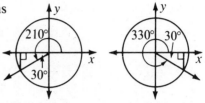

Example 1   **Determine the values of u between 0° and 360° to the nearest ten minutes that
make the equation $3 \sin u - 1 = 0$ true.**

Solve for $\sin u$, write in decimal form, and determine the reference angle.

$$3 \sin u - 1 = 0$$
$$3 \sin u = 1$$
$$\sin u = \frac{1}{3}, \text{ or } .3333$$
$$\text{Ref. } \angle: 19° \, 30'$$

◀ *Use the trigonometric table
to find closest value.*

$u = 19° \, 30' \qquad u = 180° - 19° \, 30'$
$\qquad\qquad\qquad = 160° \, 30'$

◀ *Sin is positive in the 1st and
2nd quadrants.*

Thus, $u = 19° \, 30'$ or $u = 159° \, 30'$.

Example 2   **Determine the values of θ between 0° and 360° to the nearest ten minutes that
make the equations true.**
$5 \tan \theta - 3 = -7$ $\qquad\qquad\qquad$ **$-3 \cos \theta - 4 = 5 \cos \theta + 2$**

Solve for the trigonometric function. Determine the reference angle.

$5 \tan \theta - 3 = -7$	$-3 \cos \theta - 4 = 5 \cos \theta + 2$
$5 \tan \theta = -4$	$-6 = 8 \cos \theta$
$\tan \theta = -\frac{4}{5}, \text{ or } -.8000$	$\cos \theta = -\frac{6}{8}, \text{ or } -.7500$
Ref. $\angle: 38° \, 40'$	Ref. $\angle: 41° \, 20'$
$\theta = 180° - 38° \, 40'$	$\theta = 180° - 41° \, 20'$
$\quad = 141° \, 20'$ (2nd quad.)	$\quad = 138° \, 40'$
$\theta = 360° - 38° \, 40'$	$\theta = 180° + 41° \, 20'$
$\quad = 321° \, 20'$ (4th quad.)	$\quad = 221° \, 20'$
Thus, $\theta = 141° \, 20'$ or $321° \, 20'$.	**Thus, $\theta = 138° \, 40'$ or $221° \, 20'$.**

Example 3 **Determine the values of t between $0°$ and $360°$ that make the equation $4 \sin t - 3 = 2 \sin t + 5$ true.**

$$4 \sin t - 3 = 2 \sin t + 5$$
$$2 \sin t = 8$$
$$\sin t = 4$$

Thus, there are no solutions, since $-1 \le \sin t \le 1$.

Written Exercises

Determine the values of θ between $0°$ and $360°$ to the nearest ten minutes that make each equation true.

(A) **1.** $2 \sin \theta = 1$
2. $2 \cos \theta - \sqrt{3} = 0$
3. $3 \sin \theta - 3 = 0$
4. $2 \cos \theta = -\sqrt{3}$
5. $3 \sin \theta + 2 = -1$
6. $6 \cos \theta - 2 = 0$
7. $\cot \theta = -1$
8. $3 \tan \theta - 1 = 0$
9. $\sqrt{2} \cos \theta + 1 = 0$
10. $2 \sin \theta + 1 = 0$
11. $4 \cos \theta = 6$
12. $3 \tan \theta = -\sqrt{3}$
13. $6 \cos \theta - 3 = 4 \cos \theta - 2$
14. $5 \sin \theta + 3 = 2 \sin \theta$
15. $7 \tan \theta - 1 = 2 \tan \theta + 4$
16. $-5 \sin \theta - 6 = 5 \sin \theta - 2$
17. $-8 \tan \theta - 12 = 6 \tan \theta - 1$
18. $3 \cos \theta - 4 = 5 \cos \theta - 5$

Determine the values of θ between $0°$ and $720°$ to the nearest ten minutes that make each equation true.

(B) **19.** $2 \cos \theta + \sqrt{3} = 0$
20. $\tan \theta + 1 = 0$
21. $-\cot \theta = -\sqrt{3}$
22. $5 \sin \theta - 8 = 3 \sin \theta - 9$
23. $-3 \cos \theta - 5 = -6 \cos \theta$
24. $-8 \sin \theta + 6 = -9 \sin \theta + 7$
25. $\sqrt{3} \tan \theta - 2 = 1$
26. $3 \sin \theta - 1 = -2 \sin \theta + 3$
27. $-4 \cot \theta - 7 = \cot \theta$

(C) **28.** $3 \sec \theta + 8 = 0$
29. $\sqrt{3} \csc \theta = 5\sqrt{3}$
30. $-\sec \theta - 12 = 4 \sec \theta$
31. $-4 \csc \theta = 4$
32. $-2 \sec \theta + 8.4 = 0$
33. $-3 \sec \theta + 1 = 0$
34. $\dfrac{\sec \theta}{4} - 3 = \dfrac{5 \sec \theta}{6}$
35. $\dfrac{1}{3} \csc \theta - \dfrac{3}{5} = 1$
36. $\dfrac{\csc \theta - 6}{3} = \dfrac{3}{4} \csc \theta + 1$

Determine the values of θ between $0°$ and $360°$ that make each equation true.

37. $\cos \theta = \sin \theta$
38. $-\sin \theta = \cos \theta$
39. $\tan \theta = \cot \theta$
40. $\sin \theta = \csc \theta$
41. $\cos \theta = \sec \theta$
42. $-\sec \theta = \cos \theta$

CALCULATOR ACTIVITIES

Example **Determine the values of θ between $0°$ and $360°$ to the nearest ten minutes that make the equation $\sqrt{37} \tan \theta - 8 = 0$ true.**

1. Solve for $\tan \theta$: $\tan \theta = \dfrac{8}{\sqrt{37}}$.

2. Use a calculator to find $\sqrt{37} \doteq 6.0827625$.

3. Compute $\dfrac{8}{6.0828}$ to equal 1.3151837.

4. Look in the table to find $\theta = 52°50'$ or $\theta = 232°50'$.

INTERPOLATION WITH TRIGONOMETRY   17.10

Objectives   **To use interpolation to determine functions of an angle measured to the nearest minute**
To solve an equation of the form $a \tan \theta = b$, θ to the nearest minute

You have learned how to use a table to find trigonometric values for any angle measure in multiples of 10 minutes. You will now learn how to find trigonometric values for any degree and minute measure. The process of finding values for measures not in the tables is called **interpolation.**

Example 1   **Find sin 36° 33′.**

Look in the table for known values: 36° 33′ lies between the values 36° 30′ and 36° 40′. Make a table.

	measure	sine
$10\begin{bmatrix} 3\begin{bmatrix} 36°\ 30' \\ 36°\ 33' \\ 36°\ 40' \end{bmatrix} \end{bmatrix}$		$\begin{bmatrix} .5948 \\ .5972 \end{bmatrix}n \Big].0024$

Write a proportion and solve it.

$$\frac{n}{.0024} = \frac{3}{10}$$
$$10n = .0072$$
$$n = .00072, \text{ or } n \doteq .0007$$

Add. ▶ $.5948 + .0007 = .5955$ **Thus, sin 36° 33′ = .5955.**

Example 2   **Find cos 311° 33′.**

The reference ∠ is 48° 27′.
Look in the table for known values: 48° 27′ lies between the values of 48° 20′ and 48° 30′. Make a table.

	measure	cosine
$10\begin{bmatrix} 7\begin{bmatrix} 48°\ 20' \\ 48°\ 27' \\ 48°\ 30' \end{bmatrix} \end{bmatrix}$		$\begin{bmatrix} .6648 \\ .6626 \end{bmatrix}n \Big]-.0022$

Write a proportion and solve it.

$$\frac{n}{-.0022} = \frac{7}{10}$$
$$10n = -.0154$$
$$n = .00154, \text{ or } n \doteq -.0015$$

Add. ▶ $.6648 + (-.0015) = .6633$ **Thus, cos 311° 33′ = .6633.**

The next example illustrates how to find u, if a trigonometric function of u is given.

Example 3

If tan u = .3463 ($0° < u < 90°$), find u to the nearest minute.

Look in the table for known values: .3463 lies between .3443 and .3476.

$$
10\left[n\begin{bmatrix}19°\ 0' & .3443 \\ .3463 \end{bmatrix}.0020 \\ 19°\ 10' \quad .3476 \end{bmatrix}.0033 \right.
$$

measure	tangent
19° 0'	.3443
	.3463
19° 10'	.3476

Write a proportion and solve it.

$$\frac{n}{10} = \frac{.0020}{.0033} = \frac{20}{33}$$
$$33n = 200$$
$$n = \frac{200}{33},\ \text{or } n \doteq 6$$

Add. ▶ 19° 0' + 6' = 19° 6'   **Thus, u = 19° 6'.**

Example 4

Find all values of θ ($0° < \theta < 360°$) to the nearest minute that make the equation $-4\tan\theta - 3 = -7\tan\theta + 5$ true.

Solve for tan θ. Determine the reference angle.
$$-4\tan\theta - 3 = -7\tan\theta + 5$$
$$3\tan\theta = 8$$
$$\tan\theta = \frac{8}{3},\ \text{or } 2.6667$$

Use 2.6667 to find the reference angle.

measure	tangent
69° 20'	2.6511
	2.6667
69° 30'	2.6746

Write a proportion and solve.

$$\frac{n}{10} = \frac{.0156}{.0235} = \frac{156}{235}$$
$$235n = 1560$$
$$n = \frac{1560}{235} \doteq 6$$

Add. ▶ 69° 20' + 6' = 69° 26'

Tan is positive in the 1st and 3rd quadrants. ▶ $\theta = 69°\ 26'$
$\theta = 180° + 69°\ 26' = 249°\ 26'$

Written Exercises

(A) **Find each of the following.**

1. $\sin 43° 15'$
2. $\cos 18° 32'$
3. $\tan 12° 57'$
4. $\cot 62° 38'$
5. $\cos 72° 18'$
6. $\tan 55° 8'$
7. $\cos 33° 42'$
8. $\cos 87° 53'$
9. $\tan 31° 43'$
10. $\cot 12° 24'$
11. $\sin 73° 47'$
12. $\sin 16° 18'$

Find u ($0° < u < 90°$) to the nearest minute.

13. $\cos u = .9753$
14. $\tan u = .1573$
15. $\sin u = .2875$
16. $\sin u = .9264$
17. $\cos u = .9214$
18. $\cot u = 3.5352$
19. $\cot u = .3567$
20. $\sin u = .7954$
21. $\cos u = .4535$

Find all values of θ ($0° < \theta < 360°$) to the nearest minute.

22. $2 \cos \theta + 1 = 0$
23. $5 \sin \theta - 2 = 3 \sin \theta - 1$
24. $6 \cos \theta - 4 = 3 - 3 \cos \theta$
25. $-3 \tan \theta + 5 = 4 + 3 \tan \theta$
26. $-2 \cot \theta + 8 = -5 \cot \theta$
27. $-7 \sin \theta + 5 = 7 - \sin \theta$

(B) **Find each of the following.**

28. $\cos 124° 43'$
29. $\sin 312° 24'$
30. $\tan 86° 57'$
31. $\cot 204° 36'$
32. $\sin 197° 22'$
33. $\cot 342° 53'$
34. $\cos 349° 45'$
35. $\tan 224° 27'$
36. $\sec 134° 38'$
37. $\csc 248° 17'$
38. $\csc 153° 54'$
39. $\sec 354° 48'$

Find u ($0° < u < 360°$) to the nearest minute.

40. $\tan u = .4272$
41. $\cos u = -.9177$
42. $\cot u = 1.4650$
43. $\sin u = .5987$
44. $\sin u = -.8221$
45. $\cos u = -.0650$

(C) **Find all values of θ ($0° \le \theta < 360°$).**

46. $\dfrac{\sin \theta}{\cos \theta} = -1$
47. $\dfrac{\cos \theta}{\sin \theta} = -\dfrac{\sqrt{3}}{3}$
48. $\cot \theta \cdot \sin \theta = \dfrac{\sqrt{3}}{2}$

49. $\sin \theta \sec \theta = -3.2658$
50. $\csc \theta \cos \theta = -2.4536$

51. $\dfrac{\cot \theta \cdot \tan \theta}{\sec \theta} = -.7654$
52. $\tan \theta \cdot \csc \theta \cdot (\cos \theta)^2 = -.3857$

53. $\sec^2 \theta - 2 = 0$
54. $4 \csc \theta + 6 = \csc \theta$

55. $2 \csc \theta - \dfrac{4}{\sin \theta} = -8$
56. $\dfrac{1}{\sec \theta} + 8 \cos \theta = \sec \theta \cdot \cos \theta$

CUMULATIVE REVIEW

Simplify.

1. $\dfrac{1 - \dfrac{1}{3}}{4 + \dfrac{3}{5}}$

2. $\dfrac{\dfrac{2}{3} - \dfrac{4}{5}}{\dfrac{5}{2} + \dfrac{7}{9}}$

3. $\dfrac{4 - \dfrac{5}{4x}}{7 + \dfrac{4}{3x}}$

TRIGONOMETRY OF A RIGHT TRIANGLE

Objectives
**To find trigonometric functions of acute angles in a right triangle
To simplify trigonometric expressions involving special angles**

Remember that for any angle of rotation there is an associated reference right triangle with six trigonometric ratios. You will now learn about trigonometry of right triangles. The following diagrams will help make comparisons.

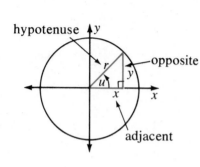

$$\sin u = \frac{y}{r} \Rightarrow \sin A = \frac{\text{length of opposite side}}{\text{length of hypotenuse}}$$

$$\cos u = \frac{x}{r} \Rightarrow \cos A = \frac{\text{length of adjacent side}}{\text{length of hypotenuse}}$$

$$\tan u = \frac{y}{x} \Rightarrow \tan A = \frac{\text{length of opposite side}}{\text{length of adjacent side}}$$

$$\cot u = \frac{x}{y} \Rightarrow \cot A = \frac{\text{length of adjacent side}}{\text{length of opposite side}}$$

Example 1   **For each right triangle, find sin 30°, cos 30°, tan 30°, and cot 30°.**

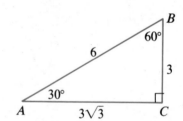

$\sin A = \dfrac{\text{opposite side}}{\text{hypotenuse}} \Rightarrow \sin 30° = \dfrac{1}{2}$

$\cos A = \dfrac{\text{adjacent side}}{\text{hypotenuse}} \Rightarrow \cos 30° = \dfrac{\sqrt{3}}{2}$

$\tan A = \dfrac{\text{opposite}}{\text{adjacent}} \Rightarrow \tan 30° = \dfrac{1}{\sqrt{3}}$, or $\dfrac{\sqrt{3}}{3}$

$\cot A = \dfrac{\text{adjacent}}{\text{opposite}} \Rightarrow \cot 30° = \dfrac{\sqrt{3}}{1}$, or $\sqrt{3}$

$\sin 30° = \dfrac{3}{6}$, or $\dfrac{1}{2}$

$\cos 30° = \dfrac{3\sqrt{3}}{6}$, or $\dfrac{\sqrt{3}}{2}$

$\tan 30° = \dfrac{3}{3\sqrt{3}}$, or $\dfrac{\sqrt{3}}{3}$

$\cot 30° = \dfrac{3\sqrt{3}}{3}$, or $\sqrt{3}$

Thus, $\sin 30° = \dfrac{1}{2}$, $\cos 30° = \dfrac{\sqrt{3}}{2}$, $\tan 30° = \dfrac{\sqrt{3}}{3}$, and $\cot 30° = \sqrt{3}$.

Example 2   **For the given triangle, find sin 60°, cos 60°, tan 60°, and cot 60°.**

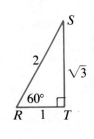

$$\sin R = \frac{\text{opp.}}{\text{hyp.}} \Rightarrow \sin 60° = \frac{\sqrt{3}}{2}$$

$$\cos R = \frac{\text{adj.}}{\text{hyp.}} \Rightarrow \cos 60° = \frac{1}{2}$$

$$\tan R = \frac{\text{opp.}}{\text{adj.}} \Rightarrow \tan 60° = \frac{\sqrt{3}}{1}, \text{ or } \sqrt{3}$$

$$\cot R = \frac{\text{adj.}}{\text{opp.}} \Rightarrow \cot 60° = \frac{1}{\sqrt{3}}, \text{ or } \frac{\sqrt{3}}{3}$$

Thus, $\sin 60° = \dfrac{\sqrt{3}}{2}$, $\cos 60° = \dfrac{1}{2}$, $\tan 60° = \sqrt{3}$, and $\cot 60° = \dfrac{\sqrt{3}}{3}$.

From Examples 1 and 2, notice the following cofunction relationship discussed on page 478.

$\sin 30° = \cos 60°$ $\tan 30° = \cot 60°$

$\sin 60° = \cos 30°$ $\tan 60° = \cot 30°$

Example 3   **Find u ($0° < u < 90°$) if $\cos 20° = \sin u$, and find u if $\tan 50° = \cot u$.**

$$\cos A = \sin (90° - A)$$
$$\cos 20° = \sin (90° - 20°)$$
$$= \sin 70°$$

Thus, $u = 70°$.

$$\tan A = \cot (90° - A)$$
$$\tan 50° = \cot (90° - 50°)$$
$$= \cot 40°$$

Thus, $u = 40°$.

The next example illustrates how to simplify trigonometric expressions.

Example 4   **Evaluate and simplify** $\dfrac{\cos 30° + \tan 45°}{\sin 30°}$.

Evaluate each trigonometric function, substitute, and simplify.

$$\frac{\cos 30° + \tan 45°}{\sin 30°} = \frac{\frac{\sqrt{3}}{2} + 1}{\frac{1}{2}} \qquad \blacktriangleleft \; \cos 30° = \frac{\sqrt{3}}{2}, \; \tan 45° = 1$$
$$\sin 30° = \frac{1}{2}$$
$$= \frac{\frac{\sqrt{3}}{2} + 1}{\frac{1}{2}} \cdot \frac{2}{2}, \text{ or } \frac{\sqrt{3} + 2}{1}$$

Thus, the value of the expression $\dfrac{\cos 30° + \tan 45°}{\sin 30°}$ is $\sqrt{3} + 2$.

Oral Exercises

For each triangle, state the measure of the side adjacent to and the side opposite the indicated angle.

1. $\angle A$ **2.** $\angle B$ **3.** $\angle H$ **4.** $\angle I$ **5.** $\angle D$ **6.** $\angle E$

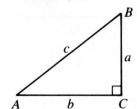 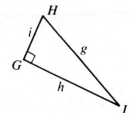 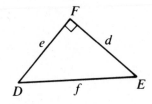

Written Exercises

(A) Find the sin, cos, tan, and cot of the measure of the indicated angle.

1. $\angle G$ **2.** $\angle H$ **3.** $\angle A$ **4.** $\angle B$ **5.** $\angle D$ **6.** $\angle E$

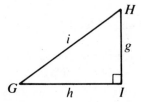 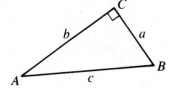 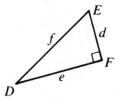

Find u, $0° < u < 90°$.

7. $\sin 34° = \cos u$ **8.** $\cos 49° = \sin u$ **9.** $\tan 73° = \cot u$ **10.** $\cot 39° = \tan u$

11. $\cot u = \tan 16°$ **12.** $\tan u = \cot 84°$ **13.** $\sin u = \cos 68°$ **14.** $\cos u = \sin 12°$

(B) **15.** $\tan 12° \, 13' = \cot u$ **16.** $\sin 63° \, 18' = \cos u$ **17.** $\cos u = \sin 74° \, 36'$

18. $\sin 46° \, 45' = \cos u$ **19.** $\tan u = \cot 74° \, 18'$ **20.** $\cot u = \tan 33° \, 19'$

Evaluate and simplify. Rationalize all denominators.

21. $\dfrac{\tan 30° + \cos 60°}{\sin 30°}$ **22.** $\dfrac{\sin 45° - \cos 30°}{\tan 60°}$ **23.** $\dfrac{\cos 45° - \sin 45°}{\tan 45°}$

24. $\dfrac{\cot 60° - \tan 45°}{\sin 60°}$ **25.** $\dfrac{\tan 60° - \tan 30°}{\tan 45°}$ **26.** $\dfrac{\sin 60° - \cot 30°}{\sin 30° + \cos 60°}$

(C) **27.** $\sin 90° \sec 60° + \dfrac{\cot 45°}{\sin 210°} - \csc 270°$ **28.** $\dfrac{\cos 270°}{\sin 90°} - \dfrac{\sin 180°}{\cos 360°} - \dfrac{\csc 225°}{\sec 315°}$

29. $\dfrac{3 \cos 60° - \tan 135° + \dfrac{4 \sec 240°}{\sin 60°}}{\dfrac{1}{2} \csc 330° - \dfrac{3}{4} \cot 315°}$ **30.** $\dfrac{\csc 30°}{\tan 135°} - \dfrac{\sec 300°}{\cot 300°} + \dfrac{3}{5} \csc 270°$

31. $\dfrac{\dfrac{1}{3} \sin (-60°) - \sqrt{3} \cos (-150°)}{\sec (-300°) + \csc (-315°)}$ **32.** $\dfrac{\sec 180° - \csc 90° + \dfrac{1}{2} \cos (-240°)}{-3 \sin (-330°)}$

USING TRIGONOMETRY OF A RIGHT TRIANGLE

Objectives   **To find the length of a side of a right triangle given the length of one side and the measure of an acute angle**
To find the measure of an acute angle of a right triangle given the lengths of two sides
To apply right triangle trigonometry

Right triangle trigonometry is often used to find the length of one side or the measure of an acute angle of a right triangle. To do this, use a function which relates that which is given and that which is to be determined. Sometimes one function is more convenient to use than another.

Example 1   **In right triangle ABC, $\angle A$ measures 43°, $c = 4.3$ m and m$\angle C = 90°$. Find a to the nearest meter.**

The *trigonometric* function that relates $\angle A$, a, and c is the sin function.

$$\sin A = \frac{\text{opp.}}{\text{hyp.}} \Rightarrow \sin 43° = \frac{a}{4.3}$$

Solve: $a = 4.3 \, (\sin 43°)$
$\doteq 4.3 \, (.6820)$, or 2.9326

Thus, $a = 3m$ to the nearest meter.

Example 2   **Find to the nearest degree the measure of each acute angle in right triangle ABC, if $b = 5$ and $c = 13$. Angle C is a right angle.**

Draw and label the diagram.
Cos A relates $\angle A$, b, and c.

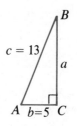

$$\cos A = \frac{\text{adj.}}{\text{hyp.}} \Rightarrow \cos A = \frac{5}{13}, \text{ or } .3846$$
$$A \doteq 67°$$

$\angle A$ and $\angle B$ are complementary.
$$A + B = 90°$$
$$67° + B = 90$$
$$B \doteq 23°$$

Thus, $\angle A$ measures 67° and $\angle B$ measures 23°.

In Examples 3 and 4 below, the angle of elevation or angle of depression is the angle between the horizontal and the line of sight. The angle of elevation equals the angle of depression.

Example 3

What is the length of a shadow cast by a 10-meter pole when the angle of elevation of the sun is 54°?

Draw and label the diagram.
The tan A or cot A relate $\angle A$, a, and b.

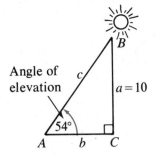

$$\tan A = \frac{a}{b}, \quad \text{or} \quad \cot A = \frac{b}{a}$$

$$\tan 54° = \frac{10}{b} \qquad\qquad \cot 54° = \frac{b}{10}$$

$$b = \frac{10}{\tan 54°} \qquad\qquad b = 10(\cot 54°)$$

$$\qquad\qquad\qquad\qquad\quad \doteq 10(.7265)$$

$$\doteq \frac{10}{1.3764} \qquad\qquad\quad \doteq 7.265$$

$$\doteq 7.265$$

Thus, the length of the shadow is approximately 7 m.

Example 4

A pilot flying at an altitude of 1200 m observes that the angle of depression of an airport is 43°. Find to the nearest meter the distance to the airport from the point on the ground directly below the plane.

Draw and label the diagram.
The tan A or cot A relate $\angle A$, a, and b.
Use cot A since the unknown will be in the numerator.

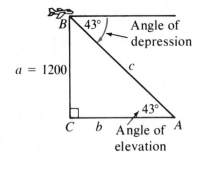

$$\cot A = \frac{b}{a}$$

$$\cot 43° = \frac{b}{1200}$$

$$b = 1200(\cot 43°)$$

$$\doteq 1200(1.0724)$$

$$\doteq 1286.88$$

Thus, the distance from C to A is 1287 m, to the nearest meter.

Oral Exercises

In right triangle ABC, which trigonometric function(s) relates the given parts with the unknown parts?

1. $A = 35°$, $a = 8$, $c = ?$
2. $B = 16°$, $b = 12$, $a = ?$
3. $A = 43°$, $b = 15$, $a = ?$
4. $A = 65°$, $a = 9$, $b = ?$
5. $B = 73°$, $b = 12$, $c = ?$
6. $B = 53°$, $a = 15$, $c = ?$
7. $A = 81°$, $a = 13$, $b = ?$
8. $B = 47°$, $b = 30$, $c = ?$

Written Exercises

(A) Find the measure of each of the other two sides of right triangle ABC, with $m\angle C = 90$, to the nearest whole number.

1. $m\angle B = 46°$, $a = 7.3$ **2.** $m\angle A = 36°$, $c = 18.3$ **3.** $m\angle B = 34°$, $b = 16.3$

4. $m\angle A = 63°$, $c = 15.1$ **5.** $m\angle B = 78°$, $a = 12.9$ **6.** $m\angle A = 53°$, $a = 38.7$

7. $m\angle B = 18°$, $b = 9.6$ **8.** $m\angle A = 15°$, $b = 37.3$ **9.** $m\angle B = 82°$, $c = 46.5$

Find the measure of each acute angle of right triangle ABC, with $m\angle C = 90$, to the nearest degree.

10. $c = 8.6$ $b = 4.9$ **11.** $b = 12.7$ $a = 9.7$ **12.** $b = 14.6$ $c = 20.3$

13. $a = 9.1$ $c = 12.3$ **14.** $c = 18.6$ $a = 14.1$ **15.** $a = 15.1$ $b = 7.6$

16. A ladder leans against a building. The top of the ladder reaches a point on the building that is 9 m above the ground. The foot of the ladder is 4 m from the building. Find to the nearest degree the measure of the angle that the ladder makes with the level ground.

17. From the top of a lighthouse 48 m high, the angle of depression of a boat at sea measures 23° 30′. Find, to the nearest m, the distance from the boat to the foot of the lighthouse.

18. From an airplane 925 m above sea level, the angle of depression of a ship measures 42° 20′. Find to the nearest meter the distance to the ship from the point at sea directly below the plane.

19. At a point on the ground 8 m from the foot of a cliff, the angle of elevation of the top of the cliff measures 38° 40′. Find the height of the cliff to the nearest meter.

(B) Find the measure of each of the other two sides to the nearest tenth and the measure of each acute angle of right triangle ABC, with $m\angle C = 90$, to the nearest minute.

20. $A = 13° 23′$, $a = 9.6$ **21.** $B = 49° 18′$, $c = 13.3$ **22.** $A = 63° 38′$, $b = 9.1$

23. $B = 74° 46′$, $b = 12.3$ **24.** $A = 11° 12′$, $b = 7.8$ **25.** $B = 16° 56′$, $c = 11.8$

26. $A = 33° 11′$, $c = 16.5$ **27.** $B = 73° 33′$, $a = 6.3$ **28.** $A = 38° 37′$, $a = 12.5$

(C) Determine BC to the nearest tenth.

29. $m\angle \theta = 60°$, $m\angle A = 45°$, $AD = 12$

30. $m\angle \theta = 45°$, $m\angle A = 30$, $AD = 10$

31. $m\angle \theta = 60°$, $m\angle A = 30°$, $AD = 16$

32. $m\angle \theta = 49° 18′$, $m\angle A = 32° 16′$, $AD = 14.3$

33. $m\angle \theta = 74° 48′$, $m\angle A = 48° 35′$, $AD = 28.6$

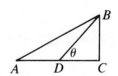

34. From point B on the ground, the angle of elevation of a kite measures 46° 18′. The kite is attached to a rope 86 m long. Find to the nearest meter the height of the kite.

35. From an airplane flying 1250 m above sea level, the angles of depression of two ships due west measure 33° 10′ and 20° 40′. Find the distance between the ships to the nearest meter.

CHAPTER SEVENTEEN REVIEW

Vocabulary

angle of rotation [17.1] sec, csc [17.5]
cofunction [17.8, 17.11] sin, cos [17.3]
quadrantral angle [17.3] tan, cot [17.4]

Sketch angles of rotation with the given measures. Determine the quadrant of each nonquadrantle angle. Identify the reference angle and reference triangle. [17.1]

1. 45° 2. −150° 3. 225° 4. −240°
5. −330° 6. 450° 7. 780° 8. −420°
9. −100° 10. −270° 11. 400° 12. −720°

★ **Find the reference angle if its terminal side intersects a circle at the following points and the center of the circle is at the origin. Determine the radius of the circle.** [17.2]

13. $(2, 2)$ 14. $(-\sqrt{3}, 1)$
15. $(a, a\sqrt{3})$ 16. $(a, -a)$

Find each of the following. Rationalize any radical denominator. [17.3–17.6]

17. cos 60° 18. tan $(-30°)$ 19. sec 150°
20. sin $(-480°)$ 21. sin 0° 22. sec 90°
23. cot 360° 24. cos $(-270°)$

Write each of the following in terms of the same trigonometric function of an acute angle. [17.3, 17.6]

25. sin 130° 26. cot 150°
27. cos $(-160°)$ 28. tan $(-220°)$

In which quadrants is each of the following true?

29. sin < 0 30. cos > 0 31. tan < 0 32. sec > 0 [17.3–17.4]

Find two values of u between 0° and 360° for which each is true. [17.3]

33. $\cos u = -\dfrac{1}{2}$ 34. $\sin u = -\dfrac{\sqrt{2}}{2}$

35. $\cot u = \dfrac{-\sqrt{3}}{3}$

★ **Find the value of each of the other trigonometric functions of u.** [17.7]

36. $\tan u = \frac{5}{12}$ and $\sin u > 0$
37. $\cos u = -\frac{3}{5}$ and $\cot u < 0$
38. $\sin u = \frac{7}{25}$ and $\sec u = -\frac{25}{24}$

[17.8]

Use the trig. table to find each of the following.

39. sin 38° 40. tan 64° 30′ 41. cos 223°
42. cos 34° 17′ 43. cot 115° 36′
44. cos 215° 33′ 45. sin $(-196° 54')$
46. tan $(-338° 28')$ 47. cot $(-134° 42')$

Find u, $u < 90°$. [17.8, 17.10, 17.11]

48. cos $u = .7790$ 49. sin $u = .9717$
50. sin 73° $=$ cos u 51. cot $u = $ tan 22°

Find all values of θ, $0° < \theta < 360°$. [17.10]

52. tan $\theta = -2.8258$ 53. cot $\theta = 6.9600$
54. cos $\theta = -.7796$ 55. sin $\theta = .4562$

Express in terms of the cofunction. [17.8]

56. sin 48° 25′ 57. tan 64° 42′

Express as a function of an angle measure less than 45°. [17.8]

58. cos 320° 59. cot 105° 30′

Determine the values of θ between 0° and 360° that make each equation true. [17.9]

60. $\sin \theta = -\frac{1}{2}$ 61. $2 \cos \theta - 1 = 5 \cos \theta + 1$
62. $5 \tan \theta - 8 = -3 \tan \theta - 5$

Evaluate and simplify. [17.11]

63. $\dfrac{\cos 45° - 3 \sin 60°}{\sec 120°}$

64. $\dfrac{\sin 45° + \cot 45°}{\cos (-120°)}$

★ 65. $\dfrac{\csc 90° + 3 \tan (-135°) - \sin 60°}{4 \cos (-315°)}$

Find the measure of each of the other two sides of right triangle ABC, with m$\angle C = 90°$, to the nearest whole number. [17.12]

66. m$\angle A = 46°$, $b = 9.2$
67. m$\angle B = 12° 18'$, $a = 12.3$

CHAPTER SEVENTEEN TEST

Sketch angles of rotation with the given measures. Determine the quadrant of each angle of rotation. Identify the reference angle and reference triangle.

1. 110°   2. −130°   3. 250°   4. −340°

Determine the lengths of the indicated sides. Rationalize all denominators.

5.

6.

Determine the point (x, y) and the radius r.

7.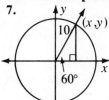

8.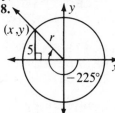

Find each of the following. Rationalize any radical denominator.

9. cos (−60°)   10. cos 180°   11. sin 420°
12. tan 135°   13. sec 180°   14. csc 420°

Write each of the following in terms of the same trigonometric function of an acute angle.

15. tan 340°   16. cos 100°   17. sin 310°

In which quadrants is each of the following true?

18. sin > 0   19. cos < 0   20. tan < 0
21. cos θ < 0 and sin θ > 0

Find two values of u between 0° and 360° for which each is true.

22. $\cos u = -\frac{\sqrt{3}}{2}$ 23. $\sin u = \frac{1}{2}$

★**Find the values of each of the other five trigonometric functions of u.**

24. $\sin u = -\frac{5}{13}$ and tan u < 0
25. $\cos u = -\frac{3}{5}$ and sin u > 0

Use the trig. table to find each of the following.

26. cos 37°   27. sin 215° 40′
28. tan (−139° 46′)   29. cos 326° 10′

Find u, u < 90°.

30. sin u = .5664   31. cos u = .2464
32. sin 65° = cos u

Find all values of θ, 0° < θ < 360°.

33. tan θ = −.6778   34. cos θ = −.4083

Express in terms of the cofunction.

35. sin 34°   36. tan 75° 26′

Express as a function of an angle measure less than 45°.

37. sin 110°   38. cos 305° 10′

Determine the values of θ between 0° and 360° that make each equation true.

39. 3 cos θ − 1 = −2 cos θ
40. −5 sin θ − 3 = 2 sin θ

Evaluate and simplify.

41. $\dfrac{\cos 60° - \sin 60°}{\cot 45°}$

Find the measure of each of the other two sides of right triangle ABC, with m∠C = 90, to the nearest whole number.

42. m∠A = 47°, b = 4.7

43. m∠B = 74° 25′, a = 10.3

44. At a point on the ground 23 m from the foot of a building, the angle of elevation of the top of the building is 48° 37′. Find to the nearest meter the height of the building.

Right Triangles

OBJECTIVE: To find the missing parts of a right triangle given two sides, or one side and an acute angle

The program below is a brief example of a menu-driven program. For each combination of known parts, the ON...GOTO statement in line 170 will branch to the appropriate portion of the program.

```
110    PRINT "ENTER THE NUMBER WHICH INDICATES WHAT YOU KNOW:"
120    PRINT   TAB( 4)"1) TWO LEGS"
130    PRINT   TAB( 4)"2) A LEG AND THE HYPOTENUSE"
140    PRINT   TAB( 4)"3) A LEG AND AN ADJACENT ANGLE"
150    PRINT   TAB( 4)"4) A LEG AND AN OPPOSITE ANGLE"
160    PRINT   TAB( 4)"5) HYPOTENUSE AND AN ACUTE ANGLE": PRINT
170    INPUT N: ON N GOTO 180,220,260,300,340
180    INPUT "ENTER THE TWO LEGS. ";A,B
190 C =   SQR (A ^ 2 + B ^ 2): PRINT "THE HYPOTENUSE = "C"
200 X =   ATN (A / B) * 180 / 3.1416
210 Y =    INT (X):R =  INT ((X - Y) * 60): GOSUB 390
220    INPUT "ENTER HYPOTENUSE, THEN LEG. ";C,A
230 B =   SQR (C ^ 2 - A ^ 2): PRINT "THE OTHER LEG = ";B
240 X =   ATN (B / A) * 180 / 3.1416
250 Y =    INT (X):R =  INT ((X - Y) * 60): GOSUB 390
260    INPUT "ENTER THE LEG, THEN THE ADJACENT ANGLE ";A,B
270 B = B * 3.1416 / 180: PRINT "THE OTHER ANGLE IS "90 - B"
       DEGREES."
280 C = A /   COS (B): PRINT "THE HYPOTENUSE = "C"."
290    PRINT "OTHER LEG IS " SQR (C ^ 2 - A ^ 2)".": GOTO 380
300    INPUT "ENTER LEG, THEN OPPOSITE ANGLE. ";A,X
310    PRINT "OTHER ANGLE IS "90 - X" DEGREES.":X = X * 3.1416
       / 180
320 C = A /   SIN (X): PRINT "THE HYPOTENUSE = "C"."
330    PRINT "OTHER LEG IS " SQR (C ^ 2 - A ^ 2)".": GOTO 380
340    INPUT "ENTER HYPOTENUSE,THEN ANGLE. ";C,X
350    PRINT "THE OTHER ANGLE IS "90 - X" DEGREES. ":X = X * 3
       .1416 / 180
360 D = C *   SIN (X): PRINT "ONE LEG IS "D"."
370    PRINT "THE OTHER LEG IS "  SQR (C ^ 2 - D ^ 2)"."
380    PRINT : PRINT : GOTO 110
390    PRINT "THE TWO ANGLES ARE ";Y;" DEGREES ";R;" MINUTES"
400    PRINT   TAB( 16);"AND ";89 - Y;" DEGREES ";60 - R;"
       MINUTES"
410    PRINT : PRINT : GOTO 110: RETURN
```

See the Computer Section beginning on page 574 for more information.

EXERCISES

1. Lines 390–410 constitute a subroutine. Rewrite the program without using a subroutine, but with the same output.

Directions: Choose the one best answer to each question or problem.

1. The lengths of the three sides of a triangle are 5, 12, and 13. A circle passes through the vertices of the triangle. Which is the length of the radius of the circle?

   (A) $\frac{13}{2}$ (B) 4   (C) 5

   (D) 6   (E) None of these

2.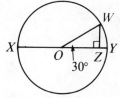

   $\overline{OW}$ is a radius of the circle with center O. If $OW = 12$, which is the value of WZ?

   (A) $6\sqrt{2}$ (B) 4   (C) 6
   (D) $6\sqrt{3}$ (E) None of these

3.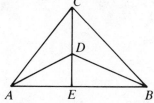

   ABC is a triangle with altitude $\overline{CE}$. $CD = 6$ and $AB = 18$. Which is the area of $ADBC$?

   (A) 108   (B) 54   (C) 90
   (D) $\frac{189}{2}$ (E) None of these

4. The average of five numbers is 5.1. Four of the numbers are 4.2, 5.7, 4.9 and 5.4. Which is the fifth number?
   (A) 5.06   (B) 5.14   (C) 5.1
   (D) 5.3   (E) None of these

5. A wheel rotates m times each minute. How many degrees does it rotate in t seconds?

   (A) $\frac{360\,m}{t}$ (B) $6\,mt$ (C) $\frac{60\,t}{m}$

   (D) $\frac{60\,m}{t}$ (E) $\frac{360\,t}{m}$

Directions: In each question you are to *compare* a Quantity I and a Quantity II. Choose your answer in the following way. Choose:
(A) if Quantity I is greater than II,
(B) if Quantity II is greater than I,
(C) if Quantity I = Quantity II,
(D) if the relationship cannot be determined.

6.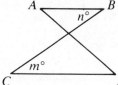

   Quantity I $x° + y°$
   Quantity II $m° + n°$

7.

   $\overline{AB}$ is parallel to $\overline{CD}$

   Quantity I m Quantity II n

8. Quantity I   area of circle with diameter
   length $\frac{2\sqrt{\pi}}{\pi}$

   Quantity II   area of a square whose side
   is 1.

18 TRIGONOMETRIC GRAPHS AND IDENTITIES

A biologist may choose a career from among many different specialties. Two examples are entomology and microbiology.

An **entomologist** is concerned with reducing the harmful species of insects that destroy crops, buildings, and clothing or cause disease in humans or wildlife. Efforts to increase and spread the many insects that are beneficial to the ''balance of nature'' are also important.

A **microbiologist** studies microorganisms such as bacteria and viruses. A career in microbiology may lead to important contributions in public health, industry, medicine, or agriculture.

Project

Write a paper on the steps that an entomologist takes to protect humans and wildlife from disease.

GRAPHING $y = \sin x$ AND $y = \cos x$ 18.1

Objectives

To graph $y = \sin x$ and $y = \cos x$
To determine the amplitude and the period of the sine and the cosine functions
To determine the quadrants in which the sin and cos functions increase or decrease

To graph the function $y = \sin x$, make a table of values by choosing special degree measures for x and finding corresponding values for y. The graph of $y = \sin x$ for $-360° \le x \le 360°$ is shown below.

x	$-360°$	$-270°$	$-180°$	$-90°$	$0°$	$90°$	$180°$	$270°$	$360°$
$\sin x$	0	1	0	-1	0	1	0	-1	0

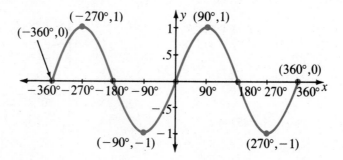

Note that the maximum value of the function is 1 and that this occurs when $x = 90°$ and when $x = -270°$. The amplitude of a trig function is its maximum value. The amplitude of the sine function is 1.

A *periodic function* is one that repeats itself. Note that the sine function is a periodic function with a period of 360°.

Example 1

Graph $y = \cos x$ for $-360° \le x \le 360°$ and determine its amplitude and period.

x	$\cos x$		x	$\cos x$
0°	1		0°	1
30°	$\frac{\sqrt{3}}{2}$ or .87		$-30°$	.87
90°	0		$-90°$	0
120°	$-\frac{1}{2}$ or $-.5$		$-120°$	$-.5$
180°	-1		$-180°$	-1
225°	$-\frac{\sqrt{2}}{2}$ or $-.71$		$-225°$	$-.71$
270°	0		$-270°$	0
330°	.87		$-330°$	.87
360°	1		$-360°$	1

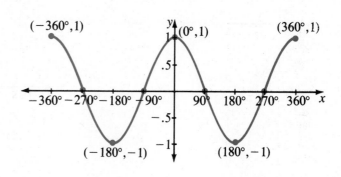

Thus, the amplitude of the cosine function is 1 and the period is 360°.

By looking at the graphs of $y = \sin x$ and $y = \cos x$, you can find where the sine increases or decreases and where the cosine increases or decreases.

1st quadrant	2nd quadrant	3rd quadrant	4th quadrant
$0° < x < 90°$	$90° < x < 180°$	$180° < x < 270°$	$270° < x < 360°$
sin increases	sin decreases	sin decreases	sin increases
cos decreases	cos decreases	cos increases	cos increases

You have learned that the values of a trig function can be found in a table. The values can also be estimated from a graph. The next example illustrates this.

Example 2   **Graph $y = \cos x$ for $0° \le x \le 90°$. Estimate cos 45° to the nearest tenth.**

x	$y = \cos x$
0°	1
30°	$\frac{\sqrt{3}}{2}$ or .9
60°	.5
90°	0

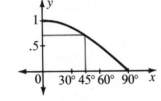

Thus, cos 45° $\doteq$ 0.7.

The next example illustrates how to solve a system of two trigonometric equations graphically.

Example 3   **Graph $y = \sin x$ and $y = \cos x$ for $-180° \le x \le 180°$ on the same set of axes. For which values of x does $\sin x = \cos x$?**

x	$y = \sin x$	$y = \cos x$
−30°	−.5	.9
−90°	−1	0
−120°	−.9	−.5
−180°	0	−1
0°	0	1
30°	.5	.9
90°	1	0
120°	.9	−.5
180°	0	−1

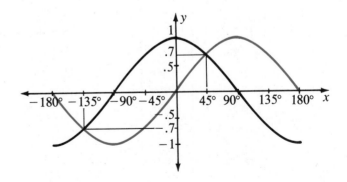

Thus, for $x = -135°$ or $x = 45°$, $\sin x = \cos x$.

Written Exercises

(A) **Complete the chart.**

	$y = \sin x$	$y = \cos x$
1. Amplitude		
2. Period		
3. Maximum value		
4. Minimum value		
5. Quadrants where function increases		
6. Quadrants where function decreases		

Sketch $y = \sin x$ for $0° \le x \le 180°$. Estimate each of the following to the nearest tenth.

7. $\sin 20°$ 　　　 8. $\sin 65°$ 　　　 9. $\sin 75°$ 　　　 10. $\sin 140°$ 　　　 11. $\sin 160°$

Sketch $y = \cos x$ for $0° \le x \le 180°$. Estimate each of the following to the nearest tenth.

12. $\cos 40°$ 　　 13. $\cos 70°$ 　　 14. $\cos 100°$ 　　 15. $\cos 130°$ 　　 16. $\cos 170°$

(B) 17. Sketch $y = \sin x$ and $y = \cos x$ for $0° \le x \le 360°$ on the same set of axes. For which values of x is $\sin x = \cos x$?

18. Sketch $y = \sin x$ and $y = \cos x$ for $-360° \le x \le 180°$ on the same set of axes. For which values of x is $\sin x = \cos x$?

19. Sketch $y = \sin x$ and $y = \cos x$ for $-180° \le x \le 360°$ on the same set of axes. For which values of x is $\sin x = \cos x$?

20. Sketch $y = \sin x$ and $y = \cos x$ for $-360° \le x \le 0°$ on the same set of axes. For which values of x is $\sin x = \cos x$?

(C) **Sketch the graph for $0° \le x \le 360°$.**

21. $y = -\cos x$ 　　　 22. $y = \sin(-x)$ 　　　 23. $y = -\sin x$ 　　　 24. $y = \cos(-x)$
25. $y = \sec x$ 　　　 26. $y = -\sec x$ 　　　 27. $y = \csc x$ 　　　 28. $y = \csc(-x)$

29. Sketch $y = \sin(-x)$ and $y = \cos(-x)$ for $-180° \le x \le 180°$ on the same set of axes. For which values of x is $\sin(-x) = \cos(-x)$?

30. Sketch $y = -\sin x$ and $y = -\cos x$ for $-180° \le x \le 180°$ on the same set of axes. For which values of x is $-\sin x = -\cos x$?

NON-ROUTINE PROBLEMS

In right triangle PRQ, the hypotenuse $\overline{QR}$ measures 20 cm and leg $\overline{PQ}$ measures 10 cm. Points S and T are on $\overline{QR}$ such that $\overline{SP}$ and $\overline{TP}$ trisect angle P. Find the length of $\overline{SP}$.

CUMULATIVE REVIEW

Find the distance between the given points. Answers should be in simplest radical form.

1. $A(3, -2)$, $B(-6, 1)$ 　　　 2. $A(-7, 8)$, $B(7, -3)$ 　　　 3. $A(12, -8)$, $B(4, -6)$

Graphing $y = \sin x$ and $y = \cos x$ 　　　　　　　　　　　　　　　　　　　　 **499**

DETERMINING AMPLITUDE

Objective

To determine the amplitude of the sine and cosine functions

You have learned that the amplitude for $y = \sin x$ and $y = \cos x$ is their maximum value, 1. You will now learn how multiplying $\sin x$ or $\cos x$ by a constant changes its amplitude.

Example 1

Sketch $y = 2 \sin x$ and $y = \sin x$ for $-180° \leq x \leq 360°$ on the same set of axes. Determine the period and amplitude for each function.

Make a table of values. Multiply $\sin x$ values by 2.

x	$y = \sin x$	$y = 2 \sin x$
0°	0	0
90°	1	2
180°	0	0
270°	−1	−2
360°	0	0
−90°	−1	−2
−180°	0	0

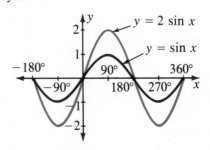

Thus, amplitude for $y = \sin x$ is 1.　Period for $y = \sin x$ is 360°.
　　　　Amplitude for $y = 2 \sin x$ is 2.　Period for $y = 2 \sin x$ is 360°.

Example 2

Sketch $y = \frac{1}{2} \cos x$ and $y = \cos x$ for $-180° \leq x \leq 360°$ on the same set of axes. Determine the period and amplitude for each function.

Make a table of values. Multiply $\cos x$ values by $\frac{1}{2}$.

x	$y = \cos x$	$y = \frac{1}{2} \cos x$
0°	1	$\frac{1}{2}$
90°	0	0
180°	−1	$-\frac{1}{2}$
270°	0	0
360°	1	$\frac{1}{2}$
−90°	0	0
−180°	−1	$-\frac{1}{2}$

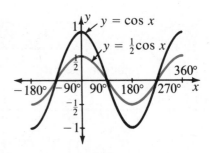

Thus, amplitude for $y = \cos x$ is 1.　Period for $y = \cos x$ is 360°.
　　　　Amplitude for $y = \frac{1}{2} \cos x$ is $\frac{1}{2}$.　Period for $y = \frac{1}{2} \cos x$ is 360°.

Example 3

Sketch $y = 3 \cos x$ and $y = \frac{1}{2} \cos x$ for $-180° \leq x \leq 360°$ on the same set of axes. Determine the period and amplitude for each function.

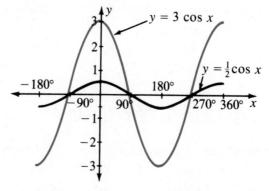

x	$y = \cos x$	$y = 3 \cos x$	$y = \frac{1}{2} \cos x$
$0°$	1	3	$\frac{1}{2}$
$90°$	0	0	0
$180°$	-1	-3	$-\frac{1}{2}$
$270°$	0	0	0
$360°$	1	3	$\frac{1}{2}$
$-90°$	0	0	0
$-180°$	-1	-3	$-\frac{1}{2}$

Thus, amplitude for $y = 3 \cos x$ is 3. Period for $y = 3 \cos x$ is $360°$.
Amplitude for $y = \frac{1}{2} \cos x$ is $\frac{1}{2}$. Period for $y = \frac{1}{2} \cos x$ is $360°$.

Amplitude and period for $y = a \sin x$ and $y = a \cos x$	If $y = a \sin x$, then the amplitude is $|a|$ and the period is $360°$, for each number a. If $y = a \cos x$, then the amplitude is $|a|$ and the period is $360°$, for each number a.

Example 4

Sketch $y = -2 \sin x$ and $y = \frac{1}{2} \sin x$ for $-180° \leq x \leq 360°$ on the same set of axes. Determine the period and amplitude for each.

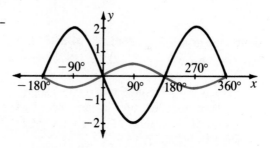

x	$y = \sin x$	$y = -2 \sin x$	$y = \frac{1}{2} \sin x$
$0°$	0	0	0
$90°$	1	-2	$\frac{1}{2}$
$180°$	0	0	0
$270°$	-1	2	$-\frac{1}{2}$
$360°$	0	0	0
$-90°$	-1	2	$-\frac{1}{2}$
$-180°$	0	0	0

Amplitude of $y = -2 \sin x$ is 2. Period of $y = -2 \sin x$ is $360°$.
Amplitude of $y = \frac{1}{2} \sin x$ is $\frac{1}{2}$. Period of $y = \frac{1}{2} \sin x$ is $360°$.

Example 5

Determine the amplitude of each trig function.

$y = 12 \sin x$ $\qquad$ $y = -6 \cos x$ $\qquad$ $y = -\frac{7}{2} \cos x$
$\downarrow$ $\qquad\qquad$ $\downarrow$ $\qquad\qquad\qquad$ $\downarrow$
Amplitude: 12 $\qquad$ Amplitude: 6 $\qquad$ Amplitude: $\frac{7}{2}$

Determining Amplitude

Oral Exercises

Determine the amplitude of each trig function.

1. $y = 5 \cos x$ **2.** $y = -2 \sin x$ **3.** $y = \frac{1}{3} \sin x$ **4.** $y = \frac{1}{5} \cos x$

5. $y = 15 \sin x$ **6.** $y = 12 \cos x$ **7.** $y = -\frac{4}{5} \cos x$ **8.** $y = -\frac{2}{3} \sin x$

Written Exercises

Ⓐ **Sketch the graph for $0° \leq x \leq 360°$. Determine the period and amplitude.**

1. $y = 3 \cos x$ **2.** $y = 2 \sin x$ **3.** $y = \frac{1}{2} \sin x$ **4.** $y = \frac{1}{3} \sin x$

5. $y = 4 \sin x$ **6.** $y = 5 \cos x$ **7.** $y = \frac{3}{4} \sin x$ **8.** $y = \frac{4}{5} \cos x$

9. Sketch $y = 2 \cos x$ and $y = \frac{1}{2} \cos x$ for $-360° \leq x \leq 360°$ on the same set of axes. Determine the period and the amplitude for each.

10. Sketch $y = 3 \sin x$ and $y = \frac{1}{2} \sin x$ for $-360° \leq x \leq 360°$ on the same set of axes. Determine the period and the amplitude for each.

Ⓑ **Sketch the graph for $-360° \leq x \leq 360°$. Determine the period and amplitude.**

11. $y = -\frac{1}{2} \cos x$ **12.** $y = -2 \sin x$ **13.** $y = -\frac{1}{3} \sin x$ **14.** $y = -3 \cos x$

15. $y = -3 \sin x$ **16.** $y = -4 \cos x$ **17.** $y = -\frac{2}{5} \cos x$ **18.** $y = -\frac{1}{2} \sin x$

19. Sketch $y = -\frac{1}{2} \sin x$ and $y = 2 \cos (-x)$ for $0° \leq x \leq 360°$ on the same set of axes. For how many values of x does $-\frac{1}{2} \sin x = 2 \cos (-x)$?

20. Sketch $y = -2 \cos x$ and $y = 2 \sin x$ for $0° \leq x \leq 360°$ on the same set of axes. For how many values of x does $-2 \cos x = 2 \sin x$?

Ⓒ **Sketch the graph for $-360° \leq x \leq 360°$. Determine the period and amplitude.**

21. $y = 2 \csc x$ **22.** $y = -\frac{1}{2} \csc x$ **23.** $y = \frac{1}{2} \sec x$ **24.** $y = -2 \sec x$

25. Sketch $y = -2 \sin x$ and $y = -\frac{1}{2} \sin x$ for $0° \leq x \leq 360°$ on the same set of axes. Estimate the values of x for which $-2 \sin x = -\frac{1}{2} \sin x$.

26. Sketch $y = -\frac{1}{2} \cos x$ and $y = -2 \cos x$ for $0° \leq x \leq 360°$ on the same set of axes. Estimate the values of x for which $-\frac{1}{2} \cos x = -2 \cos x$.

27. Sketch $y = -3 \cos x$ and $y = -2 \sin x$ for $-360° \leq x \leq 360°$ on the same set of axes. Estimate the values of x for which $-3 \cos x = -2 \sin x$.

28. Sketch $y = -2 \cos x$ and $y = \frac{1}{2} \sin x$ for $-360° \leq x \leq 360°$ on the same set of axes. Estimate the values of x for which $-2 \cos x = \frac{1}{2} \sin x$.

29. Sketch $y = \frac{1}{2} \sec x$ and $y = 2 \cos x$ for $0° \leq x \leq 360°$ on the same set of axes. Estimate the values of x for which $\frac{1}{2} \sec x = 2 \cos x$.

30. Sketch $y = \frac{1}{3} \csc x$ and $y = 2 \cos x$ for $0° \leq x \leq 360°$ on the same set of axes. Estimate the values of x for which $\frac{1}{3} \csc x = 2 \cos x$.

UNDERSEA AND UNDERGROUND EXPLORATION

Exploration below the surface of our earth, whether undersea or underground, has progressed with the invention of more and more sophisticated equipment to probe and analyze our world.

Accoustic Array, Deepstar 4000. The 18-foot-long Deepstar 4000, designed to investigate the sea bottom up to 4,000 feet below the surface, relies heavily on accoustic sensors.

Alexander the Great in glass barrel at the bottom of the sea. The first devices had no independent air supply, thus trapped air became foul and sharply limited diving time.

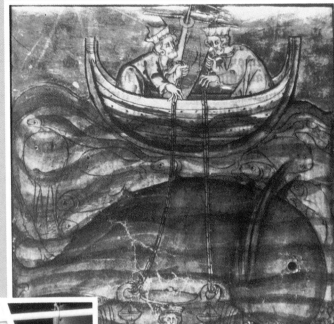

Lowering diver in "JIM," deep-sea diving suit designed by Sylvia Earle. Jacques Cousteau's invention of the "aqua-lung" rebreathing apparatus with a pressure regulator was the precursor to such advanced free-diving equipment as "JIM."

Cloche de Halley, diagram of old diving bell. Operating as an inverted tumbler, this early diving apparatus was supplied with air from the surface.

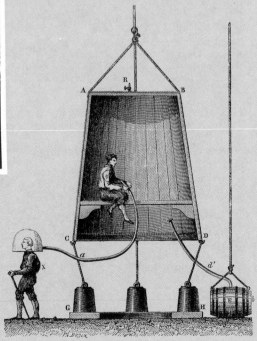

CNEXO, French deep submersible submarine. Small deep-sea submersibles give oceanographers a valuable vantage point for extended observations of marine life and conditions on the continental shelf.

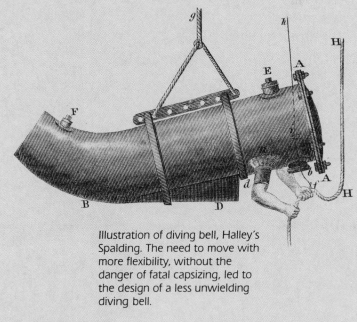

Illustration of diving bell, Halley's Spalding. The need to move with more flexibility, without the danger of fatal capsizing, led to the design of a less unwielding diving bell.

Frogmen and underwater observation chamber. During World War II frogmen were responsible for clandestine observation as well as demolition and surveying.

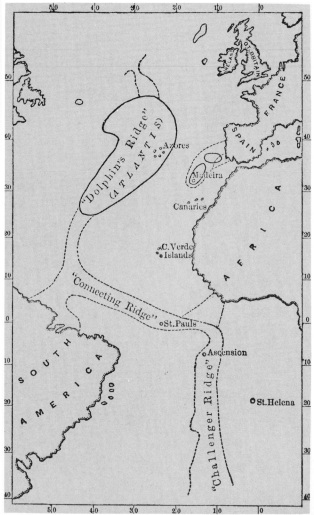

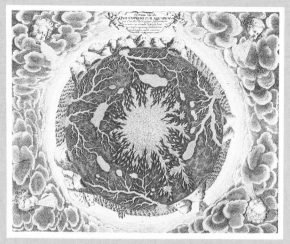

Earth cross-section with central core of fire. External geological disturbances such as volcanoes and geysers led early geologists to postulate that the earth's core was a ball of fire.

A map of islands and ridges plotted from deep sea soundings, Donnelly's Atlantis. In the 1860's, when the first trans-Atlantic cables were being laid, there was speculation that a high central plateau, discovered near the Azores, was the lost continent.

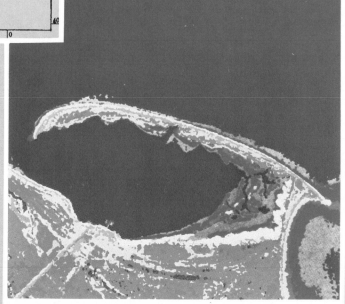

Landsat imagery assessment on the Gulf Coast. Satellites in orbit can scan an entire ocean basin, and when color patterns are analyzed, they provide data on marine life, currents, and pollutant levels.

Excavating the Gaillard (Culebra) cut for the Panama Canal. Five-hundred feet wide, this 8-mile-long cut was part of the 5-year effort to build the 51-mile canal, which shortened coast-to-coast voyages by 8,000 miles.

Coal pit worked by women. Early mining wells were not very wide because of the danger of collapse. This led to the use of women and children, who were smaller than men, as miners in the pits.

James Watt's steam-pumping engine at an English coal mine in the 1870's. His successive modifications such as the rotative shaft and double-action pistons made Watt's steam engine a vital implement in the progress of underground mining.

Illustration from *20,000 Leagues Under the Sea*. In 1870, when *20,000 Leagues Under the Sea* was published, Jules Verne had anticipated scientific inventions that would appear only after the latter half of the 20th century.

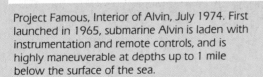

Project Famous, Interior of Alvin, July 1974. First launched in 1965, submarine Alvin is laden with instrumentation and remote controls, and is highly maneuverable at depths up to 1 mile below the surface of the sea.

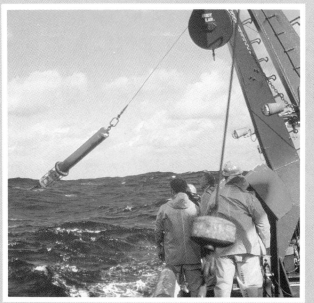

Putting a current meter into water. To examine underwater currents, instruments must be lowered to specific depths, and the information correlated to surface conditions such as sunlight and temperature.

Scientist estimating fish population. The life cycles of marine creatures suggest means of exploiting the abundance through fish farming and other aqua-agriculture.

Biology trainee. Woods Hole Oceanographic Institute in Massachusetts is a major research base for the study of all disciplines relating to the sea.

An ideal mine illustrating the types of support and central shaft. For safety and efficiency, rock in the roof and walls is shored by timbering, room and pillar constructions, and arches.

DETERMINING PERIOD

Objective   **To determine the period for y = sin bx and y = cos bx**

You have learned that the period for $y = \sin x$ and $y = \cos x$ is 360°. You will now learn how multiplying the measure of an angle by a constant changes the period.

Example 1   **Sketch $y = \cos 2x$ and $y = \cos x$ for $0° \leq x \leq 360°$ on the same set of axes. Determine the period and amplitude of each function.**

x	$2x$	$y = \cos 2x$
0°	0°	1
30°	60°	$\frac{1}{2}$
45°	90°	0
90°	180°	-1
135°	270°	0
180°	360°	1
225°	450°	0
270°	540°	-1
315°	630°	0
360°	720°	1

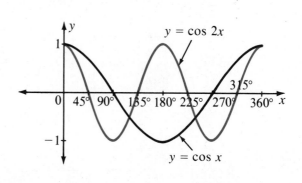

The graph of $y = \cos 2x$ completes 2 cycles from 0° to 360°. It completes one cycle every 180°. Note: $\frac{360°}{2} = 180°$.

Thus, period for $y = \cos x$ is 360°.   Amplitude for $y = \cos x$ is 1.
     Period for $y = \cos 2x$ is 180°.   Amplitude for $y = \cos 2x$ is 1.

Example 2   **Sketch $y = \sin \frac{1}{2}x$ and $y = \sin x$ for $0° \leq x \leq 720°$ on the same set of axes. Determine the period and amplitude of each function.**

x	$\frac{1}{2}x$ ·	$y = \sin \frac{1}{2}x$
0°	0°	0
60°	30°	$\frac{1}{2}$
90°	45°	$\frac{\sqrt{2}}{2}$
180°	90°	1
360°	180°	0
540°	270°	-1
720°	360°	0

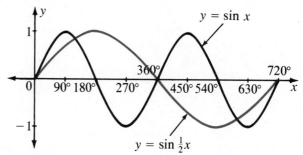

The graph of $y = \sin \frac{1}{2}x$ completes $\frac{1}{2}$ cycle from 0° to 360°. It completes one cycle every 720°. Note: $\dfrac{360°}{\frac{1}{2}} = 720°$.

Thus, period for $y = \sin x$ is 360°.   Amplitude for $y = \sin x$ is 1.
     Period for $y = \sin \frac{1}{2}x$ is 720°.   Amplitude for $y = \sin \frac{1}{2}x$ is 1.

Example 3   **Sketch $y = \sin 3x$ and $y = \sin x$ for $0° \le x \le 360°$ on the same set of axes. Determine the period and amplitude for each function.**

x	$3x$	$y = \sin 3x$
0°	0°	0
30°	90°	1
60°	180°	0
90°	270°	−1
120°	360°	0
150°	450°	1
180°	540°	0
210°	630°	−1
240°	720°	0
270°	810°	1
300°	900°	0
330°	990°	−1
360°	1080°	0

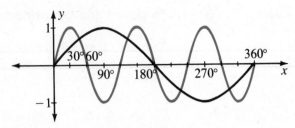

Notice that $y = \sin 3x$ completes 3 cycles from 0° to 360°. It completes one cycle every 120°.

Note: $\dfrac{360°}{3} = 120°$.

Thus, period for $y = \sin 3x$ is 120°. Amplitude for $y = \sin 3x = 1$. Period for $y = \sin x$ is 360°. Amplitude for $y = \sin x = 1$.

Period and amplitude for $y = \sin bx$ and $y = \cos bx$

If $y = \sin bx$, then the period is $\dfrac{360°}{|b|}$ and the amplitude is 1, for each *real* number $b\,(b \ne 0)$. If $y = \cos bx$, then the period is $\dfrac{360°}{|b|}$ and the amplitude is 1, for each *real* number $b\,(b \ne 0)$. Each function completes $|b|$ cycles in 360°.

Example 4   **Sketch $y = \sin\left(-\frac{1}{2}x\right)$ and $y = \sin\frac{1}{2}x$ for $0° \le x \le 720°$ on the same set of axes. Determine the period and amplitude for each function.**

x	$-\frac{1}{2}x$	$y = \sin\left(-\frac{1}{2}x\right)$
0°	0°	0
180°	−90°	−1
360°	−180°	0
540°	−270°	1
720°	−360°	0

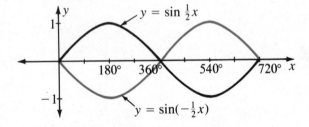

Since $\sin(-u) = -\sin u$, $\sin\left(-\frac{1}{2}x\right) = -\sin\frac{1}{2}x$. Therefore, the graph of $y = \sin\left(-\frac{1}{2}x\right)$ is symmetric to the graph of $y = \sin\frac{1}{2}x$. The x-axis is the axis of symmetry. Note: $\dfrac{360°}{\frac{1}{2}} = 720°$.

Thus, period for $y = \sin\frac{1}{2}x$ and $\sin\left(-\frac{1}{2}x\right)$ is 720°. Amplitude for $y = \sin\frac{1}{2}x$ and $\sin\left(-\frac{1}{2}x\right)$ is 1.

Example 5 Determine the period of each trig function.

$$y = \sin 6x$$
$$\text{Period: } \frac{360°}{|6|} = 60°$$

$$y = \cos\left(-\tfrac{1}{3}x\right)$$
$$\text{Period: } \frac{360°}{\left|-\tfrac{1}{3}\right|} = 1080°$$

$$y = \sin \tfrac{2}{3}x$$
$$\text{Period: } \frac{360°}{\left|\tfrac{2}{3}\right|} = 540°$$

Oral Exercises

Determine the period of each trig function.

1. $y = \sin x$
2. $y = \cos x$
3. $y = \sin \tfrac{1}{2}x$
4. $y = \cos \tfrac{1}{2}x$
5. $y = \cos 2x$
6. $y = \sin \tfrac{1}{3}x$
7. $y = \cos 9x$
8. $y = \sin 12x$

Written Exercises

(A) **Sketch the graph for $0° \le x \le 360°$. Determine the period and amplitude.**

1. $y = \cos 2x$
2. $y = \cos \tfrac{1}{2}x$
3. $y = \sin 2x$
4. $y = \sin \tfrac{1}{2}x$
5. $y = \sin \tfrac{1}{3}x$
6. $y = \sin 3x$
7. $y = \cos \tfrac{1}{3}x$
8. $y = \cos 3x$
9. $y = \cos 4x$
10. $y = \sin \tfrac{1}{4}x$
11. $y = \cos \tfrac{1}{4}x$
12. $y = \sin 4x$

(B)
13. $y = \sin(-2x)$
14. $y = \cos(-2x)$
15. $y = \cos\left(-\tfrac{1}{2}x\right)$
16. $y = \sin\left(-\tfrac{1}{3}x\right)$
17. $y = \cos\left(-\tfrac{3}{2}x\right)$
18. $y = \sin\left(-\tfrac{5}{4}x\right)$
19. $y = \sin(-3x)$
20. $y = \cos(-3x)$

21. Sketch $y = \cos 2x$ and $y = \sin 2x$ for $0° \le x \le 360°$ on the same set of axes. For how many values of x does $\cos 2x = \sin 2x$?

22. Sketch $y = \sin \tfrac{1}{2}x$ and $y = \cos \tfrac{1}{2}x$ for $0° \le x \le 360°$ on the same set of axes. For how many values of x does $\sin \tfrac{1}{2}x = \cos \tfrac{1}{2}x$?

(C) **Sketch the graph for $0° \le x \le 360°$. Determine the period and amplitude.**
23. $y = \sec 2x$
24. $y = \csc 3x$
25. $y = \sec \tfrac{1}{2}x$
26. $y = \csc(-x)$
27. Sketch the graph of $y = \sec(-2x)$ and $y = \cos(-2x)$ in the interval $0°$ to $360°$ on the same set of axes. For how many values of x does $\sec(-2x) = \cos(-2x)$?

NON-ROUTINE PROBLEMS

Solve for all real values of x and y such that:
$2 \cdot 3^x = 81y$ and $2^x = 8y$.

CUMULATIVE REVIEW

1. The height of a triangle is twice the length of its base. The area is 144 cm^2. Find the length of the base and the height.

2. The height of a triangle is 5 dm less than the length of its base. The area is 12 dm^2. Find the length of the base and the height.

CHANGE OF PERIOD AND AMPLITUDE   18.4

Objectives   **To sketch the functions defined by *y* = *a* sin *bx* and *y* = *a* cos *bx***
To solve a system of two trigonometric equations graphically

Recall that for the function defined by $y = 3 \sin x$, the amplitude is 3 and the period is 360°. In general, the amplitude for $y = a \sin x$ and $y = a \cos x$ is $|a|$. Also recall that for the function defined by $y = \cos \frac{1}{2}x$, the amplitude is 1 and the period is $\frac{360°}{\frac{1}{2}}$, or 720°. In general, the period of $y = \sin bx$ and $y = \cos bx$ is $\frac{360°}{|b|}$.

Example 1   **Sketch $y = \frac{1}{2} \cos 3x$ for 0° ≤ *x* ≤ 360°.**

Period of $y = \frac{1}{2} \cos 3x$ is $\frac{360°}{3}$, or 120°. Amplitude of $y = \frac{1}{2} \cos 3x$ is $\frac{1}{2}$.

x	$3x$	$\cos 3x$	$y = \frac{1}{2}\cos 3x$
0°	0°	1	$\frac{1}{2}$
30°	90°	0	0
60°	180°	−1	$-\frac{1}{2}$
90°	270°	0	0
120°	360°	1	$\frac{1}{2}$
150°	450°	0	0
180°	540°	−1	$-\frac{1}{2}$
210°	630°	0	0
240°	720°	1	$\frac{1}{2}$
270°	810°	0	0
300°	900°	−1	$-\frac{1}{2}$
330°	990°	0	0
360°	1080°	1	$\frac{1}{2}$

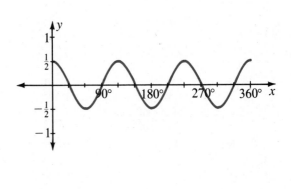

Example 2   **Sketch $y = 3 \sin \frac{1}{2}x$ for 0° ≤ *x* ≤ 720°.**

Period of $y = 3 \sin \frac{1}{2}x$ is $\frac{360°}{\frac{1}{2}}$, or 720°.

Amplitude of $y = 3 \sin \frac{1}{2}x$ is 3.

x	$\frac{1}{2}x$	$\sin \frac{1}{2}x$	$y = 3 \sin \frac{1}{2}x$
0°	0°	0	0
180°	90°	1	3
360°	180°	0	0
540°	270°	−1	−3
720°	360°	0	0

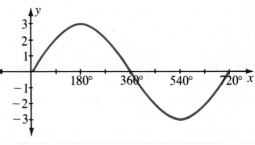

As you have learned, a system of algebraic equations may have some common solutions or no common solution. You will now study systems of two trigonometric equations. You will learn that there also may be no common solutions or many solutions. Usually, the number of points of intersection depends on the interval in which the functions are to be graphed.

Example 3 **Sketch $y = 3 \sin x$ and $y = 2 \cos \frac{1}{2}x$ for $0° \leq x \leq 720°$. For how many values of x does $3 \sin x = 2 \cos \frac{1}{2}x$?**

x	$3 \sin x$	$2 \cos \frac{1}{2}x$
0°	0	2
90°	3	$2(\frac{\sqrt{2}}{2})$, or 1.4
180°	0	0
270°	−3	$2(-\frac{\sqrt{2}}{2})$, or −1.4
360°	0	−2
450°	3	$2(-\frac{\sqrt{2}}{2})$, or −1.4
540°	0	0
630°	−3	$2(\frac{\sqrt{2}}{2})$, or 1.4
720°	0	2

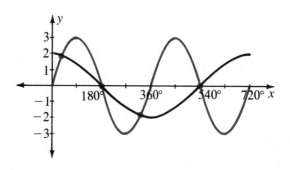

Thus, $3 \sin x = 2 \cos \frac{1}{2}x$ for 4 values of x for $0° \leq x \leq 720°$.

Example 4 **Sketch $y = -\sin 2x$ and $y = 3 \cos x$ for $-90° \leq x \leq 360°$. For how many values of x does $-\sin 2x = 3 \cos x$?**

x	$-\sin 2x$	$3 \cos 2x$
0°	0	3
45°	−1	0
90°	0	−3
135°	1	0
180°	0	3
225°	−1	0
270°	0	−3
315°	1	0
360°	0	3
−45°	1	0
−90°	0	−3

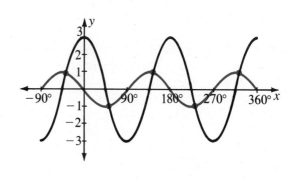

Thus, $-\sin 2x = 3 \cos x$ for 5 values of x for $-90° \leq x \leq 360°$.

Reading in Algebra

Each choice is related to the period or amplitude. Which one or ones don't belong? Defend your choices.

1. $y = 3 \sin 2x$ **a.** 3   **b.** 180°   **c.** 720°   **d.** −3
2. $y = \frac{1}{2} \cos (3x)$ **a.** 120°   **b.** 1080°   **c.** $\frac{1}{2}$ **d.** 180°
3. $y = -2 \sin \frac{1}{2}x$ **a.** 180°   **b.** 720°   **c.** 2   **d.** −2

Written Exercises

(A) **Determine the period and amplitude for each of the following.**

1. $y = -2 \cos \frac{1}{2}x$ 2. $y = 3 \sin (-2x)$ 3. $y = \frac{1}{2} \sin 2x$ 4. $y = \frac{1}{4} \cos 4x$
5. $y = \frac{1}{3} \sin 9x$ 6. $y = -3 \cos \frac{1}{3}x$ 7. $y = -2 \sin \frac{1}{2}x.$ 8. $y = 12 \cos \frac{1}{12}x$

Sketch the graph for $0° \leq x \leq 360°$. Determine the period and amplitude.

9. $y = \frac{1}{3} \cos 3x$ 10. $y = 2 \sin 2x$ 11. $y = 2 \cos 3x$ 12. $y = \frac{1}{2} \sin 4x$
13. $y = 3 \sin \frac{1}{3}x$ 14. $y = \frac{1}{2} \cos 2x$ 15. $y = 3 \sin 2x$ 16. $y = 3 \cos \frac{1}{2}x$

(B) 17. $y = -2 \sin \frac{1}{2}x$ 18. $y = -2 \cos \frac{1}{2}x$ 19. $y = -2 \sin (-\frac{1}{2}x)$
    20. $y = -\frac{1}{2} \cos 2x$ 21. $y = -\frac{1}{2} \cos (-2x)$ 22. $y = -\frac{1}{2} \sin 2x$

23. Sketch $y = \cos \frac{1}{2}x$ and $y = 3 \cos x$ for $0° \leq x \leq 360°$ on the same set of axes. For how many values of x does $\cos \frac{1}{2}x = 3 \cos x$?

24. Sketch $y = \cos 2x$ and $y = \frac{1}{2} \sin x$ for $0° \leq x \leq 360°$ on the same set of axes. For how many values of x does $\cos 2x = \frac{1}{2} \sin x$?

25. Sketch $y = -2 \cos x$ and $y = \sin (-\frac{1}{2}x)$ in the interval $-360°$ to $360°$ on the same set of axes. For how many values of x does $-2 \cos x = \sin (-\frac{1}{2}x)$?

26. Sketch $y = \sin (-2x)$ and $y = -\frac{1}{2} \cos x$ in the interval $0°$ to $360°$ on the same set of axes. For how many values of x does $\sin (-2x) = -\frac{1}{2} \cos x$?

(C) **Sketch the graph for $0° \leq x \leq 360°$. Determine the period and amplitude.**

27. $y = \frac{1}{2} \sec 2x$ 28. $y = 2 \csc 3x$ 29. $y = -2 \sec 3x$
30. $y = 2 \sec (-2x)$ 31. $y = 2 \sec 2x$ 32. $y = -2 \csc (-\frac{1}{2}x)$

33. Sketch the graph of $y = -2 \sin (-\frac{1}{2}x)$ and $y = -\frac{1}{2}\cos (-2x)$ in the interval $0°$ to $360°$ on the same set of axes. For how many values of x does $-2 \sin (-\frac{1}{2}x) = -\frac{1}{2} \cos (-2x)$?

34. Sketch the graph of $y = 2 \sec 2x$ and $y = \frac{1}{2} \sec \frac{1}{2}x$ for $0° \leq x \leq 360°$ on the same set of axes. For how many values of x does $2 \sec 2x = \frac{1}{2} \sec \frac{1}{2}x$?

CUMULATIVE REVIEW

Express each as the same function of a positive acute angle.

1. $\sin (-68°)$ 2. $\cos 130° 40'$ 3. $\sin 305° 18'$

GRAPHING $y = \tan x$ AND $y = \cot x$ 18.5

Objectives
To graph $y = \tan x$ and $y = \cot x$
To determine the amplitude and period for $y = \tan bx$ and $y = \cot bx$
To determine the quadrants in which the tan increases or decreases and in which the cot increases or decreases

To graph the function $y = \tan x$, make a table of values. After graphing, observe where the function is increasing. Also observe the period and amplitude. The graph of $y = \tan x$ for $0° \le x \le 360°$ is shown below.

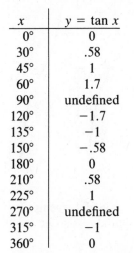

x	$y = \tan x$
0°	0
30°	.58
45°	1
60°	1.7
90°	undefined
120°	−1.7
135°	−1
150°	−.58
180°	0
210°	.58
225°	1
270°	undefined
315°	−1
360°	0

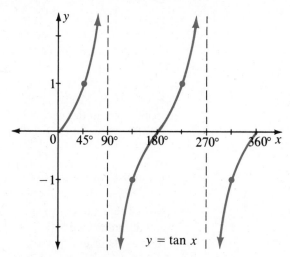

Notice that the tangent function is always increasing. It has no maximum or minimum value, hence it has no amplitude, and a period of 180°.

Example 1   **Graph $y = \tan 2x$ for $0° \le x \le 180°$.**

x	$2x$	$\tan 2x$
0°	0°	0
22.5°	45°	1
45°	90°	undefined
67.5°	135°	−1
90°	180°	0
112.5°	225°	1
135°	270°	undefined
157.5°	315°	−1
180°	360°	0

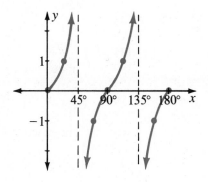

The graph of $y = \tan 2x$ completes 2 cycles from 0° to 180°. It completes one cycle every 90°. Note: $\dfrac{180°}{2} = 90°$.

Period for $y = \tan x$ is 180°. Period for $y = \tan 2x$ is 90°.

The function $y = \tan x$ has no amplitude.

The period of $y = \tan x$ is $180°$.

If $y = \tan bx$, then the period is $\dfrac{180°}{|b|}$ for each *real* number $b(b \neq 0)$.

Example 2   **Sketch $y = \cot x$ for $-180° \leq x \leq 360°$.**

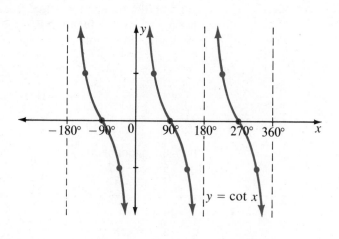

x	$y = \cot x$
$0°$	undefined
$45°$	1
$90°$	0
$135°$	-1
$180°$	undefined
$225°$	1
$270°$	0
$315°$	-1
$360°$	undefined
$-45°$	-1
$-90°$	0
$-135°$	1
$-180°$	undefined

Notice that the cotangent function is always decreasing. It has no maximum or minimum value, hence it has no amplitude. Also notice that the graph repeats itself every $180°$.

Example 3   **Sketch $y = 2 \cot \frac{1}{2}x$ for $-180° \leq x \leq 360°$.**

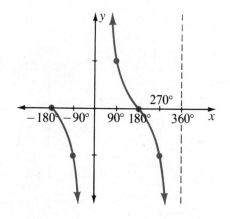

x	$\frac{1}{2}x$	$\cot \frac{1}{2}x$	$y = 2\cot\frac{1}{2}x$
$0°$	$0°$	undefined	undefined
$90°$	$45°$	1	2
$180°$	$90°$	0	0
$270°$	$135°$	-1	-2
$360°$	$180°$	undefined	undefined
$450°$	$225°$	1	2
$540°$	$270°$	0	0
$630°$	$315°$	-1	-2
$720°$	$360°$	undefined	undefined
$-90°$	$-45°$	-1	-2
$-180°$	$-90°$	0	0

The function $y = \cot x$ has no amplitude.

The period for $y = \cot x$ is $180°$.

If $y = \cot bx$, then the period is $\dfrac{180°}{|b|}$ for each real number $b(b \neq 0)$.

Written Exercises

(A) **Complete the chart.**

	$y = \tan x$	$y = \cot x$
1. Amplitude		
2. Period		
3. Maximum value		
4. Minimum value		
5. Quadrants where function increases		
6. Quadrants where function decreases		
7. Values for $-180° \le x \le 360°$ where the function is undefined		

Determine the period for each of the following.

8. $y = 2 \tan 3x$ **9.** $y = 4 \cot \frac{1}{2}x$ **10.** $y = -2 \tan 3x$ **11.** $y = -\frac{1}{2} \cot 3x$

12. $y = \frac{1}{2} \tan \frac{1}{2}x$ **13.** $y = -\frac{1}{2} \cot 2x$ **14.** $y = 3 \tan \frac{1}{3}x$ **15.** $y = -6 \cot \frac{1}{3}x$

16. $y = -6 \cot (-2x)$ **17.** $y = -\frac{1}{2} \tan (-\frac{1}{2}x)$ **18.** $y = -\frac{1}{4} \cot (-\frac{1}{3}x)$

Sketch the graph in the interval $-90° \le x \le 360°$. Determine the period.

19. $y = \tan \frac{1}{2}x$ **20.** $y = \cot 2x$ **21.** $y = \tan 3x$ **22.** $y = \cot \frac{1}{2}x$

23. $y = 2 \cot 2x$ **24.** $y = \frac{1}{2} \tan 2x$ **25.** $y = 2 \tan \frac{1}{2}x$ **26.** $y = 4 \cot 3x$

(B) **27.** $y = -2 \tan (-2x)$ **28.** $y = -\frac{1}{2} \cot (-2x)$ **29.** $y = -3 \tan (-\frac{1}{2}x)$

30. $y = -\frac{1}{2} \cot (3x)$ **31.** $y = 6 \tan (-\frac{1}{3}x)$ **32.** $y = -\cot (-\frac{1}{2}x)$

33. Sketch $y = \cot \frac{1}{2}x$ and $y = \tan 2x$ for $0° \le x \le 360°$ on the same set of axes. For how many values of x does $\cot \frac{1}{2}x = \tan 2x$?

34. Sketch $y = 2 \tan \frac{1}{2}x$ and $y = \frac{1}{2} \cot 2x$ for $0° \le x \le 360°$ on the same set of axes. For how many values of x does $2 \tan \frac{1}{2}x = \frac{1}{2} \cot 2x$?

(C) **35.** Sketch $y = -\tan (-\frac{1}{2}x)$ and $y = -2 \cot x$ for $0° \le x \le 360°$ on the same set of axes. For how many values of x does $-2 \cot x = -\tan (-\frac{1}{2}x)$?

36. Sketch $y = -\frac{1}{2} \tan (-2x)$ and $y = \cot (-2x)$ for $0° \le x \le 360°$ on the same set of axes. For how many values of x does $-\frac{1}{2} \tan (-2x) = \cot (-2x)$?

37. Sketch $y = \tan 2x$ and $y = \cot \frac{1}{2}x$ for $0° \le x \le 360°$ on the same set of axes. Estimate the values of x for which $\tan 2x = \cot \frac{1}{2}x$.

38. Sketch $y = -\tan 2x$ and $y = \sin \frac{1}{2}x$ for $0° \le x \le 360°$ on the same set of axes. Estimate the values of x for which $-\tan 2x = \sin \frac{1}{2}x$.

NON-ROUTINE PROBLEMS

$S_n = 1 - 2 + 3 - 4 + 5 - 6 + \ldots + (-1)^{n-1}n$ as $n = 1, 2, 3, 4, \ldots$.
Find $S_{25} + S_{36} + S_{60}$.

FINDING LOGARITHMS OF TRIGONOMETRIC FUNCTIONS

Objectives

To find logarithms of the sin, cos, tan, and cot of any degree measure using a log-trig table
To find the degree measure when the log-trig value is given
To interpolate using a log-trig table

In an earlier chapter you learned how to do computations with logarithms. Recall the following:

$$\log a \cdot b = \log a + \log b$$
$$\log \frac{a}{b} = \log a - \log b$$
$$\log a^n = n \log a$$
$$\log \sqrt[n]{a} = \frac{1}{n} \log a$$

Thus, to find $N = \dfrac{\sqrt[4]{(34.5) \times (7.3)^3}}{186.4}$ using logs, write

$\log N = \dfrac{1}{4}[(\log(34.5) + 3 \log 7.3)] - \log 186.4$ and solve.

The following examples illustrate the use of log-trig tables.

Example 1

Find log tan 28° 10′ and log cos 27° 40′.

The log-trig functions at the top of the table must be used with degree measures in the left-hand column. A portion of a log-trig table is shown below.

Logarithms of Trigonometric Functions

Angle	L Sin	L Tan	L Cot	L Cos	
27° 0′	9.6570	9.7072	10.2928	9.9499	**63° 0′**
10′	.6595	.7103	.2897	.9492	50′
20′	.6620	.7134	.2866	.9486	40′
30′	.6644	.7165	.2835	.9479	30′
40′	.6668	.7196	.2804	.9473	20′
50′	.6692	.7226	.2774	.9466	10′
28° 0′	9.6716	9.7257	10.2743	9.9459	**62° 0′**
10′	.6740	.7287	.2713	.9453	50′
20′	.6763	.7317	.2683	.9446	40′
30′	.6787	.7348	.2652	.9439	30′
40′	.6810	.7378	.2622	.9432	20′
50′	.6833	.7408	.2592	.9425	10′

* Subtract 10 from each entry in this table to obtain the proper logarithm of the indicated trigonometric function.

To find log tan 28° 10′ use the left-hand column and read across under L Tan (log tan) column. The log cos 27° 40′ can be found by reading across 27° 40′ under the L Cos (log cos) column. After reading the value from the table, subtract 10.

Thus, log tan 28° 10′ = 9.7287 − 10, and log cos 27° 40′ = 9.9473 − 10.

Example 2

Find log sin 46° 40' and log cot 45° 50'.

Use the right column and the bottom column heads.

43° 0'	9.8338	9.9697	10.0303	9.8641	47° 0'
10'	.8351	.9722	.0278	.8620	50'
20'	.8365	.9747	.0253	.8618	40'
30'	.8378	.9772	.0228	.8606	30'
40'	.8391	.9798	.0202	.8594	20'
50'	.8405	.9823	.0177	.8582	10'
44° 0'	9.8418	9.9848	10.0152	9.8569	46° 0'
10'	.8431	.9874	.0126	.8557	50'
20'	.8444	.9899	.0101	.8545	40'
30'	.8457	.9924	.0076	.8532	30'
40'	.8469	.9949	.0051	.8520	20'
50'	.8482	9.9975	.0025	.8507	10'
45° 0'	9.8495	10.0000	10.0000	9.8495	45° 0'
	L Cos	L Cot	L Tan	L Sin	Angle

Thus, log sin 46° 40' = 9.8618 − 10 and log cot 45° 50' = 9.9874 − 10.

Example 3

Log sin u = 9.2251 − 10   **Log tan u = .7320**
Find u, u < 90°.   **Find u, u < 90°.**

Angle	L Sin	L Tan	L Cot	L Cos	
9° 0'	9.1943	9.1997	10.8003	9.9946	81° 0'
10'	.2022	.2078	.7922	.9944	50'
20'	.2100	.2158	.7842	.9942	40'
30'	.2176	.2236	.7764	.9940	30'
40'	.2251	.2313	.7687	.9938	20'
50'	.2324	.2389	.7611	.9936	10'
10° 0'	9.2397	9.2463	10.7537	9.9934	80° 0'
10'	.2468	.2536	.7464	.9931	50'
20'	.2538	.2609	.7391	.9929	40'
30'	.2606	.2680	.7320	.9927	30'
40'	.2674	.2750	.7250	.9924	20'
50'	.2740	.2819	.7181	.9922	10'
	L Cos	L Cot	L Tan	L Sin	Angle

Thus, when log sin u = 9.2251 − 10, u = 9° 40'. When log tan u = .7320,
u = 79° 30'.

Example 4

Find log cos 36° 28'.

36° 28' lies between the values 36° 20' and 36° 30'. Make a table.

$$
10\begin{bmatrix} 8\begin{bmatrix} \begin{array}{ll} \text{Measure} & \text{log cosine} \\ 36°\,20' & .9061 \\ 36°\,28' & \\ 36°\,30' & .9052 \end{array} \end{bmatrix} n \end{bmatrix} -.0009
$$

$$\frac{8}{10} = \frac{n}{-.0009} \text{ or } 10n = -.0072$$

$$n = -.00072, \text{ or } n \doteq -.0007$$

$$.9061 + (-.0007) = .9054$$

Thus, log cos 36° 28' = 9.9054 − 10.

You will now learn how to do computations with trig functions using logs.

Example 5   **Find the value of** $\dfrac{(39.6)^3 \sin 73° 10'}{\sqrt[3]{0.315}}$ **to four significant digits.**

Use logarithms. (See pages 588–589.)

Let $s = \dfrac{(39.6)^3 \sin 73° 10'}{\sqrt[3]{0.315}}$

$$\log s = 3 \log 39.6 + \log \sin 73° 10' - \tfrac{1}{3} \log 0.315$$
$$= 3(1.5977) + 9.9810 - 10 - \tfrac{1}{3}(9.4983 - 10)$$
$$= 4.7931 + 9.9810 - 10 - \tfrac{1}{3}(29.4983 - 30)$$
$$= 4.7741 - (9.8328 - 10)$$
$$\log s = 4.9413$$

Thus, $s = 87,360.$

Oral Exercises

Find the value of each of the following.

1. log sin 48°20'
2. log cos 10°30'
3. log tan 50°10'
4. log cot 30°40'
5. log tan 40°20'
6. log sin 10°50'

Written Exercises

(A) **Find the value of each of the following.**

1. log sin 53°42'
2. log cot 76°54'
3. log cos 65°26'
4. log tan 34°18'
5. log cos 18°34'
6. log sin 22°36'
7. log tan 73°44'
8. log cot 12°58'
9. log tan 27°25'
10. log cos 46°17'
11. log sin 62°33'
12. log cos 86°42'

(B) **Find** u, $0° < u < 90°$, **for each of the following.**

13. log cos $u = 9.9446 - 10$
14. log tan $u = 9.8797 - 10$
15. log sin $u = 9.7727 - 10$
16. log tan $u = .0076$
17. log cot $u = .1795$
18. log cos $u = 9.9684 - 10$
19. log sin $u = 9.8111 - 10$
20. log cos $u = 9.4775 - 10$
21. log cot $u = .4057$

Find the value to four significant digits. Use logarithms.

22. $\dfrac{\cos 86° 10'}{(.437)^2}$

23. $\dfrac{(31.3)^3 \sqrt{0.736}}{\tan 43° 30'}$

24. $\cot 75° 50' \sqrt{\dfrac{48.2}{5.64}}$

25. $\dfrac{\sqrt[3]{46.3} \sin 35° 30'}{(1.48)^2}$

26. $\dfrac{\cos 23° 36' (63.1)^3}{\sqrt{0.487}}$

27. $\dfrac{(49.3)^3 \sqrt{107.3}}{\sin 73° 47'}$

(C) 28. $\dfrac{\sin 43° 38' \cos 63° 42'}{(\tan 17° 14')^3}$

29. $\dfrac{\sqrt[3]{\cos 33° 18'} \tan 16° 47'}{\sqrt[4]{187.6} \sin 10° 15'}$

30. $\dfrac{(12.3)^4 (\sin 73° 38')^3}{(\cos 15° 36')^2 \sqrt[3]{0.1487}}$

31. $\dfrac{(38.6)^3 \tan 23° 18'}{\sqrt[3]{(\sin 46° 37')(\cos 70° 48')}}$

32. $\dfrac{\sec 82° 42'}{\sqrt{534.3} \tan 20° 14'}$

33. $\dfrac{(713.4)^6 \csc 12° 18'}{\sqrt{0.493} \sec 46° 37'}$

QUADRATIC TRIGONOMETRIC EQUATIONS

Objective   **To solve quadratic trigonometric equations**

Solving trigonometric equations is similar to solving algebraic equations. Each example that follows will be preceded by an algebraic example.

To solve $4x^2 - 5 = 0$, add 5 to both sides, divide each side by 4, and find the square root.

$$4x^2 = 5$$
$$x^2 = \frac{5}{4}$$
$$x = \pm\frac{1}{2}\sqrt{5}$$

Example 1   **Determine the values of θ between 0° and 360° for which $4\sin^2\theta - 3 = 0$.**

Think of $\sin\theta$ as x. Note that $\sin^2\theta$ is written $(\sin\theta)^2$. Add 3 to both sides, divide each side by 4, and find the square root.

$$4(\sin\theta)^2 = 3$$
$$(\sin\theta)^2 = \frac{3}{4}$$
$$\sin\theta = \pm\sqrt{\frac{3}{4}}, \text{ or } \pm\frac{\sqrt{3}}{2}$$

$sin\ 60° = \dfrac{\sqrt{3}}{2}$ ▶ $\sin\theta = \dfrac{\sqrt{3}}{2}$ $\sin\theta = \dfrac{-\sqrt{3}}{2}$

1st quad.   2nd quad.   3rd quad.   4th quad.

Thus, $\theta = 60°, 120°, 240°$, and $300°$.

To solve the algebraic equation $2x^2 - x = 0$, factor, set each factor equal to zero, and solve.

$$x(2x - 1) = 0$$
$$x = 0 \quad \text{or} \quad 2x - 1 = 0$$
$$x = 0 \quad \text{or} \quad x = \frac{1}{2}$$

Example 2   **Determine the values of u, $0° \le u \le 360°$, for which $\tan^2 u - \sqrt{3}\tan u = 0$.**

Factor. ▶ $\tan u(\tan u - \sqrt{3}) = 0$

Set each factor ▶ $\tan u = 0$ | or $\tan u - \sqrt{3} = 0$
equal to 0. $\tan u = \sqrt{3}$

$u = 0°, u = 180°$ | Ref. angle 60°
                              1st quad. and 3rd quad.

Thus, $u = 0°, 180°, 60°$, and $240°$.

To solve the equation $3x^2 + 5x - 2 = 0$, factor, set each factor equal to zero.

$$(3x - 1)(x + 2) = 0$$
$$3x - 1 = 0 \quad \text{or} \quad x + 2 = 0$$
$$x = \frac{1}{3} \quad \text{or} \quad x = -2$$

Example 3   **Determine values of u, $0° \le u \le 360°$, for which $2 \sin^2 u + 5 \sin u - 3 = 0$.**

$$2 \sin^2 u + 5 \sin u - 3 = 0$$
$$(2 \sin u - 1)(\sin u + 3) = 0$$
$$2 \sin u - 1 = 0 \quad \text{or} \quad \sin u + 3 = 0$$
$$\sin u = \frac{1}{2} \qquad\qquad \sin u = -3$$

Ref. $\angle$: 30°   No solution, since
$\swarrow$ $\searrow$ $-1 \le \sin u \le 1$
1st quad. 2nd quad.

Thus, $u = 30°$ and $150°$.

To solve the equation $2x^2 - 5x + 1 = 0$, use the quadratic formula.

$$x = \frac{-b \pm \sqrt{b^2 - 4ac}}{2a} \qquad a = 2, b = -5, c = 1$$
$$x = \frac{-(-5) \pm \sqrt{(-5)^2 - 4(2)(1)}}{2(2)}$$
$$= \frac{5 + \sqrt{17}}{4} \text{ or } \frac{5 - \sqrt{17}}{4}$$
$$= \frac{5 + 4.123}{4} \text{ or } \frac{5 - 4.123}{4}$$
$$x = 2.281 \text{ or } .219$$

Example 4   **Determine the values of u to the nearest degree between $0°$ and $360°$ for which $3 \cos^2 u - 4 \cos u - 2 = 0$.**

Use the quadratic formula. $x = \dfrac{-b \pm \sqrt{b^2 - 4ac}}{2a}$, $a = 3, b = -4, c = -2$

$$\cos u = \frac{-(-4) \pm \sqrt{(-4)^2 - 4(3)(-2)}}{2(3)}$$
$$= \frac{4 \pm \sqrt{16 + 24}}{6}, \text{ or } \frac{4 \pm \sqrt{40}}{6}$$
$$\frac{4 + 6.32}{6} \qquad\qquad \frac{4 - 6.32}{6}$$
$$\cos u = \frac{10.32}{6} = 1.72 \quad \text{or} \quad \cos u = \frac{-2.32}{6} \doteq -.3867$$
$$\uparrow$$
No solution   Ref. $\angle$: 67°
$\swarrow$ $\searrow$
2nd quad. 3rd quad.

Thus, $u = 113°$ and $247°$.

Oral Exercises

How many values of u, $0° \leq u \leq 360°$, are there for each of the following?

1. $\sin u = -.7632$

2. $\cos u = 2.6543$

3. $\tan u = -6.3214$

4. $\tan u = .4132$

5. $\sin u = -3.1249$

6. $\cos u = -.2987$

Written Exercises

(A) **Determine all values of θ, $0° \leq \theta \leq 360°$, to the nearest degree for which the equations are true.**

1. $2 \cos^2 \theta - 5 = -4$

2. $10 \sin \theta - 5\sqrt{3} = 0$

3. $2 \cos \theta - \sqrt{3} = 0$

4. $\sin^2 \theta - 1 = 0$

5. $\cos^2 \theta = 0$

6. $\cot^2 \theta - 3 = 0$

7. $2 \sin \theta + \sqrt{3} = 0$

8. $\cot^2 \theta + \cot \theta = 0$

9. $2 \cos^2 \theta - \cos \theta = 0$

10. $2 \sin^2 \theta + \sqrt{3} \sin \theta = 0$

11. $\sqrt{3} \tan^2 \theta - \tan \theta = 0$

12. $2 \cos^2 \theta - \cos \theta - 1 = 0$

13. $\sin^2 \theta - 7 \sin \theta + 6 = 0$

14. $2 \cos^2 \theta + 3 \cos \theta + 1 = 0$

15. $\tan^2 \theta - 2 \tan \theta + 1 = 0$

(B) **16.** $\tan^2 \theta - 5 \tan \theta - 2 = 0$

17. $3 \sin^2 \theta - 8 \sin \theta - 1 = 0$

18. $4 \cos^2 \theta + 5 \cos \theta - 2 = 0$

19. $5 \cos^2 \theta - 3 \cos \theta - 1 = 0$

20. $7 \cot^2 \theta + 4 \cot \theta - 3 = 0$

21. $7 \tan^2 \theta - 3 \tan \theta - 5 = 0$

22. $2 \sin^2 \theta - 5 \sin \theta = 1$

23. $\cot^2 \theta - 7 \cot \theta + 1 = 0$

24. $3 \cos^2 \theta + 6 \cos \theta + 1 = 0$

(C) **Determine all values of θ for which the equations are true.**

25. $\sec^2 \theta - 4 = 0$

26. $4 \csc^2 \theta - 25 = 0$

27. $15 \csc^2 \theta + \csc \theta - 2 = 0$

28. $\csc^2 \theta - 7 \csc \theta + 12 = 0$

29. $2 \sec^2 \theta - \sec \theta - 15 = 0$

30. $3 \csc^2 \theta - 6 \csc \theta - 1 = 0$

31. $\sec^2 \theta - 5 \sec \theta - 3 = 0$

32. $5 \sec^2 \theta - 1 = 12$

33. $\sec^2 \theta + 7 \sec \theta - 3 = 0$

CALCULATOR ACTIVITIES

You can use a calculator to solve quadratic trigonometric equations.

Example: Solve $\sin^2 \theta - 7 \sin \theta + 5 = 0$ to the nearest ten minutes, $0 \leq \theta \leq 90°$.

1. Use the quadratic formula. $a = 1$, $b = -7$, $c = 5$.

2. Solve for $\sin \theta$: $\sin \theta = \dfrac{-(-7) \pm \sqrt{(-7)^2 - 4(1)(5)}}{2(1)}$

3. Compute $(-7)^2 - 4(5)$.

   Press ▶ $\quad -7 \otimes \ominus$; $-4 \otimes 5 \ominus$; $49 \ominus 20 \ominus$

   Display ▶ $\quad\quad 49 \quad\quad$; $\quad -20 \quad$; $\quad\quad 29$

4. Use the table to find $\sqrt{29} \doteq 5.39$.

5. Compute $\dfrac{7 + 5.39}{2}$ and $\dfrac{7 - 5.39}{2}$.

6. *Press* ▶ $\quad 7 \oplus 5.39 \oslash 2 \ominus$ and $7 \ominus 5.39 \oslash 2 \ominus$

7. *Display* ▶ $\quad 6.195$ and $.805$. Reject 6.195 since $-1 \leq \sin \theta \leq 1$.

8. Look in the table to find $\theta = 53° 40'$ to the nearest ten minutes.

Quadratic Trigonometric Equations

BASIC TRIGONOMETRIC IDENTITIES   18.8

Objective   **To verify basic trigonometric identities using reciprocal, quotient, and Pythagorean identities**

Equations that are true for all permissible values of the variable are called identities. The equation $3 + x = x + 3$ is a statement of the commutative property for addition and is true for all real numbers. It is an example of an identity.

In this lesson you will learn how to verify certain trigonometric identities. The first two examples illustrate reciprocal identities.

Example 1   **Show that $(\sin 45°)(\csc 45°) = 1$, $(\cos 60°)(\sec 60°) = 1$, and $(\tan 30°)(\cot 30°) = 1$.**

$(\sin 45°)(\csc 45°)$	1	$(\cos 60°)(\sec 60°)$	1	$(\tan 30°)(\cot 30°)$	1
$\dfrac{\sqrt{2}}{2} \cdot \sqrt{2}$	1	$\dfrac{1}{2} \cdot \dfrac{2}{1}$	1	$\dfrac{\sqrt{3}}{3} \cdot \sqrt{3}$	1
$\dfrac{2}{2}$		1		$\dfrac{3}{3}$	
1				1	

Recall that $\sin A = \dfrac{y}{r}$, $\csc A = \dfrac{r}{y}$, $\cos A = \dfrac{x}{r}$, $\sec A = \dfrac{r}{x}$, $\tan A = \dfrac{y}{x}$, and $\cot A = \dfrac{x}{y}$. You will use these in Example 2.

Example 2   **Show that for any degree measure A, $\sin A \cdot \csc A = 1$, $\cos A \cdot \sec A = 1$, and $\tan A \cdot \cot A = 1$.**

$\sin A \cdot \csc A$	1	$\cos A \cdot \sec A$	1	$\tan A \cdot \cot A$	1
$\dfrac{y}{r} \cdot \dfrac{r}{y}$	1	$\dfrac{x}{r} \cdot \dfrac{r}{x}$	1	$\dfrac{y}{x} \cdot \dfrac{x}{y}$	1
1		1		1	

Reciprocal identities

Sin and csc, cos and sec, and tan and cot are pairs of reciprocal functions.

$$\sin A \cdot \csc A = 1 \qquad\qquad \cos A \cdot \sec A = 1$$

$$\sin A = \frac{1}{\csc A} \text{ or } \csc A = \frac{1}{\sin A} \qquad \cos A = \frac{1}{\sec A} \text{ or } \sec A = \frac{1}{\cos A}$$

$$\tan A \cdot \cot A = 1$$

$$\tan A = \frac{1}{\cot A} \text{ or } \cot A = \frac{1}{\tan A}$$

Example 3   **Show that for any degree measure A, $\tan A = \dfrac{\sin A}{\cos A}$ and $\cot A = \dfrac{\cos A}{\sin A}$.**

Recall that $\sin A = \dfrac{y}{r}$, $\cos A = \dfrac{x}{r}$, $\tan A = \dfrac{y}{x}$, and $\cot A = \dfrac{x}{y}$.

$\tan A$	$\dfrac{\sin A}{\cos A}$		$\cot A$	$\dfrac{\cos A}{\sin A}$
$\dfrac{y}{x}$	$\dfrac{\frac{y}{r}}{\frac{x}{r}}$		$\dfrac{x}{y}$	$\dfrac{\frac{x}{r}}{\frac{y}{r}}$
	$\dfrac{y}{r} \cdot \dfrac{r}{x} = \dfrac{y}{x}$			$\dfrac{x}{r} \cdot \dfrac{r}{y} = \dfrac{x}{y}$

Quotient identities	$\tan A = \dfrac{\sin A}{\cos A}$ $\cot A = \dfrac{\cos A}{\sin A}$

Example 4   **Verify the quotient identities for $A = 210°$.**

$\tan 210°$	$\dfrac{\sin 210°}{\cos 210°}$		$\cot 210°$	$\dfrac{\cos 210°}{\sin 210°}$
$\dfrac{\sqrt{3}}{3}$	$\dfrac{-\frac{1}{2}}{-\frac{\sqrt{3}}{2}}$		$\sqrt{3}$	$\dfrac{-\frac{\sqrt{3}}{2}}{-\frac{1}{2}}$
	$-\dfrac{1}{2} \cdot \left(-\dfrac{2}{\sqrt{3}}\right)$			$\dfrac{-\sqrt{3}}{2}\left(-\dfrac{2}{1}\right)$
	$\dfrac{1}{\sqrt{3}}$, or $\dfrac{\sqrt{3}}{3}$			$\sqrt{3}$

Example 5   **Show that for any degree measure A, $\sin^2 A + \cos^2 A = 1$, $\tan^2 A + 1 = \sec^2 A$, and $\cot^2 A + 1 = \csc^2 A$.**

$\sin^2 A + \cos^2 A$	1	$\tan^2 A + 1$	$\sec^2 A$	$\cot^2 A + 1$	$\csc^2 A$
$\left(\dfrac{y}{r}\right)^2 + \left(\dfrac{x}{r}\right)^2$	1	$\left(\dfrac{y}{x}\right)^2 + 1$	$\left(\dfrac{r}{x}\right)^2$	$\left(\dfrac{x}{y}\right)^2 + 1$	$\left(\dfrac{r}{y}\right)^2$
$\dfrac{y^2 + x^2}{r^2}$		$\dfrac{y^2 + x^2}{x^2}$	$\dfrac{r^2}{x^2}$	$\dfrac{x^2 + y^2}{y^2}$	$\dfrac{r^2}{y^2}$
$x^2 + y^2 = r^2 \blacktriangleright \dfrac{r^2}{r^2} = 1$		$\dfrac{r^2}{x^2}$		$\dfrac{r^2}{y^2}$	

Pythagorean identities	$\sin^2 A + \cos^2 A = 1, \ \tan^2 A + 1 = \sec^2 A, \ \cot^2 A + 1 = \csc^2 A$

Example 6   **Verify the Pythagorean identities for $A = 120°$.**

The terminal side is in the 2nd quadrant. Both sin and csc are positive, all others are negative. The reference angle is $60°$.

$\sin^2 120° + \cos^2 120°$	1	$\tan^2 120° + 1$	$\sec^2 120°$	$\cot^2 120° + 1$	$\csc^2 120°$
$\left(\dfrac{\sqrt{3}}{2}\right)^2 + \left(-\dfrac{1}{2}\right)^2$	1	$(-\sqrt{3})^2 + 1$	$(-2)^2$	$\left(-\dfrac{1}{\sqrt{3}}\right)^2 + 1$	$\left(\dfrac{2}{\sqrt{3}}\right)^2$
$\dfrac{3}{4} + \dfrac{1}{4} = 1$		$3 + 1$ 4	4	$\dfrac{1}{3} + 1 = \dfrac{4}{3}$	$\dfrac{4}{3}$

Written Exercises

(A) **Express as a function of sin, cos, or both.**

1. $\csc A$ **2.** $\sec A$ **3.** $\tan A$ **4.** $\cot A$

Verify each identity for $A = 30°$.

5. $\sin^2 A + \cos^2 A = 1$ **6.** $\tan^2 A + 1 = \sec^2 A$ **7.** $1 + \cot^2 A = \csc^2 A$ **8.** $\cos A = \dfrac{1}{\sec A}$

Verify each identity for $\theta = 240°$.

9. $\sin^2 \theta = 1 - \cos^2 \theta$ **10.** $\cot \theta = \dfrac{\cos \theta}{\sin \theta}$ **11.** $\csc^2 \theta - 1 = \cot^2 \theta$ **12.** $\sin \theta = \dfrac{1}{\csc \theta}$

Verify each identity for $x = 135°$.

13. $\cos^2 x = 1 - \sin^2 x$ **14.** $\tan x = \dfrac{\sin x}{\cos x}$ **15.** $\sec^2 x = 1 + \tan^2 x$ **16.** $\csc^2 x = 1 + \cot^2 x$

Verify each identity for $B = 330°$.

17. $\sin B \cdot \csc B = 1$ **18.** $\tan B \cdot \cot B = 1$ **19.** $\cos B \cdot \sec B = 1$ **20.** $\sin B = \dfrac{1}{\csc B}$

Write an equivalent expression using only sin x.
Hint: $\sin^2 x + \cos^2 x = 1$ so, $\cos x = \pm\sqrt{1 - \sin^2 x}$.

(B) **21.** $\tan x$ **22.** $\cot x$ **23.** $\sec x$ **24.** $\csc x$

Write an equivalent expression using only cos x.
25. $\cot x$ **26.** $\sin x$ **27.** $\csc x$ **28.** $\sec x$

Prove each identity. Use the definitions of tan x, cot x, sec x, and csc x.

(C) **29.** $\tan^2 x + 1 = \sec^2 x$ **30.** $1 + \cot^2 x = \csc^2 x$

Prove each identity. Use $\sin^2 A + \cos^2 A = 1$.
31. $\tan^2 A + 1 = \sec^2 A$ **32.** $1 + \cot^2 A = \csc^2 A$

PROVING TRIGONOMETRIC IDENTITIES   18.9

Objective   **To prove trigonometric identities**

In the previous lesson you studied the reciprocal identities, the quotient identities, and the Pythagorean identities. You will now use these identities to prove other trigonometric identities.

Example 1   **Prove the identity: $\tan x = \dfrac{\sec x}{\csc x}$.**

Replace $\sec x$ with $\dfrac{1}{\cos x}$ and $\csc x$ with $\dfrac{1}{\sin x}$.

$$
\begin{array}{c|c}
\tan x & \dfrac{\sec x}{\csc x} \\[2ex]
\tan x & \dfrac{\dfrac{1}{\cos x}}{\dfrac{1}{\sin x}} \quad \blacktriangleleft \ \textit{Substitute.} \\[3ex]
 & \dfrac{1}{\cos x} \cdot \dfrac{\sin x}{1}, \text{ or } \dfrac{\sin x}{\cos x} \quad \blacktriangleleft \ \textit{Divide.} \\[2ex]
 & \tan x \quad \blacktriangleleft \ \dfrac{\sin x}{\cos x} = \tan x
\end{array}
$$

Example 2   **Prove the identity: $\csc x = \sin x + \cot x \cos x$.**

$$
\begin{array}{c|c}
\csc x & \sin x + \cot x \cos x \\[2ex]
\csc x & \sin x + \dfrac{\cos x}{\sin x} \cdot \cos x \quad \blacktriangleleft \ \cot x = \dfrac{\cos x}{\sin x} \\[3ex]
 & \dfrac{\sin^2 x + \cos^2 x}{\sin x} \\[3ex]
 & \dfrac{1}{\sin x} \quad \blacktriangleleft \ \sin^2 x + \cos^2 x = 1 \\[3ex]
 & \csc x \quad \blacktriangleleft \ \dfrac{1}{\sin x} = \csc x
\end{array}
$$

Summary   To prove a trigonometric identity:
(1) Work with each side separately.
(2) Substitute, using basic trigonometric identities, usually in terms of sin or cos.
(3) Simplify by adding, subtracting, multiplying, or dividing.
(4) When each side is the same expression, the identity has been proved.

Example 3   **Prove the identity:** $\csc x + \csc x \cdot \sec x = \dfrac{1 + \sec x}{\sin x}$.

$$
\begin{array}{c|c}
\csc x + \csc x \sec x & \dfrac{1 + \sec x}{\sin x} \\[2mm]
\hline
\csc x(1 + \sec x) & \dfrac{1 + \sec x}{\sin x} \\[2mm]
\dfrac{1}{\sin x}(1 + \sec x) & \\[2mm]
\dfrac{1 + \sec x}{\sin x} &
\end{array}
$$

Example 4   **Prove the identity:** $\dfrac{1 + \csc x}{\sec x} = \cos x + \cot x$.

$$
\begin{array}{c|c}
\dfrac{1 + \csc x}{\sec x} & \cos x + \cot x \\[2mm]
\hline
\dfrac{1 + \dfrac{1}{\sin x}}{\dfrac{1}{\cos x}} & \cos x + \cot x \\[3mm]
\dfrac{\sin x + 1}{\sin x} \cdot \dfrac{\cos x}{1} & \\[3mm]
\dfrac{\sin x \cos x}{\sin x} + \dfrac{\cos x}{\sin x} & \\[2mm]
\cos x + \cot x &
\end{array}
$$

Written Exercises

Ⓐ **Prove each identity.**

1. $\dfrac{\cot x}{\cos x} = \csc x$ **2.** $\sin x \cot x = \cos x$ **3.** $\sec x = \csc x \cdot \tan x$

4. $\cot x + \tan x = \csc x \sec x$ **5.** $\sec x = \cos x + \dfrac{\tan x}{\csc x}$ **6.** $\sin^2 x - \cos^2 x = 2 \sin^2 x - 1$

Ⓑ **7.** $\cot x - 1 = \cos x (\csc x - \sec x)$ **8.** $\sec x = \dfrac{\cos x}{1 + \sin x} + \tan x$

9. $\sin x + \dfrac{\cot x}{\sec x} = \csc x$ **10.** $\dfrac{\cot x + \cos x}{1 + \sin x} = \dfrac{\cos x}{\sin x}$

11. $\tan x - \cot x = \dfrac{1 - 2 \cos^2 x}{\sin x \cos x}$ **12.** $\dfrac{1 - \cos^2 x}{\cos x} \cdot \csc x = \tan x$

13. $2 \csc x = \dfrac{\sin x}{1 + \cos x} + \dfrac{\sin x}{1 - \cos x}$ **14.** $\cos x + \sin x = \dfrac{\sec x + \csc x}{\cot x + \tan x}$

Ⓒ **15.** $\dfrac{1 - \tan^2 x}{1 + \tan^2 x} = \dfrac{2 - \sec^2 x}{\sec^2 x}$ **16.** $\tan (180° + \theta) + \dfrac{1}{\cot (270° - \theta)} = \dfrac{\sec (180° - \theta)}{\sin (180° + \theta)}$

17. $\dfrac{\cot^2 \theta - 1}{\cot^2 \theta + 1} = \dfrac{2 - \sec^2 \theta}{\sec^2 \theta}$ **18.** $\sqrt{\dfrac{\sec \theta - \tan \theta}{\sec \theta + \tan \theta}} = \dfrac{1 - \sin \theta}{\cos \theta}$

RADIAN MEASURE

Objectives   **To express radian measure in degrees**
To express degree measure in radians
To evaluate expressions involving radian measure

You have learned that an angle of rotation can be measured in degrees. An angle of rotation can also be measured using real numbers called *radians*. If an angle of rotation, which is the central angle of a given circle, intercepts an arc with the same length as a radius, then the angle is said to have a measure of 1 radian. When it intercepts an arc whose length is twice the radius, the angle is said to have a measure of 2 radians. See the figures below.

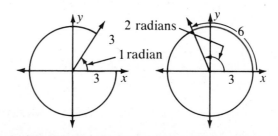

Definition of a radian	If an angle of rotation, which is the central angle of a given circle, intercepts an arc whose length is equal to the length of a radius of a circle, then the measure of the angle of rotation is 1 radian.	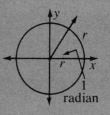

Example 1   **Find the measure of the angle of rotation.**

Since the length of the intercepted arc is 3 times the length of the radius, the measure of the angle of rotation, x, is 3 radians.

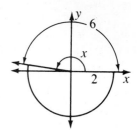

Thus, x = 3 radians.

Example 2   **Express 180° in radian measure.**

Circumference of a circle $= 2\pi r$

So, the length of $\overset{\frown}{AB} = \dfrac{1}{2}(2\pi r)$

$= \pi r$

Thus, 180° is π radians.

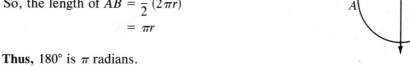

In the preceding example you learned that 180° is π radians, or 180° = π (radians). Divide both sides by 180. The result is $1° = \dfrac{\pi \, (\text{rad})}{180}$. This formula will be used to change degree measure to radian measure.

Example 3

Multiply both sides by 90°. ▶

Express each in radian measure.

| 90° | −210° |

Use $1° = \dfrac{\pi \, (\text{rad})}{180}$.

$$90° = \dfrac{\pi}{180}\,(90°)$$

$$= \dfrac{\pi}{2}$$

Use $1° = \dfrac{\pi \, (\text{rad})}{180}$.

$$-210° = \dfrac{\pi}{180}\,(-210°)$$

$$= -\dfrac{7\pi}{6}$$

Thus, 90° is $\dfrac{\pi}{2}$ radians and −210° is $-\dfrac{7\pi}{6}$ radians.

Now, divide both sides of 180° = π (radians) by π. $1(\text{rad}) = \dfrac{180°}{\pi}$. This formula will be used to change from radian measure to degree measure.

Example 4

Express in degree measure.

$$\dfrac{3}{2}\pi \qquad\qquad\qquad \dfrac{7}{6}\pi$$

Use $1(\text{rad}) = \dfrac{180°}{\pi}$.

$$\dfrac{3}{2}\pi = \dfrac{180°}{\pi}\left(\dfrac{3}{2}\pi\right)$$

$$= 90(3),\ \text{or } 270°$$

Use $1(\text{rad}) = \dfrac{180°}{\pi}$.

$$\dfrac{7}{6}\pi = \dfrac{180°}{\pi}\cdot\left(\dfrac{7}{6}\pi\right)$$

$$= 30(7),\ \text{or } 210°$$

Thus, $\dfrac{3}{2}\pi$ radians is 270° and $\dfrac{7}{6}\pi$ radians is 210°.

Example 5 illustrates how to evaluate trigonometric expressions using radian measure.

Example 5

Evaluate.

| $\sin \dfrac{\pi}{4}$ | $\cos \dfrac{2\pi}{3}$ | $\tan\left(-\dfrac{\pi}{6}\right)$ | $\cot \dfrac{7}{4}\pi$ |

$$\begin{array}{cccc}
\sin 45° & \cos 120° & \tan(-30°) & \cot 315° \\
\dfrac{\sqrt{2}}{2} & -\cos 60° & -\tan 30° & -\cot 45° \\
 & -\dfrac{1}{2} & -\dfrac{\sqrt{3}}{3} & -1
\end{array}$$

Example 6

Evaluate $\dfrac{\sin \dfrac{5\pi}{6} - \cos \dfrac{3\pi}{4}}{\tan\left(-\dfrac{\pi}{4}\right)}$.

$$\frac{\sin \dfrac{5\pi}{6} - \cos \dfrac{3\pi}{4}}{\tan\left(-\dfrac{\pi}{4}\right)} = \frac{\sin 150° - \cos 135°}{\tan(-45°)} = \frac{\sin 30° - (-\cos 45°)}{-\tan 45°} = \frac{\dfrac{1}{2} - \left(\dfrac{-\sqrt{2}}{2}\right)}{-1} = -\frac{1 + \sqrt{2}}{2}$$

Written Exercises

Express each in radians.

(A) 1. $30°$ 2. $120°$ 3. $-200°$ 4. $150°$ 5. $210°$ 6. $300°$ 7. $-60°$
    8. $-320°$ 9. $-270°$ 10. $240°$ 11. $-330°$ 12. $225°$ 13. $450°$ 14. $-360°$

Express each in degrees.

15. $\dfrac{\pi}{4}$ 16. $\dfrac{3}{2}\pi$ 17. $\dfrac{5\pi}{6}$ 18. $\dfrac{7\pi}{6}$ 19. $\dfrac{2}{3}\pi$ 20. $-\pi$ 21. $\dfrac{3}{4}\pi$

22. $-\dfrac{\pi}{8}$ 23. $-\dfrac{3}{5}\pi$ 24. $-\dfrac{2\pi}{9}$ 25. -2π 26. $\dfrac{9}{4}\pi$ 27. $\dfrac{11\pi}{6}$ 28. $-\dfrac{7\pi}{6}$

Evaluate each of the following.

(B) 29. $\sin \dfrac{\pi}{2}$ 30. $\cos \dfrac{5\pi}{2}$ 31. $\tan\left(-\dfrac{\pi}{4}\right)$ 32. $\cot \dfrac{7}{6}\pi$ 33. $\sin\left(-\dfrac{2}{3}\pi\right)$

34. $\dfrac{\tan \dfrac{5}{4}\pi - \cos \dfrac{5}{3}\pi}{\sin \dfrac{\pi}{6}}$ 35. $\dfrac{\cos \dfrac{5\pi}{4} + \tan \dfrac{7}{4}\pi}{\cot\left(\dfrac{5}{6}\pi\right)}$ 36. $\dfrac{\sin \dfrac{\pi}{9} - \cos \dfrac{3}{2}\pi}{\tan\left(-\dfrac{3\pi}{4}\right)}$

(C) 37. $\dfrac{\sec \dfrac{2\pi}{3} + \csc \dfrac{\pi}{6}}{\sec \dfrac{\pi}{4}}$

38. $\dfrac{\csc \dfrac{2\pi}{5} - \sec\left(-\dfrac{1}{3}\pi\right) + \sin \dfrac{\pi}{3}}{\tan\left(-\dfrac{\pi}{4}\right)\cos\left(-\dfrac{\pi}{3}\right)}$

39. $\dfrac{\cos \dfrac{11\pi}{6} - \sec\left(-\dfrac{\pi}{3}\right)}{\tan \dfrac{\pi}{6}\csc\left(-\dfrac{\pi}{4}\right)}$

40. $\dfrac{\sec\left(-\dfrac{\pi}{9}\right)\csc \dfrac{3\pi}{4} - \tan\left(-\dfrac{3\pi}{4}\right)}{\cos^2 \dfrac{\pi}{3} + \sin^2 \dfrac{\pi}{3}}$

If x is expressed in radians, you can use the two infinite series below to approximate $\sin x$ and $\cos x$ to several decimal places.

$$\cos x = 1 - \frac{x^2}{2!} + \frac{x^4}{4!} - \frac{x^6}{6!} + \frac{x^8}{8!} - \cdots$$

$$\sin x = x - \frac{x^3}{3!} + \frac{x^5}{5!} - \frac{x^7}{7!} + \frac{x^9}{9!} - \cdots$$

Use a calculator to approximate each of the following to 5 decimal places. Use 3.14 for π.

41. $\sin \dfrac{\pi}{4}$ 42. $\cos \dfrac{\pi}{6}$ 43. $\sin \dfrac{5\pi}{6}$ 44. $\cos \dfrac{\pi}{4}$

APPLYING RADIAN MEASURE   18.11

Objectives   **To sketch graphs of trigonometric functions using radian measure**
To write the solutions of trigonometric equations in radians

The table below lists some special degree-radian relationships.

Degrees	0°	30°	45°	60°	90°	120°	135°	150°	180°
Radians	0	$\frac{\pi}{6}$	$\frac{\pi}{4}$	$\frac{\pi}{3}$	$\frac{\pi}{2}$	$\frac{2\pi}{3}$	$\frac{3\pi}{4}$	$\frac{5\pi}{6}$	π

Degrees	210°	225°	240°	270°	300°	315°	330°	360°
Radians	$\frac{7\pi}{6}$	$\frac{5\pi}{4}$	$\frac{4\pi}{3}$	$\frac{3\pi}{2}$	$\frac{5\pi}{3}$	$\frac{7\pi}{4}$	$\frac{11\pi}{6}$	2π

To graph the function $y = \sin x$ using radian measure, label the horizontal
axis in radians and the vertical axis in real numbers. Make a table of values
by choosing special radian measures and finding the corresponding values of y.
The graph of $y = \sin x$, $0 \le x \le 2\pi$, is shown.

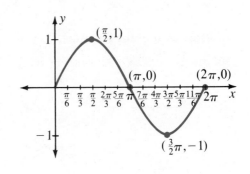

x	$y = \sin x$
0	0
$\frac{\pi}{6}$	$\frac{1}{2}$, or .5
$\frac{\pi}{3}$	$\frac{\sqrt{3}}{2}$, or .87
$\frac{\pi}{2}$	1
$\frac{2\pi}{3}$	$\frac{\sqrt{3}}{2}$, or .87
π	0
$\frac{7\pi}{6}$	$-\frac{1}{2}$, or $-.5$
$\frac{3\pi}{2}$	-1
$\frac{11\pi}{6}$	$-\frac{1}{2}$, or $-.5$
2π	0

Notice that the maximum value 1 occurs at $\frac{\pi}{2}$. The minimum value -1 occurs
at $\frac{3\pi}{2}$. The period of $y = \sin x$ is 2π.

Example 1   **Graph $y = 2 \cos x$ for $0 \le x \le 2\pi$. Determine the period and amplitude.**

x	$\cos x$	$y = 2 \cos x$
0	1	2
$\frac{\pi}{2}$	0	0
π	-1	-2
$\frac{3\pi}{2}$	0	0
2π	1	2

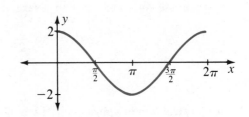

Thus, period for $y = 2 \cos x$ is 2π.
  Amplitude for $y = 2 \cos x$ is 2.

Example 2　Graph $y = \sin \frac{1}{2}x$ for $0 \le x \le 4\pi$. Determine the period and amplitude.

x	$\frac{1}{2}x$	$y = \sin \frac{1}{2}x$
0	0	0
π	$\frac{\pi}{2}$	1
2π	π	0
3π	$\frac{3\pi}{2}$	-1
4π	2π	0

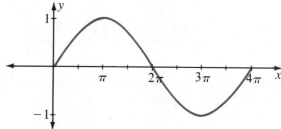

Thus, the period for $y = \sin \frac{1}{2}x$ is 4π and the amplitude is 1.

Example 3　**Sketch $y = 2\cos x$ and $y = \sin \frac{1}{2}x$ for $0 \le x \le 2\pi$ on the same set of axes. For how many values of x does $2\cos x = \sin \frac{1}{2}x$?**

Use Examples 1 and 2 to sketch the graphs.

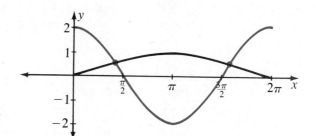

Thus, $2\cos x = \sin \frac{1}{2}x$ for 2 values of x for $0 \le x \le 2\pi$.

Example 4　**Solve $-3\tan x - \sqrt{3} = 0$ if $0 \le x \le 2\pi$.**

Solve for $\tan x$. Determine the reference angle.

$$-3\tan x = \sqrt{3}$$
$$\tan x = -\frac{\sqrt{3}}{3}$$

Reference angle: 30°

◄ *Tan is negative in quadrants 2 and 4.*

　　2nd quad.　4th quad.
　　　150°　　　330°

Thus, the solutions are $\frac{5\pi}{6}$ and $\frac{11\pi}{6}$.

Example 5　**Solve $2\cos^2 x - 5\cos x + 2 = 0$ if $0 \le x \le 2\pi$.**

$(2\cos x - 1)(\cos x - 2) = 0$　◄ *Factor.*
$2\cos x - 1 = 0$ or $\cos x - 2 = 0$　◄ *Set each factor equal to 0.*
　　　$\cos x = \frac{1}{2}$　　　　$\cos x = 2$
　　Ref. $\angle$: 60°　　　No solution, since $-1 \le \cos x \le 1$

1st quad.　4th quad.
　60°　　　300°

Thus, the solutions are $\frac{\pi}{3}$ and $\frac{5\pi}{3}$.

Written Exercises

(A) **Graph** $y = \sin 2x$ **for** $-2\pi \le x \le 2\pi$ **to find each value below.**
  1. maximum and minimum   2. period   3. amplitude

Graph $y = 3 \cos x$ **for** $-\pi \le x \le \pi$ **to find each value below.**
  4. maximum and minimum   5. period   6. amplitude

Sketch each graph for $0 \le x \le 2\pi$. **Determine the period and the amplitude.**
  7. $y = \frac{1}{2} \sin x$ 8. $y = \cos 4x$ 9. $y = 3 \cos x$ 10. $y = \sin 3x$
 11. $y = 2 \cos \frac{1}{2}x$ 12. $y = \frac{1}{2} \sin 2x$ 13. $y = \frac{1}{3} \cos 3x$ 14. $y = 4 \sin 2x$

Solve each trigonometric equation if $0 \le x \le 2\pi$.
 15. $2 \sin x + 1 = 0$ 16. $2 \cos x - \sqrt{3} = 0$ 17. $\tan x = 1$
 18. $5 \sin^2 x = 0$ 19. $2 \cos^2 x - 1 = 0$ 20. $2 \sin x + \sqrt{3} = 0$
 21. $2 \cos^2 x + \cos x = 0$ 22. $\sqrt{2} \sin^2 x - \sin x = 0$ 23. $2 \cos^2 x - 7 \cos x - 4 = 0$
 24. $\sin^2 x - \frac{1}{4} = 0$ 25. $6 \sin^2 x - 7 \sin x - 5 = 0$ 26. $4 \cos^2 - 1 = 0$

(B) 27. Sketch $y = 3 \sin x$ and $y = \cos 3x$ for $0 \le x \le 2\pi$ on the same set of axes. For how many values of x does $3 \sin x = \cos 3x$?

28. Sketch $y = \cos \frac{1}{2}x$ and $y = \sin 3x$ for $0 \le x \le 2\pi$ on the same set of axes. For how many values of x does $\cos \frac{1}{2}x = \sin 3x$?

Sketch each graph for $-\pi \le x \le 2\pi$. **Determine the period and amplitude.**
 29. $y = \tan x$ 30. $y = 3 \tan x$ 31. $y = \tan 3x$
 32. $y = -\cos x$ 33. $y = 2 \sin(-2x)$ 34. $y = -\cos 2x$
 35. $y = -2 \sin \frac{1}{2}x$ 36. $y = -3 \cos(-x)$ 37. $y = -3 \sin(-2x)$
 38. $y = -\frac{1}{2} \sin 2x$ 39. $y = \frac{1}{3} \cos 2x$ 40. $y = -\frac{1}{2} \cos 3x$

(C) 41. $y = \sec x$ 42. $y = \csc x$ 43. $y = \cot x$
 44. $y = \sec 2x$ 45. $y = -\frac{1}{2} \sec x$ 46. $y = \csc(-2x)$

Solve each trigonometric equation if $0 \le x \le 2\pi$.
 47. $\sec x = -\sqrt{2}$ 48. $\csc x = -2$ 49. $3 \sec x - 2\sqrt{3} = 0$
 50. $2 \sec x + 1 = 0$ 51. $2 \sec^2 x + 3 \sec x - 2 = 0$ 52. $\csc^2 x - 4 = 0$

53. Sketch $y = \sin 2x$ and $y = 2 \cos x$ for $-\pi \le x \le \pi$ on the same set of axes. For what values of x does $\sin 2x = 2 \cos x$?

54. Sketch $y = 2 \sin x$ and $y = \sin \frac{1}{2}x$ for $-\pi \le x \le 2\pi$ on the same set of axes. For what values of x does $2 \sin x = \sin \frac{1}{2}x$?

55. Sketch $y = -2 \sin \frac{1}{2}x$ and $y = -\frac{1}{2} \cos (-2x)$ for $0 \le x \le 2\pi$ on the same set of axes. For how many values of x does $-2 \sin \frac{1}{2}x = -\frac{1}{2} \cos(-2x)$?

56. Sketch $y = \tan(-2x)$ and $y = -2 \cos \frac{1}{2}x$ for $0 \le x \le 2\pi$ on the same set of axes. For how many values of x does $\tan(-2x) = -2 \cos \frac{1}{2}x$?

INVERSES OF TRIGONOMETRIC FUNCTIONS

Objectives
To determine the inverse of a trigonometric function
To graph inverses of the sin, cos, and tan functions
To apply inverse function notation

Recall that the inverse of a function is formed by interchanging the elements of each ordered pair of the function. The inverse of a function is not necessarily a function. The *inverse function* is formed when the inverse of a given function is also a function.

Example 1

Find the inverse of $y = \cos x$.

Interchange x and y.

$y = \cos x$

$x = \cos y$

Thus, the inverse of $y = \cos x$ is $x = \cos y$.

After finding the inverse of a function, it is customary to solve for y. Cos $y = x$ can be written in two other ways: $y = \text{arc} \cos x$, or $y = \cos^{-1} x$. Each is read "y is the angle whose cos is x." In $y = \cos^{-1} x$, -1 is *not* an exponent.

Example 2

Find u, if $u = \text{arc} \sin (-\frac{1}{2})$, $0° \leq u \leq 360°$.

Read $u = \text{arc} \sin (-\frac{1}{2})$ as u is the angle whose sin is $-\frac{1}{2}$.
Write $u = \text{arc} \sin (-\frac{1}{2})$ as $\sin u = -\frac{1}{2}$.

Ref. $\angle$: 30°

3rd quad.   4th quad.

So, $\sin 210° = -\frac{1}{2}$ and $\sin 330° = -\frac{1}{2}$.
Thus, $u = 210°$ or $330°$.

Example 3

Find θ, if $\theta = \text{arc} \cos \dfrac{\sqrt{2}}{2}$, $0 \leq \theta \leq 2\pi$.

Write $\theta = \text{arc} \cos \dfrac{\sqrt{2}}{2}$ as $\cos \theta = \dfrac{\sqrt{2}}{2}$.

Ref. $\angle$: 45°

1st quad.   4th quad.

So, $\cos 45° = \dfrac{\sqrt{2}}{2}$ and $\cos 315° = \dfrac{\sqrt{2}}{2}$.

Thus, $\theta = \dfrac{\pi}{4}$ or $\dfrac{7\pi}{4}$.

Inverses of Trigonometric Functions

In the next example you will learn how to graph the inverse of a trigonometric function.

Example 4   **Sketch y = arc sin x. Determine if the inverse is a function.**

Make a table of values for $y = \sin x$ and interchange the elements of the ordered pairs.

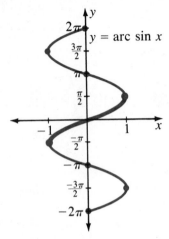

x	$y = \sin x$	$x = \sin y$	y
0	0	0	0
$\frac{\pi}{6}$	.5	.5	$\frac{\pi}{6}$
$\frac{\pi}{3}$	.87	.87	$\frac{\pi}{3}$
$\frac{\pi}{2}$	1	1	$\frac{\pi}{2}$
π	0	0	π
$\frac{3\pi}{2}$	-1	-1	$\frac{3\pi}{2}$
2π	0	0	2π
$-\frac{\pi}{2}$	-1	-1	$-\frac{\pi}{2}$
$-\pi$	0	0	$-\pi$
$-\frac{3\pi}{2}$	1	1	$-\frac{3\pi}{2}$
-2π	0	0	-2π

The graph does not pass the vertical line test, since there are many values of y for each x between -1 and 1.

Thus, the inverse of $y = \sin x$ is not a function.

The graphs of the inverses of $y = \cos x$ and $y = \tan x$ are shown below.

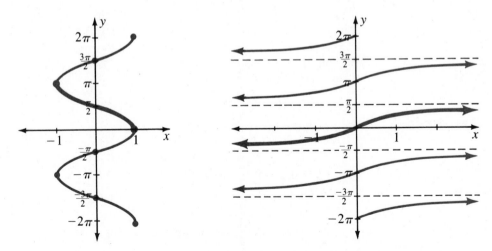

By restricting the interval of the graph of the inverse of a trig function, the inverse can be an inverse function in that interval. The shaded regions of the three graphs above are the graphs of the inverse functions.

The next three examples illustrate how to find a trig function given an inverse trig function.

Example 5   **Find cos θ if $\theta = $ Arc sin $\left(-\dfrac{3}{5}\right)$.**

Write $\theta = \text{Arc sin}\left(-\dfrac{3}{5}\right)$ as $\sin \theta = -\dfrac{3}{5}$.

The principal value of θ is

$$-\frac{\pi}{2} \le \theta \le \frac{\pi}{2} \quad \text{or} \quad -90° \le \theta \le 90°$$

$$\sin \theta = -\frac{3}{5}$$
$$\downarrow \qquad \downarrow$$

3rd quad.   4th quad.

There is no solution in the 3rd quad. since the Principal Values for sin are in the 1st and 4th quad.

Find x.
$$x^2 + y^2 = r^2$$
$$x^2 + 9 = 25$$
$$x^2 = 16$$
$$x = 4$$

Thus, $\cos \theta = \dfrac{4}{5}$.

Example 6   **Find sin $\left(\text{Arc tan }\left(\dfrac{3}{4}\right)\right)$.**

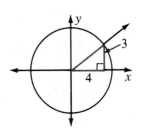

Let $A = \text{Arc tan }\dfrac{3}{4}$.

Write as $\tan A = \dfrac{3}{4}$ and $-90° \le A \le 90°$.
$$\swarrow \qquad \searrow$$

1st quad.   3rd quad.

Find r.   No solution since
$$x^2 + y^2 = r^2 \quad -90° \le A \le 90°$$
$$9 + 16 = r^2$$
$$r = 5$$

Thus, $\sin A = \dfrac{3}{5}$, so $\sin \left(\text{Arc tan }\left(\dfrac{3}{4}\right)\right) = \dfrac{3}{5}$.

Example 7   Find $\sin^{-1}\left(\tan \dfrac{5\pi}{6}\right)$.

$$\text{Tan } \dfrac{5\pi}{6} = \tan 150° = -\tan 30°, \text{ or } -\dfrac{\sqrt{3}}{3}$$

Let $A = \sin^{-1}\left(-\dfrac{\sqrt{3}}{3}\right)$.   Write as $\sin A = -\dfrac{\sqrt{3}}{3} = -\dfrac{1.732}{3}$, or $-.5773$.

Look in the table.
Ref. $\angle$: 35° to the nearest degree.

3rd quad.   4th quad.   **Thus,** $\sin^{-1}\left(\tan \dfrac{5\pi}{6}\right) = -35°$ or $-\dfrac{7\pi}{36}$.
No solution $-35°$

Reading in Algebra

**Write A if the statement is *always* true. Write S if the statement is *sometimes*
true. Write N if the statement is *never* true.**
1. Inverses of trig functions are functions.
2. The inverse for $y = \sin x$ is $x = \sin y$.
3. The inverse for a function is formed by interchanging the elements of each
   ordered pair of the function.
4. If $\theta = \text{arc tan } (-2)$, then $\tan \theta = -2$.
5. The inverse function for $y = \sin x$ is Arc sin x.
6. The Principal Values for Arc tan x are $-\dfrac{\pi}{2} \le \text{Arc tan } x \le \dfrac{\pi}{2}$.

Written Exercises

Ⓐ **Find the inverse of each function.**
1. $y = \sin \frac{1}{2}x$ 2. $y = \cos x$ 3. $y = \tan x$ 4. $y = \cot x$ 5. $y = \frac{1}{2}\cos x$

Find θ for each of the following.
6. $\theta = \text{Arc sin } 0$ 7. $\theta = \text{Arc cos } \frac{1}{2}$ 8. $\theta = \text{Arc sin } (-\frac{1}{2})$ 9. $\theta = \text{Arc sin } (-\frac{\sqrt{3}}{2})$
10. $\theta = \text{Arc cos } \frac{\sqrt{2}}{2}$ 11. $\theta = \text{Arc tan } (-1)$ 12. $\theta = \text{Arc tan } (-\frac{\sqrt{3}}{3})$ 13. $\theta = \text{Arc cos } (-1)$

14. Find $\sin A$ if $A = \text{Arc cos } (-\frac{5}{13})$.   15. Find $\cos A$ if $A = \text{Arc sin } \frac{3}{5}$.
16. Find $\tan A$ if $A = \text{Arc sin } (-\frac{12}{13})$.   17. Find $\sin A$ if $A = \text{Arc tan } (-\frac{3}{4})$.

Ⓑ **Find the value of each of the following.**
18. $\cos (\text{Arc sin } (-\frac{\sqrt{2}}{2}))$ 19. $\sin^{-1} (\tan (\frac{2}{3}\pi))$ 20. Arc sin $(\cos (-\frac{\pi}{4}))$
21. $\tan (\text{Arc cos } (\frac{\sqrt{3}}{3}))$ 22. $\cos (\text{Arc tan } (-\frac{4}{3}))$ 23. $\tan (\text{Arc cos } (-\frac{5}{13}))$

Ⓒ **Sketch each graph.**
24. $y = \text{Arc sin } 2x$ 25. $y = 2 \text{ Arc cos } \frac{1}{2}x$ 26. $y = -\frac{1}{2} \text{ Arc sin } (-2x)$
27. $y = -2 \text{ Arc tan } x$ 28. $y = -2 \text{ Arc cos } (-\frac{1}{2}x)$ 29. $y = \text{Arc cot } x$
30. $y = \text{Arc sin } x$ 31. $y = \text{Arc cos } x$ 32. $y = \text{Arc tan } x$
33. $y = 2 \text{ Arc cot } x$ 34. $y = \text{Arc sec } x$ 35. $y = \text{Arc csc } x$

COMPLEX NUMBERS AND DE MOIVRE'S THEOREM

Objectives **To write a complex number in polar form**
To write the polar form of a number as a complex number
To find a power of a complex number

The relationship shown in the figure at the right is based on the Pythagorean relation $c^2 = a^2 + b^2$. The horizontal axis is the real number axis, while the vertical axis is the imaginary axis.

Absolute Value of a Complex Number

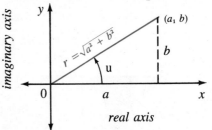

The $(a + bi)$ form for a complex number is frequently referred to as the **rectangular form** of a complex number. It is written as $r = |a + bi| = \sqrt{a^2 + b^2}$. Recall that the absolute value of a complex number like $5 - 2i$ is $r = |5 - 2i| = \sqrt{5^2 + (-2)^2}$, or $\sqrt{29}$.

According to the definition of sine and cosine, $\sin u = \dfrac{b}{\sqrt{a^2 + b^2}}$ and

$\cos u = \dfrac{a}{\sqrt{a^2 + b^2}}$ since $r = \sqrt{a^2 + b^2}$, $\sin u = \dfrac{b}{r}$ or $b = r \sin u$, and

$\cos u = \dfrac{a}{r}$ or $a = r \cos u$.

The rectangular form of a complex number can be written in trigonometric form or polar form as indicated below.

$$a + bi = r \cos u + (r \sin u)i \text{ or } r(\cos u + i \sin u)$$

Example 1   **Write the complex number $1 + i$ in polar form.**

$$\boxed{1} + \boxed{1}i$$
$$\boxed{a} + \boxed{b}i$$
▶ $a = 1$ and $b = 1$. So, $r = \sqrt{1^2 + 1^2}$, or $\sqrt{2}$.

$\sin u = \dfrac{b}{r}$ $\qquad\qquad$ $\cos u = \dfrac{a}{r}$

$\quad = \dfrac{1}{\sqrt{2}}$, or $\dfrac{\sqrt{2}}{2}$ $\qquad = \dfrac{1}{\sqrt{2}}$, or $\dfrac{\sqrt{2}}{2}$

Since both a and b are positive, the angle is in quadrant I. So, $u = \dfrac{\pi}{4}$, or $45°$.

Substitute:

$r = \sqrt{2}, u = \dfrac{\pi}{4}$ ▶ $a + bi = r \cos u + ir \sin u$

$1 + i = \sqrt{2} \left(\cos \dfrac{\pi}{4} + i \sin \dfrac{\pi}{4} \right)$

Thus, the polar form of $1 + i$ is $\sqrt{2} \left(\cos \dfrac{\pi}{4} + i \sin \dfrac{\pi}{4} \right)$.

The polar form of a complex number can also be written in rectangular form.

Example 2 **Write $2\left(\cos \dfrac{\pi}{3} + i \sin \dfrac{\pi}{3}\right)$ in rectangular form.**

$$\cos \dfrac{\pi}{3} = \dfrac{1}{2} \blacktriangleright \quad 2\left(\cos \dfrac{\pi}{3} + i \sin \dfrac{\pi}{3}\right) = 2\left(\dfrac{1}{2} + i\dfrac{\sqrt{3}}{2}\right)$$

$$\sin \dfrac{\pi}{3} = \dfrac{\sqrt{3}}{2} \blacktriangleright \qquad\qquad\qquad\qquad = 1 + i\sqrt{3}$$

Thus, the rectangular form of $2\left(\cos \dfrac{\pi}{3} + i \sin \dfrac{\pi}{3}\right)$ is $1 + i\sqrt{3}$.

De Moivre's theorem states a method for computing powers of a complex number.

De Moivre's theorem	For all real numbers r, n, and θ, $r(\cos \theta + i \sin \theta)^n = r^n(\cos n\theta + i \sin n\theta)$.

Example 3 **Compute $(-2 + 2i)^6$.**

$$a = -2, b = 2, \text{ and } r = \sqrt{4 + 4}, \text{ or } 2\sqrt{2}$$

$$\sin \theta = \dfrac{b}{r} = \dfrac{2}{2\sqrt{2}}, \text{ or } \dfrac{\sqrt{2}}{2} \qquad \cos \theta = \dfrac{a}{r} = \dfrac{-2}{2\sqrt{2}}, \text{ or } -\dfrac{\sqrt{2}}{2}$$

Let $-2 + 2i = r(\cos \theta + i \sin \theta)$
$$(-2 + 2i)^6 = [r(\cos \theta + i \sin \theta)]^6$$
$$= \left[2\sqrt{2}\left(\dfrac{-\sqrt{2}}{2} + i\dfrac{\sqrt{2}}{2}\right)\right]^6$$

$\cos \dfrac{3}{4}\pi = -\dfrac{\sqrt{2}}{2}$ and $\sin \dfrac{3}{4}\pi = \dfrac{\sqrt{2}}{2} \blacktriangleright \quad \left[2\sqrt{2}\left(\cos \dfrac{3}{4}\pi + i \sin \dfrac{3}{4}\pi\right)\right]^6$

$$\left(2\sqrt{2}\right)^6\left(\cos 6 \cdot \dfrac{3}{4}\pi + i \sin 6 \cdot \dfrac{3}{4}\pi\right)$$

$$512\left(\cos \dfrac{9}{2}\pi + i \sin \dfrac{9}{2}\pi\right)$$

$\cos \dfrac{9}{2}\pi = 0$ and $\sin \dfrac{9}{2}\pi = 1 \blacktriangleright \quad 512(0 + i(1)), \text{ or } 512i$

Thus, $(-2 + 2i)^6 = 512i$.

Written Exercises

Write each of the following in rectangular form.

Ⓐ **1.** $3\left(\cos \dfrac{\pi}{4} + i \sin \dfrac{\pi}{4}\right)$ **2.** $\left(\cos \dfrac{3\pi}{4} + i \sin \dfrac{3\pi}{4}\right)$

3. $4\left(\cos \dfrac{\pi}{6} + i \sin \dfrac{\pi}{6}\right)$ **4.** $5\left(\cos \dfrac{3\pi}{2} + i \sin \dfrac{3\pi}{2}\right)$

Compute by using De Moivre's theorem.

Ⓑ **5.** $(1 - i\sqrt{3})^4$ **6.** $(2 + i)^6$ **7.** $(-2 + 2i)^5$ **8.** $(\sqrt{3} + 2i)^3$

Vocabulary
Amplitude of a function [18.1–18.3]
Inverse of a trig function [18.12]
Periodic function [18.1–18.3]
Principal Values [18.12]
Pythagorean identities [18.8]
Quotient identities [18.8]
Radian measure [18.10]
Reciprocal identities [18.8]

Sketch each graph for $0° \leq x \leq 360°$. Determine the period and amplitude. [18.1–18.5]
1. $y = 3 \cos x$
2. $y = \sin(-2x)$
3. $y = \frac{1}{2} \sin(\frac{1}{2} x)$
4. $y = \tan 3x$
5. $y = -\frac{1}{2} \cos(-2x)$
★6. $y = -2 \sec x$

7. Sketch $y = \sin 3x$ and $y = 2 \cos x$ for $-180° \leq x \leq 180°$ on the same set of axes. For how many values of x does $\sin 3x = 2 \cos x$? [18.4]

8. Find log sin 28° 33′. [18.6]

9. Find log cot 48° 47′. [18.6]

10. Find u, $0° \leq u \leq 90°$, if log cos $u = 9.6348 - 10$.

11. Find u, $0° \leq u \leq 90°$, if log tan $u = 9.8732 - 10$. 18.6]

Find each value to four significant digits. Use logarithms. [18.6]
12. $\dfrac{(\cos 43°)^4}{\sqrt{38.7}}$
13. $\dfrac{(498)^3 \sqrt{.038}}{\tan 57° 43'}$

Solve for θ to the nearest degree ($0° \leq \theta \leq 360°$).
14. $\cos^2 \theta - 1 = 0$
15. $\sin^2 \theta - \dfrac{\sqrt{2}}{2} \sin \theta = 0$
[18.7]
16. $2 \sin^2 \theta + 5 \sin \theta - 6 = 0$

Solve for θ if $0 \leq \theta \leq 2\pi$. [18.11]
17. $\cos \theta = -\dfrac{\sqrt{2}}{2}$
18. $\tan^2 \theta + \tan \theta = 0$
19. $2 \sin^2 \theta - 3 \sin \theta + 1 = 0$
★20. $16 \sec^2 \theta - 9 = 0$

Express in terms of sin, cos, or both. [18.8]
21. tan 23°
22. csc 38°
23. cot 74° 18′
24. sec (−305°)

Verify for $A = 240°$ [18.8]
25. $\sin^2 A + \cos^2 A = 1$
26. $\tan^2 A = \sec^2 A - 1$

Write an equivalent expression using only cos x. [18.8]
27. $\cos x \cdot \dfrac{1}{\sec x}$
28. $\tan x$
29. $\sin x \cdot \sec x$

Prove each identity. [18.9]
30. $\dfrac{1}{4} \csc \theta = \dfrac{\sin \theta}{3 \sin^2 \theta - \cos^2 \theta + 1}$
31. $\dfrac{\sin \theta}{\cot \theta} + \cos \theta = \sec \theta$

Express in radians. [18.10]
32. 60°
33. −45°
34. 330°
35. −140°

Express in degrees. [18.10]
36. $\dfrac{\pi}{12}$
37. $-\dfrac{4}{3} \pi$
38. 6π
39. $-\dfrac{3}{2} \pi$

Evaluate. [18.10]
40. $\tan \dfrac{7\pi}{4}$
41. $\cos \left(-\dfrac{\pi}{6}\right)$
42. $\dfrac{\cos \dfrac{\pi}{3} - \sin \dfrac{\pi}{4}}{\tan \left(-\dfrac{\pi}{4}\right)}$
43. $\dfrac{\cot \left(-\dfrac{\pi}{3}\right) + \tan \dfrac{3\pi}{4}}{\sin \dfrac{11\pi}{6}}$

Find the inverse of each function. Is the inverse a function? [18.12]
44. $y = \cos x$
45. $y = 3 \sin x$

Find each θ. (Ex. 46–47) [18.12]
46. $\theta = \text{Arc cos} \left(-\dfrac{\sqrt{3}}{2}\right)$
47. $\theta = \text{Arc tan} \left(\dfrac{\sqrt{2}}{2}\right)$
48. Compute $(1 + i)^4$. [18.13]
49. Write $2 \left(\cos \dfrac{\pi}{3} + i \sin \dfrac{\pi}{3}\right)$ in rectangular form. [18.13]

CHAPTER EIGHTEEN TEST

Sketch each graph for $0° \leq x \leq 360°$. Determine the period and amplitude.

1. $y = 2 \sin x$ **2.** $y = \cos \left(\frac{1}{2} x \right)$

3. $y = \frac{1}{2} \sin 2x$ **4.** $y = \tan 2x$

5. Sketch $y = \sin 2x$ and $y = 3 \cos x$ for $0° \leq x \leq 360°$ on the same set of axes. For how many values of x does $\sin 2x = 3 \cos x$?

6. Find log cos 36° 45′.

7. Find u, $0° \leq u \leq 90°$, if log tan $u = 9.8685 - 10$.

Find the value to four significant digits. Use logarithms.

8. $\dfrac{(\sin 38°)^3}{\sqrt{.437}}$ **9.** $\dfrac{(137)^4 \sqrt{.0867}}{\cos 43° 36'}$

Solve for θ to the nearest degree ($0° \leq \theta \leq 360°$).
10. $2 \sin^2 \theta - 1 = 0$
11. $2 \cos^2 \theta - \cos \theta - 1 = 0$

Solve for θ if $0 \leq \theta \leq 2\pi$.

12. $\tan^2 \theta = \tan \theta$ **13.** $\sin \theta = -\dfrac{\sqrt{3}}{2}$

Express in terms of sin, cos, or both.
14. $\tan (-230°)$ **15.** $\sec 43°$

16. Verify for $A = 150°$. $\cos^2 A = 1 - \sin^2 A$

Write an equivalent expression using only cos x.
17. $\cot x$ **18.** $\tan x \cdot \csc x$

Prove the identity.
19. $\sec \theta = \dfrac{\tan \theta}{\csc \theta} + \cos \theta$

Express in radians.
20. 45°   **21.** −330°

Express in degrees.
22. $-\dfrac{5\pi}{6}$ **23.** $\dfrac{5}{3} \pi$

Evaluate.

24. $\cos \left(-\dfrac{2\pi}{3} \right)$ **25.** $\dfrac{\sin \dfrac{\pi}{6} + \cos \dfrac{7}{6} \pi}{\tan \left(-\dfrac{\pi}{3} \right)}$

Find the inverse of each function. Is the inverse a function?
26. $y = \tan x$ **27.** $y = -2 \sin x$

Find θ.
28. $\theta = \text{Arc sin} \left(-\dfrac{\sqrt{2}}{2} \right)$

Find each value.
29. $\sin \left(\text{Arc tan} \dfrac{5}{12} \right)$

30. $\tan \left(\text{Arc cos} \dfrac{5}{6} \right)$

31. Sketch $y = -\cos \left(\dfrac{1}{2} x \right)$ and $y = 3 \sin (2x)$ for $-\pi \leq x \leq 2\pi$ on the same set of axes. For how many values of x does $-\cos \dfrac{1}{2} x = 3 \sin 2x$?

Sketch each graph.
★ **32.** $y = \text{Arc cos } x$
★ **33.** $y = \text{Arc tan } x$
★ **34.** $y = \text{arc csc } x$

Solve for θ if $0 \leq \theta \leq 360°$.
★ **35.** $4 \csc^2 \theta - 9 = 0$

Evaluate.
★ **36.** $\dfrac{\sec \left(-\dfrac{\pi}{4} \right) - \csc \dfrac{3}{2} \pi}{-6 \sec \dfrac{4\pi}{3}}$

37. Compute $(-2 + 4i)^3$.

38. Write $4 \left(\cos \dfrac{3}{4} \pi + i \sin \dfrac{3}{4} \pi \right)$ in rectangular form.

COLLEGE PREP TEST

Directions: In each question you are to *compare* a Quantity I and a Quantity II. **Choose your answer in the following way. Choose:**
(A) if Quantity I is greater than II,
(B) if Quantity II is greater than I,
(C) if Quantity I = Quantity II,
(D) if the relationship cannot be determined

In questions 1 and 2 use the information given. Right triangle ABC with right angle C

1. Quantity I $(AC)^2 + (BC)^2$
   Quantity II $(AB)^2$

2. Quantity I $2(AC)^2$
   Quantity II $(AB)^2$

3. Information: Triangle ABC

   Quantity I $AC + BC$
   Quantity II AB

4. Information: Triangle ABC with $m \angle C < 90°$. Triangle RST with $m \angle T > 90°$.

   Quantity I AB
   Quantity II RS

5. Information: Right triangle ABC, C is a right angle and $m \angle B = 30°$.

   Quantity I AB
   Quantity II $2AC$

6. Quantity I $\left(\dfrac{\pi}{2} + \dfrac{\pi}{6}\right)$ radians
   Quantity II $30° + 90°$

7. Information: Right triangle ABC with right angle C. Right triangle RST with right angle T. The measure of angle S equals the measure of angle B.

   Quantity I $(AC)(RS)$
   Quantity II $(AB)(RT)$

8. Information: Right triangle ABC with right angle C and $m \angle A = 60°$. Right triangle RST with right angle T and $m \angle R = 45°$.

   Quantity I BC
   Quantity II ST

9. Information: $m \angle A > m \angle B$
   $m \angle B > m \angle C$

   Quantity I $2 \, m \angle A$
   Quantity II $m \angle B + m \angle C$

10. Information:

    Quantity I $d° + c°$
    Quantity II $a° + b° + c°$

11. Information: A triangle has angles with measures of $90°$, $y°$, and $z°$.
    Quantity I $45°$
    Quantity II $y°$

12. Information:

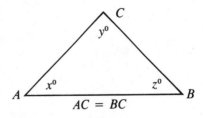

    Quantity I $x = 60°$
    Quantity II y

19 TRIGONOMETRIC LAWS AND FORMULAS

Non-Routine Problem Solving

A measure, such as a measure of life expectancy of a product, is normally distributed. If many such measures are plotted, they form a **normal (bell-shaped) curve.** The normal curve reflects the mean and the variation for a given measure. For a measure that is normally distributed, approximately 68.2% of the cases will fall between ±1 standard deviations (±1 σ) of the mean, 95.4% will fall between ±2 σ of the mean, and 99.8% will fall between ±3 σ of the mean.

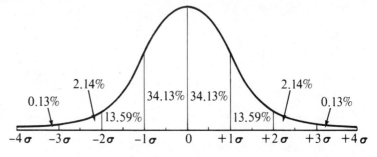

Percent of cases under portions of the normal curve

2.14%

0.13%

34.13% 34.13%

2.14%

0.13%

13.59%

13.59%

Standard deviations

−4σ −3σ −2σ −1σ 0 +1σ +2σ +3σ +4σ

Problem

The life expectancy of a certain automobile tire closely approximates the normal curve. The estimated average life is 35,000 miles, with one standard deviation of 4,000 miles. If you purchase a tire, what is the probability that it will last between 31,000 and 39,000 miles? more than 31,000 miles?

LAW OF COSINES

Objective   **To find the length of a side of a triangle using the Law of Cosines**

If two sides and the included angle of any triangle are known, the length of the third side can be found. You will now learn how to derive formulas to do this and how to apply these formulas.

For the general triangle ABC, you can show that $a^2 = b^2 + c^2 - 2bc \cos A$.
In right $\triangle ABD$, $c^2 = h^2 + x^2$, or $h^2 = c^2 - x^2$.
In right $\triangle BDC$, $a^2 = h^2 + y^2$.

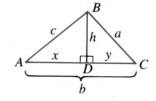

$$a^2 = h^2 + y^2$$
Since $h^2 = c^2 - x^2$ ▶ $= (c^2 - x^2) + y^2$
$= (y^2 - x^2) + c^2$
By factoring ▶ $= (y + x)(y - x) + c^2$
Since $b = x + y$, $y = b - x$ ▶ $= (b - x + x)(b - x - x) + c^2$
$= (b)(b - 2x) + c^2$
Since $\cos A = \frac{x}{c}$, $x = c \cos A$ ▶ $= b^2 - 2b(c \cos A) + c^2$
$= b^2 - 2bc \cos A + c^2$
So, $a^2 = b^2 + c^2 - 2bc \cos A$.

Law of cosines

For any triangle ABC,
$a^2 = b^2 + c^2 - 2bc \cos A$
$b^2 = a^2 + c^2 - 2ac \cos B$
$c^2 = a^2 + b^2 - 2ab \cos C$.

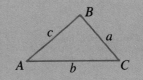

In general, the Law of Cosines states that the square of one side of a triangle equals the sum of the squares of the other two sides minus twice the product of these two sides and the cosine of the included angle.

Example 1   **In $\triangle ABC$, $b = 8$, $c = 7$, and $m \angle A = 45°$. Find a to the nearest unit.**

Use $a^2 = b^2 + c^2 - 2bc \cos A$.
$= 8^2 + 7^2 - 2(8)(7) \cos 45°$

$= 64 + 49 - 112\left(\frac{\sqrt{2}}{2}\right)$

$\doteq 64 + 49 - 56(1.414)$

$\doteq 33.8$

Thus, $a \doteq \sqrt{33.8}$, or 6.

Law of Cosines

Example 2   In $\triangle ABC$, $a = 4$, $b = 6$, and $m \angle c = 150°$. Find c to the nearest unit.

Use $c^2 = a^2 + b^2 - 2ab \cos C$.
$$\begin{aligned}
&= 4^2 + 6^2 - 2(4)(6) \cos 150° \\
&= 16 + 36 - 48(-\cos 30°) \\
&= 52 - 48\left(-\frac{\sqrt{3}}{2}\right) \\
&= 52 + 24\sqrt{3} \\
&= 52 + 24(1.732) \\
c^2 &= 52 + 41.568 \\
c &= \sqrt{93.568} \doteq 9.673
\end{aligned}$$

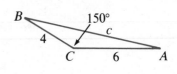

Thus, $c = 10$ to the nearest unit.

For a right triangle, the Law of Cosines becomes the Pythagorean theorem. In right triangle ABC, $m \angle c = 90°$. Use the Law of Cosines to find c.

$$\begin{aligned}
c^2 &= a^2 + b^2 - 2ab \cos C \\
&= a^2 + b^2 - 2ab \cos 90° \\
&= a^2 + b^2 - 2ab(0)
\end{aligned}$$
So, $c^2 = a^2 + b^2$.

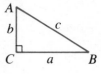

The next two examples illustrate how the Law of Cosines can be used to solve problems.

Example 3   A surveyor at point R sights two points S and T on opposite sides of a lake. Point R is 120 meters from S and 180 meters from T, and the measure of angle R is 38°. Find the distance across the lake to the nearest meter.

Use $r^2 = s^2 + t^2 - 2st \cos R$.
$$\begin{aligned}
&= (180)^2 + (120)^2 - 2(180)(120) \cos 38° \\
&= 32,400 + 14,400 - 43,200(.7880) \\
&= 46,800 - 34,041.6 \\
r^2 &= 12,758.4 \\
r &= \sqrt{12,758.4} \doteq 113
\end{aligned}$$

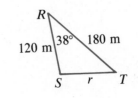

Thus, the distance across the lake $\doteq$ 113 meters.

Example 4   Two forces of 30 lb and 40 lb act on a body at an angle of 60°. Find the magnitude of the resultant to the nearest pound.

Use $a^2 = t^2 + s^2 - 2ts \cos A$.

$$\begin{aligned}
m \angle A + m \angle B &= 180° \\
m \angle A + 60° &= 180° \\
m \angle A &= 120°
\end{aligned}$$

$$\begin{aligned}
&= (40)^2 + (30)^2 - 2(40)(30) \cos 120° \\
&= 1,600 + 900 - 2,400\left(-\tfrac{1}{2}\right) \\
a^2 &= 3,700 \\
a &\doteq 60.828
\end{aligned}$$

Thus, $a = 61$, to the nearest pound.

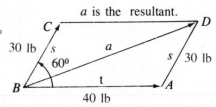

a is the resultant.

Oral Exercises

Give the form of the Law of Cosines needed to find x.

1.

A, x, 14, $140°$, B, 16, C

2.

B, x, 24, $38°$, C, 26, A

3.

C, 43, $96°$, 31, A, x, B

Written Exercises

(A) **Each exercise refers to triangle ABC. Find the length of the indicated side to the nearest unit.**

1. $a = 8$, $b = 9$, m$\angle C = 60°$. Find c.

2. $a = 10$, $c = 18$, m$\angle B = 120°$. Find b.

3. $b = 12$, $c = 15$, m$\angle A = 135°$. Find a.

4. $a = 15$, $b = 14$, m$\angle C = 150°$. Find c.

5. $a = 13$, $c = 18$, m$\angle B = 30°$. Find b.

6. $b = 17$, $c = 12$, m$\angle A = 45°$. Find a.

7. $b = 11.3$, $c = 13.1$, m$\angle A = 120°$. Find a.

8. $a = 17.3$, $c = 19.7$, m$\angle B = 135°$. Find b.

9. $a = 33.4$, $b = 22.7$, m$\angle C = 150°$. Find c.

10. $b = 51.7$, $c = 73.2$, m$\angle A = 60°$. Find a.

(B) **11.** $b = 47.3$, $c = 72.9$, m$\angle A = 18°$. Find a.

12. $a = 63.4$, $c = 27.6$, m$\angle B = 39°$. Find b.

13. $a = 12.9$, $b = 16.3$, m$\angle C = 12° \, 30'$. Find c.

14. $b = 19.6$, $c = 35.3$, m$\angle A = 45° \, 40'$. Find a.

15. $b = 19.1$, $c = 34.6$, m$\angle A = 73° \, 18'$. Find a.

16. $a = 41.3$, $b = 16.9$, m$\angle C = 81° \, 26'$. Find c.

17. $a = 49.7$, $c = 43.1$, m$\angle B = 120° \, 33'$. Find b.

18. $b = 123$, $c = 119$, m$\angle A = 105° \, 48'$. Find a.

19. Two sides of a parallelogram form an angle of 72°. The lengths of two of the sides are 18 and 24 centimeters. How long is the shorter diagonal?

20. A side and a base of an isosceles trapezoid form an angle of 54°. The length of the side is 14 centimeters and of the base is 34 centimeters. How long is a diagonal?

21. Two ships left from the same port on paths that form an angle of 68°. Ship A traveled 400 km; ship B traveled 325 km. Find to the nearest kilometer the distance between them.

22. Two airplanes left the same airport and formed a 65° angle in their flight paths. The first plane flew at 600 km/h, and the second flew at 800 km/h. How far apart were they after 2 hours?

23. Two forces of 450 lb and 600 lb act on a body at an angle of 40° 50'. Find the magnitude of the resultant to the nearest pound.

24. Two forces of 300 lb and 400 lb act on a body at an angle of 30° 40'. Find the magnitude of the resultant to the nearest pound.

(C) **25.** In $\triangle ABC$, m$\angle B < 90°$. Prove that $b^2 = a^2 + c^2 - 2ac \cos B$.

26. In $\triangle ABC$, m$\angle C < 90°$. Prove that $c^2 = a^2 + b^2 - 2ab \cos C$.

27. In $\triangle ABC$, m$\angle A > 90°$. Prove that $a^2 = b^2 + c^2 - 2bc \cos A$.

28. In $\triangle ABC$, m$\angle B > 90°$. Prove that $b^2 = a^2 + c^2 - 2ac \cos B$.

CUMULATIVE REVIEW

Use the Pythagorean relation to determine whether the following are measures of the sides of a right triangle.

1. 8, 15, 17

2. 5, 12, 13

3. $1, 2\sqrt{7}, 3\sqrt{3}$

USING THE LAW OF COSINES TO FIND ANGLE MEASURES

Objective   **To use the Law of Cosines to find the degree measure of an angle**

You can apply the Law of Cosines to find the measure of an angle of any triangle if the lengths of the three sides are known.

Using the Law of Cosines involving cos A, solve for cos A as follows.

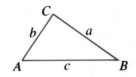

$$a^2 = b^2 + c^2 - 2bc \cos A$$
$$2bc \cos A = b^2 + c^2 - a^2$$

So, $\cos A = \dfrac{b^2 + c^2 - a^2}{2bc}$.

Finding an angle by using the law of cosines	For any triangle ABC, $\cos A = \dfrac{b^2 + c^2 - a^2}{2bc}$ $\cos B = \dfrac{a^2 + c^2 - b^2}{2ac}$ $\cos C = \dfrac{a^2 + b^2 - c^2}{2ab}$.

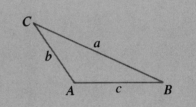

Example 1   In $\triangle ABC$, $a = 6$, $b = 7$ and $c = 4$. Find m $\angle A$ and m $\angle B$ to the nearest degree.

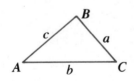

$$\cos A = \frac{b^2 + c^2 - a^2}{2bc}$$
$$= \frac{7^2 + 4^2 - 6^2}{2(7)(4)}$$
$$= \frac{49 + 16 - 36}{56}$$
$$= \frac{29}{56}$$
$$\cos A \doteq .5179$$

$$\cos B = \frac{a^2 + c^2 - b^2}{2ac}$$
$$= \frac{6^2 + 4^2 - 7^2}{2(6)(4)}$$
$$= \frac{36 + 16 - 49}{48}$$
$$= \frac{3}{48}$$
$$\cos B \doteq .0625$$

Thus, $m \angle A = 59°$ and $m \angle B = 86°$, to the nearest degree.

Example 2   In △*ABC*, the length of the sides are 3, 4, and 6. Find the measure of the smallest angle to the nearest degree.

The angle opposite the shortest side has the least measure.

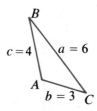

$$\cos B = \frac{a^2 + c^2 - b^2}{2ac}$$

$$= \frac{36 + 16 - 9}{2(6)(4)}$$

$$\cos B = \frac{43}{48}, \text{ or } .8958$$

$$\text{m} \angle B \doteq 26°$$

Thus, the measure of the smallest angle to the nearest degree is 26°.

Written Exercises

Find the measure of each angle to the nearest degree.

(A) 1. $a = 3$, $b = 8$, $c = 7$. Find m$\angle A$.
  2. $a = 9$, $b = 3$, $c = 9$. Find m$\angle B$.
  3. $a = 13$, $b = 12$, $c = 5$. Find m$\angle B$.
  4. $a = 2$, $b = 8$, $c = 7$. Find m$\angle C$.
  5. $a = 7$, $b = 9$, $c = 3$. Find m$\angle A$.
  6. $a = 6$, $b = 3$, $c = 4$. Find m$\angle C$.
  7. $a = 10$, $b = 2$, $c = 9$. Find m$\angle C$.
  8. $a = 10$, $b = 6$, $c = 9$. Find m$\angle A$.

(B) 9. $a = 12.3$, $b = 9.6$, $c = 7.3$. Find m$\angle A$.
  10. $a = 14.3$, $b = 19.5$, $c = 26.1$. Find m$\angle C$.
  11. $a = 17.5$, $b = 16.4$, $c = 11.7$. Find m$\angle C$.
  12. $a = 9.7$, $b = 23.4$, $c = 17.9$. Find m$\angle A$.
  13. $a = 9.6$, $b = 17.5$, $c = 12.8$. Find m$\angle B$.
  14. $a = 36.1$, $b = 44.7$, $c = 28.6$. Find m$\angle B$.

15. The measures of two sides of a parallelogram are 40 cm and 50 cm, and the length of one diagonal is 70 cm. Find the measures of the angles of the parallelogram.

16. Two straight roads $\overrightarrow{MN}$ and $\overrightarrow{PN}$ intersect at a town *N*. *M* is 62 km from *N* and *P* is 77 km from *N*. *MP* is 85 km. Find the angle that roads $\overrightarrow{MN}$ and $\overrightarrow{PN}$ make with each other.

In △*ABC*, find the measure of the smallest angle to the nearest degree.

17. $a = 6$, $b = 8$, $c = 9$
18. $a = 12$, $b = 13$, $c = 15$
19. $a = 9$, $b = 3$, $c = 8$

(C) 20. Prove: For any △*ABC*,

$$\cos B = \frac{a^2 + c^2 - b^2}{2ac}$$

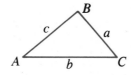

21. Prove: For any △*ABC*,

$$\cos C = \frac{a^2 + b^2 - c^2}{2ab}$$

Using the Law of Cosines to Find Angle Measures   **543**

AREA OF A TRIANGLE

Objective   **To apply formulas for the area of a triangle given the measures of two sides and the included angle**

If two sides and the included angle of any triangle are known, a formula for finding the area of the triangle can be derived as follows.

Recall that the area of a triangle is one half the product of the base and the height.

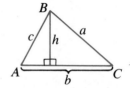

Area of $\triangle ABC = k = \frac{1}{2}bh$.

But $\sin A = \frac{h}{c}$, or $h = c \sin A$.

Now, substitute $c \sin A$ for h in the original formula.
$k = \frac{1}{2}b(c \sin A)$ or $\frac{1}{2}bc \sin A$

Area of any triangle

The area of a triangle is one half the product of the lengths of any two sides and the sine of the included angle measure.

For any triangle ABC,
$k = \frac{1}{2}bc \sin A$
$k = \frac{1}{2}ab \sin C$
$k = \frac{1}{2}ac \sin B$

Example 1   In $\triangle ABC$, $b = 10$, $c = 8$ and $m\angle A = 60°$. Find the area in simplest radical form.

$k = \frac{1}{2}bc \sin A$.
$= \frac{1}{2}(10)(8) \sin 60°$
$= 40 \cdot \dfrac{\sqrt{3}}{2}$, or $20\sqrt{3}$

Thus, the area of $\triangle ABC$ is $20\sqrt{3}$ square units.

Example 2   In $\triangle ABC$, $a = 8$, $b = 12$ and $m\angle C = 140°$. Find the area to the nearest square unit.

$k = \dfrac{1}{2} ab \sin C$.

$= \dfrac{1}{2}(8)(12) \sin 140°$

$= 48 \sin 40°$
$\doteq 48(.6428)$
$\doteq 30.8544$

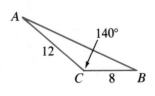

Thus, the area of $\triangle ABC$ is 31 square units to the nearest square unit.

Example 3   **In △PQR, q = 15, r = 60. Find the measure of ∠P to the nearest degree if the area is 240 square units.**

Use an area formula for q, r, and $\angle P$.

$$k = \frac{1}{2} qr \sin P$$

$$240 = \frac{1}{2} (15)(60) \sin P$$

$$\sin P = \frac{240}{450} \doteq .5333$$

Thus, the measure of ∠P is 32° to the nearest degree.

Oral Exercises

Give the form of the area formula needed to find the area of each triangle.
1. $a = 6$, $c = 8$, $m\angle B = 36°$ 2. $b = 9$, $c = 7$, $m\angle A = 73°$ 3. $a = 4$, $b = 8$, $m\angle C = 41°$

Written Exercises

(A) **In △ABC, find the area to the nearest square unit.**
1. $a = 12$, $b = 9$, $m\angle C = 43°$ 2. $b = 9$, $c = 7$, $m\angle A = 80°$ 3. $a = 6$, $c = 8$, $m\angle B = 70°$

4. $a = 7$, $b = 11$, $m\angle C = 34°$ 5. $a = 13$, $c = 12$, $m\angle B = 48°$ 6. $b = 12$, $c = 9$, $m\angle A = 38°$

In △PQR, find the measure of ∠P to the nearest degree.
7. $q = 8$, $r = 12$, $k = 38$ sq. units   8. $q = 14$, $r = 8$, $k = 54$ sq. units

9. $q = 16$, $r = 10$, $k = 74$ sq. units   10. $q = 6$, $r = 9$, $k = 22$ sq. units

(B) **In △ABC, find the area to the nearest square unit.**
11. $a = 6.2$, $b = 4.4$, $m\angle C = 150°$ 12. $a = 9.7$, $c = 4.8$, $m\angle B = 135° \, 20'$

13. $a = 9.1$, $c = 7.6$, $m\angle B = 78°$ 14. $b = 12.3$, $c = 13.7$, $m\angle A = 105° \, 30'$

15. The lengths of two sides of a parallelogram are 24 and 36 centimeters. Their included angle measures 50°. Find the area of the parallelogram.

16. The lengths of two sides of a parallelogram are 30 and 45 centimeters. Their included angle measures 130°. Find the area of the parallelogram.

(C) 17. Prove: For any triangle ABC
$k = \frac{1}{2} ab \sin C$ and $k = \frac{1}{2} ac \sin B$.

18. Prove: The area of a parallelogram is equal to the product of the lengths of two adjacent sides and the sine of their included angle.

LAW OF SINES

Objective

To apply the Law of Sines to find unknown measures in a triangle if two angles and a side of any triangle are known

If two angles and a side of any triangle are known, the measure of the third angle and the lengths of the other two sides can be found.

Recall from the previous lesson three ways to write the area of a triangle, *ABC*.

$$\frac{1}{2}\ bc\ \sin A = \frac{1}{2}\ ab\ \sin C = \frac{1}{2}\ ac\ \sin B$$

Multiplying by 2. ▶ $bc\ \sin A = ab\ \sin C = ac\ \sin B$

Dividing by abc. ▶ $\dfrac{\cancel{bc}\ \sin A}{a\cancel{bc}} = \dfrac{\cancel{ab}\ \sin C}{\cancel{ab}c} = \dfrac{\cancel{ac}\ \sin B}{\cancel{a}b\cancel{c}}$

So, $\dfrac{\sin A}{a} = \dfrac{\sin C}{c} = \dfrac{\sin B}{b}$.

Law of sines	For any triangle *ABC*, $\dfrac{\sin A}{a} = \dfrac{\sin B}{b} = \dfrac{\sin C}{c}$, or $\dfrac{a}{\sin A} = \dfrac{b}{\sin B} = \dfrac{c}{\sin C}$.

In general, the Law of Sines states that the sides of a triangle are proportional to the sines of the angles opposite them.

Example 1

In △*ABC*, *a* = 16, and m∠*A* = 40°, m∠*B* = 58°. Find *b* to the nearest unit.

Use $\dfrac{a}{\sin A} = \dfrac{b}{\sin B}$ since *a*, m∠*A*, and m∠*B* are given.

$\dfrac{16}{\sin 40°} = \dfrac{b}{\sin 58°}$

$\dfrac{16 \sin 58°}{\sin 40°} = b$

$\dfrac{16(.8480)}{.6428} = b$

$21.1076 \doteq b$

Thus, *b* = 21 to the nearest unit.

Example 2

In $\triangle ABC$, $a = 80$, $b = 60$, and $m\angle A = 52°$. Find the measure of $\angle B$ to the nearest degree.

Use $\dfrac{\sin A}{a} = \dfrac{\sin B}{b}$ since a, b, and $m\angle A$ are given.

$\dfrac{\sin 52°}{80} = \dfrac{\sin B}{60}$

$\sin B = \dfrac{60(\sin 52°)}{80}$

$\quad\quad = \dfrac{60(.7880)}{80}$

$\sin B \doteq .5910$

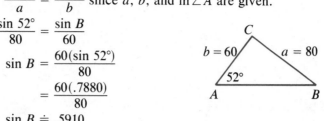

Thus, the measure of $\angle B$ is $36°$ to the nearest degree.

Example 3

In $\triangle ABC$, $b = 18$, $m\angle A = 64°$, $m\angle C = 55°$. Find c to the nearest unit.

First, find $m\angle B$.
$m\angle B + 55° + 64° = 180°$ so $m\angle B = 61°$

Use $\dfrac{b}{\sin B} = \dfrac{c}{\sin C}$.

$\dfrac{18}{\sin 61°} = \dfrac{c}{\sin 55°}$, or $c = \dfrac{18 \sin 55°}{\sin 61°}$

$c \doteq \dfrac{18(.8192)}{.8746}$, or 16.86

Thus, $c = 17$ to the nearest unit.

You can use logarithms to solve problems involving the Law of Sines.

Example 4

In $\triangle ABC$, $m\angle A = 42° 10'$, $m\angle B = 73° 18'$, $b = 36.14$. Find a to the nearest tenth.

Use $\dfrac{a}{\sin A} = \dfrac{b}{\sin B}$.

$\dfrac{a}{\sin 42° 10'} = \dfrac{36.14}{\sin 73° 18'}$, or $a = \dfrac{36.14 \sin 42° 10'}{\sin 73° 18'}$

*Take the log
of each side.* ▶ $\log a = \log 36.14 + \log \sin 42° 10' - \log \sin 73° 18'$

Interpolate. ▶ $\quad\quad = 1.5580 \quad + \quad 9.8269 - 10 \quad - (9.9813 - 10)$

$\log a = 1.4036$

Thus, $a = 25.3$.

Law of Sines

Oral Exercises

Give the form of the Law of Sines that is needed to find x.

1.

2.

3.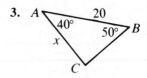

Written Exercises

(A)

1. For $\triangle ABC$, $a = 28$, $b = 59$, $m\angle B = 48°$. Find $m\angle A$ to the nearest degree.

2. For $\triangle ABC$, $a = 36$, $c = 18$, $m\angle A = 57°$. Find $m\angle C$ to the nearest degree.

3. For $\triangle PQR$, $p = 17$, $m\angle P = 75°$, $m\angle Q = 38°$. Find q to the nearest unit.

4. For $\triangle PQR$, $r = 29$, $m\angle P = 37°$, $m\angle R = 71°$. Find p to the nearest unit.

5. For $\triangle ABC$, $a = 14$, $c = 10$, $m\angle C = 30°$. Find $m\angle A$ to the nearest degree.

6. For $\triangle PQR$, $q = 5$, $r = 7$, $m\angle Q = 40°$. Find $m\angle R$ to the nearest degree.

7. For $\triangle PQR$, $p = 22$, $\sin P = \frac{1}{5}$, $\sin Q = \frac{4}{5}$. Find q to the nearest unit.

8. For $\triangle ABC$, $\sin A = .40$, $\sin B = .65$, $a = 32$. Find b to the nearest unit.

(B)

9. For $\triangle ABC$, $a = 16.3$, $m\angle B = 58°$, $m\angle C = 37°$. Find b, c, and $m\angle A$.

10. For $\triangle ABC$, $a = 12.3$, $b = 13.7$, $m\angle A = 18°\ 25'$. Find c, $m\angle B$, and $m\angle C$.

11. For $\triangle ABC$, $b = 14.8$, $c = 18.9$, $m\angle C = 42°\ 18'$. Find a, $m\angle A$, and $m\angle B$.

12. For $\triangle ABC$, $a = 15.1$, $c = 19.3$, $m\angle C = 37°\ 32'$. Find b, $m\angle A$, and $m\angle B$.

13. The distance between town B and town A is 45 kilometers. The angle formed by the roads between towns A and B and between towns A and C measures 37°, and the angle formed by $\overline{AB}$ and $\overline{BC}$ measures 110°. Find the distance from town B to town C to the nearest kilometer.

14. Two cars P and Q are parked on the same side of the street, 50 meters apart. Car R is not on the same road. The angle formed by $\overrightarrow{PQ}$ and $\overrightarrow{PR}$ is $112°\ 20'$, and the angle formed by $\overrightarrow{QR}$ and $\overrightarrow{PR}$ is $28°\ 40'$. Find the distance between P and R to the nearest meter.

(C)

15. For $\triangle RST$, show that $\dfrac{\sin R}{r} = \dfrac{\sin T}{t} = \dfrac{\sin S}{s}$.

16. For $\triangle QRS$, show that $\dfrac{q}{\sin Q} = \dfrac{r}{\sin R} = \dfrac{s}{\sin S}$.

CALCULATOR ACTIVITIES

The calculator can be an aid in solving problems involving the Law of Sines.

From Example 3, $\dfrac{18}{\sin 61°} = \dfrac{c}{\sin 55°}$ or $c = \dfrac{18 \sin 55°}{\sin 61°} = \dfrac{18(.8192)}{.8746}$.

Compute ▶ $18 \otimes .8192 \oplus .8746 \ominus 16.86$.

THE AMBIGUOUS CASE   19.5

Objective   **To determine the number of triangles that can be constructed given certain data**

In this lesson, you will compare parts of a triangle to determine the number of triangles it is possible to construct.

Example 1   **Given that $a = 18$, $b = 36$, and $m \angle A = 38°$, use the Law of Sines to find the measure of $\angle B$. How many triangles can be constructed using the given data?**

Use $\dfrac{\sin A}{a} = \dfrac{\sin B}{b}$.

$\dfrac{\sin 38°}{18} = \dfrac{\sin B}{36}$, or $\sin B = \dfrac{36 \sin 38°}{18}$

$\sin B \doteq \dfrac{36(.6157)}{18}$, or 1.2314

Thus, there is no $\angle B$ for which $\sin B = 1.2314$ since $0 \leq \sin B \leq 1$ and no triangle can be constructed with the given data.

In general, $\dfrac{\sin A}{a} = \dfrac{\sin B}{b}$

$\sin B = \dfrac{b \sin A}{a}$,

but $\sin B = \dfrac{h}{a}$, so $h = b \sin A$.

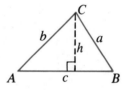

Compare h and a. When $a < h$, $(a < b \sin A)$, as in Example 1, no triangle can be constructed since $\sin B = \frac{h}{a}$ will be greater than 1. When $a = h$, $\sin B = 1$, one right $\triangle$ can be constructed.
When $a > h$, $\sin B < 1$, two $\triangle$'s can be constructed.
The following Summary is suggested.

Summary A is an acute angle and $a < b$:

1. If $a < h$, $(a < b \sin A)$,
   no triangle can be constructed.

2. If $a = h$, $(a = b \sin A)$,
   one triangle can be constructed.

3. If $a > h$, $(a > b \sin A)$,
   two triangles can be constructed.

Example 2   **Given that m∠A = 34°, a = 21, b = 32, and the height, h, is 25, can a triangle be constructed?**

A is an acute angle, $a < b$ and $a < h$.

Thus, no triangle can be constructed.

In △ABC, ∠A is an acute angle. You will now determine how many triangles can be constructed if $a = b$ and if $a > b$.

∠A is an acute angle.

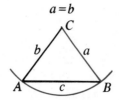

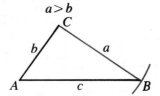

Summary   ∠A is an acute angle:
1. One triangle can be constructed if $a = b$.
2. One triangle can be constructed if $a > b$.

Example 3   **Given that m∠A = 130°, a = 12, and b = 13, find the measure of ∠C to the nearest degree. How many triangles can be constructed using the given data?**

Use $\dfrac{\sin A}{a} = \dfrac{\sin B}{b}$.

$\dfrac{\sin 130°}{12} = \dfrac{\sin B}{13}$, or $\sin B = \dfrac{13 \sin 130°}{12}$

$\sin B \doteq \dfrac{13(.7660)}{12}$, or $.8298$

So, the measure of ∠B is 56° to the nearest degree.
Find m∠C. m∠A + m∠B + m∠C = 180°
$\qquad\qquad\quad 130° + 56° + m\angle C = 180°$
$\qquad\qquad\qquad\quad 186° + m\angle C = 180°$
$\qquad\qquad\qquad\qquad\quad m\angle C = 180° - 186°$, or $-6°$

Thus, no triangle can be constructed when angle A is obtuse and $a < b$.

Example 4

Given that m∠A = 130°, $a = 12$, $b = 12$, find the measure of ∠C to the nearest degree. How many triangles can be constructed?

$$\sin B \doteq \frac{12(.7660)}{12}, \text{ or } .7660 \blacktriangleleft \sin B = \frac{b \sin A}{a}$$

So, the measure of ∠B is 50°.

$$130° + 50° + m∠C = 180°$$
$$m∠C = 0°$$

Thus, no triangle can be constructed when angle A is obtuse and $a = b$.

Example 5

Given that m∠A = 130°, $a = 12$, $b = 10$, find the measure of ∠C to the nearest degree. How many triangles can be constructed?

$$\sin B \doteq \frac{10(.7660)}{12}, \text{ or } .6383$$

So, the measure of ∠B is 40° to the nearest degree.

$$130° + 40° + m∠C = 180°$$
$$m∠C = 10°$$

Thus, one triangle can be constructed when angle A is obtuse and $a > b$.

The last three examples suggest the following.

Summary

∠A is an obtuse angle or a right angle:
1. No triangle can be constructed if $a \le b$.
2. One triangle can be constructed if $a > b$.

Example 6

How many triangles can be constructed if m∠A = 160°, $a = 20$, and $b = 60$?

∠A is an obtuse angle and $a < b$.

Thus, no triangle can be constructed when angle A is obtuse and $a < b$.

Written Exercises

How many triangles can be constructed?
1. m∠A = 36°, $a = 12$, $b = 18$
2. m∠A = 48°, $a = 6$, $b = 7$
3. m∠A = 40°, $a = 50$, $b = 20$
4. m∠A = 120°, $a = 19$, $b = 21$
5. m∠A = 90°, $a = 96$, $b = 96$
6. m∠A = 175°, $a = 71$, $b = 58$
7. m∠A = 15°, $a = 100$, $b = 100$
8. m∠A = 52°, $a = 42$, $b = 38$
9. m∠A = 133°, $a = 16$, $b = 4$
10. m∠A = 90°, $a = 14$, $b = 14$

SIN ($A \pm B$) AND SIN 2A

Objective   **To apply the formulas for sin ($A \pm B$) and sin 2A**

You can derive the formula for the sine of the sum of two angles in the following way.

Write the formulas for the areas of each of the three triangles shown at the right.

Area $\triangle ABC$ = area $\triangle ABD$ + area $\triangle BDC$

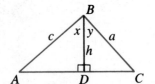

$$\frac{1}{2} ac \sin B = \frac{1}{2} ch \sin x + \frac{1}{2} ah \sin y$$

Multiply by 2. ▶ $ac \sin B = ch \sin x + ah \sin y$

Divide by ac. ▶ $\dfrac{ac}{ac} \sin B = \dfrac{ch}{ac} \sin x + \dfrac{ah}{ac} \sin y$

Substitute. ▶ $\sin B = \dfrac{h}{a} \sin x + \dfrac{h}{c} \sin y$

So, $\sin (x + y) = \cos y \sin x + \cos x \sin y$.

Sine of the sum of two angles	$\sin (A + B) = \sin A \cos B + \cos A \sin B$

Example 1   **Find sin 105° using special angle measures and the formula for sin ($A + B$). Answer may be left in simplest radical form.**

$$\sin (A + B) = \sin A \cos B \quad + \cos A \sin B$$

105° = 60° + 45° ▶ $\sin (60° + 45°) = \sin 60° \cos 45° + \cos 60° \sin 45°$

$$= \frac{\sqrt{3}}{2} \cdot \frac{\sqrt{2}}{2} \quad + \frac{1}{2} \cdot \frac{\sqrt{2}}{2}$$

$$= \frac{\sqrt{6}}{4} + \frac{\sqrt{2}}{4}$$

Thus, $\sin 105° = \dfrac{\sqrt{6} + \sqrt{2}}{4}$.

You can derive the formula for the sine of the difference of two angle measures using the formula for the sine of the sum of two angle measures. This derivation follows on the next page.

Rewrite the difference $x - y$ as the sum $x + (-y)$.

So, $\sin (x - y) = \sin [x + (-y)]$

$\qquad = \sin x \cos (-y) + \cos x \sin (-y)$

$\qquad = \sin x \cos y + \cos x(-\sin y)$, since $\begin{cases} \cos (-y) = \cos y \\ \sin (-y) = -\sin y \end{cases}$.

$\qquad = \sin x \cos y - \cos x \sin y$.

Sine of the difference of two angles	$\sin (A - B) = \sin A \cos B - \cos A \sin B$

Example 2 **Find sin 15°. Use special angle measures and the formula for sin ($A - B$). Answer in simplest radical form.**

$$\sin (A - B) = \sin A \cos B - \cos A \sin B$$
$$\sin (60° - 45°) = \sin 60° \cos 45° - \cos 60° \sin 45°$$
$$= \frac{\sqrt{3}}{2} \cdot \frac{\sqrt{2}}{2} - \frac{1}{2} \cdot \frac{\sqrt{2}}{2}, \text{ or } \frac{\sqrt{6}}{4} - \frac{\sqrt{2}}{4}$$

Thus, $\sin 15° = \dfrac{\sqrt{6} - \sqrt{2}}{4}$.

Example 3 **If sin $A = \frac{3}{5}$ and cos $B = \frac{5}{13}$, where A and B are acute angles, find the value of sin ($A + B$).**

Thus, $\sin (A + B) = \dfrac{63}{65}$.

$$\sin (A + B) = \sin A \cos B + \cos A \sin B$$
$$= \frac{3}{5} \cdot \frac{5}{13} + \frac{4}{5} \cdot \frac{12}{13}$$
$$= \frac{15}{65} + \frac{48}{65}$$
$$= \frac{63}{65}$$

Example 4 **If sin $A = \dfrac{15}{17}$ and $\angle A$ lies in the second quadrant, and if cos $B = \dfrac{4}{5}$ and $\angle B$ lies in the first quadrant, find sin ($A - B$).**

Using reference triangles, $\cos A = -\dfrac{8}{17}$ and $\sin B = \dfrac{3}{5}$.

$$\sin (A - B) = \sin A \cos B - \cos A \sin B$$
$$= \frac{15}{17} \left(\frac{4}{5} \right) - \left(\frac{-8}{17} \right) \cdot \frac{3}{5}$$
$$= \frac{60}{85} + \frac{24}{85}, \text{ or } \frac{84}{85}$$

Thus, $\sin (A - B) = \dfrac{84}{85}$.

The sine of twice (double) an angle can also be derived using the formula for the sine of the sum of two angles.

Begin by rewriting $2x$ as the sum $(x + x)$.
$$\sin (2x) = \sin (x + x)$$
$$= \sin x \cos x + \cos x \sin x$$
$$= \sin x \cos x + \sin x \cos x$$
$$= 2 \sin x \cos x.$$

Sine of twice an angle	$\sin 2A = 2 \sin A \cos A$

Example 5   **Find sin 120°. Use a special angle measure and the formula for sin 2A.**

$$\sin 2A = 2 \sin A \cos A$$
$$\sin 2(60°) = 2 \sin 60° \cos 60°$$
$$= 2\left(\frac{\sqrt{3}}{2}\right)\left(\frac{1}{2}\right), \text{ or } \frac{\sqrt{3}}{2}$$

Thus, $\sin 120° = \dfrac{\sqrt{3}}{2}$.

Example 6   **If** $\cos A = -\dfrac{5}{13}$ **and** $\angle A$ **lies in the third quadrant, find sin 2A.**

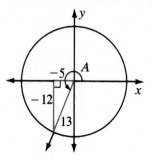

Use the reference triangle, $\sin A = -\dfrac{12}{13}$.

So, $\sin 2A = 2 \sin A \cos A$.
$$= 2\left(-\frac{12}{13}\right)\left(-\frac{5}{13}\right)$$
$$= \frac{120}{169}$$

Thus, $\sin 2A = \dfrac{120}{169}$.

Oral Exercises

Write as a function of one angle.
1. $\sin 40° \cos 30° + \cos 40° \sin 30°$
2. $\sin 50° \cos 10° - \cos 50° \sin 10°$
3. $2 \sin 15° \cos 15°$
4. $\sin 125° \cos 12° + \cos 125° \sin 12°$
5. $\cos 57° \sin 33° - \cos 33° \sin 57°$
6. $2 \cos 25° \sin 25°$
7. $\cos 70° \sin 10° + \cos 10° \sin 70°$
8. $\sin 73° \cos 62° - \cos 73° \sin 62°$

Written Exercises

Find the value of each. Use special angles and the formula for sin $(A + B)$ or sin $(A - B)$. Answers may be left in simplest radical form.

(A)   **1.** sin 75°   **2.** sin 135°   **3.** sin 60°   **4.** sin 195°   **5.** sin 285°
      6. sin 165°   **7.** sin 150°   **8.** sin 255°   **9.** sin 435°   **10.** sin 375°

Find the value of each. Use special angles and the formula for sin 2A. Answers may be left in simplest radical form.

11. sin 60°   **12.** sin 90°   **13.** sin 300°   **14.** sin 270°   **15.** sin 450°

Find the value of sin $(A + B)$ and sin $(A - B)$.

16. sin $A = \dfrac{12}{13}$ and A lies in the 2nd quad.

$\cos B = \dfrac{3}{5}$ and B lies in the 4th quad.

17. $\cos A = \dfrac{12}{13}$ and $\cos B = \dfrac{4}{5}$, A and B are both acute angles.

18. sin $A = \dfrac{15}{17}$ and A lies in the 2nd quad.

$\cos B = \dfrac{-4}{5}$ and B lies in the 3rd quad.

19. sin $A = \dfrac{-8}{17}$ and A lies in the 4th quad.

$\cos B = \dfrac{12}{13}$ and B lies in the 4th quad.

20. sin $A = \dfrac{7}{9}$ and A lies in the 1st quad.

$\cos B = \dfrac{-4}{11}$ and B lies in the 2nd quad.

21. $\cos A = \dfrac{6}{13}$ and A lies in the 4th quad.

$\sin B = \dfrac{4}{7}$ and B lies in the 1st quad.

Find the value of sin 2A.

22. $\cos A = \dfrac{-4}{5}$ and A lies in the 2nd quad.

23. $\cos A = \dfrac{12}{13}$ and A is an acute angle.

24. $\cos A = \dfrac{15}{17}$ and A lies in the 4th quad.

25. sin $A = -\dfrac{8}{17}$ and A lies in the 3rd quad.

Simplify each of the following for acute angle A.

(B)   **26.** $\dfrac{\sin^2 A}{\sin 2A}$ **27.** $\dfrac{\sin 2A}{\cos A}$ **28.** $\dfrac{\tan A}{\sin 2A}$ **29.** cot A sin 2A

Find the value of each. Use special angles, the formula for sin $(A + B)$, sin $(A - B)$, or sin 2A, and reciprocal relationships. Answers may be left in simplest radical form.

30. csc 75°   **31.** csc 135°   **32.** csc 120°   **33.** csc 15°   **34.** csc 90°

Prove each identity. Use the formula for sin $(A \pm B)$.

(C)   **35.** sin $(\pi + \theta) = -\sin \theta$ **36.** sin $(2\pi - \theta) = -\sin \theta$ **37.** sin $(\pi - \theta) = \sin \theta$

38. sin $\left(\dfrac{\pi}{2} - \theta\right) = \cos \theta$ **39.** sin $\left(\dfrac{3\pi}{2} + \theta\right) = -\cos \theta$ **40.** sin $\left(\dfrac{\pi}{2} + \theta\right) = \cos \theta$

41. sin $(180° + \theta) = -\sin \theta$ **42.** sin $(180° - \theta) = \sin \theta$ **43.** sin $(90° + \theta) = \cos \theta$
44. sin $(90° - \theta) = \cos \theta$ **45.** sin $(360° - \theta) = -\sin \theta$ **46.** sin $(270° + \theta) = -\cos \theta$

Sin $(A \pm B)$ and Sin 2A

COS $(A \pm B)$ AND COS $2A$ 19.7

Objective   **To apply the formulas for cos $(A \pm B)$ and cos $2A$**

You can derive the formula for the cosine of the sum of two angles by using the formula for the sine of the difference of two angles and the cofunction relationship of sine and cosine.

In Chapter 17, you learned the cofunction relationship $\cos u = \sin (90 - u)$. If we let $u = x + y$, it follows that,

$$\cos (x + y) = \sin [90 - (x + y)]$$
$$= \sin [(90 - x) - y] \quad \blacktriangleleft \quad 90 - (x + y) = (90 - x) - y$$
$$= \underbrace{\sin (90 - x)} \cos y - \underbrace{\cos (90 - x)} \sin y \quad \blacktriangleleft \quad \textit{using formula for sin } (A - B),$$
$$\qquad\qquad\qquad\qquad\qquad\qquad\qquad\qquad\qquad\quad A = 90 - x, B = y$$
$$\text{So, } \cos (x + y) = \qquad \cos x \quad \cos y - \quad \sin x \quad \sin y. \quad \blacktriangleleft \quad \sin (90 - x) = \cos x$$
$$\qquad\qquad\qquad\qquad\qquad\qquad\qquad\qquad\qquad\qquad\qquad\quad \textit{and cos } (90 - x) = \sin x$$

Cosine of the sum of two angles	$$\cos (A + B) = \cos A \cos B - \sin A \sin B$$

Example 1   **Find cos 75°. Use special angles and the formula for cos $(A + B)$. Answer in simplest radical form.**

$$\cos (A + B) = \cos A \cos B - \sin A \sin B$$
$75° = 45° + 30°$ ▶ $\cos (45° + 30°) = \cos 45° \cos 30° - \sin 45° \sin 30°$
$$= \frac{\sqrt{2}}{2} \cdot \frac{\sqrt{3}}{2} - \frac{\sqrt{2}}{2} \cdot \frac{1}{2}$$
$$= \frac{\sqrt{6}}{4} - \frac{\sqrt{2}}{4}$$

Thus, $\cos 75° = \dfrac{\sqrt{6} - \sqrt{2}}{4}$.

Example 2   **If cos $A = -\frac{4}{5}$ and $\angle A$ lies in the second quadrant and cos $B = \frac{12}{13}$ and $\angle B$ lies in the fourth quadrant, find cos $(A + B)$.**

Using reference triangles, $\sin A = \frac{3}{5}$ and $\sin B = -\frac{5}{13}$.
$$\cos (A + B) = \cos A \cos B - \sin A \sin B$$
$$= -\frac{4}{5} \cdot \frac{12}{13} - \frac{3}{5} \cdot \left(-\frac{5}{13}\right)$$
$$= -\frac{48}{65} + \frac{15}{65}, \text{ or } -\frac{33}{65}$$

Thus, $\cos (A + B) = -\dfrac{33}{65}$.

You can derive the formula for the cos of the difference of two angles using the formula for the cosine of the sum of two angles.

Rewrite the difference $x - y$ as the sum $x + (-y)$.
$$\cos (x - y) = \cos [x + (-y)]$$
$$= \cos x \cos (-y) - \sin x \sin (-y)$$
$$= \cos x \cos y - \sin x (-\sin y)$$
$$= \cos x \cos y + \sin x \sin y.$$

◀ $\begin{cases} \cos (-y) = \cos y \\ \sin (-y) = -\sin y \end{cases}$

Cosine of the difference of two angles	$\cos (A - B) = \cos A \cos B + \sin A \sin B$

Example 3 If $\sin A = -\dfrac{15}{17}$ and $\angle A$ lies in the fourth quadrant and if $\tan B = -\dfrac{4}{3}$ and $\angle B$ lies in the second quadrant, find $\cos (A - B)$.

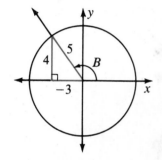
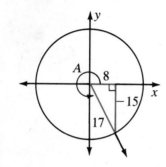

$$\cos (A - B) = \cos A \cos B + \sin A \sin B$$
$$= \left(\frac{8}{17}\right)\left(\frac{-3}{5}\right) + \left(\frac{-15}{17}\right)\left(\frac{4}{5}\right)$$
$$= \frac{-24}{85} + \frac{-60}{85}, \text{ or } \frac{-84}{85}$$

Thus, $\cos (A - B) = -\dfrac{84}{85}.$

You can derive the formula for the cosine of twice (double) an angle using the formula for the cosine of the sum of two angles.

Begin by rewriting $2x$ as the sum $(x + x)$.
So, $\cos 2x = \cos (x + x)$
$$= \cos x \cos x - \sin x \sin x$$
$$= \cos^2 x - \sin^2 x.$$

Cosine of twice an angle	$\cos 2A = \cos^2 A - \sin^2 A$

Observe that the formula for cos 2A is derived by using the formula for cos $(A + B)$ and the formula for sin 2A is derived by using the formula for sin $(A + B)$.

Cos $(A \pm B)$ and Cos 2A

Example 4   **Find cos 120°. Use a special angle measure and the formula for cos 2A.**

$$\cos 2A = \cos^2 A - \sin^2 A$$
$$\cos 120° = \cos 2(60°) = \cos^2 60° - \sin^2 60°$$
$$= \left(\frac{1}{2}\right)^2 - \left(\frac{\sqrt{3}}{2}\right)^2$$
$$= \frac{1}{4} - \frac{3}{4}, \text{ or } -\frac{1}{2}$$

Thus, cos 120° = $-\frac{1}{2}$.

Example 5   **If $\cos \theta = -\dfrac{15}{17}$ and θ lies in the third quadrant, find cos 2θ.**

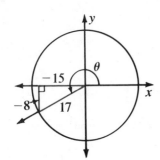

Using the reference triangle, $\cos \theta = \dfrac{-15}{17}$ and $\sin \theta = \dfrac{-8}{17}$.

$$\cos 2\theta = \cos^2 \theta - \sin^2 \theta$$
$$= \left(-\frac{15}{17}\right)^2 - \left(-\frac{8}{17}\right)^2$$
$$= \frac{225}{289} - \frac{64}{289}$$
$$= \frac{161}{289}$$

Thus, $\cos 2\theta = \dfrac{161}{289}$.

Recall from Chapter 18 that for any angle A, $\sin^2 A + \cos^2 A = 1$.
This identity can also be written as $\cos^2 A = 1 - \sin^2 A$ or $\sin^2 A = 1 - \cos^2 A$.
By using these identities, you can derive two additional formulas for cosine of twice an angle.
Begin with $\cos 2x = \cos^2 x - \sin^2 x$.

Replace $\cos^2 x$ with $1 - \sin^2 x$.
$$\cos 2x = \cos^2 x - \sin^2 x$$
$$= (1 - \sin^2 x) - \sin^2 x$$
$$= 1 - 2 \sin^2 x$$

Replace $\sin^2 x$ with $1 - \cos^2 x$.
$$\cos 2x = \cos^2 x - \sin^2 x$$
$$= \cos^2 x - (1 - \cos^2 x)$$
$$= 2 \cos^2 x - 1$$

Three formulas for cos 2A	$\cos 2A = \cos^2 A - \sin^2 A$ $\cos 2A = 1 - 2 \sin^2 A$ $\cos 2A = 2 \cos^2 A - 1$

Written Exercises

(A) Find the value of each. Use special angle measures and the formula for cos $(A + B)$ or cos $(A - B)$. Answers may be left in simplest radical form.

1. cos 105°   2. cos 15°   3. cos 135°   4. cos 195°   5. cos 285°
6. cos 165°   7. cos 150°   8. cos 255°   9. cos 435°   10. cos 375°

Find the value of each. Use special angles and the formula for cos 2A. Answer in simplest radical form.

11. cos 90°   12. cos 60°   13. cos 240°   14. cos 450°   15. cos 180°

Find the value of cos $(A + B)$ and cos $(A - B)$.

16. Cos $A = \dfrac{3}{5}$ and cos $B = \dfrac{12}{13}$.
    A and B are both acute angles.

17. Cos $A = \dfrac{15}{17}$ and cos $B = \dfrac{5}{9}$.
    A and B are both acute angles.

18. Cos $A = -\dfrac{5}{13}$ and A lies in the second quadrant. Cos $B = \dfrac{8}{17}$ and B lies in the fourth quadrant.

19. Cos $A = -\dfrac{6}{11}$ and A lies in the third quadrant. Cos $B = \dfrac{4}{5}$ and B lies in the first quadrant.

Find the value of cos 2A.

20. Cos $A = -\dfrac{4}{5}$ and A lies in the 3rd quad.

21. Cos $A = \dfrac{12}{13}$ and A lies in the 1st quad.

22. Cos $A = \dfrac{3}{7}$ and A lies in the 1st quad.

23. Cos $A = \dfrac{-7}{10}$ and A lies in the 2nd quad.

(B) Simplify.

24. $\dfrac{\cos 2A}{\sin^2 A - \cos^2 A}$

25. $\dfrac{\cos^2 A - \sin^2 A}{2 \cos^2 A - 1}$

26. $\dfrac{1 - 2 \sin^2 A}{\cos 2A}$

(C) Find the value of each. Use special angle measures and the formula for cos $(A + B)$, cos $(A - B)$, or cos 2A. Answer in simplest radical form.

27. sec 75°   28. sec 135°   29. sec 120°   30. sec 15°   31. sec 90°

Prove each identity. Use the formulas for cos $(A + B)$.

32. cos $(180° - \theta) = -\cos \theta$ 33. cos $(360° - \theta) = \cos \theta$ 34. cos $(180° + \theta) = -\cos \theta$
35. cos $(\pi + \theta) = -\cos \theta$ 36. cos $(270° + \theta) = \sin \theta$ 37. cos $(2\pi + \theta) = \cos \theta$
38. cos $(\dfrac{\pi}{2} + \theta) = -\sin \theta$ 39. cos $(\pi - \theta) = -\cos \theta$ 40. cos $(\dfrac{3}{2}\pi - \theta) = -\sin \theta$

CUMULATIVE REVIEW

1. A triangle's height and a rectangle's width are the same. The rectangle's length is three times its width. The triangle's base measures one half the rectangle's length. The rectangle's area minus the triangle's area is 9 square units. Find the height of the triangle to the nearest unit.

2. A square and a rectangle have the same width. The rectangle's length is 8 cm more than three times its width. The sum of their areas is 192 square cm. Find the width of the square to the nearest centimeter.

Cos $(A \pm B)$ and Cos 2A

TAN ($A \pm B$) AND TAN $2A$ 19.8

Objective   **To apply the formulas for tan ($A \pm B$) and tan $2A$**

You can derive the formula for the tangent of the sum of two angles using the formulas for the sine and cosine of the sum of two angles.

Recall from Chapter 18 the identity $\tan A = \dfrac{\sin A}{\cos A}$.

$$\tan (x + y) = \frac{\sin (x + y)}{\cos (x + y)}$$

$$= \frac{\sin x \cos y + \cos x \sin y}{\cos x \cos y - \sin x \sin y} \quad \blacktriangleleft \text{ Formula for } \sin (A + B)$$
$$\blacktriangleleft \text{ Formula for } \cos (A + B)$$

Divide each term by $\cos x \cos y$ and simplify.

$$\tan (x + y) = \frac{\dfrac{\sin x \cancel{\cos y}}{\cancel{\cos x} \cancel{\cos y}} + \dfrac{\cancel{\cos x} \sin y}{\cancel{\cos x} \cos y}}{\dfrac{\cancel{\cos x} \cancel{\cos y}}{\cancel{\cos x} \cancel{\cos y}} - \dfrac{\sin x \sin y}{\cos x \cos y}}$$

So, $\tan (x + y) = \dfrac{\tan x + \tan y}{1 - \tan x \tan y}$.

Tangent of the sum of two angles	$\tan (A + B) = \dfrac{\tan A + \tan B}{1 - \tan A \tan B}$

Example 1   **Find tan 75°. Use special angle measures and the formula for tan ($A + B$). Answer in simplest radical form.**

$$\tan (A + B) = \frac{\tan A + \tan B}{1 - \tan A \tan B}$$

$$\tan 75° = \tan (45° + 30°) = \frac{\tan 45° + \tan 30°}{1 - \tan 45° \tan 30°} = \frac{1 + \dfrac{\sqrt{3}}{3}}{1 - (1)\left(\dfrac{\sqrt{3}}{3}\right)}$$

$$= \frac{\dfrac{3 + \sqrt{3}}{3}}{\dfrac{3 - \sqrt{3}}{3}}, \text{ or } \frac{3 + \sqrt{3}}{3 - \sqrt{3}} \cdot \frac{3 + \sqrt{3}}{3 + \sqrt{3}} = \frac{9 + 6\sqrt{3} + 3}{9 - 3}$$

Thus, tan 75° = $2 + \sqrt{3}$.

You can derive the formula for the tangent of the difference of two angles using the formula for the tan of the sum of two angles.

Rewrite the difference $x - y$ as the sum $x + (-y)$.

$$\tan (x - y) = \tan [x + (-y)]$$
$$= \frac{\tan x + \tan (-y)}{1 - \tan x \tan (-y)}$$
$$= \frac{\tan x - \tan y}{1 + \tan x \tan y}, \text{ since } \tan (-y) = -\tan y.$$

Tangent of the difference of two angles	$\tan (A - B) = \dfrac{\tan A - \tan B}{1 + \tan A \tan B}$

Example 2   **If $\tan A = \dfrac{4}{7}$ and $\tan B = \dfrac{13}{2}$, find $\tan (A - B)$.**

$$\tan (A - B) = \frac{\tan A - \tan B}{1 + \tan A \tan B}$$
$$= \frac{\dfrac{4}{7} - \dfrac{13}{2}}{1 + \dfrac{4}{7}\left(\dfrac{13}{2}\right)}, \text{ or } \frac{-\dfrac{83}{14}}{\dfrac{66}{14}}$$

Thus, $\tan (A - B) = -\dfrac{83}{66}$.

You can derive the formula for the tangent of twice (double) an angle using the formula for the tan of the sum of two angles.

$$\tan 2x = \tan (x + x) = \frac{\tan x + \tan x}{1 - \tan x \tan x}, \text{ or } \frac{2 \tan x}{1 - \tan^2 x}$$

Tangent of twice an angle	$\tan 2A = \dfrac{2 \tan A}{1 - \tan^2 A}$

Example 3   **Find tan 120°. Use a special angle measure and the formula for tan 2x. Answer in simplest radical form.**

$$\tan 2A = \frac{2 \tan A}{1 - \tan^2 A}$$
$$\tan 120° = \tan 2(60°) = \frac{2 \tan 60°}{1 - \tan^2 60°} = \frac{2\sqrt{3}}{1 - (\sqrt{3})^2}, \text{ or } \frac{2\sqrt{3}}{-2}$$

Thus, $\tan 120° \doteq -\sqrt{3}$.

Tan $(A \pm B)$ and Tan 2A

Example 4   **If** $\tan \theta = -\dfrac{7}{6}$, **find** $\tan 2\theta$.

$$\tan 2\theta = \frac{2 \tan \theta}{1 - \tan^2 \theta} = \frac{2\left(-\dfrac{7}{6}\right)}{1 - \left(-\dfrac{7}{6}\right)^2}$$

$$= \frac{\dfrac{-14}{6}}{1 - \dfrac{49}{36}}, \text{ or } \frac{\dfrac{-14}{6}}{-\dfrac{13}{36}}, \text{ or } \frac{-14}{6} \cdot \frac{-36}{13}$$

Thus, $\tan 2\theta = \dfrac{84}{13}$.

Written Exercises

(A) Find the value of each. Use special angles and the formula for tan $(A + B)$ or tan $(A - B)$. Answer in simplest radical form.

1. $\tan 105°$ **2.** $\tan 15°$ **3.** $\tan 135°$ **4.** $\tan 120°$ **5.** $\tan 375°$

Find tan $(A + B)$ for each of the following.

6. $\tan A = \frac{5}{3}$ and $\tan B = \frac{4}{7}$

7. $\tan A = -\sqrt{3}$ and $\tan B = \frac{\sqrt{2}}{2}$

8. $\tan A = -\frac{3}{4}$ and $\tan B = \frac{5}{8}$

9. $\tan A = \frac{-5}{12}$ and $\tan B = \frac{8}{7}$

Find tan $(A - B)$ for each of the following.

10. $\tan A = -\frac{1}{3}$ and $\tan B = -\frac{\sqrt{2}}{2}$

11. $\tan A = \frac{7}{5}$ and $\tan B = -1$

12. $\tan A = \sqrt{3}$ and $\tan B = \frac{\sqrt{2}}{2}$

13. $\tan A = \frac{5}{12}$ and $\tan B = \frac{8}{17}$

Find tan $2A$ for each of the following.

14. $\tan A = \frac{4}{5}$ **15.** $\tan A = \frac{15}{17}$ **16.** $\tan A = -\frac{5}{12}$ **17.** $\tan A = -1$

18. $\tan A = -\frac{3}{4}$ **19.** $\tan A = \frac{12}{11}$ **20.** $\tan A = \frac{4}{7}$ **21.** $\tan A = -\frac{3}{8}$

(B) Tan $\theta = -\frac{5}{12}$, cos $\beta = -\frac{3}{5}$, and β lies in the second quadrant.
22. Find tan $(\theta + \beta)$.   **23.** Find tan $(\theta - \beta)$.
24. Find tan 2θ.   **25.** Find tan 2β.

Tan $\theta = -\frac{7}{3}$, sin $\beta = -\frac{5}{13}$, and β lies in the fourth quadrant.
26. Find tan $(\theta + \beta)$.   **27.** Find tan $(\theta - \beta)$.
28. Find tan 2θ.   **29.** Find tan 2β.

Find the value for each of the following.

30. $\dfrac{2 \tan 15°}{1 - \tan^2 15°}$ **31.** $\dfrac{2 \tan 22.5°}{1 - \tan^2 22.5°}$ **32.** $\dfrac{2 \tan 67.5°}{1 - \tan^2 67.5°}$

(C) Prove each identity. Use the formulas for tan $(A \pm B)$.
33. $\tan (180° + \theta) = \tan \theta$ **34.** $\tan (180° - \theta) = -\tan \theta$ **35.** $\tan (\pi + \theta) = \tan \theta$
36. $\tan (360° - \theta) = -\tan \theta$ **37.** $\tan (2\pi + \theta) = \tan \theta$ **38.** $\tan (\pi - \theta) = -\tan \theta$

HALF-ANGLE FORMULAS 19.9

Objective **To apply the formulas for sin $\frac{A}{2}$, cos $\frac{A}{2}$, and tan $\frac{A}{2}$**

You can derive the formulas for the sine, cosine, and tangent of half an angle.
You will also learn how to apply these formulas.
One of the formulas for cos $2A$ is cos $2A = 1 - 2 \sin^2 A$.
If we replace A with $\frac{x}{2}$, the following equation is true.

$$\cos 2\left(\frac{x}{2}\right) = 1 - 2 \sin^2 \left(\frac{x}{2}\right)$$

$$\cos x = 1 - 2 \sin^2 \frac{x}{2}$$

$$2 \sin^2 \frac{x}{2} = 1 - \cos x$$

$$\sin^2 \frac{x}{2} = \frac{1 - \cos x}{2} \qquad \text{So, } \sin \frac{x}{2} = \pm \sqrt{\frac{1 - \cos x}{2}}.$$

Sine of half an angle	$\sin \dfrac{A}{2} = \pm \sqrt{\dfrac{1 - \cos A}{2}}$

Example 1 **Find sin 30°. Use the formula for sin $\frac{A}{2}$.**

$$\sin \frac{A}{2} = \pm \sqrt{\frac{1 - \cos A}{2}}$$

$30° = \dfrac{60°}{2}$ ▶ $\sin \left(\dfrac{60°}{2}\right) = + \sqrt{\dfrac{1 - \cos 60°}{2}}$ ◀ *Sin 30° is positive.*

$$= + \sqrt{\frac{1 - \frac{1}{2}}{2}}, \text{ or } + \sqrt{\frac{1}{4}}$$

Thus, $\sin 30° = \dfrac{1}{2}$.

Example 2 **If cos $x = -\frac{1}{8}$ and x lies in the third quadrant, find sin $\frac{x}{2}$.**

$$\sin \frac{x}{2} = \pm \sqrt{\frac{1 - \cos x}{2}}$$

Since x lies in the third quadrant, $180° < x < 270°$. Then $\frac{x}{2}$ lies in the second quadrant, since $90° < \frac{x}{2} < 135°$.

sin $\frac{x}{2}$ is positive in the second quadrant. ▶ $\sin \dfrac{x}{2} = + \sqrt{\dfrac{1 - \cos x}{2}} = + \sqrt{\dfrac{1 - (-\frac{1}{8})}{2}}$, or $+ \sqrt{\dfrac{1 + \frac{1}{8}}{2}}$

$$= + \sqrt{\frac{9}{16}}$$

Thus, $\sin \dfrac{x}{2} = \dfrac{3}{4}$.

To derive a formula for $\cos \frac{A}{2}$, use $\cos 2A = 2 \cos^2 A - 1$ and replace A with $\frac{x}{2}$.

$$\cos 2A = 2 \cos^2 A - 1$$

$$\cos 2\left(\frac{x}{2}\right) = 2 \cos^2 \left(\frac{x}{2}\right) - 1 \qquad \cos x = 2 \cos^2 \left(\frac{x}{2}\right) - 1$$

$$2 \cos^2 \frac{x}{2} = 1 + \cos x \qquad \cos^2 \frac{x}{2} = \frac{1 + \cos x}{2}$$

So, $\cos \frac{x}{2} = \pm \sqrt{\dfrac{1 + \cos x}{2}}$

Cosine of half an angle	$\cos \dfrac{A}{2} = \pm \sqrt{\dfrac{1 + \cos A}{2}}$

Example 3

Find cos 75°. Use the formula for $\cos \frac{A}{2}$. Answer may be left in simplest radical form.

$\cos \dfrac{A}{2} = \pm \sqrt{\dfrac{1 + \cos A}{2}}$. Write 75° as $\dfrac{150°}{2}$.

cos 75° is positive. ▶ $\cos \left(\dfrac{150°}{2}\right) = + \sqrt{\dfrac{1 + \cos 150°}{2}}$

$\cos 150° = -\cos 30°$ ▶ $= \sqrt{\dfrac{1 + (-\frac{\sqrt{3}}{2})}{2}}$, or $\sqrt{\dfrac{2 - \sqrt{3}}{4}}$

$= -\frac{\sqrt{3}}{2}$.

Thus, $\cos 75° = \dfrac{\sqrt{2 - \sqrt{3}}}{2}$.

To derive a formula for $\tan \frac{x}{2}$, use $\tan A = \dfrac{\sin A}{\cos A}$ and replace A with $\frac{x}{2}$.

$$\tan \frac{x}{2} = \frac{\sin \frac{x}{2}}{\cos \frac{x}{2}} = \frac{\pm \sqrt{\dfrac{1 - \cos x}{2}}}{\pm \sqrt{\dfrac{1 + \cos x}{2}}} = \pm \sqrt{\frac{1 - \cos x}{1 + \cos x}}.$$

Tangent of half an angle	$\tan \dfrac{A}{2} = \pm \sqrt{\dfrac{1 - \cos A}{1 + \cos A}}$

Example 4

If $\cos x = \frac{1}{5}$ and x lies in the 1st quadrant, find $\tan \frac{x}{2}$.

$$\tan \frac{x}{2} = \pm \sqrt{\frac{1 - \cos x}{1 + \cos x}} = + \sqrt{\frac{1 - \frac{1}{5}}{1 + \frac{1}{5}}}, \text{ or } + \sqrt{\frac{\frac{4}{5}}{\frac{6}{5}}}$$

$$= + \sqrt{\frac{4}{6}} = \frac{2}{\sqrt{6}} = \frac{2\sqrt{6}}{6}$$

Thus, $\tan \dfrac{x}{2} = \dfrac{2\sqrt{6}}{6}$ or $\dfrac{\sqrt{6}}{3}$.

Written Exercises

(A) Find the value of each. Use special angle measures and the formula for $\sin \frac{A}{2}$, $\cos \frac{A}{2}$, or $\tan \frac{A}{2}$. Answer in simplest radical form.

1. $\sin 15°$ 2. $\cos 15°$ 3. $\sin 22\frac{1}{2}°$ 4. $\tan 22\frac{1}{2}°$ 5. $\cos 22\frac{1}{2}°$

6. $\tan 15°$ 7. $\tan 67\frac{1}{2}°$ 8. $\cos 67\frac{1}{2}°$ 9. $\sin 67\frac{1}{2}°$ 10. $\sin 75°$

Find the value of $\sin \frac{A}{2}$ and $\cos \frac{A}{2}$ given the value of $\cos A$. Answer in simplest radical form.

11. $\cos A = \frac{3}{5}$ 12. $\cos A = 0$ 13. $\cos A = -\frac{5}{13}$ 14. $\cos A = -\frac{1}{2}$

15. $\cos A = \frac{1}{4}$ 16. $\cos A = -\frac{15}{17}$ 17. $\cos A = -\frac{3}{4}$ 18. $\cos A = -1$

Find the value of $\sin \frac{A}{2}$, $\cos \frac{A}{2}$, and $\tan \frac{A}{2}$. Answer in simplest radical form.

19. $\cos A = \frac{4}{5}$ and A is in the 4th quad.   20. $\cos A = \frac{-12}{13}$ and A is in the 2nd quad.

21. $\cos A = -\frac{\sqrt{3}}{2}$ and A is in the 3rd quad.   22. $\cos A = -\frac{1}{2}$ and A is in the 2nd quad.

(B) 23. $\cos A = \frac{1}{3}$ and A is in the 1st quad.   24. $\cos A = \frac{3}{4}$ and A is in the 4th quad.

25. $\cos A = -\frac{5}{12}$ and A is in the 2nd quad.   26. $\cos A = -\frac{4}{7}$ and A is in the 3rd quad.

27. $\cos A = a$ and A is in the 1st quad.   28. $\cos A = -b$ and A is in the 2nd quad.

(C) 29. Find $\tan \frac{A}{2}$ if $\sin A = -\frac{1}{2}$ and $\cos A = -\frac{\sqrt{3}}{2}$.

30. Find $\tan \frac{\theta}{2}$ if $\cos \theta = -\frac{15}{17}$ and $90° < \theta < 180°$.

31. Find $\tan \frac{\theta}{2}$ if $\sin \theta = \frac{4}{5}$ and $\cos \theta = -\frac{5}{13}$.

32. Find $\tan \frac{B}{2}$ if $\cos B = -\frac{4}{5}$ and $180° < B < 270°$.

33. Prove: $\cot \frac{A}{2} = \pm \sqrt{\frac{1 + \cos A}{1 - \cos A}}$.

34. Prove: $\sec \frac{A}{2} = \pm \sqrt{\frac{2}{1 + \cos A}}$.

35. Prove: $\csc \frac{A}{2} = \pm \sqrt{\frac{2}{1 - \cos A}}$.

36. Find $\sec \frac{\theta}{2}$ If $\cos \theta = -\frac{1}{2}$ and $90° < \theta < 180°$.

NON-ROUTINE PROBLEMS

Quadrilateral $ABCD$ is inscribed in a circle.
Find the measure of $\angle D$ to the nearest minute.
Hint: Draw $\overline{AC}$ and use the
Law of Cosines twice, both
involving AC.

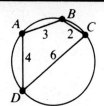

Half-Angle Formulas

DOUBLE AND HALF-ANGLE IDENTITIES   19.10

Objective   **To prove identities using double and half-angle formulas**

You have derived and applied formulas for the sine, cosine and tangent of double and half angles. You will now use these formulas to prove some trigonometric identities.

Example 1   **Prove the identity $\cos 2\theta = \dfrac{\csc^2 \theta - 2}{\csc^2 \theta}$.**

Replace $\csc^2 \theta$ with $\dfrac{1}{\sin^2 \theta}$.

$$
\begin{array}{c|c}
\cos 2\theta & \dfrac{\csc^2 \theta - 2}{\csc^2 \theta} \\[2ex]
\cos 2\theta & \dfrac{\dfrac{1}{\sin^2 \theta} - 2}{\dfrac{1}{\sin^2 \theta}} \\[3ex]
 & \dfrac{\dfrac{1 - 2\sin^2 \theta}{\sin^2 \theta}}{\dfrac{1}{\sin^2 \theta}} \\[3ex]
 & 1 - 2\sin^2 \theta \\[2ex]
 & \cos 2\theta
\end{array}
$$

Example 2   **Prove the identity $\tan 2A = \dfrac{2 \tan A}{\sec^2 A - 2\tan^2 A}$.**

Replace $\tan 2A$ with $\dfrac{\sin 2A}{\cos 2A}$, $\sec^2 A$ with $\dfrac{1}{\cos^2 A}$, and $\tan A$ with $\dfrac{\sin A}{\cos A}$.

$$
\begin{array}{c|c}
\tan 2A & \dfrac{2 \tan A}{\sec^2 A - 2\tan^2 A} \\[2ex]
\dfrac{\sin 2A}{\cos 2A} & \dfrac{2\dfrac{\sin A}{\cos A}}{\dfrac{1}{\cos^2 A} - 2\dfrac{\sin^2 A}{\cos^2 A}} \\[3ex]
\dfrac{2 \sin A \cos A}{1 - 2\sin^2 A} & \dfrac{2 \sin A}{\cos A} \cdot \dfrac{\cos^2 A}{1 - 2\sin^2 A} \\[3ex]
 & \dfrac{2 \sin A \cos A}{1 - 2\sin^2 A}
\end{array}
$$

The identity in Example 3 involves the formula for the sine of a half-angle.

Example 3   **Prove the identity $\sin^2 \dfrac{\theta}{2} = \dfrac{\sec \theta - 1}{2 \sec \theta}$.**

Replace $\sec \theta$ with $\dfrac{1}{\cos \theta}$ and $\sin \dfrac{\theta}{2}$ with $\pm \sqrt{\dfrac{1 - \cos \theta}{2}}$.

$\sin^2 \dfrac{\theta}{2}$	$\dfrac{\sec \theta - 1}{2 \sec \theta}$
$\left(\pm \sqrt{\dfrac{1 - \cos \theta}{2}} \right)^2$	$\dfrac{\dfrac{1}{\cos \theta} - 1}{2 \left(\dfrac{1}{\cos \theta} \right)}$
$\dfrac{1 - \cos \theta}{2}$	$\dfrac{1 - \cos \theta}{\cos \theta} \cdot \dfrac{\cos \theta}{2}$
	$\dfrac{1 - \cos \theta}{2}$

Written Exercises

(A) **Prove each identity.**

1. $\tan \theta = \dfrac{\sin 2\theta}{1 + \cos 2\theta}$

2. $\sin^2 \theta = \dfrac{\tan \theta \sin 2\theta}{2}$

3. $\cot \theta = \dfrac{\sin 2\theta}{2 \sin^2 \theta}$

4. $\cos^2 \dfrac{\theta}{2} = \dfrac{\sec \theta + 1}{2 \sec \theta}$

5. $\tan \dfrac{\theta}{2} = \pm \dfrac{1 - \cos \theta}{\sin \theta}$

6. $\tan \dfrac{\theta}{2} = \pm \dfrac{\sin \theta}{1 + \cos \theta}$

(B) 7. $2 \csc 2\theta = \dfrac{\cos \theta}{\sin \theta} + \dfrac{\sin \theta}{\cos \theta}$

8. $\sin^2 \theta - 2 \sin \theta \cos \theta + \cos^2 \theta = 1 - 2 \sin 2\theta$

9. $1 + \sin \theta = \left(\sin \dfrac{\theta}{2} + \cos \dfrac{\theta}{2} \right)^2$

10. $\cos \theta = \dfrac{\frac{1}{2} \sin^2 \theta}{\sin^2 \frac{1}{2} \theta} - 1$

11. $\dfrac{1 - \tan^2 \theta}{1 + \tan^2 \theta} = \cos 2\theta$

12. $\dfrac{\sin \theta + \cos \theta}{\cos \theta - \sin \theta} + \dfrac{\sin \theta - \cos \theta}{\sin \theta + \cos \theta} = 2 \tan 2\theta$

(C) 13. $\tan 2\theta = \dfrac{\sin 4\theta}{1 + \cos 4\theta}$

14. $\cos 3\theta = 4 \cos^3 \theta - 3 \cos \theta$

15. $2 \csc 2\theta = \sec \theta \csc \theta$

16. $\sec 2\theta = \dfrac{\sec^2 \theta}{2 - \dfrac{1}{\cos^2 \theta}}$

17. $\dfrac{\sec^2 \theta}{2} + \tan \theta = \dfrac{1 + \sin 2\theta}{1 + \cos 2\theta}$

18. $\dfrac{\cos 3\theta}{\sin 3\theta} = (1 + \cos 6\theta)(\csc 6\theta)$

CUMULATIVE REVIEW

Use the Pythagorean relation to determine whether the following are measures of the sides of a right triangle.

1. 6, 8, 10

2. 0.5, 0.9, 1.2

3. $\sqrt{2}, 3\sqrt{5}, 4\sqrt{3}$

CHAPTER NINETEEN REVIEW

Vocabulary
Ambiguous case [19.5]   Law of Cosines [19.1]
Area of a triangle [19.3] Law of Sines [19.4]

For triangle ABC, find the length of the indicated side to the nearest unit. [19.1]
1. $a = 8$, $c = 5$, $m\angle B = 42°$. Find b.
2. $c = 10$, $b = 6$, $m\angle A = 130°$. Find a.
3. $a = 17.3$, $b = 47.6$, $m\angle C = 67° 43'$. Find c.

4. Two forces of 50 lb and 70 lb act on a body at an angle of 72°. Find the magnitude of the resultant to the nearest pound. [19.1]

For triangle ABC, find each angle to the nearest degree. [19.2]
5. $a = 6$, $b = 9$, $c = 4$. Find $m\angle A$.
6. $a = 4$, $b = 7$, $c = 10$. Find $m\angle B$.
7. $a = 12.7$, $b = 17.9$, $c = 23.6$. Find $m\angle C$.

For △ABC, find the area to the nearest square unit. [19.3]
8. $a = 6$, $b = 8$, $m\angle C = 48°$.
9. $b = 8.3$, $c = 9.6$, $m\angle A = 63° 20'$

10. The lengths of two sides of a parallelogram are 30 and 34 centimeters. Their included angle measures 120°. Find the area of the parallelogram. [19.3]

For △PQR, find m∠P to the nearest degree. [19.3]
11. $q = 9$, $r = 16$, area = 36 square units
12. $q = 11$, $r = 9$, area = 45 square units

13. For △ABC, $a = 60$, $m\angle A = 63°$, [19.4] $m\angle B = 47°$. Find b to the nearest unit.
14. For △ABC, $a = 13.3$, $c = 18.5$, $m\angle A = 43° 40'$. Find $m\angle C$ to the nearest degree. [19.4]

How many triangles can be constructed? [19.5]
15. $m\angle A = 38°$, $a = 8$, $b = 12$
16. $m\angle A = 120°$, $a = 30$, $b = 10$

Find each. Use the formula for sin (A ± B) or sin $\frac{A}{2}$. Answer in simplest radical form. [19.6, 19.9]
17. sin 75°   18. sin 15°   19. sin $22\frac{1}{2}°$

20. If sin $A = -\frac{12}{13}$ and $\angle A$ lies in the third quadrant, find sin 2A. [19.6]

Find each. Use the formula for cos(A ± B) or cos $\frac{A}{2}$. Answer in simplest radical form. [19.7, 19.9]
21. cos 15°   22. cos $67\frac{1}{2}°$ 23. cos $22\frac{1}{2}°$

Find sin (A + B) and sin (A − B). [19.6]
24. sin $A = -\frac{4}{5}$ and $\angle A$ lies in the fourth quadrant. sin $B = \frac{5}{13}$ and $\angle B$ lies in the second quadrant.

Find cos (A + B) and cos (A − B). [19.7]
25. cos $A = -\frac{12}{13}$ and $\angle A$ lies in the third quadrant. cos $B = \frac{3}{5}$ and $\angle B$ lies in the fourth quadrant.

26. If cos $A = -\frac{1}{4}$ and $\angle A$ lies in the second quadrant, find cos $\frac{A}{2}$. [19.9]

Prove the identity. Use formula for sin (A − B).
★27. $\sin \left(\frac{3}{2} \pi - \theta\right) = -\cos \theta$. [19.6]

28. Find tan (A + B) if tan $A = -\frac{5}{12}$ and tan $B = -\frac{3}{5}$. [19.8]

29. Find tan 2A if tan $A = \frac{4}{5}$. [19.8]

30. Find tan (A − B) if tan $A = -\frac{8}{9}$ and tan $B = \frac{4}{7}$. [19.8]

31. Find tan 105°. Use the formula for tan (A + B). [19.8]

32. Prove $\cos^2 \frac{\theta}{2} = \frac{\sin \theta + \tan \theta}{2 \tan \theta}$. [19.10]

★33. Find tan $\frac{A}{2}$ if sin $A = \frac{\sqrt{3}}{2}$ and cos $A = -\frac{1}{2}$. [19.9]

★34. Prove: tan $(2\pi - \theta) = -\tan \theta$. [19.8]

For triangle *ABC*, find the length of the indicated side to the nearest unit.
1. $a = 5$, $b = 9$, m$\angle C = 50°$. Find c.
2. $c = 12$, $b = 7$, m$\angle A = 140°$. Find a.
3. $a = 23.4$, $c = 36.2$, m$\angle B = 73° 18'$. Find b.

4. Two forces of 14 lb and 16 lb act on a body at an angle of 48°. Find the magnitude of the resultant to the nearest pound.

For triangle *ABC*, find the angle measure to the nearest degree.
5. $a = 6$, $b = 5$, $c = 8$. Find m$\angle C$.
6. $a = 15$, $b = 7$, $c = 10$. Find m$\angle A$.
7. $a = 13.8$, $b = 24.6$, $c = 18.7$. Find m$\angle B$.

For $\triangle ABC$, find the area to the nearest square unit.
8. $b = 14$, $c = 8$, m$\angle A = 73°$
9. $a = 6.3$, $b = 8.5$, m$\angle C = 30° 40'$

10. The lengths of two sides of a parallelogram are 64 and 82 centimeters. Their included angle measures 70°. Find the area of the parallelogram.

For $\triangle PQR$, find m$\angle P$ to the nearest degree.
11. $q = 7$, $r = 8$, area = 22 sq. units
12. $q = 14$, $r = 10$, area = 60 sq. units

13. For $\triangle ABC$, $a = 32$, m$\angle A = 72°$, m$\angle C = 32°$. Find c to the nearest unit.

14. For $\triangle ABC$, $a = 16.4$, $c = 10.8$, m$\angle A = 38° 30'$. Find m$\angle C$ to the nearest degree.

How many triangles can be constructed?
15. m$\angle A = 28°$, $a = 7$, $b = 13$
16. m$\angle A = 130°$, $a = 16$, $b = 12$

Find each. Use the formula for sin $(A \pm B)$ or sin $\frac{A}{2}$. Answer in simplest radical form.
17. $\sin 15°$ 18. $\sin 75°$ 19. $\sin 22\frac{1}{2}°$.

20. If $\sin A = -\frac{3}{5}$ and $\angle A$ lies in the third quadrant, find sin $2A$.

Find each. Use the formula for cos $(A \pm B)$ or cos $\frac{A}{2}$. Answer in simplest radical form.
21. $\cos 135°$ 22. $\cos 22\frac{1}{2}°$ 23. $\cos 15°$

Find sin $(A + B)$ and sin $(A - B)$.
24. $\sin A = \frac{5}{7}$ and $\angle A$ lies in the second quadrant. $\sin B = -\frac{4}{5}$ and $\angle B$ lies in the third quadrant.

Find cos $(A + B)$ and cos $(A - B)$.
25. $\cos A = \frac{12}{13}$ and $\angle A$ lies in the first quadrant. $\cos B = -\frac{4}{5}$ and $\angle B$ lies in the second quadrant.

26. If $\cos A = -\frac{5}{13}$ and $\angle A$ lies in the third quadrant, find cos $\frac{A}{2}$.

27. Find tan $(A + B)$ if $\tan A = \frac{4}{5}$ and $\tan B = -\frac{2}{3}$.

28. Find tan $2A$ if $\tan A = \frac{3}{5}$.

29. Find tan $(A - B)$ if $\tan A = -\frac{8}{9}$ and $\tan B = -\frac{3}{7}$.

30. Find tan 75°. Use the formula for tan $(A + B)$.

31. Prove $\sin 2A = \dfrac{2 \cot A}{\csc^2 A}$.

32. Prove $2 \csc 2\theta = \tan \theta + \cot \theta$.

Prove the identity. Use the formula for sin $(A - B)$.
★ 33. $\sin (\frac{\pi}{2} + \theta) = \cos \theta$
★ 34. Prove $\tan (\pi - \theta) = -\tan \theta$.

★ 35. Find tan $\frac{A}{2}$ if $\sin A = -\frac{1}{2}$ and $\cos A = -\frac{\sqrt{3}}{2}$.

Law of Cosines

OBJECTIVE: **To determine the measures of angles of a triangle, given the three sides utilizing the Law of Cosines**

The Law of Cosines states: $a^2 = b^2 + c^2 - 2bc \cos A$. Solving for cos A, the equation becomes cos A $= (a^2 - b^2 - c^2)/(-2bc)$. Using this form of the Law of Cosines, you can find the measure of any angle of a triangle when you know the lengths of its three sides. The program below performs this process. First, lines 140–160 determine the cosines of all three angles of the triangle. The program then uses inverse trigonometric functions to determine the angles.

Among the inverse trigonometric functions, most forms of BASIC include only the arctangent function, symbolized by ATN. In this program, the arctangent must be expressed in terms of an angle's cosine. Line 190 does this by using the arctangent form of the following trigonometric identity:

$$\tan A = (\sqrt{1 - \cos^2 A})/\cos A.$$

All trigonometric functions in BASIC use radian measurement of angles. Lines 200, 210–230, do the necessary converting.

```
100   PRINT "WHAT ARE THE THREE SIDES OF THE TRIANGLE
      ?";: INPUT F,B,C
110   IF F + B < = B THEN 100
120   IF F + C < = B THEN 100
130   IF B + C < = F THEN 100
140 D(1) = (F ^ 2 - B ^ 2 - C ^ 2) / ( - 2 * C * B)
150 D(2) = (B ^ 2 - C ^ 2 - F ^ 2) / ( - 2 * C * F)
160 D(3) = (C ^ 2 - B ^ 2 - F ^ 2) / ( - 2 * B * F)
170   PRINT "THE ANGLES ARE "
180   FOR I = 1 TO 3
190 E =   ATN ( SQR (1 - (D(I)) ^ 2) / D(I))
200 A = E * 180 / 3.1416
210 A(I) =   INT (A)
220 S(I) =   INT ((A - A(I)) * 60)
230   PRINT  ABS (A(I));" DEGREES ";S(I);" MINUTES
240   NEXT I: PRINT : GOTO 100
```

See the Computer Section beginning on page 574 for more information.

EXERCISES

1. Use the program above to find the solutions to the odd-numbered problems 1 to 13 on page 543.

2. Write a program that INPUTs the measurement of an angle in degrees and outputs the values for all six trigonometric functions of that angle.

COLLEGE PREP TEST

DIRECTIONS: Choose the *one* best answer to each question or problem.

1.

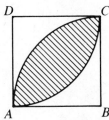

$ABCD$ is a square. $AB = 4$. B is the center of arc AC. D is the center of arc AC.
Find the area of the shaded region.
(A) $4(\pi - 2)$ **(B)** $16(\pi - 1)$ **(C)** $8(\pi - 1)$
(D) $8(\pi - 2)$ **(E)** $16\pi - 8$

2.

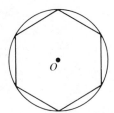

A regular hexagon is inscribed in a circle. The radius of the circle is 6 cm.
Which is the area between the circle and the regular hexagon?
(A) $36\pi - 18$ **(B)** $54\sqrt{3} - 36\pi$
(C) $36\pi - 54\sqrt{3}$ **(D)** $54\sqrt{3} - 12\pi$
(E) None of these

3.

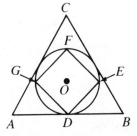

$\triangle ABC$ is equilateral with side of length s.
A circle is inscribed in the triangle.
A square is inscribed in the circle.
Which is the area of the square?
(A) $\frac{1}{6}s^2$ **(B)** s^2 **(C)** $\frac{1}{8}s^2$
(D) $\frac{1}{4}s^2$ **(E)** $\frac{1}{3}s^2$

4. For all real numbers a and b, define
$$a \star b = \frac{a + b}{ab}.$$
Find $(5 \star 5) \star 5$.

(A) 2 **(B)** 5 **(C)** $\frac{2}{5}$
(D) $\frac{27}{10}$ **(E)** $\frac{4}{25}$

5.

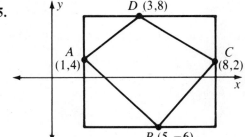

What is the area of $ABCD$?

(A) 49 **(B)** 98 **(C)** 56
(D) 43 **(E)** None of these

6.

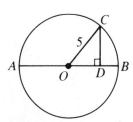

$\overline{AB}$ is a diameter of the circle with center O and radius 5.
If $\dfrac{BD}{DA} = \dfrac{1}{4}$, which is the value of OD?

(A) 4 **(B)** 12 **(C)** 13
(D) $\dfrac{5\sqrt{2}}{2}$ **(E)** 3

CUMULATIVE REVIEW

DIRECTIONS: Choose the *one* best answer to each question or problem. (Ex. 1–18)

1. Express $\sin 53° 18'$ in terms of its cofunction. [17.8]
   (A) $\csc 36° 42'$ (B) $\cos 36° 42'$
   (C) $-\cos 36° 42'$ (D) $-\csc 36° 42'$

2. In which quadrant is $\sin u < 0$ and $\cot u < 0$? [17.7]
   (A) I   (B) II   (C) III   (D) IV

3. Solve $\tan \theta = -\frac{\sqrt{3}}{3}$ for θ if $0 \le \theta \le 2\pi$. [18.11]
   (A) $\frac{5}{6}\pi$ and $\frac{11}{6}\pi$ (B) $\frac{2}{3}\pi$ and $\frac{5}{3}\pi$
   (C) $\frac{2}{3}\pi$ and $\frac{4}{3}\pi$ (D) $\frac{5}{6}\pi$ and $\frac{7}{6}\pi$

4. Express $225°$ in radians. [18.10]
   (A) $\frac{5}{4}\pi$ (B) $\frac{4}{3}\pi$ (C) $\frac{7}{6}\pi$ (D) $\frac{3}{4}\pi$

5. Express $\frac{2}{3}\pi$ radians in degrees. [18.10]
   (A) $60°$ (B) $240°$ (C) $180°$ (D) $120°$

6. Find the inverse of $y = -2 \sin x$. [18.12]
   (A) $x = \arcsin(-\frac{y}{2})$ (B) $y = \arcsin(-\frac{x}{2})$
   (C) $x = -\arcsin(\frac{y}{2})$ (D) $y = -\arcsin(\frac{x}{2})$

7. Find θ, if $\theta = \text{Arc sin } (-\frac{\sqrt{3}}{2})$. [18.12]
   (A) $240°$ (B) $300°$ (C) $-60°$ (D) $60°$

8. Find $\cos (\text{Arc tan } \frac{4}{3})$. [18.12]
   (A) $\frac{3}{5}$ (B) $-\frac{3}{5}$
   (C) $\frac{3}{5}$ and $-\frac{3}{5}$ (D) None of these

9. Evaluate $\sec (-\frac{\pi}{4}) + \csc \frac{\pi}{2}$. [18.10]
   (A) $\frac{\sqrt{2} + 2}{2}$ (B) $-\frac{\sqrt{2} + 2}{2}$
   (C) $\sqrt{2} + 1$ (D) $-\sqrt{2} + 1$

10. Write an equivalent expression for $\cot x$ using only $\sin x$. [18.8]
    (A) $\frac{1}{\sin^2 x}$ (B) $\sin^2 x$
    (C) $\pm \frac{\sin x}{\sqrt{1 - \sin^2 x}}$ (D) $\frac{\pm \sqrt{1 - \sin^2 x}}{\sin x}$

11. The lengths of two sides of a parallelogram are 20 and 30 centimeters. Their included angle measures $150°$. Find the area of the parallelogram. [19.3]
    (A) 150 cm^2 (B) 300 cm^2
    (C) $150 \sqrt{3} \text{ cm}^2$ (D) $150 \sqrt{2} \text{ cm}^2$

12. How many triangles can be constructed if $m\angle A = 46°$, $a = 16$, and $b = 14$. [19.5]
    (A) 0   (B) 1   (C) 2   (D) More than 2

13. Find $\cos 75°$. [19.6]
    (A) $\frac{\sqrt{6} + \sqrt{2}}{4}$ (B) $\frac{\sqrt{2}}{4}[\sqrt{3} - 1]$
    (C) $\frac{\sqrt{2}}{4}[1 - \sqrt{3}]$ (D) None of these

14. If $\sin A = -\frac{5}{13}$ and $\angle A$ lies in the third quadrant, find $\sin 2A$. [19.6]
    (A) $-\frac{120}{169}$ (B) $\frac{119}{169}$ (C) $-\frac{119}{169}$ (D) $\frac{120}{169}$

15. Find $\sin 15°$. [19.7]
    (A) $\frac{\sqrt{6} + \sqrt{2}}{4}$ (B) $\frac{\sqrt{2}}{2}[\sqrt{3} - 1]$
    (C) $\frac{\sqrt{6} - \sqrt{2}}{4}$ (D) $\frac{\sqrt{2} - \sqrt{6}}{4}$

16. If $\cos A = -\frac{1}{5}$ and $\angle A$ lies in the second quadrant, find $\cos \frac{A}{2}$. [19.9]
    (A) $\frac{1}{5}\sqrt{15}$ (B) $-\frac{1}{5}\sqrt{15}$
    (C) $-\frac{1}{5}\sqrt{10}$ (D) $\frac{1}{5}\sqrt{10}$

17. Find $\tan (A - B)$ if $\tan A = -\frac{5}{6}$ and $\tan B = \frac{2}{3}$. [19.8]
    (A) $-\frac{27}{8}$ (B) $\frac{3}{28}$ (C) $-\frac{27}{28}$ (D) $-\frac{3}{8}$

18. Find $\tan \frac{A}{2}$ if $\cos A = -\frac{1}{3}$ and $\angle A$ is in the 4th quadrant. [19.9]
    (A) $\sqrt{2}$ (B) $-\sqrt{2}$
    (C) $\pm\sqrt{2}$ (D) None of these

Determine the coordinates of point P and the radius r. [17.2]

19.

20.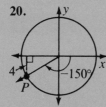

Find each of the following. Rationalize any radical denominator. [17.3–17.6]

21. $\cos(-45°)$

22. $\sin 150°$

23. $\tan 135°$

24. $\cot(-495°)$

25. $\sec 330°$

26. $\csc(-225°)$

27. $\cos 180°$

28. $\sin 270°$

29. $\tan 90°$

Write each of the following in terms of the same trigonometric function of an acute angle. [17.3–17.6]

30. $\tan 100°$

31. $\sin 340°$

32. $\cos 210°$

Find two values of u between 0° and 360° for which each is true. [17.3]

33. $\sin u = -\dfrac{\sqrt{2}}{2}$

34. $\tan u = \dfrac{\sqrt{3}}{3}$

Find the value of each of the other five trigonometric functions of u. [17.7]

35. $\cos u = -\dfrac{12}{13}$ and $\tan u > 0$.

Find all values of θ to the nearest minute ($0° < \theta < 360°$). [17.8]

36. $\tan \theta = -.1688$

Express as a function of an angle measure less than 45°. [17.8]

37. $\cos 100°$

38. $\tan 322° 20'$

Determine, to the nearest minute, the values of $0° < \theta < 360°$ that make the equation true.

39. $-6 \cos \theta + 8 = 3 \cos \theta + 4$. [17.9].

Evaluate and simplify. [17.11]

40. $\dfrac{8 \cos 30° - \sin 45°}{-2 \sec(-45°)}$

41. Find the measure of each of the other two sides of right triangle ABC, with $m\angle C = 90$, to the nearest whole number. [17.12] $m\angle A = 40° 20'$, $b = 3.86$

Sketch the graph for $0° \leq x \leq 360°$. [18.1–18.5]

42. $y = 2 \sin x$

43. $y = \cos(-2x)$

44. $y = \frac{1}{2} \sin(2x)$

45. $y = \tan 2x$

46. Sketch $y = \cos 3x$ and $y = 2 \sin x$ for $-180° \leq x \leq 180°$ on the same set of axes. For how many values of x does $\cos 3x = 2 \sin x$? [18.4]

47. Find $\log \sin 34° 37'$. [18.6]

48. Find u, $0° \leq u \leq 90°$, if $\log \cos u = 9.6487 - 10$. [18.6]

49. Find the value of the expression using logarithms. [18.6]
$$\dfrac{3.95^5 \sqrt{.079}}{\tan 32° 18'}$$

50. Solve for θ to the nearest degree: $2 \cos^2 \theta - 3 \cos \theta - 5 = 0 (0° \leq \theta \leq 360°)$. [18.7]

51. Prove the identity $2 \cos^2 \theta - 1 = \dfrac{1 - \tan^2 \theta}{1 + \tan^2 \theta}$. [18.9]

52. Sketch $y = \frac{1}{2} \sin 3x$ and $y = \cos \frac{1}{2} x$ for $-\pi \leq x \leq 2\pi$ on the same set of axes. For how many values of x does $\frac{1}{2} \sin 3x = \cos \frac{1}{2} x$? [18.11]

53. Sketch the graph of $y = \text{Arc} \sin x$. [18.12]

Each refers to triangle ABC. [19.1–19.2]

54. $a = 7$, $c = 6$, $m\angle B = 38°$. Find b.

55. $a = 5$, $b = 6$, $c = 9$. Find $m\angle B$.

56. For $\triangle ABC$, find the area to the nearest square unit. [19.3] $a = 10$, $b = 12$, $m\angle C = 52° 30'$

57. Prove the identity $\tan x = \dfrac{\sin 2x}{1 + \cos 2x}$. [19.10]

58. Prove the identity [19.10]
$$\sin^2 \dfrac{x}{2} = \dfrac{\tan \frac{x}{2} \cdot \sin x}{2}.$$

Computer Section

WRITING AND RUNNING A PROGRAM

OBJECTIVE: **To test solutions of linear equations**

For you to communicate with another human being, both of you must understand the same language. Similarly, to communicate with a computer, both you and the computer must understand the same language. Computers are designed to understand a variety of programming languages. In this text, we will use a language used by many microcomputers—BASIC (Beginner's All-purpose Symbolic Instruction Code).

The computer understands BASIC because the BASIC language has been programmed into its circuitry. In this section of the book, you will learn to use BASIC in programming your computer to solve various mathematical problems.

A computer can do many things. It can store, retrieve, process, and evaluate information and data. It performs these tasks by carrying out a computer program. A program is simply a series of numbered instructions written in a programming language the computer understands. A program can be very long and complex or as simple as the one statement PRINT 2 + 5. This instruction tells the computer to add 2 and 5 and display the answer on the monitor (screen).

PRINT is probably the most important word in the computer's vocabulary. It is used for performing mathematical computations, displaying the results, displaying messages, and performing a variety of other functions.

The computer performs mathematical calculations, using the symbols +, −, *, and / for addition, subtraction, multiplication, and division, respectively. Parentheses ( ) are used to show grouping in mathematical expressions, and the caret ∧ is used to show exponentiation (raising to a power).

When carrying out mathematical calculations, the computer follows the rules for order of operations. Type in this one-line program just as you see it.

10 PRINT 2 + 3 * 5   Press the RETURN key after each program line.
RUN

If you typed in the program exactly as shown, the computer should have displayed the result 17 on the screen. Note that the computer followed the rule for doing multiplication before addition. Now type in the following program exactly as it appears.

```
10   PRINT   3+2*8-6*4
20   PRINT   (3+2)*8-6*4
30   PRINT   3+2*(8-6)*4
40   PRINT   (3+2*8-6)*4
```

Notice that typing in the program lines does not make the computer display any results. To make the computer carry out your program and display the results, you must type in the word RUN and press the RETURN key. Before you do that, however, calculate the results yourself. Then compare your answers with the computer's. If they are not the same, there are several possible explanations.

The first explanation is that you made a mathematical error. The second explanation is that you made a typing error while entering the program. To check, type in the word LIST and press the RETURN key. The computer will display a LISTing of your program. Check it carefully. Correcting malfunctioning programs is known as "debugging."

You can program the computer to perform more useful tasks. For example, you can program the computer to check solutions of linear equations you have found.

Consider the equation $2(x + 3) = x + 11$ and its solution, 5. To check that solution, you add the value 5 into both sides of the equation. If both sides yield the same value, then the solution was correct. The following program performs this checking process for you. Before entering it, however, type in the command NEW and press the RETURN key. NEW erases the previous program stored in memory.

```
10   PRINT   2*(5+3)     Notice that an * is needed before the ( ).
20   PRINT   5+11
```

When you RUN this program, the computer will display two numbers. If they are the same, then the solution is correct.

EXERCISE: Solve equations 2, 4, 6, and 8 on page 3. Then write and run a program that tests those solutions. [*Hint:* this program should have 8 lines, 2 for each equation.]

VARIABLES IN A COMPUTER PROGRAM

OBJECTIVE: **To check solutions of equations containing variables**

In BASIC, you can use variables in almost the same way you use them in algebra. The expression $X + 6$ means the same in both. However, $2X$ in algebra must be translated to $2 * X$ in BASIC since juxtaposition does not specify multiplication in BASIC as it does in algebra. It is necessary to use $*$ for multiplication in BASIC.

You may assign a value to a variable by using the BASIC word LET. LET $X = 4$ assigns the value 4 to the variable X. Note that the variable to which the value is assigned is to the left of the $=$. LET $4 = X$ is *not* the same as LET $X = 4$. Note also the END statement in line 40, which halts execution of the program.

The solution of the equation $3X + 8 = 2(X - 5)$ is $X = -18$. A computer program that tests this result is:

```
10    LET X =  - 18
20    PRINT  3*X+8
30    PRINT  2*(X-5)
40    END
```

If the two numbers printed out by this program are the same, then the solution is correct. If the solution is not correct, you don't have to type in both expressions again. You simply have to change line 10 by assigning another value (a new possible solution) to X.

EXERCISES:

1. Solve Exercises 3, 5, and 7 on page 3. Then write a computer program that tests your solutions.

2. Instead of solving the equation in Exercise 30 on page 3, write a program that will test each side of the equation. Use the program in this lesson as an example. Then guess a possible solution and type in a LET statement that assigns that value to the variable used in your program. Keep testing numbers until you find the correct solution. [*Hint:* The correct solution is a number between 10 and 20.]

3. Solve Exercise 28 on page 3. Then demonstrate that the solution is probably correct by running the program many times, each time using a different value for the variable.

GETTING DATA INTO A PROGRAM

OBJECTIVE: **To evaluate polynomials by using INPUT, READ, and DATA statements**

In previous programs, to change the value assigned to a variable, you had to rewrite the line that assigned the value.

There are two alternatives to this cumbersome process. The first is to use the INPUT command. INPUT instructs the computer to request a variable's value from the user when the program is run. The computer requests the value by printing a question mark and a blinking cursor.

```
10   INPUT X
20   PRINT X ^ 2 + 2 * X + 3
30   END
```

This program will ask the user for a value for X. Upon receiving it, the computer will insert the value into the polynomial $X^2 + 2X + 3$ and print the result. To evaluate the expression for a different value, you can run the program again. However, by adding the line 25 GOTO 10 the computer will return on its own to line 10. It will continue requesting new values and printing results until someone stops the program.

The second way to get values into the computer is by using the READ and DATA commands. READ X instructs the computer to look for a line that begins with the word DATA. The DATA line contains the values that are to be assigned to X. The READ statement stores those DATA values one at a time for use in the program. The DATA line may appear anywhere in a program prior to an END statement.

```
10   READ X
20   PRINT X ^ 2 + 2 * X + 3
30   GOTO 10
40   DATA  1,2,3,4,5
50   END
```

After all the values have been read, the computer will display an error message indicating there are no more numbers left. At this point, you do not need to be concerned about this type of error message.

EXERCISES:

1. Write a program using INPUT that evaluates Exercises 30–33 on page 40.

2. Replace the INPUT statement in the first program above with the READ and DATA statements. Place the numbers 1–10 in the DATA line.

MULTIPLYING BINOMIALS

OBJECTIVE:

To multiply binomials by using INPUT for entering their coefficients and constants

The PRINT statement can be used to print words, instructions, or expressions by enclosing them in quotes. We will see this illustrated in the program below in which the computer outputs the product of two binomials of the form $(AX + B)(CX + D)$.

```
10   INPUT A,B
20   INPUT C,D
30   PRINT A * C"X^2 + "A * D + C * B"X +"B * D
40   GOTO 10
50   END
```

If the binomials take the form $(AX + B)(CX - D)$, a little extra planning is needed to make their product appear correctly. See if you can pick out the strategy used to do so in the program below.

```
10   INPUT A,B
20   INPUT C,D
30   PRINT A * C"X^2";
40   IF A * D + C * D < 0 THEN 60
50   PRINT "+";
60   PRINT A * D + B * C"X"B * D
70   GOTO 10
80   END
```

In this version of the program, line 40 tests the middle term to see if it is negative. If it is, the computer goes to line 60 and prints the number with its attached minus sign. If the middle term is not negative, the computer goes to line 50 and prints a plus sign before printing the number itself in line 60. This procedure is necessary because the computer does not print signs with positive numbers. Also note the semicolons in lines 30 and 50. They are used to make the output of the following PRINT statements appear on the same line with no spaces in between.

EXERCISES:

1. Write a program similar to the one above to print the product of two binomials of the form $(AX - B)(CX - D)$.

2. Write a program similar to the previous ones that will output the product of two polynomials of the form $(AX + B)(CX^2 + DX + E)$.

3. See if you can use all the above ideas to write one program that outputs the product of two binomials regardless of the signs they contain.

CONDITIONAL BRANCHING

OBJECTIVE: **To test solutions of quadratic equations by using an IF. . .THEN statement**

In previous programs, you tested the solutions of an equation by evaluating each side of the equation. If the computed values of both sides were equal, then the entered solutions were correct. By using the IF. . .THEN command, the computer can make that determination itself.

In an IF. . .THEN command, you specify a comparison test in the IF portion such as IF A = B, IF A < B, IF A > B. Following THEN, you specify a line number or another command. If the comparison test following IF is true, then the computer will branch to the statement whose line number follows THEN, or will execute the command following THEN. If the test following IF is not true, then the computer executes the next line in the program.

Use the following program for testing solutions of $X^2 - 2X - 8 = 0$. Line 10 requests your solution. Line 30 tests the solution you entered. If it is correct, the computer branches to line 60, which prints the message confirming your answer. Otherwise, the computer executes line 40.

```
10    INPUT X
20    IF X =  - 999 THEN 80
30    IF X ^ 2 - 2 * X - 8 = 0
      THEN 60
40    PRINT "THAT SOLUTION WAS
      NOT CORRECT."
50    GOTO 10
60    PRINT "THAT WAS A CORRECT
      SOLUTION."
70    GOTO 10
80    END
```

Line 20 contains a flag directive (see below).

Line 50 returns to line 10 for correct solution.
Line 70 returns to line 10 for the second correct solution or the flag.

The IF. . .THEN statement in line 20 is called a *flag directive*. It is used to end a program that otherwise would continue indefinitely. After you have entered two correct solutions, you enter − 999. This number is the *flag* that line 20 specifies as the condition for branching to line 80, which terminates the program.

EXERCISES:

1. Write programs using the READ command that test solutions of the equations in Exercises 13, 34, and 35 on page 70.

2. Choose one of the equations from Exercises 55–57 on page 70 that you have not yet solved. Then write a program that tests solutions of it. After that, see if you can find both of its solutions by randomly inserting numbers, using an INPUT statement.

LOOPING

OBJECTIVE: **To change a fraction into a decimal using a computer program containing a FOR. . .NEXT loop and the INT function**

By using a FOR. . .NEXT loop, you can make the computer carry out a particular process any specified number of times. Such a loop begins with a FOR command, such as FOR H = 1 TO 6. The number following TO specifies the number of times the loop will be executed. The NEXT command, such as NEXT H is placed at the end of the loop. As an example, the following program prints a message 25 times.

```
10   FOR H = 1 TO 25
20   PRINT "COMPUTERS ARE FUN!"
30   NEXT H
40   END
```

You can use any numerical variable in the FOR. . .NEXT statements. The computer simply uses the specified memory storage place as a counting device. In the program on the next page, the loop is set up in line 170 and terminated in line 220. The program will convert a fraction into a decimal containing any specified number, P, of decimal places. Each execution of the loop determines one digit in the fraction's decimal conversion. The last decimal place is determined by a rounding procedure in the last portion of the program.

You will also use the BASIC function, INT, in this program. INT stands for the *greatest INTeger function*. INT(X) yields the greatest integer less than or equal to the value of X. For example, INT(2.3) = 2 and INT(−1.1) = −2. The INT function has a variety of uses. In the following program, INT is used in two different ways: 1) to determine individual digits in the quotient of a long division and 2) to round the last desired digit in such a quotient.

For example, in doing the division $3 \overline{)10}$ manually, the first number you would put into the quotient is 3. When the computer divides 3 into 10, however, it finds a quotient of 3.333. . . . So, to make the computer put one digit at a time into the quotient, you instruct the computer to apply INT to the quotient, which in this example is INT(10/3) = 3. In the next program, lines 140 and 190 carry out this process.

Lines 150 and 200 determine the remainder after each division has been performed. Multiplying the remainder by 10 (lines 180 and 230) is equivalent to "bringing down" the zero in long division.

Rounding the last desired digit is accomplished in line 240 by adding .5 to the quotient before applying INT to it.

Enter and run the following program. Use it to convert several fractions into decimals of varying lengths.

```
110    PRINT "NUMERATOR";: INPUT N
120    PRINT "DENOMINATOR";: INPUT D
130    PRINT "HOW MANY DECIMAL PLACES";: INPUT P
140 A =  INT (N / D)
150 R = N - A * D
160    PRINT N"/"D" = "A".";
170    FOR I = 1 TO P - 1
180 N = 4 * 10
190 A =  INT (N / D)
200 R = N - A * D
210    PRINT A;
220    NEXT I
230 N = R * 10
240 A =  INT (N / D + .5)
250    PRINT A
260    GOTO 110
270    END
```

In lines 140, 150, 180, 190, 200, 230, and 240, note that the word LET is unnecessary in assignment statements.

See page 110 for another program that uses a FOR. . .NEXT loop and involves decimals and fractions.

EXERCISES:

1. The program above will diligently print out the decimal to the requested number of places even if the decimal terminates and you get a string of zeros. Make the necessary change in the program to make it stop printing as soon as the remainder is zero.

2. The variable in the loop need not be only a counter. It can be a value that is actually used in the program. For example, the following program will print out the multiples of 6 from 6 to 72 as the value of I goes from 1 to 12.

```
10    FOR I = 1 TO 12
20    PRINT I * 6
30    NEXT I
40    END
```

Write a program using a FOR. . .NEXT loop that will print out the squares and cubes of the integers from 1 to 25.

3. Write a program that will ask a person how old he or she is and then wish her or him a happy birthday that many times.

SQUARE ROOTS AND ABSOLUTE VALUE

OBJECTIVE: **To simplify radicals that utilize the BASIC functions SQR and ABS**

Two new BASIC functions that closely parallel mathematical operations are SQR( ), which calculates the square root of the number in parentheses, and ABS( ), which finds the absolute value of the number in parentheses. These two functions are utilized in the following program that simplifies radicals such as $\sqrt{8}$, which equals $2\sqrt{2}$.

First, line 130 of the program tests the quantity to see if it is a perfect square. If it is not a perfect square, the FOR...NEXT loop in lines 140 through 170 is executed. Line 150 tests the number to see if it contains any perfect squares as factors, beginning with 2^2 and continuing on with 3^2, 4^2, 5^2, etc. When a perfect square is found to be a factor, line 160 removes it from the radical. This process continues until all possible perfect squares have been tested.

Before continuing, a word of caution is in order. On occasion, the computer will output the result of a computation that is incorrect by a small amount. This is particularly likely to happen when performing calculations that involve exponents or the SQR function. These small errors are called *rounding errors*. Rounding errors result from the computer converting base-ten numerals into base-two numerals, performing computations on those base-two numerals, and converting the result back into base-ten numerals. For example, in computing the square root of 9, the computer may output a result like 2.99999999 or 3.00000002, instead of the exact answer, 3. There are techniques for compensating for the rounding errors. However, they will not be covered in this lesson. (See page 162 for more about rounding errors.)

Bearing in mind that rounding errors may occur in the computer's calculations, type in the following program and RUN it.

```
100   PRINT "WHAT IS THE RADICAND TO BE SIMPLIFIED?";
      : INPUT R
110 C = 1:S = R
120   IF  SQR (S) =  INT ( SQR (S)) THEN 190
130   FOR I = 2 TO  SQR (S)
140   IF S / (I ^ 2) <  >  INT (S / (I ^ 2)) THEN 160
150 C = C * I:S = S / (I ^ 2): GOTO 140
160   NEXT I
170   PRINT "SQR ("R") = "C" * SQR ("S")"
180   GOTO 100
190   PRINT "SQR ("S") = " SQR (S)
200   GOTO 100
210   END
```

EXERCISES:

1. Change the program on page 582 so that it will simplify a cube root instead of a square root. [*Hint:* There is no BASIC function for a cube root. However, cube roots can be found by using exponentiation to raise a number to the 1/3 power.]

2. Alter the program so that it will simplify a fourth root.

3. Combine all the ideas above into one program that will simplify a radical of any index. After you have attempted to write, debug, and RUN your program, compare it with the program on page 162.

It is very convenient to use the ABS function in programs that simplify odd roots. Consider line 130 in the original program. It sets loop parameters of 2 and the square root of the number in S. In a program that simplifies cube roots, however, substituting the cube root of S for its square root in line 130 will not set up the loop correctly when S is negative. In that case, line 130 must read:

$$130 \text{ FOR I} = -2 \text{ TO (S} \wedge (1/3))$$

However, this line will not work for positive numbers in S. Therefore, an easy way to set up the loop for all possible cube roots is by using ABS in the following line 130:

$$130 \text{ FOR I} = 2 \text{ TO ABS(S} \wedge (1/3))$$

4. If you didn't input negative numbers into the programs you wrote previously, go back and rewrite your programs (if necessary), using the ABS feature noted above. Then try inputting some negative numbers.

5. Write a program that makes the computer print out a table of squares, cubes, square roots, and cube roots of the numbers from 1 to 10.

6. Write a program that uses the quadratic formula to find the solutions of a quadratic equation. Use INPUT for entering values for *a*, *b*, and *c*. Use the INT function to round the answers you find to the nearest hundredth.

ON...GOTO

OBJECTIVE: **To evaluate powers of i using an ON GOTO statement**

Thus far in this text, you have examined unconditional branching, using the GOTO statement, and conditional branching using the IF...THEN statement. In this lesson, we will introduce you to another form of conditional branching using the ON...GOTO statement. ON...GOTO gives the computer the option of going to any of several program lines depending on the appearance of any of several specified conditions.

The format for the ON...GOTO statement is as follows:

ON X GOTO 100, 200, 150, 80

In this ON...GOTO statement, the computer will branch to line 100 when $X = 1$. When $X = 2$, it will branch to line 200. When $X = 3$, it will branch to line 150, and when $X = 4$, to line 80. The ON...GOTO statement requires that values of X (or any variable following ON) be natural numbers, beginning with 1. Note that the order of the line numbers following GOTO must correspond to the branching desired as $X = 1$, $X = 2$, $X = 3$, etc.

We will now use ON...GOTO in writing a program that evaluates powers of i. You know that you determine the value of a power of i by dividing the exponent by 4 and checking the remainder.

Note that the program below uses an ON...GOTO statement that places the value of the remainder in R. Because ON...GOTO requires the value of R to begin with 1, a program step was added that places 4 in R when the remainder equals 0.

```
110   PRINT "WHAT IS THE POWER OF i";: INPUT P
120 N =  INT (P / 4):R = P - N * 4
130   IF R = 0 THEN R = 4
140   PRINT "i^"P" = ";
150   ON R GOTO 160,170,180,190
160   PRINT "i": GOTO 200
170   PRINT  - 1: GOTO 200
180   PRINT "-i": GOTO 200
190   PRINT 1
200   PRINT : GOTO 110
```

EXERCISE: Write a program that multiplies two complex numbers. Use Exercises 13 to 20 on page 192 as test problems.

RANDOM NUMBERS

OBJECTIVE: **To evaluate 2 × 2 determinants generated by the computer randomly**

Programming a computer to pick random integers within specific parameters is a good example of manipulating the fundamental capabilities of a computer so that it can perform a broad range of applications. For example, the BASIC function RND generates 9–digit decimal numbers between .000000000 and .999999999. These are not very useful numbers.

However, the statement X = INT (RND(1) * 10) + 1 instructs the computer to choose whole numbers from 1 to 10. This statement modifies the RND function as follows:

1. RND(1) generates a number between .000000000 and .999999999.
2. Multiplying that number by 10 yields a number between 0.00000000 and 9.99999999.
3. Applying INT to the above product yields an integer between 0 and 9.
4. Adding 1 to that generated integer yields an integer between 1 and 10.

In general, the formula X = INT (RND(1) * (B − A + 1)) + A generates random integers from A to B inclusive.

In the program below, the computer is instructed to generate integers from −5 to +5 inclusive. (Can you follow the reasoning?) Enter and RUN the following program that tests your ability to evaluate 2 × 2 determinants.

```
110   FOR I = 1 TO 4
120   A(I) =   INT ( RND (1) * 11) - 5: NEXT I
130   PRINT "WHAT IS THE VALUE OF THE DETERMINANT?"
140   PRINT   TAB( 10)A(1) TAB( 15)A(2)
150   PRINT : PRINT   TAB( 10)A(3) TAB( 15)A(4)
160 V = A(1) * A(4) - A(2) * A(3): INPUT D
170   IF V = D THEN 190
180   PRINT "NO. THE ANSWER IS "V: GOTO 110
190   PRINT "YES. THAT WAS RIGHT."
200   FOR W = 1 TO 3000: NEXT W: GOTO 110
```

EXERCISES:

1. Rewrite the program above so that it generates integers from −9 to +9 inclusive.

2. Using the program above as a basis, write and RUN a program that evaluates 3 × 3 determinants.

EXPONENTIAL EQUATIONS

OBJECTIVE: **To solve exponential equations of the form $a^x = b$, using a FOR...NEXT loop**

You can solve any equation by simply inserting values into it until you find the value that works. Even when the solution of an equation is irrational, you can find a decimal solution that is accurate to as many decimal places as desired.

The computer program below finds solutions of equations of the form $a^x = b$ by inserting values into each given equation systematically. First, it begins by using a FOR...STEP...NEXT loop (lines 160–190) to insert integers from -10 to $+10$ into a given equation one at a time. When two consecutive integers, call them p and $p + 1$, are found such that the value of b lies between a^p and a^{p+1}, the program then uses the same FOR...STEP...NEXT loop to insert values from p to $p + 1$, incremented by .1 (as specified in line 210), into the given equation one at a time. When two successive numbers, call them q and $(q + .1)$, are found where $a^q \leq b \leq a^{q+.1}$, the loop plugs values from q to $(q + .1)$, incremented by .01, into the given equation one at a time. The program continues this process until it finds an exact solution, or a solution accurate to the number of decimal places you specified.

```
110   PRINT "WHAT IS A";: INPUT A
120   PRINT "WHAT IS B";: INPUT B
130   PRINT "HOW MANY DECIMAL PLACES (LIMIT 8 PLEASE)
      ";: INPUT N
140 L =   - 10:U = 10:C = 1:D = A ^ (L - C)
150   FOR M = 1 TO N + 1
160   FOR I = L TO U STEP C
170 E = A ^ I: IF B - D = 0 THEN 230
180   IF (B - D) * (B - E) < 0 THEN 200
190 D = E: NEXT I
200   PRINT I - C" , "I
210 L = I - C:U = I:C = C / 10: NEXT M
220   GOTO 110
230   PRINT "X = "I - C: GOTO 110
```

EXERCISES:

1. Write a program that uses logarithms to solve $a^x = b$. [The BASIC function for the natural logarithm is LOG( ).]

2. Use the technique in this lesson's program to "search for" an irrational solution to a quadratic equation.

ARITHMETIC AND GEOMETRIC PROGRESSIONS

OBJECTIVE: **To find *n* terms and the sum of *n* terms of an A.P. or a G.P.**

Below are two programs that deal with progressions. In the first program, you enter the first three terms in a progression. Then you request the number of terms you wish to have printed out and added. In the second program, the computer generates the first three terms in a progression. Then it requests that you compute a given term in the progression, and the sum of terms through the given term. Take turns with both programs.

Program 1

```
100   INPUT "ENTER THE FIRST THREE TERMS OF THE
         PROGRESSION ";A,B,C
110   INPUT "HOW MANY TERMS DO YOU WANT ? ";N
120   IF B / A = C / B THEN 150
130   PRINT "THIS IS NOT A G.P."
140   PRINT : GOTO 100
150   PRINT A" , ";
160   FOR I = 1 TO N - 1
170   PRINT A * (B / A) ^ I" , "; NEXT I
180   PRINT : PRINT "THE SUM OF "N" TERMS IS ";
190   PRINT (A - A * (B / A) ^ N) / (1 - B / A)
200   PRINT : GOTO 100: END
```

Program 2

```
110 A =   INT ( RND (1) * 30) - 15
120 D =   INT ( RND (1) * 7) - 2: IF D = 0 THEN 120
130   PRINT A" , "A + D" , "A + D + D
140   FOR I = 1 TO 5
150 N =   INT ( RND (1) * 30) + 4
160   PRINT "WHAT IS THE "N"TH TERM?": INPUT M
170   IF M = A + (N - 1) * D THEN 190
180   PRINT "NO, THE ANSWER IS "A + (N - 1) * D".":
         GOTO 200
190   PRINT "VERY GOOD"
200   PRINT "WHAT IS THE SUM OF "N" TERMS?":
         INPUT M
210   IF M = N / 2 * (2 * A + (N - 1) * D) THEN 240
220   PRINT "NO. THE SUM OF "N" TERMS IS ";
         GOTO 250
230   PRINT N / 2 * (2 * A + (N - 1) * D)".":
240   PRINT "VERY GOOD"
250   NEXT I
```

Computations Using Logarithms

Objective
To approximate products, quotients, powers, and radicals using common logarithms

If x and y are positive rational numbers and n is a positive integer, you can approximate xy, $\frac{x}{y}$, x^n, and $\sqrt[n]{x}$ to three significant digits using the laws of logarithms and a table of common logarithms. The procedure is shown in the examples below.

Example 1

Find 0.000245×3170 to three significant digits. Use logarithms.

$\log(0.000245 \times 3170)$ ◀ *log (x · y)*
$\log 0.000245 + \log 3170$ ◀ *log x + log y*
$6.3892 - 10 + 3.5011$
$9.8903 - 10$ ◀ *The mantissa .8903 is between .8899 and .8904 in the table.*

$9.8904 - 10$ ◀ *Use .8904, the nearer mantissa.*
Antilog$(9.8904 - 10) =$ antilog$(-1 + .8904) = 10^{-1} \times 7.77 = 0.777$
Thus, $0.000245 \times 3170 = 0.777$ to three significant digits.

Example 2

Find $\frac{617}{9.13}$ and $\frac{61.7}{0.0913}$ to three significant digits.

Use $\log \frac{x}{y} = \log x - \log y$.

$\log \frac{617}{9.13} = \log 617 - \log 9.13 = 2.7903 - 0.9605$ ◀ $\begin{array}{r} 2.7903 \\ (-)\ 0.9605 \\ \hline 1.8298 \end{array}$
$= 1.8298$
$\doteq 1.8299$ ◀ *Nearest mantissa is .8299.*
Antilog $1.8299 = 10^1 \times 6.76 = 67.6$

$\log \frac{61.7}{0.0913} = \log 61.7 - \log 0.0913 = 1.7903 - (8.9605 - 10)$

1.7903 − 8.9605 will not be a positive mantissa. ▶ $(-)\ \dfrac{8.9605 - 10}{}$ $1.7903 \longrightarrow 11.7903 - 10$ ◀ *Write the characteristic 1 as 11 − 10.*
$(-)\ \dfrac{8.9605 - 10}{2.8298}$

Antilog $2.8298 = 10^2 \times 6.76 = 676$
Thus, $\frac{617}{9.13} = 67.6$ and $\frac{61.7}{0.0913} = 676$ to three significant digits.

Example 3

Find $(0.0741)^3$ to three significant digits. Use logs.

log aʳ ▶ $\log(0.0741)^3$
r · log a ▶ $3 \cdot \log 0.0741 = 3(8.8698 - 10) = 26.6094 - 30$
Antilog$(26.6094 - 30) =$ antilog$(-4 + .6094) = 10^{-4} \times 4.07$
Thus, $(0.0741)^3 = 0.000407$ to three significant digits.

Example 4   Find $\sqrt{7,810}$ and $\sqrt[3]{0.0532}$ to three significant digits.

Use $\log \sqrt[n]{x} = \log x^{\frac{1}{n}} = \frac{1}{n} \cdot \log x = \frac{\log x}{n}$.

$\log \sqrt{7,810} = \frac{1}{2} \cdot \log 7,810 = \frac{3.8927}{2} = 1.94635$
Antilog $1.9465 = 10^1 \times 8.84 = 88.4$

$\log \sqrt[3]{0.0532} = \frac{1}{3} \cdot \log 0.0532 = \frac{8.7259 - 10}{3}$ ◄ *10 is not divisible*
by 3. Characteristic

$= \frac{28.7259 - 30}{3}$ ◄ *must be an integer.*
Use 28 − 30 for 8 − 10.
30 is divisible by 3.

$= 9.5753 - 10$
Antilog $(9.5753 - 10) = 10^{-1} \times 3.76 = 0.376$

Thus, $\sqrt{7,810} = 88.4$ and $\sqrt[3]{0.0532} = 0.376$.

Reading in Algebra

Given log x and log y, rewrite log x so that the mantissa of log x − log y will be positive.

1. 1.7900
$(-)$ $\underline{8.9600 - 10}$

2. 0.2200
$(-)$ $\underline{0.5100}$

3. $7.5800 - 10$
$(-)$ $\underline{7.7400 - 10}$

4. $5.4400 - 10$
$(-)$ $\underline{6.2200}$

Given log x and n, rewrite log x so that the characteristic of $\dfrac{\log x}{n}$ will be an integer.

5. $\dfrac{8.4400 - 10}{3}$

6. $\dfrac{9.3200 - 10}{4}$

7. $\dfrac{7.4800 - 10}{6}$

Written Exercises

Find the value of each expression to three significant digits. Use logs.

Ⓐ 1. $39.2 \times 53,800$

2. 0.00286×64.3

3. 0.537×0.0806

4. $\dfrac{3,080}{17.6}$

5. $\dfrac{4.52}{63.8}$

6. $\dfrac{0.0236}{0.764}$

7. $(49.4)^3$

8. $(0.723)^4$

9. $(0.0871)^2$

10. $\sqrt{5,140}$

11. $\sqrt[3]{19,400}$

12. $\sqrt[4]{82.1}$

13. $\sqrt{0.00592}$

14. $\sqrt[3]{0.0514}$

15. $\sqrt[3]{0.0736}$

Ⓑ 16. $\dfrac{4.25 \times 3,750}{543}$

17. $\dfrac{9,180}{2.37 \times 30.4}$

18. $\dfrac{0.263 \times 704}{92.2 \times 0.081}$

19. $(18.9 \times 4.04)^3$

20. $\sqrt[4]{0.263 \times 41.8}$

21. $\sqrt[3]{(502)^2}$

22. $\sqrt{\dfrac{735}{3,420}}$

23. $\left(\dfrac{0.756}{1.83}\right)^4$

24. $\sqrt[3]{\dfrac{(0.813)^2}{283 \times 0.0461}}$

Computations Using Logarithms

Squares and Square Roots

No.	Sq.	Sq. Root	No.	Sq.	Sq. Root
1	1	1.000	51	2,601	7.141
2	4	1.414	52	2,704	7.211
3	9	1.732	53	2,809	7.280
4	16	2.000	54	2,916	7.348
5	25	2.236	55	3,025	7.416
6	36	2.449	56	3,136	7.483
7	49	2.646	57	3,249	7.550
8	64	2.828	58	3,364	7.616
9	81	3.000	59	3,481	7.681
10	100	3.162	60	3,600	7.746
11	121	3.317	61	3,721	7.810
12	144	3.464	62	3,844	7.874
13	169	3.606	63	3,969	7.937
14	196	3.742	64	4,096	8.000
15	225	3.873	65	4,225	8.062
16	256	4.000	66	4,356	8.124
17	289	4.123	67	4,489	8.185
18	324	4.243	68	4,624	8.246
19	361	4.359	69	4,761	8.307
20	400	4.472	70	4,900	8.357
21	441	4.583	71	5,041	8.426
22	484	4.690	72	5,184	8.485
23	529	4.796	73	5,329	8.544
24	576	4.899	74	5,476	8.602
25	625	5.000	75	5,625	8.660
26	676	5.099	76	5,776	8.718
27	729	5.196	77	5,929	8.775
28	784	5.292	78	6,084	8.832
29	841	5.385	79	6,241	8.888
30	900	5.477	80	6,400	8.944
31	961	5.568	81	6,561	9.000
32	1,024	5.657	82	6,724	9.055
33	1,089	5.745	83	6,889	9.110
34	1,156	5.831	84	7,056	9.165
35	1,225	5.916	85	7,225	9.220
36	1,296	6.000	86	7,396	9.274
37	1,369	6.083	87	7,569	9.327
38	1,444	6.164	88	7,744	9.381
39	1,521	6.245	89	7,921	9.434
40	1,600	6.325	90	8,100	9.487
41	1,681	6.403	91	8,281	9.539
42	1,764	6.481	92	8,464	9.592
43	1,849	6.557	93	8,649	9.644
44	1,936	6.633	94	8,836	9.695
45	2,025	6.708	95	9,025	9.747
46	2,116	6.782	96	9,216	9.798
47	2,209	6.856	97	9,409	9.849
48	2,304	6.928	98	9,604	9.899
49	2,401	7.000	99	9,801	9.950
50	2,500	7.071	100	10,000	10.000

* Logarithms of Trigonometric Functions

* Subtract 10 from each entry in this table to obtain the proper logarithm of the indicated trigonometric function.

Angle	L Sin	L Tan	L Cot	L Cos	Angle
0° 0'				10.0000	90° 0'
10'	7.4637	7.4637	12.5363	.0000	50'
20'	.7648	.7648	.2352	.0000	40'
30'	7.9408	7.9409	12.0591	.0000	30'
40'	8.0658	8.0658	11.9342	.0000	20'
50'	.1627	.1627	.8373	.0000	10'
1° 0'	8.2419	8.2419	11.7581	9.9999	89° 0'
10'	.3088	.3089	.6911	.9999	50'
20'	.3668	.3669	.6331	.9999	40'
30'	.4179	.4181	.5819	.9999	30'
40'	.4637	.4638	.5362	.9998	20'
50'	.5050	.5053	.4947	.9998	10'
2° 0'	8.5428	8.5431	11.4569	9.9997	88° 0'
10'	.5776	.5779	.4221	.9997	50'
20'	.6097	.6101	.3899	.9996	40'
30'	.6397	.6401	.3599	.9996	30'
40'	.6677	.6682	.3318	.9995	20'
50'	.6940	.6945	.3055	.9995	10'
3° 0'	8.7188	8.7194	11.2806	9.9994	87° 0'
10'	.7423	.7429	.2571	.9993	50'
20'	.7645	.7652	.2348	.9993	40'
30'	.7857	.7865	.2135	.9992	30'
40'	.8059	.8067	.1933	.9991	20'
50'	.8251	.8261	.1739	.9990	10'
4° 0'	8.8436	8.8446	11.1554	9.9989	86° 0'
10'	.8613	.8624	.1376	.9989	50'
20'	.8783	.8795	.1205	.9988	40'
30'	.8946	.8960	.1040	.9987	30'
40'	.9104	.9118	.0882	.9986	20'
50'	.9256	.9272	.0728	.9985	10'
5° 0'	8.9403	8.9420	11.0580	9.9983	85° 0'
10'	.9545	.9563	.0437	.9982	50'
20'	.9682	.9701	.0299	.9981	40'
30'	.9816	.9836	.0164	.9980	30'
40'	.9945	.9966	11.0034	.9979	20'
50'	9.0070	9.0093	10.9907	.9977	10'
6° 0'	9.0192	9.0216	10.9784	9.9976	84° 0'
10'	.0311	.0336	.9664	.9975	50'
20'	.0426	.0453	.9547	.9973	40'
30'	.0539	.0567	.9433	.9972	30'
40'	.0648	.0678	.9322	.9971	20'
50'	.0755	.0786	.9214	.9969	10'
7° 0'	9.0859	9.0891	10.9109	9.9968	83° 0'
10'	.0961	.0995	.9005	.9966	50'
20'	.1060	.1096	.8904	.9964	40'
30'	.1157	.1194	.8806	.9963	30'
40'	.1252	.1291	.8709	.9961	20'
50'	.1345	.1385	.8615	.9959	10'
8° 0'	9.1436	9.1478	10.8522	9.9958	82° 0'
10'	.1525	.1569	.8431	.9956	50'
20'	.1612	.1658	.8342	.9954	40'
30'	.1697	.1745	.8255	.9952	30'
40'	.1781	.1831	.8169	.9950	20'
50'	.1863	.1915	.8085	.9948	10'
9° 0'	9.1943	9.1997	10.8003	9.9946	81° 0'
	L Cos	L Tan	L Cot	L Sin	Angle

Logarithms of Trigonometric Functions

Angle	L Sin	L Tan	L Cot	L Cos		Angle
9° 0'	9.1943	9.1997	10.8003	9.9946		81° 0'
10'	.2022	.2078	.7922	.9944		50'
20'	.2100	.2158	.7842	.9942		40'
30'	.2176	.2236	.7764	.9940		30'
40'	.2251	.2313	.7687	.9938		20'
50'	.2324	.2389	.7611	.9936		10'
10° 0'	9.2397	9.2463	10.7537	9.9934		80° 0'
10'	.2468	.2536	.7464	.9931		50'
20'	.2538	.2609	.7391	.9929		40'
30'	.2606	.2680	.7320	.9927		30'
40'	.2674	.2750	.7250	.9924		20'
50'	.2740	.2819	.7181	.9922		10'
11° 0'	9.2806	9.2887	10.7113	9.9919		79° 0'
10'	.2870	.2953	.7047	.9917		50'
20'	.2934	.3020	.6980	.9914		40'
30'	.2997	.3085	.6915	.9912		30'
40'	.3058	.3149	.6851	.9909		20'
50'	.3119	.3212	.6788	.9907		10'
12° 0'	9.3179	9.3275	10.6725	9.9904		78° 0'
10'	.3238	.3336	.6664	.9901		50'
20'	.3296	.3397	.6603	.9899		40'
30'	.3353	.3458	.6542	.9896		30'
40'	.3410	.3517	.6483	.9893		20'
50'	.3466	.3576	.6424	.9890		10'
13° 0'	9.3521	9.3634	10.6366	9.9887		77° 0'
10'	.3575	.3691	.6309	.9884		50'
20'	.3629	.3748	.6252	.9881		40'
30'	.3682	.3804	.6196	.9878		30'
40'	.3734	.3859	.6141	.9875		20'
50'	.3786	.3914	.6086	.9872		10'
14° 0'	9.3837	9.3968	10.6032	9.9869		76° 0'
10'	.3887	.4021	.5979	.9866		50'
20'	.3937	.4074	.5926	.9863		40'
30'	.3986	.4127	.5873	.9859		30'
40'	.4035	.4178	.5822	.9856		20'
50'	.4083	.4230	.5770	.9853		10'
15° 0'	9.4130	9.4281	10.5719	9.9849		75° 0'
10'	.4177	.4331	.5669	.9846		50'
20'	.4223	.4381	.5619	.9843		40'
30'	.4269	.4430	.5570	.9839		30'
40'	.4314	.4479	.5521	.9836		20'
50'	.4359	.4527	.5473	.9832		10'
16° 0'	9.4403	9.4575	10.5425	9.9828		74° 0'
10'	.4447	.4622	.5378	.9825		50'
20'	.4491	.4669	.5331	.9821		40'
30'	.4533	.4716	.5284	.9817		30'
40'	.4576	.4762	.5238	.9814		20'
50'	.4618	.4808	.5192	.9810		10'
17° 0'	9.4659	9.4853	10.5147	9.9806		73° 0'
10'	.4700	.4898	.5102	.9802		50'
20'	.4741	.4943	.5057	.9798		40'
30'	.4781	.4987	.5013	.9794		30'
40'	.4821	.5031	.4969	.9790		20'
50'	.4861	.5075	.4925	.9786		10'
18° 0'	9.4900	9.5118	10.4882	9.9782		72° 0'
	L Cos	L Cot	L Tan	L Sin		Angle

Logarithms of Trigonometric Functions

Angle	L Sin	L Tan	L Cot	L Cos		Angle
18° 0'	9.4900	9.5118	10.4882	9.9782		72° 0'
10'	.4939	.5161	.4839	.9778		50'
20'	.4977	.5203	.4797	.9774		40'
30'	.5015	.5245	.4755	.9770		30'
40'	.5052	.5287	.4713	.9765		20'
50'	.5090	.5329	.4671	.9761		10'
19° 0'	9.5126	9.5370	10.4630	9.9757		71° 0'
10'	.5163	.5411	.4589	.9752		50'
20'	.5199	.5451	.4549	.9748		40'
30'	.5235	.5491	.4509	.9743		30'
40'	.5270	.5531	.4469	.9739		20'
50'	.5306	.5571	.4429	.9734		10'
20° 0'	9.5341	9.5611	10.4389	9.9730		70° 0'
10'	.5375	.5650	.4350	.9725		50'
20'	.5409	.5689	.4311	.9721		40'
30'	.5443	.5727	.4273	.9716		30'
40'	.5477	.5766	.4234	.9711		20'
50'	.5510	.5804	.4196	.9706		10'
21° 0'	9.5543	9.5842	10.4158	9.9702		69° 0'
10'	.5576	.5879	.4121	.9697		50'
20'	.5609	.5917	.4083	.9692		40'
30'	.5641	.5954	.4046	.9687		30'
40'	.5673	.5991	.4009	.9682		20'
50'	.5704	.6028	.3972	.9677		10'
22° 0'	9.5736	9.6064	10.3936	9.9672		68° 0'
10'	.5767	.6100	.3900	.9667		50'
20'	.5798	.6136	.3864	.9661		40'
30'	.5828	.6172	.3828	.9656		30'
40'	.5859	.6208	.3792	.9651		20'
50'	.5889	.6243	.3757	.9646		10'
23° 0'	9.5919	9.6279	10.3721	9.9640		67° 0'
10'	.5948	.6314	.3686	.9635		50'
20'	.5978	.6348	.3652	.9629		40'
30'	.6007	.6383	.3617	.9624		30'
40'	.6036	.6417	.3583	.9618		20'
50'	.6065	.6452	.3548	.9613		10'
24° 0'	9.6093	9.6486	10.3514	9.9607		66° 0'
10'	.6121	.6520	.3480	.9602		50'
20'	.6149	.6553	.3447	.9596		40'
30'	.6177	.6587	.3413	.9590		30'
40'	.6205	.6620	.3380	.9584		20'
50'	.6232	.6654	.3346	.9579		10'
25° 0'	9.6259	9.6687	10.3313	9.9573		65° 0'
10'	.6286	.6720	.3280	.9567		50'
20'	.6313	.6752	.3248	.9561		40'
30'	.6340	.6785	.3215	.9555		30'
40'	.6366	.6817	.3183	.9549		20'
50'	.6392	.6850	.3150	.9543		10'
26° 0'	9.6418	9.6882	10.3118	9.9537		64° 0'
10'	.6444	.6914	.3086	.9530		50'
20'	.6470	.6946	.3054	.9524		40'
30'	.6495	.6977	.3023	.9518		30'
40'	.6521	.7009	.2991	.9512		20'
50'	.6546	.7040	.2960	.9505		10'
27° 0'	9.6570	9.7072	10.2928	9.9499		63° 0'
	L Cos	L Cot	L Tan	L Sin		Angle

Logarithms of Trigonometric Functions

Angle	L Sin	L Tan	L Cot	L Cos	Angle
36° 0'	9.7692	9.8613	10.1387	9.9080	54° 0'
10'	.7710	.8639	.1361	.9070	50'
20'	.7727	.8666	.1334	.9061	40'
30'	.7744	.8692	.1308	.9052	30'
40'	.7761	.8718	.1282	.9042	20'
50'	.7778	.8745	.1255	.9033	10'
37° 0'	9.7795	9.8771	10.1229	9.9023	53° 0'
10'	.7811	.8797	.1203	.9014	50'
20'	.7828	.8824	.1176	.9004	40'
30'	.7844	.8850	.1150	.8995	30'
40'	.7861	.8876	.1124	.8985	20'
50'	.7877	.8902	.1098	.8975	10'
38° 0'	9.7893	9.8928	10.1072	9.8965	52° 0'
10'	.7910	.8954	.1046	.8955	50'
20'	.7926	.8980	.1020	.8945	40'
30'	.7941	.9006	.0994	.8935	30'
40'	.7957	.9032	.0968	.8925	20'
50'	.7973	.9058	.0942	.8915	10'
39° 0'	9.7989	9.9084	10.0916	9.8905	51° 0'
10'	.8004	.9110	.0890	.8895	50'
20'	.8020	.9135	.0865	.8884	40'
30'	.8035	.9161	.0839	.8874	30'
40'	.8050	.9187	.0813	.8864	20'
50'	.8066	.9212	.0788	.8853	10'
40° 0'	9.8081	9.9238	10.0762	9.8843	50° 0'
10'	.8096	.9264	.0736	.8832	50'
20'	.8111	.9289	.0711	.8821	40'
30'	.8125	.9315	.0685	.8810	30'
40'	.8140	.9341	.0659	.8800	20'
50'	.8155	.9366	.0634	.8789	10'
41° 0'	9.8169	9.9392	10.0608	9.8778	49° 0'
10'	.8184	.9417	.0583	.8767	50'
20'	.8198	.9443	.0557	.8756	40'
30'	.8213	.9468	.0532	.8745	30'
40'	.8227	.9494	.0506	.8733	20'
50'	.8241	.9519	.0481	.8722	10'
42° 0'	9.8255	9.9544	10.0456	9.8711	48° 0'
10'	.8269	.9570	.0430	.8699	50'
20'	.8283	.9595	.0405	.8688	40'
30'	.8297	.9621	.0379	.8676	30'
40'	.8311	.9646	.0354	.8665	20'
50'	.8324	.9671	.0329	.8653	10'
43° 0'	9.8338	9.9697	10.0303	9.8641	47° 0'
10'	.8351	.9722	.0278	.8629	50'
20'	.8365	.9747	.0253	.8618	40'
30'	.8378	.9772	.0228	.8606	30'
40'	.8391	.9798	.0202	.8594	20'
50'	.8405	.9823	.0177	.8582	10'
44° 0'	9.8418	9.9848	10.0152	9.8569	46° 0'
10'	.8431	.9874	.0126	.8557	50'
20'	.8444	.9899	.0101	.8545	40'
30'	.8457	.9924	.0076	.8532	30'
40'	.8469	.9949	.0051	.8520	20'
50'	.8482	.9975	.0025	.8507	10'
45° 0'	9.8495	10.0000	10.0000	9.8495	45° 0'
	L Cos	L Cot	L Tan	L Sin	Angle

Logarithms of Trigonometric Functions

Angle	L Sin	L Tan	L Cot	L Cos	Angle
27° 0'	9.6570	9.7072	10.2928	9.9499	63° 0'
10'	.6595	.7103	.2897	.9492	50'
20'	.6620	.7134	.2866	.9486	40'
30'	.6644	.7165	.2835	.9479	30'
40'	.6668	.7196	.2804	.9473	20'
50'	.6692	.7226	.2774	.9466	10'
28° 0'	9.6716	9.7257	10.2743	9.9459	62° 0'
10'	.6740	.7287	.2713	.9453	50'
20'	.6763	.7317	.2683	.9446	40'
30'	.6787	.7348	.2652	.9439	30'
40'	.6810	.7378	.2622	.9432	20'
50'	.6833	.7408	.2592	.9425	10'
29° 0'	9.6856	9.7438	10.2562	9.9418	61° 0'
10'	.6878	.7467	.2533	.9411	50'
20'	.6901	.7497	.2503	.9404	40'
30'	.6923	.7526	.2474	.9397	30'
40'	.6946	.7556	.2444	.9390	20'
50'	.6968	.7585	.2415	.9383	10'
30° 0'	9.6990	9.7614	10.2386	9.9375	60° 0'
10'	.7012	.7644	.2356	.9368	50'
20'	.7033	.7673	.2327	.9361	40'
30'	.7055	.7701	.2299	.9353	30'
40'	.7076	.7730	.2270	.9346	20'
50'	.7097	.7759	.2241	.9338	10'
31° 0'	9.7118	9.7788	10.2212	9.9331	59° 0'
10'	.7139	.7816	.2184	.9323	50'
20'	.7160	.7845	.2155	.9315	40'
30'	.7181	.7873	.2127	.9308	30'
40'	.7201	.7902	.2098	.9300	20'
50'	.7222	.7930	.2070	.9292	10'
32° 0'	9.7242	9.7958	10.2042	9.9284	58° 0'
10'	.7262	.7986	.2014	.9276	50'
20'	.7282	.8014	.1986	.9268	40'
30'	.7302	.8042	.1958	.9260	30'
40'	.7322	.8070	.1930	.9252	20'
50'	.7342	.8097	.1903	.9244	10'
33° 0'	9.7361	9.8125	10.1875	9.9236	57° 0'
10'	.7380	.8153	.1847	.9228	50'
20'	.7400	.8180	.1820	.9219	40'
30'	.7419	.8208	.1792	.9211	30'
40'	.7438	.8235	.1765	.9203	20'
50'	.7457	.8263	.1737	.9194	10'
34° 0'	9.7476	9.8290	10.1710	9.9186	56° 0'
10'	.7494	.8317	.1683	.9177	50'
20'	.7513	.8344	.1656	.9169	40'
30'	.7531	.8371	.1629	.9160	30'
40'	.7550	.8398	.1602	.9151	20'
50'	.7568	.8425	.1575	.9142	10'
35° 0'	9.7586	9.8452	10.1548	9.9134	55° 0'
10'	.7604	.8479	.1521	.9125	50'
20'	.7622	.8506	.1494	.9116	40'
30'	.7640	.8533	.1467	.9107	30'
40'	.7657	.8559	.1441	.9098	20'
50'	.7675	.8586	.1414	.9089	10'
36° 0'	9.7692	9.8613	10.1387	9.9080	54° 0'
	L Cos	L Cot	L Tan	L Sin	Angle

Common Logarithms of Numbers

n	0	1	2	3	4	5	6	7	8	9
55	7404	7412	7419	7427	7435	7443	7451	7459	7466	7474
56	7482	7490	7497	7505	7513	7520	7528	7536	7543	7551
57	7559	7566	7574	7582	7589	7597	7604	7612	7619	7627
58	7634	7642	7649	7657	7664	7672	7679	7686	7694	7701
59	7709	7716	7723	7731	7738	7745	7752	7760	7767	7774
60	7782	7789	7796	7803	7810	7818	7825	7832	7839	7846
61	7853	7860	7868	7875	7882	7889	7896	7903	7910	7917
62	7924	7931	7938	7945	7952	7959	7966	7973	7980	7987
63	7993	8000	8007	8014	8021	8028	8035	8041	8048	8055
64	8062	8069	8075	8082	8089	8096	8102	8109	8116	8122
65	8129	8136	8142	8149	8156	8162	8169	8176	8182	8189
66	8195	8202	8209	8215	8222	8228	8235	8241	8248	8254
67	8261	8267	8274	8280	8287	8293	8299	8306	8312	8319
68	8325	8331	8338	8344	8351	8357	8363	8370	8376	8382
69	8388	8395	8401	8407	8414	8420	8426	8432	8439	8445
70	8451	8457	8463	8470	8476	8482	8488	8494	8500	8506
71	8513	8519	8525	8531	8537	8543	8549	8555	8561	8567
72	8573	8579	8585	8591	8597	8603	8609	8615	8621	8627
73	8633	8639	8645	8651	8657	8663	8669	8675	8681	8686
74	8692	8698	8704	8710	8716	8722	8727	8733	8739	8745
75	8751	8756	8762	8768	8774	8779	8785	8791	8797	8802
76	8808	8814	8820	8825	8831	8837	8842	8848	8854	8859
77	8865	8871	8876	8882	8887	8893	8899	8904	8910	8915
78	8921	8927	8932	8938	8943	8949	8954	8960	8965	8971
79	8976	8982	8987	8993	8998	9004	9009	9015	9020	9025
80	9031	9036	9042	9047	9053	9058	9063	9069	9074	9079
81	9085	9090	9096	9101	9106	9112	9117	9122	9128	9133
82	9138	9143	9149	9154	9159	9165	9170	9175	9180	9186
83	9191	9196	9201	9206	9212	9217	9222	9227	9232	9238
84	9243	9248	9253	9258	9263	9269	9274	9279	9284	9289
85	9294	9299	9304	9309	9315	9320	9325	9330	9335	9340
86	9345	9350	9355	9360	9365	9370	9375	9380	9385	9390
87	9395	9400	9405	9410	9415	9420	9425	9430	9435	9440
88	9445	9450	9455	9460	9465	9469	9474	9479	9484	9489
89	9494	9499	9504	9509	9513	9518	9523	9528	9533	9538
90	9542	9547	9552	9557	9562	9566	9571	9576	9581	9586
91	9590	9595	9600	9605	9609	9614	9619	9624	9628	9633
92	9638	9643	9647	9652	9657	9661	9666	9671	9675	9680
93	9685	9689	9694	9699	9703	9708	9713	9717	9722	9727
94	9731	9736	9741	9745	9750	9754	9759	9763	9768	9773
95	9777	9782	9786	9791	9795	9800	9805	9809	9814	9818
96	9823	9827	9832	9836	9841	9845	9850	9854	9859	9863
97	9868	9872	9877	9881	9886	9890	9894	9899	9903	9908
98	9912	9917	9921	9926	9930	9934	9939	9943	9948	9952
99	9956	9961	9965	9969	9974	9978	9983	9987	9991	9996

Common Logarithms of Numbers

n	0	1	2	3	4	5	6	7	8	9
10	0000	0043	0086	0128	0170	0212	0253	0294	0334	0374
11	0414	0453	0492	0531	0569	0607	0645	0682	0719	0755
12	0792	0828	0864	0899	0934	0969	1004	1038	1072	1106
13	1139	1173	1206	1239	1271	1303	1335	1367	1399	1430
14	1461	1492	1523	1553	1584	1614	1644	1673	1703	1732
15	1761	1790	1818	1847	1875	1903	1931	1959	1987	2014
16	2041	2068	2095	2122	2148	2175	2201	2227	2253	2279
17	2304	2330	2355	2380	2405	2430	2455	2480	2504	2529
18	2553	2577	2601	2625	2648	2672	2695	2718	2742	2765
19	2788	2810	2833	2856	2878	2900	2923	2945	2967	2989
20	3010	3032	3054	3075	3096	3118	3139	3160	3181	3201
21	3222	3243	3263	3284	3304	3324	3345	3365	3385	3404
22	3424	3444	3464	3483	3502	3522	3541	3560	3579	3598
23	3617	3636	3655	3674	3692	3711	3729	3747	3766	3784
24	3802	3820	3838	3856	3874	3892	3909	3927	3945	3962
25	3979	3997	4014	4031	4048	4065	4082	4099	4116	4133
26	4150	4166	4183	4200	4216	4232	4249	4265	4281	4298
27	4314	4330	4346	4362	4378	4393	4409	4425	4440	4456
28	4472	4487	4502	4518	4533	4548	4564	4579	4594	4609
29	4624	4639	4654	4669	4683	4698	4713	4728	4742	4757
30	4771	4786	4800	4814	4829	4843	4857	4871	4886	4900
31	4914	4928	4942	4955	4969	4983	4997	5011	5024	5038
32	5051	5065	5079	5092	5105	5119	5132	5145	5159	5172
33	5185	5198	5211	5224	5237	5250	5263	5276	5289	5302
34	5315	5328	5340	5353	5366	5378	5391	5403	5416	5428
35	5441	5453	5465	5478	5490	5502	5514	5527	5539	5551
36	5563	5575	5587	5599	5611	5623	5635	5647	5658	5670
37	5682	5694	5705	5717	5729	5740	5752	5763	5775	5786
38	5798	5809	5821	5832	5843	5855	5866	5877	5888	5899
39	5911	5922	5933	5944	5955	5966	5977	5988	5999	6010
40	6021	6031	6042	6053	6064	6075	6085	6096	6107	6117
41	6128	6138	6149	6160	6170	6180	6191	6201	6212	6222
42	6232	6243	6253	6263	6274	6284	6294	6304	6314	6325
43	6335	6345	6355	6365	6375	6385	6395	6405	6415	6425
44	6435	6444	6454	6464	6474	6484	6493	6503	6513	6522
45	6532	6542	6551	6561	6571	6580	6590	6599	6609	6618
46	6628	6637	6646	6656	6665	6675	6684	6693	6702	6712
47	6721	6730	6739	6749	6758	6767	6776	6785	6794	6803
48	6812	6821	6830	6839	6848	6857	6866	6875	6884	6893
49	6902	6911	6920	6928	6937	6946	6955	6964	6972	6981
50	6990	6998	7007	7016	7024	7033	7042	7050	7059	7067
51	7076	7084	7093	7101	7110	7118	7126	7135	7143	7152
52	7160	7168	7177	7185	7193	7202	7210	7218	7226	7235
53	7243	7251	7259	7267	7275	7284	7292	7300	7308	7316
54	7324	7332	7340	7348	7356	7364	7372	7380	7388	7396

TABLES

TABLES

Trigonometric Functions

x	sin x	cos x	tan x	cot x	sec x	csc x	x
0° 0'	.00000	1.0000	.00000		1.0000		90° 0'
10'	.00291	1.0000	.00291	343.77	1.0000	343.78	50'
20'	.00582	1.0000	.00582	171.88	1.0000	171.89	40'
30'	.00873	1.0000	.00873	114.59	1.0000	114.59	89° 30'
40'	.01164	.9999	.01164	85.940	1.0001	85.946	20'
50'	.01454	.9999	.01455	68.750	1.0001	68.757	10'
1° 0'	.01745	.9998	.01746	57.290	1.0002	57.299	89° 0'
10'	.02036	.9998	.02036	49.104	1.0002	49.114	50'
20'	.02327	.9997	.02328	42.964	1.0003	42.976	40'
30'	.02618	.9997	.02619	38.188	1.0003	38.202	88° 30'
40'	.02908	.9996	.02910	34.368	1.0004	34.382	20'
50'	.03199	.9995	.03201	31.242	1.0005	31.258	10'
2° 0'	.03490	.9994	.03492	28.636	1.0006	28.654	88° 0'
10'	.03781	.9993	.03783	26.432	1.0007	26.451	50'
20'	.04071	.9992	.04075	24.542	1.0008	24.562	40'
30'	.04362	.9990	.04366	22.904	1.0010	22.926	87° 30'
40'	.04653	.9989	.04658	21.470	1.0011	21.494	20'
50'	.04943	.9988	.04949	20.206	1.0012	20.230	10'
3° 0'	.05234	.9986	.05241	19.081	1.0014	19.107	87° 0'
10'	.05524	.9985	.05533	18.075	1.0015	18.103	50'
20'	.05814	.9983	.05824	17.169	1.0017	17.198	40'
30'	.06105	.9981	.06116	16.350	1.0019	16.380	86° 30'
40'	.06395	.9980	.06408	15.605	1.0021	15.637	20'
50'	.06685	.9978	.06700	14.924	1.0022	14.958	10'
4° 0'	.06976	.9976	.06993	14.301	1.0024	14.336	86° 0'
10'	.07266	.9974	.07285	13.727	1.0027	13.763	50'
20'	.07556	.9971	.07578	13.197	1.0029	13.235	40'
30'	.07846	.9969	.07870	12.706	1.0031	12.746	85° 30'
40'	.08136	.9967	.08163	12.251	1.0033	12.291	20'
50'	.08426	.9964	.08456	11.826	1.0036	11.868	10'
5° 0'	.08716	.9962	.08749	11.430	1.0038	11.474	85° 0'
10'	.09005	.9959	.09042	11.059	1.0041	11.105	50'
20'	.09295	.9957	.09335	10.712	1.0044	10.758	40'
30'	.09585	.9954	.09629	10.385	1.0046	10.433	84° 30'
40'	.09874	.9951	.09923	10.078	1.0049	10.128	20'
50'	.10164	.9948	.10216	9.7882	1.0052	9.839	10'
6° 0'	.10453	.9945	.10510	9.5144	1.0055	9.5668	84° 0'
10'	.10742	.9942	.10805	9.2553	1.0058	9.3092	50'
20'	.11031	.9939	.11099	9.0098	1.0061	9.0652	40'
30'	.11320	.9936	.11394	8.7769	1.0065	8.8337	83° 30'
40'	.11609	.9932	.11688	8.5555	1.0068	8.6138	20'
50'	.11898	.9929	.11983	8.3450	1.0072	8.4647	10'
7° 0'	.12187	.9925	.12278	8.1443	1.0075	8.2055	83° 0'
10'	.12476	.9922	.12574	7.9530	1.0079	8.0157	50'
20'	.12764	.9918	.12869	7.7704	1.0083	7.8344	40'
30'	.13053	.9914	.13165	7.5958	1.0086	7.6613	30'
	cos x	sin x	cot x	tan x	csc x	sec x	x

Trigonometric Functions

x	sin x	cos x	tan x	cot x	sec x	csc x	x
30'	.1305	.9914	.1317	7.5958	1.0086	7.6613	82° 30'
40'	.1334	.9911	.1346	7.4287	1.0090	7.4957	20'
50'	.1363	.9907	.1376	7.2687	1.0094	7.3372	10'
8° 0'	.1392	.9903	.1405	7.1154	1.0098	7.1853	82° 0'
10'	.1421	.9899	.1435	6.9682	1.0102	7.0396	50'
20'	.1449	.9894	.1465	6.8269	1.0107	6.8998	40'
30'	.1478	.9890	.1495	6.6912	1.0111	6.7655	81° 30'
40'	.1507	.9886	.1524	6.5606	1.0116	6.6363	20'
50'	.1536	.9881	.1554	6.4348	1.0120	6.5121	10'
9° 0'	.1564	.9877	.1584	6.3138	1.0125	6.3925	81° 0'
10'	.1593	.9872	.1614	6.1970	1.0129	6.2772	50'
20'	.1622	.9868	.1644	6.0844	1.0134	6.1661	40'
30'	.1650	.9863	.1673	5.9758	1.0139	6.0589	80° 30'
40'	.1679	.9858	.1703	5.8708	1.0144	5.9554	20'
50'	.1708	.9853	.1733	5.7694	1.0149	5.8554	10'
10° 0'	.1736	.9848	.1763	5.6713	1.0154	5.7588	80° 0'
10'	.1765	.9843	.1793	5.5764	1.0160	5.6653	50'
20'	.1794	.9838	.1823	5.4845	1.0165	5.5749	40'
30'	.1822	.9833	.1853	5.3955	1.0170	5.4874	79° 30'
40'	.1851	.9827	.1883	5.3093	1.0176	5.4026	20'
50'	.1880	.9822	.1914	5.2257	1.0182	5.3205	10'
11° 0'	.1908	.9816	.1944	5.1446	1.0187	5.2408	79° 0'
10'	.1937	.9811	.1974	5.0658	1.0193	5.1636	50'
20'	.1965	.9805	.2004	4.9894	1.0199	5.0886	40'
30'	.1994	.9799	.2035	4.9152	1.0205	5.0159	78° 30'
40'	.2022	.9793	.2065	4.8430	1.0211	4.9452	20'
50'	.2051	.9787	.2095	4.7729	1.0217	4.8765	10'
12° 0'	.2079	.9781	.2126	4.7046	1.0223	4.8097	78° 0'
10'	.2108	.9775	.2156	4.6382	1.0230	4.7448	50'
20'	.2136	.9769	.2186	4.5736	1.0236	4.6817	40'
30'	.2164	.9763	.2217	4.5107	1.0243	4.6202	77° 30'
40'	.2193	.9757	.2247	4.4494	1.0249	4.5604	20'
50'	.2221	.9750	.2278	4.3897	1.0256	4.5022	10'
13° 0'	.2250	.9744	.2309	4.3315	1.0263	4.4454	77° 0'
10'	.2278	.9737	.2339	4.2747	1.0270	4.3901	50'
20'	.2306	.9730	.2370	4.2193	1.0277	4.3362	40'
30'	.2334	.9724	.2401	4.1653	1.0284	4.2837	76° 30'
40'	.2363	.9717	.2432	4.1126	1.0291	4.2324	20'
50'	.2391	.9710	.2462	4.0611	1.0299	4.1824	10'
14° 0'	.2419	.9703	.2493	4.0108	1.0306	4.1336	76° 0'
10'	.2447	.9696	.2524	3.9617	1.0314	4.0859	50'
20'	.2476	.9689	.2555	3.9136	1.0321	4.0394	40'
30'	.2504	.9681	.2586	3.8667	1.0329	3.9939	75° 30'
40'	.2532	.9674	.2617	3.8208	1.0337	3.9495	20'
50'	.2560	.9667	.2648	3.7760	1.0345	3.9061	10'
15° 0'	.2588	.9659	.2679	3.7321	1.0353	3.8637	75° 0'
	cos x	sin x	cot x	tan x	csc x	sec x	x

Trigonometric Functions

x	sin x	cos x	tan x	cot x	sec x	csc x	x
30'	.3827	.9239	.4142	2.4142	1.0824	2.6131	30'
40'	.3854	.9228	.4176	2.3945	1.0837	2.5949	20'
50'	.3881	.9216	.4210	2.3750	1.0850	2.5770	10'
23° 0'	.3907	.9205	.4245	2.3559	1.0864	2.5593	67° 0'
10'	.3934	.9194	.4279	2.3369	1.0877	2.5419	50'
20'	.3961	.9182	.4314	2.3183	1.0891	2.5247	40'
30'	.3987	.9171	.4348	2.2998	1.0904	2.5078	30'
40'	.4014	.9159	.4383	2.2817	1.0918	2.4912	20'
50'	.4041	.9147	.4417	2.2637	1.0932	2.4748	10'
24° 0'	.4067	.9135	.4452	2.2460	1.0946	2.4586	66° 0'
10'	.4094	.9124	.4487	2.2286	1.0961	2.4426	50'
20'	.4120	.9112	.4522	2.2113	1.0975	2.4269	40'
30'	.4147	.9100	.4557	2.1943	1.0990	2.4114	30'
40'	.4173	.9088	.4592	2.1775	1.1004	2.3961	20'
50'	.4200	.9075	.4628	2.1609	1.1019	2.3811	10'
25° 0'	.4226	.9063	.4663	2.1445	1.1034	2.3662	65° 0'
10'	.4253	.9051	.4699	2.1283	1.1049	2.3515	50'
20'	.4279	.9038	.4734	2.1123	1.1064	2.3371	40'
30'	.4305	.9026	.4770	2.0965	1.1079	2.3228	30'
40'	.4331	.9013	.4806	2.0809	1.1095	2.3088	20'
50'	.4358	.9001	.4841	2.0655	1.1110	2.2949	10'
26° 0'	.4384	.8988	.4877	2.0503	1.1126	2.2812	64° 0'
10'	.4410	.8975	.4913	2.0353	1.1142	2.2677	50'
20'	.4436	.8962	.4950	2.0204	1.1158	2.2543	40'
30'	.4462	.8949	.4986	2.0057	1.1174	2.2412	30'
40'	.4488	.8936	.5022	1.9912	1.1190	2.2282	20'
50'	.4514	.8923	.5059	1.9768	1.1207	2.2154	10'
27° 0'	.4540	.8910	.5095	1.9626	1.1223	2.2027	63° 0'
10'	.4566	.8897	.5132	1.9486	1.1240	2.1902	50'
20'	.4592	.8884	.5169	1.9347	1.1257	2.1779	40'
30'	.4617	.8870	.5206	1.9210	1.1274	2.1657	30'
40'	.4643	.8857	.5243	1.9074	1.1291	2.1537	20'
50'	.4669	.8843	.5280	1.8940	1.1308	2.1418	10'
28° 0'	.4695	.8829	.5317	1.8807	1.1326	2.1301	62° 0'
10'	.4720	.8816	.5354	1.8676	1.1343	2.1185	50'
20'	.4746	.8802	.5392	1.8546	1.1361	2.1070	40'
30'	.4772	.8788	.5430	1.8418	1.1379	2.0957	30'
40'	.4797	.8774	.5467	1.8291	1.1397	2.0846	20'
50'	.4823	.8760	.5505	1.8165	1.1415	2.0736	10'
29° 0'	.4848	.8746	.5543	1.8040	1.1434	2.0627	61° 0'
10'	.4874	.8732	.5581	1.7917	1.1452	2.0519	50'
20'	.4899	.8718	.5619	1.7796	1.1471	2.0413	40'
30'	.4924	.8704	.5658	1.7675	1.1490	2.0308	30'
40'	.4950	.8689	.5696	1.7556	1.1509	2.0204	20'
50'	.4975	.8675	.5735	1.7437	1.1528	2.0101	10'
30° 0'	.5000	.8660	.5774	1.7321	1.1547	2.0000	60° 0'
x	cos x	sin x	cot x	tan x	csc x	sec x	x

Trigonometric Functions

x	sin x	cos x	tan x	cot x	sec x	csc x	x
15° 0'	.2588	.9659	.2679	3.7321	1.0353	3.8637	75° 0'
10'	.2616	.9652	.2711	3.6891	1.0361	3.8222	50'
20'	.2644	.9644	.2742	3.6470	1.0369	3.7817	40'
30'	.2672	.9636	.2773	3.6059	1.0377	3.7420	30'
40'	.2700	.9628	.2805	3.5656	1.0386	3.7032	20'
50'	.2728	.9621	.2836	3.5261	1.0394	3.6652	10'
16° 0'	.2756	.9613	.2867	3.4874	1.0403	3.6280	74° 0'
10'	.2784	.9605	.2899	3.4495	1.0412	3.5915	50'
20'	.2812	.9596	.2931	3.4124	1.0421	3.5559	40'
30'	.2840	.9588	.2962	3.3759	1.0430	3.5209	30'
40'	.2868	.9580	.2994	3.3402	1.0439	3.4867	20'
50'	.2896	.9572	.3026	3.3052	1.0448	3.4532	10'
17° 0'	.2924	.9563	.3057	3.2709	1.0457	3.4203	73° 0'
10'	.2952	.9555	.3089	3.2371	1.0466	3.3881	50'
20'	.2979	.9546	.3121	3.2041	1.0476	3.3565	40'
30'	.3007	.9537	.3153	3.1716	1.0485	3.3255	30'
40'	.3035	.9528	.3185	3.1397	1.0495	3.2951	20'
50'	.3062	.9520	.3217	3.1084	1.0505	3.2653	10'
18° 0'	.3090	.9511	.3249	3.0777	1.0515	3.2361	72° 0'
10'	.3118	.9502	.3281	3.0475	1.0525	3.2074	50'
20'	.3145	.9492	.3314	3.0178	1.0535	3.1792	40'
30'	.3173	.9483	.3346	2.9887	1.0545	3.1516	30'
40'	.3201	.9474	.3378	2.9600	1.0555	3.1244	20'
50'	.3228	.9465	.3411	2.9319	1.0566	3.0977	10'
19° 0'	.3256	.9455	.3443	2.9042	1.0576	3.0716	71° 0'
10'	.3283	.9446	.3476	2.8770	1.0587	3.0458	50'
20'	.3311	.9436	.3508	2.8502	1.0598	3.0206	40'
30'	.3338	.9426	.3541	2.8239	1.0609	2.9957	30'
40'	.3365	.9417	.3574	2.7980	1.0620	2.9714	20'
50'	.3393	.9407	.3607	2.7725	1.0631	2.9474	10'
20° 0'	.3420	.9397	.3640	2.7475	1.0642	2.9238	70° 0'
10'	.3448	.9387	.3673	2.7228	1.0653	2.9006	50'
20'	.3475	.9377	.3706	2.6985	1.0665	2.8779	40'
30'	.3502	.9367	.3739	2.6746	1.0676	2.8555	30'
40'	.3529	.9356	.3772	2.6511	1.0688	2.8334	20'
50'	.3557	.9346	.3805	2.6279	1.0700	2.8118	10'
21° 0'	.3584	.9336	.3839	2.6051	1.0712	2.7904	69° 0'
10'	.3611	.9325	.3872	2.5826	1.0724	2.7695	50'
20'	.3638	.9315	.3906	2.5605	1.0736	2.7488	40'
30'	.3665	.9304	.3939	2.5386	1.0748	2.7285	30'
40'	.3692	.9293	.3973	2.5172	1.0760	2.7085	20'
50'	.3719	.9283	.4006	2.4960	1.0773	2.6888	10'
22° 0'	.3746	.9272	.4040	2.4751	1.0785	2.6695	68° 0'
10'	.3773	.9261	.4074	2.4545	1.0798	2.6504	50'
20'	.3800	.9250	.4108	2.4342	1.0811	2.6316	40'
30'	.3827	.9239	.4142	2.4142	1.0824	2.6131	30'
x	cos x	sin x	cot x	tan x	csc x	sec x	x

Trigonometric Functions

x	csc x	sec x	cot x	tan x	cos x	sin x	x
52° 30'	1.6427	1.2605	1.3032	.7673	.7934	.6088	30'
20'	1.6365	1.2633	1.2954	.7720	.7916	.6111	40'
10'	1.6304	1.2662	1.2876	.7766	.7898	.6134	50'
52° 0'	1.6243	1.2690	1.2799	.7813	.7880	.6157	38° 0'
50'	1.6183	1.2719	1.2723	.7860	.7862	.6180	10'
40'	1.6123	1.2748	1.2647	.7907	.7844	.6202	20'
30'	1.6064	1.2779	1.2572	.7954	.7826	.6225	30'
20'	1.6005	1.2808	1.2497	.8002	.7808	.6248	40'
10'	1.5948	1.2837	1.2423	.8050	.7790	.6271	50'
51° 0'	1.5890	1.2868	1.2349	.8098	.7771	.6293	39° 0'
50'	1.5833	1.2898	1.2276	.8146	.7753	.6316	10'
40'	1.5777	1.2929	1.2203	.8195	.7735	.6338	20'
30'	1.5721	1.2960	1.2131	.8243	.7716	.6361	30'
20'	1.5666	1.2991	1.2059	.8292	.7698	.6383	40'
10'	1.5611	1.3022	1.1988	.8342	.7679	.6406	50'
50° 0'	1.5557	1.3054	1.1918	.8391	.7660	.6428	40° 0'
50'	1.5504	1.3086	1.1847	.8441	.7642	.6450	10'
40'	1.5450	1.3118	1.1778	.8491	.7623	.6472	20'
30'	1.5398	1.3151	1.1708	.8541	.7604	.6494	30'
20'	1.5346	1.3184	1.1640	.8591	.7585	.6517	40'
10'	1.5294	1.3217	1.1571	.8642	.7566	.6539	50'
49° 0'	1.5243	1.3250	1.1504	.8693	.7547	.6561	41° 0'
50'	1.5192	1.3284	1.1436	.8744	.7528	.6583	10'
40'	1.5142	1.3318	1.1369	.8796	.7509	.6604	20'
30'	1.5092	1.3352	1.1303	.8847	.7490	.6626	30'
20'	1.5042	1.3386	1.1237	.8899	.7470	.6648	40'
10'	1.4993	1.3421	1.1171	.8952	.7451	.6670	50'
48° 0'	1.4945	1.3456	1.1106	.9004	.7431	.6691	42° 0'
50'	1.4897	1.3492	1.1041	.9057	.7412	.6713	10'
40'	1.4849	1.3527	1.0977	.9110	.7392	.6734	20'
30'	1.4802	1.3563	1.0913	.9163	.7373	.6756	30'
20'	1.4755	1.3600	1.0850	.9217	.7353	.6777	40'
10'	1.4709	1.3636	1.0786	.9271	.7333	.6799	50'
47° 0'	1.4663	1.3673	1.0724	.9325	.7314	.6820	43° 0'
50'	1.4617	1.3711	1.0661	.9380	.7294	.6841	10'
40'	1.4572	1.3748	1.0599	.9435	.7274	.6862	20'
30'	1.4527	1.3786	1.0538	.9490	.7254	.6884	30'
20'	1.4483	1.3824	1.0477	.9545	.7234	.6905	40'
10'	1.4439	1.3863	1.0416	.9601	.7214	.6926	50'
46° 0'	1.4396	1.3902	1.0355	.9657	.7193	.6947	44° 0'
50'	1.4352	1.3941	1.0295	.9713	.7173	.6967	10'
40'	1.4310	1.3980	1.0235	.9770	.7153	.6988	20'
30'	1.4267	1.4020	1.0176	.9827	.7133	.7009	30'
20'	1.4225	1.4061	1.0117	.9884	.7112	.7030	40'
10'	1.4184	1.4101	1.0058	.9942	.7092	.7050	50'
45° 0'	1.4142	1.4142	1.0000	1.0000	.7071	.7071	45° 0'
x	sec x	csc x	tan x	cot x	sin x	cos x	x

Trigonometric Functions

x	sin x	cos x	tan x	cot x	sec x	csc x	x
30° 0'	.5000	.8660	.5774	1.7321	1.1547	2.0000	60° 0'
10'	.5025	.8646	.5812	1.7205	1.1567	1.9900	50'
20'	.5050	.8631	.5851	1.7090	1.1586	1.9801	40'
30'	.5075	.8616	.5890	1.6977	1.1606	1.9703	30'
40'	.5100	.8601	.5930	1.6864	1.1626	1.9606	20'
50'	.5125	.8587	.5969	1.6753	1.1646	1.9511	10'
31° 0'	.5150	.8572	.6009	1.6643	1.1666	1.9416	59° 0'
10'	.5175	.8557	.6048	1.6534	1.1687	1.9323	50'
20'	.5200	.8542	.6088	1.6426	1.1708	1.9230	40'
30'	.5225	.8526	.6128	1.6319	1.1728	1.9139	30'
40'	.5250	.8511	.6168	1.6212	1.1749	1.9049	20'
50'	.5275	.8496	.6208	1.6107	1.1770	1.8959	10'
32° 0'	.5299	.8480	.6249	1.6003	1.1792	1.8871	58° 0'
10'	.5324	.8465	.6289	1.5900	1.1813	1.8783	50'
20'	.5348	.8450	.6330	1.5798	1.1835	1.8699	40'
30'	.5373	.8434	.6371	1.5697	1.1857	1.8612	30'
40'	.5398	.8418	.6412	1.5597	1.1879	1.8527	20'
50'	.5422	.8403	.6453	1.5497	1.1901	1.8444	10'
33° 0'	.5446	.8387	.6494	1.5399	1.1924	1.8361	57° 0'
10'	.5471	.8371	.6536	1.5301	1.1946	1.8279	50'
20'	.5495	.8355	.6577	1.5204	1.1969	1.8198	40'
30'	.5519	.8339	.6619	1.5108	1.1992	1.8118	30'
40'	.5544	.8323	.6661	1.5013	1.2015	1.8039	20'
50'	.5568	.8307	.6703	1.4919	1.2039	1.7960	10'
34° 0'	.5592	.8290	.6745	1.4826	1.2062	1.7883	56° 0'
10'	.5616	.8274	.6787	1.4733	1.2086	1.7806	50'
20'	.5640	.8258	.6830	1.4641	1.2110	1.7730	40'
30'	.5664	.8241	.6873	1.4550	1.2134	1.7655	30'
40'	.5688	.8225	.6916	1.4460	1.2158	1.7581	20'
50'	.5712	.8208	.6959	1.4370	1.2183	1.7507	10'
35° 0'	.5736	.8192	.7002	1.4281	1.2208	1.7435	55° 0'
10'	.5760	.8175	.7046	1.4193	1.2233	1.7362	50'
20'	.5783	.8158	.7089	1.4106	1.2258	1.7291	40'
30'	.5807	.8141	.7133	1.4019	1.2283	1.7221	30'
40'	.5831	.8124	.7177	1.3934	1.2309	1.7151	20'
50'	.5854	.8107	.7221	1.3848	1.2335	1.7082	10'
36° 0'	.5878	.8090	.7265	1.3764	1.2361	1.7013	54° 0'
10'	.5901	.8073	.7310	1.3680	1.2387	1.6945	50'
20'	.5925	.8056	.7355	1.3597	1.2413	1.6878	40'
30'	.5948	.8039	.7400	1.3514	1.2440	1.6812	30'
40'	.5972	.8021	.7445	1.3432	1.2467	1.6746	20'
50'	.5995	.8004	.7490	1.3351	1.2494	1.6681	10'
37° 0'	.6018	.7986	.7536	1.3270	1.2521	1.6616	53° 0'
10'	.6041	.7969	.7581	1.3190	1.2549	1.6553	50'
20'	.6065	.7951	.7627	1.3111	1.2577	1.6489	40'
30'	.6088	.7934	.7673	1.3032	1.2605	1.6427	30'
x	cos x	sin x	cot x	tan x	csc x	sec x	x

Glossary

The explanations given in this glossary are intended to be brief descriptions of the terms listed. They are not necessarily definitions.

absolute value (of a real number) *(p. 10)* The absolute value of the real number x is written as $|x|$. $|x| = -x$ if x is a negative number, and $|x| = x$ if x is a positive number or zero.

additive identity *(p. 26)* Zero is the additive identity and has the property $x + 0 = x$, for each number x.

additive inverse *(p. 26)* The additive inverse of x is $-x$ and $x + (-x) = 0$, for each number x.

amplitude *(p. 497)* The amplitude of a trig function is its maximum value.

angle of rotation *(p. 455)* An angle of rotation is formed by rotating a ray about its end point, or vertex.

antilogarithm (base-10) *(p. 404)* Antilog $x = 10^x$, for each real number x; antilog $2 = 10^2 = 100$.

arithmetic mean *(p. 361)* The arithmetic mean of x and y is their average, $\frac{x+y}{2}$.

arithmetic means *(p. 361)* The terms between two given terms of an arithmetic progression.

arithmetic progression (A.P.) *(p. 357)* A progression formed by adding the same number to any term to obtain the next term.

arithmetic series *(p. 364)* The indicated sum of the terms of an arithmetic progression. 3, 6, 9, ... is an arithmetic progression; $3 + 6 + 9 + \ldots$ is an arithmetic series.

binomial expansion *(p. 382)* The indicated sum of the terms of the nth power of a binomial. For example, $(a + b)^3$ is $a^3 + 3a^2b + 3ab^2 + b^3$.

bounds (upper and lower) *(p. 412)* U and L are upper and lower bounds for the real zeros of a polynomial if all the real zeros are between U and L.

characteristic *(p. 403)* The integer part of a base-10 logarithm. Log $54{,}600 = 4.7372$ and log $0.0546 = 8.7372 - 10$. The characteristic of log $54{,}600$ is 4 and the characteristic of log 0.0546 is $8 - 10$, or -2.

circle *(p. 306)* The set of all points in a plane that are a given distance from a given point in the plane.

combination *(p. 435)* A selection of objects without regard to order.

combined variation *(p. 295)* A function defined by an equation of the form $y = \frac{kxz}{w}$, or $\frac{yw}{xz} = k$, where k is a nonzero constant.

complex number *(p. 187)* A number that can be written in the form $a + bi$, where a and b are real numbers and $i = \sqrt{-1}$.

conjugates *(pp. 149, 191, 419)* Two numbers of the form $x + y$ and $x - y$ are a pair of conjugates, and their product, $x^2 - y^2$, is a difference of squares. $3\sqrt{2} + 4$ and $3\sqrt{2} - 4$ are a pair of irrational conjugates, and their product, $(3\sqrt{2})^2 - 4^2$, or 2, is a rational number. $5 - 3i$ and $5 + 3i$ are a pair of imaginary conjugates, and their product, $5^2 - (3i)^2$, or 34, is a real number.

conjunction *(p. 7)* A compound sentence formed by two clauses separated by *and*. $[x > 2$ and $x < 5]$ is a conjunction.

consistent system *(p. 240)* A system of linear equations that has at least one solution.

constant function *(p. 282)* A linear function whose graph is a horizontal line, ray, or segment.

Cramer's rule *(p. 200)* Solving a system of equations using determinants.

dependent events *(p. 443)* Events in which one event depends upon the other.

dependent system *(p. 241)* A system of linear equations that has all solutions in common.

determinant *(p. 250)* A real number assigned to a matrix.

discriminant *(p. 195)* The radicand, $b^2 - 4ac$, in the quadratic formula is called the discriminant of the quadratic equation. The discriminant determines the nature of the solutions of the equation.

disjunction *(p. 7)* A compound sentence formed by two clauses separated by *or*. [$x < 2$ or $x > 5$] is a disjunction.

domain *(p. 276)* The set of all first coordinates of the ordered pairs of a relation.

ellipse *(p. 310)* The set of all points in a plane such that for each point of the set, the sum of its distance from two fixed points is constant.

exponential equation *(pp. 33, 158, 406)* Equations such as $2^x = 16$ and $5^{4x + 1} = 125$ in which the variable appears in the exponent.

factorial notation *(p. 383)* $0! = 1$ and for each positive integer k, $k! = 1 \cdot 2 \cdot 3 \cdots k$.

field *(p. 26)* A set of numbers, along with two operations, that satisfies eleven special properties.

fractional equation *(p. 116)* An equation such as $\frac{2n - 9}{n - 7} + \frac{n}{2} = \frac{5}{n - 7}$ in which a variable appears in a denominator.

function *(p. 276)* A relation, a set of ordered pairs, in which different ordered pairs have different first coordinates.

fundamental counting principle *(p. 424)* The principle used to find the number of permutations of n objects taken r at a time.

geometric mean *(p. 372)* The geometric mean (mean proportional) of x and y is $\sqrt{xy}$ or $-\sqrt{xy}$.

geometric means *(p. 372)* The terms between two given terms of a geometric progression.

geometric progression (G.P.) *(p. 368)* A progression formed by multiplying any term by the same nonzero number to obtain the next term.

geometric series *(p. 375)* The indicated sum of the terms of a geometric progression. $3 + 6 + 12 + \ldots$ is a geometric series since 3, 6, 12, … is a geometric progression.

hyperbola *(p. 319)* The set of all points in a plane such that for each point of the set, the difference of its distances from two fixed points is constant.

imaginary number *(p. 188)* Any complex number $a + bi$, where $b \neq 0$, such as $5 - 2i$ and $4i$.

inclusive events *(p. 442)* Events that can both occur at the same time.

inconsistent system *(p. 241)* A system of linear equations that has no solution.

independent events *(p. 443)* Events in which neither event depends upon the other.

independent system *(p. 240)* A system of linear equations that has exactly one solution.

inverse functions *(p. 285)* When both a relation and its inverse are functions, they are called inverse functions.

inverse relations *(p. 285)* The inverse of a relation is formed by reversing the order of the elements of each ordered pair.

inverse variation *(p. 291)* A function that is defined by an equation of the form $xy = k$ or $y = \frac{k}{x}$, where k is a nonzero constant.

irrational number *(p. 140)* A real number whose decimal numeral is nonterminating and nonrepeating.

joint variation *(p. 294)* A function that is defined by an equation of the form $y = kxz$ or $y = \frac{k}{x}$, where k is a nonzero constant.

law of cosines *(p. 539)* The law of cosines is used to find the length of a side of a triangle when the measures of the two other sides and included angle are known. In general, $a^2 = b^2 + c^2 - 2bc \cos A$, $b^2 = a^2 + c^2 - 2ac \cos B$, and $c^2 = a^2 + b^2 - 2ab \cos C$.

law of sines *(p. 546)* The law of sines is used to find the measure of an angle and length of a side of a triangle when the measures of two angles and a side are known. In general, $\frac{\sin A}{a} = \frac{\sin B}{b} = \frac{\sin C}{c}$.

linear function *(p. 282)* A function whose graph is a nonvertical line, ray, or segment. The equation $y = mx + b$ describes a linear function.

logarithm (base-n) *(p. 392)* For each positive number y and each positive number $n \neq 1$, if $y = n^x$, then $\log_n y = \log_n n^x = x$. For example, $125 = 5^3$ and $\log_5 5^3 = 3$.

logarithm (common; base-10) *(p. 403)* For each real number $y > 0$, the base-10 logarithm of y, or $\log y$, is defined as follows: If $y = 10^x$, then $\log y = \log 10^x = x$. Base-10 logarithms are also called common logarithms.

mantissa *(p. 403)* The decimal part of a base-10 logarithm. Log $54{,}600 = 4.7372$, so the mantissa of log 54,600 is .7372.

matrix *(p. 250)* An array of numbers or other elements arranged in rows and columns.

multiplicative identity *(p. 26)* One is the multiplicative identity and $x \cdot 1 = x$, for each number x.

multiplicative inverse *(p. 26)* If $x \neq 0$, the multiplicative inverse of x is $\frac{1}{x}$ and $x \cdot \frac{1}{x} = 1$, for each $x \neq 0$.

mutually exclusive events *(p. 442)* Events that cannot both occur at the same time.

parabola *(p. 327)* A set of points in a plane that are equidistant from a fixed point and a fixed line.

periodic function *(p. 497)* A function that repeats itself. If $f(x) = f(x + p)$, then f is a periodic function with a period of p.

permutation (linear) *(p. 425)* A linear arrangement of objects in a definite order.

permutation (circular) *(p. 433)* The number of circular permutations of n distinct objects is $(n - 1)!$

principal values *(p. 531)* To obtain an inverse function of a trigonometric function, restrict the range of the inverse. The restricted values are called principal values.

probability *(p. 438)* The probability of a simple event is the number of favorable outcomes divided by the total number of all possible outcomes.

progression *(p. 357)* An ordered list of numbers formed according to some pattern; a sequence.

pure imaginary number *(p. 188)* Any complex number $a + bi$, where $a = 0$, such as $4i$.

Pythagorean relation *(p. 78)* For each right triangle ABC with right angle C, $a^2 + b^2 = c^2$. And, if $a^2 + b^2 = c^2$, then triangle ABC is a right triangle with right angle C.

quadrants *(p. 209)* The two axes in a coordinate plane separate the plane into four sections called quadrants.

quadratic equation (in one variable) *(p. 68)* The standard form of a quadratic equation in one variable is $ax^2 + bx + c = 0$, where a, b, and c are numbers and $a \neq 0$.

quadratic formula *(p. 169)* The solutions of the general quadratic equation, $ax^2 + bx + c = 0$, are given by the quadratic formula,
$$x = \frac{-b \pm \sqrt{b^2 - 4ac}}{2a}.$$

quadratic inequality (in one variable) *(p. 71)* The standard form of a quadratic inequality in one variable is either $ax^2 + bx + c < 0$ or $ax^2 + bx + c > 0$, where a, b, and c are numbers and $a \neq 0$.

radian *(p. 523)* If an angle of rotation, which is the central angle of a given circle, cuts off an arc whose length is equal to the length of a radius of a circle, then the measure of the angle of rotation is 1 radian.

radical equation *(p. 178)* An equation like $\sqrt[3]{6x + 10} = -2$ in which a radicand contains a variable.

range *(p. 276)* The set of all second coordinates of the ordered pairs of a relation.

rational expression *(p. 91)* A rational expression is either a polynomial P or a quotient $\frac{P}{Q}$ of polynomials P and Q. $Q \neq 0$

rational number *(p. 89)* A rational number is one that can be written in the form $\frac{a}{b}$ where a and b are integers, $b \neq 0$. Examples of rational numbers are -8, 0, 7, $2\frac{3}{4}$, 9.62, $\sqrt{25}$, $\frac{22}{7}$, and $3.43434\ldots$

real numbers *(p. 141)* All the rational numbers together with all the irrational numbers.

rectangular hyperbola *(p. 320)* A hyperbola whose equation is $xy = k$, where k is a nonzero real number.

reference angle *(p. 455)* The positive acute angle formed between the terminal side and the x-axis.

repeating decimal *(p. 89)* A nonterminating decimal in which one block of digits continues to repeat in consecutive blocks. The numeral $3.7242424\ldots$, or $3.72\overline{24}$, is a repeating decimal.

sequence *(p. 357)* An ordered list of numbers formed according to some pattern; a progression.

slope and parallel lines *(p. 228)* If two or more different nonvertical lines have the same slope, then the lines are parallel.

slope and perpendicular lines (p. 229) If two lines have negative reciprocal slopes, the lines are perpendicular.

standard form of a linear equation (p. 216) A linear equation in two variables written in the form $ax + by = c$.

synthetic division (p. 130) A short method of dividing a polynomial in x by a binomial of the form $x - a$, where a is a rational number.

trigonometric functions

cosecant function (p. 470) For any angle of rotation u, the cosecant (csc) of u is $\frac{r}{y}$;

$$\csc u = \frac{r}{y}.$$

cosine function (p. 462) For any angle of rotation u, the cosine (cos) of u is $\frac{x}{r}$;

$$\cos u = \frac{x}{r}.$$

cotangent function (p. 466) For any angle of rotation u, the cotangent (cot) of $u = \frac{x}{y}$;

$$\cot u = \frac{x}{y}.$$

secant function (p. 470) For any angle of

rotation u, the secant (sec) of u is $\frac{r}{x}$; $\sec u = \frac{r}{x}$.

sine function (p. 462) For any angle of rotation u, the sine (sin) of u is $\frac{y}{r}$; $\sin u = \frac{y}{r}$.

tangent function (p. 466) For any angle of rotation u, the tangent (tan) of $u = \frac{y}{x}$;

$$\tan u = \frac{y}{x}.$$

trigonometric identities (p. 518) Trigonometric equations that are true for all permissible values of the variables.

value of a function (p. 279) For any ordered pair (x, y) of a function, y is the value of the function at x. For any given element of the domain of a function, there is a corresponding element in the range which is called the value of the function.

vector (p. 50) A directed line segment that can show the magnitude and direction of a force, displacement, or velocity.

zero (of a polynomial) (p. 409) A zero of a polynomial $P(x)$ is a value of x for which the value of $P(x)$ is 0. For example, $x^2 - 9$ has two zeros: 3 and -3.

Symbol List

$=$	is equal to	(x, y)	the ordered pair of numbers x, y
$<$	is less than	$\overline{AB}$	segment AB
$>$	is greater than	$\overleftrightarrow{AB}$	line AB
$\{x \mid x > 5\}$	the set of all numbers x such that x is greater than 5	AB	the distance from A to B
$\lvert x \rvert$	the absolute value of x	$\parallel$	is parallel to
$\sqrt{a}$	the principal square root of a	$\perp$	is perpendicular to
ϕ	the empty set	$f(x)$	f at x (the value of function f at x)
$\doteq$	is approximately equal to	f^{-1}	the inverse of function f
$\sqrt[n]{x}$	the nth root of x	$\log x$	the base-10 logarithm of x
$a \pm \sqrt{b}$	$a + \sqrt{b}, a - \sqrt{b}$	$\log_n x$	the base-n logarithm of x
i	$\sqrt{-1}$	$\sin^{-1}$	arc sine

Selected Answers

(This section includes answers to selected written exercises and chapter reviews.)

Page 3

1. -6 **3.** -4.5 **5.** 10 **7.** 7 **9.** -6
11. 4.6 **13.** 0.04 **15.** -14 **17.** -84.6
19. 0.05 **21.** 2.4 **23.** -4 **25.** -0.2
27. no solution **29.** 16

Page 6

1. $\{y \mid y \geq -3\}$ **3.** $\{x \mid x > 4\}$ **5.** $\{n \mid n \geq -2\}$
7. $\{x \mid x > -6\}$ **9.** $\{n \mid n \leq -3\}$ **11.** $\{x \mid x \leq -3\}$

13. $\left\{y \mid y > 3\frac{1}{2}\right\}$ **15.** $\{x \mid x \geq 10\}$

17. $\left\{a \mid a < -2\frac{1}{3}\right\}$ **19.** $\left\{y \mid y \leq 5\frac{1}{2}\right\}$

21. $\left\{x \mid x \geq -3\frac{1}{2}\right\}$ **23.** $\left\{n \mid n \geq 1\frac{1}{3}\right\}$

25. $\{y \mid y > 1\}$ **27.** $\{n \mid n \geq -21\}$ **29.** $\{y \mid y > 6\}$
31. $x < y$ given; $-1 \cdot x > -1 \cdot y$ Mult. Prop. of
Inequality; $-x > -y$ or $-y < -x$ For each
number a, $-1 \cdot a = -a$. **33.** Let $a = -2$, $b = 4$;
$-2 < 4$ but $-\frac{1}{2} \not> \frac{1}{4}$. (Answers may vary.)

Page 9

1. $\{x \mid x < -4 \text{ or } x > 4\}$ **3.** $\{x \mid x \geq -2\}$
5. $\{x \mid 3 < x < 6\}$ **7.** $\{x \mid 1 < x < 5\}$
9. $\{x \mid -3 < x < 4\}$ **11.** $\{x \mid -2 \leq x < 9\}$
13. $\{x \mid x < 3\}$ **15.** $\{x \mid 0 < x < 5\}$
17. $\{x \mid x < 1 \text{ or } x > 2\}$ **19.** $\{x \mid x \geq -5\}$
21. $\{x \mid x > 3.5\}$ **23.** $\{x \mid -3 < x < 2\}$
25. $\{x \mid -6.5 \leq x \leq 1.5\}$
27. $\{x \mid -1 < x < 2 \text{ or } x > 4\}$

29.

p	q	p or q	p and q
T	T	T	T
T	F	T	F
F	T	T	F
F	F	F	F

Page 12

1. -4, 10 **3.** -2, 10 **5.** -10, -5 **7.** -2, -1
9. $\{x \mid -3 < x < 3\}$ **11.** $\{y \mid y < -8 \text{ or } y > 8\}$
13. $\{x \mid x \leq 2 \text{ or } x \geq 4\}$ **15.** $\{a \mid -4 \leq a \leq 8\}$
17. $-2, \frac{1}{2}$ **19.** -2.8, -2 **21.** -3.5, 7
23. -4, -2.2 **25.** $\{x \mid -6 < x < 11\}$
27. $\{z \mid z \leq 3 \text{ or } z \geq 7\}$ **29.** $\{y \mid y \leq -2 \text{ or } y \geq -1\}$
31. $\{x \mid -9 < x < 5\}$ **33.** $\{x \mid 4 < x < 8\}$
35. $\{x \mid -4 < x < 8 \text{ or } x < -8 \text{ or } x > 12\}$
37. 7, -3 **39.** -13, 5 **41.** $\{x \mid 3.9 < x < 4.1\}$

Pages 16–17

1. 2 **3.** -7 **5.** smaller: x, greater: $5x + 7$
7. first: x, second: $7x$, third: $7x - 6$ **9.** 7, 11
11. 9, 45, -3 **13.** 15, 16, 17 **15.** A: 4,000,
B: 1,000, C: 3,330 **17.** 9, 12, 15, 18 **19.** 30,
45, 60 **21.** 3, 11, 21 **23.** A: 36, B: 18, C: 24
25. 70, 71, 72 **27.** -8, -6, -4
29. any three consecutive integers **31.** $\{x \mid x > -2\}$
33. $\{x \mid x \geq 4\}$

Page 19

1. length: 17 cm, width: 5 cm, area: 85 cm^2
3. a: 10 cm, b: 6 cm, c: 12 cm **5.** length:
17 dm, width: 4 dm, area: 68 dm^2 **7.** 7 m, 21 m,
21 m, 15 m, 15 m **9.** length: 9 cm, width: 3 cm
11. 24 m^2 **13.** 124 dm
15. $\{w \mid 0 \text{ cm} < w < 3 \text{ cm}\}$

Page 21

1. Selma: 5, Fred: 20 **3.** Cindy: 16, Byron: 14
5. Walt: 8, Brenda: 12, Carol: 24 **7.** falcon: 6,
eagle: 24 **9.** Cynthia: 7, Adam: 21, Fred: 5
11. Keith: 10, Phyllis: 16, Manuel: 30
13. $t + 10 + f$

Pages 24–25

1. 7% **3.** 12 yr **5.** \$5,000 **7.** \$6,000
9. 18.75 ohms **11.** 40 ohms **13.** 55 m/s
15. 315 m **17.** The first is on the way up,
and the second is on the way down.
19. 310 m/s **21.** 400 m/s **23.** 9.6 ohms
25. \$6,000 **27.** 6 yr 6 mo **29.** 12 yr 6 mo
31. \$8,500 **33.** $\frac{1}{r}$ yr **35.** 445 m
37. 45 m **39.** 120 m

Page 27

1. -3.25 **3.** -0.5 **5.** $\frac{1}{2}$ **7.** $2\frac{3}{14}$
9. 3, 7 **11.** $-5\frac{2}{3}$, 5
13. $\{x \mid x < -2 \text{ or } x > 2\}$
15. $\{z \mid z \leq -3 \text{ or } z \geq 9\}$ **17.** $\{a \mid a < 4.25\}$
19. $\{x \mid x > -3\}$ **21.** 6 **23.** a: 12 cm, b: 4 cm,
c: 10 cm **25.** pine: 5, oak: 30, spruce: 10
27. 28 ohms **29.** $\{l \mid 6 \text{ cm} < l < 16.5 \text{ cm}\}$

Page 33

1. $20a^3 b^8$ **3.** $36x^9 y^5$ **5.** $\dfrac{-2b^5}{a^5}$ **7.** $\dfrac{-2}{3m^9 p^5 q^2}$
9. y^{30} **11.** $125x^3$ **13.** $7{,}000n^{15}$ **15.** $484y^3$

17. $\dfrac{-125a^3}{64}$ **19.** $\dfrac{49c^{10}}{100d^6}$ **21.** 6 **23.** 2 **25.** 1
27. -2 **29.** $-36a^9b^{11}c^4$ **31.** $-1{,}250x^6y^9$
33. $384a^7b^9$ **35.** $-4{,}000c^6d^{18}$
37. $x^{a+2c}y^{3b+4d}$ **39.** $x^{a+bd}y^{cd}$ **41.** $x^{4a}y^{2b}$
43. $x^{2c+3}y^{3d-4}$

Page 36

1. 1 **3.** 1 **5.** $\dfrac{1}{16}$ **7.** $\dfrac{4}{9}$ **9.** 250

11. $10\dfrac{2}{3}$ **13.** 72 **15.** $\dfrac{8}{9}$ **17.** $\dfrac{1}{49}$

19. 0.027 **21.** 0.000123 **23.** 0.0468 **25.** $\dfrac{9}{x^3}$

27. $\dfrac{-10}{a^3}$ **29.** x^3 **31.** $\dfrac{7}{n^8}$ **33.** $\dfrac{x^3}{10}$ **35.** $\dfrac{5a^2}{6}$

37. x^9 **39.** $\dfrac{2n^3}{3}$ **41.** $\dfrac{1}{x^{10}}$ **43.** $\dfrac{n^{20}}{32}$ **45.** $\dfrac{1}{-125y^{12}}$

47. $\dfrac{1}{c^{12}d^{20}}$ **49.** 20,000 **51.** 0.321 **53.** $\dfrac{a^5d^6}{b^4c^3}$

55. $\dfrac{4b^5d^4}{9a^2c^3}$ **57.** $\dfrac{x^{10}}{y^{26}}$ **59.** $\dfrac{2d^{10}}{3c^3}$ **61.** $\dfrac{10}{x^5y^2}$

63. $-\dfrac{c^9}{27d^{12}}$ **65.** $\dfrac{144z^{10}}{x^6y^6}$ **67.** $\dfrac{-8y^3w^6}{27x^{12}z^9}$

69. $\dfrac{1}{x^{b-a}}$ **71.** $\dfrac{1}{z^{3n-3m}}$

Page 40

1. -58 **3.** 97, 7 **5.** 280, -40
7. $-5y + 41$ **9.** $-15a^4 + 6a^2 - 8a$
11. $2a^2 + 9$ **13.** $3x^2 + 7x + 1$
15. $-4y^2 - 2y - 3$ **17.** $12a - 9$ **19.** 1.2
21. 49.2 **23.** 6, 60, 0, 90
25. $-27, -32, 128, -7, 0$ **27.** $30x^2 + 27xy$
29. $-2n^3 + 5n^2 + 5n - 50$ **31.** 404,040
33. $-7{,}070{,}707$ **35.** $-120c^{12}$

Page 42

1. $8t^2 - 6t - 20$ **3.** $4n^2 - 64$
5. $6x^2 + 8xy - 8y^2$ **7.** $x^4 + 25x^2 + 150$
9. $10y^6 - 29y^3 + 21$ **11.** $21x^2 + 6.8x + 0.32$
13. $0.32a^2 - 4.8a - 54$ **15.** $5x^3 - 11x^2 - 2x + 8$
17. $9y^3 - 9y^2 + 14y - 8$ **19.** $8x^2 + 4x - 60$
21. $-18y^2 + 48y - 32$ **23.** $8x^4 - 4x^2y - 24y^2$
25. $9x^4 - 3x^2y^2 - 20y^4$
27. $12a^2 + 7ab + 9a - 10b^2 - 6b$
29. $3n^4 + 4n^3 - 17n^2 - 16n + 20$
31. $2.24x^2 + 1.24x + 0.12$ **33.** $12.25x^2 - 0.81$
35. $24x^3 + 76x^2 + 40x$ **37.** $50y^5 - 40y^3 + 8y$

39. $(a + b + c)(a + b + c) = a(a + b + c) +$
$b(a + b + c) + c(a + b + c) = a^2 + ab + ac +$
$ba + b^2 + bc + ca + cb + c^2 = a^2 + b^2 +$
$c^2 + ab + ab + ac + ac + bc + bc = a^2 +$
$b^2 + c^2 + 2ab + 2ac + 2bc$
41. $64x^2 + y^2 + 16z^2 - 16xy - 64xz + 8yz$
43. $3x^{4a} + 7x^{2a} + 4$ **45.** $20x^{4a} - 7x^{2a}y^b - 6y^{2b}$

Pages 45–46

1. 68 dimes, 3 quarters **3.** 4 kg at $3.50,
8 kg at $2.00 **5.** 20 L **7.** 40 L
9. 12 at 50¢, 24 at 13¢, 28 at 10¢
11. 50 kg **13.** 15 L **15.** 8 at 80¢,
16 at 60¢ **17.** 67 mL

Page 49

1. 6 h **3.** 7:20 P.M. **5.** 4 h **7.** 12:30 P.M.
9. 4 h **11.** 4:20 P.M. **13.** 12 km/h, 6 km/h,
18 km/h

Page 52

5. $\overrightarrow{HB}$ **7.** $\overrightarrow{BE}$ **9.** $\overrightarrow{BH}$

Page 53

1. $-40a^6b^7c^6$ **3.** $\dfrac{2z}{3x^4}$ **5.** $x^{5m}y^{3n+6}$ **7.** $\dfrac{a^8}{16b^{12}}$

9. $\dfrac{81x^4w^{12}}{16y^8z^{12}}$ **11.** 3,450 **13.** 7 **15.** 122 **17.** 0
19. $9xy^2 - 3xy$ **21.** $16x^4 + 24x^2y^2 + 9y^4$
23. $140c^7 + 49c^4 - 42c$ **25.** caramels: 3 kg;
mints: 5 kg **27.** 30 mL **29.** 1 h 20 min
31. 8:30 A.M.

Page 59

1. $(7x - 5)(x - 1)$ **3.** $(5y - 1)(y + 3)$
5. $(5a + 1)(a - 11)$ **7.** cannot be factored
9. $(2a + 5)(a - 2)$ **11.** $(5x - 3)(x - 5)$
13. $(3a + 4)(3a + 2)$ **15.** $(4c - 3)(c - 3)$
17. $(x + 6y)(x - 2y)$ **19.** $(x + 3y)(x - 4y)$
21. $(5c - d)(3c - d)$ **23.** $(x^2 - 2)(x^2 - 7)$
25. $(a^2 + 3)(a^2 - 6)$ **27.** $(5a^2 + 7)(3a^2 - 1)$
29. $(x + 4)(x - 18)$ **31.** $(16y + 1)(4y - 1)$
33. $(y - 2)(y + 32)$ **35.** $(4n - 3)(3n - 2)$
37. $(9a - 4)(2a - 3)$ **39.** $(8n - 7)(2n + 3)$
41. $(8a + 9)(3a - 4)$ **43.** $(5x - 6y)(x + 2y)$
45. $(8c - 5d)(3c + 4d)$ **47.** $(4x^2 - 5)(3x^2 - 2)$
49. $(3y^2 + 2)(2y^2 - 9)$ **51.** $(3a^2 + 10)(3a^2 - 2)$
53. $(3a^2b^2 + 5)(a^2b^2 - 2)$
55. $(3c^2d^2 + 4)(2c^2d^2 + 5)$
57. $(7c^2 + 6d^2)(2c^2 - 3d^2)$

59. $(x^3 - 6y^2)(x^3 - y^2)$ **61.** $(x^{3m} - 3)(x^{3m} - 4)$
63. $(3x^{2m} + 10)(3x^{2m} - 2)$
65. $(5x^a + 3y^b)(x^a + 4y^b)$
67. $(2x^{3a} - 3y^{2b})(6x^{3a} + 5y^{2b})$
69. $(x^{n-1} - 2)(x^{n-1} + 5)$
71. $(x^{a+3} + y^{a-2})(x^{a+3} + y^{a-2})$
73. $(x + m)(x + n)$

Page 61

1. $n^2 - 144$ **3.** $81x^2 - 36$ **5.** $1 - 12n + 36n^2$
7. $100 - 36a^2$ **9.** $625 - 9y^2$
11. $49c^2 + 140c + 100$ **13.** $x^3 + 8$
15. $64 - 125a^3$ **17.** $25x^2 - 64y^2$
19. $16n^4 - 1$ **21.** $9x^2 - 60xy + 100y^2$
23. $49y^4 - 16$ **25.** $100y^4 - 25z^4$
27. $16m^4 + 24m^2n^2 + 9n^4$
29. $512x^3 + y^3$ **31.** $m^6 - 27n^3$

Page 64

1. $(2b + 7)(2b - 7)$ **3.** $(y + 4)^2$
5. $(x + 3)(x^2 - 3x + 9)$ **7.** $(c^2 + 5)(c^2 - 5)$
9. $(3n - 4)^2$ **11.** $(y^2 + 9)^2$
13. $(5c - 1)(25c^2 + 5c + 1)$
15. $(x + 3 + y)(x + 3 - y)$ **17.** $(x + y + 2)^2$
19. $(x + y + 5)(x - y - 5)$
21. $(7c^2d^2 + 10)(7c^2d^2 - 10)$
23. $(5c + 2d)^2$
25. $(4m + n)(16m^2 - 4mn + n^2)$
27. $(5t - 2v)(25t^2 + 10tv + 4v^2)$
29. $(4c^3 - 5d)^2$ **31.** $(5x^3 + 6d^2)(5x^3 - 6d^2)$
33. $(0.5x + 1.1y)(0.5x - 1.1y)$
35. $(m - n + x + y)(m - n - x - y)$
37. $(x + c + d)^2$
39. $(5x - 3 + 2a - b)(5x - 3 - 2a + b)$
41. $(3x^{2m+3} + 2y^n)^2$
43. $(x^c + y^{4d})(x^{2c} - x^cy^{4d} + y^{8d})$

Page 67

1. $3(4n^2 - 5n - 1)$ **3.** $3(x - 3)(x - 4)$
5. $4(2y + 1)(2y - 1)$ **7.** $4(n + 5)^2$
9. $(y^2 + 2)(y + 3)(y - 3)$ **11.** $-1(x - 4)^2$
13. $3(y + 3)(y^2 - 3y + 9)$
15. $(4x + 3)(2y + 5)$ **17.** $-1(2n - 3)(2n + 1)$
19. $(x^2 + 6)(x + 4)$
21. $15x^2y^3(5y^2 - 2xy + 3x^2)$
23. $4a(2a - 5)(a + 2)$ **25.** $2xy(x - 1)^2$
27. $(x^2 + 1)(3x + 4)(3x - 4)$
29. $(n + 2)(n - 2)(n + 3)(n - 3)$
31. $-3(y + 3)(y - 3)$ **33.** $-a(2c + 1)^2$
35. $3(2x - 5)(4x^2 + 10x + 25)$
37. $y(y + 4)(y^2 - 4y + 16)$

39. $(3a - 7d)(4b - 3c)$ **41.** $a(a - b^2)(a - b)$
43. $(2y^2 + 3)(2y^2 - 3)(y^2 + 1)(y + 1)(y - 1)$
45. $x^4(x^{a+2} + 1)$ **47.** $y^{n+4}(y + 1)$
49. $x^2(x^{2n} + 3)^2$
51. $(x + y)(x^2 - xy + y^2)(x - y)(x^2 + xy + y^2)$
53. The factors $x + y$ and $x - y$ occur in both
answers.; $x^4 + x^2y^2 + y^4 = (x^4 + 2x^2y^2 + y^4)$
$- x^2y^2 = (x^2 + y^2)^2 - (xy)^2 =$
$(x^2 + y^2 + xy)(x^2 + y^2 - xy) =$
$(x^2 - xy + y^2)(x^2 + xy + y^2)$

Page 70

1. 5, 8   **3.** $-10, -5$ **5.** $-2, 3$ **7.** 0, 5
9. 0, 15   **11.** $-6, 6$ **13.** $-5, 5, 1, -1$
15. $-10, 10, -3, 3$ **17.** $-6, 2$ **19.** $\frac{1}{2}, 4$
21. $\frac{2}{3}, -1$ **23.** $\frac{1}{5}, \frac{1}{2}$ **25.** $-1\frac{1}{3}, 1\frac{1}{3}$
27. $-1\frac{1}{5}, 1\frac{1}{5}$ **29.** $-5, 5$ **31.** -3 **33.** 6
35. $1\frac{1}{3}, 4$ **37.** $1\frac{1}{2}, 2$ **39.** $-3\frac{1}{2}, -1$ **41.** $0, 2\frac{1}{2}$
43. $2\frac{2}{3}, 1\frac{1}{2}$ **45.** $1\frac{3}{4}, -1\frac{2}{3}$
47. $-\frac{3}{5}, \frac{3}{5}, -1, 1$ **49.** $-1\frac{1}{2}, 1\frac{1}{2}$ **51.** $-\frac{1}{2}, \frac{1}{2}$
53. $0, -6, 1$ **55.** $0, 20$ **57.** $\frac{1}{2}, 1\frac{1}{2}$
59. $\frac{2a}{3}, -2a$ **61.** $-\frac{3a}{4}, \frac{7a}{3}$
63. $\frac{-a^2}{5}, \frac{a^2}{5}, \frac{-a^2}{2}, \frac{a^2}{2}$ **65.** $4, -4, 2, -2$

Page 73

1. $\{n \mid n < 3 \text{ or } n > 5\}$ **3.** $\{c \mid -6 \leq c \leq -2\}$
5. $\{x \mid x \leq -3 \text{ or } x \geq 3\}$ **7.** $\{y \mid y < -2 \text{ or } y > 2\}$
9. $\{n \mid n < 0 \text{ or } n > 5\}$ **11.** $\{c \mid -3 \leq c \leq 0\}$
13. $\{y \mid y < -2 \text{ or } 0 < y < 3\}$
15. $\{x \mid -5 \leq x \leq 0 \text{ or } x \geq 5\}$
17. $\{x \mid -2 < x < 2.5\}$ **19.** $\{y \mid y \leq -3.5 \text{ or } y \geq 0\}$
21. $\{c \mid c < -4 \text{ or } c > 3\}$ **23.** $\{n \mid 1 < n < 3\}$
25. $\{y \mid -3 < y < -2 \text{ or } 2 < y < 3\}$
27. $\{a \mid a \leq -2 \text{ or } a \geq 2\}$
29. $\{n \mid n < -4 \text{ or } n > 4\}$ **31.** $\{x \mid x \neq 2\}$
33. $\{\text{all numbers}\}$ **35.** $\{y \mid y < -4 \text{ or } 2 < y < 4\}$
37. $\{a \mid a < -5 \text{ or } -2 < a < 2 \text{ or } a > 5\}$
39. $\{a \mid -4 < a < -2 \text{ or } 0 < a < 2\}$
41. $\{x \mid x < -4 \text{ or } -2 < x < 2 \text{ or } x > 4\}$
43. $\{x \mid -1 < x < 1 \text{ or } 3 < x < 5\}$
45. $\{x \mid x < -3 \text{ or } x > 3\}$

Pages 76–77

1. −6, −4, −2 or 4, 6, 8 **3.** −8, −7, −6 or 6, 7, 8 **5.** 5 chairs **7.** 14 boxes **9.** −3, −7 or 7, 3 **11.** $-\frac{3}{2}$, −6 or $\frac{3}{2}$, 6 **13.** −20, −15, −10 or 10, 15, 20 **15.** −16, −12, −8 or 12, 16, 20 **17.** 50 m^2 **19.** −6, −1 or 3, 8 **21.** −15, −10, −5, 0 or 10, 15, 20, 25 **23.** −0.5, 0, 0.5, 1 or 2.5, 3, 3.5, 4 **25.** Every three consecutive integers

Pages 80–82

1. 6 m, 8 m **3.** 5 cm, 12 cm, 13 cm **5.** 4 cm, 5 cm **7.** 2 m, 12 m **9.** 18 dm **11.** 25 m^2, 100 m^2 **13.** square: 9 cm^2, rectangle: 36 cm^2 **15.** 5 m **17.** 192 cm^2 **19.** rectangle: 72 m^2, triangle: 36 m^2 **21.** 27 m^2 **23.** 2 m **25.** 16 cm, 21 cm

Page 84

1. $(5c - 2)(c - 7)$ **3.** $(5a^2b^2 + 4)(5a^2b^2 - 3)$ **5.** $(3a + 4)(3a - 4)$ **7.** $(a^2 - 8)^2$ **9.** $(2c - 5d)(4c^2 + 10cd + 25d^2)$ **11.** $(5x + y - 3)(5x - y + 3)$ **13.** $16n^2 - 36$ **15.** $x^3 + 27$ **17.** $64y^4 - 36z^2$ **19.** $3x(3x - 1)(2x - 1)$ **21.** $3(2y + 3)(2y - 3)$ **23.** $(2n^2 + 3)(n + 3)(n - 3)$ **25.** $2ab(a - 5)(a + 3)$ **27.** $(x^2 + 5)(x + 4)$ **29.** $2x^{3n} \cdot (x^{2n} + 3)(x^{2n} - 3)$ **31.** $0, -2\frac{1}{3}$ **33.** −5, 5, −2, 2 **35.** $\{y \mid -4 < y < 2\}$ **37.** $\{n \mid n \le -3 \text{ or } n \ge 2.5\}$ **39.** $\{x \mid x < -6 \text{ or } -2 < x < 2 \text{ or } x > 6\}$ **41.** 9 boxes **43.** $\frac{4}{3}$, 3 or −3, −10 **45.** base: 6 cm, height: 4 cm

Page 90

1. $\frac{-9}{1}$ **3.** $\frac{-34,719}{1000}$ **5.** $\frac{8}{9}$ **7.** $\frac{71}{90}$ **9.** $\frac{43}{99}$ **11.** $\frac{51}{110}$ **13.** $\frac{40}{9}$ **15.** $\frac{500}{99}$ **17.** $\frac{97}{18}$ **19.** $\frac{229}{60}$ **21.** $\frac{571}{999}$ **23.** $\frac{2,770}{333}$ **25.** $\frac{9}{9}$ **27.** $\frac{15}{4}$

Page 93

1. 0 **3.** 3 **5.** 3, 5 **7.** $\frac{2(y - 2)}{5(y - 3)}$ **9.** $\frac{-(c + 2)}{5}$ **11.** $\frac{3(x - 1)}{5(x - 6)}$ **13.** $\frac{2}{15a^4y^2z}$ **15.** $\frac{3}{-4c}$ **17.** $\frac{2(x - 5)}{-3}$ **19.** $\frac{4(n - 3)}{3(n + 5)}$ **21.** $\frac{y + 6}{6(y + 4)}$

23. $\frac{5}{x - 3}$ **25.** $\frac{3(2n + 3)}{4a^3(n + 3)}$ **27.** $\frac{2(3c + 1)}{3c}$ **29.** $\frac{x - 2y}{4(x + y)}$ **31.** $\frac{1}{2}$ **33.** $\frac{x + x^n}{x^n(x^n - x^2)}$ **35.** $\frac{x^{4n} + x^{2n}y^n + y^{2n}}{x^{2n} + y^n}$ **37.** $\frac{x^a x^b}{x^c}$, or x^{a+b-c} **39.** $x = y, x = -y$

Page 95

1. 12 qt **3.** 45 ft **5.** $2\frac{5}{8}$ lb **7.** 20,000 mg **9.** 0.75 L **11.** 3,456 in.^{2} **13.** 5 yd^3 **15.** 44 ft/s **17.** 5 lb/gal **19.** 10 platforms **21.** \$4,462.50

Page 98

1. $\frac{3}{5}$ **3.** 2 **5.** $\frac{5}{a + 6}$ **7.** $\frac{x}{8}$ **9.** $\frac{23a - 27}{15a}$ **11.** $\frac{7}{10}$ **13.** $\frac{6x - 8}{(x + 2)(x - 2)}$ **15.** $\frac{9y + 4}{(y - 5)(y + 4)}$ **17.** 4 **19.** $\frac{8x - 1}{(x + 3)(x - 3)}$ **21.** $\frac{3x - 7}{(x + 2)(x - 2)}$ **23.** $\frac{3x^2 + 8x - 26}{x + 4}$ **25.** $\frac{y - 7}{y + 8}$ **27.** $\frac{3a + 4}{a - 4}$ **29.** $\frac{11y - 9}{6(y + 3)(y - 1)}$ **31.** $\frac{40am + 10an + 4m^2 - 4mn}{5a(m - n)(m - n)}$ **33.** $\frac{3}{x + 3}$ **35.** $\frac{9y - 51}{4(y + 3)(y - 3)}$ **37.** $\frac{2y^3 + 26}{(y + 3)(y - 1)}$ **39.** $\frac{x^c + 5}{x^c}$ **41.** $\frac{x^4y^3 + x^a y^{4n}}{x^{3a+4}y^{5n+3}}$

Page 101

1. $\frac{7}{36}$ **3.** $\frac{98}{95}$ **5.** $\frac{16b + 18a}{2b - 9a}$ **7.** $\frac{6y + 40xy}{20xy - 3x}$ **9** $\frac{42m^2n - 18mn^2 + 12}{27n + 8m - 30mn}$ **11.** $\frac{a + 2}{a - 3}$ **13.** $\frac{5y + 6}{9y - 2}$ **15.** $\frac{-3n - 4}{13n - 6}$ **17.** $\frac{2a - 36}{7a - 14}$ **19.** $\frac{5a}{5a + 13}$ **21.** 1 **23.** $\frac{a - 3b}{a + 2b}$ **25.** $\frac{x^2y^2}{y^4 - x^4}$ **27.** $\frac{-x - 1}{x^2}$

Page 105

1. 80 **3.** 298 **5.** 7 **7.** 96 **9.** 97 **11.** 98 **13.** 93.75 **15.** 13 **17.** 108 **19.** 10.4, 3.22 **21.** 497.71, 22.31 **23.** 84 **25.** 94

Page 108

1. not possible 3. -2 5. $\dfrac{3(y + 2)}{4(y + 4)}$ 7. $-\dfrac{3}{4}$

9. $\dfrac{24a^3b - 4b}{5a^2 + 18a^2b^3}$ 11. $\dfrac{x^a + 4}{x^a + 2}$ 13. $\dfrac{x + 1}{3}$

15. $\dfrac{x^a y^{3a}}{z^{5a}}$ 17. $\dfrac{5(x + y)}{6(c + 2)}$

19. 360 lb/ft^2 21. $\dfrac{5}{x - 3}$ 23. $\dfrac{7}{2n + 5}$

25. $\dfrac{3x^2 + 15x + 6}{x + 3}$ 27. $\dfrac{7x - 37}{(x + 5)(x - 5)}$

29. $\dfrac{28}{(a - 3)(a + 1)}$ 31. $\dfrac{53}{90}$ 33. $\dfrac{2,707}{990}$

Pages 112–113

1. C 3. B 5. D 7. B 9. B 11. C 13. $-\dfrac{2}{7}$

15. $\dfrac{1}{3}$, 5 17. $2\dfrac{1}{2}$, -4 19. $\{x \mid x > 7\}$

21. $\{x \mid x < -2 \text{ or } x > 6\}$ 23. $\dfrac{-32b^5c^4}{a^2}$

25. $\dfrac{3n^9}{2m^6(n - 5)}$ 27. $21m^4 - 20m^2n - 25n^2$

29. $81y^4 - 64$ 31. $\dfrac{2(a - 4)}{3(a - 6)}$

33. $(6x - 5y)(2x + 3y)$ 35. $(6a - 7b)(6a + 7b)$
37. $(x + 4y + 3)(x + 4y - 3)$ 39. $2c(5c + 2)^2$
41. $(2a - 5)(2a + 5)(a + 1)(a - 1)$
43. $(x + 2)(x^2 - 6)$ 45. a: 8 cm, b: 32 cm,
c: 26 cm 47. 12 L 49. 6, 8, 10 or -12, -10, -8

Page 118

1. 3 3. 4 5. $\dfrac{7}{2}$ 7. -6, 4 9. -5, 10

11. -8 13. 1, 6 15. 1, 4 17. $\dfrac{3}{10}$ 19. -5, 3

21. $\dfrac{5}{2}$, 3 23. 4 25. -2, $\dfrac{2}{3}$ 27. $\dfrac{3}{2}$, $\dfrac{7}{4}$

29. $-\dfrac{2}{3}$, $\dfrac{5}{2}$ 31. -2, 4 33. 1, 3 35. $\dfrac{7}{3}$

37. -3, 4 39. $\dfrac{3}{2}$

Page 121

1. 6 days 3. 15 h 5. 5 h 7. $4\dfrac{2}{7}$ wk 9. $7\dfrac{4}{5}$ h

11. Marie: 35 h, Gene: 70 h, Merv: 140 h

Page 124

1. 72 km 3. 4 km/h 5. $16\dfrac{4}{11}$ km 7. 2 km/h

9. 20 km/h 11. 48 km

Pages 126–127

1. $x = \dfrac{b + c}{a}$ 3. $x = \dfrac{b}{a - c}$ 5. $x = \dfrac{b}{4a}$

7. $x = \dfrac{ab + c}{a}$ 9. $x = \dfrac{ab - c}{2a - cd}$

11. $x = \dfrac{bc + bd}{a}$ 13. $x = \dfrac{ac - bcd}{b}$

15. $x = \dfrac{3ab}{4b + 2a}$ 17. $y = \dfrac{4x + 30}{5}$

19. $y = 4x + 10$ 21. $y = mx - bm + a$

23. $B = \dfrac{A}{1 + CD}$; 0.5 25. $x = \dfrac{ab}{a + b}$; 4.2

27. $x = \dfrac{400t - gt^2}{2}$ 29. $x = \dfrac{ab + ac + bc}{a}$

31. $x = \dfrac{ad - bc}{d - c}$ 33. $x = \dfrac{4bc - 3ac - 3ab}{bc - 2ab}$

35. $x = \dfrac{ce - b}{a - cd}$ 37. $h = \dfrac{A}{a + b + c}$; 24

39. $h = \dfrac{3V}{B}$; 1.5×10^3 or 1,500

41. $-16t^2 + 480t + 80$; 1,744

43. $-\dfrac{3a}{2}, \dfrac{3a}{2}, -\dfrac{a}{2}, \dfrac{a}{2}$ 45. $-\dfrac{b}{2}, \dfrac{b}{2}$,

$-3a$, $3a$ 47. a

Page 129

1. $2c - 1$, -1, no 3. $10x + 12$, 0, yes
5. $4y - 5$, 11, no 7. $3x^2 + 8x + 7$, 30, no
9. $2n - 1$, 10, no 11. $a^2 - 3a - 4$, 0, yes
13. $3x + 2$, -6, no 15. $a^2 - a + 1$, 0, yes
17. $2n^2 - 4n + 1$, 0, yes 19. $y^2 - y - 1$,
0, yes 21. $x^{2m} + 3$, 2, no
23. $8x^2y - 3xy^2 - 2xy$, 0, yes

Pages 132–133

1. $x^2 + 3x - 5$, -8
3. $3y^3 - 8y^2 + 10y - 20$, 51
5. $n^4 - 2n^3 + 2n^2 - 4n + 9$, -20 7. 65
9. -880 11. 209 13. 1,598 15. no
17. yes 19. yes 21. yes 23. no 25. yes
27. no 29. 536 31. 379,642
33. 72, 5, 0, 9, 8, -3, 0; $x + 1$, $x - 3$
35. 60, 0, 2, 6, 0, 20, 150; $x + 2$, $x - 1$

Page 134

1. -2 3. 16 5. $-\dfrac{4}{3}$, 1 7. 6 9. 20

11. 18 hours 13. 1 km/h 15. $x = \dfrac{ab + 3c}{a - 2c}$

17. $-a, a, -3b, 3b$ 19. $h = \dfrac{2A}{a+b}$; 15

21. $2y - 3$ 23. $n^2 + 7n + 10, R: 14$ 25. no

27. 20 29. 0, 18, 0, 0; $(x+2), (x-2), (x-5)$

Page 141
1. $-8, 8$ 3. $-4, 4$ 5. $-\sqrt{29}, \sqrt{29}$
7. $-\sqrt{70}, \sqrt{70}$ 9. $-9.33, 9.33$ 11. $-2.50,$
2.50 13. $-6.48, 6.48$ 15. $-9.27, 9.27$ 17. I
19. I 21. R 23. R 25. R 27. I 29. R
31. R 33. I 35. R 37. $-0.67, 0.67$
39. $-2.33, 2.33$ 41. $-0.07, 0.07$ 43. $-5.00,$
5.00, $-2.24, 2.24$ 45. $-1.73, 1.73$

Page 144
1. $5\sqrt{10}; 15.81$ 3. $6\sqrt{5}; 13.42$ 5. $6\sqrt{30}; 32.86$
7. $22.5\sqrt{2}; 31.82$ 9. $6\sqrt{3}$ 11. $-18\sqrt{5}$
13. $10|x^3|$ 15. $36a^7$ 17. $n^4\sqrt{n}$
19. $-10y^6\sqrt{y}$ 21. $2|c^5|d\sqrt{6d}$ 23. $24x^4y^2\sqrt{xy}$
25. $3a^2b^7|c^9|\sqrt{10b}$ 27. $56c^8\sqrt{2d}$
29. $9|x^{25}|y^{12}\sqrt{3y}$ 31. $\dfrac{a^3\sqrt{3b}}{2c^4}$ 33. x^{2m}
35. $x^{m+1}\sqrt{x}$ 37. $x^{4m}y^{5n}$ 39. $x^{3m+1}y^{4n+2}\sqrt{y}$

Page 147
1. $-5\sqrt{10} + 6\sqrt{7}$ 3. $16\sqrt{5}$ 5. $2\sqrt{n}$
7. $80\sqrt{15}$ 9. $3y^5\sqrt{2}$
11. $60 - 15\sqrt{2} + 6\sqrt{35}$ 13. $40 - 20\sqrt{11}$
15. $22 - 8\sqrt{14}$ 17. 99 19. 6,800 21. 26
23. $13 + 2\sqrt{42}$ 25. $16x - 20$
27. $36n - 72\sqrt{n} + 36$ 29. $5\sqrt{2} + 5\sqrt{3}$
31. $5c\sqrt{d}$ 33. $11mn\sqrt{3n}$ 35. $16c^2d^6\sqrt{6}$
37. $32c - 4\sqrt{7c} - 21$ 39. $3x - 8\sqrt{xy} - 3y$
41. $25c - 16d$ 43. $a + 2\sqrt{ab} + b$
45. $x + 39 + 12\sqrt{x+4}$ 47. x
49. $2n + 5 - 2\sqrt{n^2 + 5n}$
51. $(\sqrt{m} + \sqrt{n})(\sqrt{m} - \sqrt{n})$
53. $(\sqrt{c} + 6\sqrt{d})(\sqrt{c} - 6\sqrt{d})$
55. $(x + \sqrt{2}y)(x - \sqrt{2}y)$
57. $(m\sqrt{6} + 5\sqrt{n})(m\sqrt{6} - 5\sqrt{n})$

Pages 150-151
1. $\sqrt{3}$ 3. $2\sqrt{2}$ 5. a^2 7. $3n^3$ 9. $\dfrac{\sqrt{7}}{7}$

11. $\dfrac{\sqrt{66}}{12}$ 13. $\dfrac{7\sqrt{3}}{6}$ 15. $\dfrac{-5\sqrt{2}}{8}$ 17. $\dfrac{5\sqrt{x}}{x^2}$

19. $\dfrac{-5\sqrt{2d}}{4d}$ 21. $\dfrac{42 + 7\sqrt{30}}{6}$ 23. $\dfrac{\sqrt{30} + 3\sqrt{2}}{6}$
25. $\dfrac{\sqrt{55}}{11}$ 27. $\dfrac{\sqrt{46xy}}{4y}$ 29. $\dfrac{2x^3\sqrt{7xy}}{7y^2}$
31. $\dfrac{\sqrt{18c} - 30}{6}$ 33. $\dfrac{10\sqrt{22} - 10\sqrt{11}}{11}$
35. $\dfrac{10 - 3\sqrt{10}}{2}$ 37. $\dfrac{50 + 13\sqrt{10}}{54}$
39. $\dfrac{15x + 6\sqrt{xy}}{25x - 4y}$ 41. $\dfrac{3x + 3y - 6\sqrt{x+y}}{x + y - 4}$
43. $\dfrac{-10 - 6\sqrt{15} + 8\sqrt{5} - 20\sqrt{3}}{-11}$

Page 154
1. -3 3. 5 5. x^3 7. y^5 9. $2\sqrt[4]{2}$
11. $4\sqrt[3]{2}$ 13. $a\sqrt[3]{a^2}$ 15. $x^3\sqrt[3]{x}$ 17. $2\sqrt[4]{8}$
19. $5\sqrt[4]{2}$ 21. $-4x^2$ 23. $5a^2\sqrt[3]{2a}$ 25. $a\sqrt[4]{21}$
27. $2x^2\sqrt[4]{10x}$ 29. $\dfrac{2\sqrt[3]{49}}{7}$ 31. 14 33. $\dfrac{\sqrt[4]{125}}{5}$
35. $\dfrac{\sqrt[4]{22}}{2}$ 37. 6 39. -3 41. $15a^2b^4\sqrt[3]{10a^2b}$
43. $9\sqrt[3]{4}$ 45. $2\sqrt[5]{2}$ 47. $5a\sqrt[3]{a^2}$ 49. $4x^2$
51. $2n\sqrt[5]{3n^4}$ 53. $3a\sqrt[4]{2a^2}$
55. $2\sqrt[3]{49} + 2\sqrt[3]{35} + 2\sqrt[3]{25}$
57. $\dfrac{25c\sqrt[3]{c^2} - 10cd\sqrt[3]{c} + 4cd^2}{125c + 8d^3}$ 59. x^{2n}
61. x^{2n+1} 63. x^4 65. $ax^5\sqrt[n]{x}$

Page 156
1. $5^{\frac{1}{2}}$ 3. $21^{\frac{1}{4}}$ 5. $(3b)^{\frac{1}{2}}$ 7. $\sqrt{17}$
9. $6\sqrt[3]{c}$ 11. $\sqrt[4]{7n}$ 13. 5 15. 3 17. -56
19. $\dfrac{1}{2}$ 21. $\dfrac{1}{2}$ 23. 4 25. 16 27. 9
29. 135 31. $\dfrac{1}{125}$ 33. $\dfrac{1}{243}$ 35. $\dfrac{5}{16}$
37. $c^{\frac{3}{5}}$ 39. $x^{\frac{5}{6}}$ 41. $n^{\frac{7}{2}}$ 43. $5\sqrt{5}$
45. $10\sqrt[3]{x^2}$ 47. $\sqrt[4]{216t^3}$

Pages 158-159
1. $6^{\frac{7}{8}}$ 3. $y^{\frac{5}{2}}$ 5. x^4 7. n^6 9. x^2y^6
11. $16mn^3$ 13. $a^{\frac{6}{7}}$ 15. $n^{\frac{4}{3}}$ 17. $\dfrac{x^6}{y^4}$ 19. $\dfrac{5m^3}{3n^4}$

21. $\dfrac{3}{2}$ 23. $\dfrac{2}{3}$ 25. $\dfrac{1}{4}$ 27. $\dfrac{14}{15}$ 29. $\dfrac{x^{\frac{4}{5}}}{x}$

31. $210a^{\frac{37}{20}}$ 33. $\dfrac{1}{x^6}$ 35. $\dfrac{2m^3}{n^2}$ 37. $\dfrac{x^{\frac{2}{3}}y^{\frac{1}{2}}}{x}$ 39. $\dfrac{m^{\frac{2}{3}}n}{m}$

41. $\dfrac{y^2}{x^6}$ **43.** $\dfrac{16n^4}{9m^2}$ **45.** $\sqrt[10]{x^7}$ **47.** $\sqrt[4]{z^3}$

49. $\sqrt[12]{a^6 b^4 c^3}$ **51.** $\sqrt[3]{x^2} - \sqrt{x}$

53. $x\sqrt{x} - y\sqrt[4]{y}$ **55.** $9x - 12\sqrt[6]{x^3 y} + 4\sqrt[3]{y}$

57. $\sqrt[6]{x}$ **59.** $\dfrac{\sqrt[20]{x^{19}}}{x}$ **61.** $x + y$

63. $\dfrac{-2x^{\frac{2}{3}} - 2(xy)^{\frac{1}{3}} - 2y^{\frac{2}{3}}}{x - y}$

65. $\dfrac{4x^{\frac{2}{3}} - 2(3x)^{\frac{1}{3}} + 3^{\frac{2}{3}}}{8x + 3}$ **67.** $x^{\frac{1}{3}} - y^{\frac{1}{3}}$

Page 160

1. $-5.48, 5.48$ **3.** R **5.** R **7.** $2|c|\sqrt{5}$

9. $4m^5 n^2 \sqrt{2mn}$ **11.** $68 - 16\sqrt{15}$

13. $3a^2 \sqrt{a}$ **15.** $\dfrac{\sqrt{14} - \sqrt{5}}{9}$ **17.** $2c^2$

19. $7a^2 \sqrt[3]{a}$ **21.** $c\sqrt[4]{4c}$ **23.** $17^{\frac{2}{3}}$ **25.** $c^{\frac{5}{2}}$

27. $4\sqrt[5]{y^3}$ **29.** -30 **31.** $2,500$ **33.** $-18a^{\frac{1}{4}}$

35. $\dfrac{b^6}{a^3}$ **37.** $\dfrac{4x^6 y^8}{9}$ **39.** $\dfrac{5}{6}$ **41.** -1

43. $x^{3m+2} y^{5n} \sqrt{x}$ **45.** $x^2 y^3 \sqrt[n]{y^5}$

47. $x^{2a} y^a \sqrt[n]{y^{2a}}$

Page 166

1. $3, -7$ **3.** $-10, -2$ **5.** $-5 \pm 2\sqrt{5}$

7. $\dfrac{-3 \pm \sqrt{33}}{2}$ **9.** $\dfrac{1 \pm \sqrt{29}}{2}$ **11.** $4, -\dfrac{1}{2}$

13. $\dfrac{5 \pm 3\sqrt{5}}{2}$ **15.** $1, 0$ **17.** $\dfrac{7 \pm \sqrt{29}}{10}$

19. $3 \pm 2\sqrt{3}$ **21.** $\dfrac{3 \pm \sqrt{17}}{4}$ **23.** $5 \pm \sqrt{10}$

25. $2\sqrt{3}, -4\sqrt{3}$ **27.** $-3\sqrt{5}$ **29.** $3\sqrt{2} \pm \sqrt{3}$

Page 168

1. $18, x = 3$ **3.** $19, x = 3$ **5.** $1,125; x = 5$

7. 180 m, 6 s **9.** $1,800$ m^2, 30 m by 60 m

11. $\$1.50, \$450$

Page 171

1. $\dfrac{-5 \pm 2\sqrt{2}}{2}$ **3.** $-5 \pm 2\sqrt{2}$ **5.** $\dfrac{5}{2}$

7. $\dfrac{-1 \pm \sqrt{33}}{8}$ **9.** $\dfrac{1 \pm \sqrt{17}}{4}$ **11.** $-2 \pm \sqrt{5}$

13. $\dfrac{3 \pm \sqrt{2}}{2}$ **15.** $\dfrac{1 \pm 2\sqrt{2}}{3}$ **17.** $\dfrac{3 \pm \sqrt{7}}{2}$ **19.** 4

21. $\pm\sqrt{11}$ **23.** $-\dfrac{2}{3}$ **25.** $\dfrac{7}{4}, 0$ **27.** $\dfrac{5}{2}$

29. $\dfrac{9 \pm \sqrt{21}}{6}$ **31.** 5 **33.** $\dfrac{1 \pm \sqrt{7}}{2}$

35. $\dfrac{2 \pm \sqrt{58}}{3}$ **37.** $3\sqrt{2}, -\sqrt{2}$

39. $\dfrac{-\sqrt{10} \pm \sqrt{15}}{5}$ **41.** $\dfrac{5 + \sqrt{5}}{2}, \dfrac{5 - 5\sqrt{5}}{2}$

43. $\dfrac{3 \pm \sqrt{5}}{4}$ **45.** $\pm 2, \pm\sqrt{2}$

Pages 173–174

1. $1 - \sqrt{5}, -1 - \sqrt{5}$ or $1 + \sqrt{5}, -1 + \sqrt{5}$

3. $\dfrac{1 - \sqrt{7}}{2}, -1 - \sqrt{7}$ or $\dfrac{1 + \sqrt{7}}{2}, -1 + \sqrt{7}$

5. 2.8 s, 7.2 s **7.** 0.4 s, 7.6 s **9.** 7.6 s

11. $\dfrac{-5 + \sqrt{7}}{3}, -1 + \sqrt{7}$ or $\dfrac{-5 - \sqrt{7}}{3}, -1 - \sqrt{7}$

13. 1.6 s, 6.4 s **15.** 8.4 s

Pages 176–177

1. $\dfrac{6\sqrt{5}}{5}$ m, $\dfrac{12\sqrt{5}}{5}$ m **3.** 5.2 cm by 25.2 cm

5. $4\sqrt{2}$ cm **7.** 1.6 m **9.** $(4 + \sqrt{6})$ dm;

$(9 + 4\sqrt{6})$ dm^2 **11.** $(4 + 2\sqrt{2})$ m; $(12 + 8\sqrt{2})$ m^2

Page 180

1. 51 **3.** 3 **5.** 4 **7.** 5 **9.** 2 **11.** $\dfrac{8}{81}$

13. 8 **15.** 6 **17.** $\pm 2\sqrt{2}$ **19.** $2\sqrt[3]{2}$

21. $\pm 4\sqrt{15}$ **23.** 9 **25.** $-\dfrac{1}{72}$ **27.** $7, -1$

29. $\dfrac{1}{4}$ **31.** 12 **33.** 2 **35.** $2, 6$ **37.** $-4, 3$

39. $\dfrac{1}{5}, 3$ **41.** $3, 7$ **43.** 7 **45.** 6 **47.** 2

49. $l = \dfrac{T^2 g}{4\pi^2}$ **51.** 10 **53.** $V = \dfrac{4\pi r^3}{3}$

55. $T = -Mx^2 + Nx + D$

Page 182

1. $3 \pm 2\sqrt{3}$ **3.** $-6\sqrt{2}, 4\sqrt{2}$ **5.** $\dfrac{\pm\sqrt{35}}{5}$

7. $\dfrac{4 \pm \sqrt{13}}{3}$ **9.** 320 m, $t = 8$ s **11.** 3.6 s,

8.4 s **13.** $(5 + 2\sqrt{3})$ m, $(18 + 10\sqrt{3})$ m^2

15. 6 **17.** 80 **19.** $2, 3$

Page 189

1. $6i$ **3.** $8i$ **5.** $i\sqrt{22}$ **7.** $2i\sqrt{10}$ **9.** $6i\sqrt{5}$

11. $-12i\sqrt{5}$ **13.** $4 + 3i$ **15.** $8 - 5i\sqrt{2}$

17. $\pm i\sqrt{3}$ **19.** $\pm i\sqrt{10}$ **21.** $\pm 5i\sqrt{2}$

23. $4 + 2i$ **25.** $9 + 4i$ **27.** $-12 + 2i$
29. $-8 + 12i$ **31.** $7 + 32i$ **33.** $2\sqrt{15} - 4i\sqrt{15}$
35. $-1 + 16i\sqrt{5}$ **37.** $\pm 9i$ **39.** $\pm i\sqrt{65}$
41. $\pm i\sqrt{3}$ **43.** $-5 + 2i$ **45.** $5 - 3i\sqrt{3}$
47. $8 + 7i\sqrt{2}$ **49.** 5 **51.** 3 **53.** $10, 10$
55. $|a + bi| = \sqrt{a^2 + b^2}$ and $|a - bi| =$
$\sqrt{a^2 + (-b)^2} = \sqrt{a^2 + b^2}$

Page 192

1. -42 **3.** $-6i$ **5.** $12\sqrt{22}$ **7.** $-16\sqrt{3}$
9. $-10i$ **11.** $6\sqrt{3} - 6i\sqrt{15}$ **13.** $2 + 39i$
15. $-13 + 21i$ **17.** 85 **19.** $-9 + 40i$
21. $24 + 10i$ **23.** 9 **25.** $\dfrac{7i}{-4}$ **27.** $\dfrac{4i}{3}$ **29.** $\dfrac{2i}{5}$
31. $-7i$ **33.** $\dfrac{15 - 10i}{13}$ **35.** $\dfrac{1 - i}{2}$
37. $\dfrac{-6 + 17i}{25}$ **39.** $\dfrac{1 + i}{2}$ **41.** $-1,000,000$
43. $-i$ **45.** $4 + 16i\sqrt{3}$ **47.** $-46 - 12\sqrt{14}$
49. $3,969$ **51.** -8 **53.** $10\sqrt{2}, 10\sqrt{2}$
55. $(a + bi)(a - bi) = a^2 - b^2 i^2 = a^2 + b^2$,
a real number

Page 194

1. $\dfrac{3 \pm i\sqrt{3}}{2}$ **3.** $\dfrac{3 \pm i\sqrt{3}}{6}$ **5.** $\dfrac{3 \pm i\sqrt{7}}{2}$
7. $2 \pm i$ **9.** $4 \pm 2i$ **11.** $\dfrac{1 \pm i\sqrt{5}}{3}$ **13.** $\pm 3i$,
$\pm\sqrt{2}$ **15.** $\pm i, \pm 2\sqrt{2}$ **17.** $\pm i\sqrt{2}, \pm 2\sqrt{3}$
19. $\dfrac{1 \pm i\sqrt{3}}{2}, -1$ **21.** $\dfrac{3 \pm 3i\sqrt{3}}{2}, -3$
23. $-5, \dfrac{5 \pm 5i\sqrt{3}}{2}$ **25.** $\dfrac{2 \pm i\sqrt{2}}{2}$ **27.** $\dfrac{5 \pm i\sqrt{23}}{4}$
29. $\pm 2i\sqrt{2}, \pm i\sqrt{2}$ **31.** $\dfrac{\pm i\sqrt{3}}{2}, \pm 2i$
33. $\dfrac{\pm 3i\sqrt{2}}{2}, \pm i$ **35.** $3i \pm i\sqrt{17}$ **37.** -9
39. $-6i$ **41.** $-\dfrac{2}{3}, \dfrac{1 \pm i\sqrt{3}}{3}$

Page 197

R = real, Im = imaginary, Rt = rational,
Ir = irrational
1. Two R: Rt **3.** Two Im **5.** One R: Rt
7. Two R: Rt **9.** Two Im **11.** Two R: Ir
13. $10, -10$ **15.** 9 **17.** $-9, 15$ **19.** Two R: Rt
21. One R: Rt **23.** Two R: Ir **25.** Two R: Rt
27. Two R: Ir **29.** One R: Ir
31. $k < -8$ or $k > 8$

Page 200

1. sum: $-\dfrac{9}{2}$, prod.: -3 **3.** sum: 3, prod: $-\dfrac{1}{2}$
5. sum: $-\dfrac{5}{6}$, prod.: 0 **7.** sum: 6, prod.: 9
9. $x^2 - 8x + 15 = 0$ **11.** $x^2 - 4x = 0$
13. $8x^2 - 6x + 1 = 0$ **15.** $3x^2 + 16x - 12 = 0$
17. $x^2 - 14x + 49 = 0$ **19.** $4x^2 + 4x + 1 = 0$
21. $x^2 - 7 = 0$ **23.** $x^2 + 10 = 0$
25. $x^2 - 8x + 11 = 0$ **27.** $x^2 - 10x + 53 = 0$
29. $4x^2 - 8x + 1 = 0$ **31.** $3x^2 + 6x + 7 = 0$
33. sum: $-\dfrac{1}{12}$, prod.: $-\dfrac{1}{8}$
35. sum: -2, prod.: -4
37. $r + s = \dfrac{-b + \sqrt{b^2 - 4ac}}{2a} +$
$\dfrac{-b - \sqrt{b^2 - 4ac}}{2a} = \dfrac{-2b}{2a} = -\dfrac{b}{a};$
$r \cdot s = \dfrac{-b + \sqrt{b^2 - 4ac}}{2a} \cdot \dfrac{-b - \sqrt{b^2 - 4ac}}{2a} =$
$\dfrac{(-b)^2 - (\sqrt{b^2 - 4ac})^2}{4a^2} = \dfrac{b^2 - (b^2 - 4ac)}{4a^2} =$
$\dfrac{4ac}{4a^2} = \dfrac{c}{a}$ **39.** $-\dfrac{1}{2}, \dfrac{3}{2}$

Page 203

1. $10i\sqrt{3}$ **3.** $7 + 3i\sqrt{2}$ **5.** $-2,048i$
7. $6 - 8i\sqrt{10}$ **9.** $\dfrac{-6 + 10i}{17}$ **11.** $\pm 2i\sqrt{15}$
13. $-2 \pm i\sqrt{2}$ **15.** $\pm 2i, \pm 2i\sqrt{2}$
17. $3, \dfrac{-3 \pm 3i\sqrt{3}}{2}$ **19.** Two Im **21.** Two R: Ir
23. sum: $-\dfrac{3}{4}$, prod.: $-\dfrac{1}{2}$
25. $x^2 + 3x - 40 = 0$ **27.** $x^2 + 10x - 3 = 0$
29. $5, -2$

Pages 206–207

1. D **3.** A **5.** D **7.** C **9.** C **11.** B
13. $\dfrac{-22}{5}, 6$ **15.** $1, 7$ **17.** $\dfrac{5}{9}$ **19.** 2
21. $\pm 2i\sqrt{3}, \pm 2i$ **23.** $\dfrac{b - cv - at}{3a - 4c}$
25. $2(n - 3)(n^2 + 3n + 9)$ **27.** $\{x \mid x \geq -3\}$
29. $\{n \mid -5 < n < -1 \text{ or } 1 < n < 5\}$
31. $\dfrac{-y^9}{2x^8}$ **33.** $\dfrac{-2(a - 1)}{3(a + 3)}$ **35.** $\dfrac{xy(2x + 3y)}{2y + 3x}$
37. $-5c\sqrt{2cd}$ **39.** $3c\sqrt[4]{2c^2}$ **41.** $-9 + 5i\sqrt{3}$

43. $5x^2 - 20x + 3$, $R: -4$ **45.** 17 **47.** R

49. I **51.** $\dfrac{4}{125}$ **53.** $\sqrt[4]{125y^3}$

55. $13\dfrac{1}{3}$ min **57.** $(25 + 10\sqrt{5})$ m,

$(450 + 200\sqrt{5})$ m²

Page 211

1. left 3, up 2 **3.** up 1 **5.** right 4, up 3
7. left 1, up 4 **9.** left 3 **11.** $A(3, 0)$; $B(-5, 3)$;
$C(-1, -3)$; $D(5, 2)$; $E(0, 1)$; $F(-3, 2)$; $G(-4, -1)$;
$H(2, -2)$; $I(1, -4)$ **13.** I: D; II: B, F; III: C, G;
IV: H, I (In Ex. 15–47 the coordinates of three
points are given. The points you use may differ.)
15. $(0, 3)$, $(2, -1)$, $(-2, 7)$ **17.** $(0, -2)$,

$\left(\dfrac{1}{2}, 0\right)$, $(1, 2)$ **19.** $(0, 4)$, $(1, 6)$, $(-1, 2)$

21. $(0, -5)$, $(5, 0)$, $(2, -3)$ **23.** $(0, -5)$, $(1, -3)$,
$(-1, -7)$ **25.** $(0, -3)$, $(1, -1)$, $(-1, -5)$
27. $(0, 3)$, $(1, 1)$, $(-1, 5)$ **29.** $(0, 4)$, $(2, 0)$, $(1, 2)$
31. $(0, 6)$, $(4, 0)$, $(2, 3)$ **33.** $(0, 3)$, $(3, -2)$,
$(-3, 8)$ **35.** $(-3, 0)$, $(-1, 1)$, $(1, 2)$ **37.** $(2, 0)$,
$(5, 4)$, $(-1, -4)$ **39.** $(-10, 0)$, $(-7, 2)$, $(-4, 4)$
41. $(6, 2)$, $(24, 0)$, $(-3, 3)$ **43.** $(0, 4)$, $(6, 0)$,
$(3, 2)$ **45.** $(0, 3)$, $(6.4, 0)$, $(1, 2.5)$ **47.** $(0, -1.4)$,
$(-1.2, 0)$, $(1, -2.6)$

Page 215

1. 5 **3.** 5 **5.** 4 **7.** 10 **9.** $\dfrac{3}{10}$, up, right

11. -4, down, right **13.** undefined, vertical

15. $-\dfrac{1}{2}$, down, right **17.** 0, horizontal

19. $\dfrac{15}{19}$, up, right **21.** 3, up, right

23. 0, horizontal **25.** $2 = 2 = 2 = m$

27. $\dfrac{2}{3} = \dfrac{2}{3} = \dfrac{2}{3} = m$ **29.** $\dfrac{16}{9}$ **31.** $\dfrac{5}{4}$ **33.** $\dfrac{5}{2}$

35. $-\dfrac{b}{a}$ **37.** $\dfrac{7b}{a}$ **39.** $\dfrac{s}{t}$ **41.** 2 **43.** $-\dfrac{1}{6}$

Page 219

1. $2x - 3y = 19$ **3.** $5x - 6y = -41$
5. $4x + y = -4$ **7.** $4x + 3y = 6$
9. $5x + 8y = -3$ **11.** $4x + y = -4$
13. $x - y = -4$ **15.** $7x + 2y = 8$
17. $3x + 10y = 51$ **19.** $y = 5$ **21.** $x = 8$
23. $9x - 5y = -45$ **25.** $y = 12$ **27.** $y = -3$
29. $\dfrac{3}{5}$, -2 **31.** $\dfrac{5}{3}$, -5 **33.** $-\dfrac{3}{4}$, 10

35. $-\dfrac{2}{3}$, 2 **37.** -8, -15 **39.** 0, 2

41. $-\dfrac{23}{10}$, $\dfrac{7}{10}$ **43.** $-\dfrac{5}{17}$, $\dfrac{6}{17}$ **45.** $-\dfrac{29}{4}$, -1

47. $-\dfrac{248}{75}$, $-\dfrac{148}{75}$ **49.** $x - y = 4$

51. $Ax + By = C$

Page 222

(In Ex. 1–19, 33–37, the coordinates of three points
are given.) **1.** $(0, 3)$, $(3, 5)$, $(-3, 1)$ **3.** $(0, 1)$,
$(-3, 0)$, $(3, 2)$ **5.** $(0, -1)$, $(1, -3)$, $(-1, 1)$
7. $(0, -2)$, $(1, -5)$, $(-1, 1)$ **9.** $(0, 4)$, $(1, 2)$,
$(-1, 6)$ **11.** $(0, -3)$, $(1, 1)$, $(-1, -7)$ **13.** $(0, 2)$,
$(2, -1)$, $(-2, 5)$ **15.** $(0, 5)$, $(1, 5)$, $(-1, 5)$
17. $(0, 4)$, $(4, 0)$, $(2, 2)$ **19.** $(-4, 0)$, $(-4, 1)$,
$(-4, -1)$ **21.** yes **23.** no **25.** no **27.** yes
29. yes **31.** yes **33.** $(1, 3)$, $(-3, 4)$, $(5, 2)$

35. $\left(0, -\dfrac{1}{2}\right)$, $\left(1, -\dfrac{7}{6}\right)$, $\left(-1, \dfrac{1}{6}\right)$ **37.** $\left(0, \dfrac{20}{3}\right)$, $\left(\dfrac{20}{7}, 0\right)$,

$\left(3, -\dfrac{1}{3}\right)$ **39.** $(2, 4)$ **41.** $x - y = 0$; $x + y = 0$

Page 225

1. $AB = \sqrt{82}$; $BC = 1$; $AC = 9$ **3.** $AB = \sqrt{85}$;
$BC = 7$; $AC = 6$ **5.** $AB = 5$; $BC = 3$; $AC = 4$
7. $2\sqrt{5}$ **9.** $\sqrt{97}$ **11.** $3\sqrt{13}$ **13.** $\sqrt{149}$
15. $\sqrt{41}$ **17.** $2\sqrt{5}$ **19.** $\sqrt{10}$ **21.** $2\sqrt{2}$

23. $\dfrac{1}{6}\sqrt{229}$ **25.** $\dfrac{1}{6}\sqrt{221}$ **27.** $\dfrac{17}{20}$

29. $\sqrt{b^2 + 4a^2}$ **31.** $\sqrt{9p^2 + 16s^2}$ **33.** a

35. $\dfrac{1}{12}\sqrt{b^2 + 16a^2}$ **37.** $\dfrac{1}{8}\sqrt{25b^2 + 16a^2}$

39. yes **41.** yes **43.** no **45.** yes **47.** yes
49. 6, -2 **51.** $1 \pm 4\sqrt{2}$
53. $c + b + \sqrt{e^2 + d^2} + \sqrt{e^2 + (c - d + b)^2}$

55. $\dfrac{1}{30}\sqrt{6481} + \dfrac{7}{30}\sqrt{349} + \dfrac{3}{2}\sqrt{10}$

Page 227

1. $(5, 3)$ **3.** $\left(-4, -\dfrac{1}{2}\right)$ **5.** $(3, 2)$ **7.** $(2, 1)$

9. $(2, 1)$ **11.** $(-4, -3)$ **13.** $\left(\dfrac{7}{2}, 8\right)$

15. $\left(-\dfrac{13}{2}, -\dfrac{13}{2}\right)$ **17.** $\left(\dfrac{5}{12}, -\dfrac{3}{20}\right)$

19. $\left(-\dfrac{2}{15}, \dfrac{13}{24}\right)$ **21.** $\left(-\dfrac{2}{15}, -\dfrac{1}{15}\right)$

23. $(4a, 3b)$ **25.** $(m + 2, n - 3)$ **27.** $(9, 8)$
29. $(-1, -2)$ **31.** $(4, -2)$ **33.** $(8, 1)$

35. $\left(-\dfrac{5}{3}, -\dfrac{19}{6}\right)$ **37.** $\left(\dfrac{33}{5}, \dfrac{7}{2}\right)$ **39.** $4\sqrt{5}$
41. Midpt. $M(-1, 2)$; $AM = MC = 5$;
$BM = MD = 10$

Page 230

1. $x + 4y = 33$ **3.** $2x - 5y = 14$ **5.** $x = -3$
7. parallel **9.** perpendicular **11.** neither
13. $y = -2$ **15.** $x - y = 16$ **17.** $2x - 3y = 14$
19. $9x - 7y = 68$ **21.** $2x + 3y = 22$
23. $2x - 3y = 24$ **25.** $x + y = 2$
27. $x + y = 1$ **29.** $3x - 4y = 8$
31. $bx + ay = 2ab$ **33.** $(a - c)x + (b - d)y$
$= a^2 - c^2 - d^2 + b^2$

Page 233

1. $m(\overline{AB}) = \dfrac{5 - 2}{2 - 1} = 3$; $m(\overline{BC}) = \dfrac{6 - 5}{-1 - 2}$
$= -\dfrac{1}{3}$; $AB \perp BC$ **3.** $m(\overline{AB}) = \dfrac{3 + 5}{7 - 5} = \dfrac{8}{2} = 4$;
$m(\overline{BC}) = \dfrac{5 - 3}{-1 - 7} = \dfrac{2}{-8} = -\dfrac{1}{4}$; $\overline{AB} \perp \overline{BC}$
5. $m(\overline{AB}) = \dfrac{-8 - 2}{-1 - 1} = \dfrac{-10}{-2} = 5$; $m(\overline{BC})$
$= \dfrac{-3 + 8}{0 + 1} = 5$; $m(\overline{AB}) = m(\overline{BC})$
7. $AC = \sqrt{(5 - 1)^2 + (4 - 2)^2} = \sqrt{16 + 4}$
$= \sqrt{20} = 2\sqrt{5}$ Midpoint D of $\overline{AB} = (5, 2)$;
Midpoint E of $\overline{BC} = (6, 4)$; $DE =$
$\sqrt{(4 - 2)^2 + (6 - 5)^2} = \sqrt{4 + 1} = \sqrt{5}$;
$DE = \dfrac{1}{2}AC$ **9.** $m(\overline{AB}) = 0$; $m(\overline{BC}) = \dfrac{b}{0}$,
undefined, $AB \perp BC$; $m(\overline{AD}) = \dfrac{b}{0}$, undefined,
$\overline{AB} \perp \overline{AD}$; $m(\overline{CD}) = 0$, $\overline{CD} \perp \overline{AD}$; $\overline{CD} \perp \overline{BC}$

Page 236

1. left 2, down 3 **3.** left 7 **5.** right 5, down 5
7. $A(-2, -3)$ **9.** $C(1, 3)$ **11.** $E(1, 0)$
13. $\left(0, 1\dfrac{1}{2}\right), \left(-1\dfrac{1}{2}, 0\right), \left(1, 2\dfrac{1}{2}\right)$ **15.** 4 **17.** 2
19. $\dfrac{-y}{x}$ **21.** $2x - 3y = -11$
23. $x - 2y = -11$ **25.** $-4, 13$ **27.** yes
29. $2\sqrt{17}$ **31.** $\dfrac{5}{6}\sqrt{2}$ **33.** $\left(\dfrac{11}{2}, \dfrac{1}{2}\right)$
35. $Q(5, -2)$ **37.** $2x + 3y = 5$
39. $x - 2y = 16$

Page 242

1. $(2, -1)$; consistent, independent **3.** $(-1, -1)$;
consistent, independent **5.** $(2, 3)$; consistent,
independent **7.** no solution; inconsistent
9. $(0, 3)$; consistent, independent **11.** $(5, -3)$;
consistent, independent **13.** no solution;
inconsistent **15.** $(1, -2)$; consistent, independent
17. any two equations with $m_1 = m_2$, $b_1 \neq b_2$
19. 3 **21.** 6

Page 246

1. $(-1, -2)$ **3.** $(-4, -7)$ **5.** all (x, y) such that
$x = 2y - 3$ **7.** $\left(-1, \dfrac{3}{4}\right)$ **9.** $\left(-1, \dfrac{5}{2}\right)$
11. $\left(\dfrac{3}{10}, \dfrac{1}{10}\right)$ **13.** $(2, 3)$ **15.** $(0, 3)$
17. 20 km/h **19.** Alicia: 14; Bill: 10

Page 249

1. $(7, -3, 6)$ **3.** $(1, -3, 3)$ **5.** $(2, 1, -1)$
7. $(4, -5, 8)$ **9.** $(2, -3, 4)$ **11.** 8, 10, 12
13. nickels: 40, dimes: 30, quarters: 10
15. $\left(\dfrac{14}{65}, \dfrac{2}{17}, \dfrac{14}{29}\right)$

Page 252

1. 16 **3.** 1 **5.** -38 **7.** 51 **9.** 22 **11.** $-\dfrac{1}{4}$
13. $2m + 8n$ **15.** $-6p - 16r$ **17.** $(-4, -2)$
19. $(-1, 2)$ **21.** $(-1, 0)$ **23.** $\left(-\dfrac{19}{7}, \dfrac{8}{7}\right)$
25. $(0.05, -0.15)$ **27.** p^2 **29.** $-2n$
31. $\left(\dfrac{nr + ms}{n + 2m}, \dfrac{s - 2r}{n + 2m}\right)$
33. $\left(\dfrac{c_1 b_2 - c_2 b_1}{a_1 b_2 - a_2 b_1}, \dfrac{a_1 c_2 - a_2 c_1}{a_1 b_2 - a_2 b_1}\right)$

Page 256

1. -75 **3.** 15 **5.** -63 **7.** -28 **9.** $(1, -1, 2)$
11. $(3, -1, 0)$ **13.** $(1, -1, 2)$ **15.** $(4, -8, 3)$
17. $(5, 4, 10)$ **19.** $(-1, 2, 3)$

Page 259

1. yes **3.** no **5.** $\begin{bmatrix} 7 & 0 & 10 \\ 4 & -2 & -1 \\ 6 & -6 & -3 \end{bmatrix}$
7. $\begin{bmatrix} 0 & 0 & 0 \\ 0 & 0 & 0 \end{bmatrix}$ **9.** $\begin{bmatrix} -9 & -15 \\ 3 & -6 \end{bmatrix}$
11. $\begin{bmatrix} 8 & -3 \\ -7 & -6 \end{bmatrix} + \begin{bmatrix} 0 & 0 \\ 0 & 0 \end{bmatrix} = \begin{bmatrix} 8 & -3 \\ -7 & -6 \end{bmatrix}$

13. $\begin{bmatrix} 12 & 0 \\ -2 & -21 \end{bmatrix} = \begin{bmatrix} 12 & 0 \\ -2 & -21 \end{bmatrix}$

15. $\begin{bmatrix} 12 & 0 \\ -2 & -21 \end{bmatrix} = \begin{bmatrix} 12 & 0 \\ -2 & -21 \end{bmatrix}$

Page 263

1. yes **3.** no **5.** $\begin{bmatrix} 6 & -9 & -15 \\ 18 & -30 & -18 \\ -24 & -36 & 12 \end{bmatrix}$

7. $\begin{bmatrix} -7 & -6 & -9 \\ -8 & 1 & 5 \end{bmatrix}$ **9.** $\begin{bmatrix} 5 & 1 & 8 \\ 11 & 20 & 0 \end{bmatrix}$

11. $\begin{bmatrix} 8 & -8 & 0 \\ -4 & 6 & -2 \end{bmatrix}$ **13.** $\begin{bmatrix} 1 & 0 \\ 0 & 1 \end{bmatrix}; \begin{bmatrix} -\dfrac{1}{20} & -\dfrac{1}{10} \\ \dfrac{3}{10} & -\dfrac{2}{5} \end{bmatrix}$

15. $\begin{bmatrix} 1 & 0 & 0 \\ 0 & 1 & 0 \\ 1 & 0 & 1 \end{bmatrix} \begin{bmatrix} -\dfrac{2}{27} & \dfrac{-8}{27} & \dfrac{11}{54} \\ -\dfrac{19}{27} & \dfrac{-22}{27} & \dfrac{37}{54} \\ -\dfrac{8}{27} & \dfrac{-5}{27} & \dfrac{17}{54} \end{bmatrix}$

17. $\begin{bmatrix} -5 & -33 \\ 17 & -19 \end{bmatrix} \neq \begin{bmatrix} -2 & -36 \\ 17 & -22 \end{bmatrix}$

19. $\begin{bmatrix} -19 & 30 \\ -36 & 6 \end{bmatrix} = \begin{bmatrix} -19 & 30 \\ -36 & 6 \end{bmatrix}$

21. $\begin{bmatrix} 1 & -6 \\ 3 & -2 \end{bmatrix} \cdot \begin{bmatrix} -\dfrac{1}{8} & \dfrac{3}{8} \\ -\dfrac{3}{16} & \dfrac{1}{16} \end{bmatrix} = \begin{bmatrix} 1 & 0 \\ 0 & 1 \end{bmatrix}$

23. True **25.** True **27.** True

Page 267
1. $(1, -1)$ **3.** $(-1, -1)$ **5.** $(3, -1)$
7. $(1, -2, 1)$ **9.** $(2, -1, -3)$
11. $(-2, -1, 3)$ **13.** $(-1, -1, 2)$
15. $(-2, -2, 2)$ **17.** $(4, -1, -2)$
19. $(1, -1, 2, -2)$ **21.** $(2, -1, 2, 1)$
25. $\dfrac{1}{5}\begin{bmatrix} 3 & -4 \\ 2 & -1 \end{bmatrix}$ **27.** $\dfrac{1}{cf + de}\begin{bmatrix} f & d \\ -e & c \end{bmatrix}$

Page 271

1. 72 **3.** 52 **5.** brand A: 6 kg; brand B: 9 kg
7. 785 **9.** Ralph: 10; Lisa: 18 **11.** \$4,000 at 9%,
\$2,400 at 8% **13.** 25%: 6 L; 40%: 5 L; $37\frac{1}{2}$%: 4 L

Page 272

1. $(4, 1)$; consistent, independent **3.** all (x, y)
such that $y = -\dfrac{3}{4}x + 2$; consistent, dependent

5. $(1, 1)$ **7.** $(1, -2)$ **9.** no solution
11. $(-1, 2, 3)$ **13.** 32 **15.** $-n^2$ **17.** 55
19. $(-2, 3)$ **21.** $(0.6, 0.5)$
23. $\begin{bmatrix} -9 & 12 & 2 \\ 11 & -12 & 13 \end{bmatrix}$
25. $\begin{bmatrix} 1 & 0 & 0 \\ 0 & 1 & 0 \\ 0 & 0 & 1 \end{bmatrix}; \begin{bmatrix} -3 & 1 & -2 \\ -2 & -1 & -2 \\ 3 & -2 & 1 \end{bmatrix};$
$\begin{bmatrix} -\dfrac{5}{3} & 1 & -\dfrac{4}{3} \\ -\dfrac{4}{3} & 1 & -\dfrac{2}{3} \\ \dfrac{7}{3} & -1 & \dfrac{5}{3} \end{bmatrix}$ **27.** 38

Page 278

1. $D = \{3, 2, -1, -2, 0\}; R = \{1, -3, 5, -2, 2\};$ Yes
3. $D = \{0, 8, -3, 4, 5\}; R = \{0, 7, 6, -1\};$ Yes
5. $D = \{3, 4, -4, -3, 2\}; R = \{2, -2, -3\};$ Yes
7. $D = \{1, -1, 2, -2, 3\}; R = \{-2, 0, -4, 3\};$ Yes
9. $D = \{4, -1, -3\}; R = \{-1, 4, 3, -3, 0\};$ No
11. $G = \{(-3, 2), (-1, -2), (2, -1), (3, 1)\};$
$D = \{-3, -1, 2, 3\}; R = \{2, -2, -1, 1\}$
13. $G = \{(-3, -3), (-2, -2), (-1, -1), (0, 0),$
$(1, -1), (2, -2), (3, -3)\};$
$D = \{-3, -2, -1, 0, 1, 2, 3\}; R = \{-3, -2, -1, 0\}$
15. $D = \{x \mid x \le 3\}; R = \{y \mid y \text{ is a real number}\};$ No
17. $D = \{x \mid -3 \le x < 2\}; R = \{y \mid -2 \le y < 2\};$
Yes **19.** $D = \{x \mid 0 \le x < 4\};$
$R = \{y \mid -2 < y < 2\};$ No **21.** $D = \{x \mid 0 \le x\};$
$R = \{y \mid y \text{ is a real number}\};$ No
23. $D = \{\text{all even integers}\}; R = \{\text{all integers}\};$ Yes
25. $D = \{\text{squares of all integers} > 1\};$
$R = \{\text{all integers} > 1\};$ Yes

Page 281

1. -3 **3.** -2 **5.** -1 **7.** 1 **9.** 5 **11.** -5
13. $-2b + 5$ **15.** -3 **17.** $a^2 + 2ab + b^2$
$+ 3a + 3b - 1$ **19.** $\{4, 5, 7\}$ **21.** $\{-3, -1, 3\}$
23. $\{1, -3\}$ **25.** $\{3, 2, 1, 0\}$ **27.** 3 **29.** -8
31. {all real numbers except 0}; none
33. {all real numbers except 3 and 4}; 2
35. {all real numbers except -6 and 2}; 2
37. -6 **39.** -7 **41.** $2h$ **43.** $2a - 2t$
45. $2a^2 - 7$ **47.** $36a^2 + 36a + 7$ **49.** 2
51. 2 **53.** $x - 4, x + 4$, No **55.** 46, -3

Page 284

1. Linear **3.** Constant, Linear **5.** Neither
7. Neither **9.** Linear **11.** Neither

13. Constant, Linear 15. Linear 17. Linear
19. Constant, Linear 21. Neither 23. Constant, Linear

Page 287

1. $\{(3, -2), (3, -1), (-1, 3)\}$, Not a function
3. $\{(-1, 4), (-2, 5), (-3, 6)\}$, Function
5. $\{(11, 7), (10, 8), (9, 9)\}$, Function 7. Function
9. Function 11. Function 13. Not a function
15. Function 17. Not a function
19. $y = \frac{9}{10}x + 15$, Function
21. $y = \frac{2}{21}x + \frac{8}{7}$, Function
23. $y = \pm\sqrt{x^2 + 4}$, Not a function
25. $y = \pm\sqrt{x - 5}$, Not a function 27. $y = \frac{9}{x}$,
Function 29. 2 31. 0 33. 2.5 35. 10 37. a

Page 290

1. Yes; $-\frac{1}{3}$; $y = -\frac{1}{3}x$ 3. Yes; 5; $A = 5r$
5. Yes; 1; $M = 1N$ 7. Yes; 3 9. Yes; 2
11. No 13. Yes; 6 15. Yes; $\frac{2}{3}$ 17. -63
19. 10 21. 6 23. $\frac{5}{3}$ 25. 25 27. 250
29. 150 31. 22.5 lb 33. B is multiplied by c.

Page 293

1. No 3. No 5. Yes; $\frac{1}{12}$; $ab = \frac{1}{12}$
7. Yes; 12 9. No 11. No 13. Yes; $\frac{2}{3}$
15. Yes; 10 17. 6 19. -27 21. 18 lb/in.^{2}
23. $4\frac{1}{2}$ 25. $\frac{24}{125}$

Page 296

1. Neither 3. Combined; 6 5. Neither
7. Joint; $\frac{5}{2}$ 9. 72 11. 112 13. 40 15. 2
17. 240 19. $\frac{7,203}{2}$ 21. The value of V is
multiplied by 8. 23. The value of t is multiplied
by $\frac{16}{25}$. 25. The value of z is multiplied by $\frac{4}{3}$.

Page 298

1. $D = \{-2, 0, 8, 4\}$; $R = \{3, -7, -3\}$; Function

3. $D = \{$all integers$\}$; $R = \{$squares of all integers$\}$;
Function 5. $D = \{x \mid -2 \le x < 3\}$;
$R = \{y \mid 0 \le y \le 2\}$; Function 7. -2 9. -3
11. 14 13. $\{3, 5, -1\}$ 15. $\{$all real numbers
except 1$\}$; 2 17. -26 19. 15 21. $2c + h$
23. Constant, Linear 25. Linear
27. $\{(1, 3), (6, 3), (9, 3)\}$; Function 29. Not a
function 31. Not a function 33. Direct
variation; -3 35. 30 37. 2 39. 120

Page 305

1. origin 3. y-axis 5. y-axis 7. x-axis, origin
9. none of these 11. $(-4, 5), (4, -5), (4, 5)$
13. $(-3, -5), (3, 5), (3, -5)$ 15. $(-2, -5)$,
$(2, 5), (2, -5)$ 17. $(5, -4), (-5, 4), (-5, -4)$
19. $(-3, 4), (3, -4), (3, 4)$ 21. $(-8, 8), (8, -8)$,
$(8, 8)$ 23. $\left(-\frac{2}{3}, -\frac{4}{5}\right), \left(\frac{2}{3}, \frac{4}{5}\right), \left(\frac{2}{3}, -\frac{4}{5}\right)$
25. $\left(\frac{5}{6}, \frac{1}{3}\right), \left(-\frac{5}{6}, -\frac{1}{3}\right), \left(-\frac{5}{6}, \frac{1}{3}\right)$
27. $\left(-1, \frac{2}{3}\right), \left(1, -\frac{2}{3}\right), \left(1, \frac{2}{3}\right)$
29. $(-a, b), (a, -b), (a, b)$ 31. $(x, y), (-x, -y)$,
$(-x, y)$ 33. $(-x, y), (x, -y), (x, y)$ 35. $(-5, 2)$,
$(5, -2), (5, 2)$ 37. $(3.3, 2.3), (-3.3, -2.3)$,
$(-3.3, 2.3)$ 39. y-axis 41. y-axis 43. even
45. even 47. odd

Page 309

1. $x^2 + y^2 = 1$ 3. $x^2 + y^2 = 9$
5. $x^2 + y^2 = 4$ 7. $x^2 + y^2 = 2$
9. $x^2 + y^2 = 3$ 11. 5 13. 9 15. 6
17. $2\sqrt{3}$ 19. $(x - 5)^2 + (y - 6)^2 = 49$
21. $(x - 11)^2 + (y + 7)^2 = \frac{49}{4}$ 23. $x^2 + y^2 = 5$
25. $(x - 2)^2 + (y + 3)^2 = 85$
27. $(x + 6)^2 + (y + 3)^2 = 104$
29. $x^2 + y^2 = 64$ 31. $x^2 + (y - 3)^2 = 16$;
$(0, 3)$; 4 33. $(x + 4)^2 + (y + 2)^2 = 25$;
$(-4, -2)$; 5
(For Ex. 35–37, the center and radius of
the circle are given.) 35. $(0, -2)$, 4
37. $(-5, 3)$, 3 39. $(x - 1)^2 + (y - 4)^2 = 65$
41. $(x - 4)^2 + y^2 = 25$
43. $(x - 2)^2 + (y + 5)^2 = 25$
45. $\left(-\frac{c}{2}, -\frac{d}{2}\right), \frac{1}{2}\sqrt{c^2 + d^2 - 4e}$

Page 313

1. $\dfrac{x^2}{9} + \dfrac{y^2}{4} = 1$ 3. $\dfrac{x^2}{4} + \dfrac{y^2}{3} = 1$

5. $\dfrac{x^2}{64} + \dfrac{y^2}{36} = 1$ 7. $a = \pm 4,\ b = \pm 3$

9. $a = \pm 5,\ b = \pm 2$ 11. $a = \pm 2,\ b = \pm 4$
13. $a = \pm 3,\ b = \pm 10$ 15. $a = \pm 5,\ b = \pm 10$
17. $a = \pm 6,\ b = \pm 2$

19. $\dfrac{(x + 1)^2}{16} + \dfrac{(y - 3)^2}{25} = 1$

21. $\dfrac{(x + 2)^2}{4} + \dfrac{(y + 3)^2}{16} = 1$

23. $\dfrac{(x + 5)^2}{7} + \dfrac{(y - 2)^2}{36} = 1$

25. $(-2, 3);\ 10, 4$ 27. $(4, 2);\ 4, 8$

29. $\dfrac{x^2}{16} + \dfrac{y^2}{7} = 1$ 31. $\dfrac{x^2}{25} + \dfrac{y^2}{9} = 1$

33. $\{x \mid -3 \le x \le 3\};\ \{y \mid -4 \le y \le 4\}$
35. $\{x \mid -2 \le x \le 2\};\ \{y \mid -9 \le y \le 9\}$
37. $\{x \mid -4 \le x \le 4 \text{ and } y \mid -2 \le y \le 2\}$
39. 25π 41. 10π

Page 318

1. 1 unit left   3. 1 unit right
5. 1 unit up   7. 4 units down   9. 2 right,
1 up   11. 1 right, 2 down   13. $x = 1;\ (1, 2)$

15. $x = -\dfrac{3}{2};\ \left(-\dfrac{3}{2}, -\dfrac{25}{2}\right)$ 17. $x = -\dfrac{1}{8};\ \left(-\dfrac{1}{8}, \dfrac{113}{16}\right)$

19. 5.0   21. 4.0   23. 0.3, 3.7   25. 3.0
27. 32.0   29. 3.0   31. -22.0
33. 20.0   35. $x = 0;\ (0, -2)$
37. $x = 0;\ (0, 1)$ 39. $x = y^2 + 2y - 6;$ no;
$y = -1$ 41. $x = -2y^2 - 4y + 5;$ no; $y = -1$

43. $x = 4y^2 - 5y - 2;$ no; $y = \dfrac{5}{8}$

Page 323

1. $x = \pm 4$ 3. $x = \pm 2$ 5. $x = \pm 3$ 7. $x = \pm 7$
9. $x = \pm 2$ 11. $x = \pm 10$ 13–15. Neither
x- nor y-intercept exists.   17. $y = \pm 4$
19. $y = \pm 9$ 21. $y = \pm 3$ 23. $y = \pm 4$

25. $y = \dfrac{3}{5}x;\ y = -\dfrac{3}{5}x$ 27. $y = \dfrac{5}{2}x;\ y = -\dfrac{5}{2}x$

29. $y = \dfrac{6}{5}x;\ y = -\dfrac{6}{5}x$ 31. $y = \dfrac{2}{3}x;\ y = -\dfrac{2}{3}x$

33. $x = 0;\ y = 0$ 35. $x = 0;\ y = 0$
37. $x = 0;\ y = 0$ 39. $x = 0;\ y = 0$

41. $y = 2x;\ y = -2x$ 43. $y = \dfrac{4}{3}x;\ y = -\dfrac{4}{3}x$

45. $y = 2x;\ y = -2x$ 47. $y = \dfrac{1}{2}x;\ y = -\dfrac{1}{2}x$

49. Use the basic graph of $\dfrac{x^2}{16} - \dfrac{y^2}{9} = 1$ for

reference. The hyperbola defined by
$\dfrac{(x - 1)^2}{16} - \dfrac{(y + 2)^2}{9} = 1$ has center at

$C(1, -2)$. Move the basic graph 1 unit to
the right and 2 units down.   51. Use the

basic graph of $\dfrac{x^2}{9} - \dfrac{y^2}{4} = 1$ for reference.

The hyperbola defined by $\dfrac{(x - 2)^2}{9} - \dfrac{(y - 3)^2}{4} =$

1 has center at $C(2, 3)$. Move the basic
graph 2 units to the left and 3

units down.   53. $\dfrac{(x + 4)^2}{9} - \dfrac{y^2}{25} = 1$

55. $\dfrac{(y + 5)^2}{4} - \dfrac{(x - 3)^2}{4} = 1$ 57. $\dfrac{x^2}{64} - \dfrac{y^2}{36} = 1$

59. $\dfrac{x^2}{81} - \dfrac{y^2}{19} = 1$ 61. $\dfrac{(x - 4)^2}{25} - \dfrac{(y + 1)^2}{9} = 1$

63. $\dfrac{(x + 8)^2}{100} - \dfrac{(y + 2)^2}{64} = 1$

Page 326

(In Ex. 1–57, P = parabola, C = circle,
E = ellipse, H = hyperbola.)   1. E   3. H   5. H
7. C   9. E   11. P   13. C   15. P   17. P   19. H
21. E   23. C   25. H   27. E   29. H   31. H
33. C   35. C   37. H   39. P   41. P   43. C   45. H
47. H; x-axis, y-axis, origin   49. C;
x-axis, y-axis, origin   51. C; x-axis, y-
axis, origin   53. H; x-axis, y-axis, origin
55. H; origin   57. H; origin

Page 330

1. $y^2 = 8x$ 3. $y^2 = -8x$ 5. $x^2 = -4y$
7. $x^2 = 8y$ (In Ex. 9–15, the coordinates of
three points of each parabola are given.)
9. $(0, 0), (2, -1), (4, -4)$ 11. $(0, 0), (1, 2),$
$(4, 4)$ 13. $(0, 0), (5, 10), (20, 20)$ 15. $(0, 0),$

$(3, -3),\ (9, -27)$ 17. $\dfrac{x^2}{25} + \dfrac{y^2}{9} = 1$

19. $\dfrac{x^2}{16} + \dfrac{y^2}{12} = 1$ 21. $\dfrac{x^2}{11} + \dfrac{y^2}{36} = 1$

23. $\dfrac{x^2}{25} + \dfrac{y^2}{9} = 1$ 25. $\dfrac{x^2}{49} + \dfrac{y^2}{81} = 1$

27. $\dfrac{x^2}{25} + \dfrac{y^2}{9} = 1$ or $\dfrac{x^2}{9} + \dfrac{y^2}{25} = 1$

29. $\dfrac{x^2}{9} + \dfrac{y^2}{4} = 1$ or $\dfrac{x^2}{4} + \dfrac{y^2}{9} = 1$

Page 332

1. origin **3.** x-axis **5.** $(-5, -1)$, $(5, 1)$, $(5, -1)$
7. $(-3, 8)$, $(3, -8)$, $(3, 8)$ **9.** (m, n), $(-m, -n)$, $(-m, n)$ **11.** even **13.** $x^2 + y^2 = 169$
15. $x^2 + y^2 = 8$ **17.** $(x - 3)^2 + (y + 2)^2 = 16$
19. $(x - 1)^2 + (y + 8)^2 = 8$ **21.** circle with center at $C(4, -3)$ and radius $\sqrt{14}$
23. $a = \pm 4$, $b = \pm 2$ **25.** $a = \pm 2$, $b = \pm 36$
27. $\dfrac{x^2}{25} + \dfrac{y^2}{9} = 1$ **29.** $(0, 0)$; $x = 0$
31. parabola with vertex at $(-1, 4)$ and passing through $(-2, 5)$ and $(1, 5)$
33. -0.1, 3.6 **35.** $x = 0$, $y = 0$
37. hyperbola with center at $C(1, 2)$
39. circle **41.** ellipse **43.** circle
45. $y^2 = -16x$ **47.** $\dfrac{x^2}{36} + \dfrac{y^2}{85} = 1$

Page 335

1. C **3.** B **5.** B **7.** C **9.** C **11.** A **13.** B
15. D **17.** $3, -\dfrac{1}{4}$ **19.** $\dfrac{-5 \pm \sqrt{61}}{2}$
21. $(6a - 7b)(6a + 7b)$ **23.** $\dfrac{4b^8}{a^6}$
25. $\dfrac{n - 3}{-2(n + 4)}$ **27.** $-7 + i\sqrt{2}$ **29.** $10, 36$
31. Line; passes through $(0, 6)$, $(4, 9)$, $(-4, 3)$
33. $2x - 3y = 3$ **35.** Slopes of consecutive sides of figure formed are 5, 1, 5, and 1.
37. $(1, -1, 2)$ **39.** 11 **41.** 3 **43.** $(-6, -1)$, $(6, 1)$, $(6, -1)$ **45.** $(x - 3)^2 + (y + 1)^2 = 29$
47. parabola; 2 left, 3 down **49.** circle; 2 left, 2 up

Page 340

1. $(-1.2, -2.8)$, $(3.9, 0.6)$ **3.** $(3.6, -1.6)$, $(-1.7, 3.7)$ **5.** $(2, 0)$, $(4.5, 1.3)$ **7.** $(1.2, -1.6)$, $(0, -2)$ **9.** $(-4, -2)$, $(2, 4)$ **11.** $(4, 5)$, $(6, 15)$
13. $(4, 3)$, $(2, -3)$ **15.** $(-4, -3)$, $(0, 5)$
17. $\left(\dfrac{13 + 2\sqrt{46}}{3}, \dfrac{-5 - \sqrt{46}}{3} \right)$,
$\left(\dfrac{13 - 2\sqrt{46}}{3}, \dfrac{-5 + \sqrt{46}}{3} \right)$

19. $\left(\dfrac{13 + \sqrt{137}}{8}, \dfrac{-25 + 3\sqrt{137}}{16} \right)$,
$\left(\dfrac{13 - \sqrt{137}}{8}, \dfrac{-25 - 3\sqrt{137}}{16} \right)$
21. $\left(-\dfrac{11}{19}, \dfrac{63}{19} \right)$, $(1, -3)$ **23.** $(7, 5)$, $(0, -2)$
25. $\left(\dfrac{-4 + \sqrt{431}}{5}, \dfrac{-8 + 2\sqrt{431}}{5} \right)$,
$\left(\dfrac{-4 - \sqrt{431}}{5}, \dfrac{-8 - 2\sqrt{431}}{5} \right)$
27. $\left(\dfrac{31 + 3\sqrt{42}}{11}, \dfrac{-5 + 2\sqrt{42}}{11} \right)$,
$\left(\dfrac{31 - 3\sqrt{42}}{11}, \dfrac{-5 - 2\sqrt{42}}{11} \right)$

Page 343

1. $(-4, 0)$, $(4, 0)$ **3.** $(3.6, 1.4)$, $(-3.6, -1.4)$
5. $(4, 0)$, $(-4, 0)$ **7.** $(5, 3.6)$, $(5, -3.6)$, $(-5, +3.6)$, $(-5, -3.6)$ **9.** $(2\sqrt{5}, 2)$, $(2\sqrt{5}, -2)$, $(-2\sqrt{5}, 2)$, $(-2\sqrt{5}, -2)$ **11.** $\left(\dfrac{\sqrt{3}}{3}, 4 \right)$, $\left(-\dfrac{\sqrt{3}}{3}, 4 \right)$,
$\left(\dfrac{\sqrt{3}}{3}, -4 \right)$, $\left(-\dfrac{\sqrt{3}}{3}, -4 \right)$
13. $\left(\sqrt{7}, \dfrac{3}{2} \right)$, $\left(\sqrt{7}, -\dfrac{3}{2} \right)$, $\left(-\sqrt{7}, \dfrac{3}{2} \right)$, $\left(-\sqrt{7}, -\dfrac{3}{2} \right)$
15. $(3, 2)$, $(3, -2)$, $(-3, 2)$, $(-3, -2)$
17. $(1.0, -0.5)$ **19.** $(1.0, 0.6)$, $(-1.9, -3.0)$

Page 348

1. above, including the line containing $(2, 0)$, $(4, 3)$
3. below the line containing $(0, 2)$, $(5, 0)$
5. below, including the line containing $(0, -7)$, $(3, 0)$ **7.** above, including the line containing $(0, 8)$, $(1, 6)$ **9.** below the line containing $(0, 8)$, $(1, 5)$
11. below, including the line containing $(0, -4)$, $(2, -3)$ **13.** right, including the line containing $(8, 0)$, $(8, 5)$ **15.** below the line containing $(0, -4)$, $(5, -4)$ (For Ex. 17–35, the coordinates of two points of each line are given. The points of a given line may or may not be part of the graph.)
17. below, including the line containing $(0, 3)$, $(1, 1)$ *and* left of the line containing $(0, -1)$, $(1, 3)$ **19.** above, including the line containing $(0, 2)$, $(3, 0)$ *and* below the line containing $(0, -2)$, $(4, 0)$
21. above, including the line containing $(0, 3)$, $(3, 1)$ *and* above, including the line containing $(0, -1)$, $(4, 2)$ **23.** below, including the line containing

$(0,3)$, $(-5,-1)$ *and* above, including the line containing $(0,3)$, $(4,8)$ **25.** below, including the line containing $(0,-2)$, $(3,-4)$ *and* below the line containing $(0,2)$, $(2,3)$ **27.** below, including the line containing $(0,-3)$, $(3,-1)$ *and* below the line containing $(0,-3)$, $(2,-2)$ **29.** above the line containing $(0,3)$, $(1,1)$ *and* above, including the line containing $(0,0)$, $(3,3)$ **31.** above, including the line containing $(0,-3)$, $(5,-3)$ *and* left, including the line containing $(-2,0)$, $(-2,5)$ **33.** below, including the line containing $(0,3)$, $(1,2)$ *and* below, including the line containing $(0,-2)$, $(5,2)$ *and* above the line containing $(0,2)$, $(1,4)$ **35.** below the line containing $(4,3)$, $(7,4)$ *and* below, including the line containing $(0,-3)$, $(2,0)$ *and* below the line containing $(0,3)$, $(1,7)$

Page 351

1. inside, including the circle containing $(0,3)$, $(3,0)$ **3.** inside the circle containing $(5,0)$, $(-5,0)$ **5.** inside, including the two branches of the hyperbola containing $(2,3)$, $(3,2)$ and $(-2,-3)$, $(-3,-2)$ **7.** inside the two branches of the hyperbola containing $(3,1)$, $(1,3)$ and $(-3,-1)$, $(-1,-3)$ **9.** inside, including the ellipse containing $(5,0)$, $(0,2)$ **11.** inside the ellipse containing $(10,0)$, $(0,3)$ **13.** outside, including the two branches of the hyperbola containing $(3,0)$, $(4,1.8)$, $(-3,0)$, $(-4,1.8)$ **15.** outside the two branches of the hyperbola containing $(2,0)$, $(6,8.5)$ and $(-2,0)$, $(-6,8.5)$ **17.** outside, including the ellipse containing $(3,0)$, $(0,2.4)$ **19.** outside the two branches of the hyperbola containing $(2,0)$, $(6,8.5)$ and $(-2,0)$, $(-6,8.5)$ **21.** outside, including the ellipse containing $(1.7,0)$, $(0,3)$ **23.** inside, including the two branches of the hyperbola containing $(1.7,0)$, $(3,2.8)$ and $(-1.7,0)$, $(-3,-2.8)$ **25.** inside, including the circle containing $(0,4)$, $(4,0)$ *and* above, including the line containing $(0,-6)$, $(3,-4)$ **27.** inside the two branches of the hyperbola containing $(1,1)$, $(2,0.5)$ and $(-1,-1)$, $(-2,-0.5)$ *and* above, including the line containing $(0,3)$, $(5,-1)$ **29.** outside, including the ellipse containing $(4,0)$, $(0,2)$ *and* below, including the line containing $\left(0,-\frac{2}{3}\right)$, $\left(3,-\frac{8}{3}\right)$

31. inside, including the ellipse containing $(5,0)$, $(0,9)$ *and* below, including the line containing $(0,-0.75)$, $(-1.5,0)$ **33.** outside, including the circle containing $(0,5)$, $(5,0)$ *and* outside the two branches of the hyperbola containing $(1,-5)$, $(5,-1)$, $(-1,5)$, $(-5,1)$ **35.** outside, including the ellipse containing $(2,0)$, $(0,5)$, $(-2,0)$, $(0,-5)$ *and* inside, including the circle containing $(6,0)$, $(0,6)$ **37.** outside, including the ellipse containing $(4,0)$, $(0,2)$ *and* outside, including the two branches of the hyperbola containing $(3,0)$, $(6,3)$, $(-3,0)$, $(-6,3)$ **39.** outside, including the ellipse containing $(0,2)$, $(2.4,0)$ *and* inside, including the circle containing $(3,0)$, $(0,3)$

Page 352

1. $(0.5,2.0)$, $(-1.5,-1.9)$ **3.** $(4,2)$, $(-5,0)$ **5.** $(2.9,0.6)$, $(2.9,-0.6)$, $(-2.9,0.6)$, $(-2.9,-0.6)$ **7.** $\left(\dfrac{-15+\sqrt{249}}{4},\dfrac{3-\sqrt{249}}{8}\right)$, $\left(\dfrac{-15-\sqrt{249}}{4},\dfrac{3+\sqrt{249}}{8}\right)$ **9.** $(2,0)$ $(1,-1)$ **11.** $(3,1)$, $(-3,-1)$, $(1,3)$, $(-1,-3)$ **13.** above, including the line containing $(0,-3)$, $(1,-1)$ **15.** above, including the line containing $(0,-4)$, $(2,-1)$ **17.** inside, including the circle containing $(2,0)$, $(0,2)$ **19.** outside, including the two branches of the hyperbola containing $(-1,5)$, $(-5,1)$, $(1,-5)$, $(5,-1)$ **21.** outside, including the two branches of the hyperbola containing $(6,0)$, $(7,4)$, $(6,0)$, $(-7,4)$ **23.** above, including the line containing $(0,3)$, $(1,2)$ *and* below the line containing $(0,-4)$, $(1,-1)$ **25.** below the line containing $(0,5)$, $(1,8)$ *and* above, including the line containing $\left(-\frac{1}{2},4\right)$, $\left(\frac{1}{2},3\frac{1}{2}\right)$ **27.** above, including the line containing $(0,-2)$, $(1,0)$ *and* above, including the line containing $(0,4)$, $(1,3)$ *and* below, including the line containing $(0,2)$, $(3,4)$ **29.** outside, including the two branches of the hyperbola containing $(1,6)$, $(6,1)$, $(-1,-6)$, $(-6,-1)$ *and* above, including the line containing $(0,0)$, $(3,3)$ **31.** outside the two branches of the hyperbola containing $(5,0)$, $(6,1.3)$, $(-5,0)$, $(-6,1.3)$ *and* below, including the line containing $(0,2)$, $(2,1)$

Page 359

1. 26.8, 28, 29.2 **3.** $\dfrac{7}{12}, \dfrac{2}{3}, \dfrac{3}{4}$

5. 9, 3, −3, −9 **7.** $5.00, $7.40, $9.80, $12.20
9. 15 kg, 11.6 kg, 8.2 kg, 4.8 kg **11.** −200

13. 54 **15.** $-22\dfrac{1}{5}$ **17.** $-1 + 3\sqrt{3}$,

$-3 + 4\sqrt{3}, -5 + 5\sqrt{3}$ **19.** Not an A.P.
21. $5\sqrt{2}, 6\sqrt{2}, 7\sqrt{2}$ **23.** $-x + 8y, -4x + 11y,$
$-7x + 14y$ **25.** $-3, \sqrt{6}, 3 + 2\sqrt{6}, 6 + 3\sqrt{6}$
27. $x^2 - y, 2x^2, 3x^2 + y, 4x^2 + 2y$
29. $25 - 21\sqrt{2}$ **31.** $-17x + 22y - 19$
33. $19,100 **35.** 75 m **37.** 32 **39.** 21 **41.** 26

43. $c = 3$ or $c = -\dfrac{1}{2}$ **45.** $a = -4$

47. 46, 41, 36

Page 362

1. 19, 26 **3.** 2, 6, 10, 14 **5.** $6\dfrac{1}{2}, 7, 7\dfrac{1}{2}$

7. 5.7, 8 **9.** 49 **11.** 2.8 **13.** 7, 23
15. $12,200, $13,000, $13,800, $14,600, $15,400,
$16,200, $17,000, $17,800, $18,600, $19,400
17. 42, 9 **19.** $7\sqrt{7}, 11\sqrt{7}$ **21.** $-2 - 2i,$
$-4 + i, -6 + 4i, -8 + 7i$ **23.** $-5a + 6b,$
$-2a + 3b, a$ **25.** $4x$ **27.** $5 - 3\sqrt{2}$ **29.** a

31. $\dfrac{2x + y}{3}, \dfrac{x + 2y}{3}$ **33.** $\dfrac{(2x - 1) + (2x + 1)}{2} = \dfrac{4x}{2}$

$= 2x$, an even integer

Page 367

1. 1,550 **3.** 2,500 **5.** 959 **7.** 4,500
9. 871 **11.** 63 **13.** 1,325 **15.** −134

17. $\displaystyle\sum_{k=1}^{4} 6k$ **19.** $\displaystyle\sum_{k=7}^{26} k$ **21.** $\displaystyle\sum_{k=4}^{8} 2k$ **23.** $\displaystyle\sum_{k=1}^{12}$

$(14 - 5k)$ **25.** $4,100\sqrt{3}$ **27.** 170.3 **29.** 1,262.5

31. 610 **33.** $\displaystyle\sum_{k=1}^{50} 10kx = 10x + 20x + 30x +$

$\cdots + 500x = 10x(1 + 2 + 3 + \cdots + 50) =$

$10x \cdot \displaystyle\sum_{k=1}^{50} k$ **35.** $\displaystyle\sum_{c=1}^{30} (c + t) =$

$(1 + t) + (2 + t) + (3 + t) + \cdots + (30 + t)$
$= 30t + (1 + 2 + 3 + \cdots + 30) = 30t +$

$\displaystyle\sum_{c=1}^{30} c$ **37.** $\displaystyle\sum_{n=1}^{k} (2n - 1) = 1 + 3 + 5 + 7 +$

$\cdots + (2k - 1) = \dfrac{k}{2}[1 + (2k - 1)] = \dfrac{k}{2} \cdot 2k = k^2$;

the sum of the first k positive odd
integers is k^2.

Page 370

1. 240, 480, 960 **3.** 9, −27, 81 **5.** $-\dfrac{1}{4}$,

$-1, -4, -16$ **7.** 32, 16, 8, 4 **9.** 6, 0.6, 0.06,

0.006 **11.** $64, -24, 9, -\dfrac{27}{8}$ **13.** −810

15. 34,000,000 **17.** −0.0096 **19.** 2,048

21. −128 **23.** $\dfrac{1}{64}$ **25.** 0.000037

27. 0.000064 **29.** $-\dfrac{5}{4}$ **31.** $768x^8$

33. $8x^{15}y^{22}$ **35.** $-100\sqrt{5}$ **37.** $128x^{11}$,
$512x^{14}, 2,048x^{17}$ **39.** $24 + 16\sqrt{5}$,
$48 + 32\sqrt{5}, 96 + 64\sqrt{5}$ **41.** $24x^9, -48x^{11}$,
$96x^{13}$ **43.** 0.25 dm **45.** $10\dfrac{2}{3}$ m **47.** 8

49. $-\dfrac{2}{25}$ **51.** $2, \dfrac{1}{2}$ **53.** 11th term

Page 374

1. 3, 9, 27 or −3, 9, −27 **3.** −10 **5.** 6, −12
7. 12 **9.** −8, 4 **11.** −15, −45, −135, or 15,
−45, 135 **13.** 2 **15.** 3, 1 **17.** 22, 44, 88, 176
19. $3\sqrt{2}$ **21.** −10 **23.** 3, 12 or −3, −12
25. 2, 8 or −8, −2 **27.** 0.5 g, 1 g, 2 g, 4 g, 8 g,
16 g, 32 g **29.** 0.04 **31.** $3\sqrt{7}, 21$ **33.** $10x^5$,
$20x^7, 40x^9$ or $-10x^5, 20x^7, -40x^9$ **35.** 6
37. $2i$ or $-2i$ **39.** $4i\sqrt{3}$ or $-4i\sqrt{3}$ **41.** $\sqrt{2}$ or
$-\sqrt{2}$ **43.** $xy\sqrt{xy}$ or $-xy\sqrt{xy}$ **45.** x, g, y

implies $y = xr^2$; $r^2 = \dfrac{y}{x}$; $r = \dfrac{\pm\sqrt{xy}}{x}$;

$g = x \cdot \dfrac{\pm\sqrt{xy}}{x} = \pm\sqrt{xy}$

Page 377

1. 66,666 **3.** −684 **5.** $31\dfrac{7}{8}$ **7.** 333,333.3

9. 102.2 **11.** 5,115 **13.** −340 **15.** 66,666.66
17. 4,368 **19.** 24.9984 **21.** 222,222.2 **23.** 126

25. $\dfrac{31}{32}$ **27.** 170 **29.** $-85\dfrac{1}{4}$ **31.** $-21\dfrac{3}{8}$

33. $728\dfrac{2}{3}$ **35.** −364 **37.** 2,084 **39.** 6,825

41. 382 m **43.** 159.375 cm **45.** $\displaystyle\sum_{k=1}^{4} ax^{k-1}$

$= ax^0 + ax^1 + ax^2 + ax^3 =$

$a \cdot 1 + a(x^1 + x^2 + x^3) = a + a \cdot \displaystyle\sum_{k=1}^{3} x^k$

Page 381

1. Does not exist **3.** $6\frac{3}{4}$ **5.** 1 **7.** $\frac{3}{4}$

9. $77.\overline{7}$ or $77\frac{7}{9}$ **11.** $\frac{7}{9}$ **13.** $\frac{200}{9}$

15. $\frac{14}{3}$ **17.** $\frac{563}{90}$ **19.** 128 m^2 **21.** 405 m

23. $-9\frac{1}{7}$ **25.** $40\frac{1}{2}$ **27.** 1 **29.** $\frac{6,680}{99}$

31. $\frac{1,700}{99}$ **33.** $\frac{6,290}{99}$ **35.** $x < -1$ or $x > 1$

Page 383

1. $a^8 + 8a^7b + 28a^6b^2 + 56a^5b^3 + 70a^4b^4$
$+ 56a^3b^5 + 28a^2b^6 + 8ab^7 + b^8$
3. $x^6 + 12x^5 + 60x^4 + 160x^3 + 240x^2$
$+ 192x + 64$ **5.** $x^{16} + 8x^{14}y + 28x^{12}y^2$
$+ 56x^{10}y^3 + 70x^8y^4 + 56x^6y^5 + 28x^4y^6$
$+ 8x^2y^7 + y^8$ **7.** $16r^4 + 160r^3t + 600r^2t^2$
$+ 1,000rt^3 + 625t^4$ **9.** $64x^6 + 19.2x^5 + 2.4x^4$
$+ 0.16x^3 + 0.006x^2 + 0.00012x + 0.000001$
11. $-20x^6y^3$ **13.** $720x^9y^2$ **15.** $160x^9$
17. $32c^{15} + 240c^{12}d^2 + 720c^9d^4 + 1,080c^6d^6$
$+ 810c^3d^8 + 243d^{10}$ **19.** $16x^4 - 32x^3\sqrt{y}$
$+ 24x^2y - 8xy\sqrt{y} + y^2$ **21.** $\frac{1}{x^5} + \frac{5}{x^4y} + \frac{10}{x^3y^2}$
$+ \frac{10}{x^2y^3} + \frac{5}{xy^4} + \frac{1}{y^5}$ **23.** $x^{11} - 22x^{10}y$
$+ 220x^9y^2 - 1,320x^8y^3 + \cdots$
25. $1 + 0.07 + 0.0021 + 0.000035 + \cdots$
27. $1 + 0.18 + 0.0135 + 0.00054 + \cdots$

29. $\sqrt[3]{x^2} + \frac{6\sqrt[3]{x^2}}{x} - \frac{9\sqrt[3]{x^2}}{x^2} + \frac{36\sqrt[3]{x^2}}{x^3} - \cdots$

Page 385

1. 0.841 **3.** 2.7183 **5.** $\frac{517}{720} \doteq 0.7180555\ldots$

7. 0.2955 **9.** 0.9950 **11.** 1.0000

13. 1.6487 **15.** $S = \frac{a}{1-r} = \frac{1}{1-x}$ **17.** $\frac{1}{3-z}$

Page 386

1. 14, 6, -2, -10 **3.** 125.8 **5.** 3.9 **7.** 195

9. 16 **11.** $\sum_{k=1}^{11}(4k+14)$ or $\sum_{k=4}^{14}(4k+2)$

13. $-32x^4, -64x^5, -128x^6$ **15.** -324 **17.** -6,
$-12, -24$, or $6, -12, 24$ **19.** $127\frac{3}{4}$ **21.** $21\frac{1}{4}$

23. $-2,605$ **25.** Does not exist **27.** $32\sqrt{3}$
29. $x^4 + 12x^3y + 54x^2y^2 + 108xy^3 + 81y^4$

31. $\frac{1}{m^4} + \frac{4}{m^3n} + \frac{6}{m^2n^2} + \frac{4}{mn^3} + \frac{1}{n^4}$
33. $x^{10} - 20x^9y^2 + 180x^8y^4 - 960x^7y^6$
35. 41, 7

Page 391

	Domain	Range	Constantly	$(x, 0)$ or $(0, y)$
1.	reals	$y > 0$	increasing	$(0, 1)$
3.	reals	$y > 0$	increasing	$(0, 1)$
5.	reals	$y > 0$	decreasing	$(0, 1)$
7.	reals	$y > 0$	increasing	$(0, 5)$
9.	reals	$y > 0$	increasing	$(0, 3)$
11.	reals	$y > 0$	decreasing	$(0, 2)$
13.	reals	$y > 4$	increasing	$(0, 5)$
15.	reals	$y > 6$	decreasing	$(0, 9)$

17. $-4 \le x \le 2$ $\frac{1}{16} \le y \le 4$ decreasing $\left(0, \frac{1}{4}\right)$

Page 393

1. 2 **3.** 4 **5.** 1 **7.** -1 **9.** -4 **11.** $\frac{1}{5}$

13. 10 **15.** 2 **17.** $\frac{1}{8}$ **19.** -3 **21.** $\frac{4}{5}$ **23.** $\frac{3}{2}$

25. 4 **27.** -2 **29.** d **31.** 2 **33.** 5 **35.** 4

Page 396

1. $y = \log_5 x$ **3.** $y = 8^x$

	Domain	Range	Constantly	$(x, 0)$ or $(0, y)$
5.	$x > 0$	reals	increasing	$(1, 0)$
7.	$x > 0$	reals	increasing	$(1, 0)$
9.	$x > 0$	reals	increasing	$(1, 0)$
11.	$x > 0$	reals	increasing	$(1, 0)$
13.	$x > 0$	reals	decreasing	$(1, 0)$
15.	$x > 0$	reals	increasing	$\left(\frac{1}{9}, 0\right)$

17. $y = \log_{\frac{1}{5}} x$ **19.** $y = \left(\frac{1}{5}\right)^x$ **21.** Graph

passes through $(11, 3)$, $(5, 1)$, $(4, 0)$, $\left(3\frac{1}{4}, -2\right)$.

Page 399

1. $\log_2 5 + \log_2 a$ **3.** $\log_b 2 + \log_b m + \log_b n$
$+ \log_b p$ **5.** $\log_3 4 - \log_3 a$ **7.** $\log_b r = \log_b d$
$- \log_b 2$ **9.** $\log_5 B = \log_5 2 + \log_5 V - \log_5 3$

11. $\log_3 \frac{7}{t}$ **13.** $\log_5 3c$ **15.** $\log_b 12c$ **17.** 5

19. 6 **21.** 6 **23.** 6 **25.** $\log_b 5 + \log_b c -$

$\log_b 3 - \log_b d$ **27.** $\log_4 \frac{21}{mn}$

29. $\log_b h = \log_b L - \log_b 2 - \log_b \pi - \log_b r$
31. $\log_5 r = \log_5 4 + \log_5 A - \log_5 3 - \log_5 t$

Page 402

1. $5 \log_2 a$ **3.** $2(\log_5 m + \log_5 n)$
5. $3(\log_4 v - \log_4 h)$ **7.** $\log_3 v =$
$\frac{1}{2}(\log_3 2 + \log_3 g + \log_3 h)$ **9.** $\log_4 r =$
$\frac{1}{2}(\log_4 A - \log_4 \pi)$ **11.** $\log_2 \sqrt[3]{6c}$ **13.** $\log_5 \sqrt{2cd}$
15. $\log_3 \sqrt[4]{\frac{6y}{5z}}$ **17.** 6 **19.** 2 **21.** 3 **23.** 5
25. 24 **27.** $3(\log_5 m + 2 \log_5 n - 3 \log_5 p)$
29. $\log_{10} A = \log_{10} P + 12 \log_{10} 1.09$
31. $\log_2 r = \frac{1}{2}(\log_2 3 + \log_2 V - \log_2 \pi - \log_2 h)$
33. $\log_4 \frac{\sqrt[3]{10ab}}{a^2 b}$ **35.** $\log_b \sqrt[4]{\frac{3x^3}{10y}}$ **37.** 5 **39.** 5, 8
41. $\frac{2}{5} \log_b 3 - \frac{1}{5} \log_b 2 - \frac{2}{5} \log_b 4 + \frac{1}{5} \log_b 27 =$
$\frac{2}{5} \log_b 3 - \frac{1}{5} \log_b 2 - \frac{2}{5} \log_b 2^2 + \frac{1}{5} \log_b 3^3 =$
$\log_b 3^{\frac{2}{5}} + \log_b 3^{\frac{3}{5}} - (\log_b 2^{\frac{1}{5}} + \log_b 2^{\frac{4}{5}}) =$
$\log_b 3^{\frac{5}{5}} - \log_b 2^{\frac{5}{5}} = \log_b 3 - \log_b 2$

Page 405

1. 4.2967 **3.** $9.3160 - 10$ **5.** $7.7875 - 10$
7. 7.5132 **9.** 83,300 **11.** $\sqrt[3]{10}$
13. 2, 3 **15.** -3, 3

Page 408

1. 2.32 **3.** 3.32 **5.** 0.686 **7.** 0.387
9. 6.15 **11.** 1.71 **13.** $14,600; $6,600
(by logs) or $14,624; $6,624 (by calculator)
15. 3,070,000 (by logs) or 3,072,000 (by
calculator) **17.** 7.01×10^{18} (by logs) or
7.04×10^{18} (by calculator) **19.** 1.82
21. 3.42 **23.** 15.4

Page 411

1. $-1, 4, \frac{1}{3}$; $(x + 1)(x - 4)(3x - 1)$ **3.** 2,
$-\sqrt{7}, \sqrt{7}$; $(x - 2)(x + \sqrt{7})(x - \sqrt{7})$ **5.** $3, -5, \frac{1}{2}$;
$(x - 3)(x + 5)(2x - 1)$ **7.** $2, -3, \frac{5}{2}$ **9.** $4, -4,$
-5 **11.** $1, -1, -2, \frac{3}{2}$ **13.** $3, -3, i\sqrt{3}, -i\sqrt{3}$
15. $4, -7, \sqrt{2}, -\sqrt{2}$ **17.** $1, 2, -2, \frac{-1 + i\sqrt{3}}{2},$
$\frac{-1 - i\sqrt{3}}{2}$ **19.** $1, 3, -3, \frac{-1 + i\sqrt{3}}{2}, \frac{-1 - i\sqrt{3}}{2}$

Page 415

1. 4, -2 **3.** 2, -2 **5.** $\frac{2}{3}, \pm\sqrt{3}$ **7.** $1, \frac{2}{3},$
$\pm\sqrt{5}$ **9.** $2, \frac{3}{2}, \pm i$ **11.** $\frac{1}{2}, \frac{1}{3}, \pm\sqrt{2}$ **13.** $2, -\frac{1}{3},$
$\pm\frac{1}{2}$ **15.** $-\frac{1}{2}, \frac{1}{4}, \pm\sqrt{3}$ **17.** 3, -3 **19.** 5, -2
21. $\frac{1}{2}$, mult. 2; $\pm 2i$ **23.** $\frac{1}{2}$, mult. 2; $-\frac{1}{3}$,
mult. 2 **25.** $-1, \frac{1}{2}, \frac{1}{3}, \pm 2i$

Page 417

1. horizontal line, y-intercept: 3
(In Ex. 3–11, three points are given for
each graph.) **3.** $(-5, 0), (0, -5), (2, 7)$;
turning point: $(-2, -9)$ **5.** $(-1, 0), (0, 12)$,
$(5, -18)$ **7.** $(0, 15), (3, 3), (5, 35)$; turning
point: $(2, -1)$ **9.** $(-4, 0), (0, 48), (3, -84)$
11. $(-3, 0), (0, -9), (3, 0)$

Page 420

1. domain: reals; range: $y > 0$; constantly
decreasing; through: $(0, 1)$ **3.** 4 **5.** $\frac{3}{4}$
7. $y = \left(\frac{1}{4}\right)^x$ **9.** $\frac{1}{3}(\log_b 3 + 2 \log_b v - \log_b k)$
11. $\log_4 c^2 \sqrt[3]{a}$ **13.** 6 **15.** 32 **17.** 0.0167
19. 5.73 **21.** (1): 3, -3;
(2): $(x - 1)(x + 2)(x - \sqrt{3})(x + \sqrt{3})$; (3): 1, -2,
$\sqrt{3}, -\sqrt{3}$ **23.** (1): 1, -1;
(2): $12\left(x - \frac{1}{3}\right)\left(x + \frac{1}{2}\right)\left(x + \frac{1}{2}\right)$; (3): $\frac{1}{3}$; $-\frac{1}{2}$, mult. 2
25. $3a$ **27.** 7

Page 427

1. 120 **3.** 720 **5.** 6 **7.** 210 **9.** 2,600,000
11. 30 **13.** 36 **15.** 360 **17.** 125
19. 15,600,000

Page 429

1. 72 **3.** 1,080 **5.** 120 **7.** 17,280 **9.** 3,250
11. 64 **13.** 360 **15.** 172,800 **17.** 1,304
19. 48 **21.** 6

Page 432

1. 840 **3.** 495 **5.** 990 **7.** 168 **9.** 2,520
11. 20 **13.** 1,680 **15.** 90 **17.** 3,780
19. 12,600

Page 434

1. 720 **3.** 3,628,800 **5.** 362,880 **7.** 144 **9.** 3

Page 437

1. 21 **3.** 816 **5.** 63 **7.** 715 **9.** 45
11. 120 **13.** 84 **15.** 45 **17.** 31

Page 440

1. $\frac{1}{6}, \frac{1}{6}, \frac{1}{2}$ **3.** $\frac{4}{7}, \frac{3}{7}$ **5.** $\frac{1}{13}, \frac{1}{52}, \frac{12}{13}$

7. $\frac{1}{3}, \frac{1}{4}, \frac{5}{12}, \frac{2}{3}, \frac{7}{12}$ **9.** 16, 50, 33 **11.** 6

13. $\frac{1}{2}$, or 1 to 2 **15.** $\frac{5}{6}$

Page 444

1. $\frac{1}{8}$ **3.** $\frac{3}{8}$ **5.** $\frac{1}{8}$ **7.** $\frac{1}{8}$ **9.** $\frac{7}{8}$ **11.** $\frac{4}{8}$ or $\frac{1}{2}$

13. $\frac{3}{8}$ **15.** $\frac{1}{6}$ **17.** $\frac{1}{3}$ **19.** $\frac{1}{18}$ **21.** $\frac{1}{6}$ **23.** $\frac{1}{36}$

25. $\frac{7}{12}$ **27.** $\frac{8}{9}$ **29.** $\frac{2}{9}$ **31.** $\frac{2}{9}$ **33.** $\frac{1}{18}$

35. $\frac{1}{9}$ **37.** $\frac{1}{6}$ **39.** $\frac{1}{5}$ **41.** $\frac{7}{9}$ **43.** $\frac{2}{3}$ **45.** $\frac{2}{9}$

47. $\frac{7}{13}$ **49.** $\frac{9}{13}$ **51.** GGG, GGB, GBG, GBB, BGG, BGB, BBG, BBB

Page 447

1. $\frac{1}{6}$ **3.** $\frac{33}{391,510}$ **5.** $\frac{1}{56}$ **7.** $\frac{15}{56}$ **9.** $\frac{15}{28}$ **11.** $\frac{7}{54}$

13. $\frac{1}{17}$ **15.** $\frac{13}{102}$ **17.** $\frac{13}{51}$ **19.** $\frac{1}{8}$ **21.** $\frac{1}{4}$

23. $\frac{1}{2}$ **25.** $\frac{7}{52}$ **27.** $\frac{73}{153}, \frac{80}{153}, \frac{28}{153}, \frac{5}{17}$

Page 448

1. 1,980 **3.** 1 **5.** 180 **7.** 60 **9.** 5,040 **11.** 60
13. 70 **15.** 144 **17.** $\frac{1}{13}, \frac{12}{13}$ **19.** $\frac{2}{3}$ **21.** $\frac{1}{6}$

23. $\frac{2}{15}, \frac{4}{29}$

Page 452

1. D **3.** C **5.** B **7.** C **9.** B **11.** A **13.** C
15. A **17.** $-6a^8b$ **19.** base: 2 cm; height: 20 cm

21. $5a\sqrt[3]{2a^2}$ **23.** $3\sqrt[3]{x^2}$ **25.** $\frac{4i}{3}$
27. $(-3, 0), (0, -2)$ **29.** $\sqrt{65}$ **31.** $(-1, 2, 1)$
33. {all real numbers except -5 and 3}
35. $(x + 3)^2 + (y - 2)^2 = 97$ **37.** $1\frac{1}{2}$ **39.** 5

Page 457

1. I **3.** III **5.** IV **7.** Quad. $\angle$ **9.** Quad. $\angle$
11. IV **13.** II **15.** I **17.** II **19.** I **21.** III
23. IV **25.** II **27.** II **29.** III **31.** 50° **33.** 45°
35. 45° **37.** 45° **39.** 60° **41.** Quad. $\angle$

Page 461

1. $6\sqrt{3}, 6$ **3.** $3\sqrt{2}, 3\sqrt{2}$ **5.** 16, $16\sqrt{2}$
7. $(-3\sqrt{3}, -3), 6$ **9.** $(-3, 3\sqrt{3}), 6$
11. 8, $8\sqrt{3}$ **13.** $4\sqrt{3}, 2\sqrt{3}$ **15.** $12\sqrt{2}, 12$
17. 45°, $3\sqrt{2}$ **19.** 60°, 8 **21.** 30°, 4
23. 60°, 10 **25.** 30°, 8 **27.** 45°, $a\sqrt{2}$
29. 60°, $2r$ **31.** 45°, $r\sqrt{2}$

Page 465

1. $\frac{1}{2}\sqrt{3}$ **3.** $\frac{\sqrt{2}}{2}$ **5.** $\frac{1}{2}$ **7.** 0 **9.** 0 **11.** 0

13. $\frac{\sqrt{2}}{2}$ **15.** $-\frac{1}{2}\sqrt{3}$ **17.** -1 **19.** $-\frac{\sqrt{2}}{2}$

21. $\frac{1}{2}\sqrt{3}$ **23.** $\frac{1}{2}$ **25.** $-\cos 45°$ **27.** $-\cos 30°$
29. $\sin 45°$ **31.** $\sin 30°$ **33.** $-\sin 45°$
35. $-\cos 30°$ **37.** 45°, 135° **39.** 30°, 150°
41. 135°, 225° **43.** 150°, 210° **45.** II, III
47. III, IV **49.** 120°, 480° **51.** 330°, 690°
53. 300°, 660°

Page 469

1. 1 **3.** $\sqrt{3}$ **5.** $\frac{\sqrt{3}}{3}$ **7.** undefined **9.** -1

11. undefined **13.** -1 **15.** $-\frac{\sqrt{3}}{3}$ **17.** $\frac{\sqrt{3}}{3}$

19. 0 **21.** $\frac{\sqrt{3}}{3}$ **23.** undefined **25.** $-\cot 70°$
27. $\cot 50°$ **29.** $-\tan 20°$ **31.** $\tan 70°$
33. $-\tan 30°$ **35.** $\cot 70°$ **37.** $-\frac{3}{4}$ **39.** $-\frac{8}{17}$

41. $-\frac{4}{3}, -\frac{3}{4}$ **43.** $\frac{15}{4}, \frac{4}{15}$ **45.** 150°, 330°

47. 45°, 225° **49.** 150°, 330° **51.** 45°, 225°
53. 120°, 480° **55.** 300°, 660°

Page 472

1. $\sqrt{2}$ **3.** 2 **5.** $\dfrac{2\sqrt{3}}{3}$ **7.** -2 **9.** $-\sqrt{2}$ **11.** 2

13. 2 **15.** $\sqrt{2}$ **17.** $\dfrac{2\sqrt{3}}{3}$ **19.** $-\sec 60°$

21. $-\csc 60°$ **23.** $-\sec 60°$ **25.** $\csc 60°$

27. $-\sec 10°$ **29.** $\csc 10°$ **31.** -2

33. $\dfrac{3}{2}$ **35.** $-\dfrac{6}{5}$ **37.** $180°$ **39.** $225°, 315°$

41. $60°, 120°$ **43.** $60°, 300°$ **45.** $180°, 540°$

47. $240°, 600°$

Page 474

1. $-\dfrac{1}{2}$ **3.** $-\dfrac{\sqrt{3}}{3}$ **5.** $\dfrac{2\sqrt{3}}{3}$ **7.** $-\dfrac{\sqrt{3}}{2}$ **9.** 2

11. $-\dfrac{\sqrt{3}}{2}$ **13.** $-\dfrac{1}{2}$ **15.** $\sqrt{3}$ **17.** 2 **19.** $\dfrac{1}{2}$

21. $\dfrac{\sqrt{2}}{2}$ **23.** 1 **25.** $\sqrt{2}$ **27.** $-\tan 10°$

29. $\cot 85°$ **31.** $-\sin 80°$ **33.** $\sec 18°$

35. $\cos 35°$ **37.** $\cos 10°$ **39.** $\sin 10°$

41. $\csc 15° 35'$ **43.** $\cos 59° 15'$ **45.** $\cot (-u)$

$= \dfrac{b}{-a} = -\dfrac{b}{a} = -\cot u,\ a > 0,\ b > 0$

[Quad. II, IV] **47.** $\csc (-u) = \dfrac{r}{-a} = -\dfrac{r}{a}$

$= -\csc u,\ a > 0,\ b > 0$ [Quad. III, IV]

Page 476

1. $\sin u = \dfrac{12}{13}$, $\cos u = -\dfrac{5}{13}$, $\cot u = -\dfrac{5}{12}$,

$\csc u = \dfrac{13}{12}$, $\sec u = -\dfrac{13}{5}$ **3.** $\sin u = -\dfrac{12}{13}$,

$\tan u = \dfrac{12}{5}$, $\cot u = \dfrac{5}{12}$, $\csc u = -\dfrac{13}{12}$,

$\sec u = -\dfrac{13}{5}$ **5.** $\sin u = -\dfrac{4}{5}$, $\cos u = \dfrac{3}{5}$,

$\tan u = -\dfrac{4}{3}$, $\csc u = -\dfrac{5}{4}$, $\sec u = \dfrac{5}{3}$

7. $\sin u = \dfrac{5}{13}$, $\tan u = -\dfrac{5}{12}$, $\cot u = -\dfrac{12}{5}$,

$\csc u = \dfrac{13}{5}$, $\sec u = -\dfrac{13}{12}$ **9.** $\sin u = -\dfrac{4}{5}$,

$\cos u = -\dfrac{3}{5}$, $\cot u = \dfrac{3}{4}$, $\csc u = -\dfrac{5}{4}$,

$\sec u = -\dfrac{5}{3}$ **11.** $\sin u = \dfrac{\sqrt{2}}{2}$, $\cos u = -\dfrac{\sqrt{2}}{2}$,

$\cot u = -1$, $\csc u = \sqrt{2}$, $\sec u = -\sqrt{2}$

Page 480

1. .4067 **3.** .9205 **5.** 1.0977 **7.** 2.3183

9. .8718 **11.** .8342 **13.** 1.3190 **15.** .8342

17. $-.7431$ **19.** -2.1445 **21.** 5.3093

23. -1.1988 **25.** .9863 **27.** $15° 10'$ **29.** $70° 10'$

31. $47° 30'$ **33.** $22° 10'$ **35.** $75° 30'$ **37.** $76° 20'$

39. $\sin 65°$ **41.** $\tan 80°$ **43.** $\sin 39° 50'$

45. $\tan 34° 40'$ **47.** $\sin 55° 30'$

Page 482

1. $30°, 150°$ **3.** $90°$ **5.** $270°$ **7.** $135°, 315°$

9. $135°, 225°$ **11.** none **13.** $60°, 300°$ **15.** $45°$,

$225°$ **17.** $141° 50', 321° 50'$ **19.** $150°, 210°, 510°$,

$570°$ **21.** $30°, 210°, 390°, 570°$ **23.** no values

25. $60°, 240°, 420°, 600°$ **27.** $144° 30', 324° 30'$,

$504° 30', 684° 30'$ **29.** $11° 30', 168° 30', 371° 30'$,

$528° 30'$ **31.** $270°, 630°$ **33.** no values

35. $12°, 168°, 372°, 528°$ **37.** $45°, 225°$

39. $45°, 135°, 225°, 315°$ **41.** $0°, 180°$

Page 485

1. .6852 **3.** .2300 **5.** .3040 **7.** .8320

9. .6180 **11.** .9602 **13.** $12° 46'$ **15.** $16° 43'$

17. $22° 52'$ **19.** $70° 22'$ **21.** $63° 2'$ **23.** $30°, 150°$

25. $9° 28', 189° 28'$ **27.** $199° 28', 340° 32'$

29. $-.7384$ **31.** 2.1842 **33.** -3.2472

35. .9810 **37.** -1.0764 **39.** 1.0042

41. $156° 35', 203° 25'$ **43.** $36° 47', 143° 13'$

45. $93° 44', 266° 16'$ **47.** $120°, 300°$ **49.** $107° 2'$,

$287° 2'$ **51.** $139° 57', 220° 3'$

Page 488

1. $\dfrac{g}{i}, \dfrac{h}{i}, \dfrac{g}{h}, \dfrac{h}{g}$ **3.** $\dfrac{a}{c}, \dfrac{b}{c}, \dfrac{a}{b}, \dfrac{b}{a}$ **5.** $\dfrac{d}{f}, \dfrac{e}{f}, \dfrac{d}{e}, \dfrac{e}{d}$

7. $56°$ **9.** $17°$ **11.** $74°$ **13.** $22°$ **15.** $77° 47'$

17. $15° 24'$ **19.** $15° 42'$ **21.** $\dfrac{2\sqrt{3} + 3}{3}$ **23.** 0

25. $\dfrac{2\sqrt{3}}{3}$ **27.** 1 **29.** $\dfrac{30 - 64\sqrt{3}}{-3}$

31. $\dfrac{18 - 2\sqrt{3} - 9\sqrt{2} + \sqrt{6}}{12}$

Page 491

1. $b = 8, c = 11$ **3.** $a = 24, c = 29$ **5.** $b = 61$,

$c = 62$ **7.** $a = 30, c = 31$ **9.** $a = 6, b = 46$

11. $m\angle A = 37, m\angle B = 53$ **13.** $m\angle A = 48$,

$m\angle B = 42$ **15.** $m\angle A = 63, m\angle B = 27$

17. 110 m **19.** 6 m **21.** $a = 8.7, b = 10.1$,

$m\angle A = 40° 42'$ **23.** $a = 3.3, c = 12.7$,

$m\angle A = 15° 14'$ **25.** $a = 11.3, b = 3.4$,

$m\angle A = 73° 4'$ **27.** $b = 21.3, c = 22.1$,

$m\angle A = 16° 27'$ **29.** 28.4 **31.** 13.9

33. 46.9 **35.** 1,401 m

Page 492

1. I, 45° **3.** III, 45° **5.** I, 30° **7.** I, 60°
9. III, 80° **11.** I, 40° **13.** 45°, $2\sqrt{2}$

15. 60°, $2a$ **17.** $\frac{1}{2}$ **19.** $-\frac{2\sqrt{3}}{3}$ **21.** 0

23. undefined **25.** sin 50° **27.** $-\cos 20°$
29. III, IV **31.** II, IV **33.** 120°, 240°

35. 120°, 300° **37.** $\sin u = \frac{4}{5}$, $\tan u = -\frac{4}{3}$,

$\cot u = -\frac{3}{4}$, $\sec u = -\frac{5}{3}$, $\csc u = \frac{5}{4}$ **39.** .6157

41. $-.7314$ **43.** $-.4792$ **45.** .2907
47. .9896 **49.** 76°20′ **51.** 68° **53.** 8°11′,
188°11′ **55.** 27°9′, 152°51′ **57.** cot 25°18′
59. $-\tan 15°30′$ **61.** 131°49′, 228°11′

63. $\frac{\sqrt{2} - 3\sqrt{3}}{-4}$ **65.** $\frac{8\sqrt{2} - \sqrt{6}}{8}$

67. $b = 3$, $c = 13$

Page 499

1. 1, 1 **3.** 1, 1 **5.** I, IV; III, IV **7.** 0.3
9. 0.9 **11.** 0.3 **13.** 0.3 **15.** -0.6
17. 45°, 225° **19.** $-135°$, 45°, 225°

Page 502

1. 360°, 3 **3.** 360°, $\frac{1}{2}$ **5.** 360°, 4 **7.** 360°, $\frac{3}{4}$

9. 360°, 2; 360°, $\frac{1}{2}$ **11.** 360°, $\frac{1}{2}$ **13.** 360°, $\frac{1}{3}$

15. 360°, 3 **17.** 360°, $\frac{2}{5}$ **19.** 2 **21.** 360°, none

23. 360°, none **25.** 0°, 180°, 360° **27.** 56°, 236°,
$-124°$, $-304°$ **29.** 60°, 120°, 240°, 300°

Page 505

1. 180°, 1 **3.** 180°, 1 **5.** 1,080°, 1 **7.** 1,080, 1
9. 90°, 1 **11.** 1,440°, 1 **13.** 180°, 1 **15.** 720°, 1
17. 240°, 1 **19.** 120°, 1 **21.** 4 **23.** 180°, none
25. 720°, none **27.** 5

Page 508

1. 720°, 2 **3.** 180°, $\frac{1}{2}$ **5.** 40°, $\frac{1}{3}$ **7.** 720°, 2

9. 120°, $\frac{1}{3}$ **11.** 120°, 2 **13.** 1,080°, 3 **15.** 180°, 3

17. 720°, 2 **19.** 720°, 2 **21.** 180°, $\frac{1}{2}$ **23.** 2

25. 4 **27.** 180°, none **29.** 120°, none
31. 180°, none **33.** 0

Page 511

1. none; none **3.** none; none **5.** I, II, III, IV,
none **7.** $-90°$, 90°, 270°; $-180°$, 0°, 180°, 360°
9. 360° **11.** 60° **13.** 90° **15.** 540° **17.** 360°
19. 360° **21.** 60° **23.** 90° **25.** 360° **27.** 90°
29. 360° **31.** 540° **33.** 5 **35.** 2 **37.** 36°,
108°, 180°, 252°, 324°

Page 514

1. $9.9063 - 10$ **3.** $9.6188 - 10$ **5.** $9.9768 - 10$
7. $10.5350 - 10$ **9.** $9.7150 - 10$
11. $9.9481 - 10$ **13.** 28°20′ **15.** 36°20′
17. 33°29′ **19.** 40°20′ **21.** 21°27′ **23.** 27,710
25. 0.9520 **27.** 1,292,000 **29.** 0.1182
31. 39,910 **33.** 7.303×10^{16}

Page 517

1. 45°, 135°, 225°, 315° **3.** 30°, 330° **5.** 90°, 270°
7. 240°, 300° **9.** 90°, 270°; 60°, 300° **11.** 0°, 180°;
30°, 210° **13.** 90° **15.** 45°, 225° **17.** 187°, 353°
19. 33°, 327°; 104°, 256° **21.** 47°, 227°, 147°, 327°
23. 8°, 188°; 82°, 262° **25.** 60°, 120°; 240°, 300°
27. no values **29.** 114°, 246°; 71°, 289°
31. 80°, 280° **33.** 98°, 262°

Page 520

1. $\frac{1}{\sin A}$ **3.** $\frac{\sin A}{\cos A}$ **5.** $\left(\frac{1}{2}\right)^2 + \left(\frac{\sqrt{3}}{2}\right)^2 = 1$

7. $1 + (\sqrt{3})^2 = 2^2$ **9.** $\left(-\frac{\sqrt{3}}{2}\right)^2 = 1 - \left(-\frac{1}{2}\right)^2$

11. $\left(-\frac{2}{\sqrt{3}}\right)^2 - 1 = \left(\frac{1}{\sqrt{3}}\right)^2$

Page 525

1. $\frac{\pi}{6}$ **3.** $-\frac{10\pi}{9}$ **5.** $\frac{7\pi}{6}$ **7.** $-\frac{\pi}{3}$ **9.** $-\frac{3\pi}{2}$

11. $-\frac{11\pi}{6}$ **13.** $\frac{5\pi}{2}$ **15.** 45° **17.** 150°

19. 120° **21.** 135° **23.** $-108°$ **25.** $-360°$

27. 330° **29.** 1 **31.** -1 **33.** $-\frac{\sqrt{3}}{2}$

35. $\frac{\sqrt{6} + 2\sqrt{3}}{6}$ **37.** 0 **39.** $\frac{-3\sqrt{2} + 4\sqrt{6}}{4}$

Page 528

1. 1, -1 **3.** 1 **5.** 2π **7.** 2π, $\frac{1}{2}$ **9.** 2π, 3

11. 4π, 2 **13.** $\frac{2\pi}{3}$, $\frac{1}{3}$ **15.** $\frac{7\pi}{6}$, $\frac{11\pi}{6}$

17. $\frac{\pi}{4}$, $\frac{5\pi}{4}$ **19.** $\frac{\pi}{4}$, $\frac{5\pi}{4}$, $\frac{3\pi}{4}$, $\frac{2\pi}{4}$

21. $\dfrac{\pi}{2}, \dfrac{3\pi}{2}, \dfrac{2\pi}{3}, \dfrac{4\pi}{3}$ **23.** $\dfrac{2\pi}{3}, \dfrac{4\pi}{3}$ **25.** $\dfrac{7\pi}{6}, \dfrac{11\pi}{6}$

23. $\dfrac{120}{169}$ **25.** $\dfrac{240}{289}$ **27.** $2\sin A$ **29.** $2\cos^2 A$

Page 532

1. $y = 2\sin^{-1} x$ **3.** $y = \tan^{-1} x$
5. $y = \cos^{-1}(2x)$ **7.** $60°$ **9.** $-60°$ **11.** $-45°$

13. $180°$ **15.** $\dfrac{4}{5}$ **17.** $-\dfrac{3}{5}$ **19.** none

21. $\sqrt{2}$ **23.** $-\dfrac{12}{5}$

Page 535

1. $360°, 3$ **3.** $720°, \dfrac{1}{2}$ **5.** $180°, \dfrac{1}{2}$ **7.** 2
9. $9.9425 - 10$ **11.** $36°\,45'$ **13.** $15,210,000$
15. $0°, 45°, 135°, 180°, 360°$ **17.** $\dfrac{3\pi}{4}, \dfrac{5\pi}{4}$

19. $\dfrac{\pi}{6}, \dfrac{5\pi}{6}, \dfrac{\pi}{2}$ **21.** $\dfrac{\sin 23°}{\cos 23°}$ **23.** $\dfrac{\cos 74°\,18'}{\sin 74°\,18'}$

25. $\left(-\dfrac{\sqrt{3}}{2}\right)^2 + \left(-\dfrac{1}{2}\right)^2 = 1$ **27.** $\cos^2 x$

Page 541

1. 9 **3.** 25 **5.** 9 **7.** 21 **9.** 54 **11.** 32 **13.** 5
15. 34 **17.** 81 **19.** 25 **21.** 410 km **23.** 985 lb

Page 543

1. $22°$ **3.** $67°$ **5.** $41°$ **7.** $55°$ **9.** $92°$ **11.** $40°$
13. $102°$ **15.** $102°, 78°$ **17.** $41°$ **19.** $19°$

Page 545

1. 37 **3.** 23 **5.** 58 **7.** $52°$ **9.** $68°$ **11.** 7
13. 34 **15.** 662 cm^2

Page 548

1. $21°$ **3.** 11 **5.** $44°$ **7.** 88 **9.** $13.9, 9.8, 85°$
11. $27, 105°\,54', 31°\,48'$ **13.** 50 km

Page 551

1. 2 **3.** 1 **5.** 0 **7.** 1 **9.** 1

Page 555

1. $\dfrac{\sqrt{6} + \sqrt{2}}{4}$ **3.** $\dfrac{\sqrt{3}}{2}$ **5.** $-\left(\dfrac{\sqrt{2} + \sqrt{6}}{4}\right)$ **7.** $\dfrac{1}{2}$

9. $\dfrac{\sqrt{6} + \sqrt{2}}{4}$ **11.** $\dfrac{\sqrt{3}}{2}$ **13.** $-\dfrac{\sqrt{3}}{2}$ **15.** 1

17. $\dfrac{56}{65}, -\dfrac{16}{65}$ **19.** $-\dfrac{171}{221}, -\dfrac{21}{221}$

21. $\dfrac{-\sqrt{4,389} + 24}{91}, \dfrac{-\sqrt{4,389} - 24}{91}$

Page 559

1. $\dfrac{\sqrt{2} - \sqrt{6}}{4}$ **3.** $-\dfrac{\sqrt{2}}{2}$ **5.** $\dfrac{-\sqrt{2} + \sqrt{6}}{4}$

7. $-\dfrac{\sqrt{3}}{2}$ **9.** $\dfrac{-\sqrt{2} + \sqrt{6}}{4}$ **11.** 0 **13.** $-\dfrac{1}{2}$

15. -1 **17.** $\dfrac{75 - 16\sqrt{14}}{153}, \dfrac{75 + 16\sqrt{14}}{153}$

19. $\dfrac{-24 + 3\sqrt{85}}{55}, \dfrac{-24 - 3\sqrt{85}}{55}$ **21.** $\dfrac{119}{169}$

23. $-\dfrac{1}{50}$ **25.** 1 **27.** $\sqrt{6} + \sqrt{2}$ **29.** -2

Page 562

1. $-2 - \sqrt{3}$ **3.** -1 **5.** $2 - \sqrt{3}$
7. $3\sqrt{3} - 4\sqrt{2}$ **9.** $\dfrac{61}{124}$ **11.** -6 **13.** $-\dfrac{11}{244}$

15. $\dfrac{255}{32}$ **17.** undefined **19.** $-\dfrac{264}{23}$ **21.** $\dfrac{-48}{55}$

23. $\dfrac{33}{56}$ **25.** $\dfrac{24}{7}$ **27.** $-\dfrac{69}{71}$ **29.** $-\dfrac{120}{119}$ **31.** 1

Page 565

1. $\dfrac{1}{2}\sqrt{2 - \sqrt{3}}$ **3.** $\dfrac{1}{2}\sqrt{2 - \sqrt{2}}$ **5.** $\dfrac{1}{2}\sqrt{2 + \sqrt{2}}$

7. $\sqrt{3 + 2\sqrt{2}}$ **9.** $\dfrac{1}{2}\sqrt{2 + \sqrt{2}}$

11. $\pm\dfrac{1}{5}\sqrt{5}, \pm\dfrac{2}{5}\sqrt{5}$ **13.** $\pm\dfrac{3}{13}\sqrt{13}, \pm\dfrac{2}{13}\sqrt{13}$

15. $\pm\dfrac{1}{4}\sqrt{6}, \pm\dfrac{1}{4}\sqrt{10}$ **17.** $\pm\dfrac{1}{4}\sqrt{14}, \pm\dfrac{1}{4}\sqrt{2}$

19. $\dfrac{1}{10}\sqrt{10}, -\dfrac{3}{10}\sqrt{10}, -\dfrac{1}{3}$

Page 568

1. 5 **3.** 44 **5.** $32°$ **7.** $100°$ **9.** 36 **11.** $30°$
13. 49 **15.** 2 **17.** $\dfrac{\sqrt{6} + \sqrt{2}}{4}$ **19.** $\dfrac{1}{2}\sqrt{2 - \sqrt{2}}$

21. $\dfrac{1}{2}\sqrt{2 + \sqrt{3}}$ **23.** $\dfrac{1}{2}\sqrt{2 + \sqrt{2}}$

25. $-\dfrac{56}{65}, -\dfrac{16}{65}$ **27.** $\sin\left(\dfrac{3}{2}\pi - \theta\right)$

$= \sin\dfrac{3}{2}\pi \cos\theta - \sin\theta \cos\dfrac{3\pi}{2}$

$= (-1)\cos\theta - \sin\theta\,(0) = -\cos\theta$ **29.** $\dfrac{40}{9}$

31. $-2 - \sqrt{3}$ **33.** $\sqrt{3}$

Index

Abscissa, 209
Absolute value
  of a complex number, 189, 192, 533
  equation property for, 10
  equations involving, 10, 70
  inequalities involving, 11, 73
  inequality properties for, 11
  of a real number, 10
Addition
  of complex numbers, 188, 201
  of logarithms, 397
  of matrices, 258–259
  of polynomials, 38
  properties of matrices, 258–259
  of radicals, 145, 153
  of rational expressions, 96
  in solving systems of two or three linear equations, 244–249
  of vectors, 50–51
Additive identity, 26, 201, 258
Additive identity matrix, 258
Additive inverse (opposite), 26, 96–97, 188, 202, 258
  of complex numbers, 188, 202
  of matrices, 258
Ambiguous case, 549–551
Amplitude, 497, 503–504, 509–510
Angle(s)
  acute, 463
  cosecant of, 470–471
  cosine of, 462–464
  cosine of sum or difference of two, 556–557
  cotangent of, 466–469
  coterminal, 456
  of depression, 490
  double, 554, 557, 561, 566
  of elevation, 490
  half, 563–564, 566
  initial side of, 455
  and law of cosines, 542–543
  measure of, 455–456
  negative functions of, 473–474
  of negative measure, 455
  obtuse, 551
  of positive measure, 455
  quadrantal, 456
  reference, 456
  right, 551
  of rotation, 455–456, 475
  secant of, 470–471

  sine of, 462–464
  sine of sum or difference of two, 552–553
  tangent of, 466–469
  tangent of sum or difference of two, 560–561
  terminal side of, 455
Antilogarithm, 404, 407, 588–589
Applications, 26, 83, 106–107, 181, 201–202, 234, 297, 344–345, 418–419
Approximately equal ($\doteq$), 140
Arc cosine, 529–531
Arc sine, 529–531
Arc tangent, 531
Arithmetic mean(s), 361–362
Arithmetic progressions, 357–358
Arithmetic sequences, 357–358
Arithmetic series, 364–366
Arrangements, probability and, 445–446
Associative properties, 26, 106, 202, 259, 262
Asymptotes, 320–322
Augmented matrix, 265–266
Averages, 102–105, 361
  mean, 102–105
  median, 103
  mode, 102
Axes, 209
Axis of symmetry, 303, 314–317

Base(s)
  of exponents, 31, 35, 390
  of natural logarithms, 385
Base-b logarithms, 392, 406
Base-10 antilogarithms, logarithms, 403–404, 406, 588–589
Binomial(s), 37, 57–58, 60, 128, 130–131, 146
Binomial expansions, 382–383
Binomial theorem, 382, 437
Biography, 186, 356, 454
Bounds, upper and lower, 412–414

Calculator activities, 82, 90, 127, 133, 174, 177, 242, 360, 371, 378, 482, 517, 548
Career(s), 138, 208, 275, 496
Chapter Test, see Tests
Characteristic, 403
Circle, 306–309, 324–325, 354

Circular permutations, 433–434
Clockwise rotation, 455
Closure properties, 26, 107, 201–202
Coefficients, 38, 115, 129, 195, 382, 409, 413
Cofunctions, 478
College Prep Test, see Tests
Combinations, 435–436, 445
Combined variation, 295
Common binomial factor, 66
Common difference, 357, 364
Common monomial factor, 65–66
Common ratio, 368, 375, 379
Commutative properties, 26, 201–202, 209
Complementary functions, 478
Completing the square, 165–167
Complex fraction, 99
Complex numbers, 187–188, 190–191, 195, 201–202, 409–410, 419, 533–534
Complex rational expressions, 99
  simplifying, 99–100
Computations using logarithms, 588–589
Computer activities, 86, 110, 136, 162, 184, 235, 300, 331, 354, 450, 494, 570
Computer section, 574–587
Conditional permutations, 428–429
Conic section(s), 302–333
  circle, 306–309
  ellipse, 310–313
  hyperbola, 319–323
  parabola, 314–318
Conjugate zero theorem, 419
Conjugates, property of, 149, 191, 199, 419
Conjunctions, 7–9, 11, 71
Consecutive integers, 15, 74
Consistent systems, 240
Constant functions, 282–284
Constant of proportionality, 289
Constant term, 409, 413
Constant of variation, 288, 290, 291, 293
Converse of the Pythagorean relation, 224
Conversion factor, 94
Coordinate geometry, 209–237

Coordinate plane, 209, 303–304
Coordinates of a point, 209
Cosecant, 470–471
Cosine(s), 462–464, 556–558, 564
  as cofunction of sine, 478
  as complementary to sine, 478
  as infinite series, 384, 525
  law of, 539–540
Cotangent, 466–469
Coterminal angle, 456
Counterclockwise rotation, 455
Cramer's rule, 250, 254
Cube roots, 152, 193–194
Cubes, 61, 193
  difference of two, 61, 159
  factoring difference of two, 63, 154, 159
  factoring sum of two, 63, 154, 159, 194
  sum of two, 61, 159
Cumulative Reviews, see Reviews

Decimal
  as irrational number, 140
  nonrepeating, 140
  nonterminating, 140
  part of logarithm, 403
  rational number as, 89–90
  repeating, 89, 140, 380
  terminating, 89, 380
Degree of a polynomial, 37, 409–411
Degrees, radians, 523–524
De Moivre's theorem, 534
Denominator(s), 96–97, 99–100, 115–117
  of a complex rational expression, 99–100
  least common multiple (LCM) of, 97
  rationalizing, 148–150, 153–154, 191
Dependent events, 443
Dependent systems, 241
Descartes' (rule of signs), 418
Descending order of exponents, 39, 129
Determinants, 250–256
  Cramer's rule, 250, 254
  minor's, 255
Differences, 96, 100
  of complex numbers, 187–188
  of radicals, 145
  of two cubes, 61, 63, 154, 159
  of two squares, 60, 62, 146
Digits, 268–269
Dimensional analysis, 94–95

Direct variation, 288–289
Directrix of a parabola, 327
Discriminant, 195–196
Disjunction, 7–9, 11, 71
Distance, 47–48, 223–224
Distance formula(s), 47–48, 122–123, 224
Distinguishable permutations, 431–432
Distributive property, 2, 26, 41
Dividend, 128, 130
Division
  of complex numbers, 191
  of imaginary numbers, 191
  of polynomials, 128, 131
  of positive numbers using logarithms, 397–398
  of powers, 31, 35
  of radicals, 148, 153, 159
  of rational expressions, 92
  synthetic, 130–132
Divisor, 128, 130
Domain
  of an exponential function, 390–391
  of a function, 279
  of a logarithmic function, 392, 394–396
  of a rational function, 280
  of a relation, 276
Double angle identities, 566
Double angles, 554, 557–558, 561
Dry mixtures, 43–44

e, e^x, 385
Ellipse, 310–313, 324, 328–330
  equation of, 328–330
  identifying of, 324–325
  properties of, 329
Equality for
  complex numbers, 201
  exponentials, 33, 158
  logarithms, 398
  matrices, 257
Equation(s)
  absolute value and, 10, 70
  of the asymptotes of a hyperbola, 320–322
  of axis of symmetry of a parabola, 315–317
  of a circle, 306–308, 324
  derived, 116
  of an ellipse, 310–312, 324, 328–330
  exponential, 33, 158, 390, 393, 406
  fourth-degree, 69, 193, 410

  fractional, 115–117
  general form of conic section, 324
  of a higher degree, 409–411
  of a hyperbola, 319–321, 324
  of a hyperbola with x-intercepts, 319–320
  of a hyperbola with y-intercepts, 321
  with imaginary number solutions, 193
  of a line, see Linear equation(s)
  linear, see Linear equation(s)
  of linear-quadratic systems, 338–339
  linear trigonometric, 481–482
  literal, 125–126
  logarithmic, 393, 398, 401
  of a parabola, 314–315, 317, 324
  polynomial, 410, 414
  properties, 1
  quadratic, see Quadratic equations
  of quadratic systems, 341–342
  quadratic trigonometric, 515–516
  radical, 178–179
  of a rectangular hyperbola, 321
Equation property for absolute value, 10
Event(s)
  compound, 441–444
  dependent, 443
  exclusive, 442
  inclusive, 442
  independent, 443
  mutually exclusive, 442
  simple, 438–439
Exponent(s), 31, 35, 143
  irrational number as, 390–391, 408
  rational number as, 155–156
  zero as, 34
Exponential equations, 33, 158, 390, 393, 406
Exponential expression(s)
  in radical form, 143, 155–156
  simplifying, 31–32, 34–35, 157–159
Exponential form, 155
Expression(s)
  exponential, 31–32, 34–35, 155–157
  logarithmic, 392, 397–398, 400–401
  radical, 142–143, 145–146, 148–150, 152–156
Extraneous solution(s), 116, 178, 398, 401
Extremes of a proportion, 117

Factor(s), 57, 62–63, 65–66
  greatest common, 65–66
  perfect square, 142–143
  of a perfect square trinomial, 62
  of a trinomial, 57–58, 86
Factor theorem, 128
Factorials, 383–384, 426, 431–433, 435–436, 450
Factoring, 57–58, 65–66
  completely, 65–66
  a difference of two squares, 62
  a perfect square trinomial, 62
  a polynomial into its first-degree factors, 409
  and quadratic equations, 68–69
  and quadratic inequalities, 71–72
  a sum or difference of two cubes, 63
Field, 26, 106, 201
Foci
  of an ellipse, 328–329
  of a hyperbola, 319
  of a parabola, 327
FOIL method, 41, 146
Formula(s)
  for area of a rectangle, 79
  for area of a triangle, 79, 544–545
  for the arithmetic mean of two numbers, 362
  bacteria, for counting, 408
  binomial theorem, 382
  for circular permutations, 433
  compound interest, 407
  for $\cos x$ as infinite series, 384, 525
  for cosine of the difference of two angles, 557
  for cosine of the sum of two angles, 556
  for cosine of twice an angle, 557
  distance, 47–48, 122–123, 224
  for distinguishable permutations, 431
  for e as infinite series, 385
  for e^x as infinite series, 385
  for expected value, 83
  fundamental counting principle, 424
  for geometric mean of two numbers, 373
  $h = vt - 5t^2$, 23, 173
  for half-angles, 563–564
  as a logarithmic equation, 398, 400
  for $\log_b x \cdot y$ and $\log_b \frac{x}{y}$, 397
  for $\log_b x^r$ and $\log_b \sqrt[n]{x}$, 400
  midpoint, 226, 232

  for number of combinations, $\binom{n}{r}$; 435
  for nth term of A.P., 358
  for nth term of G.P., 369
  for perimeter of a rectangle, 18, 125
  Pythagorean, 78, 115
  quadratic, 169–170, 172–173, 175
  rth term of $(a + b)^n$, 383
  for simple interest, 22
  simple pendulum, 181
  for $\sin x$ as infinite series, 384, 525
  for sine of the difference of two angles, 553
  for sine of the sum of two angles, 552
  for sine of twice an angle, 554
  slope, 231–232
  special product, 60–61
  for the sum of finite geometric series, 375, 376
  for the sum of the first n terms of arithmetic series, 364–366
  for the sum of infinite geometric series, 379
  for tangent of the difference of two angles, 561
  for tangent of the sum of two angles, 560
  for tangent of twice an angle, 561
  for total amount, 22
  for total amount A
    including compound interest, 407
    including simple interest, 22
  for total resistance, 23
  volume, 176, 181
Fourth-degree equations, 69, 193, 410
Fraction(s), 99, 115
Frequency table, 103
Function(s), 276–296
  amplitude of, 497, 500–501, 503–504, 506–507, 509–510
  composite, 280, 300
  constant, 282–283
  cosecant, 470–471
  cosine, 462–464
  cotangent, 466–469
  determining whether the inverse is a, 286
  exponential, 390
  greatest integer, 284
  inverse, 285–286, 392, 394–395
  linear, 282
  logarithmic, 392
  of negative angles, 473–474

  periodic, 497
  polynomial, 409–414, 416–417
  reciprocal, 468
  as a relation, 276
  secant, 470–471
  sine, 462–464
  tangent, 466–469
  trigonometric, 462–475, 529–532
  values of, 279–281
  values of trigonometric, 475
Fundamental counting principle, 424–426, 445–446
Fundamental theorem of algebra, 410

Geometric mean, 372–373
Geometric progressions, 368–369
Geometric series, 375–377, 379–380
Geometry, coordinate, 209–237
Graph(s), 277, 279, 282–283, 286
  absolute value inequality, 11
  compound linear inequalities, 7–8
  conjunctions, 7–8
  disjunctions, 7–8
  of the equation of a circle, 306–308
  of the equation of an ellipse, 310–312
  of the equation of a hyperbola, 319–322
  of the equation of a parabola, 314–318
  of exponential functions, 390–391
  of inverses of trigonometric functions, 529–530
  of a linear equation in two variables, 210
  of linear inequalities, 4–5, 7–8
  of linear inequality systems, 346–347
  of linear-quadratic inequality systems, 349–350
  of linear-quadratic systems, 338
  of logarithmic functions, 394–396
  of an ordered pair in a coordinate plane, 209
  of a quadratic function (parabola), 314–318, 328
  quadratic inequality in one variable, 71–72
  of quadratic systems, 341
  of a relation, 276–277
  slope-intercept method for, 220–221
  of systems of two linear equations, 240–241
  of systems of two trigonometric

equations, 507
of trigonometric functions, 497–498, 509–510
of trigonometric functions applying radian measure, 526–527
Greatest common factor, 65–66

Half-angle formula, 563–564
Half-angle identities, 567
Hexagon, 19
Horizontal line(s), 214
Hyperbola, 319–322
Hypotenuse, 78, 175

Identity (Identities)
  additive, 26, 201
  double angle, 566
  half-angle, 567
  multiplicative, 26, 202
  Pythagorean, 520
  quotient, 519
  reciprocal, 518
  trigonometric, 518–522, 566–567
Imaginary numbers, 188, 193–195, 201–202, 418, 419
Inconsistent systems, 241
Independent events, 443
Independent systems, 240
Index, 152
Inequality (inequalities)
  absolute value and, 11, 73
  compound linear, 7–8
  linear, see Linear inequalities
  polynomial, 72
  properties of, 4
  properties for absolute value, 11
  quadratic, 71–72
  solutions of, see Solution set(s)
  systems of linear, 346–347
Infinite geometric series, 379–380
Infinite series, special, 384–385
Integer(s), 15, 26, 74, 89
Integral zero theorem, 409
Interpolation, 483–484, 512–513
Inverse
  additive, 26, 96, 188, 201
  of exponential function, 394
  of log function, 394–395
  multiplicative, 26, 92, 106, 202
Inverse functions, 285, 392, 394–395, 529–532
Inverse relations, 285
Inverse variation, 291–292
Irrational numbers, 140–141, 175
  as exponents, 390–391, 408

Joint variation, 294–295

Kovalevsky, Sonya, 186

Law of cosines, 539–540, 542–543, 570
Law of sines, 546–547
Laws of logarithms, 397–398, 400–401
Least common denominator (LCD), 97, 99, 115
Least common multiple (LCM), 97
Like radicals, 145
Like terms, 38, 145
Line(s)
  determining whether a point is on, 220–221
  equation of, 216–218
  horizontal, 214
  negative slope of, 213–214
  parallel, 228
  perpendicular, 229–230
  positive slope of, 213–214
  slope of, 212–214
  undefined slope of, 214
  vertical, 214
  y-intercept of, 217
  zero slope of, 214
Linear equation(s)
  consistent system of, 240
  dependent system of, 241
  inconsistent system of, 241
  independent system of, 240
  in one variable, 1–2
  in two variables, 210, 216
  point-slope form of, 216–217
  properties of, 1
  slope-intercept form of, 217–218, 235
  solving problems using systems of, 268–270
  solving a system of three, 247–248
  solving systems of three using matrix transformations, 265–266
  solving by using Cramer's rule, 250, 254
  solving by using determinants, 250–251, 253–254
  solving by using minors, 255
  system of two, 240–241
  system of two solved by addition, 244–246
  system of two solved by inverse of a matrix, 264–265
  system of two solved by substitution, 243
Linear functions, 282
Linear inequalities, 4–5, 7–8

Linear inequality systems, 346–347
Linear polynomials, 37
Linear-quadratic systems, 338–339
Linear-quadratic inequality systems, 349–350
Linear trigonometric equations, 481–482
Literal equations, 125–126
Logarithm(s), 392–405, 588–589
  base-b, 392, 406
  base of natural, 385
  base-10, 403–404
  characteristic of, 403
  common, 403
  computations using, 588–589
  decimal part of, 403
  integer part of, 403
  mantissa of, 403
  of a power, 400–401, 588–589
  of a product, 397–398
  of a quotient, 397–398
  of a radical, 400–401
  of trigonometric functions, 512–514
Logarithmic equations, 397–398, 400–401
Logarithmic expressions, 399, 401–402
Logarithmic function, inverse of, 394–395
Log-trig table, 590–592
Lower bound, 412

Major axis, of an ellipse, 329
Mantissa, 403
Matrix (matrices), 257–267
  addition properties of, 258–259
  additive identity, 258
  additive inverse, 258
  associative addition property of, 259
  augmented, 265–266
  commutative addition property of, 259
  equality of, 257
  multiplication properties of, 260–262
  multiplicative identity, 261
  multiplicative inverse, 261–262
  product of scalar and, 260
  product of two, 260–261
  properties of, 257–262
  property of sum of two, 258
  square, 257
  transformations, 265–266
Maximum value, 167–168
Mean (average), 102–105
Mean(s)

arithmetic, 361–362
average, 102–105
geometric, 372–373
of a proportion, 117
proportional, 372–373
Measure
of central tendency, 102
degree, 523–524
radian, 523–527
Median, 102–103
Midpoint formula, 226, 232
Minor axis, of an ellipse, 329
Minors, 255
Mixture(s), 43–46
dry, 43–44
wet, 44–45
Mode, 102
Monomial(s), 37
Motion problems, 47–48, 122–123
Multiplication
of complex numbers, 190–191, 202
of exponentials, 31, 35, 157
FOIL method, 41, 146
of polynomials, 41–42, 60–61
properties of matrices, 260–262
of rational expressions, 92
of a scalar and a vector, 52
zero product, property of, 68–69
Multiplicative identity, 26, 202
Multiplicative identity matrix, 261
Multiplicative inverse, 26, 92, 106, 202
Multiplicative inverse matrix, 261
Multiplicity, 411, 418
Mutually exclusive events, 442

Natural logarithms, 385
Nature of solutions of quadratic equations, 195
Negative angles, functions of, 473–474
Negative exponents, 34–35
Negative measure of angle of rotation, 455
Negative slope of a line, 213–214
Nonrepeating decimal, 140
Non-routine problems, 3, 42, 46, 77, 82, 88, 101, 114, 118, 144, 151, 164, 177, 197, 222, 239, 287, 302, 323, 337, 340, 389, 405, 423, 430, 432, 434, 437, 472, 499, 505, 511, 538, 565
Nonterminating decimal, 140
nth root, 152–154
nth term, 358, 369
Number(s)
averages of, 102–105

complex, 187–188, 190–192, 195, 201–202, 410, 419, 533–534
conjugate complex, 191, 419
conjugate irrational, 149, 199
finding in word problems, 13–15, 74–75, 172
imaginary, 188, 190–194, 201–202
irrational, 140–141, 175–176, 195
line, 4, 7, 10–11, 71–72
pure imaginary, 188, 190
rational, 26, 89–90, 106–107, 115, 140–141, 195
real, 141, 188, 195, 390
Number line, 4, 7, 10–11, 71–72

Obtuse angle, 551
Opposites, 26, 96, 188, 201
Ordered pair, 209, 416
Ordered triple, 247, 254
Ordinate, 209
Origin, 10, 209, 306, 310, 319

Parabola, 314–318, 324–325, 327–328, 331, 417
axis of symmetry, 314–317
Parallel lines, 228
Pascal, Blaise, 356
Pentagon, 19
Perfect cube, 193
Perfect square, 139
Perfect square factor, 142–143
Perfect square trinomial, 60–61, 165
Perimeter, 18, 125
Period of a function, 500–501, 503–507, 509–510
Period of a simple pendulum, 181
Periodic function, 497
Permutation(s), 425–426, 428–429, 431–433, 445
Perpendicular lines, 229–230
Plane, coordinate, 209, 223–224
Point(s)
circle as set of, 306
collinear, 231
coordinates of, 209–210
determining whether on a line, 220–221
distance between two, 223–224
ellipse as set of, 310, 328
hyperbola as set of, 319
parabola as set of, 327
symmetric, 303–304
turning, 314, 416–417
Point-slope form of an equation of a line, 216–217
Polynomial(s), 37–39, 409–419
binomial factor of, 128–131

degree of, 37, 410, 416
division of, 128–131
equation, 409
factoring completely, 65–66, 409–411, 413–414
function, 416
integral, 409
multiplication of, 41, 60–61
quadratic, 37
subtraction of, 38–39
zero of, 409
Positive measure of angle of rotation, 455
Positive slope of a line, 213–214
Power(s)
of a complex number, 534
logs of, 400–401, 588–589
of a power, 32, 157
of a product, 32, 157
product of, 31–32, 157
of a quotient, 32, 157
quotient of, 31–32, 157
Power series, 385
Principal value, 531
Probability, 438–447
Problem solving
age, 20
applying formulas, 22–25
area, 78–80
compound interest, 407–408
digit, 268–271
electrical circuit, 23
geometric, 78–81, 175–176
height of object sent upward, 23, 25, 167, 173
maximum income, area, height, 167–168
mixture, 43–45
motion, 47–49, 122–123
non-routine problems, 3, 42, 46, 77, 82, 88, 101, 114, 118, 144, 151, 164, 177, 197, 222, 239, 287, 302, 323, 337, 340, 389, 405, 423, 430, 432, 434, 437, 472, 499, 505, 511, 538, 565
number, 13–17, 74–76, 172
perimeter, 18
quadratic equations, 74–75, 78–80, 172–173, 175–176
simple interest, 22
solving by using combined variation, 295
solving by using direct variation, 289
solving by using inverse variation, 292
solving by using joint variation, 295

solving by using the quadratic
formula, 172–173, 175–176
solving by using a system of
linear equations, 268–271
steps in solving word, 15, 172
work, 119–121
writing word phrases in algebraic
terms, 13, 15
Problem-solving method
formulating a situation, 56
four steps, x
planning how to solve, 30
Product(s)
of complex numbers, 190–191,
201–202
formula for special, 60–61
logarithms of, 397–398, 588–589
of the means and extremes of a
proportion, 117
nth root of, 142–143, 152–153
of a number and a logarithm, 401
of powers, 31–32, 157
in proportions, 117
of pure imaginary numbers, 190
of radicals, 145, 153, 159
of rational expressions, 92
of a scalar and a matrix, 260
of a scalar and a vector, 52
of solutions of a quadratic
equation, 198–200
square root of, 142
of square roots, 145–146
of the sum and difference of two
terms, 60
of two binomials, 41
of two matrices, 260–261
Progression(s), 357–358, 368–369
Proof(s), 106–107, 201–202
Property (properties)
absolute value, 10–11
additive identity, 26, 201, 258
additive inverse, 26, 96–97,
188–202, 258
associative, 26, 106, 202, 259,
262
binomial expansions, 382–383
closure, 26, 107, 201–202
commutative, 26, 201–202, 259
complex numbers, 187–192,
201–202
of conjugates, 149, 191, 199, 419
conjunctions, 7–9
discriminant, 195
disjunctions, 7–9
distributive, 2, 26, 41
of an ellipse, 329
equality for,
complex numbers, 201

exponentials, 33, 158
logarithms, 398
matrices, 257
equation for absolute value, 10
equation, linear, 1
exponential functions, 390–391
of exponents, 31–32, 34, 155–156
field, 26
i and $-i$, 187
inequality
of absolute value, 11
compound linear, 7–8
polynomial, 72
quadratic, 71
logarithmic functions, 395–396
of matrices, 257–262, 266
of matrix addition, 257–259
of matrix multiplication, 260–262
multiplicative identity, 26, 148,
202, 261
multiplicative inverse, 1, 26, 92,
202, 261
negative product of two factors,
71
nth root of a product, 142–143,
152
nth root of a quotient, 150, 153
positive product of two factors,
71
power
of a power, 32
of a product, 32
of a quotient, 32
powers
of i, 187, 191–192
of a binomial, 382–383
product
of powers, 31
of radicals, 145, 152
of scalar and matrix, 260
of scalar and vector, 52
of two matrices, 260
proof(s), 106–107, 201–202
proportion, 117
quotient of powers, 31
quotient of radicals, 148, 153
rational numbers, 89, 106–107
sum and product of solutions of
quadratic equation, 198
of sum of two matrices, 258
vector, 50–52
zero-product, 68
Proportion, 117, 126
Pure imaginary numbers, 188, 190
Pythagorean identities, 520
Pythagorean relation, 78, 175,
223–224

Quadrant(s), 209
Quadrantal angle, 456
Quadratic equations, 68–70
coefficients of, 195–196
discriminant of, 195–196
factoring, 68–69
with imaginary solutions,
187–188, 193–194
nature of solutions, 195–196
product of solutions of, 198–200
solving by completing the square,
165–166
solving by using the quadratic
formula, 169–170, 184
solving by using square root,
139–140
standard form of, 68
sum of solutions of, 198–200
and word problems, 74–75,
78–80, 172–173, 175–176
Quadratic formula, 169–170,
172–173, 175, 193, 195
Quadratic inequality (inequalities)
factoring, 71–72
graphing the solutions of, 71–72
solution set of, 71–72
Quadratic polynomial, 37
Quadratic systems, 341–342
Quadratic trigonometric equations,
515–516
Quadratic trinomial, 57
Quotient(s)
of complex numbers, 190–191
identities, 519
logarithm of, 397–398, 588–589
nth root of, 153
of numbers in factorial notation,
383–384, 431–432, 435–436
of polynomials, 128–131
power of, 32
of powers, 31, 157
of radicals, 148–150, 153–154
of rational expressions, 92

Radian measure, 523–527
Radical(s), 139, 152–153
combining, 145
cube roots, 152, 193–194
exponents as, 390–391, 408
fifth roots, 152
fourth roots, 152
logarithms of, 400–401, 588–589
nth root, 152
products of, 145–146, 153
quotients of, 148–150, 153–154
simplifying, 142–143, 145–146,
148–150, 152–154, 162
Radical equations, 178–179, 181, 373

INDEX